Methodologies in Amine Synthesis

Methodologies in Amine Synthesis

Challenges and Applications

Edited by Alfredo Ricci and Luca Bernardi

WILEY-VCH

Editors

Prof. Alfredo Ricci
University of Bologna
Department of Industrial Chemistry
"Toso Montanari"
Viale del Risorgimento 4
40136 Bologna
Italy

Prof. Luca Bernardi
University of Bologna
Department of Industrial Chemistry
"Toso Montanari" & INSTM
Research Unit Bologna
Viale del Risorgimento 4
40136 Bologna
Italy

Library of Congress Card No.:
applied for

British Library Cataloguing-in-Publication Data
A catalogue record for this book is available from the British Library.

Bibliographic information published by the Deutsche Nationalbibliothek
The Deutsche Nationalbibliothek lists this publication in the Deutsche Nationalbibliografie; detailed bibliographic data are available on the Internet at <http://dnb.d-nb.de>.

Print ISBN: 978-3-527-34739-1
ePDF ISBN: 978-3-527-82617-9
ePub ISBN: 978-3-527-82619-3
oBook ISBN: 978-3-527-82618-6

Typesetting SPi Global, Chennai, India
Printing and Binding CPI Group (UK) Ltd, Croydon CR0 4YY

Printed on acid-free paper

10 9 8 7 6 5 4 3 2 1

Contents

Preface *xi*

1 **Substitution-type Electrophilic Amination Using Hydroxylamine-Derived Reagents** *1*
Zhe Zhou and László Kürti
1.1 Introduction *1*
1.2 Cu-Catalyzed Reactions *2*
1.3 Electrophilic Amination Reactions Catalyzed by Other Transition Metals *15*
1.4 Electrophilic Amination with Hydroxylamine-derived Metallanitrenes *18*
1.5 Transition-Metal-Free Electrophilic Amination Reactions *22*
1.6 Conclusion *28*
References *28*

2 **Remote Functionalizations Using Nitrogen Radicals in H-Atom Transfer (HAT) Reactions** *31*
Ji Hye Kim, Elizabeth M. Dauncey and Daniele Leonori
2.1 Introduction *31*
2.2 Intramolecular 1,5-H-Atom Transfer (1,5-HAT) *31*
2.3 Photoinduced Strategies *34*
2.3.1 Reductive Strategies *34*
2.3.1.1 1,5-HAT via Iminyl Radicals *34*
2.3.1.2 1,5-HAT via Amidyl and Sulfamidyl Radicals *37*
2.3.2 Oxidative Strategies *39*
2.3.2.1 1,5-HAT via Iminyl Radicals *39*
2.3.2.2 1,5-HAT via Amidyl and Sulfamidyl Radicals *40*
2.3.3 Photoinduced Bond Homolysis *43*
2.4 Thermal Strategies *46*
2.5 Summary and Conclusions *50*
References *50*

3 **Radical-Based C–N Bond Formation in Photo/Electrochemistry** *53*
Binbin Huang, Yating Zhao and Wujiong Xia
3.1 Introduction *53*
3.2 C—N Bond Formation via N-radical Species Addition *54*
3.2.1 Radical Addition to C—C Double/Triple Bonds *55*
3.2.1.1 Amidyl Radical Addition *55*
3.2.1.2 Hydrazonyl Radical Addition *62*
3.2.1.3 Aminium Radical Cation Addition *64*
3.2.2 Radical Species Addition to Aromatic Rings *71*
3.3 Amination via N-atom Nucleophilic Addition *77*
3.3.1 Aromatic $C(sp^2)$—H Bond Amination *77*
3.3.2 Olefinic $C(sp^2)$—H Bond Amination *82*
3.3.3 Activated $C(sp^3)$—H Bond Amination *85*
3.3.3.1 Benzylic $C(sp^3)$—H Bond Amination *85*
3.3.3.2 N-α-$C(sp^3)$—H Bond Amination *87*
3.4 Amination via Radical Cross-coupling *90*
3.4.1 Aryl $C(sp^2)$—N Bond Formation via Radical Cross-coupling *90*
3.4.1.1 Aryl $C(sp^2)$—N Bond Formation Using Diarylamines *91*
3.4.1.2 Aryl $C(sp^2)$—N Bond Formation Using Azoles *94*
3.4.2 Other C—N Bond Formation via Radical Cross-coupling *96*
3.5 Summary and Conclusions *98*
References *100*

4 **Propargylamines: Recent Advances in Asymmetric Synthesis and Use as Chemical Tools in Organic Chemistry** *103*
Fei Zhao, Seong-Heun Kim and Daniele Castagnolo
4.1 Introduction *103*
4.2 Metal-Catalyzed Asymmetric Synthesis of Propargylamines *104*
4.2.1 Enantioselective A^3 Coupling *104*
4.2.1.1 Enantioselective A^3 Coupling Involving Primary Amines *106*
4.2.1.2 Enantioselective A^3 Coupling Involving Secondary Amines *108*
4.2.2 Enantioselective Propargylic Amination of Propargylic Esters with Amines *112*
4.2.3 Cu-Catalyzed Enantioselective Ring Opening of Alkynyl-Substituted Epoxides/Lactones/Carbonates *116*
4.2.4 Enantioselective Addition of Terminal Alkynes to Enamines/Enamides *118*
4.2.5 Rh/Ru-Catalyzed Enantioselective Hydrogenation of Alkynyl-Substituted Enamides/Imines *120*
4.2.6 Enantioselective C–H Activation: Synthesis of Cyclic Propargylamines *123*
4.3 Enzymatic Synthesis of Propargylamines *126*
4.4 Photoredox Synthesis of Propargylamines *130*
4.5 Organocatalyzed Asymmetric Synthesis of Propargylamines *134*
4.6 Propargylamines as Building Blocks in the Synthesis of Heterocycles *137*
4.6.1 Synthesis of Pyrroles from Propargylamines *138*
4.6.2 Synthesis of Pyrrolines from Propargylamines *139*

4.6.3 Synthesis of Pyridines from Propargylamines *140*
4.6.4 Synthesis of Quinolines from Propargylamines *142*
4.6.5 Synthesis of Oxazoles from Propargylamines *145*
4.6.6 Synthesis of Thiazoles from Propargylamines *148*
4.7 Conclusions *150*
References *150*

5 Transition-Metal-Catalyzed Chiral Amines Synthesis *155*
Anton Cunillera, Carmen Claver, Cyril Godard, Martine Urrutigoïty and Philippe Kalck
5.1 Introduction *155*
5.2 Asymmetric Reductive Amination *156*
5.3 Asymmetric Hydroamination *162*
5.4 Asymmetric Hydroaminoalkylation *168*
5.5 Asymmetric Hydroaminomethylation *171*
5.6 Coupling on a Chiral Metal Center *176*
5.7 Conclusion *177*
References *178*

6 Industrial Relevance of Asymmetric Organocatalysis in the Preparation of Chiral Amine Derivatives *187*
Luca Bernardi, Armando Carlone and Francesco Fini
6.1 Introduction *187*
6.2 Organocatalysis in Manufacture: Representative Examples *189*
6.3 Case Studies *200*
6.3.1 Pregabalin *200*
6.3.1.1 Pathway A: Desymmetrization of Glutaric Anhydride 53 *202*
6.3.1.2 Pathway B: Addition of an Amino α-Carbanion 55 to Michael Acceptors *204*
6.3.1.3 Pathway C: Addition of Acetate Enolate Equivalents to Nitroalkene 56 *206*
6.3.2 Bicyclic α-Amino Acid Core of Telaprevir *214*
6.3.3 5-(Trifluoromethyl)-2-isoxazolines as Antipest Agents *223*
6.4 Summary and Conclusions *230*
References *231*

7 Biocatalytic Synthesis of Chiral Amines Using Oxidoreductases *243*
Sebastian C. Cosgrove, Jeremy I. Ramsden, Juan Mangas-Sanchez and Nicholas J. Turner
7.1 Introduction *243*
7.2 Amine Oxidases *243*
7.2.1 Introduction *243*
7.2.2 (*S*)-Selective Amine Oxidases *244*
7.2.2.1 Monoamine Oxidase from *Aspergillus niger* *244*
7.2.2.2 Directed Evolution of MAO-N *244*
7.2.2.3 Synthetic Applications and Cascades *247*
7.2.2.4 Monoamine Oxidase from *Pseudomonas monteilii* ZMU-T01 *249*

7.2.2.5 Cyclohexylamine Oxidase from *Brevibacterium oxydans* (CHAO) *249*
7.2.3 (*R*)-Selective Amine Oxidases *252*
7.2.3.1 D-Amino Acid Oxidase (pkDAO) *252*
7.2.3.2 6-Hydroxy-D-nicotine Oxidase (6-HDNO) from *Arthrobacter nicotinovorans* *252*
7.3 Amine Dehydrogenases *252*
7.3.1 Introduction *252*
7.3.2 Discovery and Engineering of AmDH *253*
7.3.2.1 Leucine Dehydrogenase *253*
7.3.2.2 Phenylalanine Dehydrogenase and Chimeric Amine Dehydrogenase *254*
7.3.2.3 Native Amine Dehydrogenase *254*
7.3.3 Synthetic Applications of AmDH *256*
7.3.3.1 Primary Amine Synthesis with Engineered AmDH *256*
7.3.3.2 Primary Amine Synthesis with Natural AmDH *258*
7.3.3.3 Substrate Promiscuity in AmDH *259*
7.3.3.4 Cascade Reactions that Use AmDH *260*
7.4 Imine Reductases *262*
7.4.1 From Biosynthesis to Biocatalysis *262*
7.4.2 Biocatalytic Application of Imine Reductases *264*
7.4.2.1 IREDs in Cascade and Chemoenzymatic Synthesis *265*
7.4.3 IRED Engineering *267*
7.4.4 Imine Reductases Catalyzing Reductive Amination *267*
7.4.5 Imine Reductase-Catalyzed Amine Alkylation Cascades *270*
7.4.6 Engineering of Reductive Aminases *270*
7.5 Engineered Cytochrome P450s *270*
7.6 Conclusions and Perspectives *274*
References *274*

8 Engineering Functional Nanomaterials Through the Amino Group *285*
Giacomo Filippini, Paolo Pengo, Susanna Bosi, Giulio Ragazzon, Lucia Pasquato and Maurizio Prato
8.1 Abbreviations *285*
8.1 Introduction *287*
8.2 Quantification of Nanomaterial-Bound Amino Groups *288*
8.3 Exploiting Amino Compounds for the Functionalization of Carbon-Based Nanomaterials *290*
8.3.1 Historical Backgrounds: Allotropes of Carbon *290*
8.3.2 Use of Amines for the Functionalization of Carbon Nanostructures *290*
8.3.3 Other Functionalization Procedures of Common Carbon Nanostructures *297*
8.3.4 Exfoliation of Graphite with Melamine *299*
8.3.5 Other Carbon Nanomaterials *301*
8.3.5.1 Carbon Nanohorns *301*
8.3.5.2 Carbon Nanodiamonds *305*
8.3.5.3 Carbon Nano-onions *307*

8.3.6 Amino-Functionalized Carbon-Based Nanomaterials for Analytical Applications *308*
8.4 Amines in the Synthesis and Functionalization of Carbon Dots *309*
8.4.1 Amines as CD Constituents *310*
8.4.2 Amine-Rich CDs from Arginine and Ethylenediamine (NCDs) *312*
8.4.2.1 One-Pot Functionalization of NCDs *312*
8.4.2.2 Postfunctionalization of NCDs *313*
8.4.2.3 Use of CD-Supported Amines in Organocatalysis *315*
8.5 Amines for the Engineering of Hybrid Organic–Inorganic Nanomaterials *316*
8.5.1 Amines as Head Groups or End Groups on Self-assembled Monolayers on Flat Surfaces *316*
8.5.2 Alkylamines in the Preparation of Semiconductor Quantum Dots *319*
8.5.2.1 Sulfur–Amine and Selenium–Amine Systems *319*
8.5.2.2 Capping Ligands for Quantum Dots and Ligand Exchange by Amines *321*
8.5.3 Alkylamines as Reagents for the Synthesis and Passivation of Metal Nanoparticles *322*
8.5.3.1 Alkylamines as Capping Agents for Metal Nanoparticles *322*
8.5.3.2 Displacement of Amines from the Surface of Metal Nanoparticles *325*
8.5.4 Amines on the Outer Surface of Organic–inorganic Hybrid Nanoparticles *325*
8.5.5 Postfunctionalization of Amine-Terminated Organic–Inorganic Hybrid Nanoparticles *328*
References *329*

9 Recent Advances in the Synthesis of Nitrogen Compounds from Biomass Derivatives *341*
Ana C. Fernandes
9.1 Introduction *341*
9.2 Synthesis of Nitrogen Compounds from Chitin and Its Derivatives *341*
9.3 Synthesis of Amines and Formamides from α-Amino Acids *345*
9.4 Synthesis of Nitrogen Compounds from Cellulosic Biomass Derivatives *348*
9.5 Synthesis of Nitrogen Compounds from Lignin Derivatives *366*
9.6 Synthesis of Nitrogen Compounds from Triglycerides and Fatty Alcohols *370*
9.7 Conclusion *373*
References *373*

10 Recent Advances in the Synthesis of Arylamines in the Light of Application in Pharmaceutical and Chemical Industry *377*
Dino Berthold, Alexander M. Haydl, Joyce C. Leung, Ulrich Scholz, Qing Xiao and Zhibin Zhu
10.1 Modern Approaches to Transition-Metal-Catalyzed C–N Coupling in Industry *377*
10.1.1 Introduction *377*
10.1.2 Transition-Metal-Catalyzed C—N-Bond Formation *377*
10.1.2.1 Ullmann-Type Amination *378*
10.1.2.2 Buchwald–Hartwig Amination *383*

10.2 New Methodologies in the Synthesis of Arylamines on the Brink of Industrial Application 388
10.2.1 Introduction 388
10.2.2 Catalytic C–H Amination 389
10.2.2.1 Catalytic C–H Amination under Standard Conditions 389
10.2.2.2 Photoredox Catalysis 399
10.2.2.3 Electrochemical Approaches 402
10.2.3 Decarboxylative Aryl Amination 405
10.2.4 Nickel-Catalyzed C–N Coupling 407
10.2.5 Other Metal-Catalyzed Cross-Couplings 414
10.2.6 Reductive Amination 417
10.2.7 Hydroamination 419
10.2.8 Summary and Conclusions 424
10.3 Advances to Arylamine Formation Using Intensified and More Sustainable Process Technologies 424
10.3.1 Introduction 424
10.3.2 Flow Chemistry 425
10.3.2.1 Pd-Catalyzed C—N Bond Forming Reaction 425
10.3.2.2 Nucleophilic Aromatic Substitution 426
10.3.2.3 Telescoped Sequence of Nitration and Hydrogenation in Flow Synthesis 428
10.3.2.4 Chan–Lam Coupling 428
10.3.3 Immobilization of Catalysts/Supported Catalysts 429
10.3.4 Personal Accounts from Contract Manufacturing Companies on Utility of Modern Flow Amination Methods 431
10.4 Miscellaneous Aspects of Aromatic Amination Reactions in the World of Active Pharmaceutical Ingredients 431
10.4.1 Cohort of Concerns and its Regulatory Impact on Amine-Based Active Pharmaceutical Compounds 431
10.4.2 Role of Control of Elemental Impurities in Human Pharma Applications 432
10.4.3 Transition Metal Accounting 433
10.4.4 Recycling of Metals, Ligands, and Other Cost Drivers of Aromatic Amination 434
10.4.5 Regulatory Requirements for the Submission of a Catalytic Reaction in a New Drug Application 434
References 435

Index 445

Preface

The books "Modern Amination Methods" and "Amino Group Chemistry – From Synthesis to the Life Sciences" edited by one of us for Wiley-VCH in 2000 and in 2007, respectively, were intended to provide the reader with exhaustive overviews of the most advanced methodologies for the C—N bond formation and with the role played by the amino function in those processes that are closely related to the life sciences.

These books were well received by the chemical community and highlighted the importance of keeping scientists to be continuously aware of the progress in the field of amino group chemistry. In 2018, setting up to discuss the prospect of coediting a new book on amino group chemistry, we asked ourselves if the past decade witnessed sufficient breakthroughs to make such update worth to be read by academicians or industrialists. It did not take too much time to convince ourselves that such endeavor would have been worthwhile and timely. In the past years, most of the breakthrough discoveries in synthetic organic chemistry embed efficient preparations of nitrogen-containing compounds. Arguably, amino group chemistry lies at the core of recent methodological trends. Furthermore, the disclosure of new materials, from medicine to nanoscience, has often been grounded on nitrogen derivatives. Thus, we embarked in this adventure, approaching the selection of the topics with an unbiased and wide-range attitude. In this new book, we aimed at providing not only an overview over specific aspects of amino chemistry but also a unique journey through modern chemistry.

With the very useful insights into the topic selection provided by Wiley anonymous reviewers, we commenced to approach several prospective authors. We have been very lucky to find contact authors among the most authoritative of their fields, who, together with their coworkers, assembled chapters characterized by clear structures, schemes, mechanisms, and figures accompanying the text together with exhaustive and up-to-date reference sections.

Throughout the 10 chapters of the book, amines are discussed with respect to the most advanced methodologies for their preparation, from design to scope surveys to applications to biologically active targets or precursors thereof, including large-scale industrial settings. Because the pharmaceutical and agrochemical industries are nowadays mostly away from the development of racemic compounds, several of the book chapters refer to enantioselective synthesis of chiral amines. It is worth noting that the contributions to this book present interlinks that, as a fil rouge crossing the whole book, allow the reader to face some of the topics under different perspectives.

The book starts with a topic that constitutes a continuum between the 2000s monographies and this one: electrophilic amination. Indeed, substitution-type electrophilic amination has gained a tremendous progress in the decade. This unconventional, yet synthetically appealing approach to C—N bond formation is nowadays performed under transition metal (TM)- and metal-free catalysis aiming to employ bench-stable and easily obtainable reagents for forging new C—N bonds. In this context, Chapter 1 summarizes the most recent achievements complying with the requirements for an ideal simple and versatile synthetic methodology.

The following part of the book brings the reader in the domain of nitrogen radicals as reactive intermediates, a reactivity stream displaying high complementarity to classical ionic and TM mediated processes. The use of photo- and electrochemical induced catalytic strategies is widely discussed (Chapters 2 and 3), overviewing the recent dramatic advances enabling to couple challenging C–H activation processes with C—N bond forming events. Although aminations with high enantioselectivity are yet rarely accessed, these radical methodologies targeting the synthesis of a wide range of heterocyclic systems and an array of complex N-containing structures outline a new and highly promising frontier.

As an enabling methodology, the development and application of asymmetric catalysis is now an integral part of the work of any major research unit in academia and industry. In the past decade, organo- and biocatalysis have flanked TM catalysis as key players for the enantioselective synthesis of chiral amino compounds, as overviewed in the ensuing Chapters 4–7. Worth noting is the possibility that these various catalytic approaches might concur toward attaining the same targets. This is exemplified in Chapter 4 devoted to the synthesis of chiral propargylamines. Many of these compounds possessing pharmaceutical properties have found applications in the treatment of diseases and can be considered among the most efficient building blocks for the synthesis N-containing heterocycles.

The relevance of high atom efficiency and very limited waste production (high atom economy) is of paramount importance for the development of methodologies with industrial prospects. Reactions such as TM-catalyzed asymmetric hydroamination, reductive amination, and hydroaminomethylation, reported in Chapter 5, meet these goals, an additional advantage being the simplicity and the easy availability of the starting materials, which is only partially counterbalanced by the complexity of the used ligands. Because of their relevance, these reactions are occasionally mentioned elsewhere in this book, although in the frame of different contexts, thus providing the reader with an exhaustive set of complementary information.

In the following chapters, metal-free stereoselective catalytic methodologies at the service of amine synthesis, represented by organo- and biocatalysis, are treated with a greater focus on industrial applications reflecting the dynamic nature of these catalytic streams. The central goal of organocatalysis (Chapter 6) is herein addressed to the emerging utilization of this methodology as an exceedingly useful synthetic tool for the large-scale preparation of active pharmaceutical ingredients (APIs) in industrial settings. An emphasis is placed on three case studied in which organocatalysis stands out with respect to other enantioselective strategies, for its versatility and potential for scale-up. The revolutionary approach of enzyme engineering, in the context of oxidoreductase enzymes applied to the synthesis of N-containing biologically relevant intermediates and products, is described in Chapter 7.

The intrinsic selectivity of enzymes provides an obvious advantage in the key challenge of developing new synthetic platforms amenable to industrialization.

The following two chapters stand out for their uniqueness, targeting topics that were not even touched in the previous two books on amino chemistry but of exceptional prominence and timeliness. Chapter 8 deals with the use of amines in the synthesis, stabilization, and functionalization of organic–inorganic hybrid nanomaterials, highlighting through a large number of case studies the pivotal role played by the amino group to unlock practical applications in the fields of biology, medicine, and energy production. Conversely, the synthesis of a number of valuable amino compounds is reported in Chapter 9, from the transformation of renewable biomass resources that already incorporate the amino groups such as chitin, chitosan, and amino acids or by modifying bio-based compounds, followed by amination. Rare aminosugars, precursors of medicinal compounds, and a wide range of heterocycles are obtained avoiding the use of fossil-based feedstocks, thus providing a remarkable step forward in the ongoing shift from depleting to renewable resources.

The final Chapter 10 addresses current applications of TM-catalyzed aromatic amination in industrial settings by discussing a large number of case studies related to the manufacturing process of pharmaceutical compounds. In addition, with reference to the seminal work by Ullmann, Buchwald, and Hartwig, this contribution points on new concepts still at academic level, but either further extending the applicability of new methodologies, or on the brink of being industrially used. Also approaches with a focus on process intensification and sustainability (flow chemistry and catalyst immobilization) are presented, together with a view of the accompanying questions when applying the methodology of aromatic amination in the pharmaceutical industry. To make aware the reader about these challenges, themes such as the control of elemental impurities, the TM accounting, and the metal recycling are treated as well. Besides being a highly useful and up-to-date source of information on the TM-catalyzed aromatic amination in industry, this contribution will hopefully provide inspiration for academic research in developing new methodologies amenable to industrialization.

We warmly thank all the distinguished scientists and their coauthors for their rewarding and highly instructive contributions. Without their effort, even more valuable considering it partially coincided with a difficult period at the international level, this volume would have not been possible. Grateful acknowledgments are also addressed to the Wiley-VCH editorial staff, and in particular to Anne Brennführer, Aruna Pragasam, Elke Maase, and Katherine Wong, who encouraged us at project outset and helped us in a very competent manner in all the phases of the preparation of this book.

Bologna
07 April 2020

Alfredo Ricci and Luca Bernardi

1

Substitution-type Electrophilic Amination Using Hydroxylamine-Derived Reagents

Zhe Zhou and László Kürti

Rice University, Department of Chemistry, 6500 Main Street, Houston, TX 77030, USA

1.1 Introduction

Electrophilic amination is a class of organic reactions where C—N bonds are formed via the use of electrophilic aminating reagents [1–4]. Depending on the specific reaction pathway, electrophilic amination reactions can be classified either as substitution or addition. This chapter focuses on substitution reactions. Aminating reagents for the substitution-type electrophilic aminations are essentially NR_2^+ synthons, and the electrophilicity on the nitrogen atom is generally achieved by attaching a more electronegative functionality (X) to the nitrogen atom that can serve as a leaving group. Common structural motifs for this class of reagents include chloramines, hydroxylamines, and oxaziridines (i.e. cyclic hydroxylamine derivatives). Because of the safety hazards associated with the use of chloramines, recent developments in this area have been focused on the use of more stable hydroxylamine-type reagents (Scheme 1.1).

Substitution-type electrophilic amination reactions can operate under either uncatalyzed or catalyzed conditions. The majority of catalytic substitution-type electrophilic amination reactions are catalyzed by complexes of transition metals (TMs). In the uncatalyzed reactions, the carbon nucleophile directly attacks the electrophilic nitrogen atom and a new C—N bond is formed. In the TM-catalyzed reactions, the transition metal first enters into the N—X bond via an oxidative addition, and the new C—N bond is formed after sequential ligand exchange and reductive elimination (Scheme 1.2). The major difference between TM-catalyzed substitution-type electrophilic amination reactions and TM-catalyzed C–N cross-coupling reactions (i.e. Buchwald–Hartwig coupling) is the role of the nitrogen source: it acts as an electrophile in the former while as a nucleophile in the latter.

The majority of the literature in this area concerns the TM-catalyzed versions of substitution-type electrophilic amination. Therefore, they will be discussed first in this chapter.

Methodologies in Amine Synthesis: Challenges and Applications, First Edition.
Edited by Alfredo Ricci and Luca Bernardi.

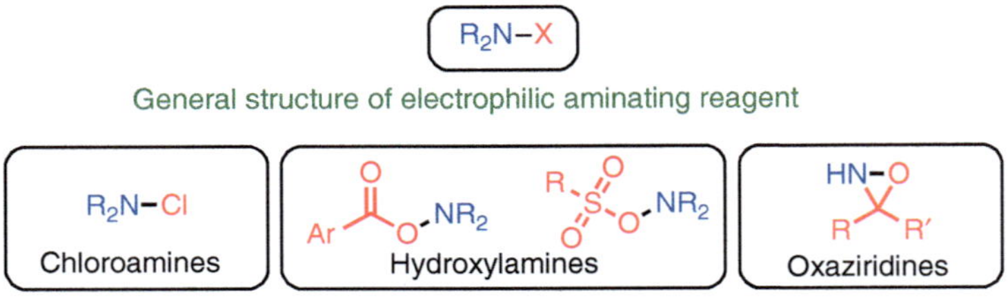

Scheme 1.1 General structure of electrophilic aminating reagent.

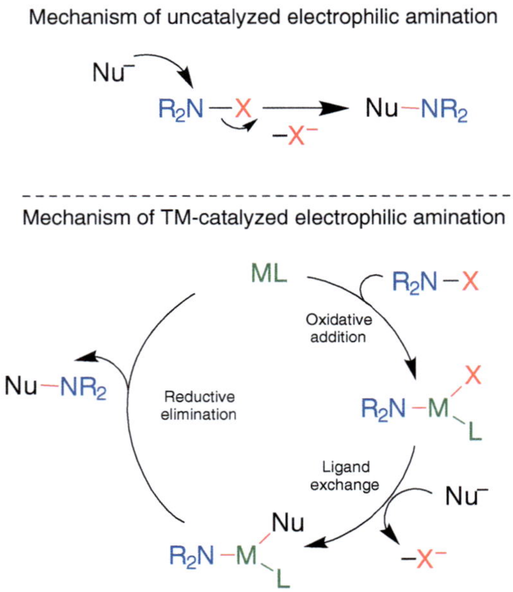

Scheme 1.2 Mechanisms of two main types of electrophilic amination.

1.2 Cu-Catalyzed Reactions

Narasaka and coworkers reported an early iteration of Cu-catalyzed substitution-type electrophilic amination of Grignard reagents utilizing *O*-sulfonyloximes as aminating reagents. The reactions can also give products without copper catalysis, albeit with much lower yields. Because of the nature of the aminating reagents, subsequent acidic hydrolysis is needed to convert the imine products to the desired amines [5, 6]. A similar amination process of organozinc reagents was reported by the Erdik group around the same time. In this case, in addition to the *O*-sulfonyloximes, the authors showed that the reaction can also proceed with methoxyamine when excess of the organometallic reagent was used (Scheme 1.3) [7, 8]. These early reactions have several drawbacks that affect their utilizations, including a very limited substrate scope, low conversion rate due to many side reactions, and, most importantly, the need to use a strong acid to hydrolyze the initial imine

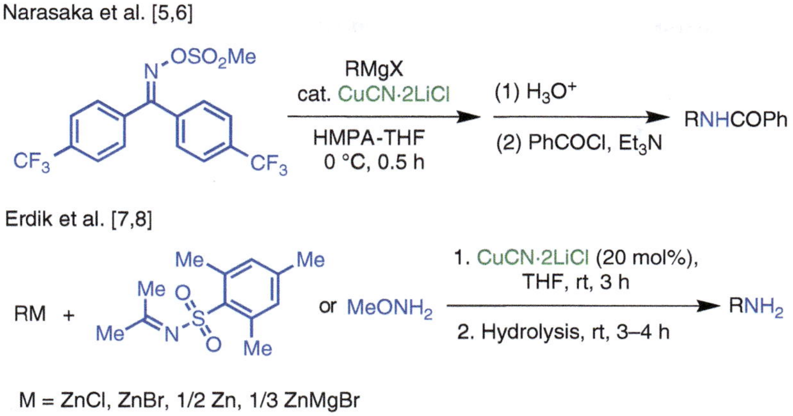

Scheme 1.3 Early examples of Cu-catalyzed electrophilic amination. Source: Erdik and Ay [2] and Tsutsui et al. [5].

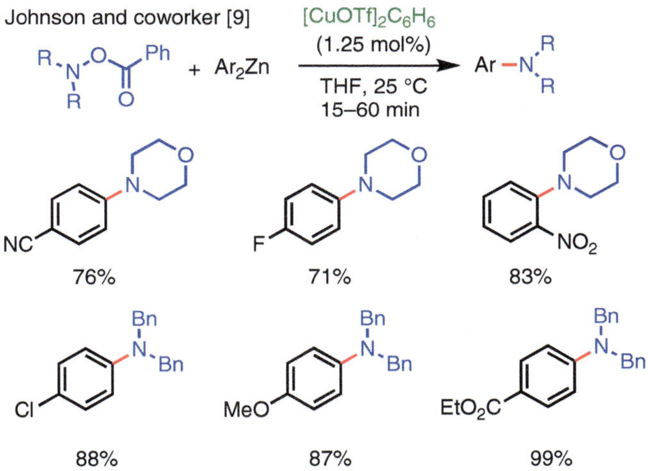

Scheme 1.4 Cu-catalyzed electrophilic amination of organozinc reagents.

products. The use of strong acidic conditions in the hydrolysis makes these procedures unsuitable for substrates containing acid-labile functionalities.

In 2004, the Johnson group at UNC reported the Cu-catalyzed electrophilic amination of diorganozinc reagents using acyl hydroxylamines as aminating reagents [9]. This is the first report of Cu-catalyzed electrophilic amination reactions that give tertiary amines as products (Scheme 1.4). Compared to the *O*-sulfonyloximes, *O*-acyl hydroxylamines are more synthetically accessible and have better atom economy. A limitation, however, is that the nitrogen must be fully substituted (i.e. no acidic N—H bond is tolerated). In place of the diorganozinc substrate, this reaction can also use a Grignard reagent as the nucleophile, which is arguably more accessible and convenient to use [10].

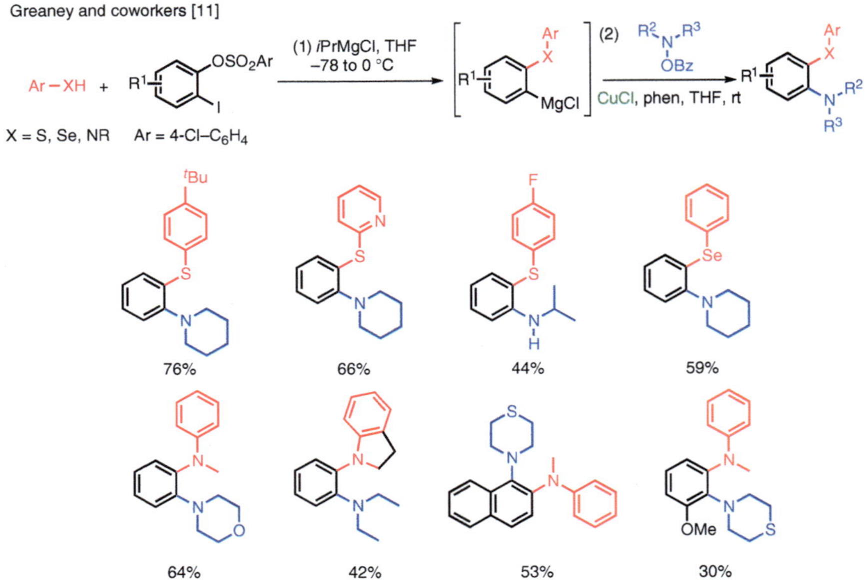

Scheme 1.5 Cu-catalyzed electrophilic amination via aryne intermediate.

After the initial disclosure of Johnson and coworkers, subsequent research showed that other organometallic reagents can also serve as the nucleophile.

A unique reaction was reported by Greaney and coworkers at the University of Manchester in the UK [11]. In this case, an aryne intermediate is first generated in situ via iodine–magnesium exchange, followed by elimination. A nucleophile subsequently attacks the aryne, and the resulting arylmetal undergoes Cu-catalyzed electrophilic amination with the hydroxylamine-derived electrophilic aminating reagent to furnish aryl amines as the final products (Scheme 1.5). This reaction gives access to unique aromatic products with a 1,2-bis substitution pattern. The nucleophiles that can be used in this reaction include arylamines, thiophenols, and arylselenols. The overall transformation exhibits excellent regioselectivity; however, the isolated yield is lowered in those cases in which the reactants are sterically encumbered.

One important consideration in these types of reactions is that the nucleophile has to be able to coexist with the aminating reagent in order for the catalytic cycle to be established. Otherwise, the nucleophile may directly react (i.e. protonation and/or uncatalyzed nucleophilic attack) with the aminating reagent and may lead to the formation of undesired by-products. This puts limitations on both the nucleophilicity and basicity of the organometallic nucleophiles. Organozinc compounds, Grignard reagents, and organoaluminum [12] reagents have been shown to be suitable nucleophiles for these types of reactions, while organolithium reagents are generally unsuitable because of their strong basicity and tendency to participate in side reactions that involve electron transfer pathways.

To address this issue and provide a solution for the direct electrophilic amination of organolithium reagents via copper catalysis, the group of A. B. Smith III (University of

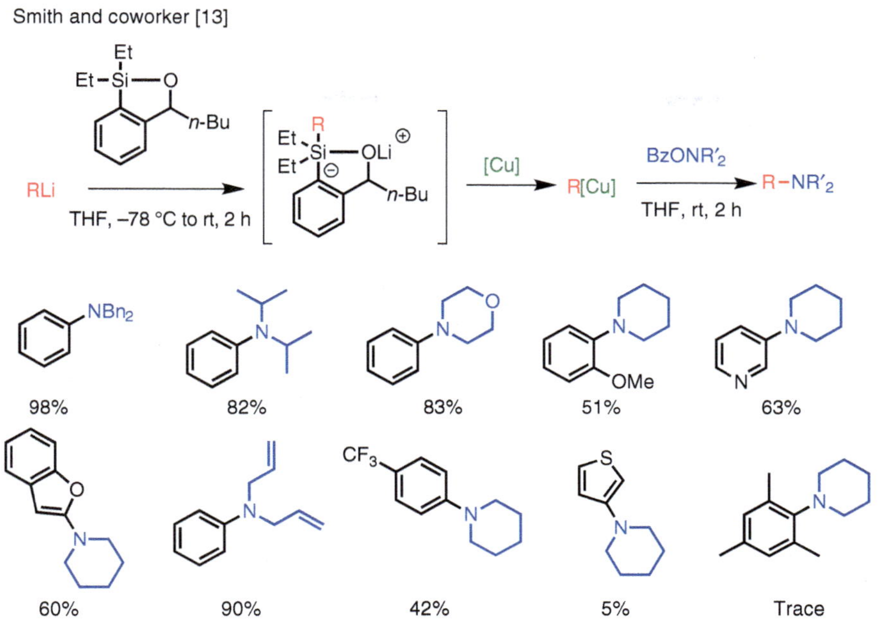

Scheme 1.6 Cu-catalyzed electrophilic amination of organolithium reagents.

Pennsylvania, 2013) developed a protocol in which a siloxane transfer reagent is used to modulate the reactivity of the organolithium reagents (Scheme 1.6) [13]. The siloxane reagent acts as an "attenuator" for the more reactive organolithium reagents and lowers its reactivity by forming a less basic complex that can more readily participate in the catalytic cycle and, in the meantime, prevents the direct attack on the aminating reagent by the organolithium.

This type of C—N bond-forming reaction can also take place directly between cuprates and electrophilic aminating reagents. One reason to justify the use of stoichiometric amounts of a copper salt (i.e. full transmetalation) is that the low basicity of the resulting cuprate can tolerate the presence of N–H protons in the aminating reagent, thereby enabling the synthesis of primary and secondary amine products without the use of excess organometallic reagent. The Kürti group has demonstrated this possibility with the use of sterically hindered N–H oxaziridines (Scheme 1.7) [14]. Compared to the tertiary amines, unprotected primary amines are more versatile building blocks for further functionalization.

These sterically hindered N–H oxaziridines can be readily synthesized on multigram scale from the corresponding N–H imines and *meta*-chloroperbenzoic acid (*m*CPBA). The N–H oxaziridines are bench-stable compounds and can also be readily purified via flash column chromatography and using regular silica gel as the stationary phase. The steric hindrance created by the bulky alkyl groups reduces the kinetic acidity of the oxaziridine N—H bond, thus allowing the cuprates to be aminated as opposed to suffering unproductive proton transfer.

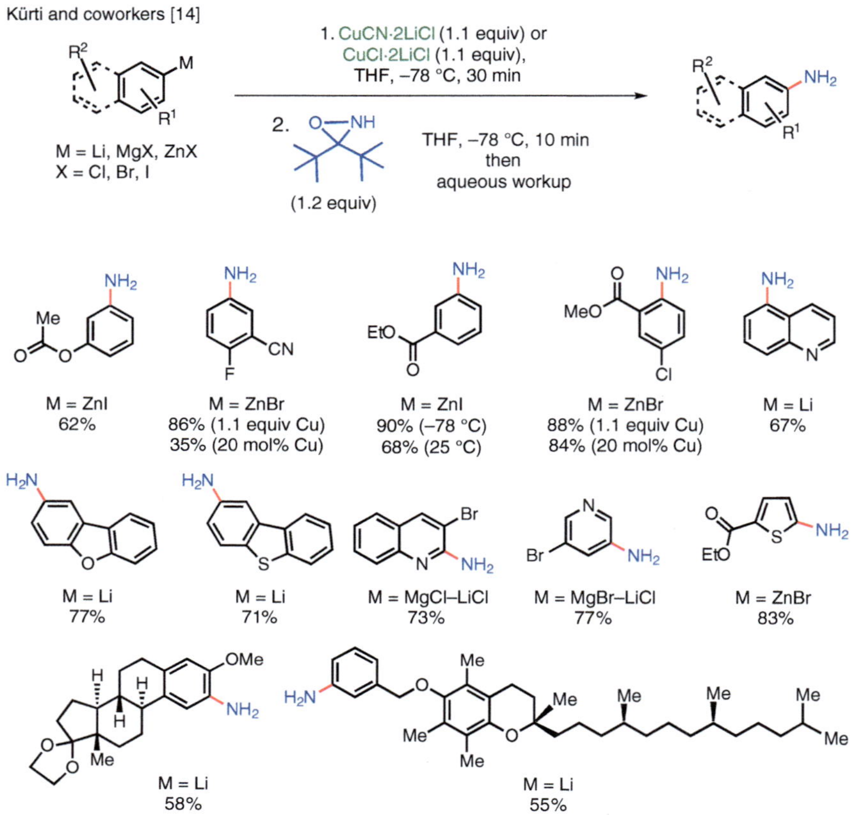

Scheme 1.7 Electrophilic amination of arylcuprates using a NH-oxaziridine.

Since the advent of direct C–H cupration of arenes, it is now also possible to utilize cuprate nucleophiles without first going through a separate transmetalation step. Uchiyama and coworkers (at RIKEN, Japan) showed that it was indeed possible to directly aminate aryl cuprates with *O*-benzyl hydroxylamine. The directed C–H cupration of arenes was achieved using a strong base $(TMP)_2Cu(CN)Li_2$, which can selectively deprotonate at the ortho position of the amide directing group. The resulting aryl cuprates can be directly primary aminated with benzyl hydroxylamine (Scheme 1.8) [15]. The hygroscopic nature of *O*-benzyl hydroxylamine necessitates its use as a stock solution in an organic solvent.

The instability of organometallic reagents used in the aforementioned reactions imposes limits on their widespread utilization, especially in an industrial setting. Efforts have been made to replace the air- and moisture-sensitive organometallic reagents with more stable alternatives that can be conveniently stored and used. Miura and coworkers (Osaka University, Japan) have shown that both arylboronates (Scheme 1.9) [16] and arylsilanes (Scheme 1.10) [17] can serve as starting materials in Cu-catalyzed electrophilic amination reactions. These reactions can proceed under ambient temperature and furnish the corresponding anilines in good to excellent isolated yields. The enhanced stability of arylboronates and arylsilanes reduces the complexity of the operation and, at the same

Uchiyama and coworkers [15]

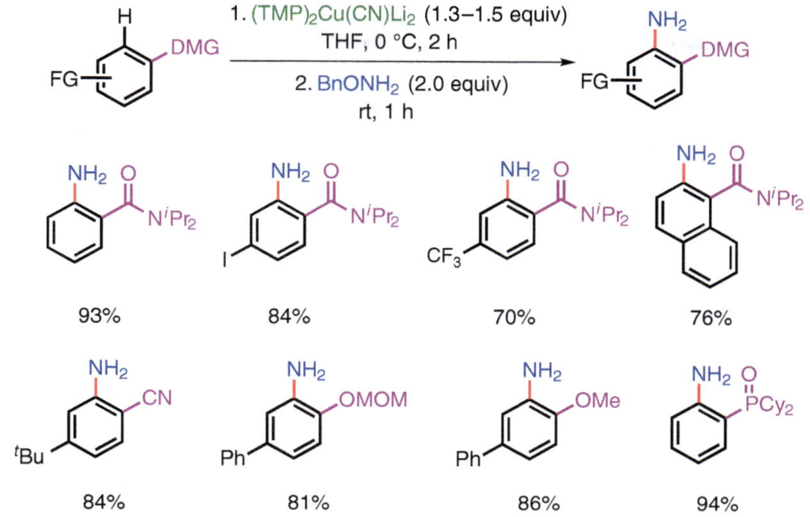

Scheme 1.8 Electrophilic amination via directed C–H cupration.

Miura and coworkers [16]

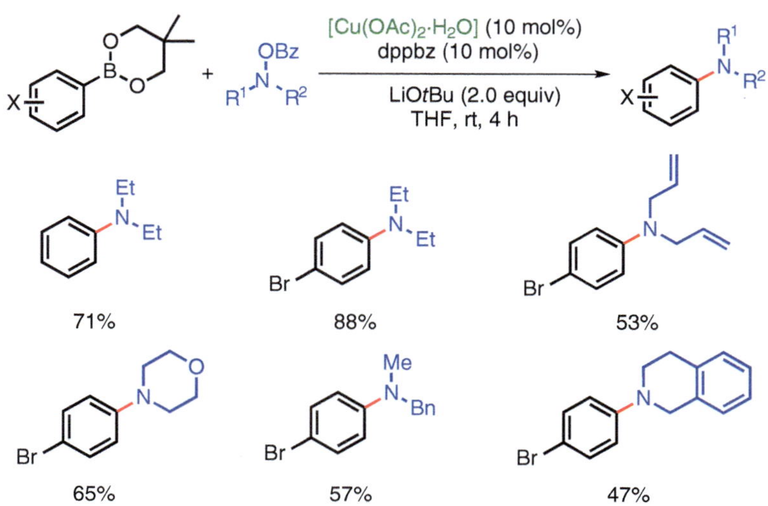

Scheme 1.9 Cu-catalyzed electrophilic amination of arylboronates.

time, the wide commercial availability of arylboronates also adds convenience to these types of reactions.

Hirano and Miura discovered that ambident nucleophiles such as silyl ketene acetals are also suitable nucleophiles for the Cu-catalyzed electrophilic amination reactions. These reactions result in the formation of alpha-amino esters as products. The first generation of this reaction uses chloramines as aminating reagents [18], while the second generation

Miura and coworkers [17]

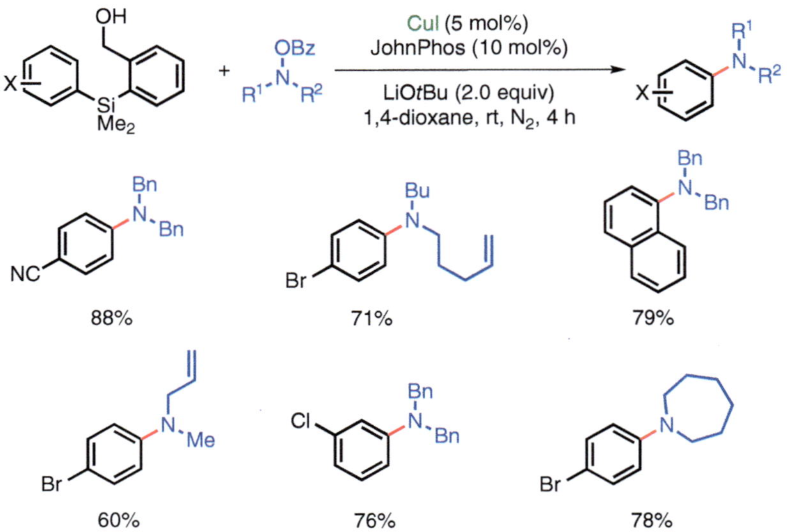

Scheme 1.10 Cu-catalyzed electrophilic amination of aryl silanes. Source: Modified from Miki et al. [17].

can proceed with the more stable and much safer *O*-benzoyl hydroxylamines (Scheme 1.11) [19]. This method provides a potential route for the syntheses of unnatural as well as modified natural amino acids.

Weak nucleophiles such as styrenes and some electron-deficient heterocycles can also participate in Cu-catalyzed electrophilic amination reactions.

In the case of styrenes, the substrates can undergo hydroamination or aminoboration depending on the specific reaction conditions. Hirano, Miura, and coworkers have demonstrated that styrenes can be stereoselectively functionalized with benzoyl hydroxylamine and bis(pinacolato)diboron under Cu catalysis (Scheme 1.12) [20]. The resulting products can further participate in transition-metal-catalyzed cross-coupling reactions.

When polymethylhydrosiloxane (PMHS) is used instead of bis(pinacolato)diboron, hydroamination products can be obtained under similar reaction conditions (Scheme 1.13). In these cases, it is proposed that the reaction proceeds with an initial CuH addition across the C—C bond of the olefins, followed by the electrophilic amination of the resulting cuprates [21].

With chiral ligands, the hydroamination reactions can give enantiomerically enriched products. Both the Miura (Scheme 1.14) and Buchwald (Scheme 1.15) groups have developed conditions using chiral phosphine ligands [21, 22].

Buchwald and coworkers have also reported the hydroamination of aryl acetylenes. The reaction is highly stereoselective, giving *E*-enamines as the major products. The enamine products can be further reduced to give alkyl amines, which are important building blocks in organic synthesis (Scheme 1.16) [23].

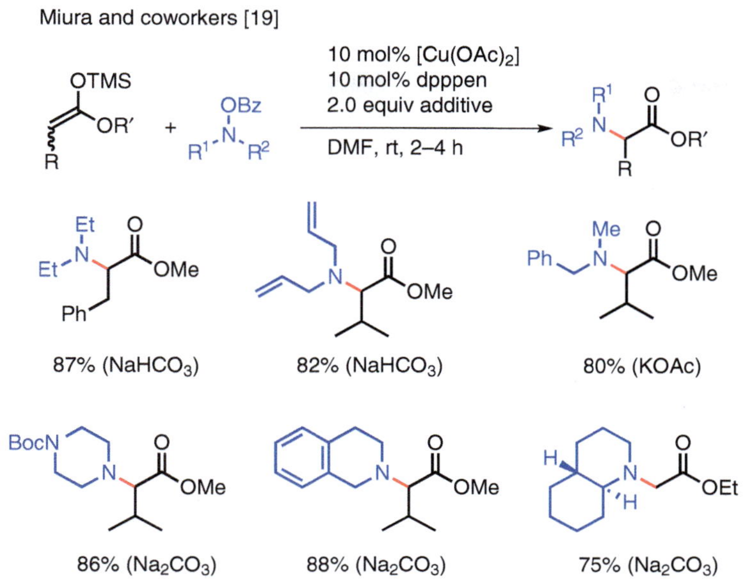

Scheme 1.11 Cu-catalyzed electrophilic amination of silyl enol ethers. Source: Modified from Matsuda et al. [19].

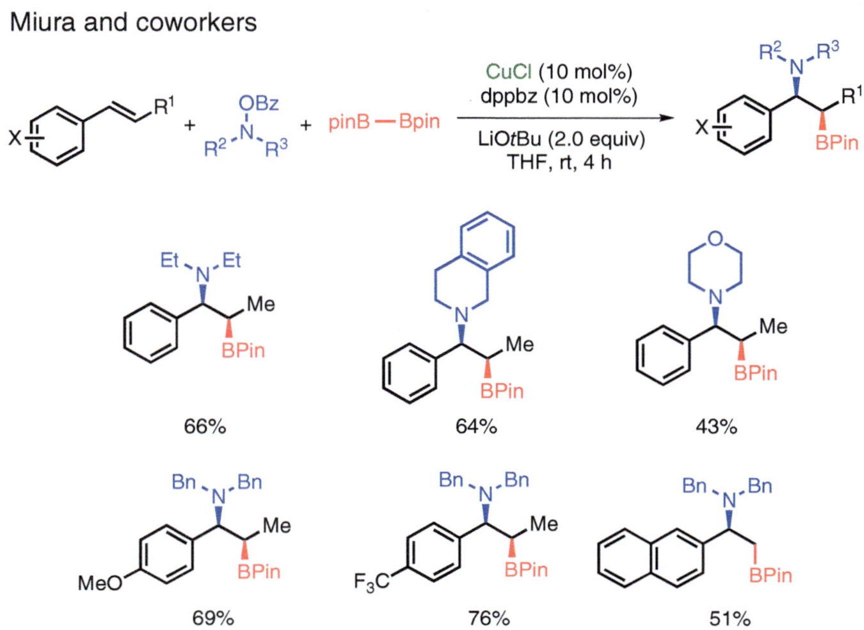

Scheme 1.12 Cu-catalyzed electrophilic catalyzed aminoboration of styrenes. Source: Modified from Matsuda et al. [20].

Miura and coworkers [21]

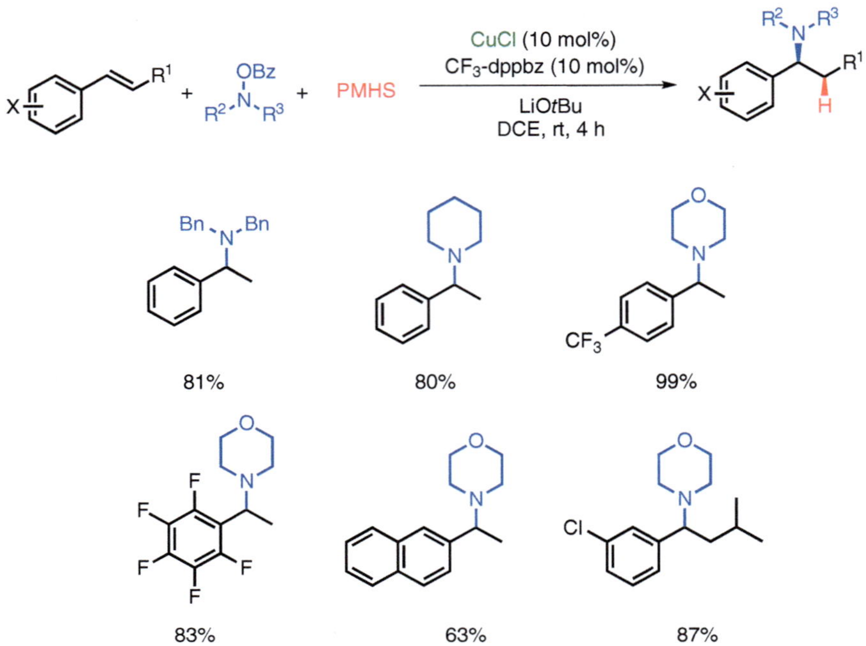

Scheme 1.13 Cu-catalyzed electrophilic hydroamination of styrenes. Source: Modified from Miki et al. [21].

Miura and coworkers [21]

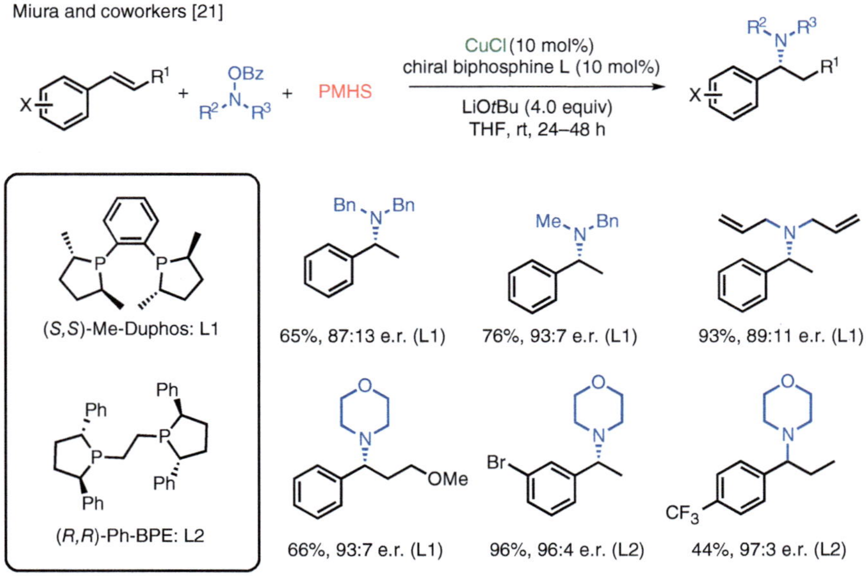

Scheme 1.14 Enantioselective Cu-catalyzed electrophilic hydroamination of styrenes. Source: Miki et al. [21].

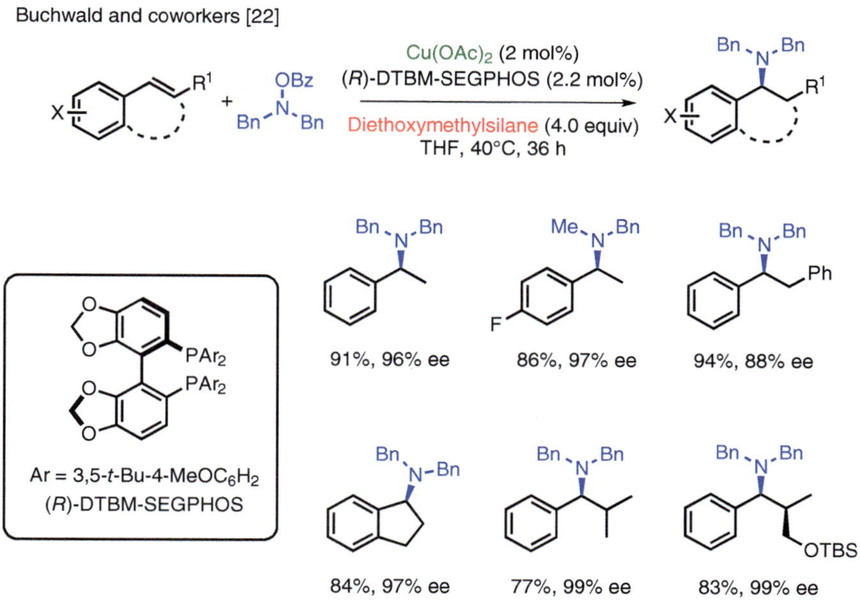

Scheme 1.15 Enantioselective Cu-catalyzed electrophilic hydroamination of styrenes. Source: Modified from Zhu et al. [22].

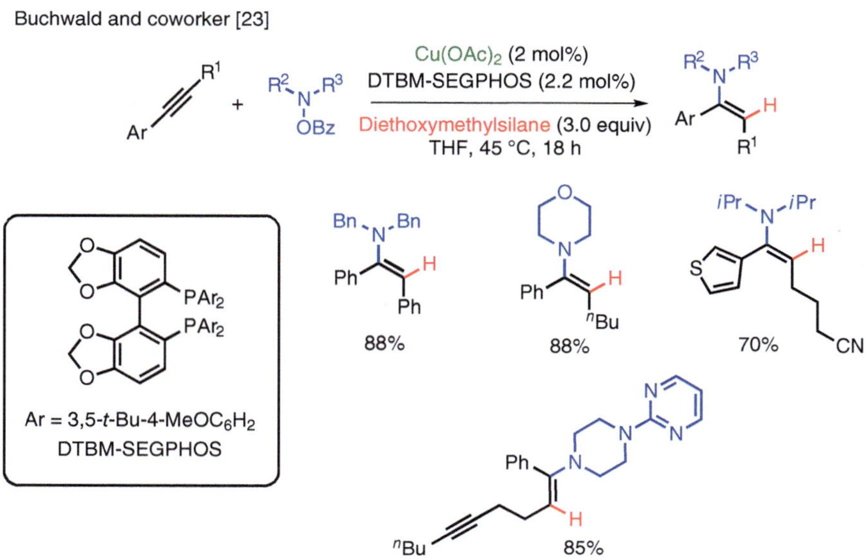

Scheme 1.16 Cu-catalyzed electrophilic amination of alkynes. Source: Shi and Buchwald [23].

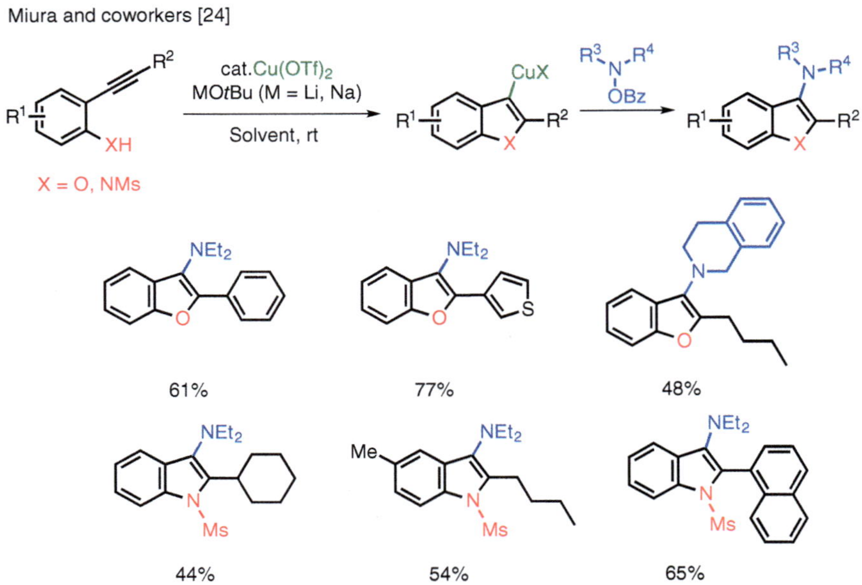

Scheme 1.17 Cu-catalyzed annulative electrophilic amination. Source: Modified from Matsuda et al. [24].

ortho-Alkynyl phenols and anilines can also undergo annulative amination with electrophilic aminating reagents under Cu catalysis. Miura and coworkers have developed conditions for the synthesis of aminated benzofurans and indoles (Scheme 1.17) [24]. The transformation is operationally simple and proceeds at room temperature. The mechanism was probed and the authors concluded that the most plausible pathway is a nonradical electrophilic amination of the heteroarylcuprate species in the C—N bond-forming step.

Similar intramolecular reactions can also take place with substrates containing unactivated terminal alkenes. In 2015, the Wang group (Duke University) reported the copper-catalyzed vicinal diamination of unactivated alkenes with hydroxylamines that is both regio- and stereoselective. The first iteration of this reaction takes place on unsaturated amides and gives 4-amino-2-pyrrolidones as the products (Scheme 1.18) [25]. This transformation is considered to be the first metal-catalyzed alkene 1,2-diamination that enables the direct incorporation of an electron-rich amino group.

In 2016, Wang and coworkers successfully expanded the substrate scope to include unsaturated carboxylic acids, which undergo amino-lactonization under the reaction conditions (Scheme 1.19) [26]. The overall transformation allows the practitioner to access quickly and efficiently a wide range of amino-substituted γ- and δ-lactones as well as 1,2-amino alcohol derivatives, which are of significant value in the synthesis of natural products and active pharmaceutical ingredients.

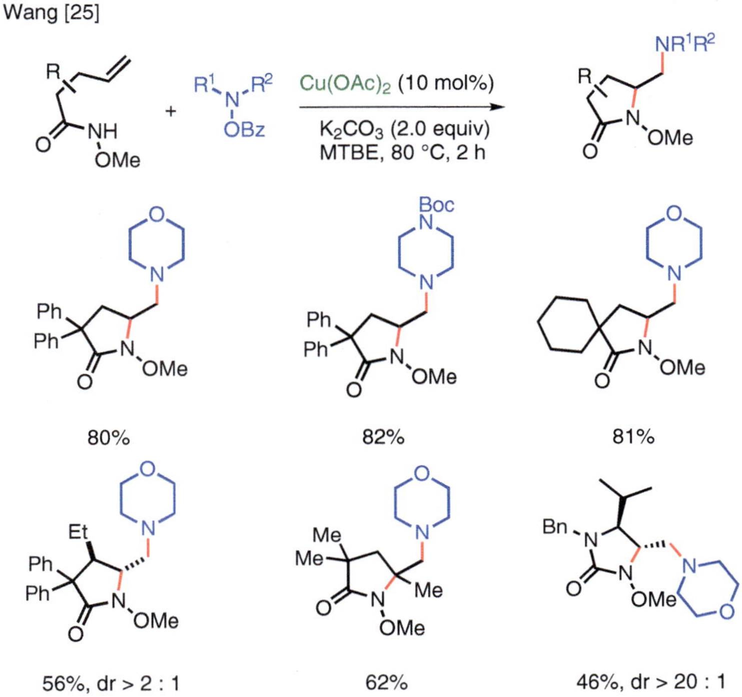

Scheme 1.18 Cu-catalyzed electrophilic diamination. Source: Modified from Shen and Wang [25].

An unusual case of ring-opening amination of cyclopropanols has been reported by the Dai group [27]. In this reaction, a base-initiated ring-opening of cyclopropanol generates a carbanion nucleophile, which participates in the Cu-catalyzed electrophilic amination and affords β-aminoketones as products (Scheme 1.20). The catalytic cycle involves the oxidation of the Cu(I) complex to the corresponding Cu(III) species by the hydroxylamine reagent. Next, the Cu(III) intermediate promotes the ring-opening of the cyclopropanol substrate and the resulting copper-homoenolate undergoes reductive elimination to form the new C—N bond and to regenerate the catalytically active Cu(I) species. Overall, the transformation proceeds under mild reaction conditions and it is also compatible with a number of sensitive functionalities such as esters, epoxides, and unsaturated carbonyl compounds.

With electron-deficient arenes, direct C–H amination is also possible. Miura and coworkers have reported the Cu-catalyzed direct C–H amination using benzoyl hydroxylamines as aminating reagents (Scheme 1.21) [28]. Electron-deficient aromatic substrates such as fluoroarenes, oxadiazoles, and thiazoles can be directly aminated to furnish the corresponding aryl and heteroaryl amines. The Yotphan group later expanded the substrate scope to include benzoxazoles [29], while the Li group applied the reaction to enable the C–H amination of quinoline *N*-oxide [30].

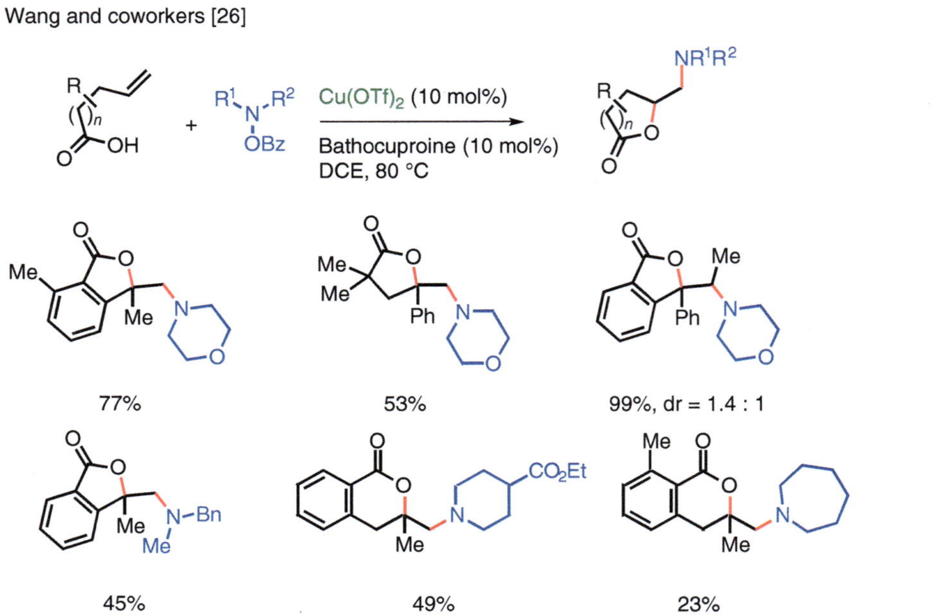

Scheme 1.19 Cu-catalyzed electrophilic amino-lactonization. Source: Modified from Hemric et al. [26].

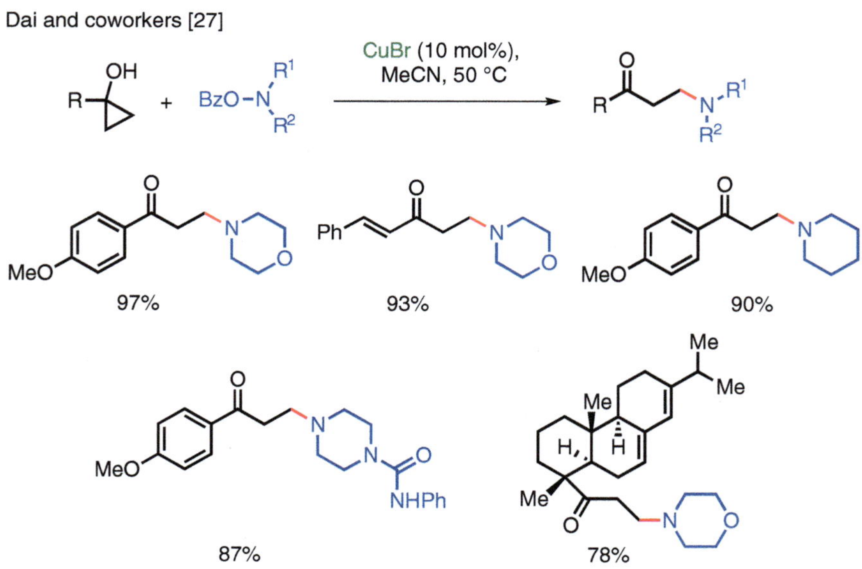

Scheme 1.20 Cu-catalyzed ring-opening amination. Source: Modified from Ye and Dai [27].

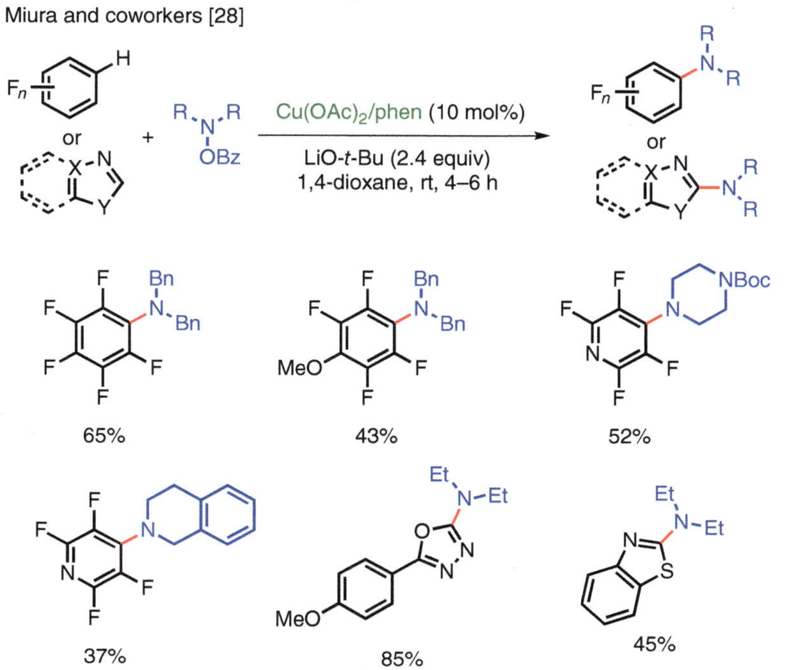

Scheme 1.21 Cu-catalyzed C–H amination of heterocycles. Source: Matsuda et al. [28].

1.3 Electrophilic Amination Reactions Catalyzed by Other Transition Metals

Although the current focus of TM-catalyzed electrophilic amination is copper catalysis, other transition metal complexes are also capable of serving as catalysts in these reactions.

Complexes of both Ni and Co can catalyze the electrophilic amination of organozinc reagents. The reaction mechanism is similar to the Cu-catalyzed reactions; that is, the N—O or N—Cl bond of the aminating reagent undergoes cleavage when the metal enters into it via oxidative addition. The Johnson [31], Jarvo [32], and Knochel [33] groups reported their findings in several publications (Scheme 1.22).

J.-Q. Yu and coworkers (The Scripps Research Institute) have successfully combined the Pd-catalyzed C–H activation and electrophilic amination. Both sp^2 (Scheme 1.23) and sp^3 (Scheme 1.24) C—H bonds can be aminated with this process. Mechanistically, these reactions proceed via a Pd(II)/Pd(IV) cycle: after the C–H activation, the resulting Pd(II) complex oxidatively inserts into the N—O bond of the aminating reagent, which is followed by a reductive elimination to form the new C—N bond. In some cases, it is necessary to add a Ag(I) salt as a sacrificial oxidant to help establish the catalytic cycle [34, 35].

The Yu group also demonstrated that rhodium and ruthenium complexes can also catalyze these types of reactions via similar mechanisms (Scheme 1.25) [36].

Apart from using a directing group in the C–H functionalization process, Catellani-type reactions can also achieve directed C–H activation without the use of amide-type directing

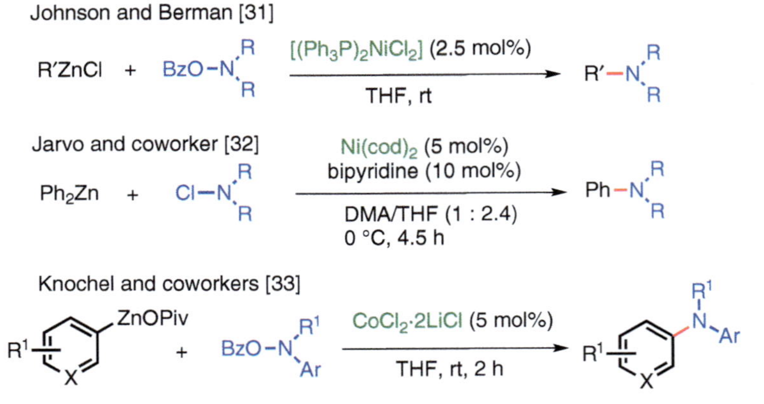

Scheme 1.22 Electrophilic amination catalyzed by other transition metals. Source: Johnson and Berman [31], Barker and Jarvo [32], and Lutter et al. [33].

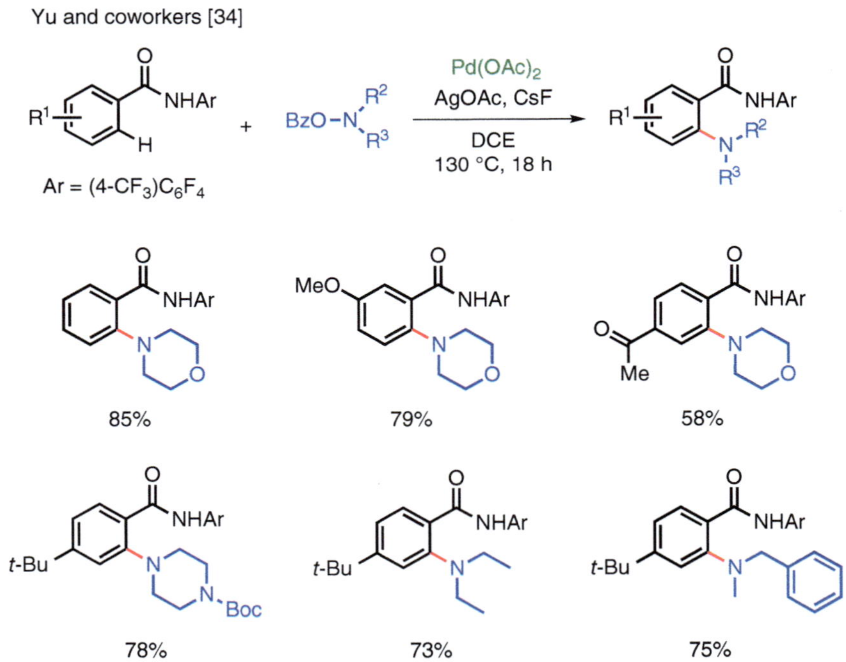

Scheme 1.23 Pd-catalyzed aromatic C–H amination via electrophilic amination. Source: Modified from Yoo et al. [34].

groups. Dong and coworkers applied this strategy in the development of Pd-catalyzed electrophilic amination reactions (Scheme 1.26) [37]. The main difference between this transformation and the standard Buchwald–Hartwig protocol is that the C–H group adjacent to the aryl halide moiety (i.e. the ortho position) gets aminated, while in the Buchwald–Hartwig reactions, the C—X bond itself gets aminated (i.e. the ipso position).

Yu and coworkers [35]

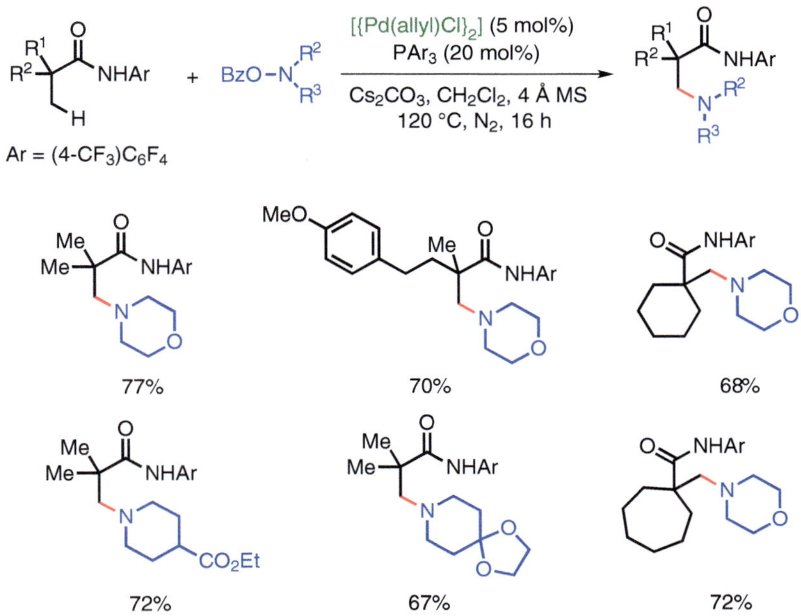

Scheme 1.24 Pd-catalyzed aliphatic C–H amination via electrophilic amination. Source: Modified from He et al. [35].

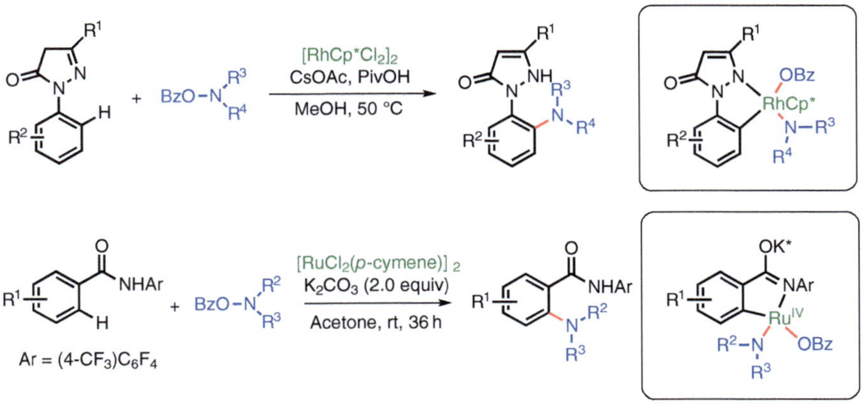

Scheme 1.25 Ru-catalyzed C–H amination. Source: Modified from Shang et al. [36].

A unique Fe-catalyzed reaction was reported by the Yang group in which styrenes undergo formal hydroamination to afford tertiary amines with Markovnikov regioselectivity [38]. In this case, an organoferrates initially formed from a Grignard reagent and the iron(II) catalyst. This organoferrate intermediate engages the styrene substrate to form a hydride complex, which hydrometallates the styrene double bond and subsequently undergoes electrophilic amination (Scheme 1.27).

Dong and Dong [37]

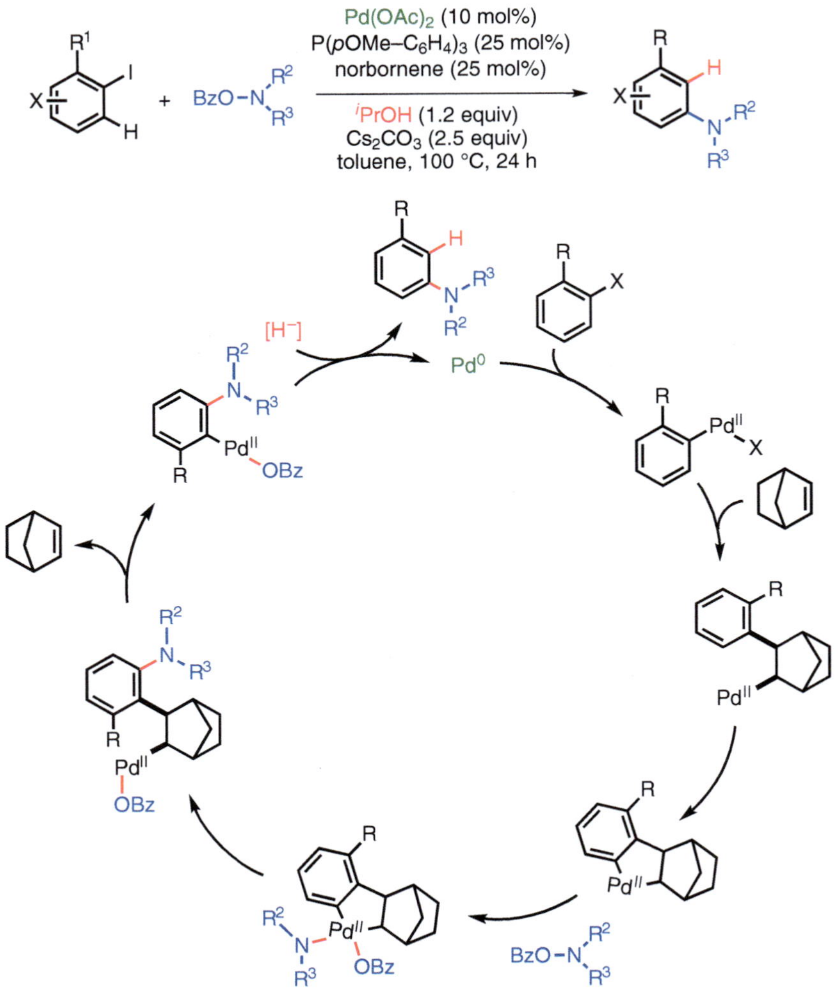

Scheme 1.26 Pd-catalyzed Catellani-type C–H electrophilic amination. Source: Modified from Dong and Dong et al. [37].

1.4 Electrophilic Amination with Hydroxylamine-derived Metallanitrenes

A nitrene is the nitrogen analog of a carbene, which is a six-electron electron-deficient species. Although it is formally uncharged and univalent, it can be considered to be an electrophile because of the unsatisfied octet. These highly reactive intermediates are involved in many chemical reactions, including electrophilic aminations [39–41].

Nitrene intermediates can be generated from various precursors. Pioneering studies by Khan and Kwart in the 1960s showed that elemental copper can catalyze the decomposition

Yang and coworkers [38]

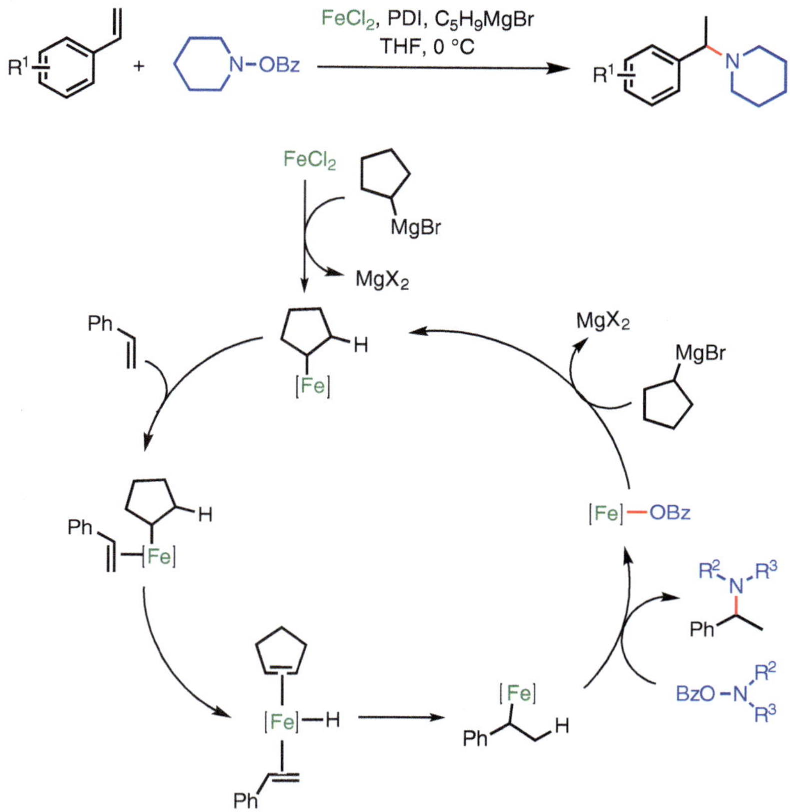

Scheme 1.27 Fe-catalyzed electrophilic amination of styrenes. Source: Modified from Huehls et al. [38].

of a sulfonyl azide, and the resulting reaction mixture can aminate cyclohexene and give aziridine as one of the products [42]. This result is consistent with the participation of metallanitrene intermediates. However, because of the instability and toxicity of organic azides, this procedure is not widely adapted by the organic chemistry community. In later years, (*N*-(sulfonyl)imino)phenyliodinane was found to be a suitable alternative nitrogen source. In addition to Cu, Mn and Fe are also capable of catalyzing the formation of metallanitrenes. These “tamed” nitrenes are more stable and have longer lifetimes, making them uniquely suitable for synthetic studies. However, the *N*-sulfonyl groups are hard to remove, limiting the usefulness of the resulting products.

Recently, studies have shown that hydroxylamines can also serve as the nitrogen sources for metallanitrenes. Using hydroxylamines, it is now possible to generate N–H and N–alkyl amine products without sulfonyl groups. This section focuses on reactions that involve hydroxylamine-derived metallanitrenes.

Metallanitrenes are generated from O-activated hydroxylamines, in which the nitrogen atom bears a strong leaving group and a transition metal catalyst. In 2014, Kürti, Falck,

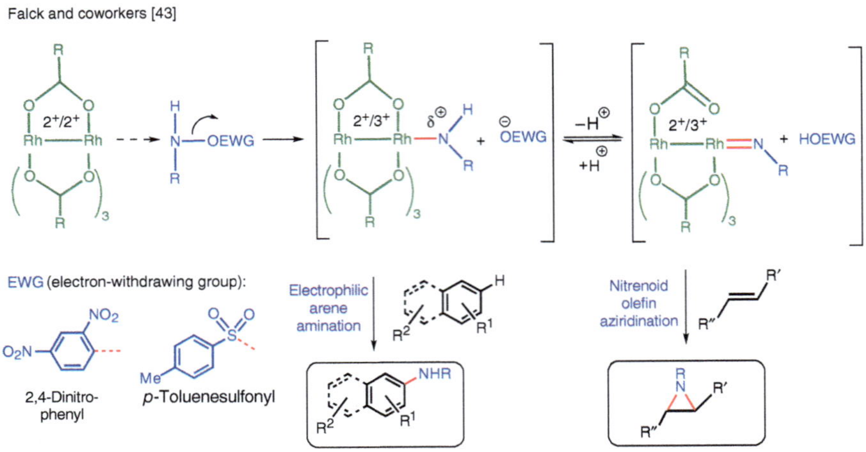

Scheme 1.28 Rh-catalyzed formation of metallanitrenes.

and coworkers developed an N–H aziridination procedure for unactivated olefins using *O*-(2,4-dinitrophenyl)hydroxylamine (DPH) as the nitrogen source and a dirhodium carboxylate catalyst. Computational studies conducted by the Ess group support the intermediacy of a Rh–nitrene pathway (Scheme 1.28) [43].

As the smallest nitrogen-containing heterocycles, aziridines are important synthetic building blocks for the synthesis of amines. Recently, it has been shown that natural product analogs containing aziridine functionalities can possess improved biological activities than their natural counterparts [44]. Despite their importance, the majority of literature has focused on the synthesis of aziridines bearing electron-withdrawing sulfonyl groups on the nitrogen atom (i.e. *N*-tosylaziridines). The stability of the sulfonyl group renders the deprotection and functionalization of these aziridines difficult. In comparison, unprotected N–H aziridines can be further functionalized with relative ease.

The transformation is operationally simple as neither the aminating reagent nor the catalyst is air- or moisture-sensitive. The functional group tolerance is excellent as hydroxyl groups, epoxides, and esters in the substrates are unaffected. The NH-aziridination is stereospecific and no scrambling of the olefin stereochemistry is observed even in sensitive styrene-type substrates. When more than one C—C double bond is present in the substrate, the more electron-rich one undergoes aziridination preferentially. The aziridination of terminal double bonds requires a slightly higher catalyst loading. The triple bond of alkynes and electron-deficient double bonds (i.e. α,β-unsaturated carbonyl compounds) remain unchanged under the reaction conditions. The choice of solvent is important because the presence of trifluoroethanol is required for aziridination.

The first iteration of this reaction uses the nitro group-containing DPH as the aminating reagent (Scheme 1.29). Because of its thermal instability, there were some concerns about its safety in industrial-scale settings. To address this issue, the Kürti group developed an improved version of this reaction in 2017 [45]. Instead of the nitro-containing DPH, the new version can proceed with the stable and inexpensive hydroxylamine-*O*-sulfonic acid (HOSA) (Scheme 1.30). Because HOSA is a zwitterionic compound, its solubility in

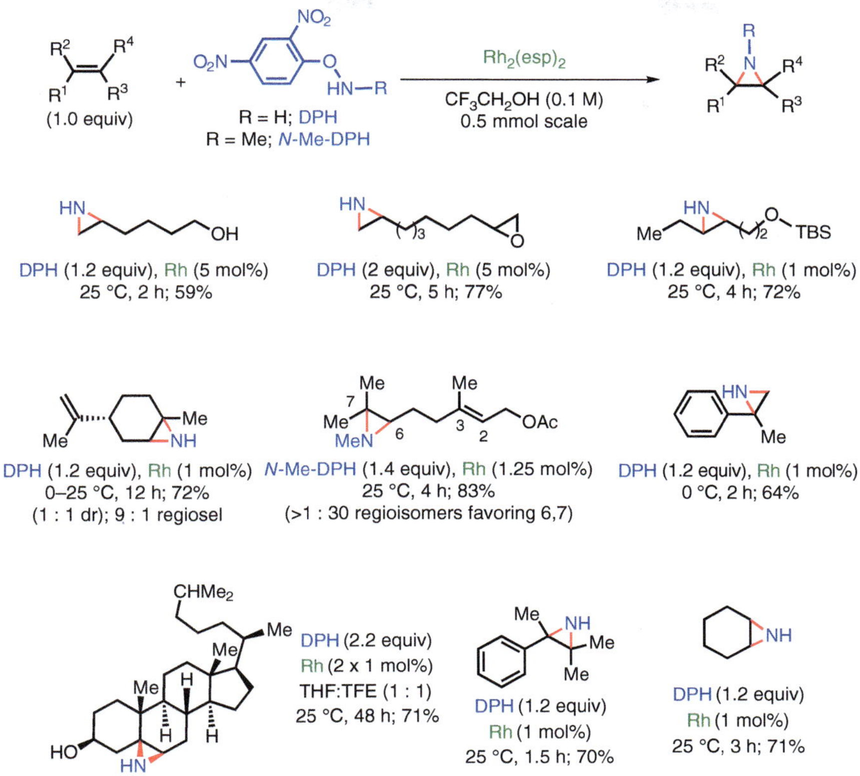

Scheme 1.29 Rh-catalyzed NH-aziridination of unactivated olefins using DPH.

common organic solvents is very low. However, through the addition of an equivalent of a mild base (i.e. pyridine), HOSA can be solubilized. Hexafluoroisopropanol (HFIP) was identified as the optimal solvent, and under these modified conditions, the olefin aziridination reaction can proceed at room temperature and with generally higher rates compared to the original DPH process. The scope of substrates is further expanded and now includes nitrogen heterocycles with basic nitrogen atoms. A further advantage is that the inorganic sulfate by-product is water-soluble and nontoxic, which greatly simplifies the purification.

Because *N*-alkyl HOSA derivatives can be readily prepared on multigram scale [46], it was demonstrated that *N*-Me as well as *N*-isopropyl units could be stereospecifically transferred to olefins, leading to the corresponding *N*-alkylaziridines.

Further studies on this catalytic system by the three groups showed that the Rh–nitrene intermediate can also facilitate the C–H amination of electron-rich aromatic rings (Scheme 1.31) [47]. In 2016, the Falck, Kürti, and Ess groups published the direct C–H amination of arenes. By using a more reactive aminating reagent and modifying the acidity of the reaction, Rh–nitrene intermediate can be directly inserted into the aromatic π-system of electron-rich arenes. Using this protocol, both inter- and intramolecular arene C–H amination can be achieved.

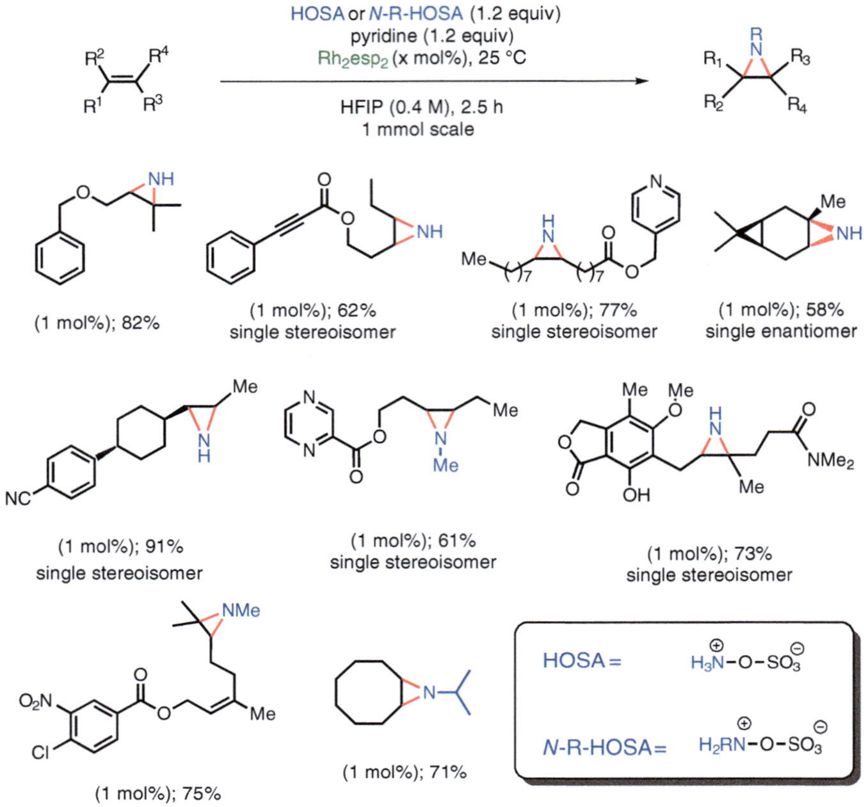

Scheme 1.30 Rh-catalyzed NH-aziridination of unactivated olefins using HOSA.

1.5 Transition-Metal-Free Electrophilic Amination Reactions

Although the mechanism of uncatalyzed electrophilic amination reactions seems straightforward, in practice, these uncatalyzed reactions often suffer from poor efficiency and low yields. A common issue with uncatalyzed electrophilic amination is the side reaction between the nucleophile and the highly reactive aminating reagent. It is especially difficult to directly synthesize primary or secondary amines under uncatalyzed conditions because the unmasked NH protons will usually quench the strongly basic nucleophiles used in the reactions (Scheme 1.32).

In recent years, several approaches have been developed to address these issues and finally achieve practical TM-free electrophilic aminations.

One approach is to use a mild nucleophile to avoid quenching and/or side reactions. In 2012, the Kürti group reported a TM-free primary amination of arylboronic acids (Scheme 1.33) [48].

This strategy exploits the fact that hydroxylamines can act as both electrophiles and nucleophiles. The nitrogen atom in the DPH aminating reagent first acts as a nucleophile to attack the boronic acid, and subsequently, it acts as an electrophile to accept an

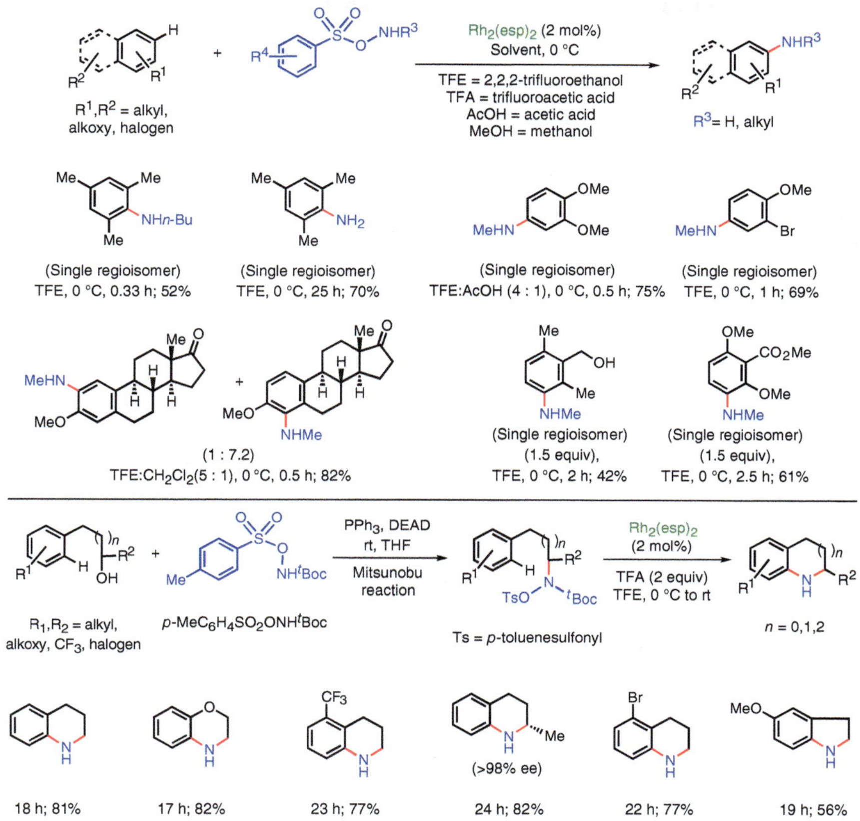

Scheme 1.31 Rh-catalyzed aromatic C–H amination. Source: Modified from Paudyal et al. [47].

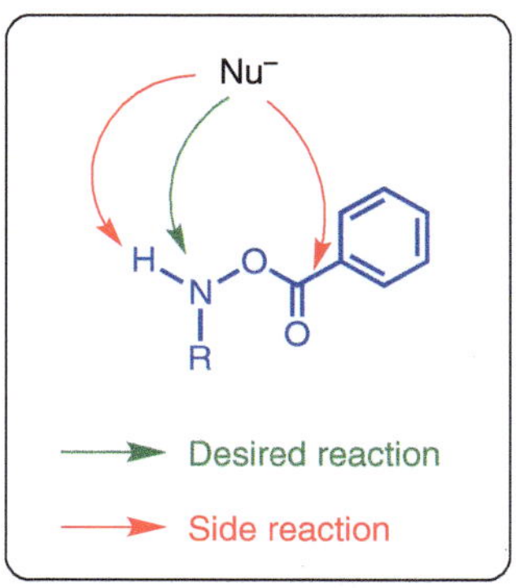

Scheme 1.32 Problems with uncatalyzed electrophilic amination.

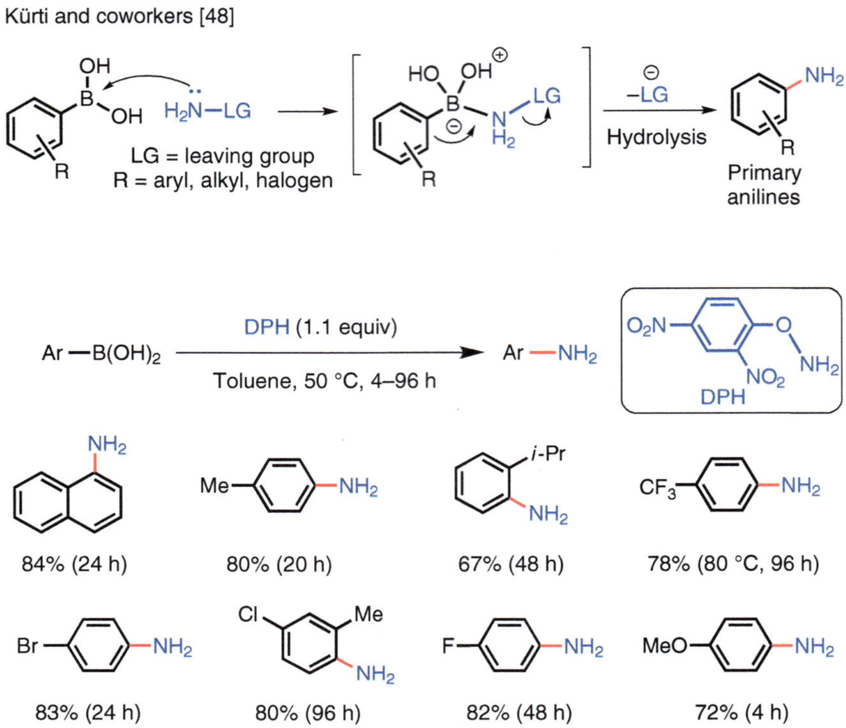

Scheme 1.33 TM-free electrophilic amination of arylboronic acids. Source: Modified from Zhu et al. [48].

intramolecular nucleophilic attack to furnish the aniline products. Before this method, the direct conversion of arylboronic acids to the corresponding primary arylamines under mild conditions was not possible. Now, even halogenated primary arylamines may be readily prepared under transition-metal-free conditions.

Another approach to achieve TM-free electrophilic amination while avoiding undesired quenching of the nucleophile is to modify the structure of the aminating reagent. In 2017, the Kürti group reported the use of sterically hindered oxaziridines to achieve the TM-free direct primary amination of aryl Grignard reagents and aryl lithiums (Scheme 1.34) [49]. Subsequent studies carried out in the Kürti group successfully expanded the substrate scope to include alkyl organometallic reagents [50].

The backbone of these aminating reagents can be readily recovered and reused for the preparation of more reagents (i.e. NH-oxazirdines). Remarkably, the N—H bond does not undergo deprotonation as this pathway is sterically inhibited (i.e. because of the kinetic decrease of the N—H bond acidity). It is intriguing that the *N*-alkyl derivatives of these oxaziridines transfer the oxygen atom to aryl and alkyl Grignard reagents at low temperatures with complete chemoselectivity (i.e. no N-transfer occurs with these reagents). Both the direct primary amination and hydroxylation of Grignard reagents are currently considered to be still extremely challenging, and they represent unmet synthetic needs. The camphor-derived *N*-benzyl oxaziridine is significantly less reactive and considerably more

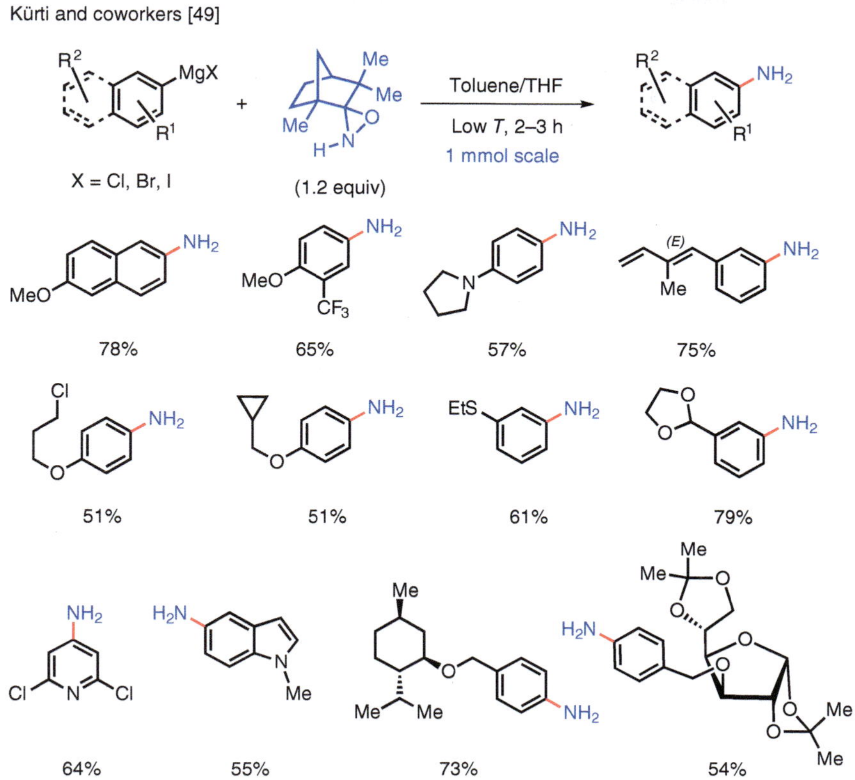

Scheme 1.34 TM-free electrophilic amination of arylmetals using NH-oxaziridines. Source: Modified from Gao et al. [49].

stable than Davis's oxaziridine; thus, it allows for highly chemoselective hydroxylations in the presence of multiple sensitive functionalities such as sulfides, amines, and alkenes.

For weaker nucleophiles such as olefins and arenes, successful TM-free electrophilic aminations rely on highly reactive aminating reagents. Pioneering works in this area by the Bower group have shown that it is possible to achieve TM-free intramolecular electrophilic amination using highly reactive sulfonyl hydroxylamines under acid catalysis (Scheme 1.35). The unprotected tosylhydroxylamines are too unstable to be isolated, so they are instead generated *in situ* from the Boc-protected precursors under acidic conditions. These reactions are proposed to proceed via the "Butterfly Mechanism" akin to the Prilezhaev reaction [51].

The α-aminoketone moiety is commonly found in biological molecules, natural products, and active pharmaceutical ingredients. Despite their apparent importance, synthetic access to these compounds is not always straightforward. This is especially evident when the desired amino ketones have a primary amino group attached to a fully substituted α-carbon atom. The most common strategy for synthesizing these compounds involves a two-step approach, in which either an azido, a nitro, or a hydroxylamino group is first installed at the α-position. A subsequent hydrogenation, usually catalyzed by a transition

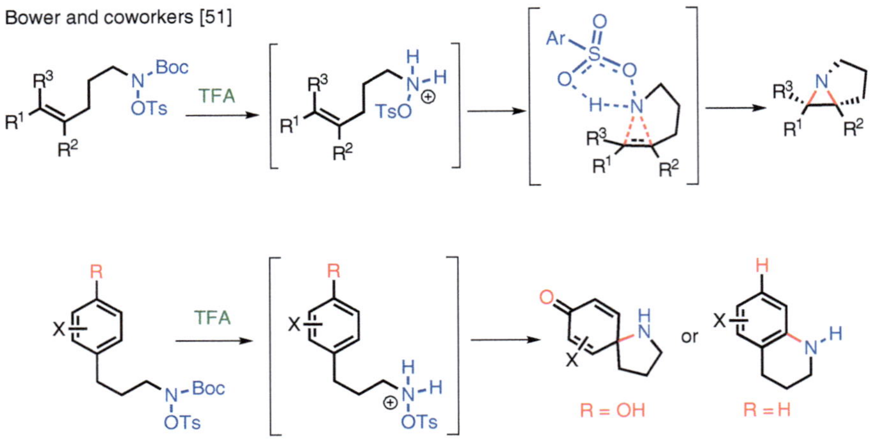

Scheme 1.35 TM-free Prilezhaev reaction. Source: Modified from Farndon et al. [51].

metal such as Pd, Pt, or Raney Ni, is then carried out to obtain the corresponding primary α-aminoketone. Aside from the added steps, this approach also has limitations such as the lack of chemoselectivity during the reduction and the potential instability of the azido- or nitro-intermediates.

In 2019, the Kürti group reported the intermolecular TM-free Aza-Rubottom reaction [52]. Electron-rich silyl enol ethers can be directly converted to the corresponding α-primary amino ketones in one step. This reaction also proceeds via the "Butterfly Mechanism," and the reactivity of the hydroxylamine aminating reagent is enhanced by cooperative hydrogen bonding interactions provided by the HFIP solvent molecules (Scheme 1.36).

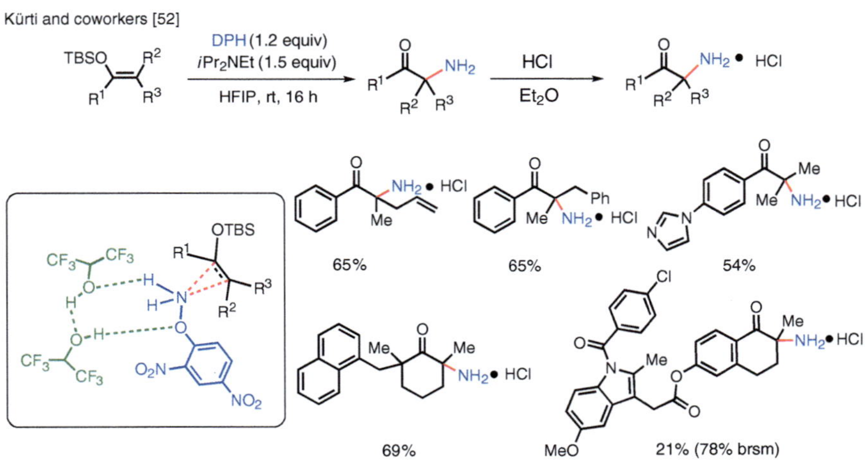

Scheme 1.36 TM-free Rubottom oxidation.

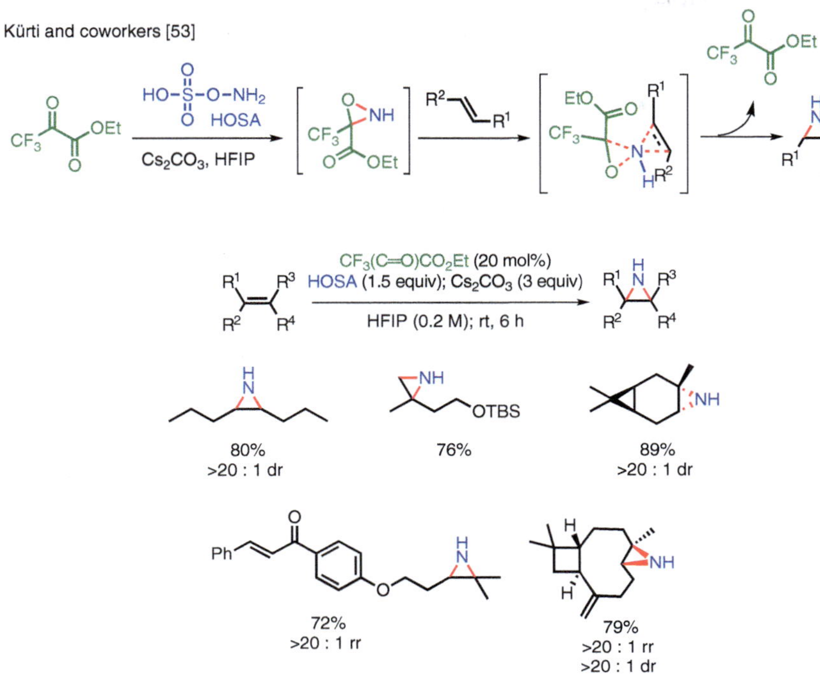

Scheme 1.37 TM-free NH-aziridination of unactivated olefins. Source: Modified from Cheng et al. [53].

Two sets of conditions for the synthesis of primary α-aminoketones from silyl enol ethers were developed. Electron-rich substrates can be α-aminated without transition metal catalysis, while substrates bearing electron-withdrawing substituents can undergo α-amination with the help of Rh or Cu catalysts.

Further studies in the Kürti group have shown that it is possible to achieve TM-free NH-aziridination of unactivated olefins with highly reactive NH-oxaziridines via the "Butterfly Mechanism" [53]. These unique NH-oxaziridines bear one or more strongly electron-withdrawing group(s), which greatly enhance the electrophilicity of the nitrogen atom. With the further enhancement provided via the hydrogen bonding interactions by the HFIP solvent molecules, these highly reactive intermediates are capable of TM-free transfer of the NH onto the unactivated olefins (Scheme 1.37).

However, because of their high reactivities, these highly electron-deficient NH-oxaziridines cannot be isolated. The Kürti group solved this issue by forming these intermediates *in situ* from HOSA and ketones bearing electron-withdrawing groups. Mass spectrometric studies confirmed the existence of the highly reactive oxaziridine intermediates.

By using a chiral nonracemic ketone as the organocatalyst, the reaction can give enantiomerically enriched products.

1.6 Conclusion

In the past two decades, a number of C—N bond-forming reactions have been developed that take advantage of hydroxylamine-based electrophilic aminating agents as sources of nitrogen. A great deal of structural diversity has been achieved in terms of the products. Olefins, substituted aromatic systems, as well as organometallic compounds have been successfully aminated. Although the vast majority of reported methods utilize transition metal complexes as catalysts, metal-free and even organocatalytic methods have also emerged during the past decade. The two emerging trends are to incorporate unprotected amino groups directly and to use inexpensive and nontoxic transition metal catalysts such as iron complexes. We are confident that during the next decade, we will see further innovation in electrophilic amination chemistries.

References

1 (a) Canè, F., Brancaleoni, D., Dembech, P. et al. New developments on organocopper-mediated electrophilic amination. In: *New Horizons in Organic Synthesis*, 118–129. New Age International Publishers. (b) Bernardi, P., Dembech, P., Ricci, A., and Seconi, G. (1999). *The Journal of Organic Chemistry* 64: 641–643, and references therein. (c) Corpet, M. and Gosmini, C. (2014). *Synthesis-Stuttgart* 46: 2258–2271.
2 Erdik, E. and Ay, M. (1989). *Chemical Reviews* 89: 1947–1980.
3 Dong, X., Liu, Q., Dong, Y., and Liu, H. (2017). *Chemistry (Easton).* 23: 2481–2511.
4 Zhou, Z. and Kürti, L. (2019). *Synlett* 30: 1525–1535.
5 Tsutsui, H., Ichikawa, T., and Narasaka, K. (1999). *Bulletin of the Chemical Society of Japan* 72: 1869–1878.
6 Tsutsui, H., Hayashi, Y., and Narasaka, K. (1997). *Chemistry Letters* 26: 317–318.
7 Erdik, E. and Daşkapan, T. (1999). *Synthetic Communications* 29: 3989–3997.
8 Erdik, E. and Daşkapan, T. (1999). *Journal of the Chemical Society, Perkin Transactions 1* 21: 3139–3142.
9 Berman, A.M. and Johnson, J.S. (2004). *Journal of the American Chemical Society* 126: 5680–5681.
10 Berman, A.M. and Johnson, J.S. (2006). *The Journal of Organic Chemistry* 71: 219–224.
11 Garcia-Lopez, J.A., Cetin, M., and Greaney, M.F. (2015). *Angewandte Chemie (International Ed. in English)* 54: 2156–2159.
12 Zhou, S., Yang, Z., Chen, X. et al. (2015). *The Journal of Organic Chemistry* 80: 6323–6328.
13 Nguyen, M.H. and Smith, A.B. (2013). *Organic Letters* 15: 4872–4875.
14 Zhou, Z., Ma, Z., Behnke, N.E. et al. (2017). *Journal of the American Chemical Society* 139: 115–118.
15 Tezuka, N., Shimojo, K., Hirano, K. et al. (2016). *Journal of the American Chemical Society* 138: 9166–9171.
16 Matsuda, N., Hirano, K., Satoh, T., and Miura, M. (2012). *Angewandte Chemie (International Ed. in English)* 51: 3642–3645.
17 Miki, Y., Hirano, K., Satoh, T., and Miura, M. (2013). *Organic Letters* 15: 172–175.

18 Miura, T., Morimoto, M., and Murakami, M. (2012). *Organic Letters* 14: 5214–5217.

19 Matsuda, N., Hirano, K., Satoh, T., and Miura, M. (2012). *Angewandte Chemie (International Ed. in English)* 51: 11827–11831.

20 Matsuda, N., Hirano, K., Satoh, T., and Miura, M. (2013). *Journal of the American Chemical Society* 135: 4934–4937.

21 Miki, Y., Hirano, K., Satoh, T., and Miura, M. (2013). *Angewandte Chemie (International Ed. in English)* 52: 10830–10834.

22 Zhu, S., Niljianskul, N., and Buchwald, S.L. (2013). *Journal of the American Chemical Society* 135: 15746–15749.

23 Shi, S.L. and Buchwald, S.L. (2015). *Nature Chemistry* 7: 38–44.

24 Matsuda, N., Hirano, K., Satoh, T., and Miura, M. (2012). *The Journal of Organic Chemistry* 77: 617–625.

25 Shen, K. and Wang, Q. (2015). *Chemical Science* 6: 4279–4283.

26 Hemric, B.N., Shen, K., and Wang, Q. (2016). *Journal of the American Chemical Society* 138: 5813–5816.

27 Ye, Z. and Dai, M. (2015). *Organic Letters* 17: 2190–2193.

28 Matsuda, N., Hirano, K., Satoh, T., and Miura, M. (2011). *Organic Letters* 13: 2860–2863.

29 Yotphan, S., Beukeaw, D., and Reutrakul, V. (2013). *Tetrahedron* 69: 6627–6633.

30 Zhu, C., Yi, M., Wei, D. et al. (2014). *Organic Letters* 16: 1840–1843.

31 Johnson, J.S. and Berman, A.M. (2005). *Synlett* 11: 1799–1801.

32 Barker, T.J. and Jarvo, E.R. (2009). *Journal of the American Chemical Society* 131: 15598–15599.

33 Lutter, F.H., Graßl, S., Grokenberger, L. et al. (2019). *ChemCat Chem* 11: 5188–5197.

34 Yoo, E.J., Ma, S., Mei, T.S. et al. (2011). *Journal of the American Chemical Society* 133: 7652–7655.

35 He, J., Shigenari, T., and Yu, J.Q. (2015). *Angewandte Chemie International Edition* 54: 6545–6549.

36 Shang, M., Zeng, S.-H., Sun, S.-Z. et al. (2013). *Organic Letters* 15: 5286–5289.

37 Dong, Z. and Dong, G. (2013). *Journal of the American Chemical Society* 135: 18350–18353.

38 Huehls, C.B., Lin, A., and Yang, J. (2014). *Organic Letters* 16: 3620–3623.

39 Dequirez, G., Pons, V., and Dauban, P. (2012). *Angewandte Chemie (International Ed. in English)* 51: 7384–7395.

40 Shimbayashi, T., Sasakura, K., Eguchi, A. et al. (2019). *Chemistry (Easton).* 25: 3156–3180.

41 Starkov, P., Jamison, T.F., and Marek, I. (2015). *Chemistry (Easton).* 21: 5278–5300.

42 Kwart, H. and Khan, A.A. (1967). *Journal of the American Chemical Society* 89: 1951–1953.

43 Jat, J.L., Paudyal, M.P., Gao, H. et al. (2014). *Science* 343: 61–65.

44 Nicolaou, K.C., Rhoades, D., Wang, Y. et al. (2017). *Journal of the American Chemical Society* 139: 7318–7334.

45 Ma, Z., Zhou, Z., and Kürti, L. (2017). *Angewandte Chemie, International Edition* 56: 9886–9890.

46 Strom, A.E. and Hartwig, J.F.J. (2013). *Organic Chemistry* 78: 8909–8914.

47 Paudyal, M.P., Adebesin, A.M., Burt, S.R. et al. (2016). *Science* 353: 1144.

48 Zhu, C., Li, G., Ess, D.H. et al. (2012). *Journal of the American Chemical Society* 134: 18253–18256.

49 Gao, H., Zhou, Z., Kwon, D.-H. et al. (2017). *Nature Chemistry* 9: 681–688.

50 Behnke, N.E., Kielawa, R., Kwon, D.-H. et al. (2018). *Organic Letters* 20: 8064–8068.

51 Farndon, J.J., Young, T.A., and Bower, J.F. (2018). *Journal of the American Chemical Society* 140: 17846–17850.

52 Zhou, Z., Cheng, Q.-Q., and Kürti, L. (2019). *Journal of the American Chemical Society* 141: 2242–2246.

53 Cheng, Q.-Q., Zhou, Z., Jiang, H. et al. (2020). *Nature Catalysis* 3: 386–392.

2

Remote Functionalizations Using Nitrogen Radicals in H-Atom Transfer (HAT) Reactions

Ji Hye Kim, Elizabeth M. Dauncey and Daniele Leonori

University of Manchester, Department of Chemistry, Oxford Road, Manchester M13 9PL, UK

2.1 Introduction

Nitrogen-containing molecules represent a central class of compounds with applications spanning therapeutic agents, agrochemicals, food additives, and materials. Chemical methodologies able to streamline the synthesis and modification of these molecules underpin the development of high-value products central to the well-being of our society [1].

Nitrogen radicals are reactive intermediates with powerful applications to the synthesis of nitrogenated molecules [2]. Their reactivity can be used for direct C—N bond formation via exo-trig cyclization (Scheme 2.1a) and addition to electron-rich π-system (Scheme 2.1b) or for remote sp^3 carbon functionalization, exploiting their ability to partake in radical transpositions [3, 4]. From this perspective, there are two types of processes that can be used, β-fission (Scheme 2.1c) and intramolecular 1,5-H-atom transfer (1,5-HAT) (Scheme 2.1d). The former reactivity normally involves radical ring opening of small cyclic systems, while the latter enables the effective functionalization of N-containing alkyl chains.

This chapter will discuss selected recent advances in the application of nitrogen radicals in the field of 1,5-HAT. We will arrange the content highlighting the key mechanistic features of the HAT process and will then divide the discussion on the basis of the chemical process used to access the key nitrogen radicals. In each case, we aim to highlight the application of this radical reactivity in the assembly or the late-stage functionalization of complex and bioactive materials.

2.2 Intramolecular 1,5-H-Atom Transfer (1,5-HAT)

Nitrogen radicals are among the most efficient open-shell intermediates known to undergo selective transpositions via intramolecular HAT [5, 6]. This reactivity can be used for the functionalization of nitrogenated alkyl chains at specific sp^3 carbons that can be difficult to target by either ionic or transition-metal-mediated strategies [3]. The HAT process displays

Methodologies in Amine Synthesis: Challenges and Applications, First Edition.
Edited by Alfredo Ricci and Luca Bernardi.

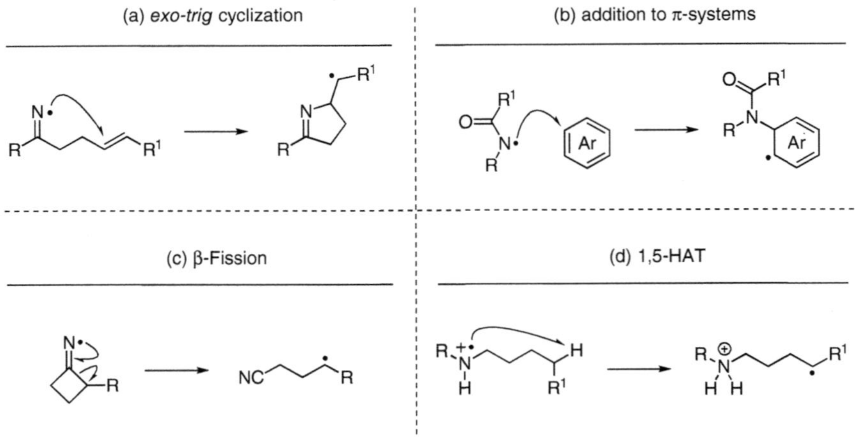

Scheme 2.1 Radical transposition processes using nitrogen radicals.

a strong 1,5-selectivity, which means that the sp^3 carbon δ to the nitrogen radical is the one that will be functionalized (Scheme 2.2a). This high fidelity has played a fundamental role in the application of these processes in synthetic organic chemistry.

The ability of nitrogen radicals to undergo very efficient and effective transposition is also of high historical relevance to radical chemistry and organic synthesis as a whole. Indeed, this elementary step is the key event in the named reaction developed over 100 years ago by Hoffman, Löffler, and Freitag (HLF) [7]. Interestingly, the HLF reaction represents the first known example of radical transposition process and a very effective strategy for the construction of pyrrolidine heterocycles [5, 6]. In general terms, this reactivity is based on the conversion of a protonated *N*-haloamine (**1**) into an aminium radical (**2**) that undergoes very facile 1,5-HAT, resulting in a nitrogen-to-carbon radical relay (Scheme 2.2b). The incipient δ-carbon radical (**3**) is then halogenated (S_H2) (**4**) as part of an efficient radical chain propagation. A final basic treatment induces an ionic cyclization reaction, leading to the construction of a pyrrolidine scaffold (**5**). Although this reactivity has been established over 100 years ago, there is still continuous interest in further developments and innovations especially in terms of expanding the types of functionalizations and also control the overall stereochemical outcome.

Mechanistically, 1,5-HAT reactions of nitrogen radicals are strongly polarity- and enthalpy-driven processes with stringent geometrical constraints. This means that three main factors need to be considered when approaching the design of these transformations.

Geometrical factors. The 1,5-HAT process goes through a six-membered ring transition state, which, depending on the 3D structure of the nitrogen radical, might favor or thwart reactivity [8]. This is seen in a classical example of HLF reactivity for the assembly of a complex-bridged azabicyclic system (Scheme 2.3a) [9]. Although the intermediate *N*-chloroamine **6** contains four different sites with 1,5-relationship to the aminium radical **7**, only the geometrically proximal methyl group is functionalized (**8**).

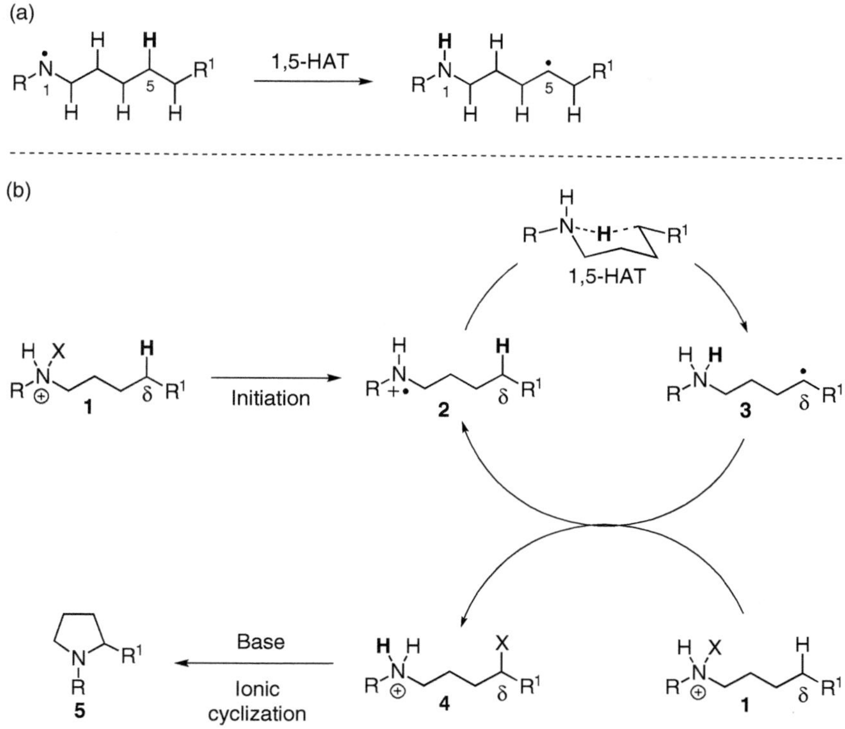

Scheme 2.2 1,5-HAT process in HLF reaction.

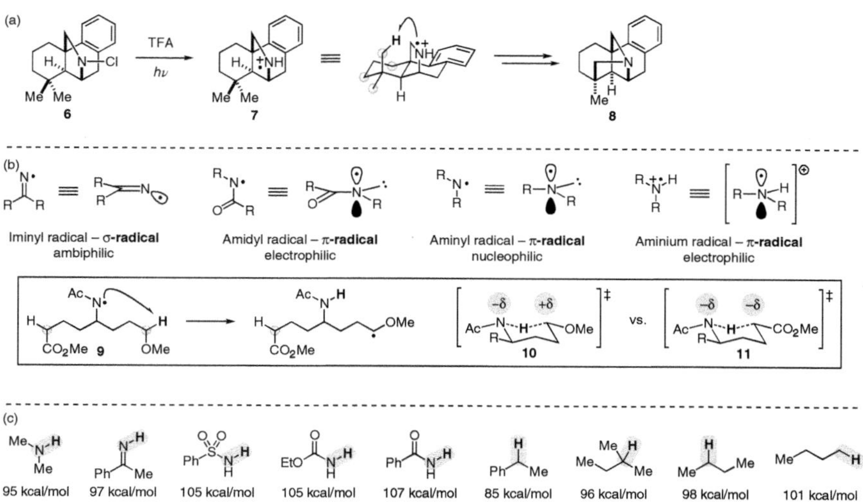

Scheme 2.3 Geometrical factors, polar and enthalpic effects in 1,5-HAT process. Source: (a) Katohgi et al. [9] and (c) Luo [10].

Polar effects. Nitrogen radicals are reactive intermediates with distinct and different philicities based on their substitution patterns. There are four general classes: (i) iminyl, (ii) amidyl (including carbamoyls, sulfamidyls, and phosphoramidyls), (iii) aminyl, and (iv) aminium radicals (Scheme 2.3b). The philicity and the hybridization of these species change significantly: iminyl radicals [11] are σ-radicals with ambiphilic character, while all the others are π-radicals with the amidyls [12] and the aminiums [13] being electrophilic and the aminyls [14] slightly nucleophilic.

Although the 1,5-HAT process goes via a sterically favorable six-membered ring transition state, polar effects have a strong interplay in its stabilization [15]. This means that when the innate polarity of the H-atom matches the one of the nitrogen radical, then a charge transfer stabilization takes place, leading to a more facile 1,5-HAT process. As an example, the general amidyl radical **9** would undergo selective abstraction from the α-*O*-methylene group rather than the α-ester one, as transition state **10** enables a better stabilization of polar effects than transition state **11**.

Enthalpic effects. The exothermicity, hence the feasibility, of 1,5-HAT reactions is dictated by the relative energy difference between the sp^3 C—H bond broken and the N—H bond formed [16]. In general, N—H bonds have higher bond-dissociation energy (BDEs) than sp^3 C—H bonds, which facilitate 1,5-HAT reactivities (Scheme 2.3c) [10]. This is particularly evident for the reaction of amidyl and aminium radicals, which lead to the formation of very strong N—H bonds. Iminyl radicals have witnessed a more limited application in radical relay strategies owing to the weaker nature of their corresponding N—H bonds, which makes 1,5-HAT possible mostly from activated or tertiary sp^3 centers. Aminyl radicals are the only class of nitrogen radicals that are not used in these processes owing to their weaker N—H bonds and also to negative polar effects.

2.3 Photoinduced Strategies

2.3.1 Reductive Strategies

2.3.1.1 1,5-HAT via Iminyl Radicals

Iminyl radicals have been identified as powerful reactive intermediates for a number of reactions such as intramolecular cyclization but also radical ring opening and 1,5-HAT. In this case, the pioneering work of Forrester [17–20] has clearly demonstrated their ability to partake into radical transposition, but the harsh chemical methods for their generation limited applications in mainstream organic synthesis.

The Leonori group recently reported that iminyl (and amidyl radicals) can be generated upon visible light irradiation from easy-to-make *O*-aryl oximes [21] via oxidative quenching photoredox cycles and also via the formation of electron donor–acceptor (EDA) complex with tertiary amines [22–24]. These strategies have been adopted by others in the subsequent development of interrupted HLF-type reactions.

Fu and coworkers described a photochemical intramolecular α-*N* sp^3 C–H imination based on *O*-aryl oximes **12** for the preparation of functionalized imidazoles **13** (Scheme 2.4) [25]. In this case, the authors exploited the known ability of these derivatives to undergo EDA complex formation with a tertiary amine in an intramolecular setting (**12**). This EDA

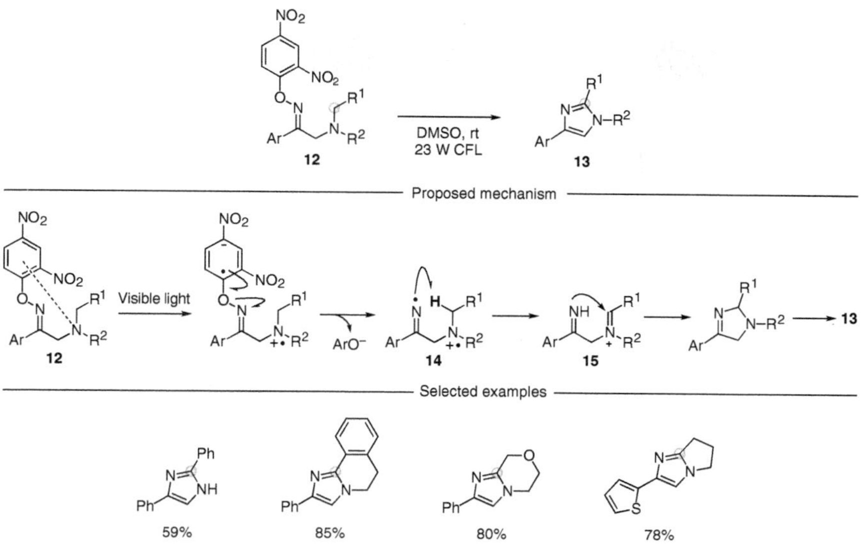

Scheme 2.4 Photochemical intramolecular C(sp^3)–H imination of tertiary aliphatic amines containing β-*O*-aryl oximes. Source: Modified from Li et al. [25].

complex underwent photoinduced single-electron transfer (SET) and fragmentation across the weak N—O bond to give the iminyl **14**. This slightly nucleophilic radical was then able to trigger a polarity-matched 1,5-HAT from the α-N sp^3 carbon providing **15**. An ionic intramolecular cyclization, followed by aromatization, delivered the desired products **13**. This methodology can only be used to assemble five-membered ring systems and is limited on having an aryl substituent on the oxime. However, the site of 1,5-HAT could be embedded into a further cyclic system, thus leading to interesting bicyclic structures.

More recently, Wu and coworkers have demonstrated that similar *O*-aryl oximes **16** can also undergo EDA complex formation with 1,4-diazabicyclo[2.2.2]octane (DABCO)·$(SO_2)_2$ **17** and that this can be used for iminyl radical generation upon visible light irradiation (Scheme 2.5) [26]. This step was followed by a 1,5-HAT process, and the resulting carbon radical was able to intercept SO_2 en route to products **18**. In this way, a rare example of amino-sulfonylation of unactivated sp^3 C—H bonds was achieved, leading to the assembly of biologically relevant 1,2-thiazine dioxides.

The ability of iminyl radicals to undergo 1,5-HAT has been recently employed for the conversion of activated acyclic *O*-acyl oximes **19** into a variety of cyclic ketones **20** (Scheme 2.6). In this work, Nevado and coworkers described a redox-neutral process, where the SET reduction of radical precursor **19** was performed by a visible-light-excited Ir(III) photocatalyst [27]. In this case, the authors used *O*-acyl oximes that have a reduction potential accessible by many photoexcited catalysts [28]. This SET led to the corresponding iminyl radical **21** that underwent selective 1,5-HAT. The distal carbon radical **22** was exploited in an intramolecular cyclization onto the (hetero)aromatic substituent. Oxidation of intermediate **23** by the Ir(IV) catalyst gave the imine **24**, which, upon hydrolysis, delivered a variety of fused bicyclic ketones in high yields. The six-membered ring radical cyclization enabled excellent control of relative stereochemistry when substituents were present.

Scheme 2.5 Photoinduced aminosulfonylation of C(sp^3)−H bonds with sulfur dioxide. Source: Modified from Li et al. [26].

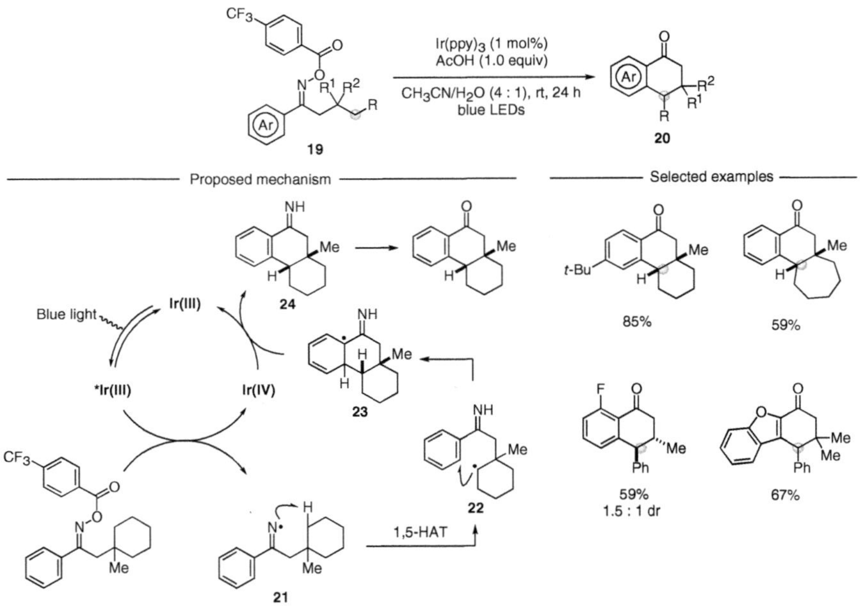

Scheme 2.6 Redox neutral remote functionalization of C(sp^3)−H bonds via iminyl radicals. Source: Modified from Shu and Nevado [27].

The direct installation of vinyl groups onto unactivated sp^3 C—H bonds is a transformation with powerful applications as the olefin functionality can then be elaborated in a number of different ways. An interesting solution to this challenge has been provided by Yu and coworkers that have developed cascade reactions between electron-poor *O*-acyl oximes **25** and styrenyl boronic acids **26** (Scheme 2.7) [29]. This reactivity was also based on the generation of an iminyl radical by reductive SET from a photoexcited catalyst, its following transposition by 1,5-HAT, and the reaction of the corresponding distal carbon radical with the olefin. Although the methodology was only used to install styrenyl-type substituents, it enabled the fast assembly of small-molecule building blocks **27** difficult to target by other methodologies.

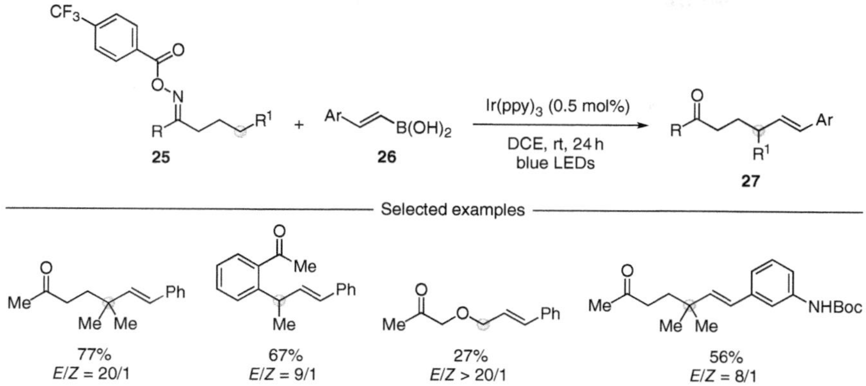

Scheme 2.7 Intermolecular remote C(sp^3)–H and C–C vinylations via iminyl radicals under photoredox-catalysis. Source: Modified from Shen et al. [29].

Shortly after, Duan and coworkers demonstrated that a related cascade reaction could be implemented using directly styrenes [30]. In this case, a broad range of substitution pattern was tolerated.

2.3.1.2 1,5-HAT via Amidyl and Sulfamidyl Radicals

As discussed in Section 2.2, the strong electrophilic character of amidyl and sulfamidyl radicals means that very effective 1,5-HAT transpositions can be achieved.

A powerful example based on SET reductions can be found in the work by Wang, who reported the remote sp^3 C–H allylation of amides (Scheme 2.8) [31]. In this case, the authors used the highly electron-poor *O*-aryl hydroxyamides **28** ($E^{red}_{1/2} = -0.9$ V vs. saturated calomel electrode [SCE]) that were previously identified as powerful electrophores for SET by the photoexcited state of the organic dye eosin Y ($^*E^{red}_{1/2} = -1.1$ V vs. SCE). This event led to a fragmentation across the weak N—O bond, thus delivering the key amidyl radical **29**. A very facile 1,5-HAT process was used to generate the distal carbon radical **30**, which underwent allylation by reaction with a variety of sulfone reagents (e.g. **31**). This step provided the product **32** with concomitant release of $PhSO_2^{\cdot}$. The authors proposed a redox-neutral photoredox process, whereby *N,N*-diisopropylethylamine (DIPEA) closed the cycle by SET with the oxidized eosin Y. However, as quantum yield determination was not performed, a potential radical chain mechanism cannot be ruled out [32]. This reaction enabled the modification of secondary and tertiary centers and also displayed broad functional group compatibility.

The ability of amidyl radicals to undergo 1,5-HAT has also been exploited as part of a strategy, leading to remote sp^3 C–H arylated products **35**. Yu and coworkers demonstrated that a site-selective Minisci-type reaction [33] can be achieved using the electron-poor *O*-acyl hydroxy-amides **33** as amidyl radical precursors and several *N*-heteroarenes **34** (Scheme 2.9) [34]. Upon visible light irradiation, the excited dye 3CzClIPN promoted SET reduction of **33** (in analogy to the *O*-acyl oximes), thus enabling access to the corresponding nitrogen radicals **36**. A 1,5-HAT process gave the carbon radical **37**, which underwent addition to an activated heteroarene to give **38**. The authors suggested that under basic conditions, a proton and electron transfer took place, thus leading to the desired product **39**. This reactivity

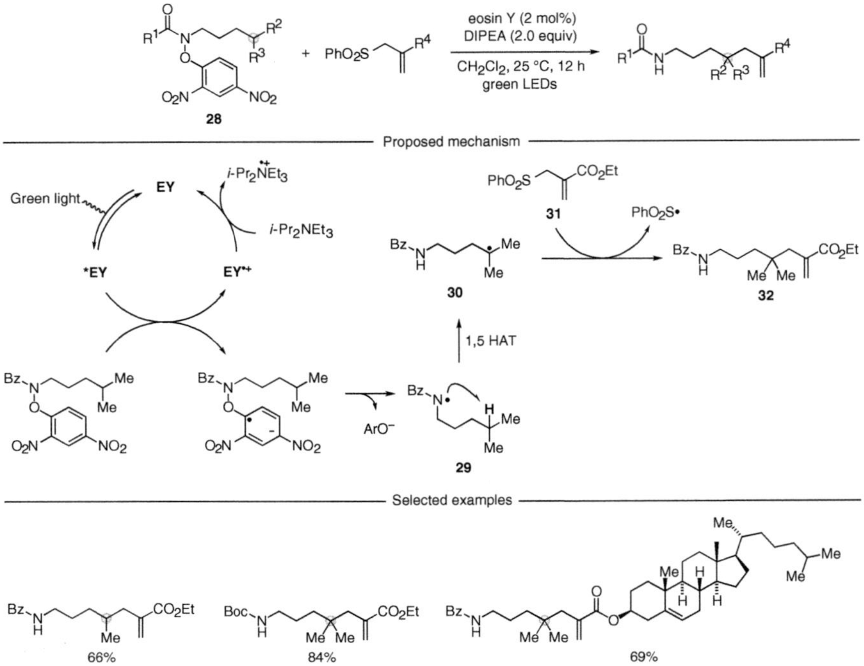

Scheme 2.8 Remote C(sp^3)–H allyation of amides using organic photocatalyst. Source: Modified from Wu et al. [31].

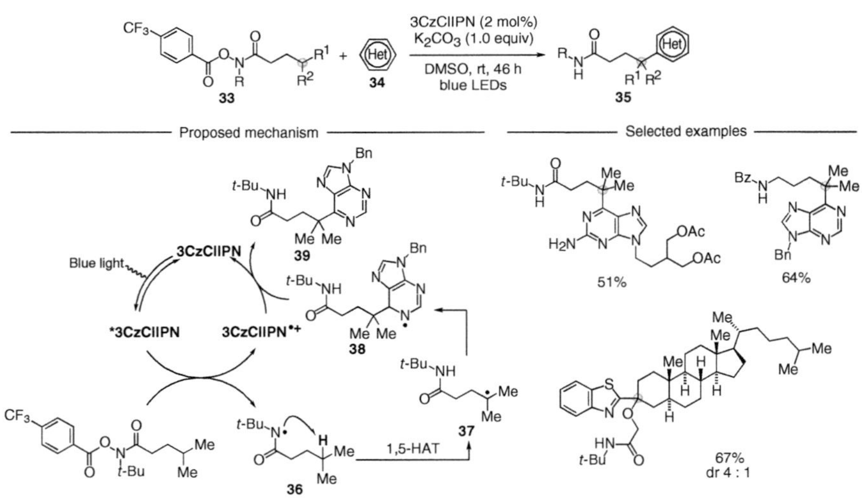

Scheme 2.9 Site-selective remote C(sp^3)–H heteroarylation of amides. Source: Modified from Chen et al. [34].

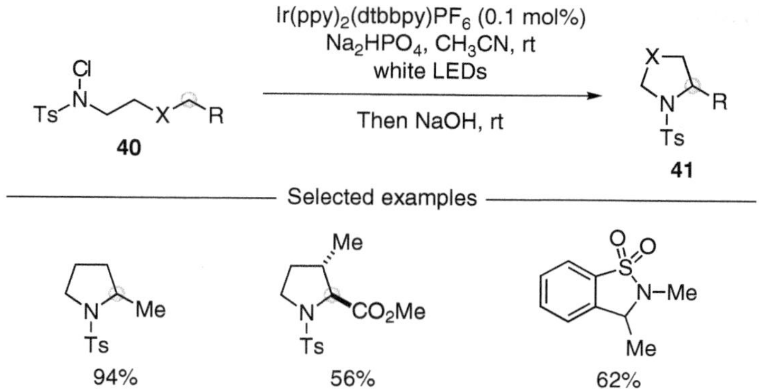

Scheme 2.10 Visible-light-promoted C(sp^3)–H amidation and chlorination of *N*-chlorosulfonamides. Source: Modified from Quin and Yu [35].

displayed remarkable functional group tolerance in terms of the compatible heterocycles and was used for the modification of biologically active systems.

The generation of amidyl radicals through photoinduced SET reduction is not restricted to electrophores embedded into a N–O systems but can also be performed on *N*-Cl derivatives. This was demonstrated by Yu and coworkers in the HLF-type reactivity of *N*-Cl–*N*-Ts amines **40** by using a heteroleptic Ir(III) photocatalyst under white light irradiation (Scheme 2.10) [35]. In this case, after generation of a very electrophilic sulfamidyl radical, the 1,5-HAT took place, and then the corresponding carbon radical was chlorinated, most likely, as part of a radical chain propagation. This protocol has also been used in the assembly of many pyrrolidines **41** and also in late-stage functionalizations of some bioactive materials, demonstrating the strong functional group compatibility.

2.3.2 Oxidative Strategies

2.3.2.1 1,5-HAT via Iminyl Radicals

Inspired by the pioneering work of Forrester [17–20], the Leonori and the Studer group took carboxylic acid-containing oximes **42** and used them to generate functionalized imine and ketone derivatives **43–45** (Scheme 2.11) [36–38]. These processes were all based on a reductive quenching photoredox cycle and required basic conditions to trigger the SET oxidation of the carboxylate functionality (**46**). This event led to two β-scissions extruding CO_2 and acetone and forming the corresponding iminyl radical **47**. This species then underwent 1,5-HAT and the incipient carbon radical **48** was intercepted and diversified with several radical trapping agents (X–Y) such as *N*-chlorosuccinimide (NCS; chlorination, **43**), selectfluor (fluorination, **44**), and Michael acceptors (Giese addition, **45**). A final electron transfer between the reduced photocatalyst and the Y· intermediate, generated upon atom transfer/addition, led to an overall redox-neutral process. This approach for remote ketone functionalization tolerated important functionalities such as protected amines, heteroaromatics, free alcohols, and ester groups among others. Interestingly, in the case of NCS, the intermediate NH imine was directly intercepted by NCS before hydrolysis, leading to the corresponding N–Cl imines that showed remarkable stability.

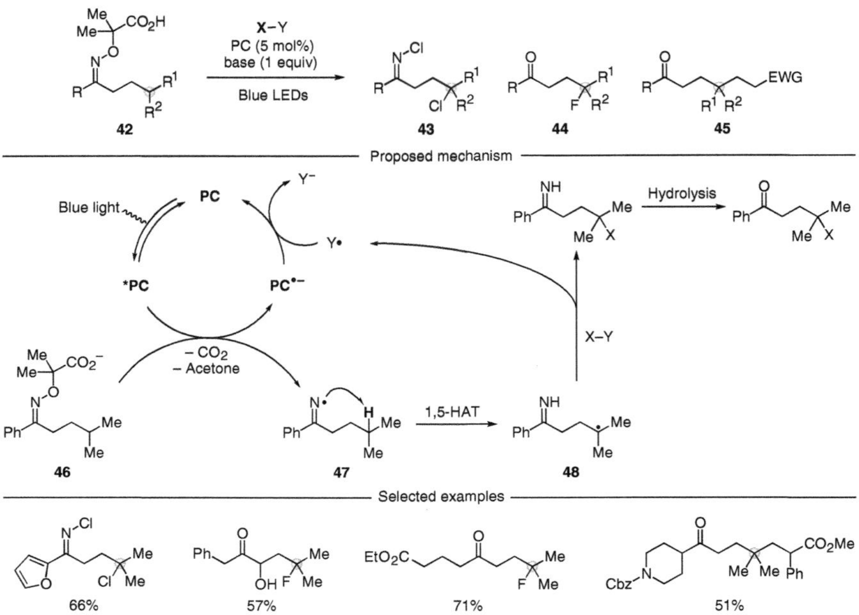

Scheme 2.11 Site-specific intermolecular γ-C(sp^3)–H functionalization of ketones via iminyl radicals. Source: Jiang and Studer [36], Morcillo et al. [37], and Davies et al. [38].

2.3.2.2 1,5-HAT via Amidyl and Sulfamidyl Radicals

A related approach for nitrogen radical generation was then used by the Leonori group to introduce a range of functionalities at the γ-position of amides and at the δ-position of protected amines using carboxylic acid-containing hydroxyamides **49** (Scheme 2.12) [37]. In analogy to their previous work on iminyl radicals, photoinduced oxidation of **49** under basic conditions enabled amidyl radical generation, followed by 1,5-HAT and distal carbon radical functionalization. In this case, halogenating agents such as NCS and selectfluor allowed introduction of γ chlorine and fluorine atoms, while S-containing phthalimides enabled direct thioetherification. More importantly, 2-iodoxybenzoic acid (IBX) reagents were successful to achieve remote C—C bond formation and introduce CN as well as alkyne functionalities. This strategy enabled the selective modification of tertiary, secondary, and even primary centers and was also used for the late-stage functionalization of small dipeptides.

A very powerful approach for distal functionalization of amides has been possible from the corresponding N–H substrates using photoinduced proton-coupled electron transfer (PCET). As initially demonstrated by Knowles and coworkers, this activation mode bypasses the need to install a stoichiometric electrophore on the substrates and therefore streamlines the preparation of the desired reaction products. This strategy has been initially used to trigger cyclization reactions [39], and also used in interrupted HLF-type reactivity as demonstrated by the groups of Knowles and Rovis (Scheme 2.13) [40, 41]. Following a reductive-quenching photoredox cycle and in the presence of a suitable base, **50** was effectively oxidized by photoinduced PCET generating the amidyl radical **51**. This

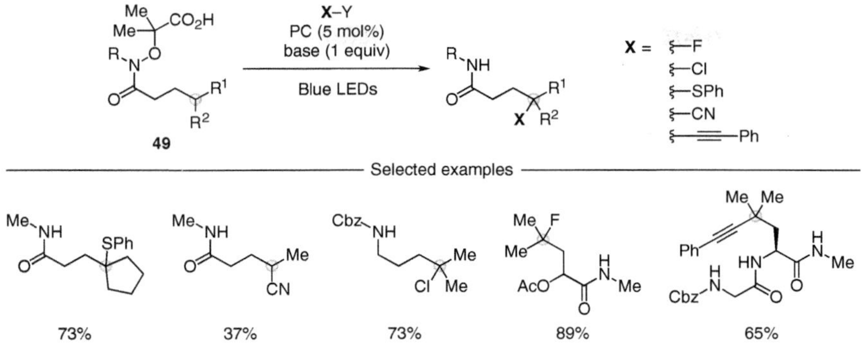

Scheme 2.12 Nitrogen radical mediated remote functionalization of amides and amines. Source: Modified from Morcillo et al. [37].

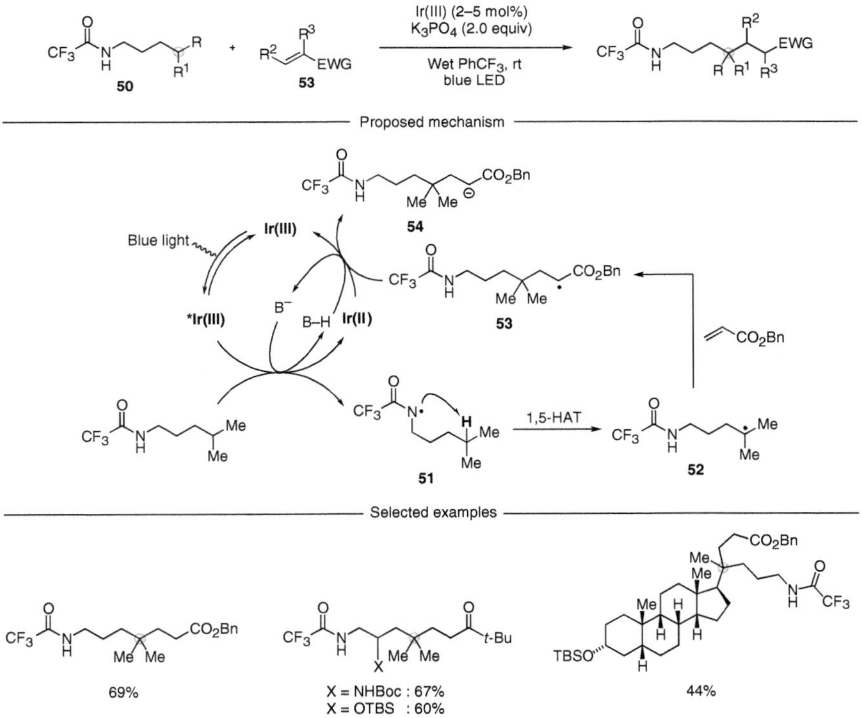

Scheme 2.13 Amide-directed selective C—C bond formation at C(sp^3)—H bonds. Source: Choi et al. [40] and Chu and Rovis [41].

species was exploited in a 1,5-HAT transposition and the resulting carbon radical **52** reacted with a Michael acceptor (**53**), thus assembling a distal sp^3–sp^3 C—C bond. Owing to electron-poor nature of the intermediate α-ester radical **53**, the photoredox cycle was closed by SET with the reduced Ir(II) photocatalyst that delivered the stabilized anion **54** that was protonated. The design of the amide substrates was critical to the feasibility and the overall success of the PCET as the carbonyl functionality must be able to acidify the N–H proton enough to encourage the deprotonation step. This somewhat controlled the nature of the amidyl radicals that could be generated. Nevertheless, this methodology displayed strong functional group tolerance as demonstrated by the number of Michael acceptors and functionalized amides that could be used.

This PCET reactivity for amidyl radical generation and 1,5-HAT has also been exploited by Tamber to achieve remote amide allylation using allyl chlorides and a Ni(0) cocatalyst (Scheme 2.14) [42].

A powerful expansion of this PCET strategy for amidyl radical generation and 1,5-HAT has been recently introduced by Rovis and coworkers, who merged the photoredox manifold with nickel catalysis (Scheme 2.15) [43]. This dual catalytic process used alkyl bromides as coupling partners and enabled a site-selective C–H alkylation delivering products **55**. Mechanistically, the trifluoroacetic acid (TFA)-amide was oxidized through PCET by the excited Ir(III) catalyst upon visible light irradiation to form the amidyl radical **56**. This

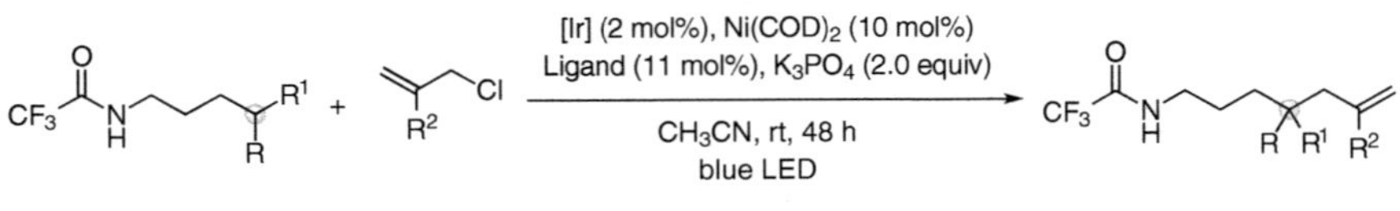

Scheme 2.14 Remote amide allylation using allyl chlorides and Ni(0) cocatalyst. Source: Modified from Xu and Tambar [42].

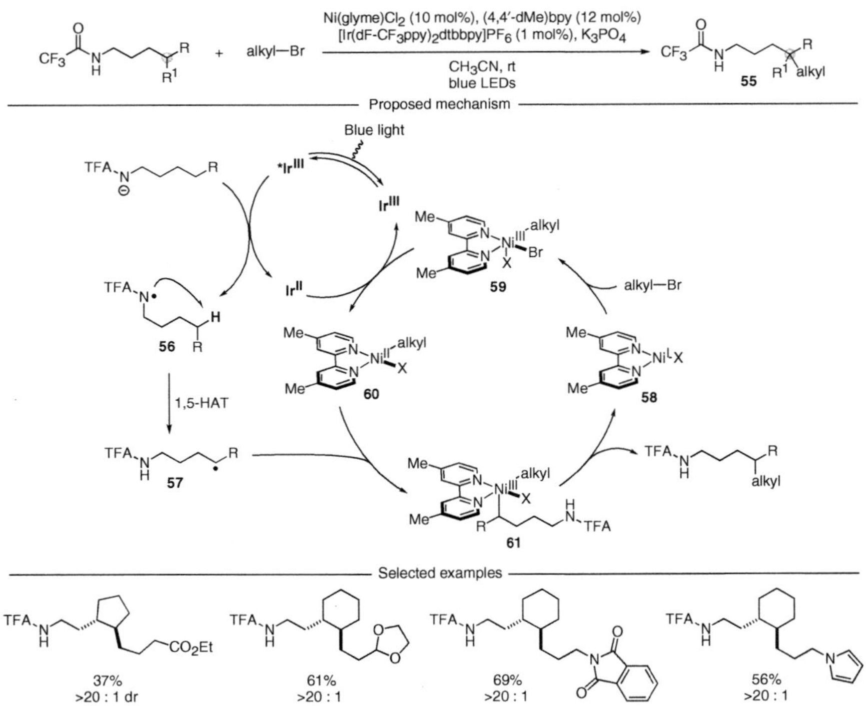

Scheme 2.15 Regioselective cross-coupling of C(sp^3)–H bonds and alkyl bromides under a combination of photoredox and nickel catalysis.

species underwent 1,5-HAT, delivering the carbon radical **57**. At the same time, oxidative addition of a nickel(I) complex **58** to the alkyl bromide resulted in an alkyl–Ni(III) complex **59** that was proposed to undergo SET with the reduced photocatalyst. This event would close the photoredox cycle and generate an alkyl–Ni(II) species **60** that could be intercepted by **57**. This radical transmetalation enabled access to a dialkyl–Ni(III) intermediate **61** from which reductive elimination is facile. This final step would forge the key sp^3–sp^3 C—C bond and close the nickel cycle. Although this reactivity is restricted to the use of primary alkyl bromides as coupling partners, the authors demonstrated its ability to construct a broad range of remotely alkylated products.

2.3.3 Photoinduced Bond Homolysis

The weak nature of N—X (X = halogen) bonds has historically been of fundamental importance for the development of HLF reactions. However, the requirement for prior synthesis of *N*-iodo or *N*-bromo amines, which are sometimes unstable, represented a significant limitation.

In a series of elegant reports Muñiz and coworkers developed conditions for HLF conversion of unfunctionalized *N*-Ts amines **62** into the corresponding pyrrolidines **63** by using substoichiometric amounts of I_2 and the I(III) reagent PhI(mCBA)$_2$ as the oxidant (Scheme 2.16) [44–46]. These processes are believed to be initiated by the reaction of I_2

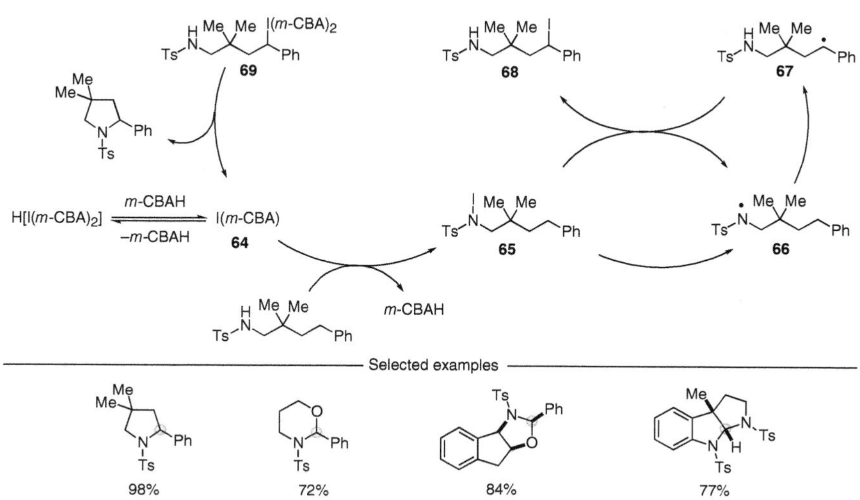

Scheme 2.16 Iodine-catalyzed visible-light mediated C–H aminations. Source: Zhang and Muniz [44], Becker et al. [45], Martinez and Muniz [46].

with PhI(*m*CBA)$_2$ that leads to the formation of I(*m*CBA) **64**, the active catalyst. Following detailed mechanistic studies, the authors suggested that this species promoted the in situ formation of an *N*-iodo–*N*-Ts amine **65**, which underwent facile N—I bond homolysis upon visible light irradiation. Following 1,5-HAT from the sulfonyl radical **66**, the distal carbon radical **67** was iodinated to give **68** as part of a radical-chain propagation. In classical HLF strategies, a basification is required to trigger the ionic cyclization; however, in this case, the PhI(*m*CBA)$_2$ was able to directly oxidize **68** to the corresponding I(III) intermediate **69**. The very activated nature of this species means that an immediate nucleophilic displacement was possible, leading to the assembly of the desired N-heterocycle. This step would also regenerate the active I(*m*CBA) catalyst. The authors have showcased this reactivity mostly for the functionalization of benzylic centers and the preparation of pyrrolidines. However, careful juxtaposition of substituents has also enabled access to the six-membered ring (piperidine) analogs.

More recently, this reactivity could also be triggered by catalytic generation of the corresponding *N*-bromo–*N*-Ts amine, which represents a catalytic version of the original HLF reactivity [47].

A different strategy based on N—I bond homolysis was recently introduced by Nagib and coworkers to generate imidate radicals, which could be used in HLF settings for C–H amination (Scheme 2.17) [48]. This methodology started with simple aliphatic alcohols, which upon addition of Cl_3C–CN formed imidates **70**. These species were iodinated by the in situ formation of I_2 (KI + PhI(OAc)$_2$) and under visible light irradiation underwent N—I bond homolysis to generate the corresponding iminyl-like nitrogen radicals. Upon 1,5-HAT, the carbon radicals were iodinated, and then an ionic cyclization was used to access heterocycles, which, upon hydrolysis, gave the *β*-amino alcohol products **71**.

The same authors later published a thermal, catalytic variant of the radical chaperone strategy, allowing them to generate analogous amino alcohols, however with less oxidant, which suppressed unwanted side reactions [49]. The side reactions occur from radicals derived from homolysis of excess AcOI, which can oxidize α-aminyl C—H bonds, thereby competing with the desired 1,5-HAT process.

Zhu and coworkers developed a protocol relying on the in situ formation of an *N*-iodo–*N*-Ts amine followed by photochemical N—I bond homolysis. While pioneering work by Gomez-Suarez and coworkers [50–54] in the area focused on cyclization to form

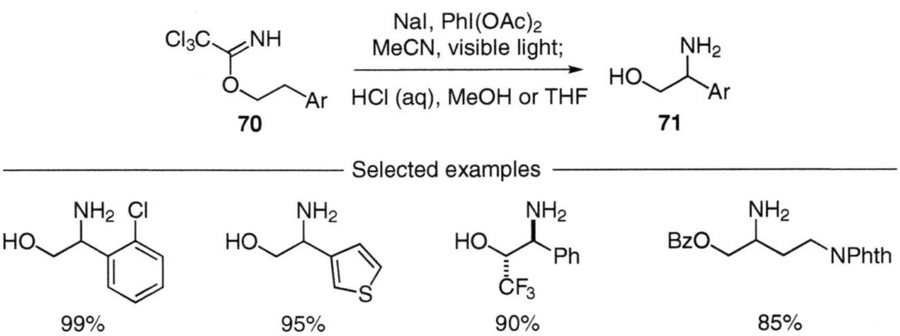

Scheme 2.17 Imidate radical mediated β-C–H amination of alcohols. Source: Modified from Wappes et al. [48].

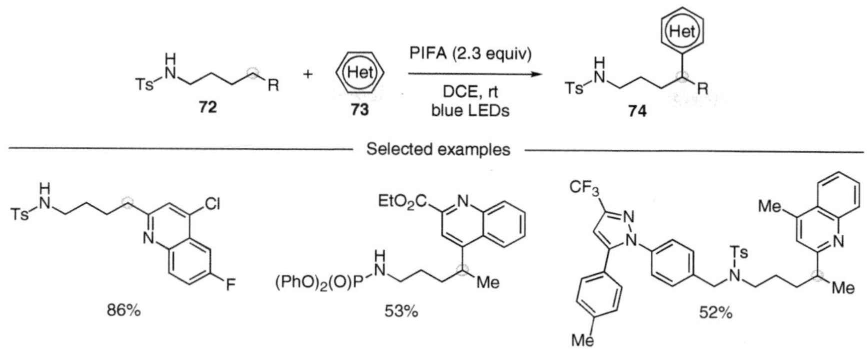

Scheme 2.18 Remote heteroarylation of amides via sulfonamidyl radicals [55]. Source: Gomez-Suarez et al. [50], Hernandez et al. [51], Francisco et al. [52], Francisco et al. [53], and Martin et al. [54].

pyrrolidine derivatives, this new strategy sought to achieve an interrupted reactivity for selective sp^3 C–H arylation of acyclic sulfonamides at distal positions (Scheme 2.18). By taking simple reagents **72**, in the presence of phenyl-iodine bis(trifluoroacetate) (PIFA), they efficiently formed the corresponding *N*-iodo-sulfonamide species. Upon visible light irradiation, N—I bond homolyses took place, furnishing the electrophilic sulfonamidyl radical. This species underwent efficient 1,5-HAT, and the incipient carbon radical was intercepted with a protonated N-heterocycle **73** in a Minisci-type reaction. Further oxidation delivered the desired γ-arylated products **74** in high yields. This methodology proved general with respect to the heteroaromatic coupling partner and could tolerate, for example, aryl halides, esters, and ethers. Furthermore, with respect to the amide portion, a number of different sulfonamides, carboxamides, and phosphoramides could be used, including a celecoxib derivative.

Alexanian and coworker developed a site-specific C–H functionalization of amides using *N*-dithiocarbamates **75** as precursors for amidyl radicals (Scheme 2.19) [56]. Owing to the weak nature of N—S bonds, visible light irradiation (or standard radical initiation) enabled

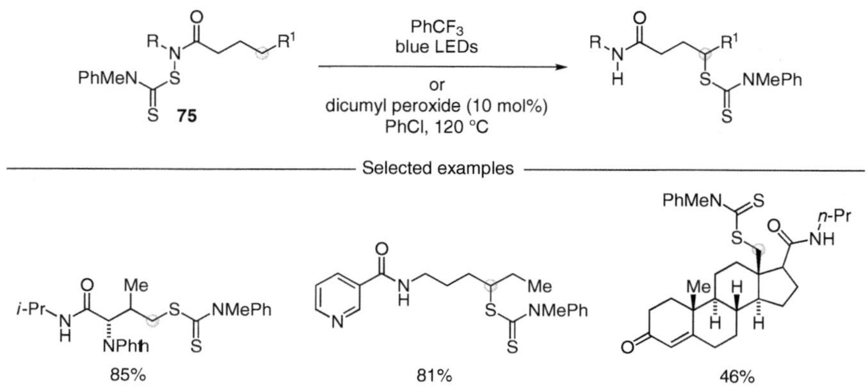

Scheme 2.19 Site-specific C–H functionalization via HAT using *N*-dithiocarbamates. Source: Modified from Na and Alexanian [56].

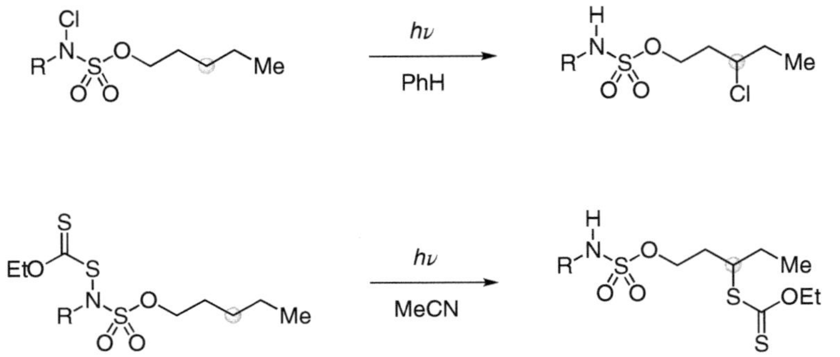

Scheme 2.20 γ-C(sp^3)–H chlorination and xanthylation of sulfamate esters via 1,6-HAT process. Source: Short et al. [57] and Ayer and Roizen [58].

the generation of a nitrogen radical that underwent 1,5-HAT, followed by xanthylation as part of a very efficient radical chain propagation. A wide range of complex molecules has been modified with this procedure, including complex natural products and pharmaceutical derivatives, thus showcasing the broad functional group compatibility.

So far, all HAT strategies discussed have been mostly based on 1,5-abstractions.The Roizen group has recently demonstrated how γ-C(sp^3)–H chlorination and xanthylation of sulfamate esters is possible via a very selective 1,6-process (Scheme 2.20) [57, 58]. While this chemistry is mechanistically related to other procedures previously discussed, it has a peculiar selectivity because of the elongated N—S—O bond system that stabilizes a unique seven-membered ring transition state.

2.4 Thermal Strategies

The generation of nitrogen radicals and their utilization in radical transpositions is not restricted to photochemical approaches, and many powerful and applicable methods based on thermal processes have been reported.

Iminyl radicals can be conveniently accessed from electron-poor *O*-acyl oximes by SET reduction as mentioned before (see Section 2.3.1.1). Similar starting materials **76** have also been used in combination with Fe-catalysis to access remote azidation of ketones (Scheme 2.21) [59]. The proposed mechanism relied on an initial reduction of an Fe(III) species by $TMSN_3$ to generate the active Fe(II) catalyst. This species was postulated to reduce the *O*-acyl oxime **76** and give, after fragmentation, the corresponding iminyl radical **77**. As these reactions were run under acidic conditions (AcOH), protonation of the iminyl radical would generate a more electrophilic species **78**, thus reinforcing the polar effect and resulting in a more favorable 1,5-HAT process. The resulting carbon radical **79** was proposed to undergo an Fe(III)-mediated azide transfer to give the product **80** and regenerate the active Fe(II) catalyst. This methodology tolerated several functionalities including alkenes, heterocycles, nitriles, and substituted aromatics. Furthermore, by using

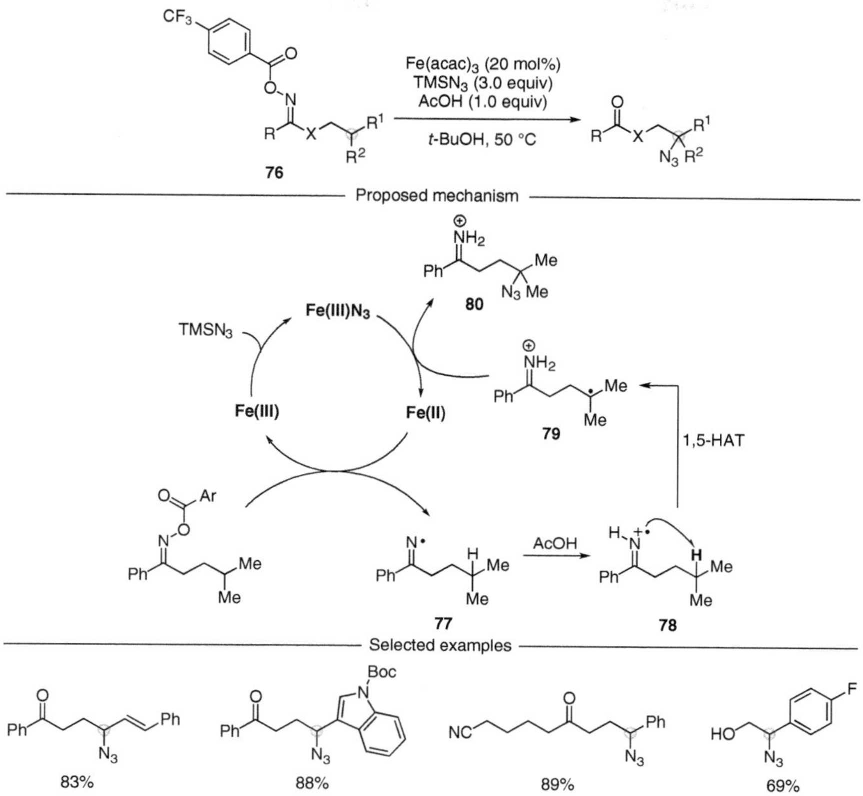

Scheme 2.21 Iron-catalyzed remote C(sp^3)–H azidation of ketones via iminyl and imidate radicals. Source: Modified from Torres-Ochoa et al. [59].

N-acyloxy imidates, valuable β-azido alcohols (after further hydrolysis) could also be obtained.

Cook and coworkers reported the direct Csp^3–H fluorination of *N*-fluoroamides **81**, employing $Fe(OTf)_2$ as the catalyst (Scheme 2.22) [60]. The mechanistic studies suggest that Fe(II) could directly cleave the N—F bond to generate an amidyl radical, which underwent nitrogen-to-carbon radical relay. Fe-mediated fluorine transfer gave the products in generally high yields, also showing significant functional group tolerance.

Using a similar strategy, Zhu and coworkers designed a copper-catalyzed variant, which allowed for a remote arylation using preformed *N*-fluoroamides **82** (Scheme 2.23) [61]. Mechanistically, the authors proposed that Cu(I)-mediated reductive cleavage of the amide N—F bond generated the amidyl radical **83**, which underwent 1,5-HAT from the benzylic position. Simultaneously, the Cu(II) species underwent transmetalation with the arylboronic acid **84**, generating an Ar–Cu(II) species. This intermediate was intercepted by the carbon radical **85** to give an Ar, alkyl–Cu(III)complex **86** from which reductive elimination is facile. This step generated the desired arylated product **87** along with the

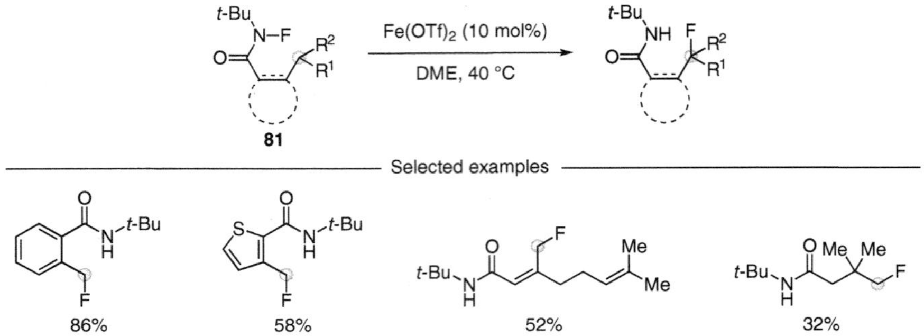

Scheme 2.22 Amide-directed fluorination of C–H bonds catalyzed by iron. Source: Modified from Groendyke et al. [60].

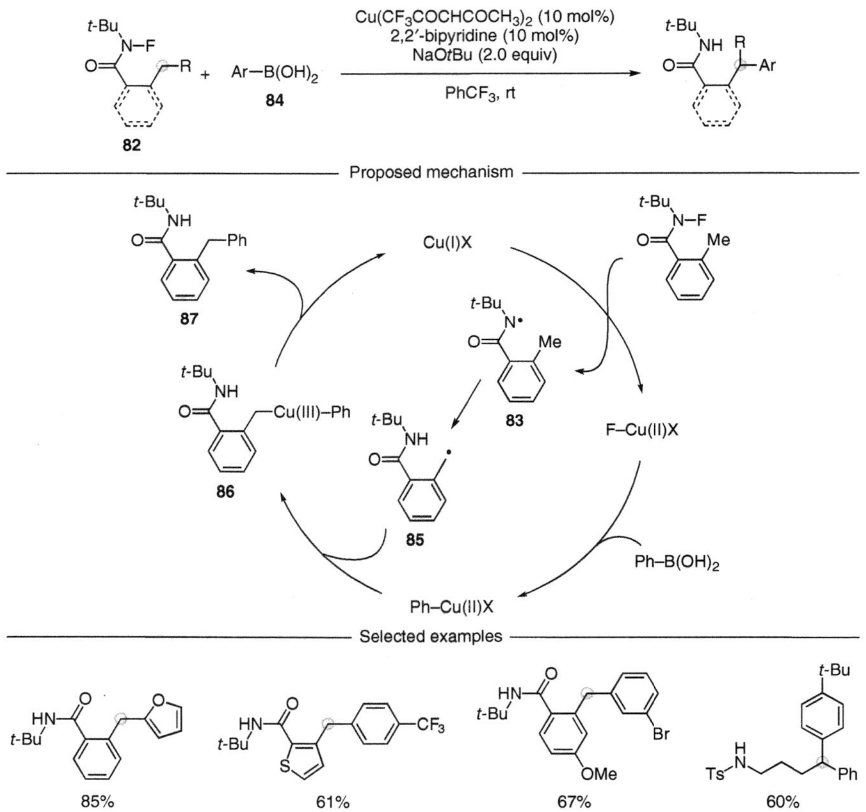

Scheme 2.23 Copper-catalyzed remote arylation of C(sp^3)–H bonds via amidyl radicals. Source: Modified from Li et al. [61].

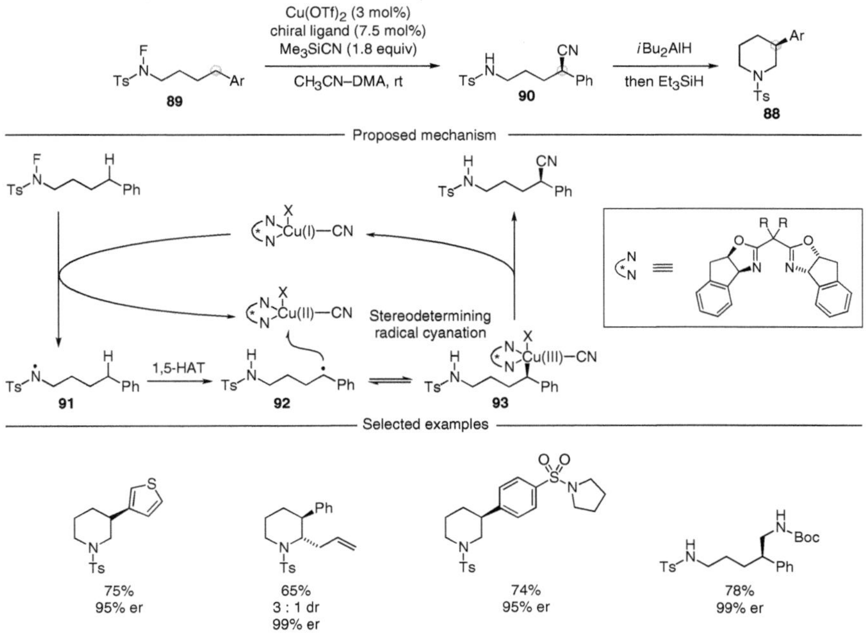

Scheme 2.24 Enantioselective remote C–H cyanation of amines via copper-catalyzed radical relay.

active Cu(I) catalyst. The methodology showed excellent functional group compatibility around the aromatic rings and could also be applied to heteroaromatic substrates. This Cu(I)/(II)/(III) manifold was also extended to the functionalization of nonbenzylic positions. However, these reactions generally gave lower yields owing to the increased flexibility of the chain when compared to the substrates with an embedded aromatic, which have a rigid and defined conformation.

Nagib and coworkers have recently provided a solution to the asymmetric synthesis of 3-aryl substituted piperidines **88** using an interrupted HLF reaction (Scheme 2.24) [62]. By employing a Cu(I) catalyst in the presence of a chiral Box ligand, the authors achieved a rare example of enantioselective remote cyanation of *N*-fluoroamides **89** in high yields and excellent enantioselectivities. These cyanated products **90** were then converted in just two steps into the corresponding piperidines. The authors proposed a mechanism where an initial transmetalation between the Cu(I) catalyst and Me_3SiCN provided a Cu(I) intermediate capable of reducing the N—F bond in the amide starting materials. This step gave the amidyl radical **91** that underwent N-to-C radical relay via 1,5-HAT. The incipient carbon radical **92** was intercepted by the Cu(II)CN species to give an alkyl–Cu(III)CN **93** intermediate, from which a stereoselective reductive elimination provided the product and regenerated the Cu(I) catalyst. Although this chemistry is only compatible with the functionalization of benzylic positions, the authors applied it to the preparation of a range of cyanated substrates, all synthesized in good yields and enantioselectivities. Next, they subjected a number of products to a two-step reduction with *i*-Bu_2AlH (to form a hemiaminal), followed by Et_3SiH to furnish the desired C-3 arylated piperidine products.

2.5 Summary and Conclusions

Nitrogen radicals are powerful intermediates in synthetic chemistry. Their ability to undergo selective H-atom transfer reactions represents a valuable tool for the assembly and functionalization of organic molecules. Although the large body of work and examples are already available, several challenges are still unsolved. For example, the ability to control in a general sense the stereochemical outcome of these transformations is currently not possible but would be highly desirable. Furthermore, these processes have rarely been used in large-scale settings, especially at an industrial level, so further work would be required to identify reaction protocols that can be translated into process development. Finally, an area where continuous development and application is required involves the use of these strategies for the late-stage modification of complex and bioactive materials. The ability to selectively target specific and unactivated sp^3-centers, where a functionality can be introduced, would represent a powerful tool in order to better explore chemical space around lead molecules.

References

1 Ricci, A. (2008). *Amino Group Chemistry*. Weinheim, Germany: Wiley-VCH.
2 Zard, S.Z. (2008). *Chemical Society Reviews* 37: 1603–1618.
3 Stateman, L.M., Nakafuku, K.M., and Nagib, D.A. (2018). *Synthesis* 50: 1569.
4 Chen, J.-R., Hu, X.-Q., Lu, L.Q., and Xiao, W.-J. (2016). *Chemical Society Reviews*: 2044–2056.
5 Stella, L. (1983). *Angewandte Chemie, International Edition* 22: 337.
6 Wolff, M.E. (1963). *Chemical Reviews* 63: 55.
7 Hofmann, A.W. (1879). *Berichte der Deutschen Chemischen Gesellschaft* 12: 984–990.
8 Nechab, M., Mondal, S., and Bertrand, M.P. (2014). *Chemistry - A European Journal* 20: 16034.
9 Katohgi, M., Togo, H., Yamaguchi, K., and Yokoyama, M. (1999). *Tetrahedron* 55: 14885–14900.
10 Luo, Y.-R. (2003). *Handbook of Bond Dissociation Energies in Organic Compounds*. CRC Press.
11 Walton, J.C. (2014). *Accounts of Chemical Research* 47: 1406.
12 Lessard, J., Griller, D., and Ingold, K.U. (1980). *Journal of the American Chemical Society* 102: 3262.
13 Chow, Y.L., Danen, W.C., Nelsen, S.F., and Rosenblatt, D.H. (1978). *Chemical Reviews* 78: 243.
14 Angelini, L., Davies, J., Simonetti, M. et al. (2019). *Angewandte Chemie, International Edition* 58: 5003.
15 Bietti, M. (2018). *Angewandte Chemie, International Edition* 57: 16618–16637.
16 Hioe, J., Šakić, D., Vrček, V., and Zipse, H. (2015). *Organic & Biomolecular Chemistry* 13: 157.
17 Forrester, A.R., Fullerton, J.D., and McConnachie, G. (1983). *Journal of the Chemical Society, Perkin Transactions 1*: 1759.

18 Forrester, A.R., Napier, R.J., and Thomson, R.H. (1981). *Journal of the Chemical Society, Perkin Transactions 1*: 984.

19 Forrester, A.R., Gill, M., Sadd, J.S., and Thomson, R.H. (1979). *Journal of the Chemical Society, Perkin Transactions 1*: 612.

20 Forrester, A.R., Gill, M., Sadd, J.S., and Thomson, R.H. (1975). *Journal of the Chemical Society, Chemical Communications*: 291.

21 Narasaka, K. and Kitamura, M. (2005). *European Journal of Organic Chemistry*: 4505.

22 Davies, J., Booth, S.G., Essafi, S. et al. (2015). *Angewandte Chemie, International Edition* 54: 14017.

23 Davies, J., Svejstrup, T.D., Reina, D.F. et al. (2016). *Journal of the American Chemical Society* 138: 8092.

24 Davies, J., Morcillo, S.P., Douglas, J.J., and Leonori, D. (2018). *Chemistry - A European Journal* 24: 12154.

25 Li, J., Zhang, P., Jiang, M. et al. (2017). *Organic Letters* 19: 1994.

26 Li, Y., Mao, R., and Wu, J. (2017). *Organic Letters* 19: 4472.

27 Shu, W. and Nevado, C. (2017). *Angewandte Chemie, International Edition* 56: 1881.

28 Jiang, H., An, X., Tong, K. et al. (2015). *Angewandte Chemie, International Edition* 54: 4055.

29 Shen, X., Zhao, J.-J., and Yu, S. (2018). *Organic Letters* 20: 5523.

30 Chen, L., Guo, L.-N., Ma, Z.-Y. et al. (2019). *The Journal of Organic Chemistry* 84: 6475–6482.

31 Wu, K., Wang, L., Colón-Rodríguez, S. et al. (2019). *Angewandte Chemie, International Edition* 58: 1774–1778.

32 Cismenia, M.A. and Yoon, T.P. (2015). *Chemical Science* 6: 5426.

33 Duncton, M.A.J. (2011). *Medicinal Chemistry Communications* 2: 1135.

34 Chen, H., Fan, W., Yuan, X.-A., and Yu, S. (2019). *Nature Communications* 10: 4743.

35 Quin, Q. and Yu, S. (2015). *Organic Letters* 17: 1894.

36 Jiang, H. and Studer, A. (2018). *Angewandte Chemie, International Edition* 57: 1692.

37 Morcillo, S.P., Dauncey, E.M., Kim, J.H. et al. (2018). *Angewandte Chemie, International Edition* 57: 12945.

38 Davies, J., Sheikh, N.S., and Leonori, D. (2017). *Angewandte Chemie, International Edition* 56: 13361.

39 Miller, D.C., Choi, G.J., Orbe, H.S., and Knowles, R.R. (2015). *Journal of the American Chemical Society* 137: 13492.

40 Choi, G.J., Zhu, Q., Miller, D.C. et al. (2016). *Nature* 539: 268–271.

41 Chu, J.C.K. and Rovis, T. (2016). *Nature* 539: 272–275.

42 Xu, B. and Tambar, U.K. (2019). *ACS Catalysis* 9: 4627–4631.

43 Thullen, S.M., Treacy, M.S., and Rovis, T. (2019). *Journal of the American Chemical Society* 141: 14062–14067.

44 Zhang, H. and Muniz, K. (2017). *ACS Catalysis* 7: 4122.

45 Becker, P., Duhamel, T., Stein, C.J. et al. (2017). *Angewandte Chemie, International Edition* 56: 8004.

46 Martinez, C. and Muniz, K. (2015). *Angewandte Chemie, International Edition* 54: 8287.

47 Bergès, J., García, B., and Muñiz, K. (2018). *Angewandte Chemie, International Edition* 57: 15891–15895.

48 Wappes, E.A., Nakafuku, K.M., and Nagib, D.A. (2017). *Journal of the American Chemical Society* 139: 10204.

49 Nakafuku, K.M., Twumasi, R.K., Vanitcha, A. et al. (2019). *The Journal of Organic Chemistry* 84: 13065–13072.

50 Gomez-Suarez, A., Gasperini, D., Vummaleti, S.V.C. et al. (2014). *ACS Catalysis* 4: 2701.

51 Hernandez, R., Rivera, A., Salazar, J.A., and Suarez, E. (1980). *Journal of the Chemical Society, Chemical Communications*: 958.

52 Francisco, C.G., Herrera, A.J., and Suarez, E. (2003). *The Journal of Organic Chemistry* 68: 1012.

53 Francisco, C.G., Herrera, A.J., Martin, A. et al. (2007). *Tetrahedron* 48: 6384.

54 Martin, A., Perez-Mertin, I., and Suarez, E. (2009). *Tetrahedron* 65: 6147.

55 Tang, N., Wu, X., and Zhu, C. (2019). *Chem. Sci.* 10: 6915–6919.

56 Na, C.G. and Alexanian, E.J. (2018). *Angewandte Chemie, International Edition* 57: 13106–13109.

57 Short, M.A., Blackburn, J.M., and Roizen, J.L. (2017). *Angewandte Chemie, International Edition* 57: 296.

58 Ayer, S.K. and Roizen, J.L. (2019). *The Journal of Organic Chemistry* 84: 3508–3523.

59 Torres-Ochoa, R.O., Leclair, A., Wang, Q., and Zhu, J. (2019). *Chemistry - A European Journal* 25: 9477–9484.

60 Groendyke, B.J., AbuSalim, D.I., and Cook, S.P. (2016). *Journal of the American Chemical Society* 138: 12771.

61 Li, Z., Wang, Q., and Zhu, J. (2018). *Angewandte Chemie, International Edition* 57: 13288.

62 Zhang, Z., Zhang, X., and Nagib, D.A. (2019). *Chem* 5: 3127.

3

Radical-Based C–N Bond Formation in Photo/Electrochemistry

Binbin Huang[1], Yating Zhao[2] and Wujiong Xia[1]

[1] *Harbin Institute of Technology, State Key Lab of Urban Water Resource and Environment, School of Science, Shenzhen 518055, China*
[2] *Quzhou University, College of Chemical and Material Engineering, Quzhou 324000, China*

3.1 Introduction

As ubiquitous fundamental linkages, C—N bonds prevalently exist in various value-added compounds such as natural products, pharmaceutical agents, functional materials, synthetic intermediates, and coordinating ligands. Thus, developing efficient methodologies for the construction of C—N bonds has always been a hot research goal in synthetic chemistry [1]. Among the existing approaches toward C—N bonds, cross-dehydrohalogenative couplings of C—H/N—X (X = halide or *pseudo*-halide) bonds or C—X/N—H bonds are generally well established, highly efficient, and consequently widely used, including the classical Buchwald–Hartwig amination and Ullman amination. However, from the perspective of green chemistry, these methods are not the ideal choices because the requirement for prefunctionalization of the coupling substrates considerably lowers their atom and step economy.

During the past few decades, cross-dehydrogenative coupling (CDC) has emerged as an attractive technique, in which preinstallation of a reactive group in either of the coupling precursors is avoided, thus featuring higher atom and step economy in synthetic chemistry. Accordingly, CDC amination, which furnishes C—N bonds directly from unfunctionalized C—H bonds and N—H bonds, is acknowledged as one of the most straightforward approaches for C—N bond construction. In this field, transition-metal-mediated CDC aminations have been more extensively investigated, as summarized in the reviews by Patureau and coworker [2] and Chang and coworker [3]. Generally, these protocols depend on the interaction of the metal catalysts with the preinstalled directing groups on the substrates to facilitate the following amination. However, both the installation and the removal of such groups are tedious work that makes these strategies practically compromised. While, in other C–H aminations without chelation, the substrate scopes of carbon moieties are mostly limited to certain acidic C—H bonds that are more prone to activation. Nevertheless, these limitations have been overcome, wherein the direct conversion of unfunctionalized N—H and C—H bonds into C—N bonds have been achieved via reactive metal-nitrenoid species [4]. Apart from metal chemistry, metal-free direct oxidative C–H aminations have also been

Methodologies in Amine Synthesis: Challenges and Applications, First Edition.
Edited by Alfredo Ricci and Luca Bernardi.

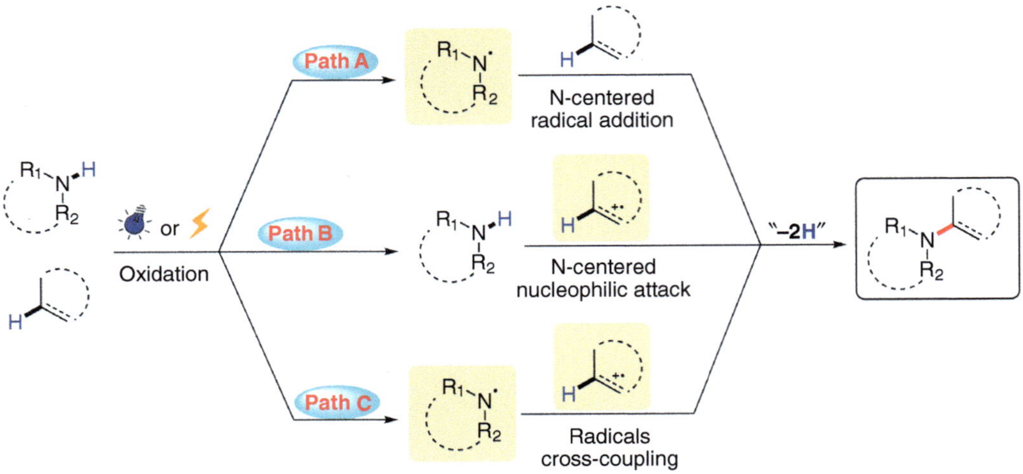

Scheme 3.1 C–N bond formation applying N–H bonds via photo/electrochemical methods.

realized by the groups of Antonchick, Deboef, Chang, etc., via the mediation of hypervalent iodine reagents [5].

Radical chemistry is reviving in organic synthesis owing to its inherently advantageous properties and has provided various alternative approaches for C—N bond formation. As is documented in one review on nitrogen-centered radical amination, apart from some metal-coordinated aminations, considerable progress has been advanced on photocatalytic strategies in the past years [6]. Moreover, remarkable achievements in direct electrochemical amination have also been witnessed more recently. On the basis of the active radical intermediates involved, pathways of oxidative radical aminations can be classified into three major patterns: N-radical addition (Scheme 3.1, path A), N-atom nucleophilic addition (Scheme 3.1, path B), and radical cross-coupling (Scheme 3.1, path C), which have enriched the methodology of C—N bond formation and broadened the substrate scope to some extent.

Accordingly, this chapter will highlight the recent advances in the photo/electrochemically mediated, radical-involved cross-coupling reactions of unfunctionalized N—H bonds with diverse C—H bonds, among which several cyclization or hydroamination examples with certain unsaturated carbon-centered moieties are also included. To stimulate the design and development of more versatile and practical amination protocols, extra emphasis is put on the reaction mechanisms. The key radical intermediates involved in these transformations are highlighted in yellow boxes.

3.2 C–N Bond Formation via N-radical Species Addition

The most familiar pathway for C—N bond formation is via the addition of in situ generated N-radical species to unsaturated carbon-centered moieties. Visible light photoredox catalysis has provided a large array of methodologies for C—N bond construction, and in major cases, the active N-radical species are produced by the reductive cleavage of N—X (X = N,

O, S, Cl, and Br) bonds via single-electron transfer (SET) processes, which have already been comprehensively summarized by Xiong and Zhang [6], Xiao and coworkers [7], and Kärkäs [8]. However, unlike the reductive cleavage of weak N—X bonds, direct oxidative scission of strong N—H bonds to form N-radical species is far more challenging because of their higher bond dissociation energies, which usually requires harsh conditions or strong stoichiometric oxidants.

3.2.1 Radical Addition to C–C Double/Triple Bonds

3.2.1.1 Amidyl Radical Addition

In the successful examples directly employing N—H bonds as the radical precursors, the N—H bonds of most aminating reagents possess relatively low pK_a values to facilitate their effective oxidation. Besides, electron-deficient functional groups such as acyl groups and sulfonyl groups (–Ts) adjacent to the N-atom can also significantly stabilize the generated N-radical species. Thus, amidyl or sulfonamidyl radicals often appear as the key intermediates in the N-radical-based C—N bond formation.

During the past few years, proton-coupled electron transfer (PCET) strategy has been adopted in synthetic chemistry by Knowles' group, particularly for the homolytic activation of strong N—H bonds to form N-radical species that could further engage in various C—N bonds constructing transformations. Different from the straightforward hydrogen atom transfer (HAT) process, in a typical PCET-enabled amination, the most crucial step is the formation of a discrete hydrogen bond complex between the N—H bond of the substrate and the catalytic Brønsted base to modulate the redox potential, which enables the following concerted exchange step of an electron and a proton to access the N-radical species.

In 2015, employing alkene **1** with a remote amidyl substituent as the substrate, Knowles and coworkers sequentially developed PCET-based intramolecular carbo- and hydroamination using a combination of an iridium complex photocatalyst and a catalytic Brønsted base (Scheme 3.2) [9]. As is proposed by the authors, the N—H bond of **1** first interacts with the phosphate base, forming a hydrogen bond to lower the requirement of redox potential for the following electron transfer process. The concerted process of a one-electron oxidation by photoexcited $Ir^{III}*$ species and a proton abstraction by the base provides the key amidyl radical **2**, which then cyclizes onto the attached C=C bonds to form a new C—N bond with an adjacent C-centered radical (**3**). Subsequently, the nucleophilic radical **3** can further add onto an electrophilic olefin acceptor to deliver the carboamination product **4** after a tandem electron/proton transfer process. On the other hand, in the presence of catalytic thiophenol, radical **3** can also undergo a HAT process with the hydrogen donor to furnish the hydroamination product **5**, while the resulting $PhS^·$ is then engaged in a SET process with the Ir^{II} species to afford PhS^- and regenerate ground-state photocatalyst Ir^{III}. The following protonation of PhS^- by the conjugate acid of the phosphate base indicates the closure of both catalytic cycles.

Later in 2016, Knowles' group employed the same strategy in the selective alkylation of remote $C(sp^3)$—H bonds, which engages a 1,5-HAT process with the aid of the amidyl radicals produced via PCET [10]. Simultaneously, a similar work was reported by Rovis' group [11]. Therefore, PCET activation has been proven as a powerful reaction mode for direct

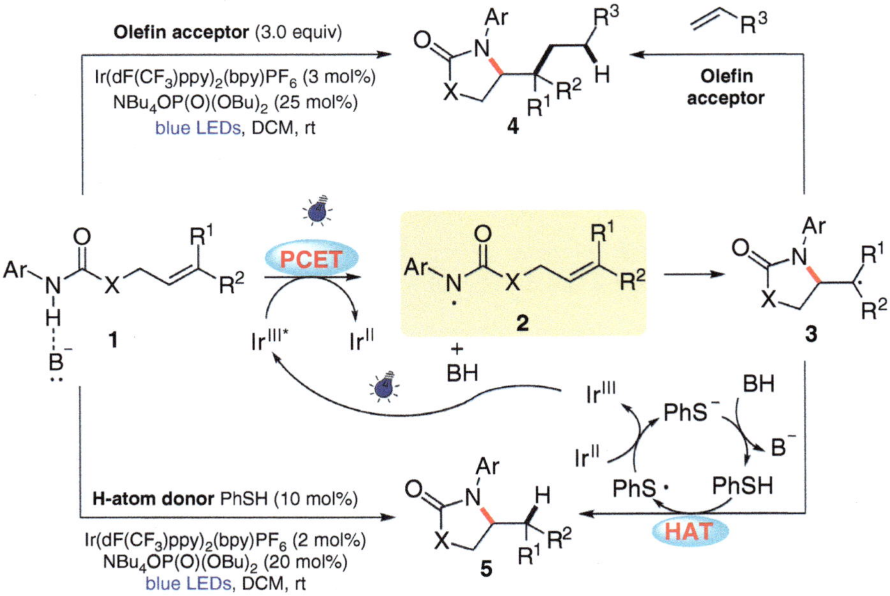

Scheme 3.2 PCET-mediated intramolecular carbo/hyrdroamination of alkenes. Source: Modified from Gentry and Knowles [9].

N-radical formation from strong N—H bonds. Furthermore, it is also a promising platform to broadly explore this strategy in the construction of diversified C—N bonds.

More recently, Knowles and coworkers reported a sulfonamidyl-based hydroamination strategy, again via PCET activation (Scheme 3.3) [12]. Hydroamination products **7** and **10** could be smoothly furnished from intramolecular cyclization of **6** and intermolecular reaction between **8** and **9**, respectively, employing 2,4,6-triisopropyl-thiophenol (TRIP thiol) as the HAT catalyst via the same pathway as depicted in Scheme 3.2. Apart from the broad substrate scope demonstrated, a series of tandem amination/C–H alkylation sequences were performed to highlight the synthetic versatility (Scheme 3.3b). The terminal alkene **11** was first subjected to the intermolecular anti-Markovnikov hydroamination with *p*-methoxyphenyl (PMP) sulfonamide **12** to afford the alkylated sulfonamide **13**, and the newly installed secondary sulfonamide was then activated in a second oxidative PCET event using the same Ir/phosphate pair, leading to the site-selective abstraction of the δ-C—H bond to afford a carbon-centered radical that can be further trapped by an electron-deficient olefin **14** to afford the final product **15**.

The reviving synthetic organic electrochemistry has been providing environmentally benign alternatives to the traditional synthetic methods because of its generally high atom economy and good functional group compatibility. In the electrochemical C—N bond formation from unfunctionalized N–H/C–H precursors, a series of advances have been achieved by Xu's group in the recent years. In 2016, Xu's group disclosed an electrocatalytic hydroamination of alkenes, in which an inexpensive organometallic reagent ferrocene was employed as a redox catalyst to enable the direct generation of amidyl radicals from *N*-aryl amides **16** (Scheme 3.4) [13]. Based on the optimization of the reaction conditions, a mixed

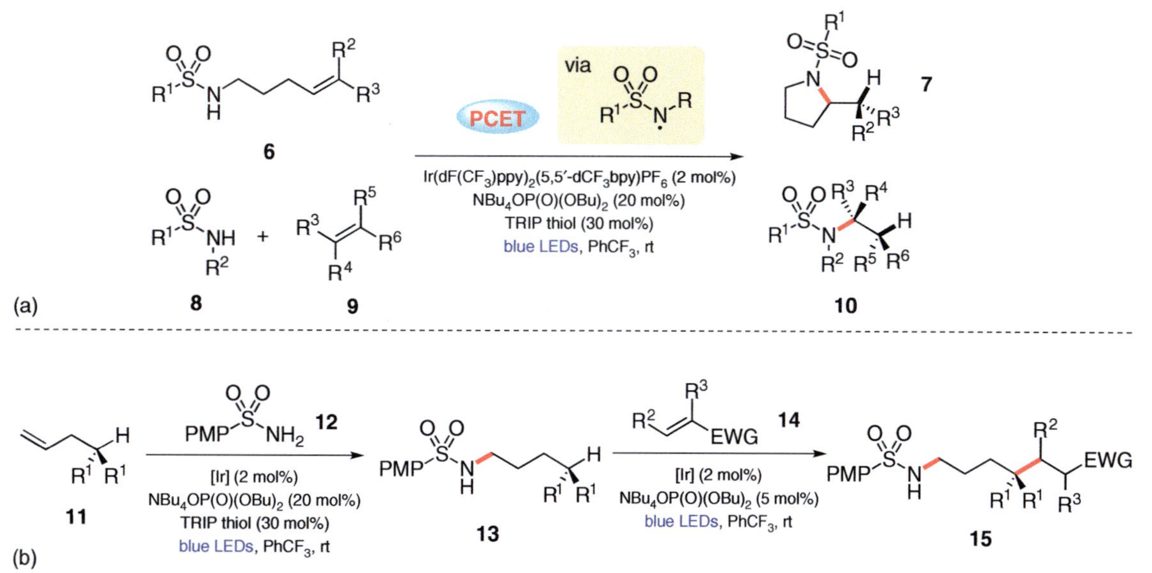

Scheme 3.3 PCET-mediated intra/intermolecular amination of alkenes. (a) Hydroamination of alkenes with primary or secondary sulfonamides. (b) Tandem amination/C–H alkylation. Source: Modified from Zhu et al. [12].

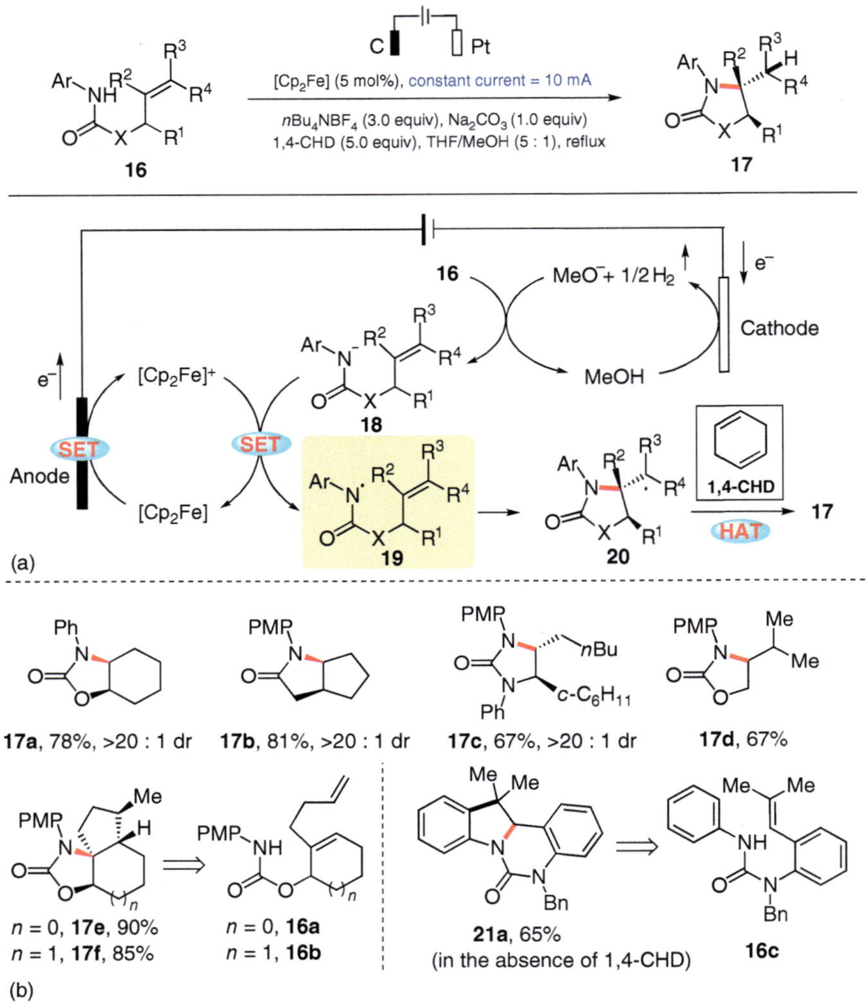

Scheme 3.4 Electrocatalytic intramolecular hydroamination of alkenes. Source: Modified from Zhu et al. [13].

solvent tetrahydrofuran (THF)/MeOH in 5 : 1 ratio is demonstrated as the optimal choice, while no reaction takes place in the single solvent MeOH, which is also supported by the cyclic voltammetry (CV) observations.

As depicted in their mechanistic proposal (Scheme 3.4a), this reaction is supposed to begin with the anodic oxidation of ferrocene ([Cp_2Fe]) and the simultaneous cathodic reduction of cosolvent methanol. Subsequently, the electrochemically generated base MeO^- deprotonates the amide group of substrate **16** to afford anionic intermediate **18**, which can be easily oxidized by [Cp_2Fe]$^+$ to provide the key amidyl radical **19**, along with the regenerated mediator [Cp_2Fe]. Radical **19** then cyclizes onto its tethered alkene to furnish intermediate **20**, which further acquires a hydrogen atom from the H-atom donor

1,4-cyclohexadiene (1,4-CHD) to yield the desired product **17**. The scope exploration reveals that various carbamates, ureas, and amides can serve as viable substrates, providing the corresponding products in good to high yields under the standard conditions (Scheme 3.4b, **17a–17d**). Notably, in the reactions of diene substrates **16a** and **16b**, tandem cyclization processes occurred to furnish polycyclic products **17e** and **17f** with high efficiency. Moreover, under standard conditions but in the absence of 1,4-CHD, substrate **16c** finally turns into indoline **21a** after an oxidative termination step.

Again, Xu and coworkers employed the same electrochemical N-radical formation strategy in the synthesis of highly functionalized (aza)indoles **23** through intramolecular N-radical species addition of **22** to their remotely attached alkynyl moieties (Scheme 3.5)

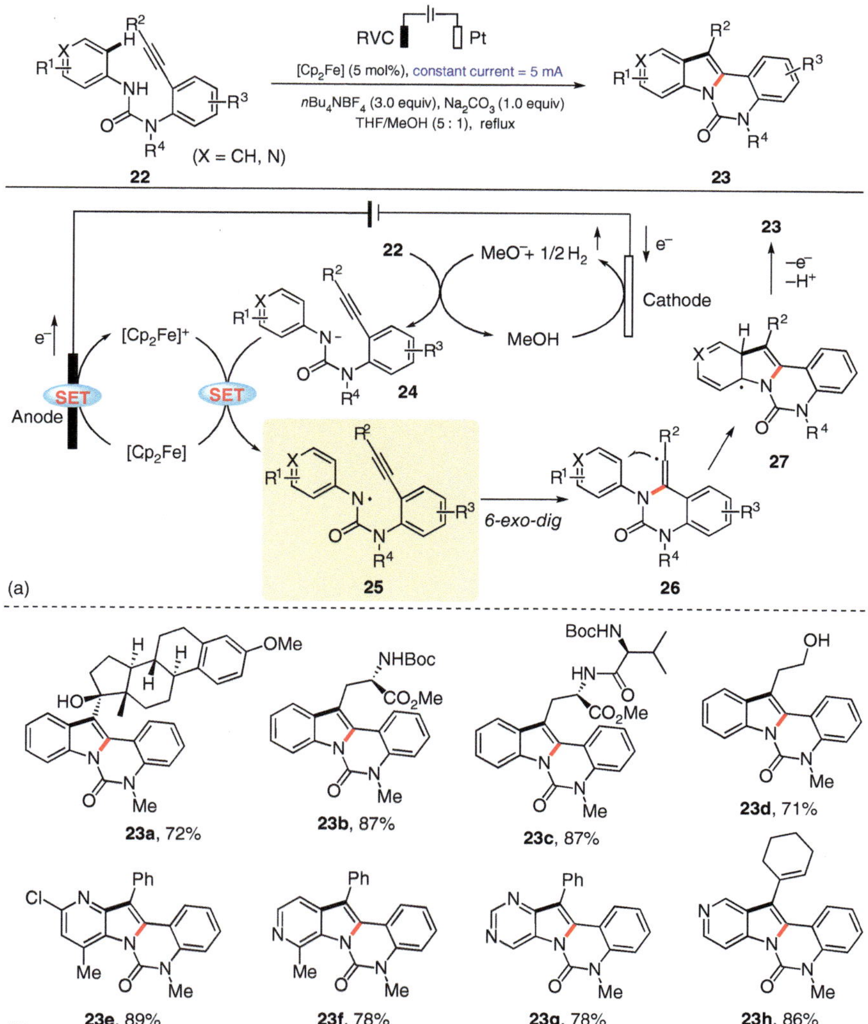

Scheme 3.5 Electrochemical synthesis of multifunctionalized (aza)indoles. RVC, reticulated vitreous carbon. Source: Modified from Hou et al. [14].

[14]. In this transformation, ferrocene is also selected as the redox mediator upon anodic oxidation, and the simultaneous cathodic reduction converts the cosolvent methanol into MeO^- base and H_2 gas. Subsequently, a SET process between the oxidized $[Cp_2Fe]^+$ and anion **24** from deprotonation of **22** produces an electron-deficient, N-centered radical **25** and meanwhile regenerates $[Cp_2Fe]$. Radical **25** would then preferentially undergo a 6-exo-dig cyclization to vinyl radical **26**, followed by a second cyclization to give the delocalized radical **27**, as also supported by the density functional theory (DFT) calculations. Finally, rearomatization of **27** after an oxidation/deprotonation sequence delivers the final product **23**. A broad substrate scope is demonstrated under the standard conditions, exhibiting high functional group tolerance (Scheme 3.5b). Notably, the late-stage modification of ethinyl estradiol proceeds smoothly to deliver the indole-functionalized estradiol **23a**. The acid/base-sensitive chiral amino esters (**23b**, **23c**) as well as a free alcohol (**23d**) are also well tolerated under the electrolysis.

In 2017, Xu and coworkers adopted substrate **28** with a polysubstituted alkene moiety to achieve an electrochemical aza-Wacker-type cyclization, wherein the key amidyl radical **29** is generated by direct anodic oxidation of the amidyl N—H bond without the aid of a base (Scheme 3.6) [15]. The following intramolecular cycloaddition of radical **29** to its pendant alkenyl group furnishes C-centered radical **30**, which is more prone to suffer further oxidation rather than H-atom abstraction, to deliver the corresponding cation **31**. The desired product **32** is finally generated from cation **31** after deprotonation. This

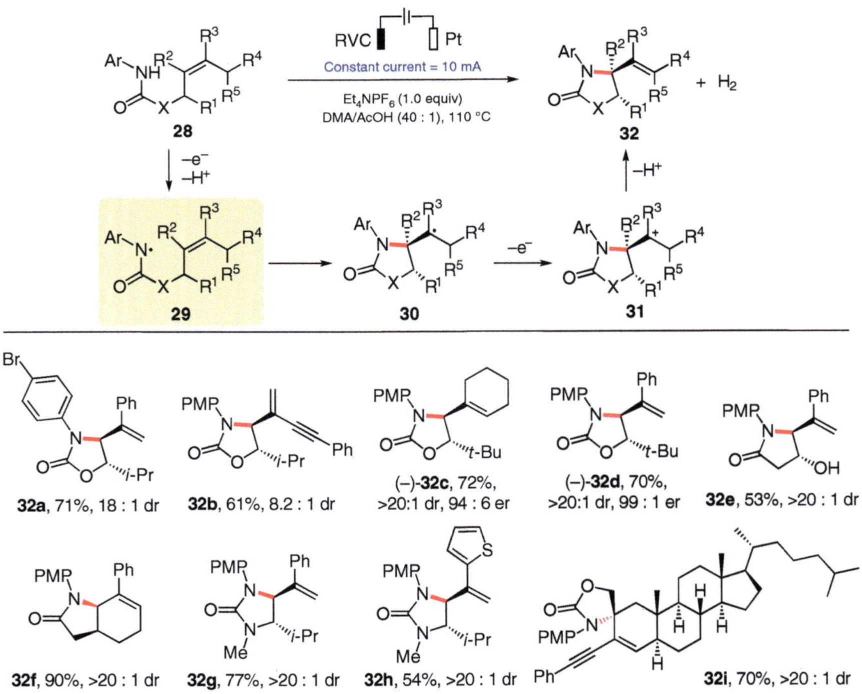

Scheme 3.6 Electrochemical intramolecular oxidative amination of tri- and tetra-substituted alkenes. Source: Modified from Xiong et al. [15].

electrochemical protocol achieves the challenging intramolecular oxidative amination of sterically demanding alkenes in the absence of a metal catalyst and is proven compatible with a wide range of carbamates (**32a–32d**), amides (**32e**, **32f**), and ureas (**32g**, **32h**). Moreover, product **32i** with a steroid-based core is smoothly afforded in 70% yield from a tetrasubstituted alkene under electrolysis, demonstrating the extra synthetic potential of this protocol.

A regiospecific electrochemical [3+2] annulation for the preparation of various imidazo-fused *N*-heteroaromatic compounds was more recently described by Xu's group (Scheme 3.7) [16]. Employing a novel tetra-arylhydrazine (**34**) as the catalyst in a mixed solvent of acetonitrile and water, this electrosynthesis smoothly converts substrate **33** into

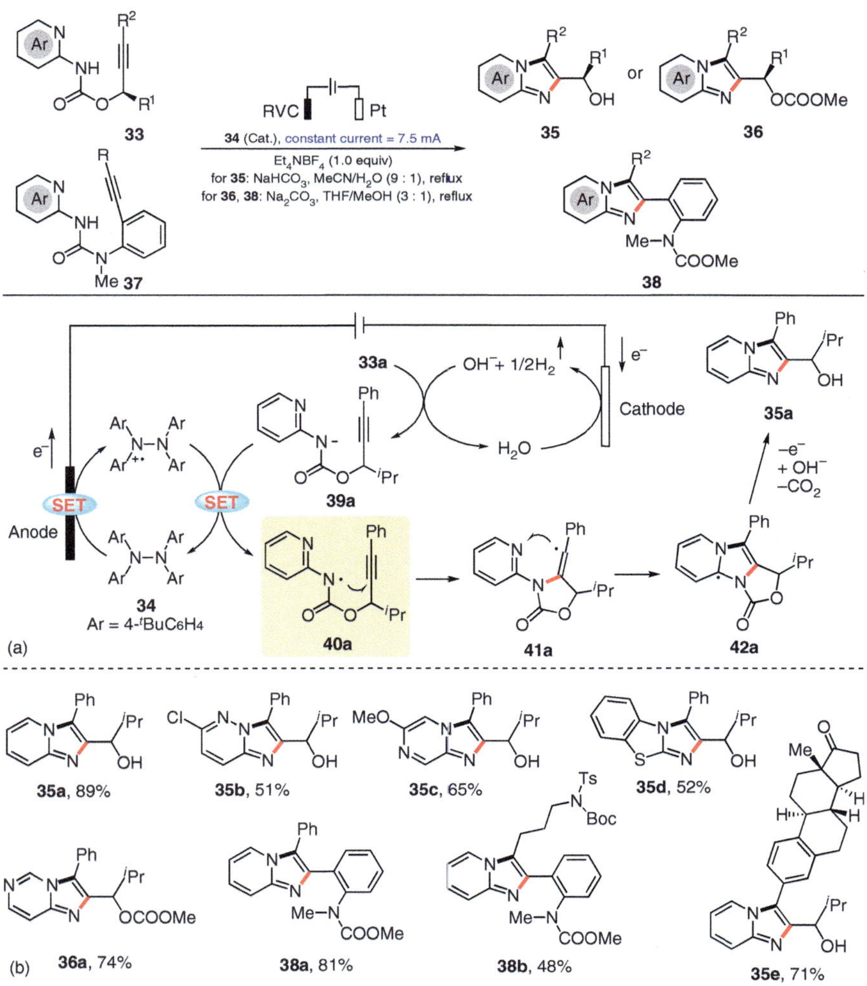

Scheme 3.7 Electrochemical synthesis of imidazo-fused N-heteroaromatic compounds. Source: Modified from Hou et al. [16].

35 through a cyclization/C—N bond cleavage sequence and the release of one molecule of CO_2. Conversely, another product **36**, wherein the carbonyl survived, could be obtained from certain substrates under another set of reaction conditions using THF/methanol as the solvent. Further attempts subjecting urea-linked **37** under the second conditions also resulted in the corresponding imidazopyridines **38** with methoxycarbonyl substituents installed. The mechanism of this transformation is quite similar to the aforementioned ones, and the DFT studies reveal that the C—N bond-forming pathway of **41a** is both thermodynamically and kinetically favored over the alternative C=C bond-forming pathway.

3.2.1.2 Hydrazonyl Radical Addition

Hydrazones are also competent N-radical precursors under basic conditions. Since 2014, β,γ-unsaturated hydrazone **43** has been chosen by Xiao's group as a substrate to study a series of visible-light-induced intramolecular cyclization reactions, for the synthesis of various dihydropyrazole or dihydropyridazine derivatives (Scheme 3.8a). Mechanistically, the acidic N—H bond of **43** can be readily deprotonated by a base such as NaOH or K_2CO_3, and the resulting anionic species **44** is prone to SET oxidation to generate the crucial radical intermediates **45**. Next, there are two R^4-dependent scenarios, 5-exo-trig or 6-endo-trig cyclization, for the further transformation of N-radical **45** to access the corresponding radical intermediate **46** or **47**, which can finally transform into diversified five- or six-membered products after variable processes such as HAT, radical addition, etc.

The first work in this series is an intramolecular hydroamination of **43** bearing terminal alkenes, which was developed in 2014 (Scheme 3.8b) [17]. The desired product **48** is produced from the five-membered intermediate **46** via a HAT process with the reaction solvent $CHCl_3$, which is demonstrated by a deuterium-labeling experiment using $CDCl_3$ as the solvent. Moreover, radical **46** can also be intercepted by scavengers such as 2,2,6,6-tetramethylpiperidine-1-oxyl (TEMPO) and diphenyltelluride (PhTeTePh) to form the corresponding adducts. In the second work published in 2016, the substituent R^4 of **43** is an aromatic group, instead of a hydrogen atom, giving access to 1,6-dihydropyridazines **49** under photoredox conditions, which is facilitated by a TEMPO-mediated HAT process [18]. As calculated by DFT, the additional Ar-group significantly lowers the activation free energy of a 6-endo-trig cyclization pathway and simultaneously stabilizes the formed intermediate **47**. On the basis of these previous findings, Xiao and coworkers next set out to expand this N-radical formation strategy of diversely substituted β,γ-unsaturated hydrazones to other cascade transformations, including the synthesis of dihydropyrazole-fused benzosultams **50** through the addition of radical **46** to the aromatic ring of the pendant –Ts group [19], the synthesis of –OH-functionalized pyrazoline **51a** or pyridazine **51b** through trapping of oxygen by the radical intermediate **46** or **47** [20], and the synthesis of **52a** or **52b** via the addition of radical **46** or **47** to allyl sulfones [21].

In 2016, another hydrazonyl radical-based photoredox amination followed by a Smiles rearrangement was reported by Belmont's research group, wherein the N—H bond activation strategy is similar to that in Xiao's work (Scheme 3.9) [22]. Phthalazine derivatives **54** with varied substituents are obtained from alkynyl-substituted sulfonyl hydrazones **53** under basic conditions with the extrusion of SO_2 (Scheme 3.9a). In light of the mechanistic studies including CV and fluorescence quenching experiments, it is proposed that the photoexcited catalyst Ru^{II*} undergoes one-electron reduction with anion **55**, generated by

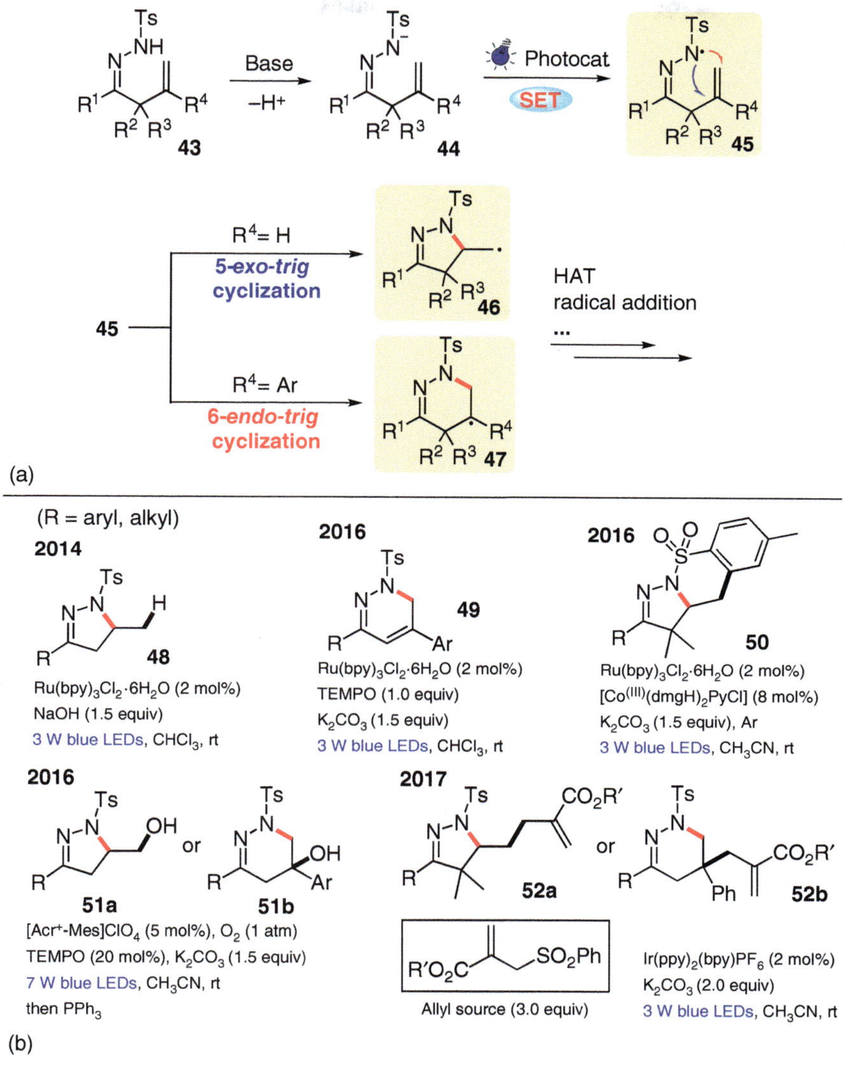

Scheme 3.8 Applying hydrazonyl radical for C–N bond formation in photochemistry. Source: Modified from Hu et al. [17].

deprotonation of substrate **53** by the base NaOH. Subsequently, the produced N-radical **56** cyclizes onto its tethered alkyne to form vinylic radical **57**, which is further engaged in a radical Smiles rearrangement to yield a delocalized radical species **58** after the loss of one molecule of SO_2. After acquiring an electron from Ru^I species to complete the photoredox catalytic cycle, radical **58** is reduced to anionic species **59**, which is then protonated by the protic solvent EtOH to deliver the final product **54** (Scheme 3.9b). Furthermore, the quantum yield of this reaction is measured as $\Phi = 0.33$, which excludes a radical chain propagation process during the transformation. Additionally, attempts to realize one-pot amination/Smiles rearrangement sequences by preparing **53** in situ from the corresponding

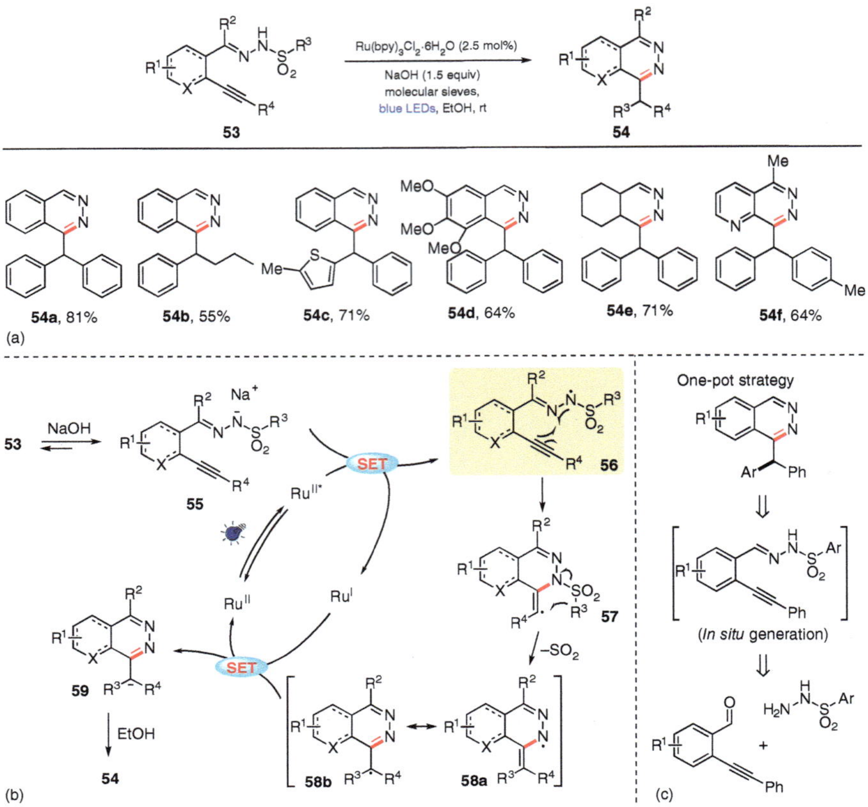

Scheme 3.9 Visible-light-enabled access to phthalazine derivatives via an amination/Smiles rearrangement cascade. Source: Modified from Brachet et al. [22].

aldehydes and sulfonyl hydrazides also succeeded, with the same efficiency compared with the original two-step protocol (Scheme 3.9c).

3.2.1.3 Aminium Radical Cation Addition

Electrophilic addition of aminium radical cations to unsaturated C=C bonds represents another attractive approach for the C—N bond construction. Generally, in visible-light-induced reactions, direct SET oxidation of the N—H bonds by the excited photocatalysts affords the crucial aminium radical cation intermediates, which are further converted into α-amino radicals after addition to unsaturated C=C bonds to allow the C—N bond formation. In this field, representative work has been demonstrated independently by the research groups of Zheng and Knowles.

In 2012, Zheng's group disclosed a visible-light-induced intramolecular C—N bond formation/cyclization reaction of styryl diarylamines **60** and **62** to prepare *N*-arylindoles **61** and **63** under the irradiation of a white light-emitting diode (LED) and ambient conditions (Scheme 3.10) [23]. Interestingly, a 1,2-carbon shift process is observed in the reaction of substrates **62** with their C2 position fully occupied, affording R^1-shifted indole products **63** in moderate yields, wherein aryl groups are preferentially transferred over alkyl

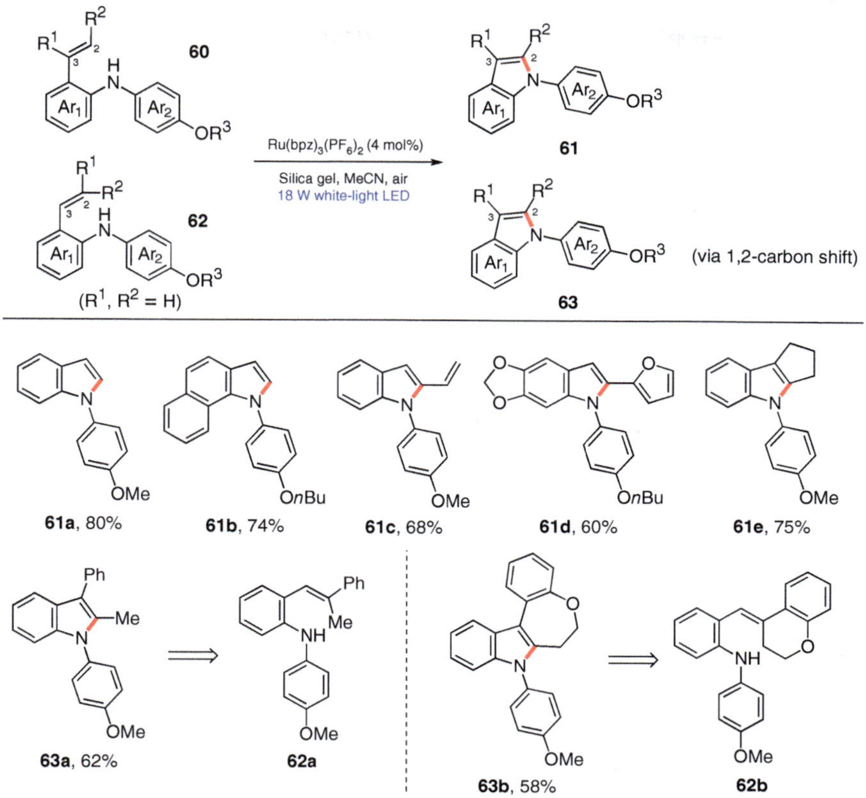

Scheme 3.10 Photocatalytic C–N bond formation for the preparation of *N*-arylindoles. Source: Modified from Maity and Zheng [23].

groups (**63a**). A reasonable reaction scope is presented, covering various substrates with electron-donating or electron-withdrawing substituents on Ar_1, as well as a wide range of substituents on C2 and C3 (**61a–61e**). Moreover, the R^1-migration process represents an elegant strategy for ring expansion (**63b**). Nevertheless, functional groups on Ar_2 are limited to *para*-alkoxy ones to facilitate their effective oxidation.

A plausible mechanism for the above photocatalytic transformation is proposed in Scheme 3.11. The reaction occurs with the oxidation of substrate **60** by the excited state photocatalyst Ru^{II*} to generate the key aminium radical cation **64**. The subsequent intramolecular electrophilic addition of **64** to its tethered C=C bonds affords intermediate **65**, which then turns into benzylic radical **66** after a proton abstraction. The oxidation into the corresponding benzylic cation **67** and the further aromatization of **67** upon deprotonation finally result in the desired *N*-arylindole **61**. As for substrate **62**, the reaction proceeds through the same pathway before reaching benzylic cation **68**. Then, a 1,2-carbon shift event follows, in which one of the substituents on C2 migrates onto C3, providing cationic intermediate **69**, which is further deprotonated to deliver the final product **63**.

Thereafter, the authors applied this intramolecular C—N bond formation strategy and the same privileged structure to another photoredox cyclization protocol, in which two

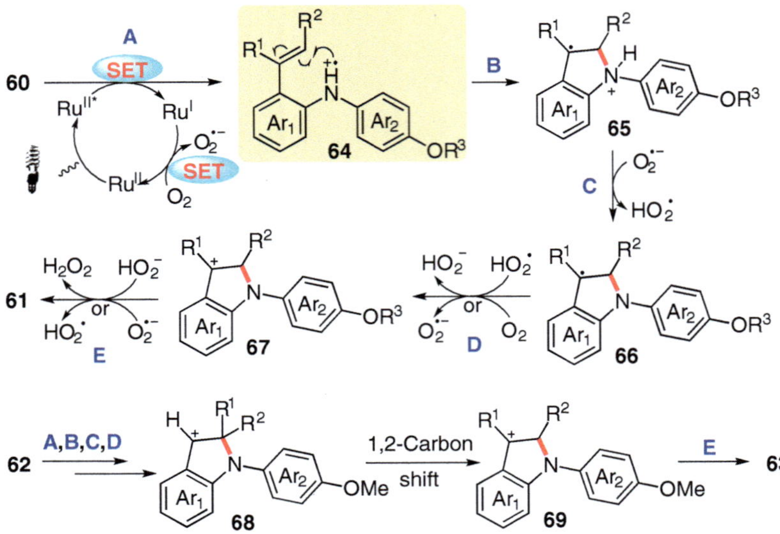

Scheme 3.11 Mechanistic proposal for the photocatalytic synthesis of *N*-arylindoles.

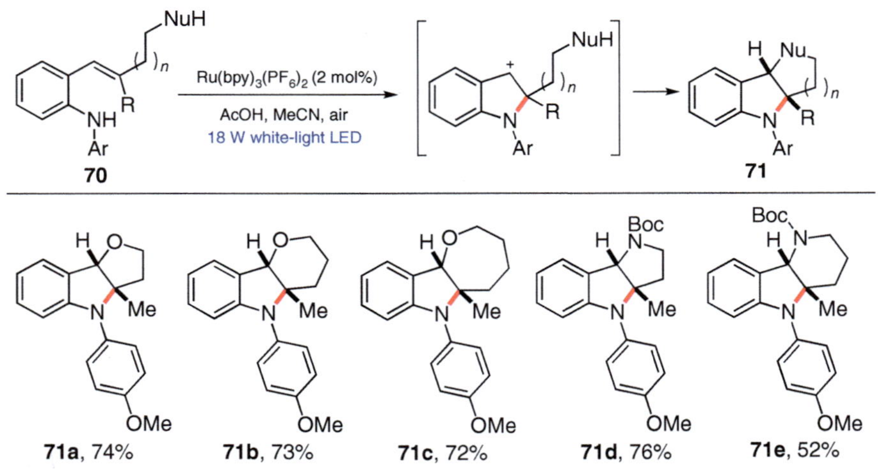

Scheme 3.12 Visible-light-initiated tandem reaction for the synthesis of fused *N*-arylindolines. Source: Modified from Morris et al. [24].

aliphatic rings are formed in one step to furnish fused indolines **71** (Scheme 3.12) [24]. Herein, remote nucleophilic functional groups such as –OH and –NHBoc are installed at the C2 position of the 2-styryl anilines **70** to attack the in situ generated benzylic carbocations, and the desired cyclization products **71** are successfully obtained under photocatalytic conditions similar to their previous work.

In 2014, an intramolecular anti-Markovnikov hydroamination protocol of aryl olefins was disclosed by Knowles' group, for the construction of structurally diverse *N*-aryl heterocycles **73** in a simple reaction system under mild photocatalytic conditions (Scheme 3.13) [25].

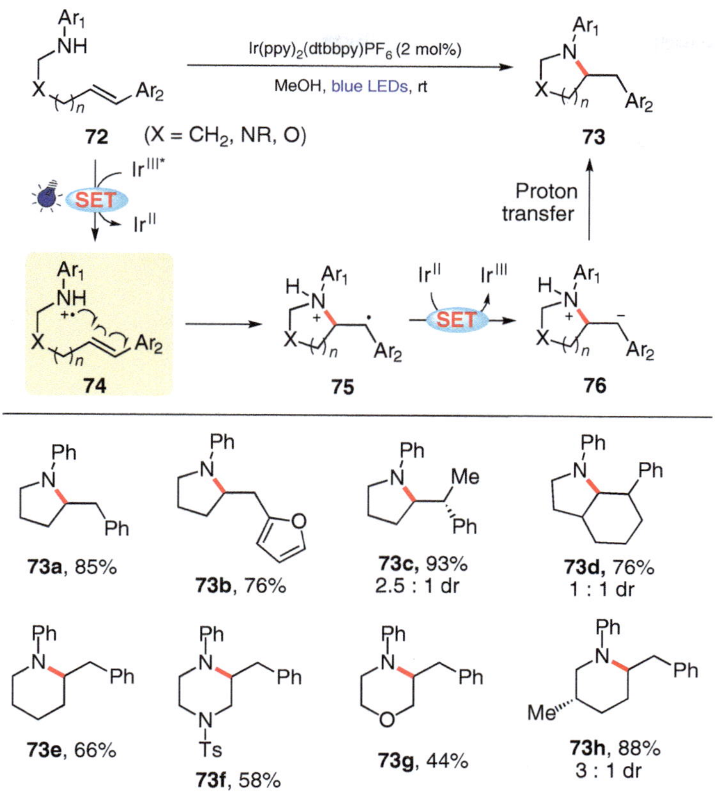

Scheme 3.13 Photocatalytic intramolecular hydroamination of olefins with aminium radical cations. Source: Modified from Musacchio et al. [25].

The reaction is proposed to begin with the direct oxidation of aniline **72** by the photoexcited catalyst Ir^{III*} to generate the crucial aminium radical cation **74**, as supported by both CV data and luminescence quenching experiments. The following sequence of the intramolecular addition to the tethered alkene of intermediate **74**, the one-electron reduction of benzylic radical **75** by Ir^{II}, and the proton transfer from alcoholic solvent to **76** delivers the final product **73**. A wide range of substrates bearing various substituents on both aromatic groups (Ar^1 and Ar^2) are proven compatible with this method, providing cyclization products in good to excellent yields, including five- or six-membered, heterocylic, and bicyclic ones (**73a–73h**). However, application of this protocol to intermolecular couplings turned out to be unsuccessful, which is likely due to the fact that the bimolecular C—N bond formation fails to outcompete the favorable electron back-transfer from the Ir^{II} complex to the aminium intermediate, as speculated by the authors.

With their further exploration into photocatalytic amination of alkenes, Knowles and coworkers next developed an intermolecular anti-Markovnikov hydroamination of unactivated alkenes **78** with secondary aliphatic amines **77**, which is not accessible using their previous protocol (Scheme 3.14) [26]. Various structurally diverse alkenes, including terminal and di/tri/tetra-substituted internal ones, smoothly undergo this redox-neutral

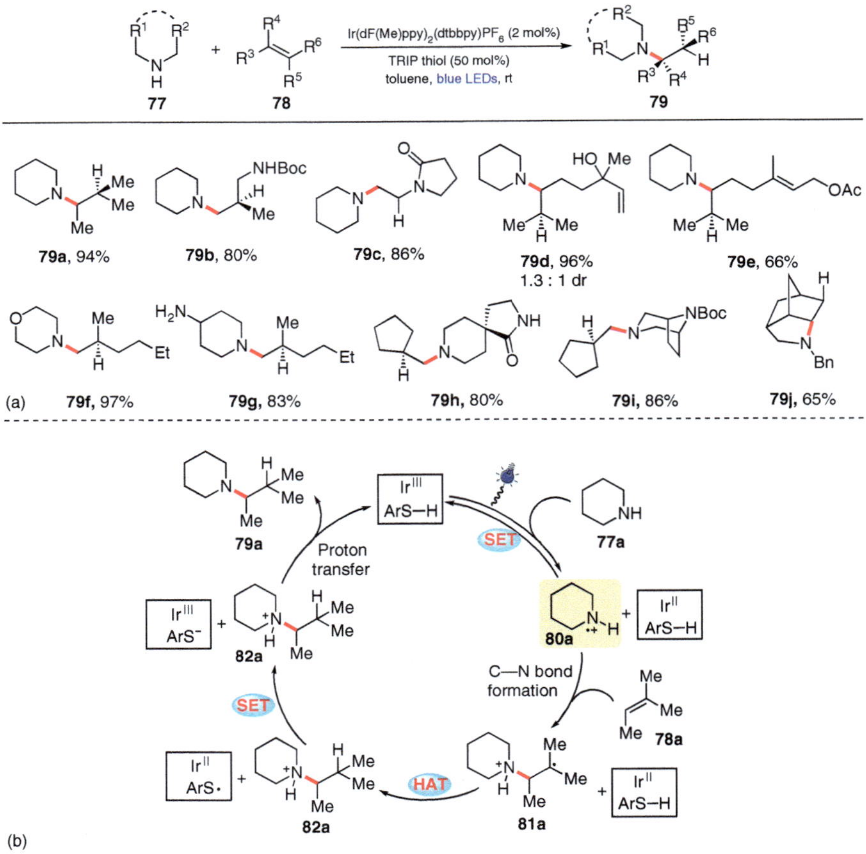

Scheme 3.14 Photocatalytic intermolecular hydroamination of unactivated olefins with secondary alkyl amines. Source: Modified from Musacchio et al. [26].

hydroamination with a wide range of secondary alkyl amines under facile conditions (**79a–79i**). Specifically, the aminium radical cation intermediate would preferentially add to the more electron-rich olefin when two electronically differentiated C=C bonds exist in one substrate (**79d**, **79e**). Notably, this direct hydroamination method can also be applied to the intramolecular C—N bond construction (**79j**). Nevertheless, aromatic amines, α-amino acids, and tetramethylpiperidine have been proven as inert amine partners in this reaction.

The mechanism proposed by the authors is depicted in Scheme 3.14b. As supported by Stern–Volmer experiments, piperidine **77a** can effectively quench the photoexcited $Ir^{III}*$ to produce the N-centered radical cation species **80a**, which then adds onto an alkene acceptor, e.g. **78a**, to form a new C—N bond along with an adjacent C-radical (**81a**). Differently from their previous amination protocol (Scheme 3.13), intermediate **81a** then abstracts a hydrogen atom from the TRIP thiol cocatalyst to deliver a closed-shell ammonium ion **82a** and a transient thiyl radical ($ArS^{\cdot}$). Subsequently, a SET process takes place between $ArS^{\cdot}$ and the reduced Ir^{II} species to provide ArS^- and the regenerated ground-state photocatalyst Ir^{III}. The final hydroamination product **79a** is furnished after the deprotonation of **82a** by anion ArS^-. According to the obtained redox potentials, the final products **79** could possibly be further oxidized by the excited $Ir^{III}*$, but high yields of **79** have still been obtained in these transformations without meaningful amounts of decomposition. The authors speculate that this outcome may result from the protective action of the thiol cocatalyst by reducing any α-amino radical that could be engaged in further deleterious side reactions.

Again in 2019, an intermolecular anti-Markovnikov hydroamination of unactivated olefins **84** was achieved by Knowles' group, using simple primary alkyl amines **83** under mild photoredox conditions (Scheme 3.15) [27]. An identical mechanism to the former one via an aminium radical cation intermediate **86** is proposed for this transformation employing a different iridium photocatalyst with divergent reactivity. Despite the presence of excess olefin, high selectivities, generally over 20 : 1, are observed for secondary over tertiary amine products.

In the typical visible-light-induced reactions described above, the photoredox catalysts harvest visible light to reach their excited states, which then undergo further single-electron redox events with reactants. Recently, a new photoactivation paradigm has emerged, wherein the transition-metal-based photocatalysts also directly engage in the bond formation/cleavage processes. In this area, a growing number of photoinduced C—N bonds forming methodologies using Cu-based unconventional photocatalysts have lately been developed by Fu, Hwang, Kobayashi, etc., and their progresses have been summarized by Gevorgyan in 2017 [28]. As an example, Hwang and coworkers reported a visible-light-induced, Cu-catalyzed reaction for the preparation of α-ketoamides from anilines **87** and alkynes **88** under ambient conditions (Scheme 3.16) [29]. Using oxygen as a green oxidant, this ligand- and base-free method features high atom economy and is compatible with a wide range of aniline and alkyne substrates. Notably, bioactive epoxide hydrolase inhibitors **89i** and **89j** are smoothly prepared within a single step from commercial starting materials by means of this strategy.

A plausible pathway is proposed on the basis of a series of detailed mechanistic studies (Scheme 3.17). Via a weak Cu^I–aniline complex formed by catalyst CuCl and aniline **87**,

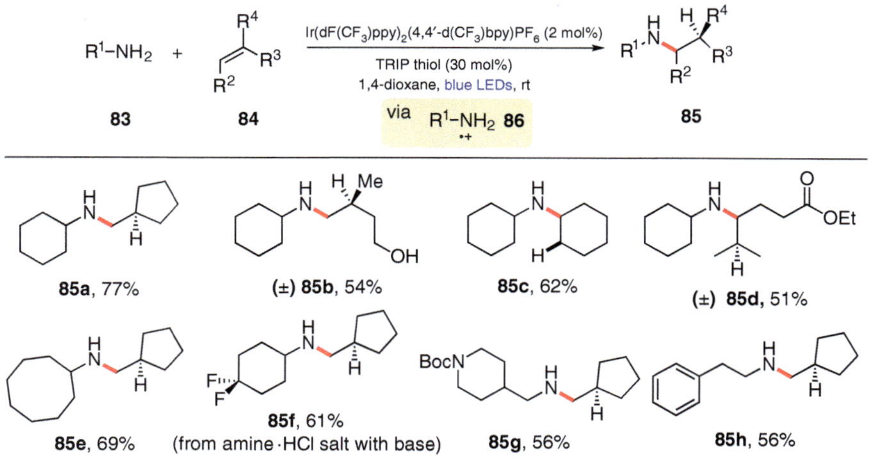

Scheme 3.15 Photocatalytic intermolecular anti-Markovnikov hydroamination of unactivated olefins with primary alkyl amines. Source: Miller et al. [27].

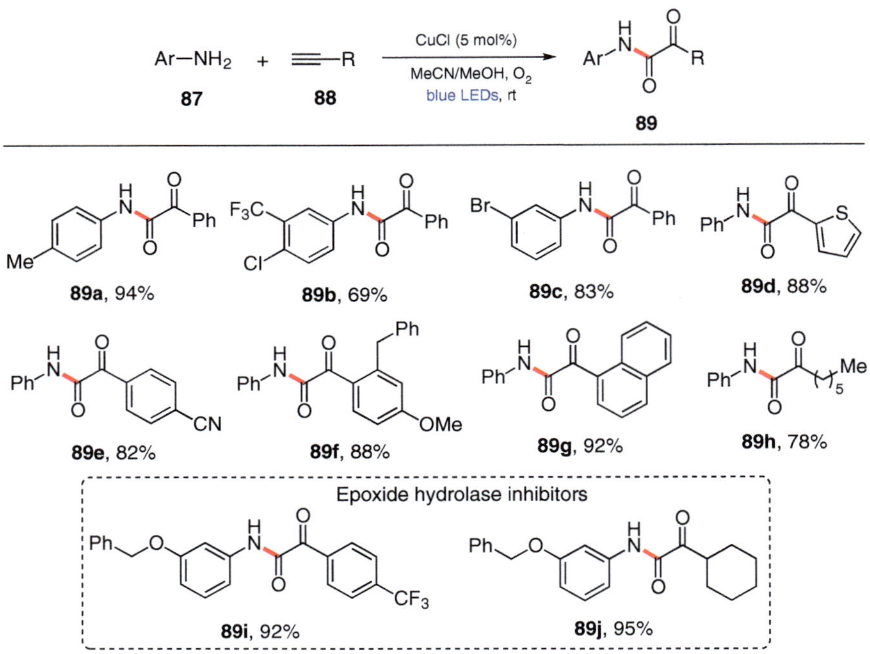

Scheme 3.16 Photoinduced, CuCl-catalyzed oxidative C–N coupling of anilines with terminal alkynes. Source: Modified from Sagadevan et al. [29].

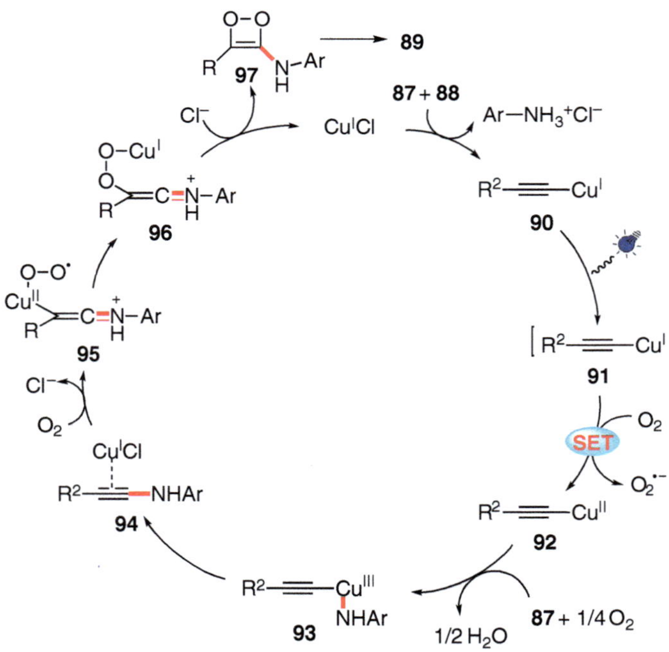

Scheme 3.17 Proposed mechanism for the photoinduced, CuCl-catalyzed C–N coupling of anilines with terminal alkynes.

Cu^I-phenylacetylide **90** is generated upon the addition of alkyne **88**, which then reaches its excited state **91** under blue light irradiation. The following SET oxidation of **91** by O_2 provides an electron-deficient Cu^{II}-phenylacetylide **92** as well as a superoxide. The following nucleophilic addition of aniline **87** to complex **92** results in the Cu^{III} species **93**, which next undergoes reductive elimination to furnish the highly reactive Cu^I-coordinated ynamine **94**. The intermediate **94** readily reacts with O_2 to generate the Cu^{II} peroxo-complex **95**, which then tautomerizes into the Cu^I species **96**. Finally, complex **96** is attacked by Cl^- to regenerate the catalyst CuCl, and the subsequent ring cleavage of the resulting intermediate **97** furnishes the final α-ketoamide product **89**.

3.2.2 Radical Species Addition to Aromatic Rings

In the early years, the direct oxidative aminations of unfunctionalized arenes were achieved using iodine-based reagents by the research groups of DeBoef, Chang, and Antonchick, wherein the employed amine partners were mainly sulfonamides and phthalimides. With the development of photoredox and electrochemical methodologies, more and more sustainable synthetic alternatives have emerged for the direct $C(sp^2)$—N bond formation of simple arenes via N-radical species addition pathways.

In 2016, Yu and Zhang and coworkers reported a visible-light-induced CDC amination of heteroarenes **98** with sulfonamides **99** using bleach (aqueous NaClO solution) as the oxidant (Scheme 3.18) [30]. Multiple heteroarenes such as indoles, pyrroles, and benzofurans

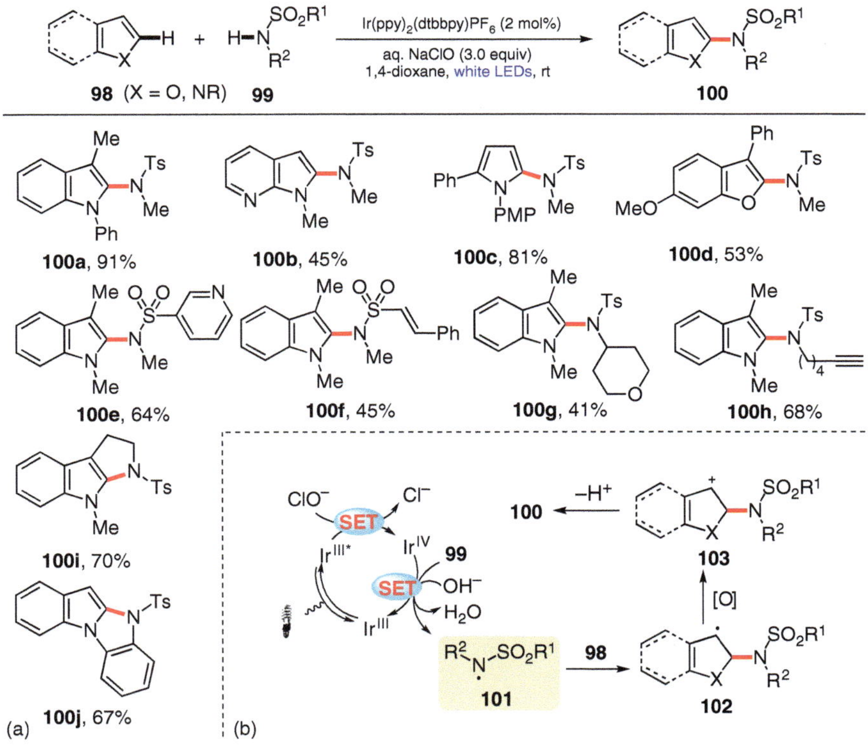

Scheme 3.18 Visible-light-induced direct C–H amination of heteroarenes with sulfonamides. Source: Modified from Tong et al. [30].

can undergo this transformation with *N*-methyl-*para*-toluene sulfonamide to afford their corresponding C2-amide products (**100a–d**). The scope of sulfonamides is also explored using 1,3-dimethyl-1*H*-indole as the heteroarene partner, with R^1 and R^2 being aromatic or aliphatic moieties (**100e–100h**). Furthermore, attempts to apply this protocol to intramolecular C2-amination/cyclization of indoles succeeded in building fused indole derivatives (**100i, 100j**). The proposed mechanism is depicted in Scheme 3.18b. After excitation of the photocatalyst Ir^{III} under visible light irradiation, the excited Ir^{III*} is then oxidized by NaClO to its higher valence state Ir^{IV}, as demonstrated by fluorescence quenching experiments. The resulting Ir^{IV} is capable of directly oxidizing sulfonamide **99** into N-centered radical **101**, meanwhile regenerating Ir^{III} to complete the photocatalytic cycle. The subsequent addition of N-radical **101** onto the C2 position of heteroarene **98** forms benzylic radical **102**, which undergoes further oxidation/deprotonation sequence to deliver the desired product **100** via cationic intermediate **103**.

In the following year, another CDC amination of heteroarenes **104** using phthalimide **105** as the N–H source was then disclosed by Itoh's group, employing 2-*tert*-butylanthraquinone (2-*t*-Bu-AQN) as the photocatalyst and aerobic oxygen as the external oxidant (Scheme 3.19) [31]. Multiple heteroarenes are smoothly aminated under the metal-free conditions, including diverse N-protected indoles and pyrroles as well as benzo[*b*]thiophene (**106a–106d**).

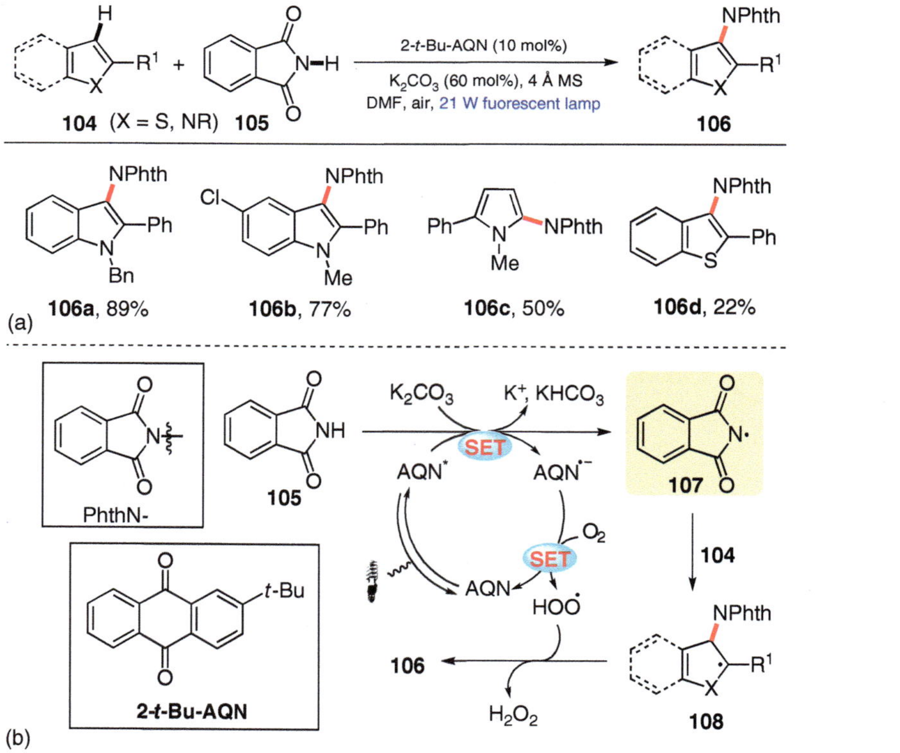

Scheme 3.19 Visible-light-mediated C–H amination of heteroarenes. Source: Modified from Yamaguchi et al. [31].

Based on several control experiments, the authors postulate that, after the deprotonation by K_2CO_3 and the single-electron oxidation by the photoexcited AQN*, phthalimide **105** is converted into the key N-centered radical **107**, and the resulting $AQN^{\cdot-}$ is then oxidized back to the original AQN by aerobic oxygen to complete the photocatalytic cycle. Subsequently, the desired amination product **106** is furnished from intermediate **107** after a radical addition/SET oxidation/aromatization sequence.

In the same year, Itami and coworkers disclosed a CDC amination of aromatic C(sp^2)–H enabled by visible light photoredox catalysis, using sulfonimides **110** as the N–H sources and a widely used Ru-complex as the photocatalyst (Scheme 3.20) [32]. The substrate scopes of both coupling partners are reasonable, covering a variety of arylsulfonimides bearing various substituents and different arenes and heterocycles (**111a–111f**). The mechanism proposed by the authors is presented in Scheme 3.20b. First, the hypervalent iodine oxidant 1-butoxy-1λ^3-benzo[*d*][1,2]iodaoxol-3(1*H*)-one (IBB) undergoes a SET reduction by the photoexcited catalyst Ru^{II}* to generate a radical intermediate **112** along with a Ru^{III} species. The subsequent single-electron oxidation of sulfonimide **110** by Ru^{III} yields the key imidyl radical **113**, which then adds onto arene **109** to afford radical intermediate **114**. Next, cationic intermediate **116** formed via a SET process between **112** and **114**, further aromatizes to afford the desired amination product **111**.

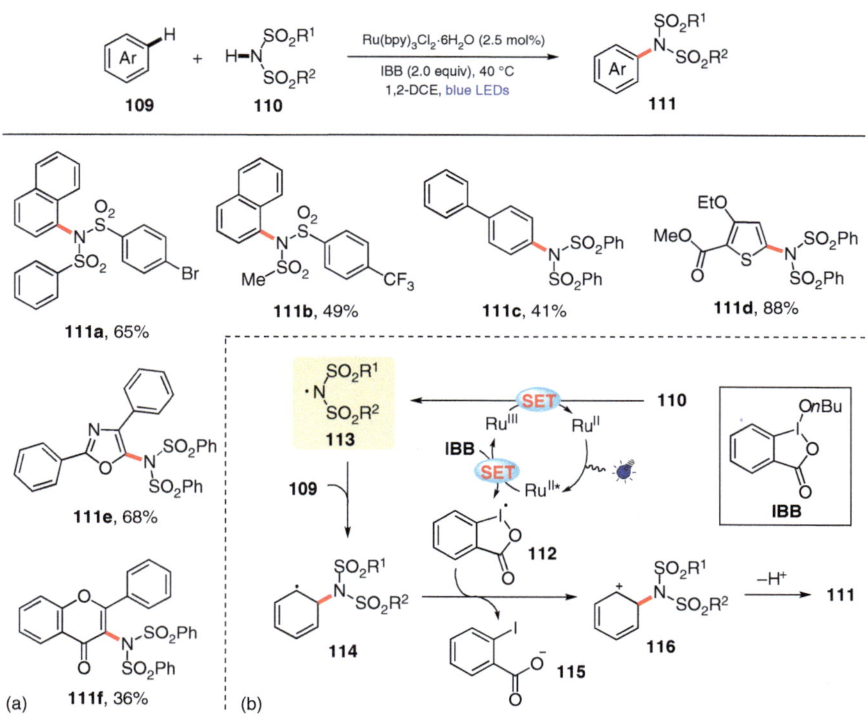

Scheme 3.20 Photocatalytic dehydrogenative C–H imidation of arenes with sulfonimides. Source: Modified from Ito et al. [32].

Remarkably, aliphatic amines have also been successfully applied as aminating reagents for the CDC amination of arenes. In 2017, Nicewicz's group reported a challenging intermolecular C—N bond formation between simple arenes **117** and primary aliphatic amines **118** by means of photoredox catalysis (Scheme 3.21) [33]. This transformation, which employs a mixed solvent of 1,2-dichloroethane (1,2-DCE) and phosphate buffer (pH = 8), accommodates a wide range of electron-rich aromatic and heteroaromatic compounds and diverse primary aliphatic amines including amino acids. According to fluorescence quenching experiments, both the arenes and amines are able to quench the excited photocatalyst; thus, both of their radical cations are potential reaction intermediates. However, even the less electron-rich arenes such as benzene and toluene, which are unlikely to be oxidized by the excited photocatalyst, can undergo smooth amination to afford the corresponding aniline products (**119d**, **119e**). Consequently, an arene radical cation pathway can be excluded in such cases. Based on all mechanistic studies, a plausible mechanism is presented by the authors as shown in Scheme 3.21b. For the reactions of

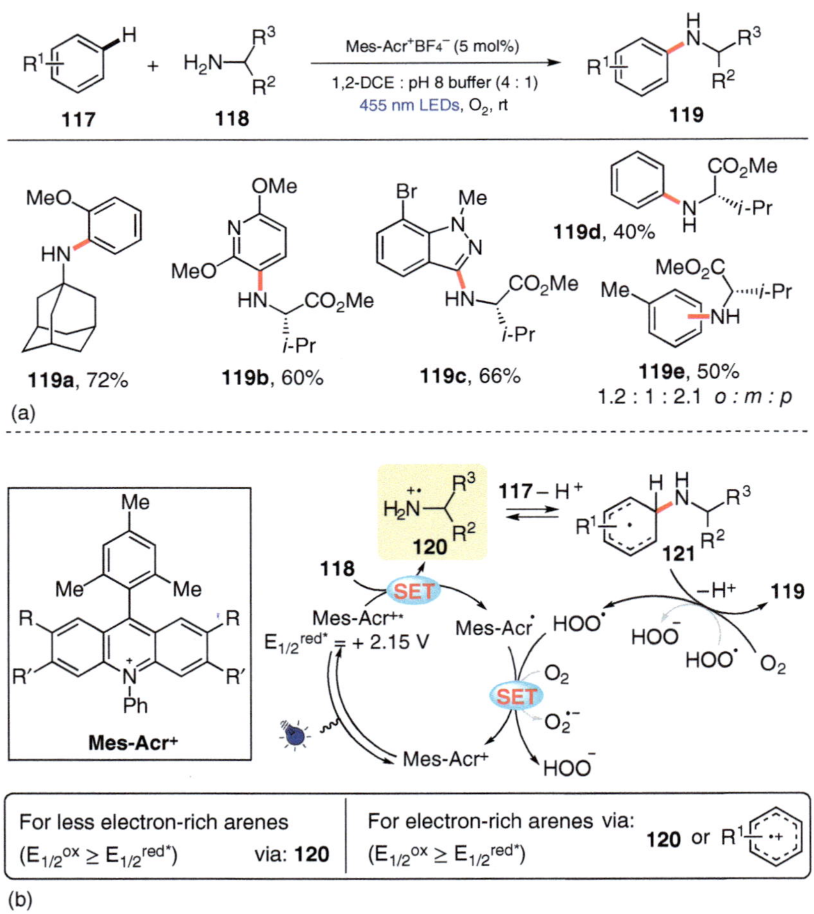

Scheme 3.21 Photocatalytic aryl C–H amination using primary aliphatic amines. Source: Modified from Margrey et al. [33].

electron-poor/neutral arenes, radical cation **120** produced from SET oxidation of amine **118** by the photoexcited Mes-Acr+* is the only possible intermediate, which subsequently adds onto arene **117** to form a cyclohexadienyl radical **121**. Upon oxidation by O_2 and deprotonation, **121** aromatizes to form the desired amination product **119**. As for the reactions of electron-rich arenes, both amine radical cation and arene radical cation can serve as potential intermediates, and the mechanism via the latter will be discussed in Section 3.3.

Organic electrochemistry has also been applied to generate N-radicals directly from N—H bonds for addition to aromatic moieties. In 2017, Xu and coworkers described an electrochemical approach to access amidinyl radical **123** via the anodic N—H bond cleavage of substrate **122** (Scheme 3.22) [34]. Through the intramolecular cyclization of N-radicals **123** onto their arene and heteroarene moieties, various tetracyclic benzimidazoles or pyridoimidazoles **124** are furnished with high efficiency. The reaction can also easily be scaled-up by simply switching up the constant current (**124a–124d**). Additionally, substrate **122e** with two potential cyclization sites is selected as an example to investigate the cyclization tendency of amidinyl radical **123e**. Based on both experimental and theoretical results, 6-endo-trig cyclization (path a) to form a six-membered ring takes precedence over the alternative five-membered ring formation (path b).

In 2018, Lei's group developed a metal- and oxidant-free electrochemical protocol for the oxidative C-H amination of unprotected phenols (Scheme 3.23) [35]. Various phenothiazine derivatives **126** generally couple at the ortho-position of electron-rich phenols

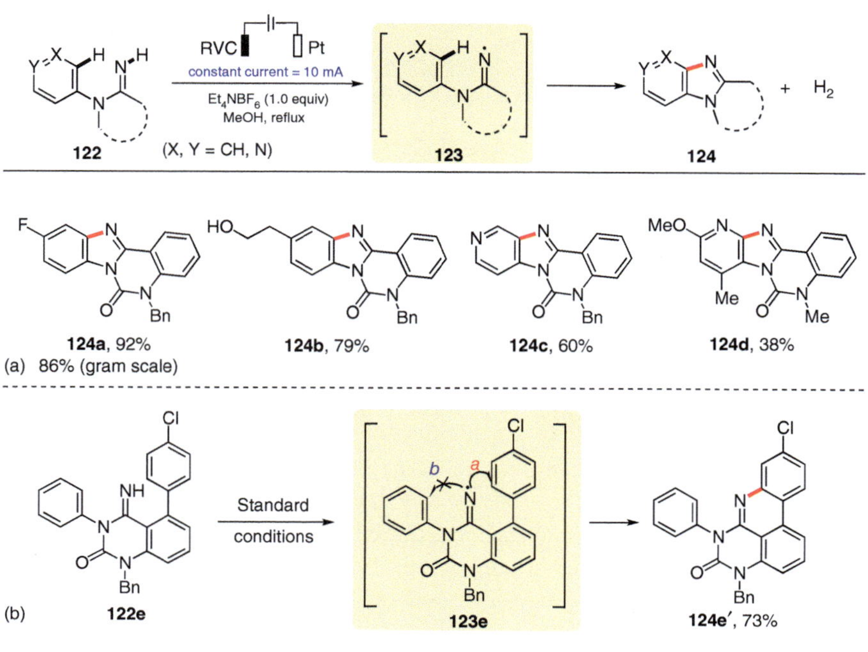

Scheme 3.22 Anodic N—H bond cleavage for aromatic C—H bond amination. Source: Modified from Zhao et al. [34].

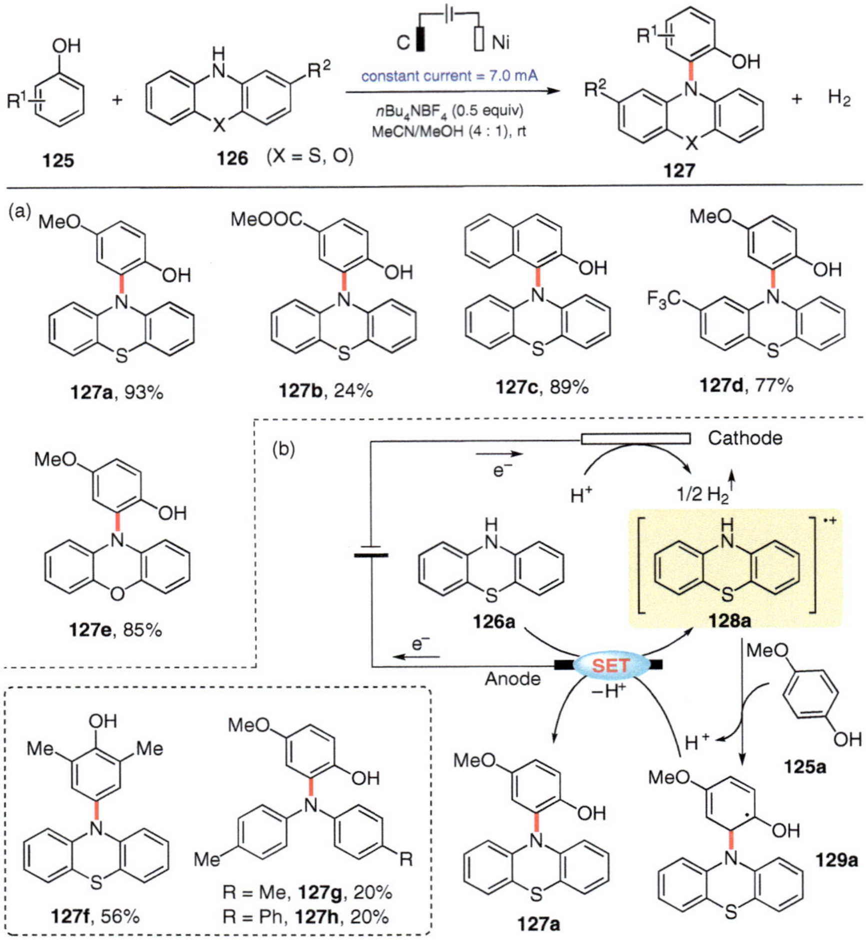

Scheme 3.23 Electrochemical oxidative C–H amination of phenols. Source: Modified from Tang et al. [35].

125, affording N-aryl phenothiazines **127** in good to excellent yields. Low yields of **127** are achieved using phenols with an electron-withdrawing group (**127b**) or electron-neutral diaryl amines (**127g**, **127h**). When both ortho-positions of the phenol are occupied, the amination occurs at the para-position instead (**127f**). The reaction is believed to proceed via an N-radical cation addition to the aromatic ring of the phenol (Scheme 3.23b). Taking **127a** as an example, phenothiazine **126a** is first oxidized at the anode to generate a radical cation intermediate **128a**, which then undergoes electrophilic addition to **125a**, affording the hydroxyl carbon radical **129a**. Subsequent anodic oxidation and deprotonation of **129a** gives the final C(sp^2)—N bond formation product **127a**. Simultaneously, the reduction of protons at the cathode releases hydrogen gas during the process. In the following year, these authors have also applied the protocol to the modification of biomolecules [36].

3.3 Amination via N-atom Nucleophilic Addition

As introduced in Section 3.2, construction of C—N bonds via N-radical addition pathway relies on N-centered radical (or radical cation) species, generated by oxidation of N—H bonds, as key intermediates. On the contrary, aminations via N-atom nucleophilic addition pathway proceed through the oxidation of the C–H partners, and the amines usually serve as nucleophiles to attack the generated C-centered cationic intermediates during the C—N bonds forming processes.

3.3.1 Aromatic C(sp^2)—H Bond Amination

As documented in the literature precedents, direct oxidation of (hetero)arenes is feasible under photo- or electrochemical conditions, and the resulting radical cation intermediates could then be attacked by nucleophiles such as –NHR, –OH, and $-CN^-$. Yet, among the electrochemical amination reactions via arene radical cation intermediates, methods directly employing simple N–H nucleophiles are still not common. Nevertheless, notable progress has been advanced via photoredox catalysis, wherein strongly nucleophilic azoles usually serve as the amine partners.

In 2015, a photocatalytic site-selective CDC amination of unfunctionalized arenes with azoles was described by Nicewicz's group using an organic catalyst system consisting of an acridinium photocatalyst and a nitroxyl radical (Scheme 3.24) [37]. Upon irradiation with visible light, the photoexcited acridinium photocatalyst Mes-Acr^{+*} is able to directly oxidize arene **130** into the corresponding radical cation **134**, which is subsequently attacked by the amine nucleophile **131** to form cationic radical **135**. Deprotonation of **135** affords radical intermediate **136**, which further undergoes aromatization via a TEMPO-mediated HAT process to furnish the final product **133**. Alternatively, intermediate **136** can also be trapped by O_2 to form 1,3-cyclohexadienyl peroxyl radical **137**, which then converts into product **133** after the internal elimination of the hydroperoxyl radical $HOO^{\cdot}$. The closure of two catalytic cycles is indicated by the regeneration of ground-state photocatalyst Mes-Acr^+ from the SET oxidation of Mes-$Acr^{\cdot}$ by O_2 or $HOO^{\cdot}$, and TEMPO from a HAT process between TEMPO-H and O_2 or O-radical species such as $O_2^{\cdot-}$ and $HOO^{\cdot}$. In the scope exploration, a variety of aromatic compounds including complex drug-like structures are firstly investigated, with the corresponding amination products obtained in moderate to good yields. A series of heterocyclic nucleophiles including pyrazoles, triazoles, tetrazoles, imidazoles, and benzimidazoles are proven as suitable coupling partners for the CDC amination with arenes. Particularly, reactions between a commercially available ammonium carbamate $H_4N^+H_2NCO_2^-$ (**132**) and various (hetero)arenes directly furnish primary aniline products.

The direct oxidative C–N cross-coupling between arenes and azoles has also been realized with other photocatalytic systems. In 2017, Lei's group employed a dual catalytic system combining an acridinium photocatalyst with a Co-based cocatalyst for the C—N bond formation to access *N*-arylazoles (Scheme 3.25) [38]. This sustainable protocol does not require any sacrificial oxidant, and the released H_2 gas is the only by-product. Various aromatics including substituted benzenes, biphenyls, and anisole derivatives are smoothly aminated under the standard conditions. Based on the kinetic isotope effect (KIE) experiments and

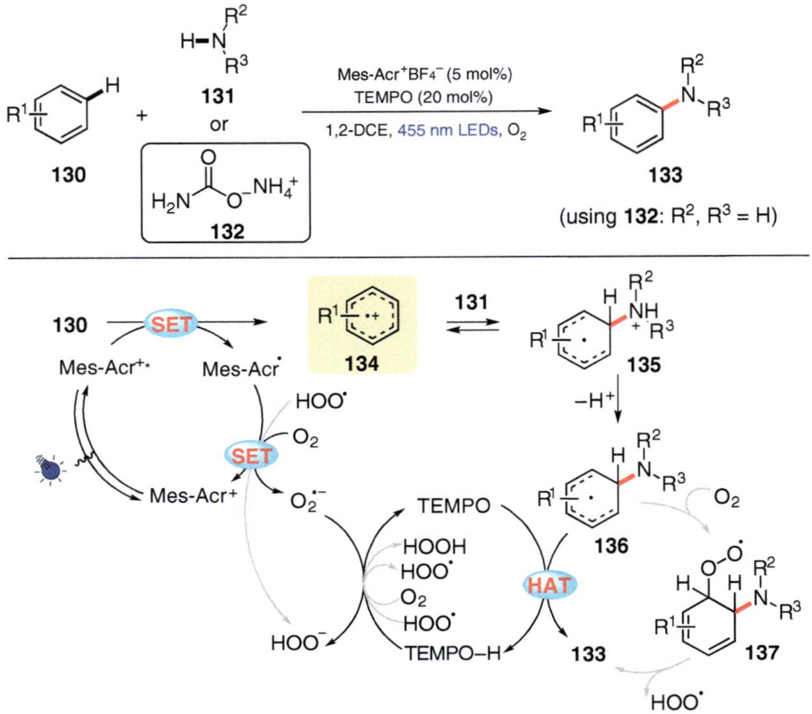

Scheme 3.24 Site-selective aromatic C–H amination via photoredox catalysis. Source: Modified from Romero et al. [37].

other mechanistic studies, the reaction begins with the oxidation of **138** by the photoexcited Mes-Acr^{+*}, which leads to the formation of Mes-Acr$^{\cdot}$ and radical cation **141**. The ground-state photocatalyst Mes-Acr^{+} is then regenerated by single-electron oxidation of Mes-Acr$^{\cdot}$ by Co^{III} catalyst. On the other hand, nucleophilic attack of amine **139** to radical cation **141** forms radical intermediate **142** after deprotonation, which subsequently undergoes a single-electron oxidation by Co^{II} to form cation **143**. The final product **140** is delivered after deprotonation of **143**.

In the same year, Pandey's group reported a regioselective method for the direct C(sp^2)–H amination of anisoles with different azoles under visible light conditions using Selectfluor as an external oxidant (Scheme 3.26) [39]. As a strong oxidant, Selectfluor is capable of oxidizing the photoexcited Ru^{II*} to its higher valence state Ru^{III} and meanwhile furnishing radical cation **147**. The resulting Ru^{III} then readily undergoes a SET reduction with the electron-rich arene **144** to complete the photocatalytic cycle and simultaneously generates radical cation **148**, which is further attacked by the amine nucleophile **145** to give radical intermediate **149**. After deprotonation of **149** by 2,6-lutidine and H-atom abstraction of **150** by radical **147**, the desired amination product **146** is finally obtained.

In 2019, an electrochemical alternative was disclosed by Wu and coworkers, wherein exclusive ortho-selectivity of amination was achieved in the most cases, under metal- and

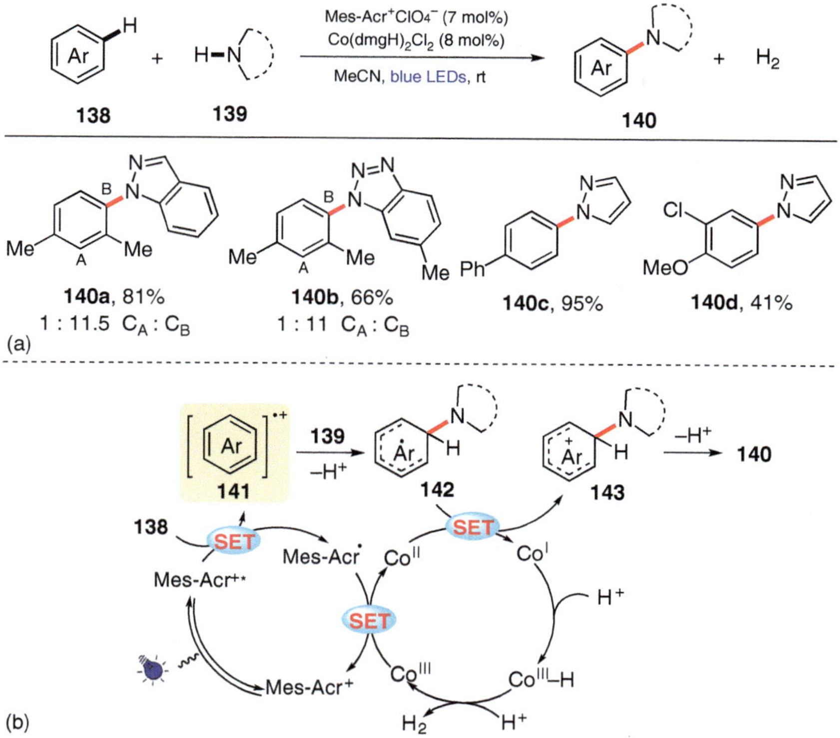

Scheme 3.25 Photoinduced oxidant-free C–H amination of arenes with azoles. Source: Modified from Niu et al. [38].

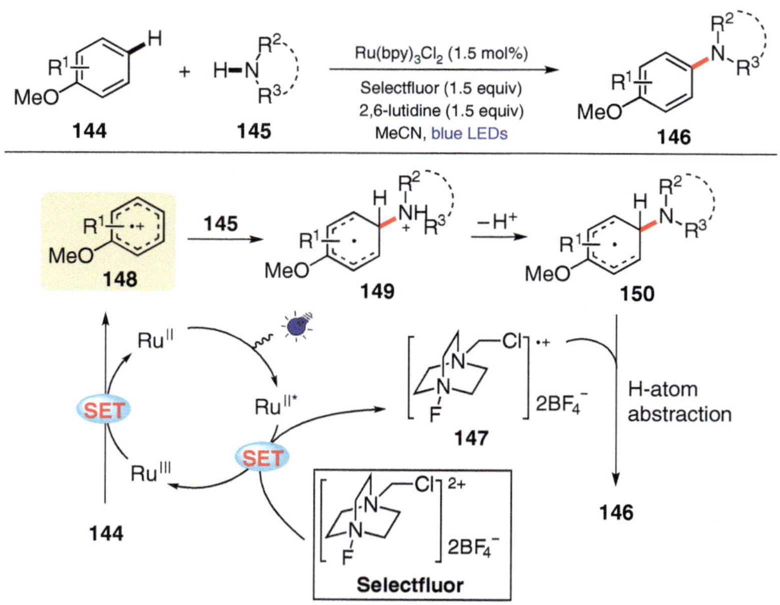

Scheme 3.26 Selective C–H amination of electron-rich arenes via photoredox catalysis. Source: Modified from Pandey et al. [39].

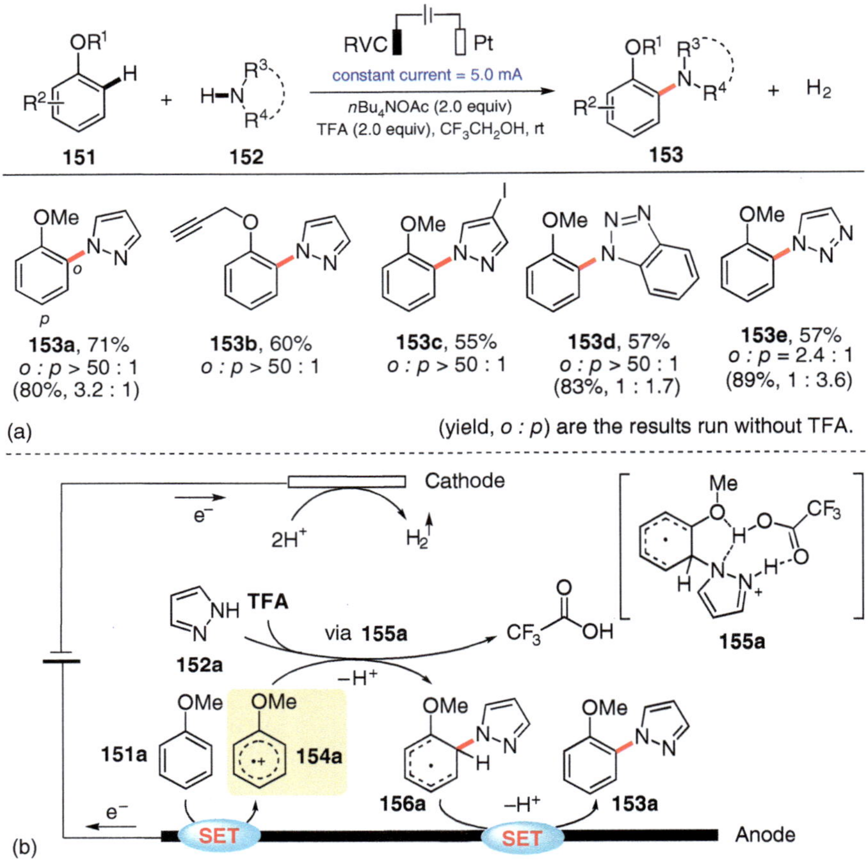

Scheme 3.27 Electrochemical ortho-amination of aromatic C–H bonds with azoles. Source: Modified from Wang et al. [40].

oxidant-free conditions (Scheme 3.27) [40]. During the optimization of the reaction conditions using anisole **151a** and pyrazole **152a**, the authors found that the addition of trifluoroacetic acid (TFA) could effectively enhance the ortho-regioselectivity. Similarly, in the subsequent scope evaluation, it was also observed that reactions with TFA provided significantly higher *o*:*p* selectivity than the corresponding ones without it (**153a**, **153d**, and **153e**). The proposed mechanism is depicted in Scheme 3.27b. Anisole **151a** is preferentially oxidized to provide the radical cation **154a** at the anode, after which pyrazole **152a** as a nucleophile attacks **154a** to form the ortho-aminated intermediate **156a** via a putative intermediate **155a** with TFA-assisted hydrogen bond interaction. Final product **153a** is formed after deprotonation and rearomatization of **156a**.

Electron-rich pyrroles and thiophenes can easily be oxidized into their corresponding radical cations to undergo further C—N bond formation processes via N-atom nucleophilic addition pathway. In 2016, König and coworkers realized a photocatalytic C(sp^2)–H sulfonamidation of pyrroles to reach a range of *N*-(2-pyrrole)sulfonamides **159** (Scheme 3.28) [41]. In this transformation, N-substituted pyrrole **157** is first oxidized by the photoexcited

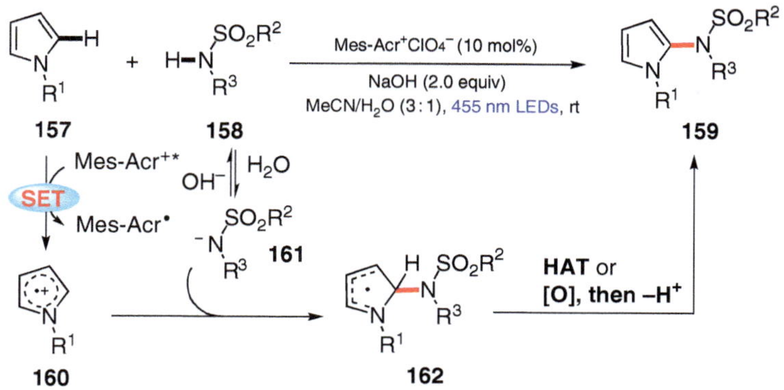

Scheme 3.28 Direct C2-sulfonamidation of pyrroles via visible-light photoredox catalysis. Source: Modified from Meyer et al. [41].

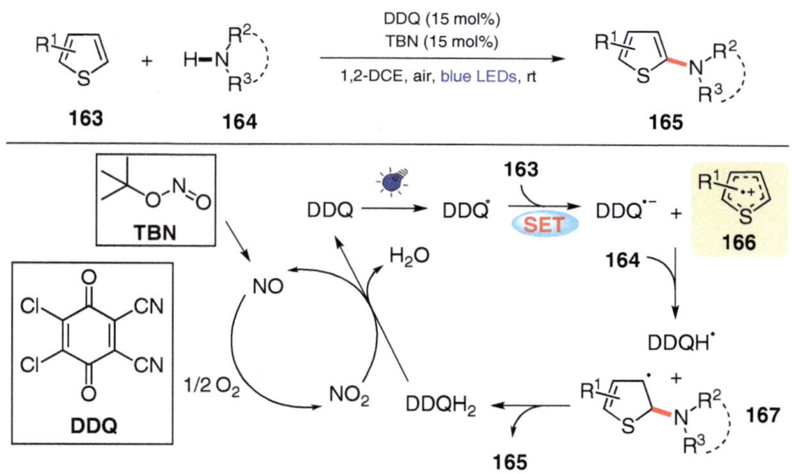

Scheme 3.29 DDQ-mediated C2-amination of thiophenes via visible-light photoredox catalysis. Source: Modified from Song et al. [42].

organic dye Mes-Acr$^+$ into its corresponding radical cation **160**, which is attacked by strong anionic nucleophile **161**, resulting from deprotonation of sulfonamide **158** to generate radical intermediate **162**. The final amination product **159** is afforded via a HAT process with $O_2^{\cdot-}$ or alternatively through further oxidation and deprotonation sequence of **162**.

In the following year, Lei and coworkers disclosed a photocatalytic C(sp^2)–H amination of thiophenes with azoles, employing 2,3-dicyano-5,6-dichlorobenzoquinone (DDQ) as an organic photocatalyst, *tert*-butyl nitrite (TBN) as an electron transfer mediator, and aerobic oxygen as the oxidant (Scheme 3.29) [42]. The photoexcited catalyst DDQ* upon visible light irradiation possesses a high oxidation potential (E_{red} = 3.18 V vs. saturated calomel electrode [SCE]) and thus can easily oxidize thiophene **163** into radical cation **166**. Nucleophilic addition by azole **164** and the following proton donation to DDQ$^{\cdot-}$ convert radical cation **166** into radical intermediate **167**, which undergoes further HAT process with the

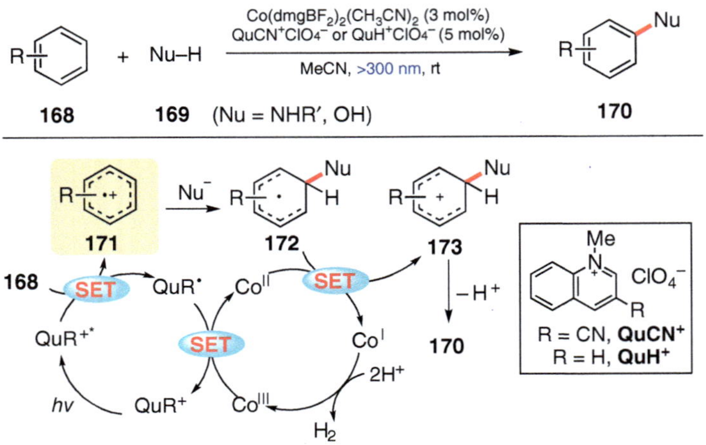

Scheme 3.30 Photocatalytic benzene C–H amination and hydroxylation with hydrogen evolution. Source: Modified from Zheng et al. [43].

formed DDQH· to yield the desired products **165** and $DDQH_2$. To complete the photocatalytic cycle, $DDQH_2$ then participates in another cycle involving TBN to regenerate the original photocatalyst DDQ.

Apart from the most widely used N–H nucleophile azoles, ammonia as an abundantly available and cheap reagent has also attracted attention of researchers, and has been exploited in nucleophilic CDC amination. In 2016, Wu and Tung and coworkers developed a dual catalytic cross-coupling between aromatic C—H bonds and ammonia or water under visible-light conditions to furnish anilines or phenols with the evolution of hydrogen gas (Scheme 3.30) [43]. *N*-Methylquinolinum salts $QuCN^+ClO_4^-$ and $QuH^+ClO_4^-$ are employed as the photocatalysts (QuR^+), whose the excited states possess enough oxidizing power (E_{red}* = 2.72 and 2.46 V vs. SCE, respectively) to accept an electron from benzene (E_{ox} = 2.48 V vs. SCE) to furnish the reduced photocatalyst QuR· along with the crucial radical cation **171**. Subsequently, the former is oxidized by Co^{III} catalyst to regenerate the ground-state photocatalyst QuR^+; meanwhile, the latter is captured by a nucleophile to form radical **172**, which further donates an electron to Co^{II} to produce cationic species **173**. The final product **170** is formed after further deprotonation of **173**. On the basis of CV studies, the redox potential values demonstrate that product **170** is also capable of quenching the excited photocatalyst QuR^{+*}, which may result in its further functionalization. However, mono-functionalized benzene is obtained as the only product, and no multiaminated or hydroxylated products can be detected even after prolonged reaction time. A hypothesis for this result, advanced by the authors, is that the rapid back-electron transfer from the reduced QuR· to the formed radical cation of **170** protects product **170** from overoxidation.

3.3.2 Olefinic C(sp^2)—H Bond Amination

As good radical acceptors, alkenes are capable of trapping various radicals to deliver the corresponding addition products. Nonetheless, the direct nucleophilic amination of alkenes depends on the prior oxidation of the C=C bonds to form the crucial C-centered radical cation intermediates, which is followed by subsequent nucleophilic addition of N–H partners.

During the past decade, Nicewicz's group has developed a variety of visible-light-induced anti-Markovnikov hydrofunctionalization protocols for alkenes via the addition of nucleophiles to the in situ generated C-centered radical cations, employing the strongly oxidizing acridinium salts as photocatalysts. Based on their previous work on the intramolecular hydroamination of alkenes **174** using a combination of catalytic Mes-Acr$^+$BF$_4^-$ and thiophenol as the photocatalyst and H-atom donor, respectively (Scheme 3.31a) [44], they later reported an intermolecular anti-Markovnikov hydroamination of alkenes by replacing the thiophenol cocatalyst with a combination of phenyl disulfide and 2,6-lutidine (Scheme 3.31b) [45]. Therein, trifluoromethanesulfonamide (TfNH$_2$) and azoles (**177**) are employed as the N–H nucleophiles to approach a variety of phenethylamine derivatives **178**. In line with their previous work on nucleophilic alkene functionalization, the authors suggest that the key step is the SET process between alkene **176** and the photoexcited acridinium (Mes-Acr^{+*}), which results in the generation of key radical cation intermediate **179**. Subsequently, nucleophiles TfNH$_2$ or **177** undergo reversible addition to the least-substituted side of **179** to form the stabilized C-radical **180**, which may further undergo a HAT process with a hydrogen atom donor, presumably thiophenol, to deliver the desired product **178**. On the other hand, the putative HAT reagent thiophenol is supposed to come from phenyl disulfide after a homolysis/single-electron reduction/protonation sequence.

In 2017, Lei and coworkers developed an external oxidant-free photocatalytic CDC amination and etherification of alkenes with azoles and alcohols using a dual catalytic system

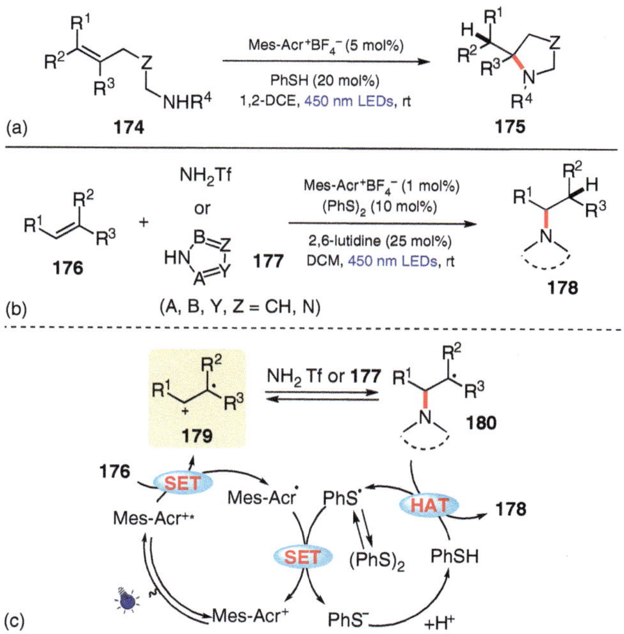

Scheme 3.31 Anti-Markovnikov hydroaminations of alkenes by two component organic photoredox systems. Source: Modified from Nguyen and Nicewicz [44].

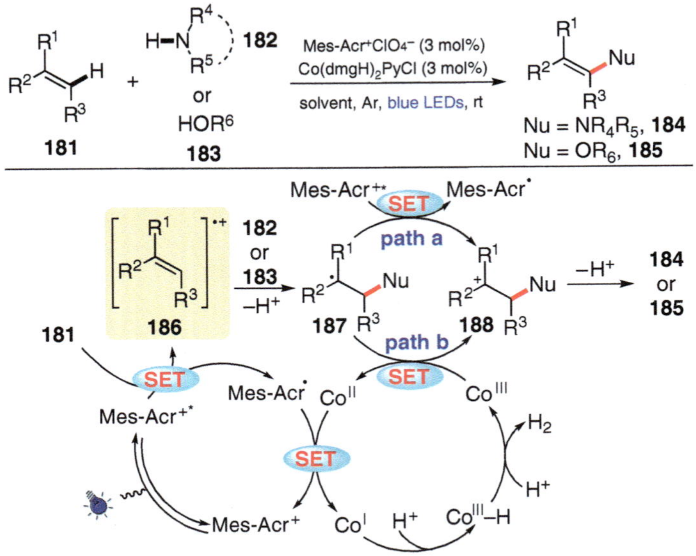

Scheme 3.32 Photocatalytic dehydrogenative cross-coupling of alkenes with azoles and alcohols via a dual catalytic system. Source: Modiifed from Yi et al. [46].

combining photocatalyst Mes-Acr$^+$ClO$_4^-$ and a Co-based cocatalyst (Scheme 3.32) [46]. The single-electron oxidation of alkene **181** by the photoexcited Mes-Acr^{+*} is considered as the key initial step, providing Mes-Acr$^\cdot$ and radical cation species **186**. Subsequently, the employed nucleophile **182** (or **183**) reacts with **186** to form the more stable radical intermediate **187**, which is further oxidized into the corresponding cationic species **188** by either Mes-Acr^{+*} or Co^{III}. Then, the final anti-Markovnikov product **184** (or **185**) is delivered after deprotonation of **188**. As for the Co catalytic cycle, the reduced Co^{II} undergoes further reduction by Mes-Acr$^\cdot$ to yield Co^I species, which is protonated twice and then releases one molecule H_2 to regenerate the Co^{III} species.

In 2018, again in Lei's laboratory, a photoinduced oxidative [4+2] annulation of imines and alkenes was developed by utilizing a dual photoredox/cobaloxime catalytic system in a mixed solvent of 1,2-DCE and hexafluoroisopropanol (HFIP) (Scheme 3.33) [47]. This method not only obviates the application of stoichiometric oxidants but also exhibits excellent atom economy by generating hydrogen gas as the only by-product, achieving high regioselectivity and *trans*-diastereoselectivity even if alkenes are employed in *Z*/*E* mixtures. On the basis of Stern–Volmer studies and CV experiments, a plausible mechanism is proposed. Firstly, single-electron oxidation of **190** by the excited-state photosensitizer leads to the generation of the alkene radical cation intermediate **192** and to the reduced Mes-Acr$^\cdot$. Subsequently, the nucleophilic attack of **189** furnishes the radical **193** after deprotonation, in which the C=C single bond prefers a more stable trans-configuration upon rotation as proposed by the authors. Next, the radical cyclization of **193** gives the intermediate **194**, which is further oxidized to furnish **195**. After elimination of a proton, the desired 3,4-dihydroisoquinoline product **191** is afforded. On the other hand, the reduced photosensitizer is oxidized by the Co^{III} species to complete the photoredox

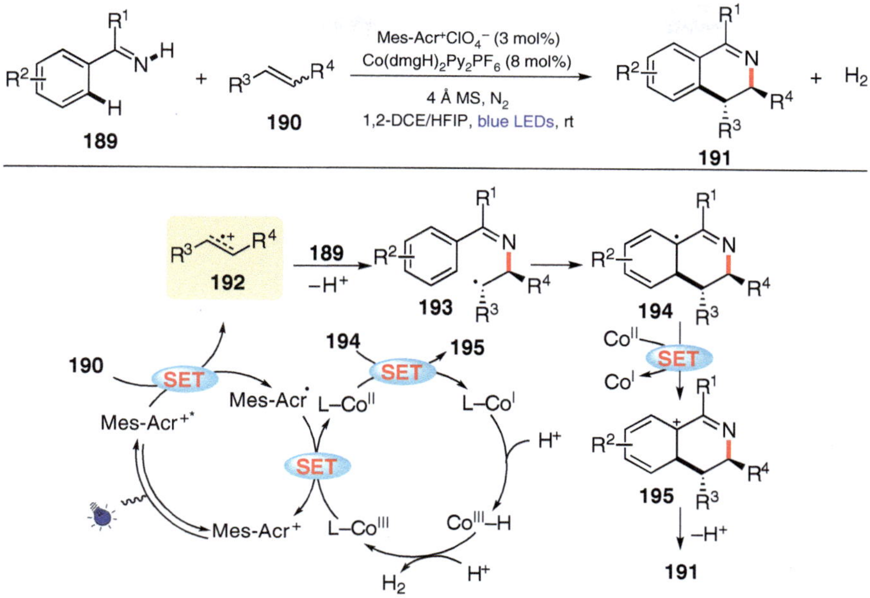

Scheme 3.33 Photoinduced oxidative [4+2] imine/alkene annulation with H_2 liberation. Source: Modified from Hu et al. [47].

catalytic cycle. As for the cobalt side, the Co^{II} species is reduced to form a Co^{I} intermediate, which can be protonated to produce Co^{III}-hydride. The Co^{III}-hydride species would then release H_2 upon the interaction with protons.

3.3.3 Activated C(sp^3)–H Bond Amination

In the early years, C(sp^3)–H amination reactions were mainly achieved via transition metal catalysis or with hypervalent iodine reagents. Recently, quite a few methods for CDC amination of activated C(sp^3)—H bonds, such as benzylic C(sp^3)–H and C(sp^3)–H in α-position to a nitrogen atom, have been realized via the formation of the corresponding cationic species and subsequent addition of nitrogen nucleophiles by means of photo/electrochemistry.

3.3.3.1 Benzylic C(sp^3)–H Bond Amination

As an example for the direct amination of benzylic C(sp^3)—H bonds, Pandey's group disclosed a photocatalytic CDC amination of alkylbenzenes **196** with *N*-methoxyamides **197** under metal-free and external oxidant-free conditions (Scheme 3.34) [48]. The 410 nm wavelength visible light source is obtained by using a combination of Pyrex and a $CuSO_4$:NH_3 solution filter from a 450 W Hanovia medium pressure lamp. Compound 9,10-dicyanoanthracene (DCA) is employed as a photosensitizer, which facilitates an array of C(sp^3)–H amination reactions between **196** and **197** with good functional group tolerance (**198a–198f**). Notably, this protocol is also applicable to intramolecular C—N bond construction to provide five-membered rings (**198g**). The mechanistic proposal is depicted in Scheme 3.34b. First, the single-electron oxidation of **197** by the photoexcited

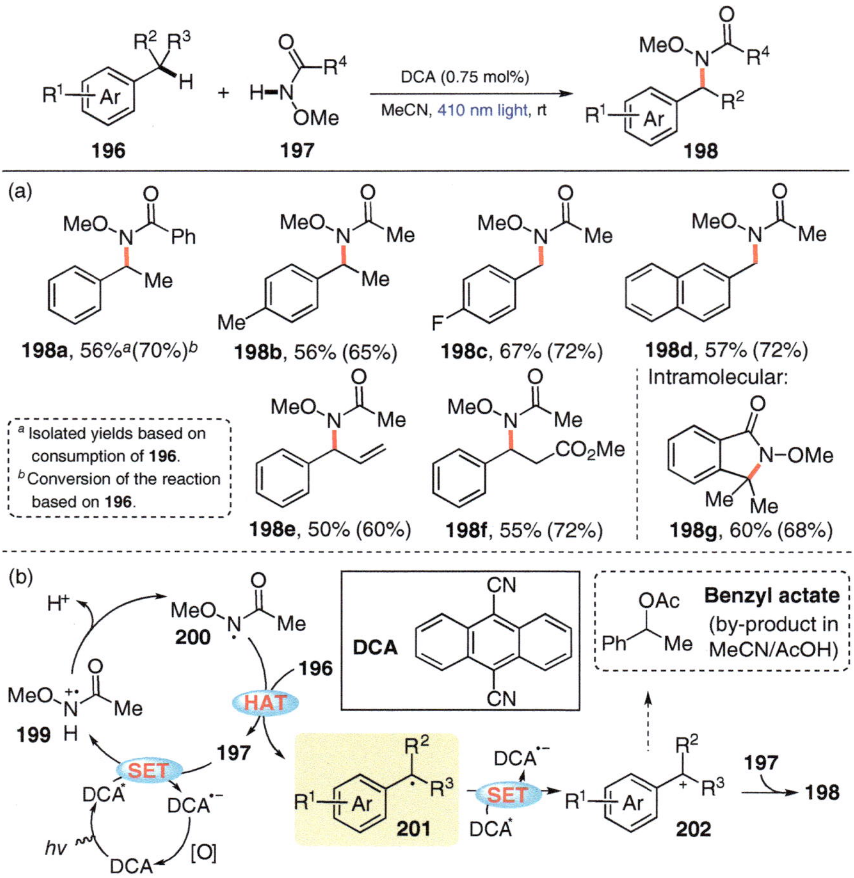

Scheme 3.34 Visible-light-enabled direct benzylic C(sp^3)−H amination. Source: Modified from Pandey and Laha [48].

DCA* yields the *N*-methoxyamide radical cation **199**, which then converts into radical intermediate **200** after deprotonation. The following HAT process from the benzylic position of alkylbenzene **196** to radical **200** regenerates **197** and meanwhile gives benzylic radical **201**. The subsequent oxidation of **201** generates benzylic cation **202**, which further reacts with **197** to deliver the desired amination product **198**. The results of control experiments and some experimental observations add credit to this putative mechanism. For example, the existence of N-centered radical intermediate **200** is suggested by the formation of its dimeric product, while benzylic radical **201** is captured by TEMPO in a radical trapping experiment. To demonstrate the intermediacy of benzylic cation **202**, the authors performed the model reaction using a mixture of MeCN/AcOH (9 : 1) as the solvent, which provides the desired **198** in diminished yield, along with the by-product benzyl acetate in 10% yield.

Alkyl-substituted arenes with available benzylic C(sp^3)—H bonds tend to undergo direct C–H functionalization at the benzylic position via the preliminarily formed arene radical cation intermediate upon single-electron oxidation. In 2016, Pandey et al. developed a

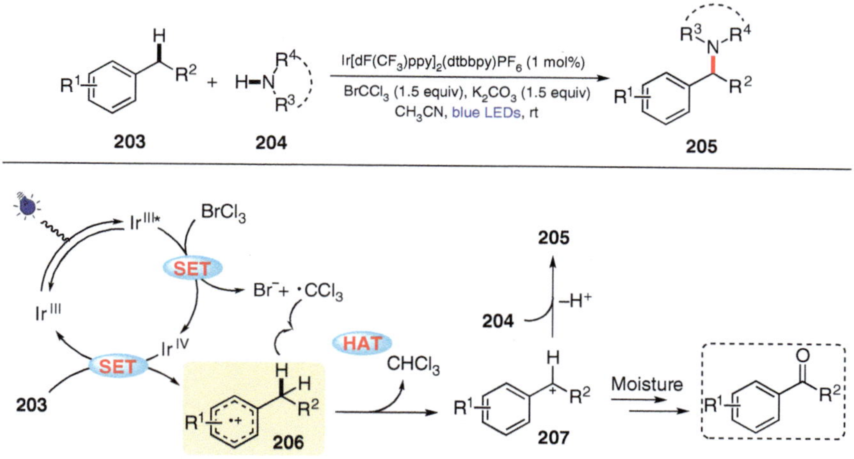

Scheme 3.35 Benzylic C–H amination via visible-light photoredox catalysis. Source: Modified from Pandey et al. [49].

visible-light-mediated cross-dehydrogenative amination of benzylic C(sp^3)—H bonds with azole derivatives, employing an Ir-based photocatalyst and bromotrichloromethane ($BrCCl_3$) as an oxidative quencher (Scheme 3.35) [49]. Upon photoexcitation, the excited catalyst (Ir^{III*}) first undergoes a single-electron oxidation by $BrCCl_3$, affording the high-valence catalyst Ir^{IV} and a trichloromethyl radical $^{\cdot}CCl_3$. Subsequently, arene substrate **203** is oxidized by the Ir^{IV} species to generate radical cation **206**, along with the regenerated ground-state Ir^{III} photocatalyst. Then, radical $^{\cdot}CCl_3$ as a competent H-atom abstractor engages in a HAT process with intermediate **206** to provide benzylic cation **207**, which further reacts with nucleophile **204** to deliver the final product **205**. Moreover, it is possible to trap cation **207** with moisture to in situ generate the benzylic alcohol intermediate, which is prone to further oxidation to furnish the corresponding carbonyl compound.

3.3.3.2 N-α-C(sp³)—H Bond Amination

The activated C(sp^3)–H adjacent to a nitrogen atom is also prone to photo/electrochemical oxidation and to subsequent nucleophilic amination. In 2019, Zeng and coworkers reported an electrochemical dehydrogenative imidation of *N*-methyl benzylamines **208** with phthalimides **209** to obtain various phthalimide-protected *gem*-diamines **210** (Scheme 3.36) [50]. Notably, the amination occurs regioselectively at the methyl group rather than at the benzylic position of each *N*-methyl benzylamine. Apart from multiple phthalimides with different substituents (**210a–210e**), certain triazoles are also viable amine sources (**210f**). Based on CV studies, revealing that benzylamine **208a** is easier to be oxidized than phthalimide **209a**, a plausible mechanism is proposed as shown in Scheme 3.36b. The reaction is initiated by anodic oxidation of **208a** to give the radical cation **211a**, which undergoes deprotonation and further oxidation to generate the iminium intermediate **212a**. Meanwhile, MeOH is reduced at the cathode to give MeO^- and H_2. Simultaneously, **209a** is deprotonated

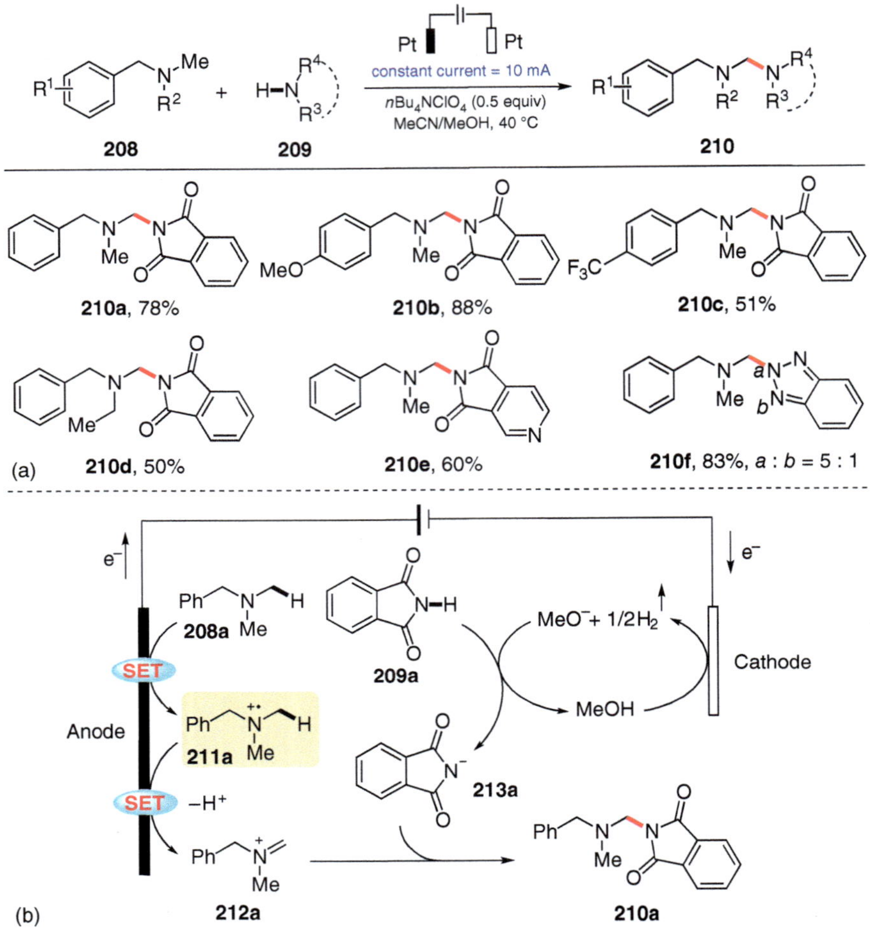

Scheme 3.36 Electrochemical dehydrogenative imidation of *N*-methyl-substituted benzylamines. Source: Modified from Lian et al. [50].

by MeO^- to give the anionic intermediate **213a**, which then undergoes a Mannich-type addition to **212a** generating product **210a**.

Almost simultaneously, Lei's group also reported an electrochemical direct imidation of *N*-methyl anilines with hydrogen evolution under acidic conditions (Scheme 3.37) [51]. Acetic acid plays a key role in this reaction, other acids such as HCOOH, H_2SO_4, and TsOH could not enable the reaction to proceed, while *n*PrCOOH led to reduced efficiency. Cyclic amides, heteroatom-containing amides, and succinimides are well tolerated in this transformation and afford products in moderate to good yields (**216a–216e**). The proposed reaction mechanism is nearly identical to the previous one.

Recently, an elegant visible-light-driven intramolecular C–N cross-coupling reaction was disclosed by Zheng's research group under mild and metal-free conditions (Scheme 3.38) [52]. Various polycyclic quinazolinone derivatives (**221a–221f** and **223a–221d**), including

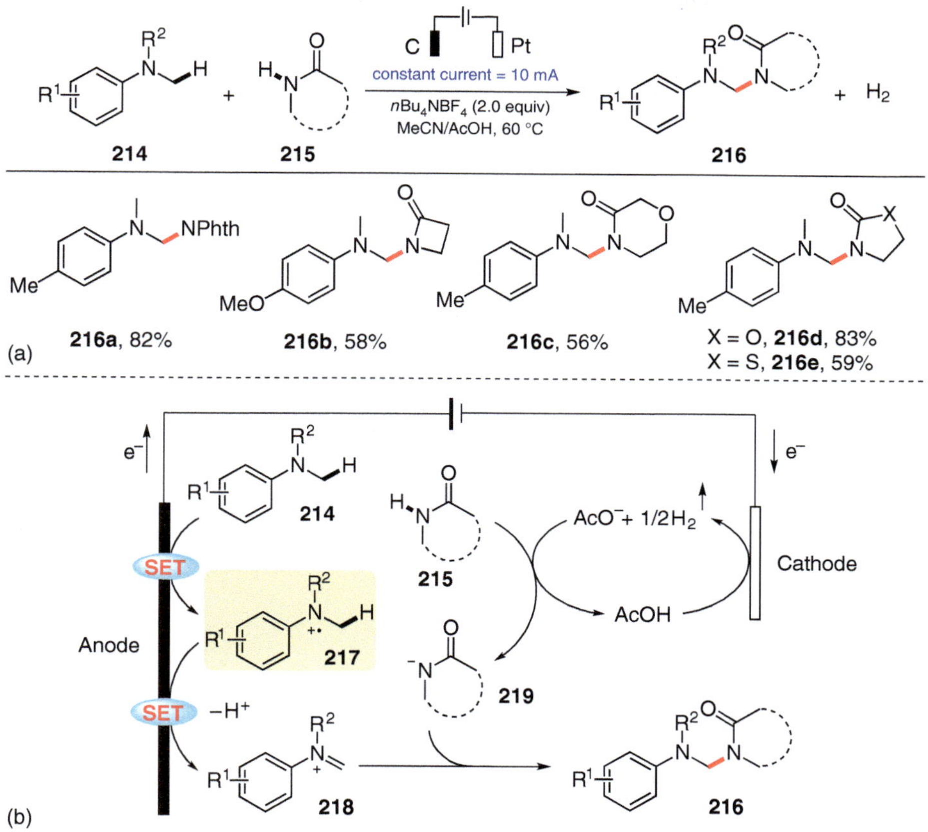

Scheme 3.37 Electrochemical dehydrogenative imidation of *N*-methyl substituted anilines. Source: Modified from Wang et al. [51].

natural products tryptanthrin, rutaecarpine, and their analogs (**223b–221d**), could be smoothly furnished via this protocol. Further demonstration of gram-scale synthesis and solar-driven transformation proves the potential of this strategy for practical applications.

The mechanistic proposal, on the basis of control experiments, DFT calculations, UV–vis spectroscopy, and electron paramagnetic resonance (EPR) studies, as well as X-ray characterization of the key intermediate, is outlined in Scheme 3.39 with **221a** taken as the example. The reaction is initiated by a photoisomerization of **220a** upon irradiation with visible light to produce the photoisomer **222a**, which displays absorbance around 395 nm and has a short lifetime. The long-lived photoisomer complex **223a** is then generated by the coordination of (*R*)-(-)-1,1'-binaphthyl-2,2'-diyl hydrogenphosphate [(*R*)-BPA] through hydrogen bonding, which can induce the SET process under irradiation with visible light to generate the radical pair **224a**. The resulting superoxide radical anion abstracts a hydrogen atom from **224a** to produce amino radical **225a**, which undergoes intramolecular 1,6-HAT process to generate radical **226a**. Then, the radical recombination of **226a** and HOO$^{\cdot}$ forms the key intermediate hydroperoxide **227a**, which is easily converted into iminium ion **228a**

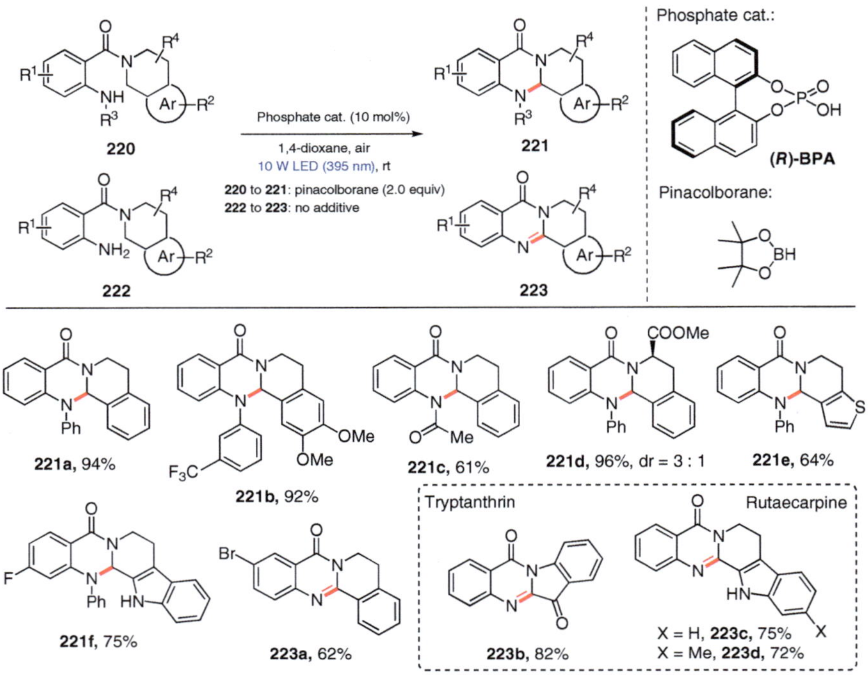

Scheme 3.38 Light-driven direct intramolecular C−N cross-coupling. Source: Modified from Jing et al. [52].

under acidic conditions provided by the Brønsted acid (*R*)-BPA. Finally, the polycyclic quinazolinone product **221a** is formed through intramolecular nucleophilic addition with release of H_2O_2. As an alternative pathway, the oxidation by-product **229a**, formed from the hydroperoxide **227a**, can be efficiently converted into **221a** under the standard reaction conditions, as proven by a control experiment.

3.4 Amination via Radical Cross-coupling

The last pattern for the radical-involved C—N bond formation is the cross-coupling of C- and N-radicals, which is relatively undeveloped compared with the aforementioned two pathways, presumably because of the challenge to provide an appropriate environment for the simultaneous generation of both C- and N-radicals. Moreover, the generated C- and N-radical in the same system are supposed to obey to the principle of persistent radical effect (PRE) to furnish the desired C–N cross-coupling product rather than the homocoupling ones.

3.4.1 Aryl C(sp^2)−N Bond Formation via Radical Cross-coupling

Among the reported C—N bond construction methods directly from aromatic C(sp^2)—H bonds and N—H bonds via radical cross-coupling pathway, both of the coupling partners are required to be redox active to produce their corresponding radical intermediates upon

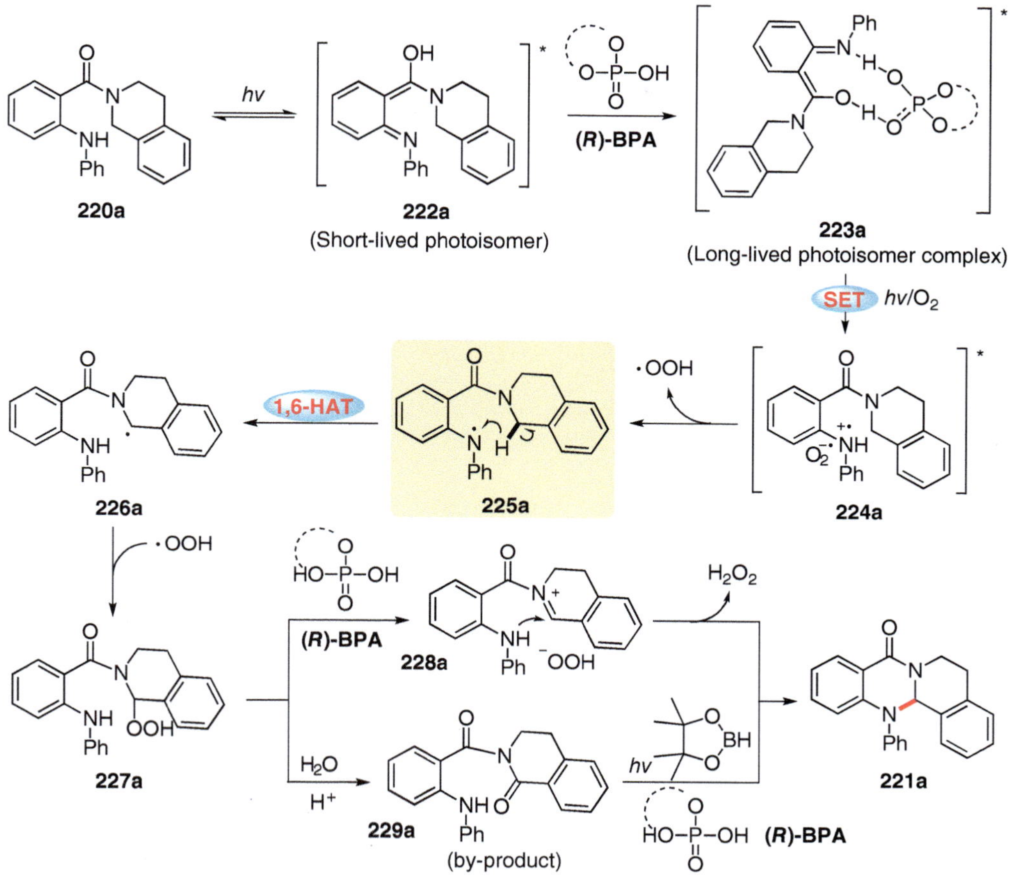

Scheme 3.39 Mechanistic proposal for the light-driven direct intramolecular C–N cross-coupling.

oxidation. Therefore, in most cases, electron-rich arenes such as phenols and anilines are chosen as the aryl radical precursors. As for the amine partner, electron-rich diarylamines and azoles are commonly used. Sulfonamides after deprotonation can also turn into the corresponding radicals upon oxidation, as previously described. For these amine sources, the stabilizing effect for the N-radicals also appears to be crucial for the success of the radical cross-coupling.

3.4.1.1 Aryl C(sp^2)–N Bond Formation Using Diarylamines

In 2016, Xia and coworkers developed a cross-dehydrogenative C–N coupling between phenols **230** including trimethylsilyl (TMS)-protected ones and phenothiazines **231** to afford a variety of ortho-aminated phenols **232** employing $K_2S_2O_8$ as the external oxidant under mild visible light photoredox conditions (Scheme 3.40) [53]. This protocol requires no additional photocatalyst, which is probably because of the inherent properties of phenothiazines (**231**) that can absorb visible light to enable the following energy transfer process. Notably, instead of inhibiting the amination or lower the reaction efficiency, additional

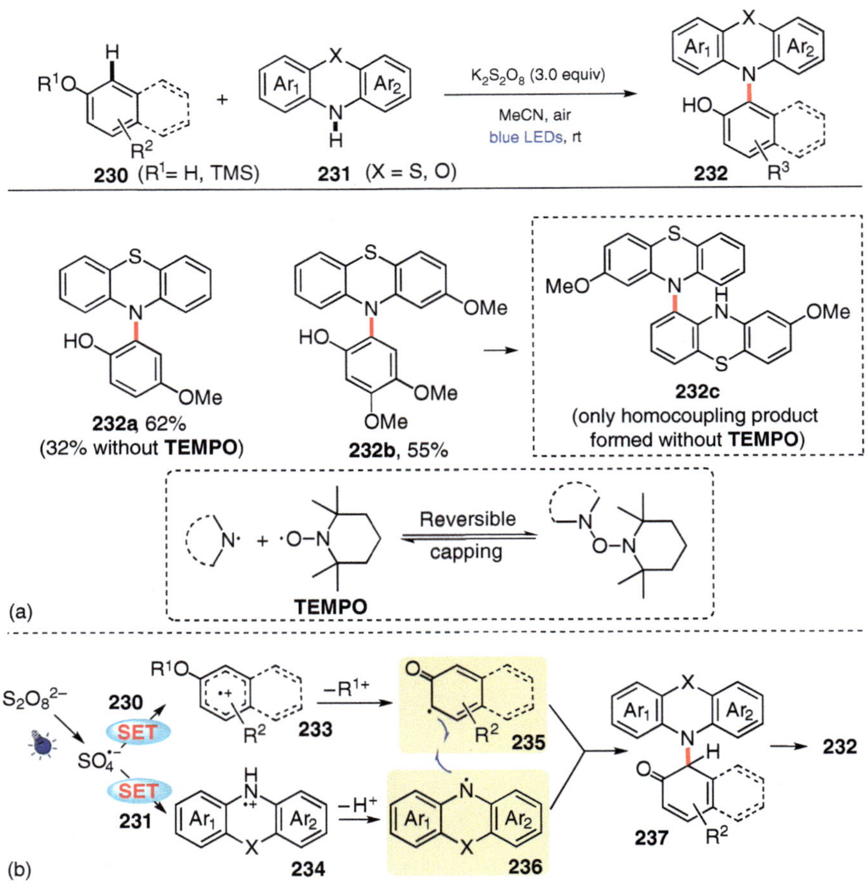

Scheme 3.40 Visible-light-promoted CDC amination of phenols and phenothiazines. (a) Reactions with additional 2.0 equiv of TEMPO. (b) The mechanistic proposal. Source: Modified from Zhao et al. [53].

radical scavenger TEMPO dramatically improves the yield of desired product **232a**. Moreover, the reaction system of 3,4-dimethoxyphenol (**230b**) and 2-methoxylphenothiazine (**231b**), which only yields **232c** as a homocoupling product of **231b** under the original conditions, does furnish the desired C–N cross-coupling product **232b** with extra 2.0 equiv of TEMPO added. On the basis of some relevant literature, a hypothesis that TEMPO can covalently and reversibly prolong the lifetime of the transient N-radicals is reasonably proposed by the authors. Mechanistically, under the irradiation of visible light, oxidant $K_2S_2O_8$ is capable of oxidizing both phenol and phenothiazine to provide C-radical **235** and N-radical **236**, respectively. The following radical–radical cross-coupling process generates intermediate **237**, which further tautomerizes into the final amination product **232**.

However, the amine scope of this catalyst-free amination is limited to S- or O-containing cyclic diarylamines, which does not include acyclic ones because they are incapable of absorbing visible light and possess higher oxidation potentials. To accommodate these

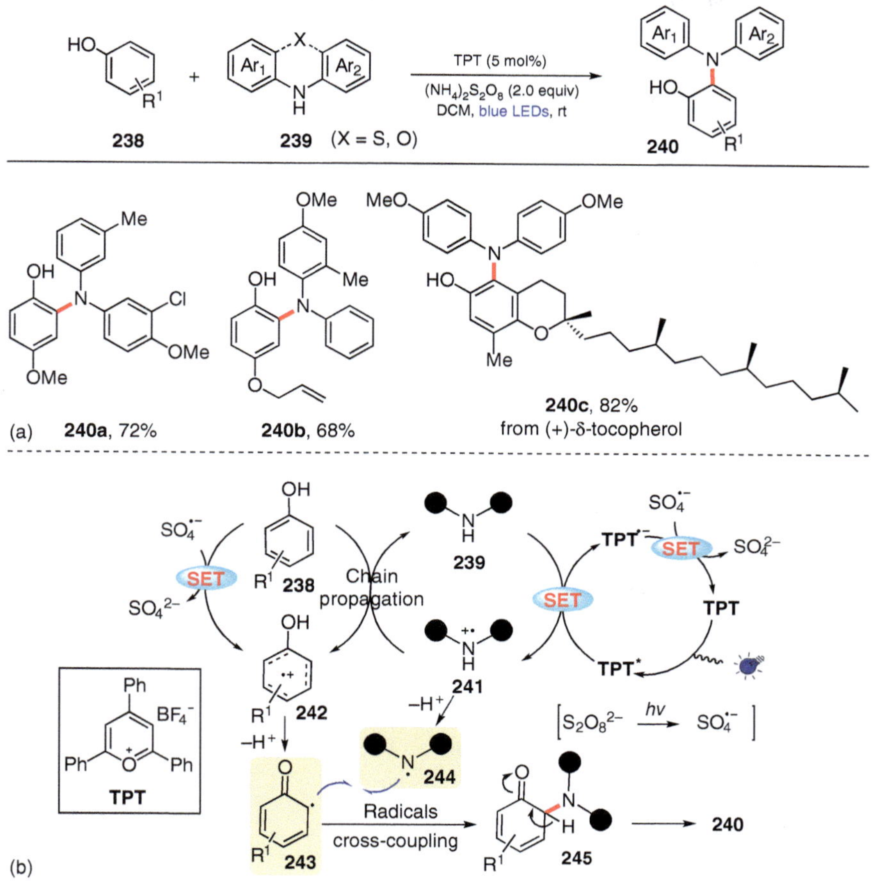

Scheme 3.41 Visible-light-mediated CDC amination of phenols and acyclic diarylamines. Source: Modified from Zhao et al. [54].

substrates, Xia's group subsequently developed another photocatalytic strategy to enable the CDC amination of phenols **238** with acyclic diarylamines **239** employing the organic salt 2,4,6-triphenylpyrylium (TPT) as a photocatalyst (Scheme 3.41) [54]. This protocol is proven compatible with a wide range of electron-rich acyclic diarylamines and phenols including the natural product (+)-δ-tocopherol (**240a–240c**). Notably, phenothiazine or phenoxazine substrates, which require prolonged time to complete the reaction using the previous method, convert much faster under these conditions. As demonstrated by a series of mechanistic studies including EPR spectroscopy and fluorescence quenching experiments, diaryl amine **239** is able to reductively quench the photoexcited catalyst TPT*, generating the corresponding radical cation **241** along with reduced $TPT^{\cdot-}$. A radical chain propagation process is indicated by the determined quantum yield ($\Phi = 19$), which is proposed to take place between phenol **238** and N-centered radical cation **241** to provide phenol radical cation **242** regenerated amine **239**. An alternative pathway to reach **242** is

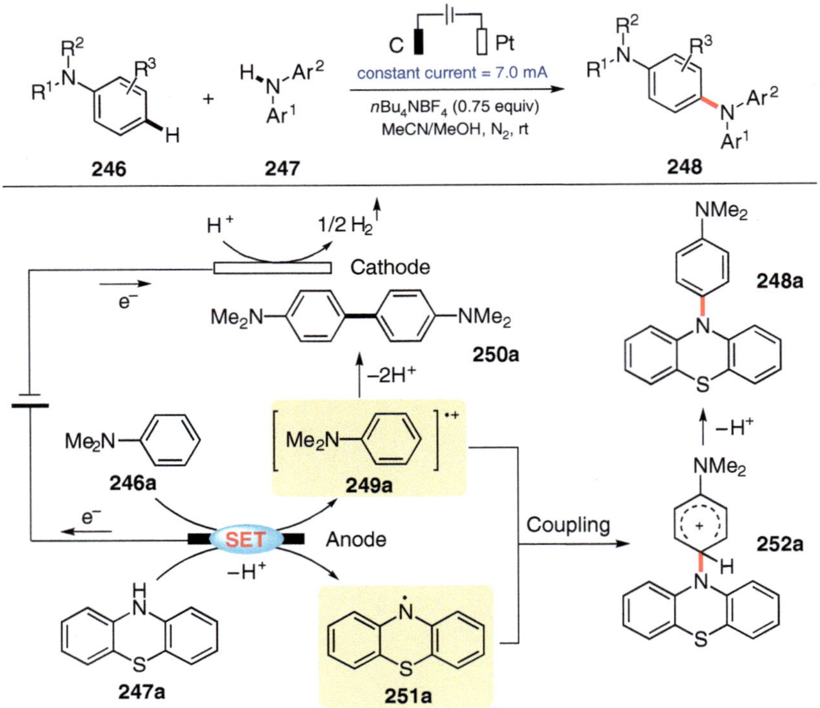

Scheme 3.42 Electro-oxidative para-selective C–H/N–H cross-coupling with hydrogen evolution. Source: Modified from Liu et al. [55].

through the SET oxidation of **238** by persulfate. Finally, radicals **244** and **243** formed from deprotonation couple with each other to deliver the cross-coupling product **240**.

An electrochemical protocol via a similar mechanism was described by Lei's group in early 2019, through which various triarylamines can be synthesized by coupling electron-rich arenes and diarylamine derivatives (Scheme 3.42) [55]. More specifically, a series of para-selective amination products **248** are furnished from the reactions between aniline derivatives **246** and diarylamines **247**. Both coupling partners are found to be redox active during the mechanistic investigation, as exemplified by a plausible mechanism related to **248a**. Therefore, **246a** and **247a** are simultaneously oxidized at the anode to afford radical cation **249a** and radical **251a** after deprotonation. The subsequent radical/radical cross-coupling produces cationic species **252a**, which turns into the final product **248a** after further deprotonation. Shortly after the publication of this work, Song and Li and coworkers also disclosed their independent research, which is very close to Lei's work [56].

3.4.1.2 Aryl C(sp^2)–N Bond Formation Using Azoles

In 2017, Adimurthy's group developed a regioselective C–H amination of heteroarenes with azoles using acridinium salt Mes-Acr$^+$ClO$_4$$^-$ as the photocatalyst under visible light irradiation (Scheme 3.43) [57]. Quinoline amides **253** and imidazopyridines **254** serve as eligible substrates, which undergo regioselective amination at C5 and C3 position, respectively,

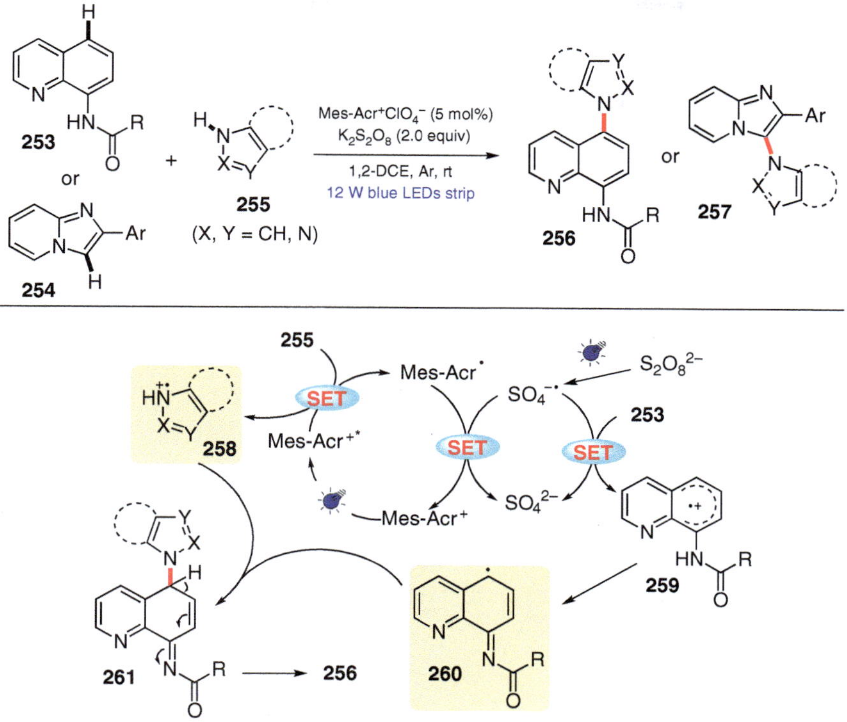

Scheme 3.43 Selective C–H amination of heteroarenes with azoles via an organic photoredox system. Source: Modified from Samanta et al. [57].

with various azoles **255**. Taking the amination of **253** as an example, it is proposed that the transformation proceeds through cross-coupling of N-radical cation **258** and C-radical species **260** on the basis of a series of mechanistic investigations.

More recently, Lei and coworkers disclosed an electrochemical C-3 amination of imidazo[1,2-*a*]pyridines **262** with amines **263** via an oxidative C–H/N–H cross-coupling (Scheme 3.44) [58]. The reaction is supposed to proceed via a cross-coupling process of electrochemically generated C- and N-radicals (Scheme 3.44b). Additionally, an intramolecular C–N coupling through a Cp_2Fe-mediated sequence of N-radical formation and addition to the adjacent unsaturated moiety was also reported in the same article for the synthesis of 10*H*-benzo[4,5]imidazo[1,2-*a*]indole derivatives **266** from **265**.

One more electrochemical C–H azolation was next developed by Feng and Chen and coworkers in 2019, wherein phenol and aniline derivatives are regioselectively aminated with azoles via a radical–radical cross-coupling pathway (Scheme 3.45) [59]. During the investigation of the substrate scope, high ortho-selectivity to the –XH group of this azolation is observed, and the sterically less-hindered ortho-position is preferred when asymmetrical phenols are applied (**273b**). In light of the mechanistic studies and reported work, the reaction mechanism is believed to involve a radical/radical cross-coupling of **275** and **276**, which are generated from anodic oxidation of substrates **271** and **272**.

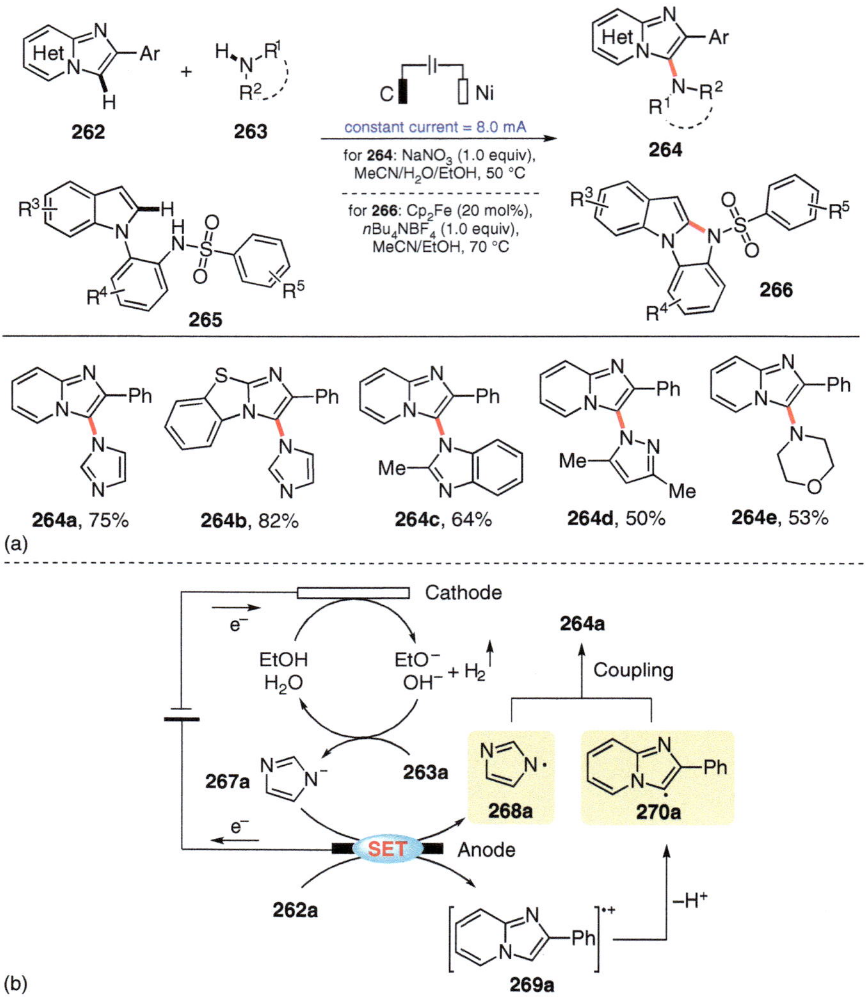

Scheme 3.44 Electrochemical oxidative C–H/N–H cross-couplings for C–N bond formation with hydrogen evolution. Source: Modified from Yu et al. [58].

3.4.2 Other C–N Bond Formation via Radical Cross-coupling

The conjugate amination of α,β-unsaturated carbonyl compounds represents a fundamental strategy to construct highly valuable β-aminocarbonyl moieties. However, the direct addition of N-centered radicals to electron-deficient C=C bonds is unfavorable, at least as an intermolecular processes. To address this challenge, Meggers and coworkers realized a visible-light-induced enantioselective β-amination of α,β-unsaturated 2-acyl imidazoles **278** with *N*-aryl carbamates **279** using an iridium-based photocatalyst in combination with a chiral-at-rhodium Lewis acid (Δ-RhO), through a PCET-enabled N–H activation with the aid of a weak phosphate base, followed by a C–N radical cross-coupling (Scheme 3.46) [60]. The reaction mechanism is proposed in Scheme 3.46b, which begins with a PCET

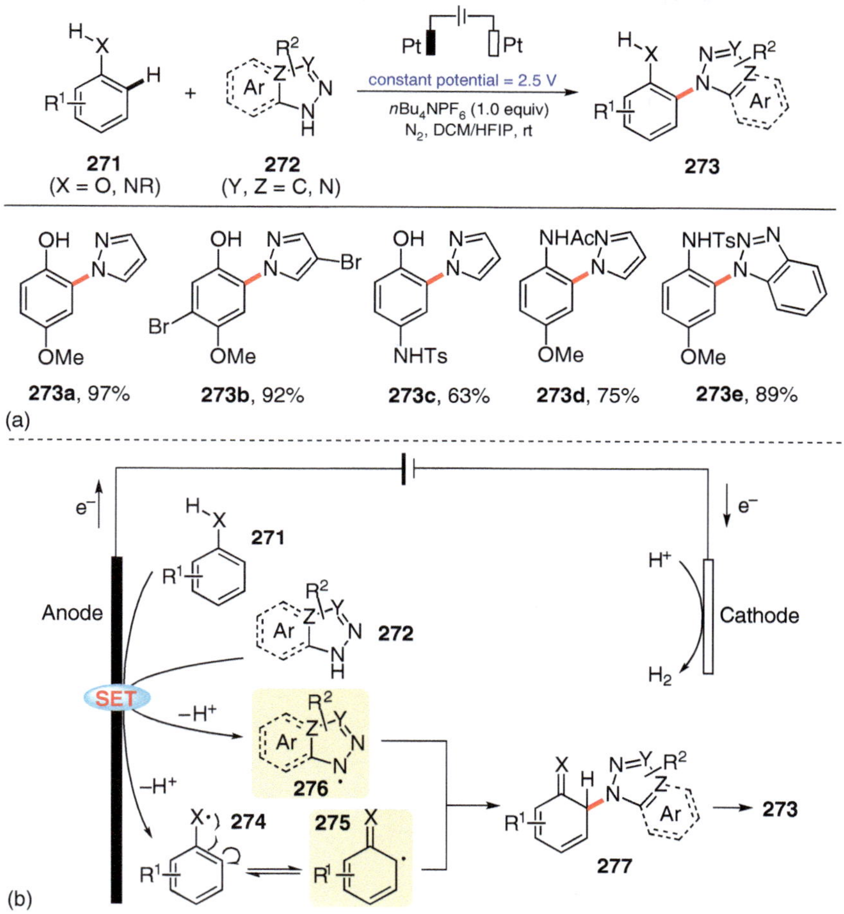

Scheme 3.45 Electro-oxidative C–H azolation of phenol and aniline derivatives. Source: Modified from Feng et al. [59].

process between the photoexcited Ir^{III}* and substrate **279** with the aid of the Brønsted base, affording the carbamoyl radical **281** and the reduced Ir^{II} species. On the other hand, chiral catalyst Δ-RhO binds the α,β-unsaturated compound **278** to form the Rh-bonded intermediate **282**,which then undergoes single-electron reduction with Ir^{II} to provide the Rh-enolate radical intermediate **283** and meanwhile regenerate photocatalyst Ir^{III}. Subsequently, the cross-coupling of radical intermediates **281** and **283** furnishes the C—N bond of **284**. Alternative pathways via the addition of carbamate radical **281** or the corresponding anion to the C=C bonds of Rh-coordinated intermediate **282** have been discussed and excluded on the basis of the experimental observations.

Direct amination of C(sp^3)—H bonds via C/N-radical cross-coupling pathway often employs a HAT strategy (mainly 1,5-HAT) and occurs intramolecularly. This topic has been discussed in detail within another chapter. As a distinct example, in 2017, Lei's group reported a metal- and oxidant-free electrochemical protocol for the direct intermolecular

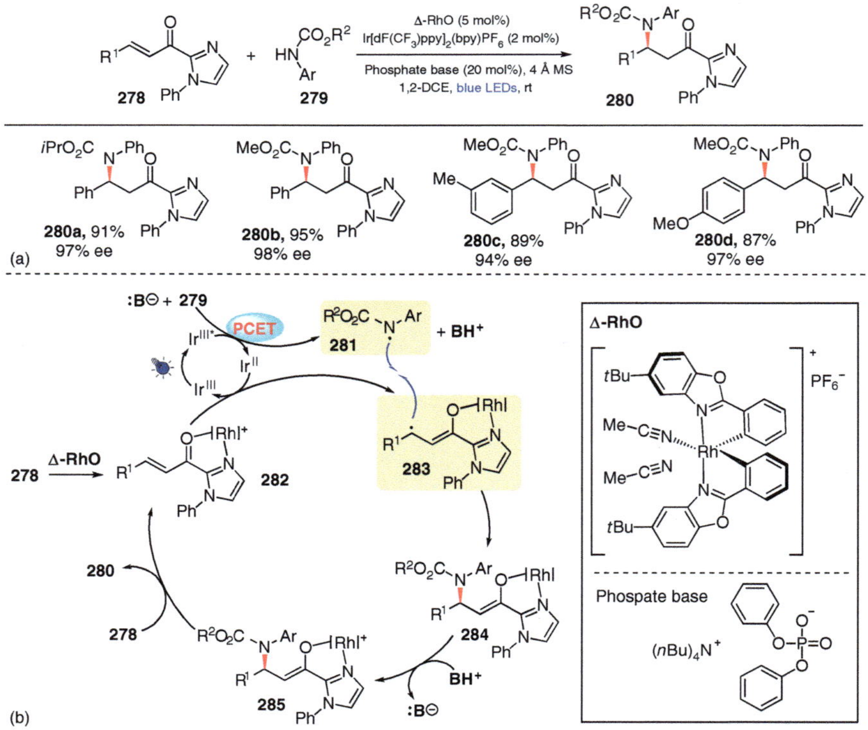

Scheme 3.46 Enantioselective amination via PCET followed by stereo-controlled radical cross-coupling. Source: Modified from Zhou et al. [60].

C(sp^3)—N bond construction via a radical cross-coupling pathway (Scheme 3.47) [61]. The activated C(sp^3)—H bonds on benzylic or allylic sites, as well as those on the α-positions to O, S, and N atoms (**286**), can be aminated with various azoles **287** using this method (**288a–288f**). A tentative mechanism is proposed by the authors on the basis of the mechanistic studies and literature reports, as shown in Scheme 3.47b, in which N-radical **289** serves as both a HAT reagent to produce C-radical **290** and as a coupling partner to react with **290** to afford C–N cross-coupling product **288**.

3.5 Summary and Conclusions

This chapter focuses on the recent progress in direct C—N bond construction from unfunctionalized C–H and N–H partners via photo- or electrochemical methods. The reactions highlighted feature higher atom and step economy compared with the well-established cross-dehydrohalogenative couplings between C—H and N—X or C—X and N—H bonds. Amino functionalities, ranging from secondary to primary amines, are efficiently inserted into various C-centered moieties. Particularly, the conceptually literal CDC amination, providing the amination products with the evolution of H_2 gas, has been achieved by dual photocatalytic systems as well as catalyst-free electrochemical methods without the requirement for any external oxidant. Furthermore, as summarized in this chapter, the radical

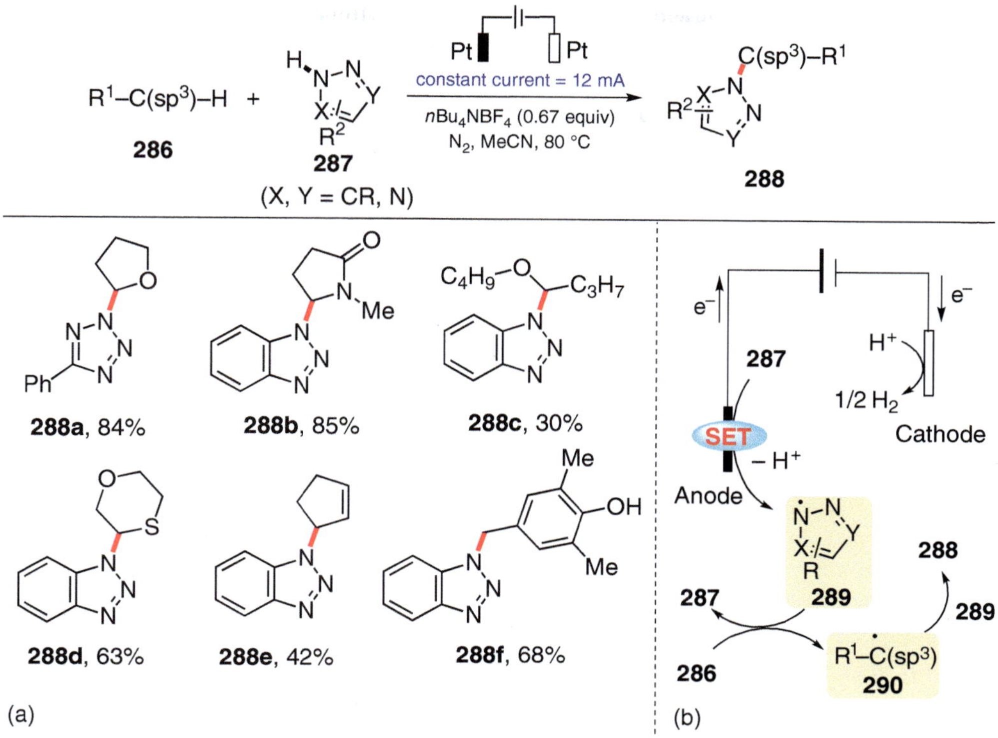

Scheme 3.47 Electrochemical intermolecular oxidative C(sp^3)–H/N–H cross-coupling. Source: Modified from Wu et al. [61].

intermediates engaged in these protocols have expanded the mechanistic patterns to N—H bonds nucleophilic addition, radical cross-coupling, and others such as HAT, which are beyond the familiar N-radical species addition.

Although an array of notable achievements has already been advanced in the recent years, there still remain certain challenges/opportunities in this field: (i) The N–H sources are generally limited to azoles, sulfonamides, secondary amides, carbamates, and diaryl amines. Only few examples have been developed that make use of primary amines or secondary amines. Therefore, developing a versatile strategy with a broader amine scope is still highly desirable; (ii) the direct amination of C(sp^3)–H has only been realized for the aliphatic ones on the activated sites such benzylic or α-positions to a heteroatom, and the HAT-facilitated amination is mostly limited to the functionalization at the remote δ-H to a nitrogen substitution. The exploration of more versatile C(sp^3)–H aminations is also very attractive; (iii) highly enantioselective aminations as well as site-specific aminations on general aromatic rings are still rarely achieved in the simple radical-involved processes, which needs further investigation.

In the near future, further efforts in this area may focus on the development of new catalysts or combined catalytic systems to broaden the amination generality, both in photo- and electrochemical protocols. Comprehensive mechanistic investigations are also necessary for the elucidation of the activation mode of the direct aminations, which can provide a better understanding of the reaction pathway and conversely inspire the design of more elegant reactions with a broader scope. Moreover, application of chiral catalysts to coordinate with the radical intermediates or directing groups installed on the substrates would potentially address the issue of enantioselectivity. Further disclosures are anticipated to make the direct amination strategy one of the most valuable approaches for C—N bond construction.

References

1 Shin, K., Kim, H., and Chang, S. (2015). *Accounts of Chemical Research* 48: 1040–1052.
2 Louillat, M.L. and Patureau, F.W. (2014). *Chemical Society Reviews* 43: 901–910.
3 Kim, H. and Chang, S. (2016). *ACS Catalysis* 6: 2341–2351.
4 Park, Y., Kim, Y., and Chang, S. (2017). *Chemical Reviews* 117: 9247–9301.
5 Selected example and references cited therein:Manna, S., Serebrennikova, P.O., Utepova, I.A. et al. (2015). *Organic Letters* 17: 4588–4591.
6 Xiong, T. and Zhang, Q. (2016). *Chemical Society Reviews* 45: 3069–3087.
7 Chen, J.R., Hu, X.Q., Lu, L.Q., and Xiao, W.J. (2016). *Chemical Society Reviews* 45: 2044–2056.
8 Kärkäs, M.D. (2017). *ACS Catalysis* 7: 4999–5022.
9 Gentry, E.C. and Knowles, R.R. (2016). *Accounts of Chemical Research* 49: 1546–1556.
10 Choi, G.J., Zhu, Q., Miller, D.C. et al. (2016). *Nature* 539: 268–271.
11 Chu, J.C.K. and Rovis, T. (2016). *Nature* 539: 272–275.
12 Zhu, Q., Graff, D.E., and Knowles, R.R. (2018). *Journal of the American Chemical Society* 140: 741–747.
13 Zhu, L., Xiong, P., Mao, Z.Y. et al. (2016). *Angewandte Chemie, International Edition* 55: 2226–2229.
14 Hou, Z.W., Mao, Z.Y., Zhao, H.B. et al. (2016). *Angewandte Chemie, International Edition* 55: 9168–9172.
15 Xiong, P., Xu, H.H., and Xu, H.C. (2017). *Journal of the American Chemical Society* 139: 2956–2959.
16 Hou, Z.W., Mao, Z.Y., Melcamu, Y.Y. et al. (2018). *Angewandte Chemie, International Edition* 57: 1636–1639.
17 Hu, X.Q., Chen, J.R., Wei, Q. et al. (2014). *Angewandte Chemie, International Edition* 53: 12163–12167.
18 Hu, X.Q., Qi, X., Chen, J.R. et al. (2016). *Nature Communications* 7: 11188.
19 Zhao, Q.Q., Hu, X.Q., Yang, M.N. et al. (2016). *Chemical Communications* 52: 12749–12752.
20 Hu, X.Q., Chen, J., Chen, J.R. et al. (2016). *Chemistry - A European Journal* 22: 14141–14146.
21 Zhao, Q.Q., Chen, J., Yan, D.M. et al. (2017). *Organic Letters* 19: 3620–3623.
22 Brachet, E., Marzo, L., Selkti, M. et al. (2016). *Chemical Science* 7: 5002–5006.
23 Maity, S. and Zheng, N. (2012). *Angewandte Chemie, International Edition* 51: 9562–9566.

24 Morris, S.A., Nguyen, T.H., and Zheng, N. (2015). *Advanced Synthesis and Catalysis* 357: 2311–2316.

25 Musacchio, A.J., Nguyen, L.Q., Beard, G.H., and Knowles, R.R. (2014). *Journal of the American Chemical Society* 136: 12217–12220.

26 Musacchio, A.J., Lainhart, B.C., Zhang, X. et al. (2017). *Science* 355: 727–730.

27 Miller, D.C., Ganley, J.M., Musacchio, A.J. et al. (2019). *Journal of the American Chemical Society* 141: 16590–16594.

28 Parasram, M. and Gevorgyan, V. (2017). *Chemical Society Reviews* 46: 6227–6240.

29 Sagadevan, A., Ragupathi, A., Lin, C.-C. et al. (2015). *Green Chemistry* 17: 1113–1119.

30 Tong, K., Liu, X., Zhang, Y., and Yu, S. (2016). *Chemistry - A European Journal* 22: 15669–15673.

31 Yamaguchi, T., Yamaguchi, E., and Itoh, A. (2017). *Organic Letters* 19: 1282–1285.

32 Ito, E., Fukushima, T., Kawakami, T. et al. (2017). *Chem* 2: 383–392.

33 Margrey, K.A., Levens, A., and Nicewicz, D.A. (2017). *Angewandte Chemie, International Edition* 56: 15644–15648.

34 Zhao, H.B., Hou, Z.W., Liu, Z.J. et al. (2017). *Angewandte Chemie, International Edition* 56: 587–590.

35 Tang, S., Wang, S., Liu, Y. et al. (2018). *Angewandte Chemie, International Edition* 57: 4737–4741.

36 Song, C., Liu, K., Wang, Z. et al. (2019). *Chemical Science* 10: 7982–7987.

37 Romero, N.A., Margrey, K.A., Tay, N.E., and Nicewicz, D.A. (2015). *Science* 349: 1326–1330.

38 Niu, L., Yi, H., Wang, S. et al. (2017). *Nature Communications* 8: 14226.

39 Pandey, G., Singh, D., and Laha, R. (2017). *Asian Journal of Organic Chemistry* 6: 469–474.

40 Wang, J.H., Lei, T., Nan, X.L. et al. (2019). *Organic Letters* 21: 5581–5585.

41 Meyer, A.U., Berger, A.L., and König, B. (2016). *Chemical Communications* 52: 10918–10921.

42 Song, C., Yi, H., Dou, B. et al. (2017). *Chemical Communications* 53: 3689–3692.

43 Zheng, Y.W., Chen, B., Ye, P. et al. (2016). *Journal of the American Chemical Society* 138: 10080–10083.

44 Nguyen, T.M. and Nicewicz, D.A. (2013). *Journal of the American Chemical Society* 135: 9588–9591.

45 Nguyen, T.M., Manohar, N., and Nicewicz, D.A. (2014). *Angewandte Chemie, International Edition* 53: 6198–6201.

46 Yi, H., Niu, L., Song, C. et al. (2017). *Angewandte Chemie, International Edition* 56: 1120–1124.

47 Hu, X., Zhang, G., Bu, F., and Lei, A. (2018). *Angewandte Chemie, International Edition* 57: 1286–1290.

48 Pandey, G. and Laha, R. (2015). *Angewandte Chemie, International Edition* 54: 14875–14879.

49 Pandey, G., Laha, R., and Singh, D. (2016). *The Journal of Organic Chemistry* 81: 7161–7171.

50 Lian, F., Sun, C., Xu, K., and Zeng, C. (2019). *Organic Letters* 21: 156–159.

51 Wang, P., Yang, Z., Wu, T. et al. (2019). *ChemSusChem* 12: 3073–3077.

52 Jing, D., Lu, C., Chen, Z. et al. (2019). *Angewandte Chemie, International Edition* 58: 14666–14672.

53 Zhao, Y., Huang, B., Yang, C., and Xia, W. (2016). *Organic Letters* 18: 3326–3329.

54 Zhao, Y., Huang, B., Yang, C. et al. (2017). *ACS Catalysis* 7: 2446–2451.

55 Liu, K., Tang, S., Wu, T. et al. (2019). *Nature Communications* 10: 639.

56 Wu, Y.C., Jiang, S.S., Song, R.J., and Li, J.H. (2019). *Chemical Communications* 55: 4371–4374.

57 Samanta, S., Ravi, C., Rao, S.N. et al. (2017). *Organic & Biomolecular Chemistry* 15: 9590–9594.

58 Yu, Y., Yuan, Y., Liu, H. et al. (2019). *Chemical Communications* 55: 1809–1812.

59 Feng, P., Ma, G., Chen, X. et al. (2019). *Angewandte Chemie, International Edition* 58: 8400–8404.

60 Zhou, Z., Li, Y., Han, B. et al. (2017). *Chemical Science* 8: 5757–5763.

61 Wu, J., Zhou, Y., Zhou, Y. et al. (2017). *ACS Catalysis* 7: 8320–8323.

4

Propargylamines: Recent Advances in Asymmetric Synthesis and Use as Chemical Tools in Organic Chemistry

Fei Zhao, Seong-Heun Kim and Daniele Castagnolo

School of Cancer and Pharmaceutical Sciences, King's College London, 150 Stamford Street, SE1 9NH London, UK

4.1 Introduction

Propargylamines are a wide class of organic compounds presenting peculiar chemical properties, mainly because of their unique structure composed of an amine group in β-position to an alkyne moiety. Propargylamines can behave as electrophiles or nucleophiles in organic transformations, thus proving to be exceptional building blocks in the construction of diverse pharmaceutical ingredients, chemicals, and natural products [1]. The carbon–carbon triple bond of propargylamines is electrophilic and can react with nucleophilic species. In addition, propargylamines possess an amine moiety that can undergo a variety of nucleophilic reactions. The combination of a nucleophilic amine with an alkyne moiety in the same molecule offers a unique mix of reactivities that can be exploited in many chemical reactions and in the construction of challenging organic compounds.

Historically, the first paper describing the use of propargylamines as building blocks in chemical synthesis goes back to the 1940s–1950s, and since then, their use in organic synthesis has grown exponentially [2–4]. Nowadays, propargylamines can be considered as the best choice of substrates for the synthesis of many heterocycles such as oxazoles, imidazoles, thiazoles, and pyrroles, as well as in bioconjugation reactions and click chemistry.

Not less important, many propargylamines also possess pharmaceutical properties and have found direct applications in the treatment of the neurodegenerative Parkinson's and Alzheimer's diseases (Figure 4.1) [5, 6]. Pargyline, a *N*-benzyl-*N*-methylpropargylamine, is an irreversible selective monoamine oxidase-B (MAO-B) inhibitor, which found applications in the treatment of hypertension, while the analog clorgiline, an irreversible and selective inhibitor of monoamine oxidase A (MAO-A), has been used as antidepressant. Similarly, the two stereo-defined chiral propargylamines rasagiline and selegiline have found applications in the treatment of Parkinson's and Alzheimer's diseases, respectively, by inhibiting the action of MAO-B enzymes.

Because of their unique chemical properties, many synthetic methods have been developed in the past decades to access propargylamines, especially in enantiomerically pure form. Structurally simple propargylamines can be synthesized by alkylation reactions

Methodologies in Amine Synthesis: Challenges and Applications, First Edition.
Edited by Alfredo Ricci and Luca Bernardi.

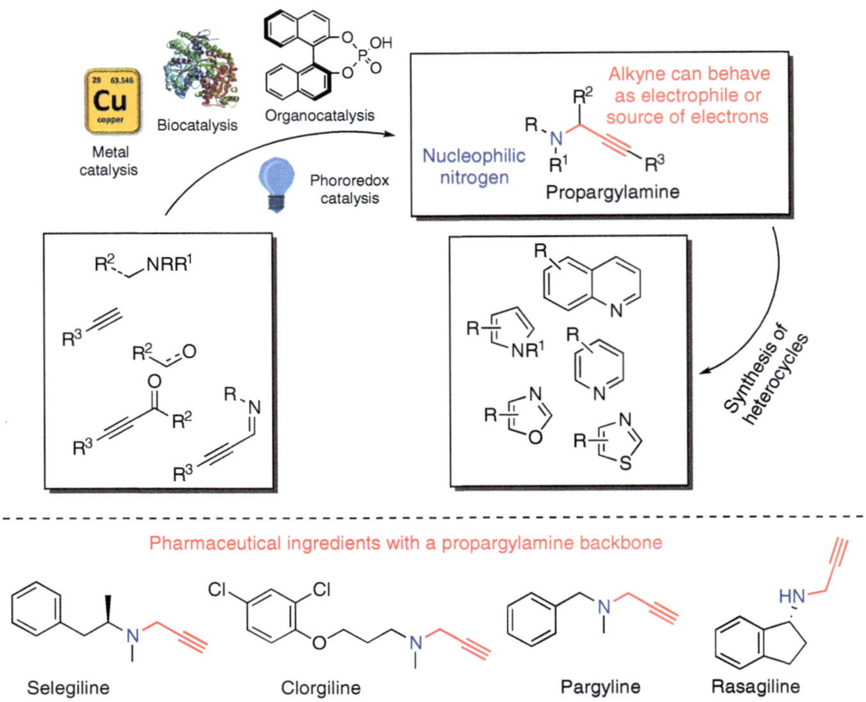

Figure 4.1 Propargylamines: synthetic methods and drugs containing a propargylamine backbone.

of amines with propargylic halides, propargylic phosphates, or propargylic triflates or by functionalization of commercially available propargylamines with alkyl halides [7–9]. However, many propargylamines of chemical and pharmaceutical interest are chiral, and their synthesis has often represented a major challenge in organic chemistry.

This chapter describes the recent advances made in the asymmetric synthesis of propargylamines in the past decade. A detailed review covering all the different methodologies used to synthesize propargylamines has been recently reported by our research group [10]. Thus, in this chapter, the most recent advances in the enantioselective synthesis of propargylamines via metal-catalyzed reactions will be presented. In addition, recent catalytic and green methodologies to access propargylamines via biocatalyzed, photoredox, and organocatalyzed reactions are described. Finally, in the last part of the chapter, an overview of the use of propargylamines as building blocks in the synthesis of diverse heterocycles and other chemicals is presented.

4.2 Metal-Catalyzed Asymmetric Synthesis of Propargylamines

4.2.1 Enantioselective A^3 Coupling

The metal-catalyzed, asymmetric, three-component coupling of aldehydes, amines, and alkynes (A^3 coupling) which features readily available inputs and simple operation, is considered as one of the most straightforward strategy for the synthesis of chiral propargylamines. This coupling involves the asymmetric addition of the alkyne nucleophiles to the imines/iminiums prepared in advance or formed in situ from aldehydes and amines in the presence of a metal catalyst and a chiral ligand.

Among the metals used, copper complexes are the most frequently used catalysts because of their outstanding coordination properties with a variety of chiral ligands as well as their high affinity with alkynes. With regard to the chiral ligands employed, bidentate and tridentate nitrogen-based ligands such as bis(oxazolinyl)pyridine (Pybox) ligands, bis(imidazoline) (Pybim) ligands, binaphthyl bis(imine), N-tosylated aminoimines, and di(2-aminoaryloxy)1,1′-binaphthyl ligands are those normally used in asymmetric A^3 coupling involving primary amines, while the axially chiral P,N-ligands such as Quinap, Pinap, StackPhos, StackPhim, and UCD-Phim were the optimal ligands for the enantioselective A^3 couplings with secondary amines (Scheme 4.1).

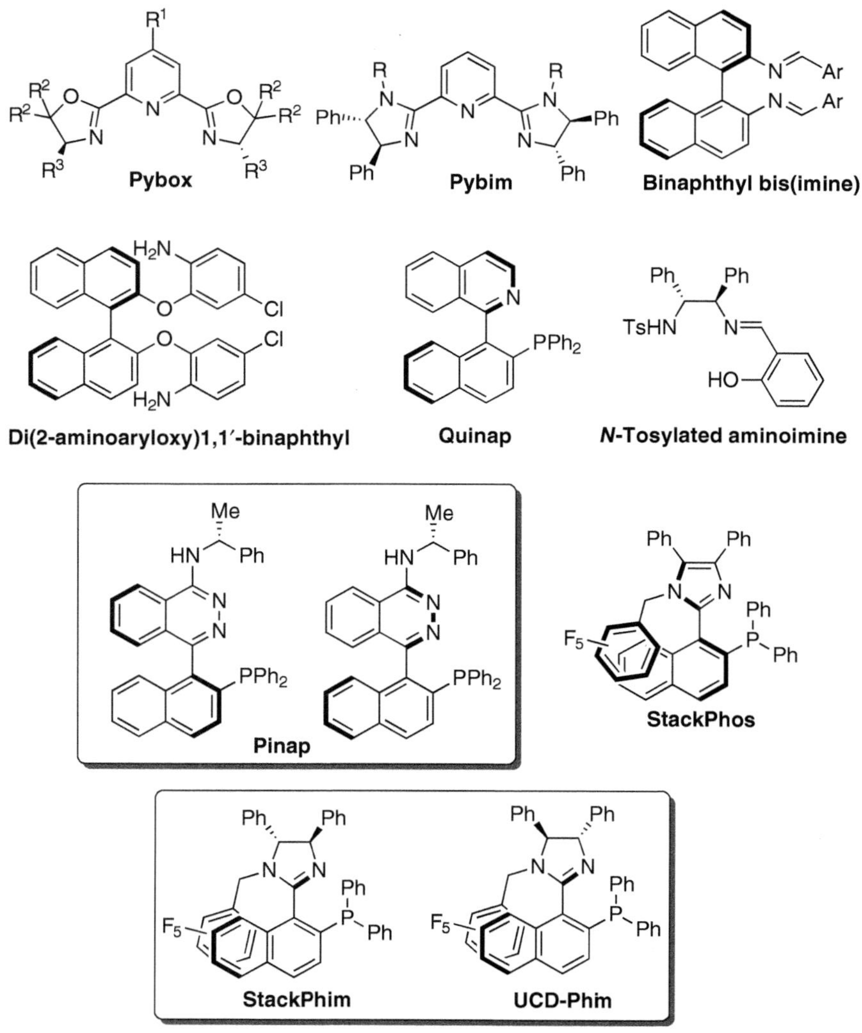

Scheme 4.1 Chiral ligands used in asymmetric A^3 coupling reactions.

Very recently, Guiry and coworkers published a comprehensive review on the asymmetric A^3 couplings [11]. Therefore, this chapter will focus on the most representative examples on enantioselective A^3 couplings.

4.2.1.1 Enantioselective A^3 Coupling Involving Primary Amines

In 2002, the group of Li described the first example of catalytic enantioselective A^3 coupling exploiting a CuOTf catalyst and a Pybox ligand **L1** (Scheme 4.2) [12]. This process was relatively sluggish because it took four days to complete the reactions at 22 °C; the increase of the reaction temperature to 35 °C halved the reaction time but at the cost of enantioselectivity. As demonstrated by the authors, both toluene and water turned out to be suitable solvents, although higher yields and enantioselectivities were observed in the former. This simple process tolerates a variety of aromatic aldehydes and anilines, and they could couple with phenyl acetylene to provide a diverse range of chiral propargylic amines **1** in good to high yields and enantiomer excesses (ees) under mild reaction conditions.

Later, Li and coworkers also examined various aliphatic alkynes as the reaction component; however, the ees of the corresponding propargylamines were lower (<86%) [13]. This disappointing result was attributed to the lower steric hindrance of alkyl alkynes as compared with the aryl-substituted ones. After Li's pioneering work, numerous modifications and variants of asymmetric A^3 coupling using Pybox-type ligands have been reported, with the aim of broadening the substrate scope, improving the enantioselectivities and exploiting other transition metal catalysts. These reports made the asymmetric A^3 coupling to be a practical, powerful, and straightforward strategy for the synthesis of enantiomerically pure propargylamines.

A Pybim ligand **L2**, which could be regarded as the analog of Pybox ligands, was employed together with a copper catalyst by the group of Nakamura to catalyze the asymmetric A^3 coupling of aldehydes, 4-methoxyaniline, and alkynes (Scheme 4.3a) [14]. Notably, the Pybim-type ligand **L2** was highly efficient in stereocontrol in the coupling reactions of aromatic alkynes as well as alkyl alkynes. These reagents react smoothly with a broad range of aryl/alkyl aldehydes and 4-methoxyaniline to give the desired propargylamines **2** in high ees.

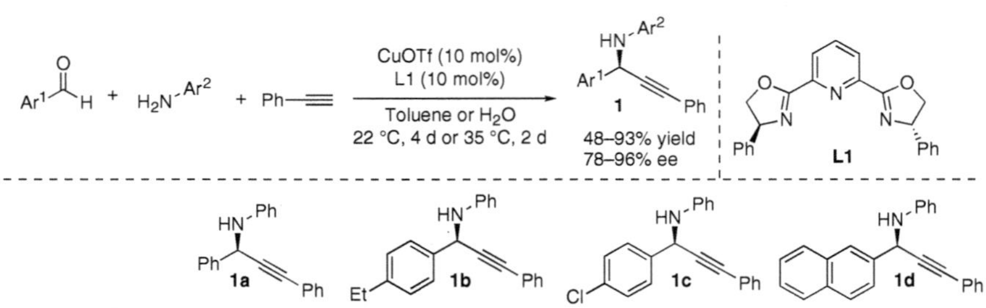

	1a	1b	1c	1d
Toluene: 22 °C, 4 d	78% yield, 96% ee	70% yield, 96% ee	85% yield, 94% ee	63% yield, 88% ee
Toluene: 35 °C, 2 d	83% yield, 93% ee	73% yield, 95% ee	90% yield, 92% ee	67% yield, 82% ee
H_2O: 22 °C, 4 d	71% yield, 84% ee	68% yield, 89% ee	70% yield, 87% ee	57% yield, 86% ee
H_2O: 35 °C, 2 d	77% yield, 80% ee	68% yield, 78% ee	74% yield, 85% ee	65% yield, 78% ee

Scheme 4.2 Li's approach to the enantioselective synthesis of propargylamines. Source: Based on Wei and Li [12].

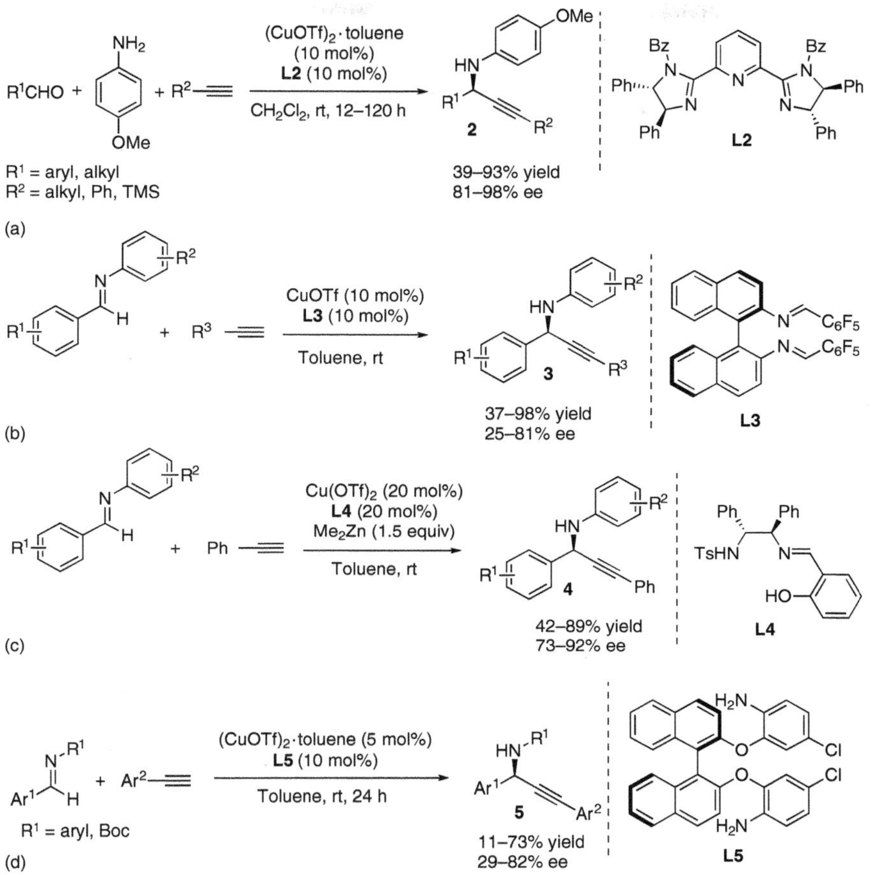

Scheme 4.3 Enantioselective A^3 coupling approaches to propargylamines. Source: (a) Modified from Nakamura et al. [14], (b) modified from Benaglia et al. [15], (c) modified from Liu et al. [16], and (d) modified from Hatano et al. [17].

In 2004, a chiral binaphthyl bis(imine) ligand **L3**, which could be synthesized from (*S*)-binaphthyl diamine and pentafluorobenzaldehyde, was used by Benaglia and coworkers to catalyze the asymmetric addition of aryl and alkyl acetylenes to *N*-aryl arylimines (Scheme 4.3b) [15]. Interestingly, alkyl acetylenes took part in this A^3 coupling to deliver the products in higher ees than aryl acetylenes. However, only moderate to good ees were observed with all the desired propargylamines **3** in the presence of **L3**, indicating that the binaphthyl bisimine-type ligands are less efficient in stereocontrol in A^3 coupling reactions than Pybox-type ligands and Pybim-type ligands.

In 2007, Chan's group found that a chiral tridentate N-tosylated aminoimine ligand **L4** was effective in chiral induction in copper-catalyzed alkynylation of *N*-aryl arylimines (Scheme 4.3c) [16]. A broad range of *N*-aryl arylimines was reacted with phenylacetylene to afford the corresponding propargylamines **4** in good to high yields and ees. Later, in 2008, Ishihara's group reported the enantioselective alkynylation of *N*-aryl aldimines with arylacetylenes using a catalyst system consisting of copper triflate and a di(2-aminoaryloxy)1,1′-binaphthyl ligand **L5**. Low to good yields and moderate to good ees were observed with the desired products **5** (Scheme 4.3d) [17].

4.2.1.2 Enantioselective A³ Coupling Involving Secondary Amines

The A^3 coupling of secondary amines with aldehydes and alkynes involves the formation of iminium ion intermediates rather than imine species. Although the iminiums are more electrophilic as compared with imines, the coordination of a chiral catalyst to iminiums is more difficult because of the lack of a lone electron pair on nitrogen atom. Therefore, it is much more challenging to achieve the asymmetric A^3 coupling involving secondary amines. The usage of axially chiral P,N-ligands such as Quinap, Pinap, StackPhos, Stack-Phim, and UCD-Phim well addresses this issue.

In 2003, Knochel and coworkers described the first example of enantioselective A^3 coupling involving secondary amines employing a catalytic amount of CuBr and (*R*)-(+)-Quinap **L6** (Scheme 4.4) [18]. Despite the requirement of a long reaction time, various aldehydes and alkynes underwent this asymmetric A^3 coupling with dibenzylamine or diallylamine to produce the desired propargylamines **6** in good to high yields and ees under mild reaction conditions. In their catalytic system, alkyl aldehydes were found to offer higher enantioselectivities than the aromatic ones, and trimethylsilylacetylene was found to give higher enantioselectivities than other alkynes. As shown in Scheme 4.5, the proposed reaction mechanism indicates that the dimeric chiral copper complex **7** coordinates with alkynes first to produce the copper-alkyne complex **8**, which reacts with the aminal formed from amines and aldehydes to give a copper-alkyne-aminal complex **9**. Deprotonation of the coordinated alkyne and the elimination of H_2O provide a copper-alkyne-iminium intermediate, and the following addition of the acetylide to the iminium ion under a chiral sphere affords the chiral propargylamines **6** and regenerates the catalyst **7**.

Following Knochel's breakthrough work on asymmetric A^3 coupling with secondary amines, the group of Carreira adopted the copper complexes of the axially chiral P,N-ligands

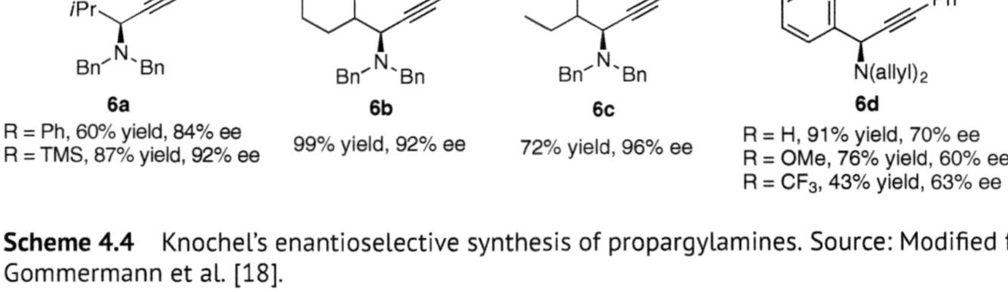

Scheme 4.4 Knochel's enantioselective synthesis of propargylamines. Source: Modified from Gommermann et al. [18].

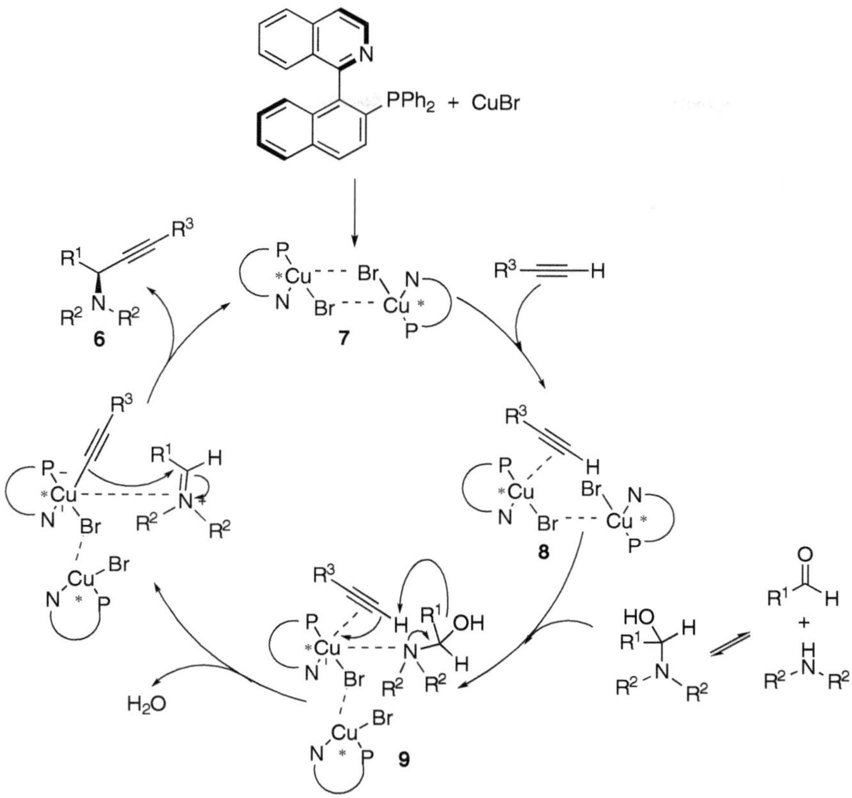

Scheme 4.5 Proposed mechanism for the (*R*)-(+)-Quinap-catalyzed A^3 coupling.

(Pinap **L7** and **L8**), which feature a chiral axis as well as a chiral center, to catalyze the asymmetric A^3 coupling of alkyl aldehydes, dibenzylamine, and alkynes (Table 4.1) [19].

All the desired propargylamines **10** were obtained in high yields and excellent ees, and both enantiomers could be prepared in high ees by changing the axial configuration of the Pinap ligand. From a chemical point of view, Pinap **L7** and **L8** are structurally similar to Quinap. However, it should be noted that the Pinap-type ligands **L7** and **L8** are superior to the Quinap-type ligand **L6** in stereocontrol in the asymmetric A^3 coupling involving secondary amines. In 2013, the group of Aponick reported a novel imidazole-based chiral biaryl P,N-ligand **L9** ((*S*)-StackPhos) for the synthesis of chiral propargylamines **11** (Scheme 4.6) [20]. The new ligand was designed with the aim to increase the rotational barrier in biaryl systems and was then employed with CuBr in A^3 coupling reactions. Amazingly, not only aliphatic aldehydes but also aromatic aldehydes could react well with dialkyl amines and trimethylsilylacetylene to deliver the desired propargylamines in high yields and excellent ees. Compared with Quinap- or Pinap-mediated approaches, in which a long reaction time (several days) is required to achieve good to high yields, the StackPhos **L9**-catalyzed A^3

Table 4.1 Carreira's asymmetric approach to propargylamines.

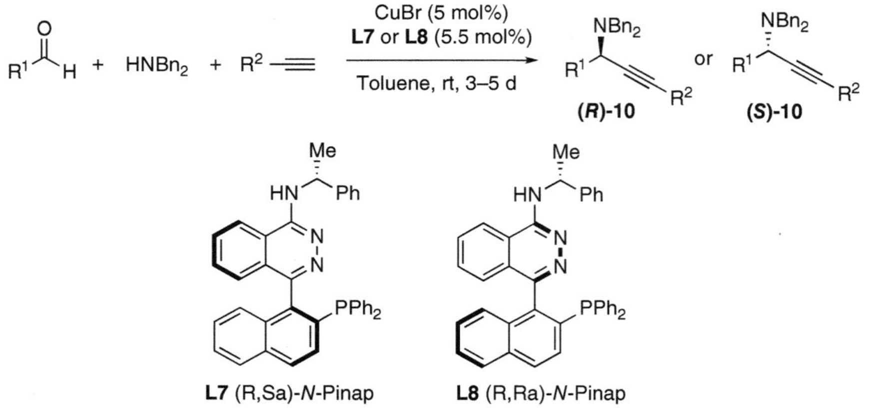

Compound	R¹	R²	Ligand	Yield (%)	ee (%)
10a	*i*Pr	TMS	**L7**	84	98 (*R*)
			L8	82	99 (*S*)
10b	*i*Pr	Ph	**L7**	88	90 (*R*)
			L8	82	95 (*S*)
10c	*i*Bu	*n*Bu	**L7**	74	91 (*R*)
			L8	72	94 (*S*)

couplings are generally completed within a shorter time (24 hours). Therefore, considering the reactivity and efficiency in stereocontrol, the StackPhos **L9** is a better ligand than Quinap **L6** and Pinap **L7**/**L8** for the asymmetric A^3 couplings involving secondary amines.

Later, using a similar strategy, the same group achieved the synthesis of chiral 3-amino-skipped diynes **12** from alkynylic aldehydes with StackPhos **L9** (Scheme 4.7) [21].

In 2017, Aponick and Guiry almost simultaneously reported the StackPhos-derived chiral P,N-ligands StackPhim **L10** and UCD-Phim **L11**, respectively (Schemes 4.8 and 4.9) [22, 23].

Compared with StackPhos, both **L10** and **L11** possess central and axial chirality elements, and they were proved to be highly efficient in catalyzing the asymmetric A^3 coupling reactions. It should be noted that the axial chirality element in **L10** and **L11** is crucial for chiral induction, as the employment of corresponding ligands without axial chirality but carrying the central chirality causes remarkably reduced enantioselectivity.

Aponick and coworkers described an A^3 coupling/cyclization sequence with CuBr/StackPhim (**L10**) to prepare chiral arylmethanamines **14** with high yields and ees (Scheme 4.8) [22], while Guiry and coworker reported CuBr/UCD-Phim (**L11**)-catalyzed asymmetric A^3 couplings to furnish the corresponding propargylamines **15** in good to excellent yields and ees (Scheme 4.9) [23].

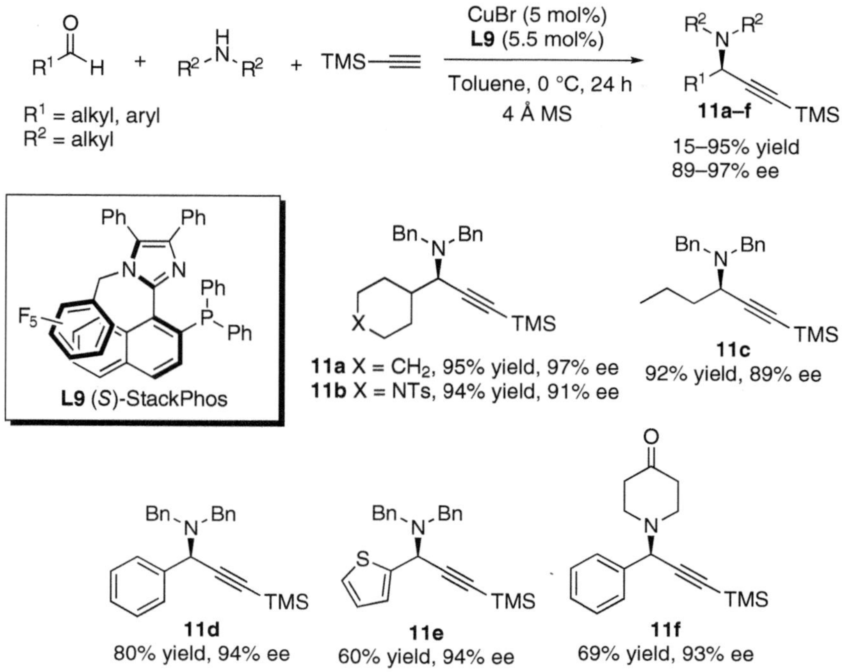

Scheme 4.6 Synthesis of propargylamines using (*S*)-StackPhos ligand. Source: Modified from Cardoso et al. [20].

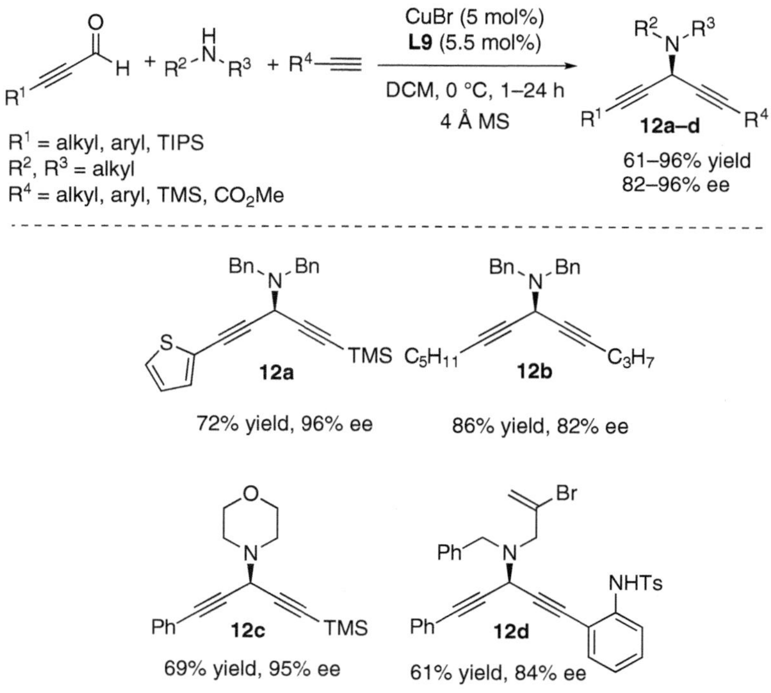

Scheme 4.7 Aponick's strategy for the asymmetric synthesis of 3-amino skipped diynes. Source: Paioti et al. [21].

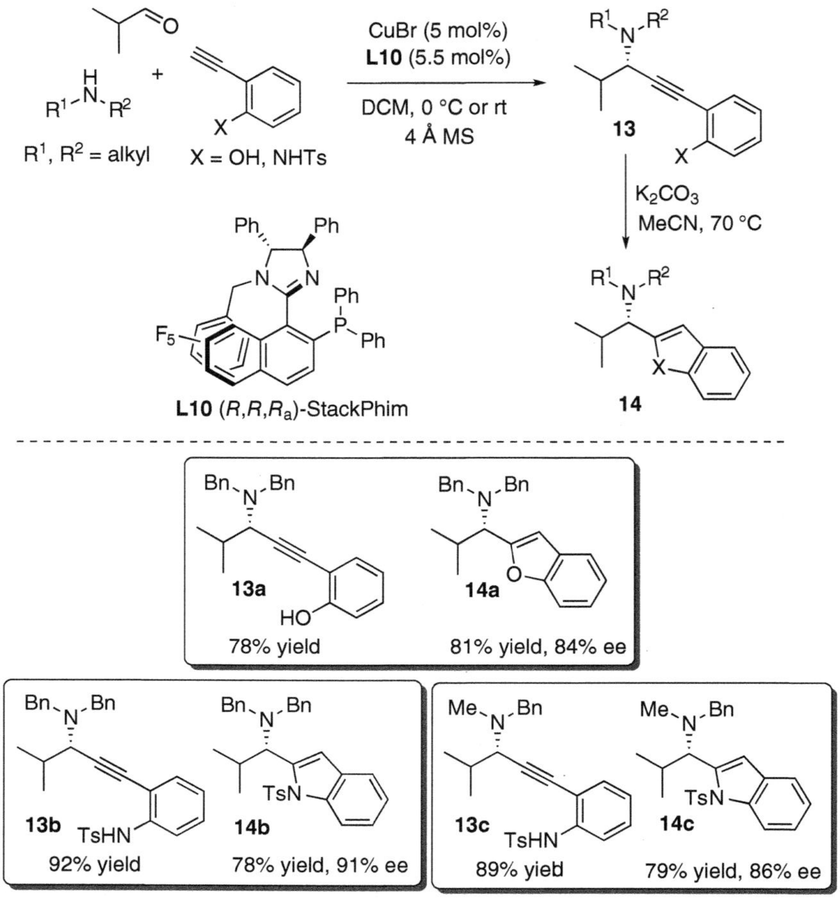

Scheme 4.8 Asymmetric synthesis of propargylamines using the StackPhim ligand **L10**. Source: Paioti et al. [22].

4.2.2 Enantioselective Propargylic Amination of Propargylic Esters with Amines

The first example of asymmetric propargylic amination reaction to access enantiomerically pure propargylamines was reported by van Maarseveen et al. in 2008 (Scheme 4.10) [24].

In this work, the combination of CuI, Pybox **L12**, and $i\mathrm{Pr_2NEt}$ was found to be effective for the enantioselective synthesis of propargylamines **17** from propargylic acetates **16** and primary anilines. It should be noted that the base $i\mathrm{Pr_2NEt}$ was indispensable in terms of both the yields and enantioselectivities, and the terminal acetylene group was also crucial for the success of this reaction. Besides, the aryl groups at the R¹ position were necessary for high enantioselectivities as the introduction of alkyl substituents caused poor enantioselectivities. With the CuI/Pybox/$i\mathrm{Pr_2NEt}$ system, a diversity of chiral propargylamines **17** was obtained in good to high yields and ees.

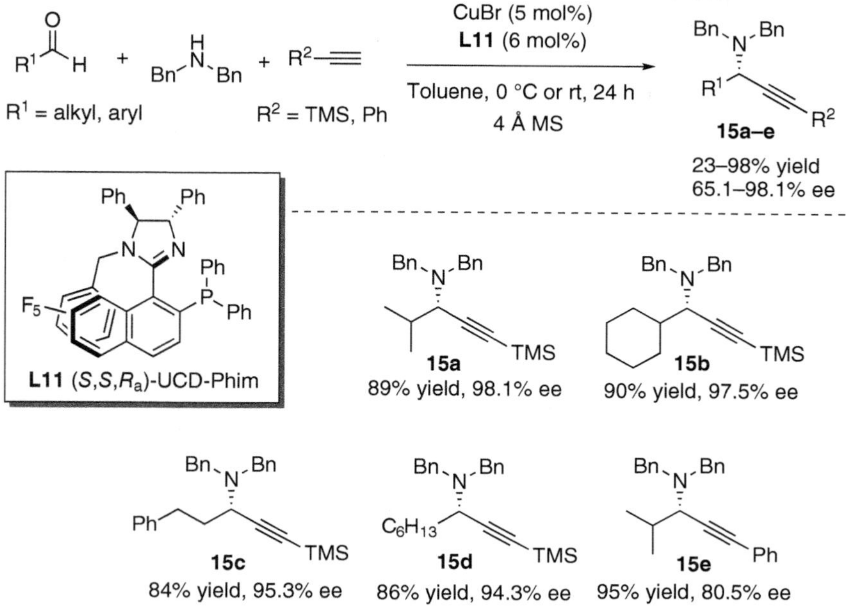

Scheme 4.9 Asymmetric synthesis of propargylamines using the UCD-Phim **L11**. Source: Rokade and Guiry [23].

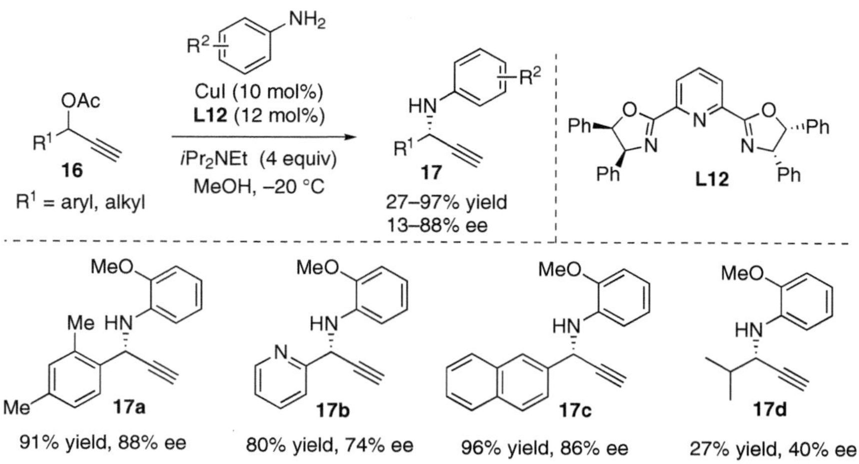

Scheme 4.10 Asymmetric synthesis of propargylamines using ligand **L12**. Source: Detz et al. [24].

Later, van Maarseveen and coworkers identified another two Pybox ligands **L13** and **L14**, with which even diverse propargylic esters **18** bearing alkyl groups at the propargylic position could be aminated in good to high ees (Scheme 4.11) [25].

A proposed mechanism (Scheme 4.12) involves the formation of the copper-allenylidene intermediates, which were then attacked by the amines to give the products. Finally, the

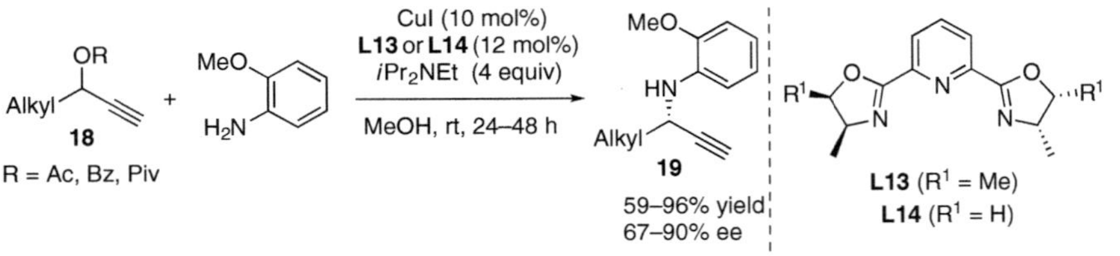

Scheme 4.11 Synthesis of propargylamines using Pybox ligands **L13** and **L14**. Source: Detz et al. [25].

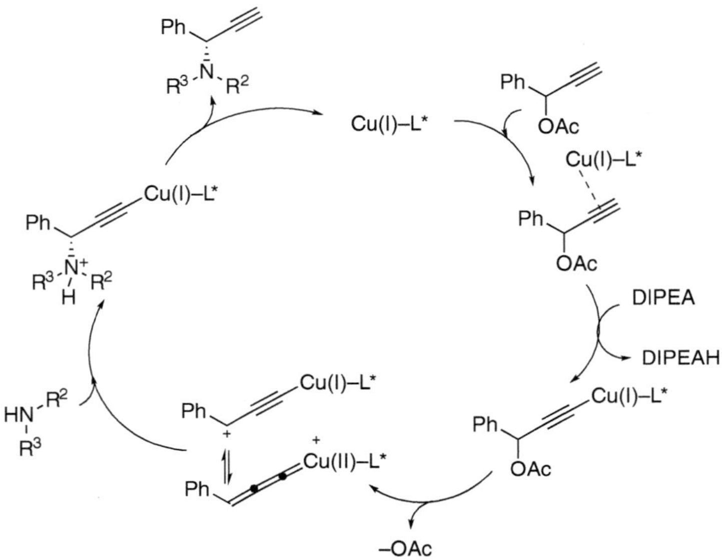

Scheme 4.12 Proposed mechanism for the synthesis of propargylamines using Pybox ligands **L13** and **L14**.

utility of this reaction was demonstrated in the synthesis of the optically active ClickPhine ligand, the antibiotic (+)-anisomycin, and the cytokine modulator (−)-cytoxazone.

Since van Maarseveen's pioneering work, several variations on the asymmetric propargylic amination appeared in the literature. For example, soon after van Maarseveen's first report, Nishibayashi's group reported a similar enantioselective propargylic amination of acetates **20** with secondary amines catalyzed by CuOTf·$(C_6H_6)_{0.5}$/**L15** (Scheme 4.13a) [26, 29].

Likewise, the aryl group at the propargylic position was essential to achieve high enantioselectivities. However, much higher enantioselectivities were generally observed with secondary amines such as *N*-alkyl anilines and *N*,*N*-dialkyl amines rather than with primary anilines. This contrasts sharply with van Maarseveen's CuI/Pybox system, in which primary anilines produced the highest enantioselectivity of the products.

Subsequently, Nishibayashi and coworkers reported a Cu/**L16**-catalyzed asymmetric propargylic amination of propargylic pentafluorobenzoates **22** carrying an alkyl group at

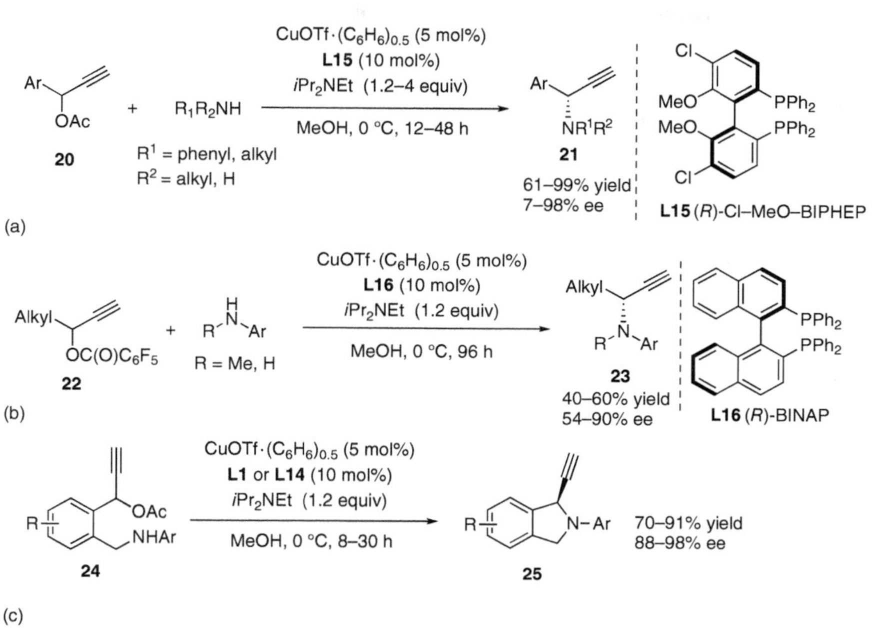

Scheme 4.13 Nishibayashi's approaches to propargylamines. Source: (a) Modified from Hattori et al. [26], (b) modified from Yoshida et al. [27], and (c) modified from Shibata et al. [28].

the propargylic position by amines (Scheme 4.13b) [27]. In addition, they also developed an intramolecular version of this enantioselective propargylic amination using Cu/Pybox, which afforded the optically active 1-ethynyl-isoindolines **25** in high yields and ees (Scheme 4.13c) [28].

Afterward, in 2012, the group of Hu improved van Maarseveen's work by employing chiral tridentate P,N,N ligands **L17** and **L18** (Scheme 4.14) [30]. The system of CuCl/**L17** was demonstrated to be effective for the asymmetric propargylic amination of aryl-substituted propargylic acetates **26** with primary anilines, secondary phenyl alkyl anilines, and even dialkyl amines, and all the aminated products were produced in good to high yields and excellent ees. Notably, the combination of CuCl or $Cu(OAc)_2{\cdot}H_2O$ with **L17** or **L18** worked in a highly enantioselective manner on the amination of the more challenging aliphatic propargylic acetates **28**, affording the propargylamines **29** in good to high yields and ees.

Recently, Tsuji and Kawatsura also disclosed an improved procedure that uses a $Ni(cod)_2$/**L19** complex to catalyze the asymmetric propargylic amination of propargylic carbonates **30** bearing an internal alkyne group (Scheme 4.15) [31]. A variety of primary and secondary aryl amines could undergo this reaction to deliver the desired products **31** in good to high yields and excellent ees. As suggested by the authors, an allenyl nickel intermediate **32** is proposed as the key intermediate. This Ni-catalyzed protocol provides an additional complementarity to the pre-existing Cu-catalyzed methods, in which propargylic esters bearing an internal alkyne moiety do not react at all.

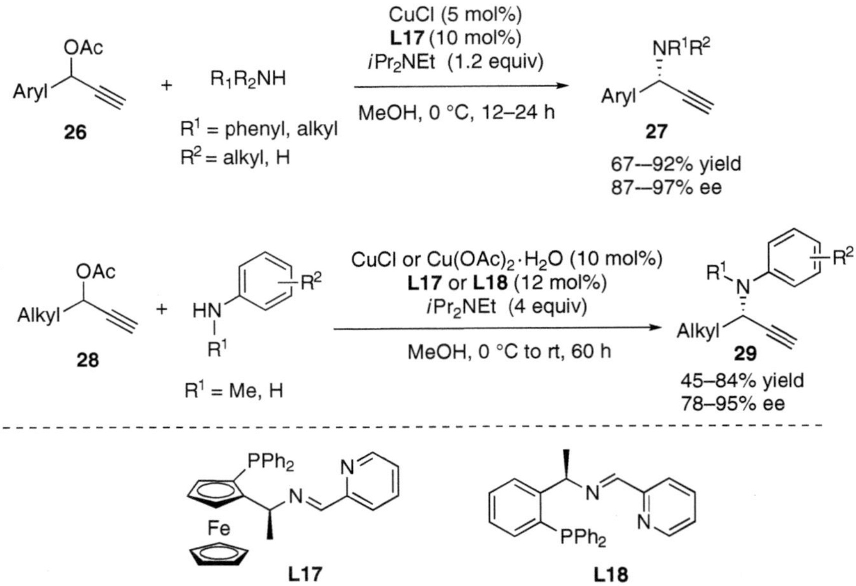

Scheme 4.14 Hu's synthesis of propargylamines using chiral tridentate P,N,N ligands **L17** and **L18**. Source: Modified from Zhang et al. [30].

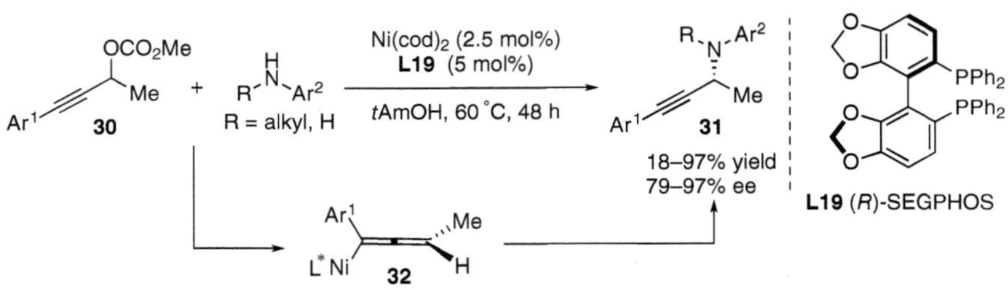

Scheme 4.15 Use of a $Ni(cod)_2$/**L19** complex in the asymmetric synthesis of propargylamines. Source: Modified from Watanabe et al. [31].

4.2.3 Cu-Catalyzed Enantioselective Ring Opening of Alkynyl-Substituted Epoxides/Lactones/Carbonates

Prompted by the copper-allenylidene intermediates proposed in the copper-catalyzed enantioselective propargylic substitutions of propargylic acetates with amines, the group of Nishibayashi envisaged that ethynyl epoxides could also be employed as the substrates to generate similar intermediates, which could be attacked by the amines to afford the chiral propargylamines. In 2009, they reported a $Cu(OTf)_2$-catalyzed enantioselective ring-opening reaction of ethynyl epoxides **33** by amines using the chiral ligand **L20** to give the β-OH chiral propargylamines **34** with up to 98% yield and 94% ee (Scheme 4.16) [32]. Overall, much higher enantioselectivities were observed with anilines as compared with monoalkyl- and diakylamines. The electronic nature and position of the substituents

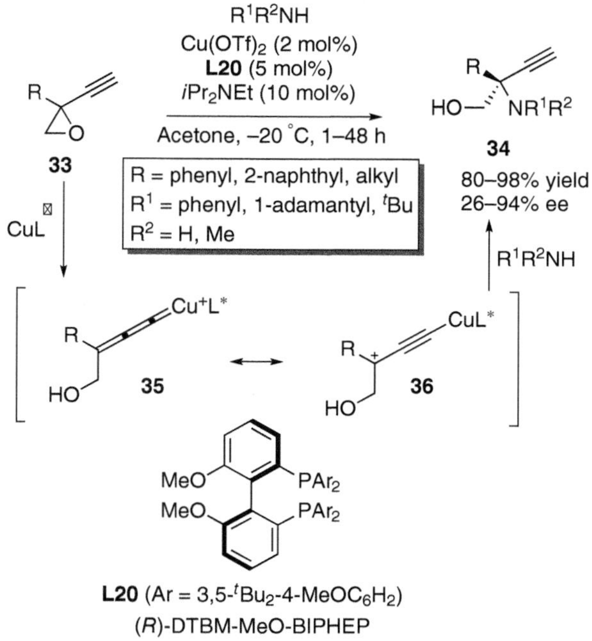

Scheme 4.16 Synthesis of propargylamines from ethynyl epoxides. Source: Modified from Hattori et al. [32].

in the phenyl ring of anilines significantly affected the enantioselectivities, and the highest enantioselectivities (93–94% ee) were observed with 4-trifluoromethylaniline and 4-methoxycarbonylaniline. Besides, the reactions of ethynyl epoxides bearing aryl groups at the R position generally furnished the corresponding propargylamines with much higher enantioselectivities than substrates carrying alkyl groups at the same position. Control experiments showed that ethynyl epoxides bearing an internal alkyne moiety could not undergo this reaction, which indicated that the copper-allenylidene complexes **35–36** were likely to be involved as the key intermediates.

Following a similar strategy, in 2017, Kleij and Guo disclosed a $Cu(CH_3CN)_4(OTf)$-catalyzed enantioselective synthesis of γ-COOH chiral propargylamines **38** bearing a quaternary stereocenter via the asymmetric ring-opening reaction of lactones **37** with amines (Scheme 4.17a) [33]. Various lactones reacted well with diverse primary amines to produce the desired chiral γ-ethynyl-γ-amino acids **38** with up to 98% yield and 98 : 2 enantiomer ratio (er) in the presence of Cu/**L21** catalytic system. Likewise, Zhang and coworkers also achieved in 2018 the asymmetric synthesis of β-OH chiral propargylamines **40** from propargyl-substituted cyclic carbonates **39** and amines under the catalysis of CuI/**L22** (Scheme 4.17b) [34].

A variety of cyclic carbonates with electron-rich, electron-deficient, or halogenated aromatic rings as well as anilines and alkyl amines were well tolerated to provide the β-amino-β-ethynyl alcohols **40** with up to 99% yield and 94 : 6 er.

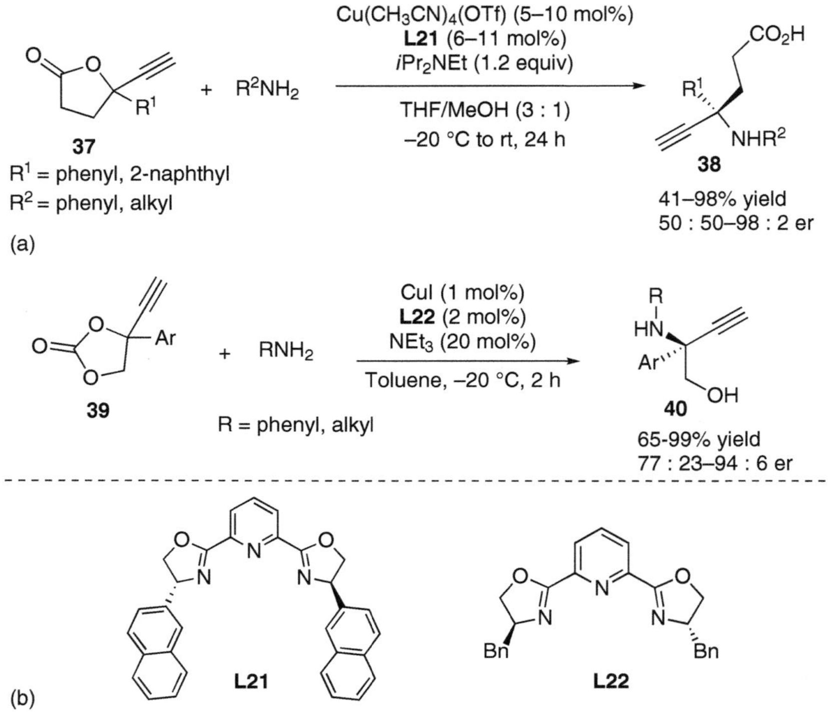

Scheme 4.17 Asymmetric synthesis of propargylamines from lactones and cyclic carbonates. Source: (a) Modified from Gómez et al. [33] and (b) modified from Tian et al. [34].

4.2.4 Enantioselective Addition of Terminal Alkynes to Enamines/Enamides

In contrast to the numerous examples of enantioselective addition of alkynes to imines (A^3 couplings), relatively less attention has been paid to the asymmetric addition of alkynes to the structurally similar enamines/enamides.

This is likely due to the fact that enamines/enamides face the challenge of achieving high regioselectivity and enantioselectivity. In 2002, Knochel and coworkers were the first to exploit the asymmetric addition of terminal alkynes to enamines to synthesize chiral propargylamines **42** [35].

As shown in Scheme 4.18, a CuBr/Quinap **L6**-catalyzed addition of various alkynes to diverse enamines **41** took place regioselectively and enantioselectively, leading to the formation of propargylamines **42** in good to high yields and ees. Interestingly, a dimeric complex $[BrCu(Quinap)]_2$ was isolated as an air-stable solid. Based on this, the authors proposed a mechanism (Scheme 4.19) that commences with the dissociation of dimeric copper complex to yield the monomeric copper species. Then, the successive coordination of the alkyne and enamine to the copper complex gives the zwitterionic intermediate, which undergoes a transfer of the alkyne moiety to the immonium ion to deliver the products and regenerate the catalyst. Recently, the group of Li employed a similar strategy to realize the asymmetric assembly of chiral propargylamines **44** by Rh/**L23**-catalyzed enantioselective hydroalkynylation of enamides **43** (Scheme 4.20) [36]. In particular, the stereochemistry of the C=C

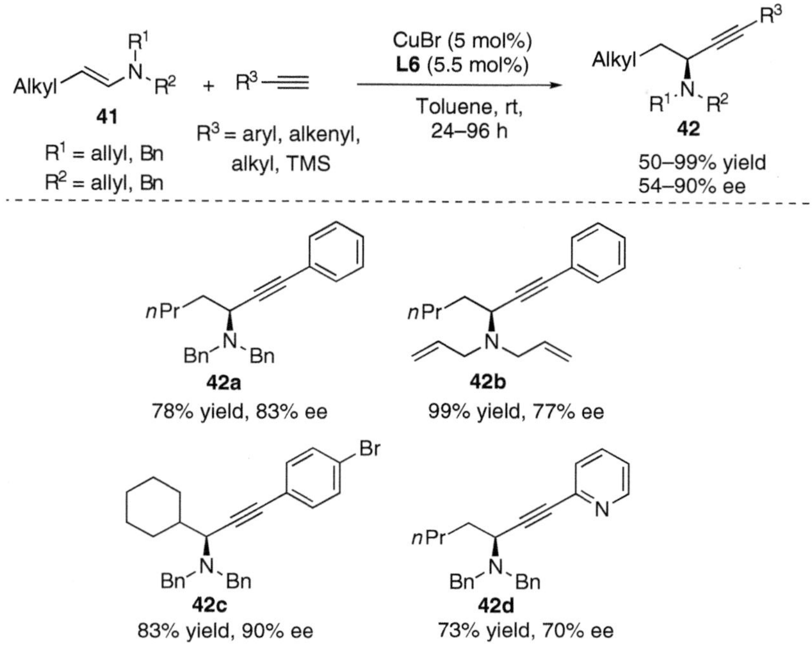

Scheme 4.18 Asymmetric synthesis of propargylamines from enamines. Source: Modified from Koradin et al. [35].

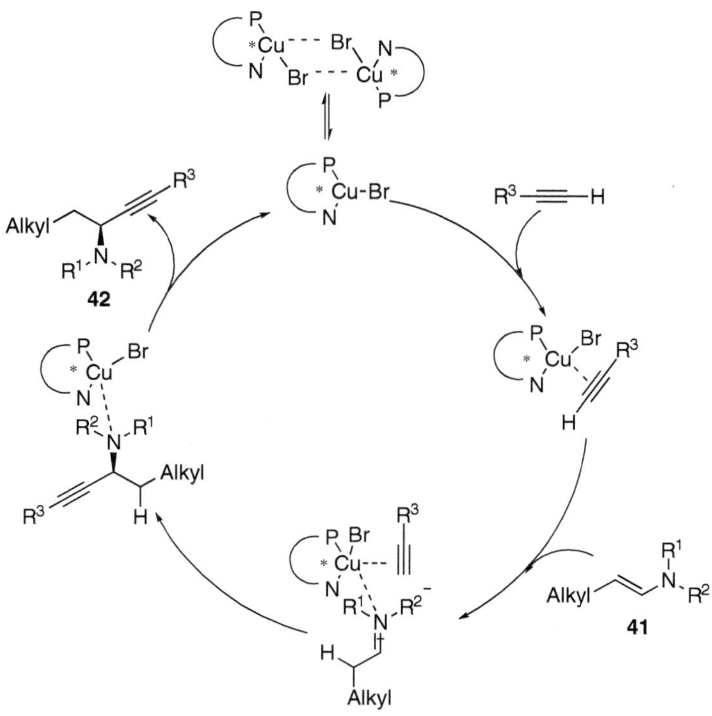

Scheme 4.19 Proposed mechanism for the CuBr/Quinap **L6**-catalyzed synthesis of propargylamines.

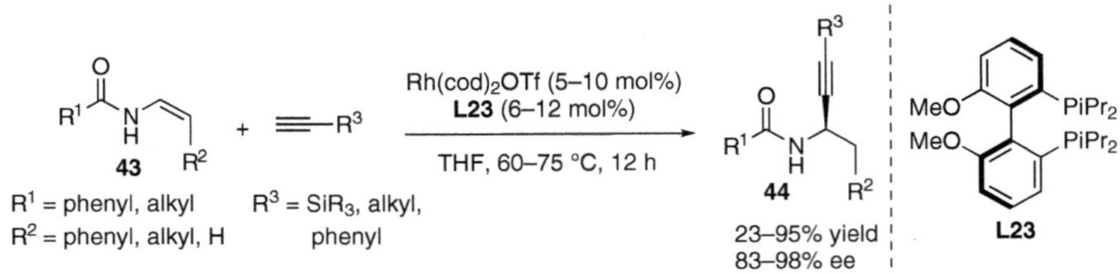

Scheme 4.20 Synthesis of propargylamines via Rh/**L23**-catalyzed enantioselective hydroalkynylation of enamides. Source: Modified from Bai et al. [36].

double bond of enamides **43** has an impact on the reaction yield and enantioselectivity, and the Z-configuration was proved to be beneficial for achieving high yields and enantioselectivities. It is noteworthy that this method features an extraordinary broad substrate scope and excellent enantioselectivities. A variety of Z-enamides could react with diverse terminal alkynes to afford the desired propargylamines **44** in good to high yields and excellent ees.

4.2.5 Rh/Ru-Catalyzed Enantioselective Hydrogenation of Alkynyl-Substituted Enamides/Imines

Rh/Ru-catalyzed enantioselective hydrogenation of alkynyl-substituted enamides/imines, which normally features a low catalyst loading and high chiral induction, represents an efficient method for the generation of chiral propargylamines. The main challenge of this hydrogenation process is to achieve high levels of chemoselectivity toward the carbon–carbon bonds and to identify an appropriate chiral ligand. In 2013, Zhang and Wang reported a [Rh(cod)(S_C,R_P)-DuanPhos]BF_4-catalyzed enantioselective hydrogenation of the alkynyl-substituted enamide **45** to afford the corresponding propargylamine **47** with an excellent yield and ee (Scheme 4.21) [37]. It should be noted that the TMS group at the alkyne moiety seems essential to keep the alkyne group untouched, as the hydrogenation of substrate **46** carrying a phenyl group at the alkyne moiety led to the formation of the corresponding chiral Z-allylic amine **48**.

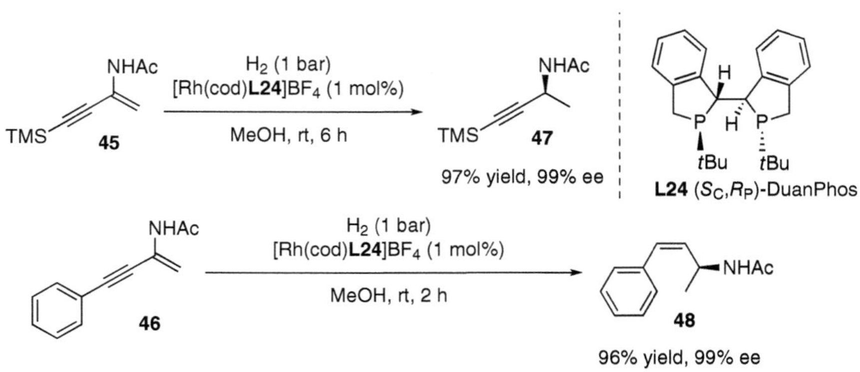

Scheme 4.21 Enantioselective hydrogenation of the alkynyl-substituted enamide. Source: Modified from Liu et al. [37].

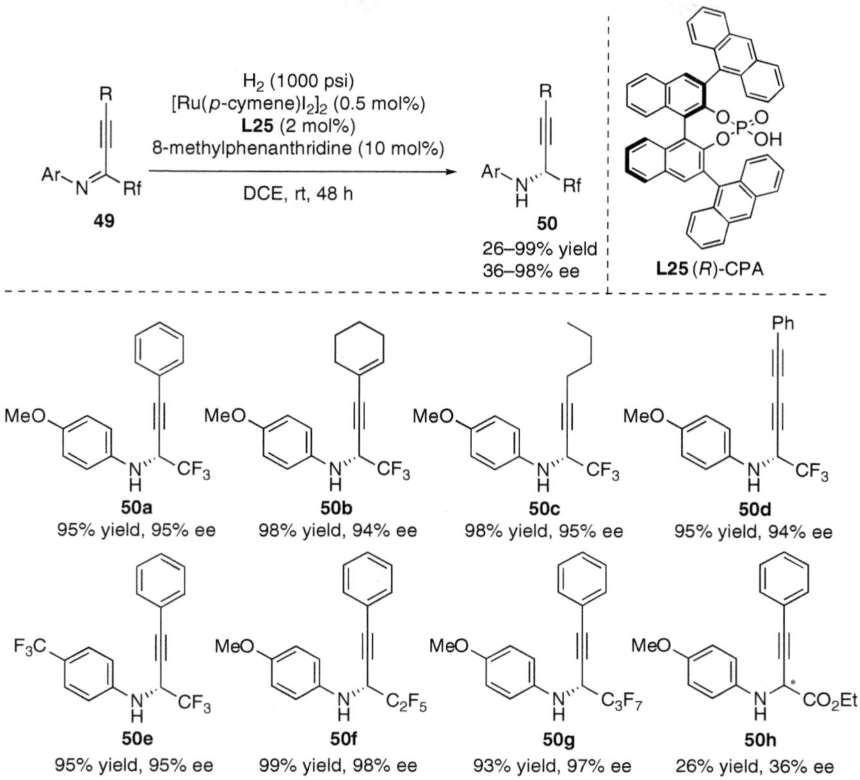

Scheme 4.22 Chemoselective biomimetic hydrogenation of fluorinated alkynyl-ketimines. Source: Modified from Chen et al. [38].

Inspired by the natural NAD(P)$^+$/NAD(P)H redox cofactor couples, in 2016, Zhou and coworkers developed a [Ru(*p*-cymene)I_2]$_2$/**L25**/8-Me-PD system, which achieved the enantioselective synthesis of fluorinated propargylamines **50** through the highly chemoselective biomimetic hydrogenation of the carbon–nitrogen double bond of fluorinated alkynyl-ketimines **49** in the presence of a carbon–carbon triple bond (Scheme 4.22) [38].

This protocol shows a broad substrate scope, excellent chemoselectivity, and high enantioselectivity. In general, the reactions of various substrates worked well to deliver the desired chiral fluorinated propargylamines **50** with excellent yields and ees, irrespective of the electronic nature of the substituents. The proposed mechanism (Scheme 4.23) begins with the ruthenium-catalyzed in situ generation of 8-Me-DHPD from 8-Me-PD. Then, the asymmetric hydrogen transfer from 8-Me-DHPD to the C=N bonds in the presence of **L25** occurs to provide the chiral products. The very slow background reaction, in which the C=N bonds were directly hydrogenated by Ru/H_2, and the high chiral induction by **L25** are responsible for the excellent enantioselectivity observed.

In 2018, Jiang and Zhang and coworkers disclosed a rhodium-catalyzed chemo- and enantioselective hydrogenation of alkynyl-aryl hydrazones **51** using **L26** as the chiral ligand (Scheme 4.24) [39]. The directing group *N*-NH(*p*-NO_2Bz) and the additive $PhCO_2H$

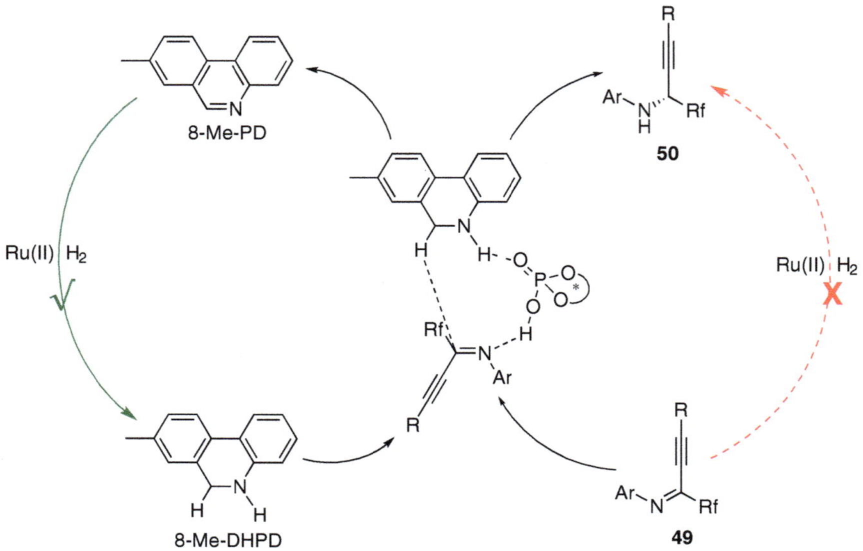

Scheme 4.23 Proposed mechanism for the reduction of ketimines. Source: Modified from Chen et al. [38].

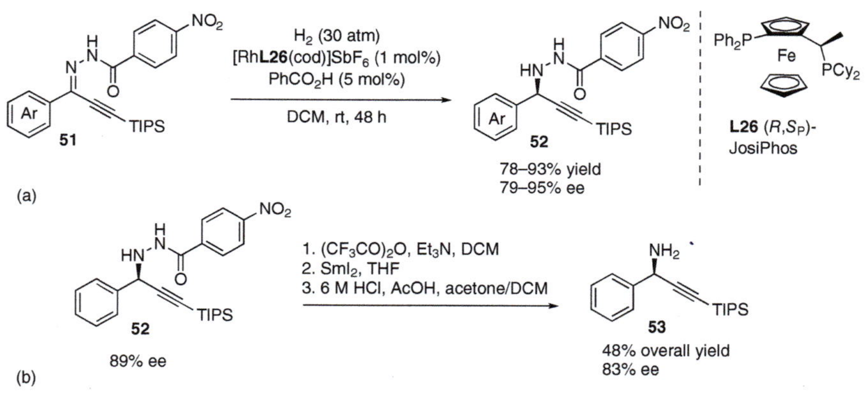

Scheme 4.24 Rhodium-catalyzed hydrogenation of alkynyl–aryl hydrazones. Source: Modified from Fan et al. [39].

are required for achieving high enantioselectivities and high conversions, respectively. Similarly, the bulky silyl group linked to the C≡C bonds improves the chemoselectivity.

As a result, a variety of alkynyl-aryl hydrazones **51** bearing different aryl substituents like phenyl and naphthyl groups underwent this enantioselective hydrogenation smoothly to produce the desired chiral propargyl hydrazines **52** with good to high yields and ees and excellent chemoselectivity. More importantly, the chiral propargyl hydrazines **52** obtained could be converted into the corresponding chiral propargylamines **53** through deprotection.

4.2.6 Enantioselective C–H Activation: Synthesis of Cyclic Propargylamines

In 2018, the group of Zhang presented an example of 2-(phenylethynyl)-pyrrolidine synthesis from aliphatic diazo compounds via cobalt-catalyzed intramolecular alkylation of the propargylic C(sp^3)—H bond (Scheme 4.25) [40]. Although a complex chiral **[Co(P1)]** catalyst was required and only a moderate yield and ee were obtained for the desired product, this process provides an alternative route to the chiral propargylamine motif. This reaction represents an enantioselective radical cyclization for the construction of pyrrolidine ring **55** by metalloradical C–H activation.

As shown in Scheme 4.26, the reaction mechanism begins with the radical activation of the diazo compound by the Co(II)-metalloradical to afford the α-Co(III)-alkyl radical, which undergoes the H-abstraction to give the ε-Co(III)-alkyl radical.

Then, the intramolecular radical substitution happens to yield the product and regenerates the catalyst. Subsequently, using the same strategy, they achieved the synthesis of chiral 2-alkynyl-substituted indoline **57** from the conformationally rigid aryldiazomethane **56** (Scheme 4.27a) [41]. A different cobalt catalyst **[Co(P2)]** was found to be efficient in chiral induction, producing the desired product with a 87% ee.

Soon afterward, the same group reported the synthesis of six-membered chiral heterocyclic sulfamides **59** carrying the propargylamine substructure from sulfamoyl azides through Co(II)-based metalloradical 1,6-amination of C(sp^3)—H bonds (Scheme 4.27b) [42].

A variety of alkynyl-substituted heterocyclic sulfamides **59** were prepared in high yields and excellent enantioselectivities. Zhang and coworkers developed a new strategy for the enantioselective construction of C—C and C—N bonds via asymmetric metalloradical catalysis, which were proved to be efficient in the assembly of chiral N-heterocycles containing a propargylamine motif.

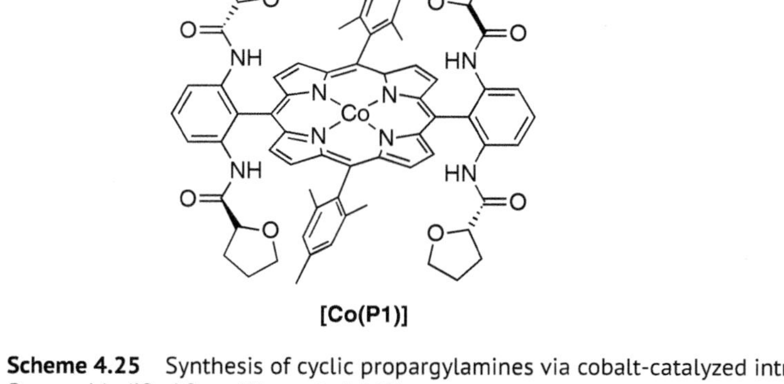

Scheme 4.25 Synthesis of cyclic propargylamines via cobalt-catalyzed intramolecular alkylation. Source: Modified from Wang et al. [40].

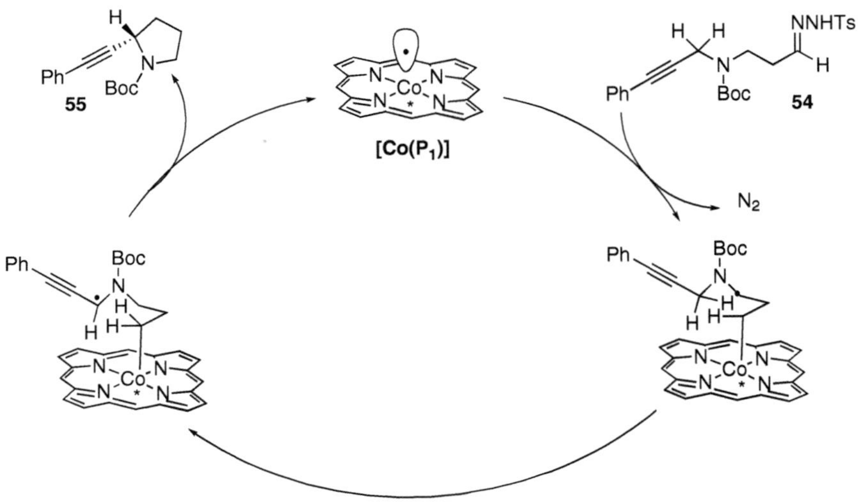

Scheme 4.26 Mechanism of the cobalt-catalyzed intramolecular cyclization.

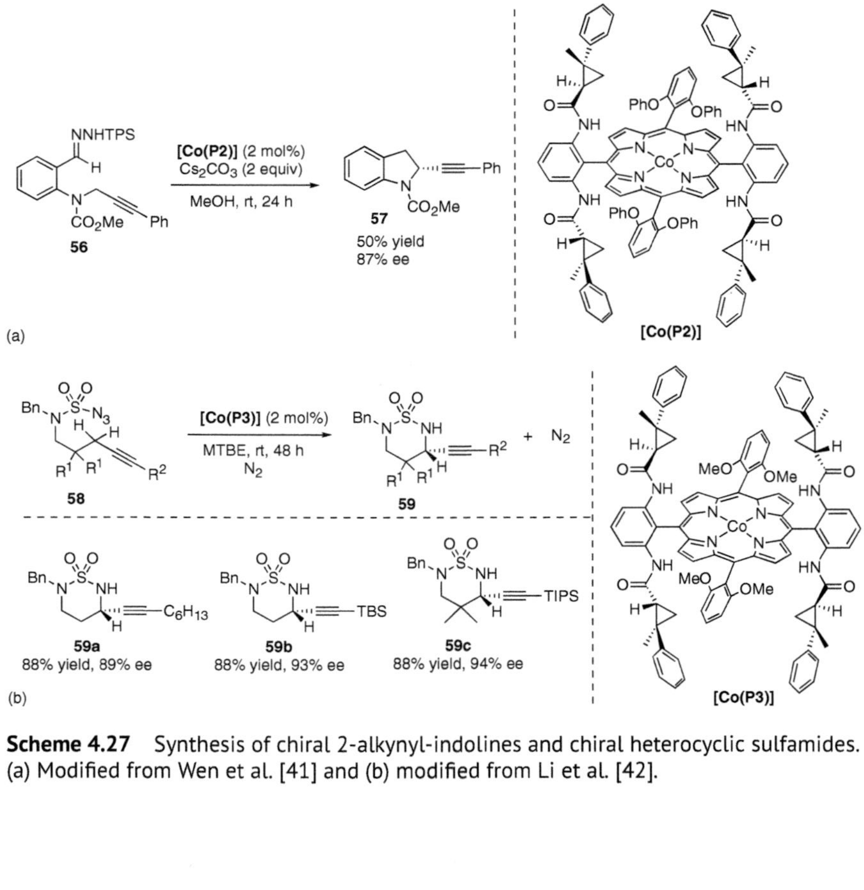

Scheme 4.27 Synthesis of chiral 2-alkynyl-indolines and chiral heterocyclic sulfamides. Source: (a) Modified from Wen et al. [41] and (b) modified from Li et al. [42].

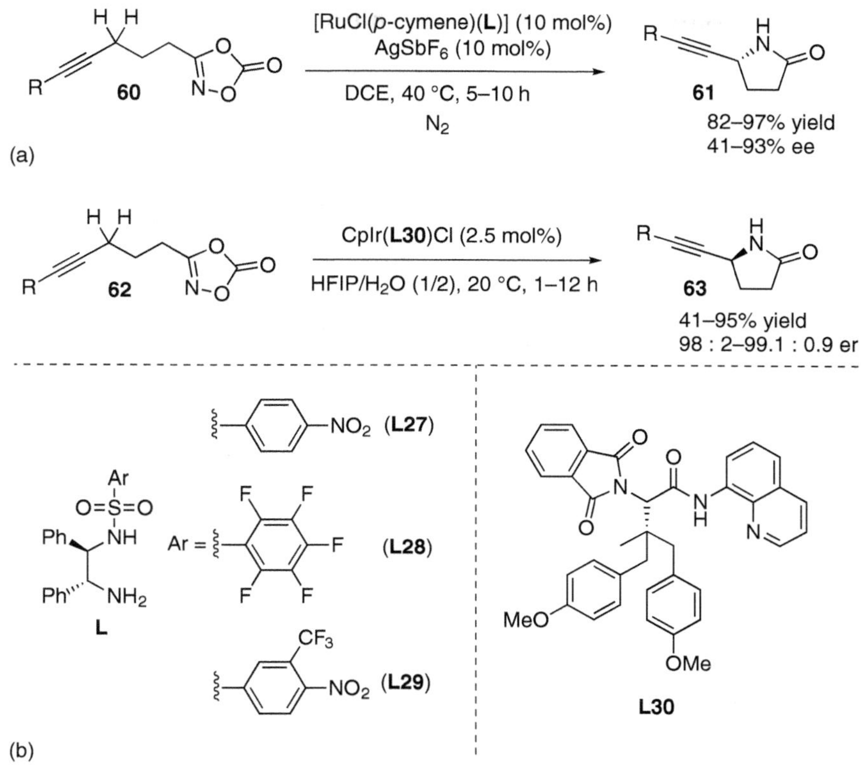

Scheme 4.28 Synthesis of cyclic propargylamines via intramolecular nitrene insertion to propargylic C—H bonds. Source: (a) Modified from Xing et al. [43] and (b) modified from Wang et al. [44].

Very recently, Yu and coworkers reported the Ru-catalyzed asymmetric synthesis of 5-alkynyl-substituted pyrrolidin-2-ones **61** from the corresponding 1,4,2-dioxazol-5-ones **60** via the intramolecular nitrene insertion to propargylic C—H bonds (Scheme 4.28a) [43]. Successful enantiocontrol of this annulation has been achieved by employing chiral diphenylethylene diamines with electron-withdrawing arylsulfonyl substituents as the ligands. The coordination of the Ru catalyst to the chiral diamine ligands and the nitrogen atom of dioxozalone substrates, and the hydrogen bond between the oxygen atom of dioxozalone substrates and the NH_2 group of the diamine ligands are both responsible for the chiral induction. The desired five-membered lactams containing the propargylamine nucleus were synthesized in high yields and moderate to high ees.

Almost simultaneously, Chen and coworkers reported a similar transformation through iridium catalysis with an α-amino-acid-based chiral ligand **L30** (Scheme 4.28b) [44]. The authors disclosed that the ligand could form a well-defined groove-type chiral pocket around the Ir center, which allows stereocontrolled binding of the dioxozalone substrates **62** to undergo the intramolecular $C(sp^3)$–H amidation affording the desired γ-alkynyl γ-lactams **63** in good to excellent yields and outstanding enantiomeric ratios.

4.3 Enzymatic Synthesis of Propargylamines

The tremendous progresses made in the past decades in the fields of DNA cloning, expression, and purification have boosted the use of enzymes in synthetic chemistry. Enzyme catalysis, or biocatalysis, has undergone a huge development in the past few years, and it is nowadays an established and robust technique in the synthesis of many pharmaceutical ingredients, organic building blocks, and synthetic precursors of drugs and fine chemicals [45, 46]. The use of enzymes (biocatalysts) in organic synthesis offers clear economic and environmental benefits over standard chemical syntheses [47]. In fact, the intrinsic chirality of enzymes can be exploited to carry out enantioselective transformations under mild conditions (ambient temperature and atmospheric pressure), thus avoiding the requirement of more extreme conditions, which could lead to isomerization and racemization or rearrangement of the products. Moreover, the enzymes for biocatalysis can be produced from readily available and inexpensive renewable resources and can replace the use of scarce precious metals in synthesis. Enzymes are also essentially nonhazardous, nontoxic, and biodegradable and, especially in the case of supported enzymes, can be recycled and reused several times in the same transformations.

Enzymatic reactions to access propargylamines have been developed in the past two decades. The first enzymatic synthesis of enantiopure 1-aryl-propargylamines has been described by Botta's research group in 1999 (Scheme 4.29) [48]. The enzymatic kinetic resolution of racemic 1-aryl-proprgylamines with *Candida antarctica lipase B* (CAL-B), using EtOAc as the acetyl source, led to **(*S*)-66** in enantiopure form, and to **(*R*)-66** with excellent ees (>88%). CAL-B selectively catalyzed the acetylation of the (*R*)-enantiomer, leaving the (*S*)-enantiomer unreacted. Deprotection of **67** with 6 N HCl finally afforded **(*R*)-66**. The substituents on the aryl group of the propargylamine influence the enzymatic resolution probably because of steric factors. In fact, poor ees were obtained for compounds bearing substituents on the ortho position. The methodology was later adopted for the synthesis of the paclitaxel side chain [49] and antifungal drugs [50, 51].

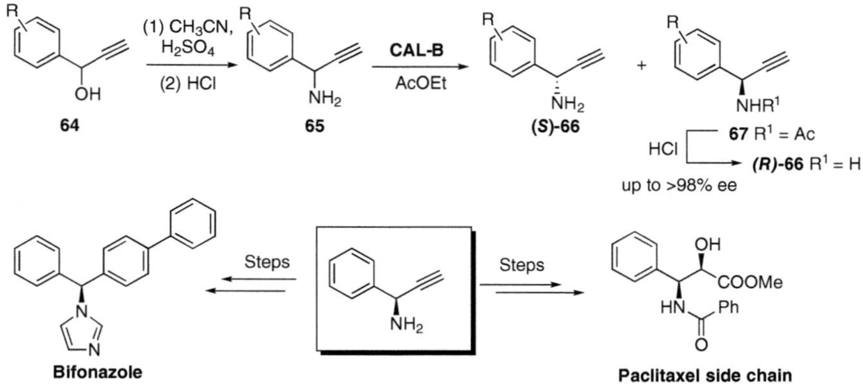

Scheme 4.29 CAL-B-catalyzed enzymatic kinetic resolution of 1-aryl-propargylamines. Source: Modified from Messina et al. [48].

Another enzymatic kinetic resolution approach to synthesize enantiopure propargylamines **69** from the acyl precursor **68** was reported in 1999 by Landis and coworkers using immobilized penicillin G amidohydrolase (PGA) in phosphate buffer solution (pH = 7.4) [52]. The enzyme PGA promotes the selective deacylation of racemic **68**, leading to propargylamine **(*S*)-69** while leaving the (*R*)-enantiomer unreacted. However, treating **(*R*)-68** with an excess of PGA at a slightly higher temperature led to the deacylation of the substrate to give the propargylamine enantiomer **(*R*)-69** (Scheme 4.30).

In both previous examples, an enzymatic kinetic resolution of racemic propargylamines was carried out, thus leading to a maximum 50% conversion of the substrates into each enantiomer. More recently, biocatalytic approaches to synthesize enantiopure propargylamines from prochiral carbonyls have been developed.

Kroutil and coworkers described the conversion of prochiral propargyl ketones **70** into enantiopure propargylamines **71** using ω-transaminases (ω-TAs) (Scheme 4.31) [53]. Both the (*R*) and (*S*) enantiomers can be easily synthesized using either (*R*)-selective ω-TAs originating from *Arthrobacter sp.* and *Aspergillus terreus* or an (*S*)-selective ω-TA from *Chromobacterium violaceum*. Alanine was used as the source of nitrogen while alanine dehydrogenase (AlaDH) was used as a recycling enzyme to convert the pyruvate into alanine exploiting ammonium formate as a sacrificial substrate. The resulting propargylamines **71** were obtained with high conversions (up to 99%) and excellent ees (>99%).

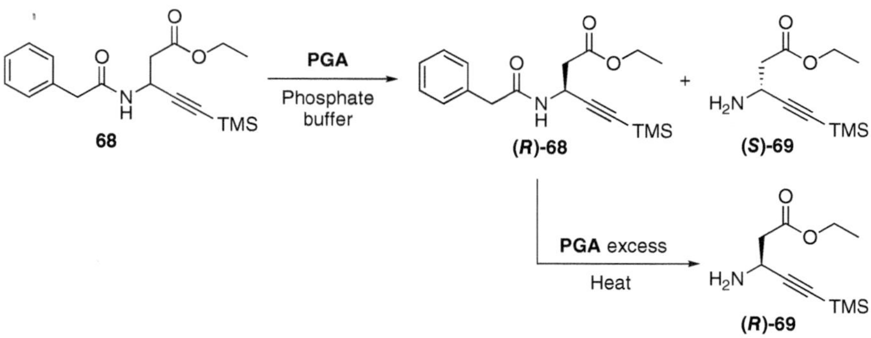

Scheme 4.30 Enzymatic kinetic resolution of propargylamines using penicillin G amidohydrolase (PGA).

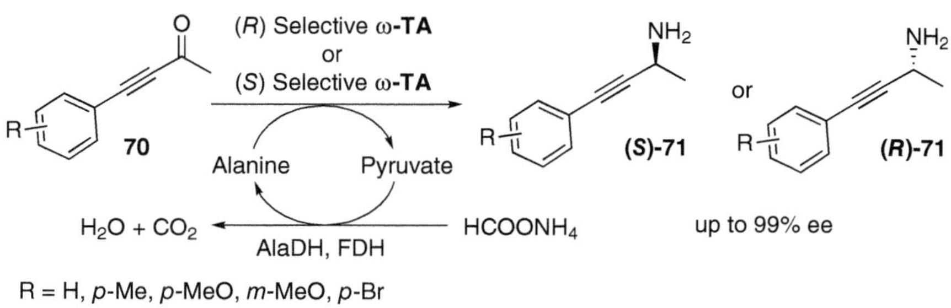

Scheme 4.31 Synthesis of propargylamines using ω-transaminases. Source: Modified from Schmidt et al. [53].

Table 4.2 Synthesis of propargylamines using AspRedAm.

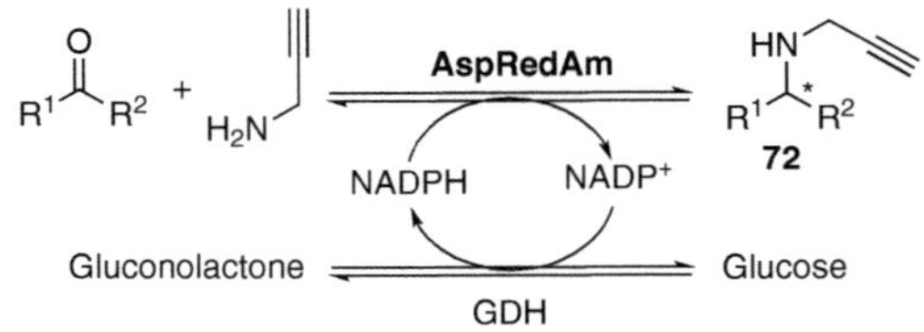

Entry	Propargylamine 408	Conversions (%)	ee (%)
1.	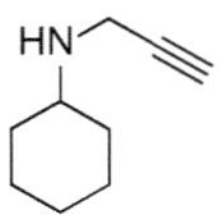	94	—
2.	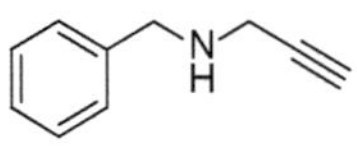	92	—
3.	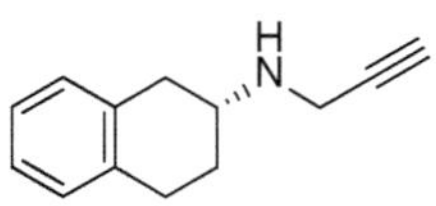	45	97
4.	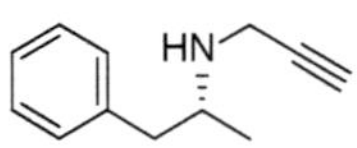	>97	>98
5.	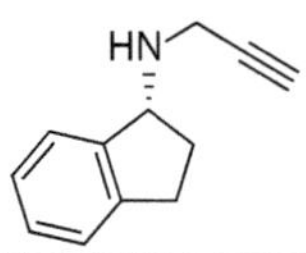	64	95

Turner and coworkers recently developed an NADP(H)-dependent reductive aminase enzyme from *Aspergillus oryzae* (AspRedAm), which is able to catalyze the reductive coupling of a broad set of carbonyl compounds with a variety of primary and secondary amines to afford chiral amine derivatives [54]. The AspRedAm enzyme is able to catalyze a reductive amination reaction, allowing the formation of an imine intermediate, followed by its reduction to afford **72** as shown in Table 4.2. Propargylamines were obtained with up to >98% conversion and >98% ee.

Propargylamines can also be accessed from alcohols through biocatalytic hydrogen-borrowing [55], by combining an alcohol dehydrogenase (ADH) with an amine dehydrogenase (AmDH) and catalytic NAD^+ (Scheme 4.32). Turner's group combined nonselective NADPH-dependent ADHs with the *Asp*RedAm enzyme to generate a redox-neutral cycle, which is able to catalyze the conversion of alcohols into N-alkylated propargylamines **75** [56]. Three different ADH enzymes were combined with *Asp*RedAm, namely, the variant of the ADH from *Thermoanaerobacter ethanolicus* (TeSADH W110A), ADH-150, and the ADH from *Sphingobium yanoikuyae* (SyADH). The three ADH enzymes resulted in similar conversions for **75b**, while the variant SyADH showed the best conversion for **75a**. The

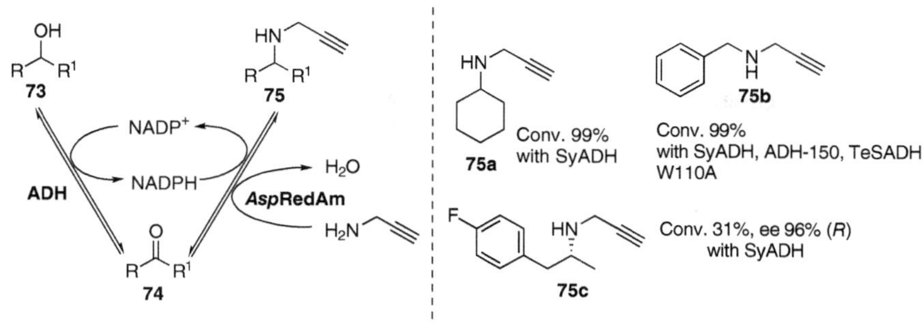

Scheme 4.32 Synthesis of propargylamines through the combination of alcohol dehydrogenase (ADH) and amine dehydrogenase (AmDH) enzymes.

enantiopure propargylamine **75c** was also obtained from the appropriate ketone with an excellent ee (96%).

Imine reductases (IREDs) are enzymes that catalyze the reduction of imines into amines [57]. In addition to this standard reactivity, some IREDs can also catalyze the reductive amination of ketones into chiral amine products [58]. Höhne and coworkers synthesized the two enantiomers of the drug rasagiline **77** with high yields and good ees (up to 90%) using two IRED enzymes (IR-14 and IR-Sip) provided by Roche [59, 60] (Scheme 4.33a).

A panel of diverse *N*-methylamino acid dehydrogenases (NMAADHs), ketimine reductases (KIREDs), and Δ^1-pyrroline-5-carboxylate reductases (P5CRs) has also been studied by Watson and Roiban and coworkers to access a variety of chiral N-functionalized amino acids (Scheme 4.33b). The *N*-propargyl amino acid **79** was obtained by treating the ketone **78** with different reductases (En01–En07, from the three different classes of enzymes) in

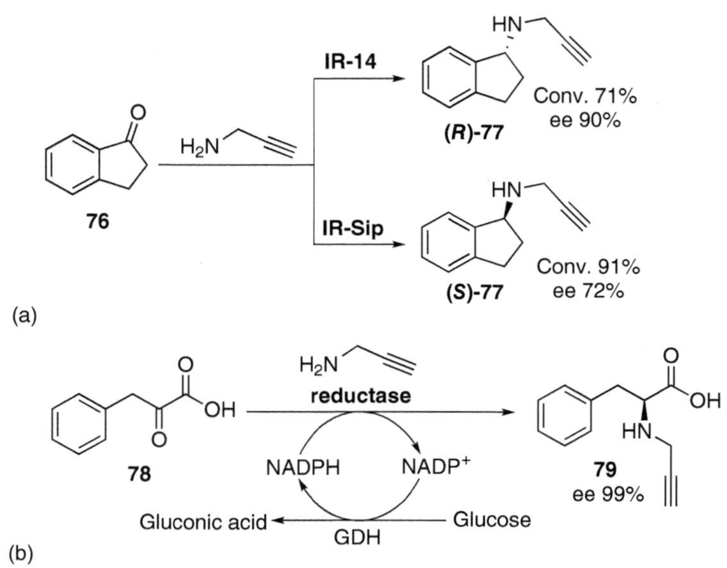

Scheme 4.33 Reductase-biocatalyzed synthesis of propargylamines.

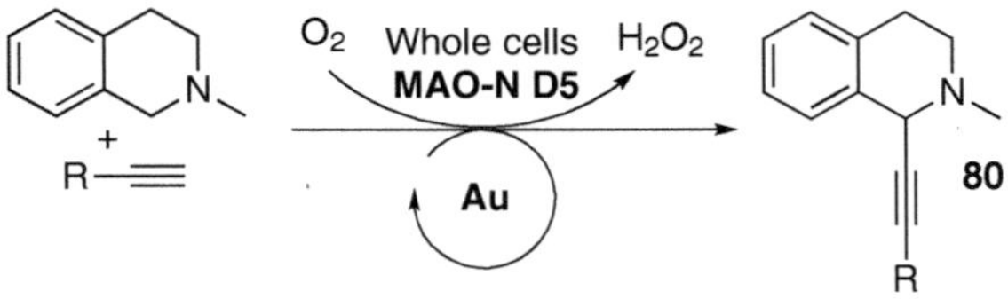

Scheme 4.34 MAO-N/Au chemoenzymatic synthesis of cyclic propargylamines.

the presence of glucose dehydrogenase (GDH). All the enzymes led to the formation of **79** with excellent ees (>99%) but poor conversions [61].

Finally, an elegant chemoenzymatic approach to cyclic propargylamines **80** was reported by Turner and Greaney through the combination of MAO-N biocatalysts and the gold chemocatalyst $HAuCl_4 \cdot H_2O$ (Scheme 4.34). The MAO-N enzyme, variant D5, oxidizes the tetrahydroisoquinoline substrate into the corresponding iminium, to which a gold-activated terminal alkyne is attacked, leading to products **80** in good yields [62].

4.4 Photoredox Synthesis of Propargylamines

Photoredox catalysis has emerged in recent years as a powerful and sustainable strategy for the synthesis of a variety of molecules [63]. Photoredox synthetic approaches rely on the ability of metal complexes and organic dyes to convert visible light into chemical energy by engaging in single-electron transfer (SET) processes with organic substrates, thus enabling the generation of reactive radical intermediates. Among the many, a few photoredox methods to synthesize propargylamines have also been developed.

Rueping's group first showed how photoredox catalysis could be exploited to synthesize the cyclic propargylamine derivatives **81** [64]. The *N*-phenyl-tetrahydroisoquinoline was initially treated with phenylacetylene in the presence of different photoinitiators such as $[Ru(bpy)_3](PF_6)_2$ or $[Ru(bpy)_2(dtbbpy)](PF_6)_2$ and metal catalysts such as silver or copper salts. A dual photoredox and metal catalytic cycle was in fact hypothesized for this reaction as shown in Scheme 4.35. The photoinitiators react with isoquinoline, leading to an iminium intermediate **83** and are regenerated by oxygen. The iminium **83** can then be transformed into the propargylamine derivative **81** via the second metal catalytic cycle. The best reaction conditions were found when the reaction was performed with $[Ru(bpy)_2(dtbbpy)](PF_6)_2$, $(MeCN)_4CuPF_6$, and a 5 W fluorescent bulb as the light source, leading to the formation of **81** in 88% yield in 16 hours. The scope of the reaction was expanded to a number of isoquinoline substrates bearing various substituents on the aromatic ring as well as the nitrogen.

In 2015, Waser and coworkers reported a photoredox catalytic decarboxylative alkynylation of carboxylic acids, leading to cyclic propargylamines (Scheme 4.36a) [67]. An *N*-Cbz-proline was decarboxylated by the Ir-photocatalyst $Ir[dF(CF_3)ppy]_2(dtbbpy)PF_6$ activated by blue LED, generating a proline radical intermediate **84**. The latter then reacted with the ethynylbenziodoxolone (EBX), leading to propargylamine derivative **85**.

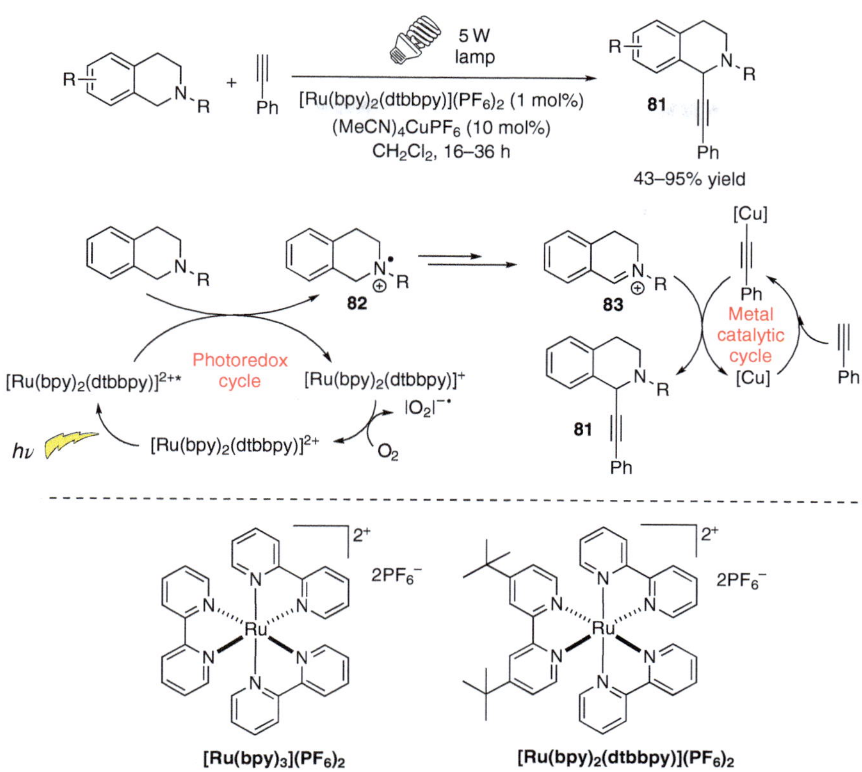

Scheme 4.35 Rueping's photoredox approach to propargylamines.

In parallel, Xiao and Lu and coworkers reported a similar methodology to access linear propargylamines **86** (Scheme 4.36b) [65]. Later, Fu and coworkers described a decarboxylative alkynylation approach of *N*-(acetoxy)phthalimides of α-amino acids by merging photoredox with copper catalysis [66]. The methodology allows the synthesis of a variety of propargylamines **87** with assistance of $[Ru(bpy)_3]Cl_2$ photocatalyst and visible light (Scheme 4.36c). The ruthenium photocatalysis promotes the formation of an iminium intermediate by decarboxylation, while CuI reacts with phenylacetylene, exploiting the phtalimide anion generated during the decarboxylation as a base, and forms a copper-acetylene intermediate. This latter finally reacts with the iminium ion, leading to the propargylamines **87**.

Hashmi and coworkers described the synthesis of propargylamines through an α-C(sp^3)–H alkynylation of unactivated tertiary aliphatic amines using 1-iodoalkynes as the alkynylating reagents and $[Au_2(m\text{-dppm})_2]^{2+}$ as the photocatalyst [68]. The reaction proceeds under sunlight to afford a variety of propargylamines **88** in 50–81% yields. The use of 1-iodoalkynes instead of EBX reagents proved to be beneficial, leading to an increase

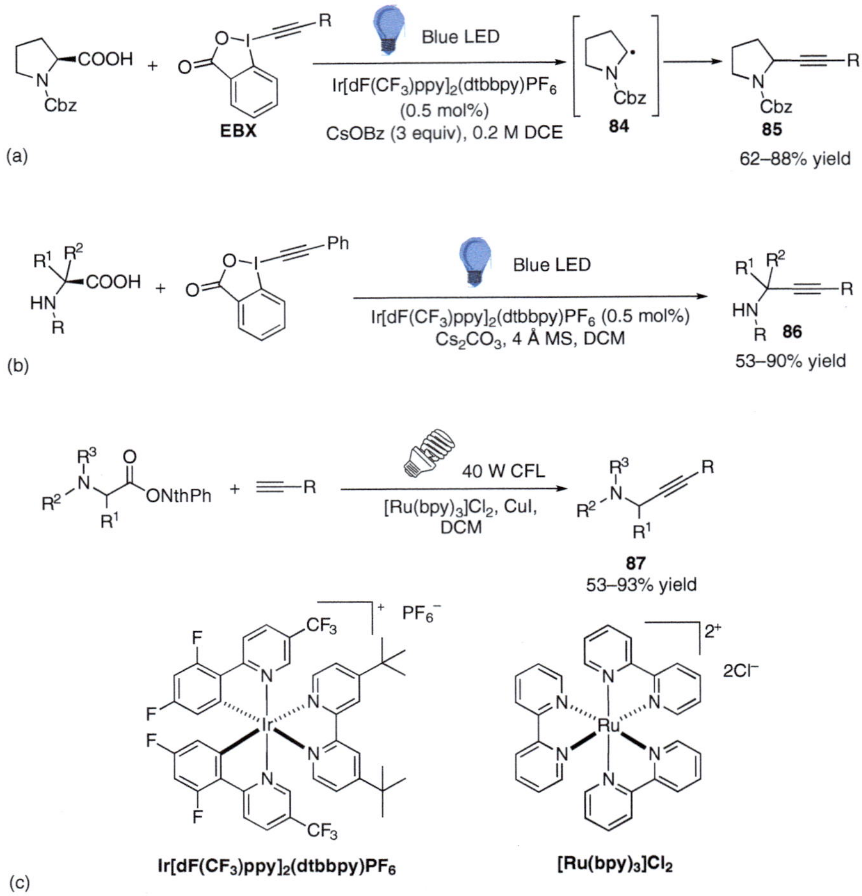

Scheme 4.36 Photoredox synthesis of propargylamines using ethynylbenziodoxolones. (a) Waser's approach. Source: Modified from Zhou et al. [65]. (b) Xiao and Lu's approach. Source: Modified from Zhang et al. [66]. (c) Fu's approach.

of the yields. The authors hypothesized that the photocatalysts can activate both the amine and the alkyne substrates into radical intermediates as shown in Scheme 4.37.

A new photocatalytic methodology has been recently developed to access propargylamines via copper photoredox-catalyzed A^3 coupling of arylamines, terminal alkynes, and alcohols [69]. Blue LED irradiation of *N*-alkylanilines, phenylacetylene, and benzoquinone (BQ) with 5 mol% CuCl in CH_3OH can provide propargylamines **89** in good to excellent yields. Chiral amines **90** can also be obtained when different alcohol substrates were used (Scheme 4.38).

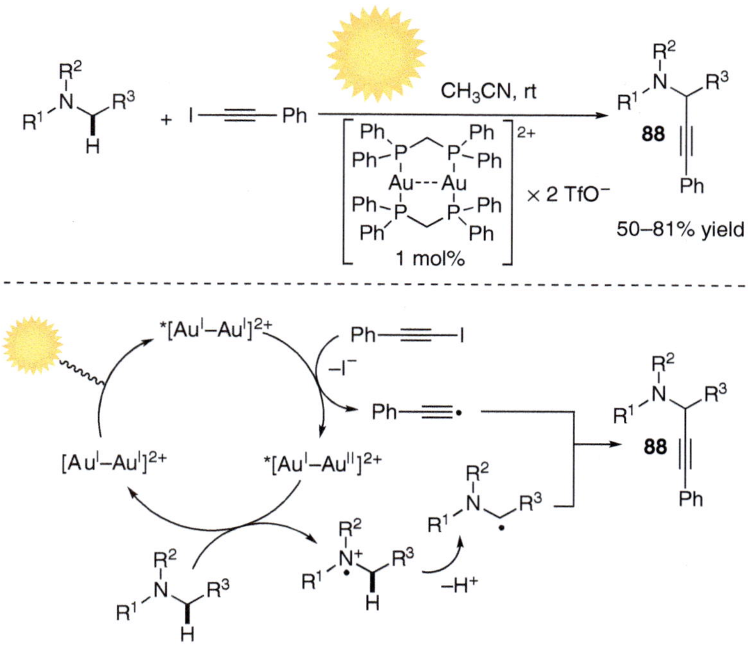

Scheme 4.37 Synthesis of propargylamines via photoredox decarboxylative alkynylation and α-C(sp^3)–H alkynylation of unactivated tertiary aliphatic amines.

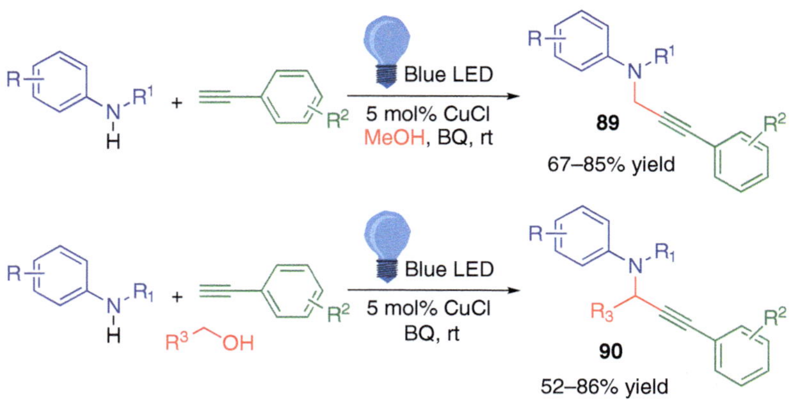

Scheme 4.38 Copper photoredox-catalyzed A^3 coupling of arylamines, terminal alkynes, and alcohols.

4.5 Organocatalyzed Asymmetric Synthesis of Propargylamines

The organocatalytic stereoselective synthesis of chiral propargylamines has been described by a number of authors, mainly exploiting Brønsted acids as catalysts. Maruoka and coworkers developed a Mannich-type reaction to synthesize enantioenriched propargylamines **92–93** through the in situ generation of C-alkynyl imines **95** from aminals **91** catalyzed by chiral Brønsted acids **94** [70]. The in situ generated imine intermediates react with aldehydes, β-ketoesters, or acetylacetone to afford the desired propargylamines with high enantioselectivity. It was found that the reaction of aminals **91** proceeded through the reversible and gradual generation of imine **95** (or the corresponding iminium salt) via acid-catalyzed elimination of $BocNH_2$ and the subsequent carbon–carbon bond formation with aldehydes/ketones (Scheme 4.39). The acid has a dual role, not only accelerating the reaction but also promoting the formation of imine intermediate. In the case of aldehydes, the reaction proceeds via the formation of *N*-Boc-imines (or *N*-Boc-iminum salt) and enecarbamates, giving anti-Mannich adducts as major diastereomers.

Palomo and coworkers also developed an organocatalyzed asymmetric Mannich-type reaction of aldehydes with inactivated imines **96** by combining a proline-based catalyst and *p*-nitrobenzoic acid (PNBA) cocatalyst (Scheme 4.41) [71]. The dialkylprolinol ether **97** was found as the best among the proline-based catalysts giving rise to *anti*-Mannich adducts with a good yield of 70–75% when used with PNBA. Propargylamines **98** were obtained with an anti-syn ratio higher than 90 : 10 and with ee > 95%. The authors

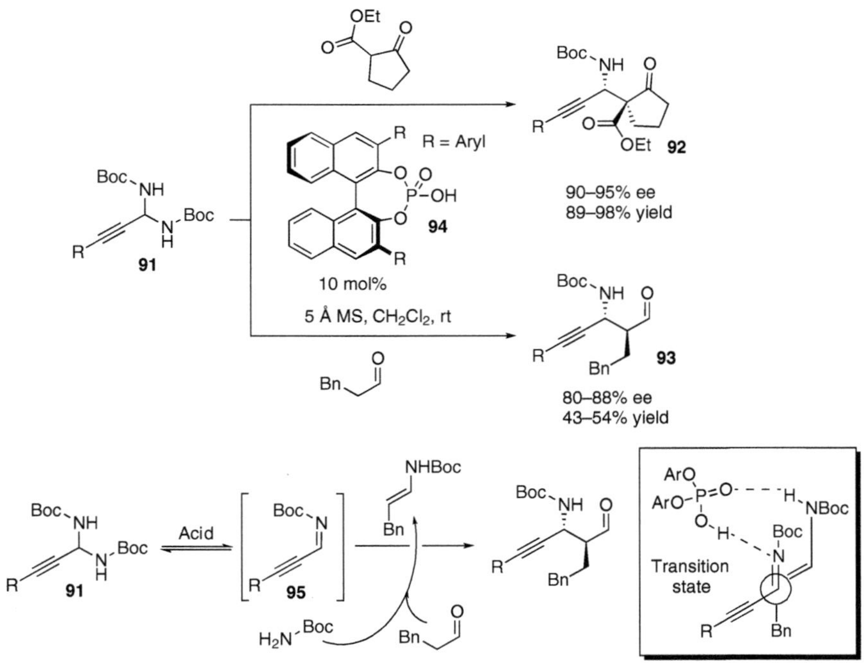

Scheme 4.39 Brønsted acid-catalyzed Mannich-type synthesis of propargylamines.

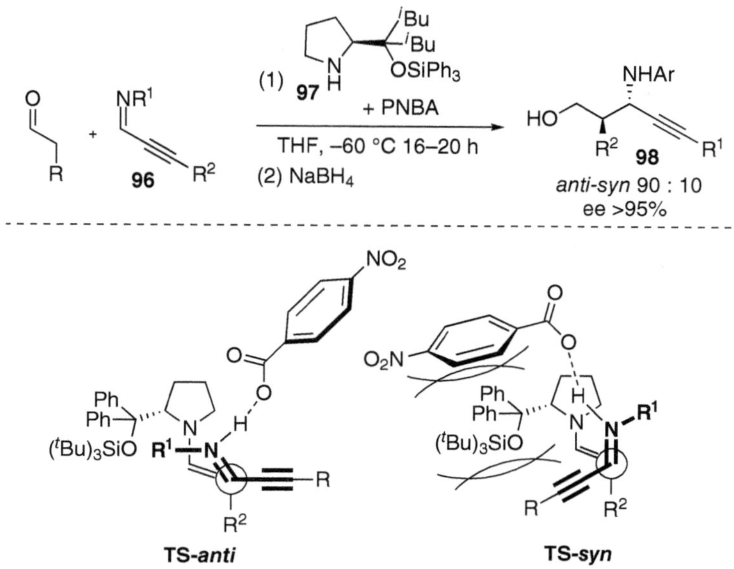

Scheme 4.40 Organocatalyzed asymmetric Mannich-type reaction of aldehydes with inactivated imines.

attributed the preference for the formation of the anti-product to unfavorable interactions in the syn-transition state between the bulky substituents on pyrrolidine ring and both aromatic system of benzoic acid and alkynyl moiety of the imine (Scheme 4.40). The Mannich-type reaction of aldehydes with alkynyl imines favoring the syn configuration was also described by the same group [72]. Instead of using Brønsted acids, the urea Brønsted base cocatalyst **99** was used in combination with proline. Under these reaction conditions, the propargylamine **100** was obtained with a syn/anti ratio up to 99 : 1 and ee up to 99%.

The syn configuration was favored through Houk–List type model in which the tertiary amine–urea cocatalyst further stabilizes the transition state as illustrated in Scheme 4.41.

The asymmetric synthesis of propargylamines using a Brønsted base was also described by Wang et al. [73]. The authors reacted *N*-Boc-*N*,*O*-acetal **101** with malonate and β-ketoesters in the presence of different organocatalysts such as **102a,b**. When the acetal was reacted with a malonate (thio)ester in the presence of catalyst **102a**, the propargylamine **103** was obtained with excellent er but poor dr. On the other hand, when **101** was reacted with the β-ketoester in the presence of catalyst **102b**, both excellent er and dr were observed. The proposed mechanism involves the cleavage of the C—O bond of the *N*,*O*-acetal, which is promoted by the mild noncovalent hydrogen bond interactions with catalysts **102**. The interaction also leads to the stabilization of the *N*-Boc-C-alkynyl imine intermediate, which is generated in situ (Scheme 4.42).

Recently, an arylation strategy to synthesize 1-aryl-propargylamines was described by Shao and coworkers. The methodology is based on the asymmetric aza-Friedel–Crafts-type arylation of alkynyl imines with aryl nucleophiles catalyzed by the chiral acid **106** [74]. The scope of the reaction was investigated and electron-deficient, electron-neutral,

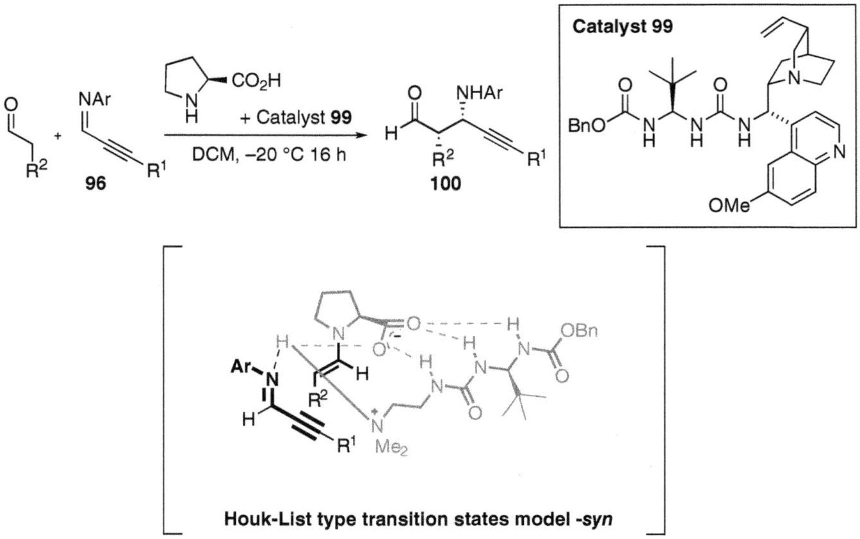

Scheme 4.41 syn-Selective Mannich synthesis of propargylamines. Source: Modified from Gómez-Bengoa et al. [71].

electron-rich phenols, as well as 1- and 2-naphthols, gave the desired propargylamines **107** in good yields and high enantioselectivities. The mechanistic investigation reveals that the chiral phosphoric acid **106** can simultaneously activate both the phenol and the in situ generated C-alkynyl imine intermediate through hydrogen bonding interactions. This creates a chiral environment where the phenol preferentially attacks the Re-face of imine intermediate affording propargylamines **107** with high enantioselectivity (Scheme 4.43). More recently, the reaction of C-alkynyl *N*-Boc *N,O*-acetals such as **105** with acetylenamides in the presence of a chiral phosphoric acid catalyst has also been described, leading to β-keto-propargylamines with up to 95% ee [75].

On the other hand, enantioenriched propargylamines can be obtained via organocatalyzed kinetic resolution of the racemates (Scheme 4.44). Seidel and coworkers firstly demonstrated that the organocatalyst **110** could be used to deracemize propargylamines **108** [76]. The kinetic resolution reaction was conducted via a dual catalytic method, which involves a thiourea-amide anion binding catalyst **110** in combination with DMAP. The approach affords enantioenriched acyl-propargylamines **109** with a high selectivity factor (s-factor), up to 56.

The key step of Seidel's approach is the formation of the ion pairs **111** and **112**. The achiral acyl pyridinium salt **111** (ion pair I) is formed from DMAP and an acylating reagent. The salt **111** reacts with the chiral thiourea catalyst **110**, leading to the chiral ion pair II **112**. Cossy and coworkers reported a similar approach using the chiral catalyst **114**, which afforded the corresponding enantioenriched N-acetylated propargylamines **115** with s-factors up to 193 [77].

Finally, Peng and coworkers recently reported the asymmetric synthesis of chiral fluorinated propargylamines **119** via organocatalytic reduction of alkynyl ketimine substrates **116** using chiral phosphoric acid **117** as a catalyst and benzothiazoline **118** as a hydride

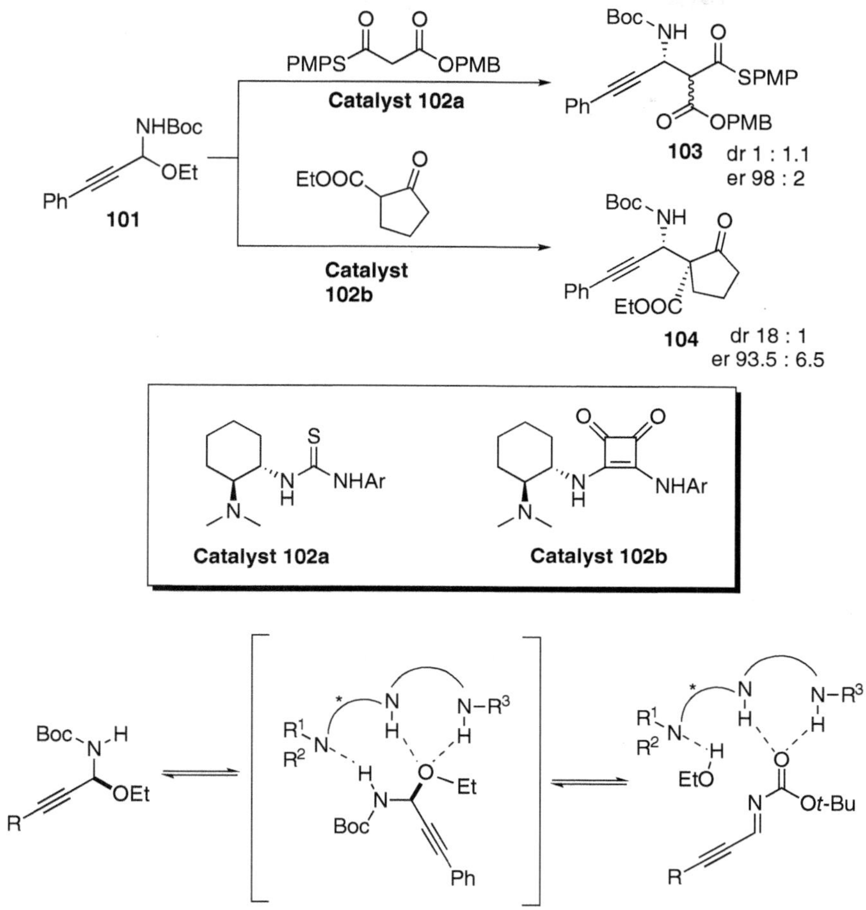

Scheme 4.42 Asymmetric synthesis of *syn*-propargylamines under Brønsted base catalysis.

source. The benzothiazoline promotes a Re-face hydrogen transfer attack as shown in Scheme 4.45 [78].

4.6 Propargylamines as Building Blocks in the Synthesis of Heterocycles

Propargylamines possess a unique chemical structure consisting of a nucleophilic amine group and an electrophilic carbon–carbon triple bond combined in the same molecule, which makes them versatile substrates for the synthesis of a variety of heterocyclic organic compounds such as pyrroles, pyridines, quinolines, oxazoles, and thiazoles. In the last part of this chapter, the use of propargylamines as building blocks in the synthesis of heterocyclic compounds is highlighted.

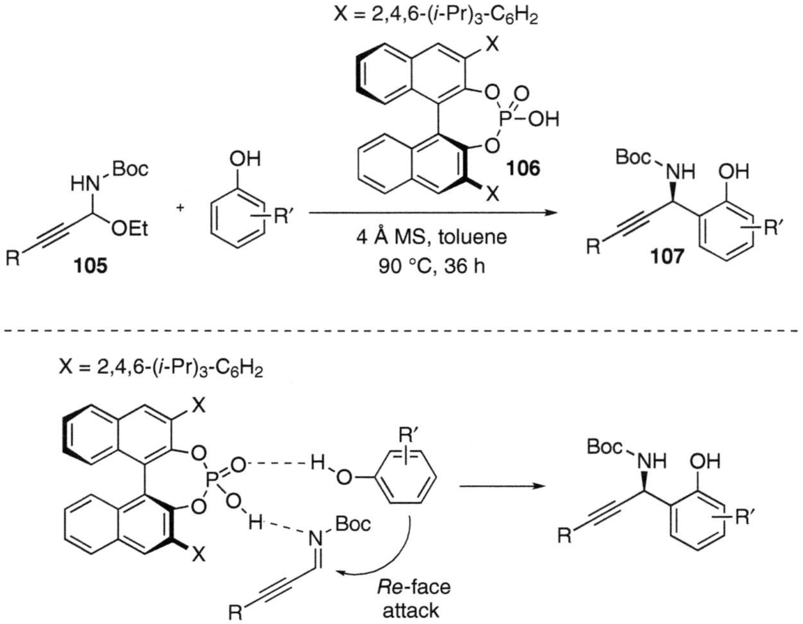

Scheme 4.43 Synthesis of propargylamine via Friedel–Crafts-type arylation.

4.6.1 Synthesis of Pyrroles from Propargylamines

Palladium catalysts are known to have good affinity for carbon–carbon triple bonds and thus can be used to activate propargylamines to afford various heterocycles including pyrroles. Trost et al. reported a Pd-catalyzed synthesis of pyrroles **121** through the addition of alkynes to propargylamines **120**, followed by a 5-endo-dig-cyclization and tautomerization of the ynenoate intermediate. Interestingly, the reaction was not influenced by the electronic character of the substituents, allowing a wide range of the substrates to be converted efficiently to the corresponding pyrroles under optimized conditions (Scheme 4.46a) [79]. In addition to Pd, other metal catalysts have also been used to convert propargylamines into pyrroles. Yamamoto and coworkers developed an iridium-catalyzed synthesis of the polycyclic pyrrole **125**. The *N*-allyl-propargylamine **122**, obtained via Cu-catalyzed Mannich condensation, was converted into cyclic diene **123** through Ir-catalyzed cycloisomerization. The diene **123** then reacted with a dienophile to provide intermediate **124**, which is finally converted into pyrrole **125** by dehydrogenative aromatization (Scheme 4.46b) [80].

The alkene–alkyne metathesis reaction between propargylamines and ethyl vinyl ether (EVE) catalyzed by Grubbs' catalyst has also emerged as an effective method for the construction of pyrroles. Our group described a one-pot microwave-assisted alkene-alkyne cross-metathesis/cyclization sequence to access pyrroles from propargylamines **126**. The propargylamine was reacted with EVE at 120 °C in the presence of Grubb's catalyst and $CuSO_4$ as the additive. The reaction proceeds via the formation of cross-metathesis diene intermediate **127**, which then cyclized in the presence of $CuSO_4$ to afford a variety of 1,2,3-substituted pyrroles **128** (Scheme 4.46c) [81].

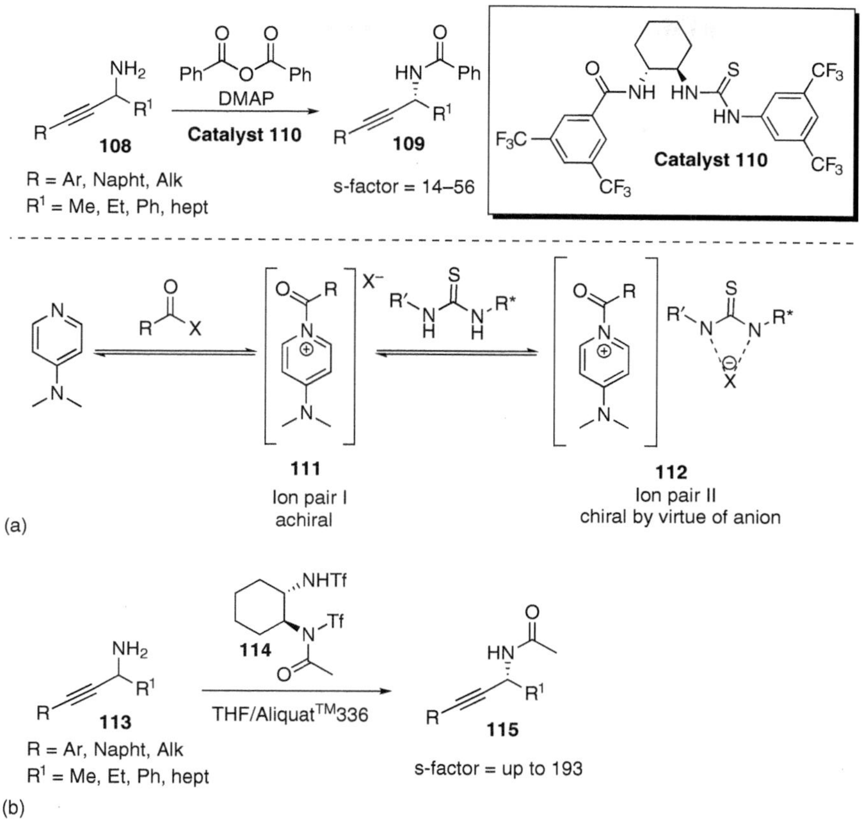

Scheme 4.44 Organocatalytic kinetic resolution of propargylamines. (a) Seidel's approach and (b) Cossy's approach.

Finally, Meng et al. reported an interesting approach for the synthesis of pyrroles through a metal free base-catalyzed intramolecular cyclization reaction of *N*-cyanopropargylamines (Scheme 4.46d). The reaction occurs through a cyclization–decyanation–aromatization cascade of propargylamine **129**. When the reaction was carried out at 140 °C, the pyrrole **130** was obtained as the only product, while at 110 °C, the 2-cyano-substituted pyrrole **131** was recovered as the major product [82].

4.6.2 Synthesis of Pyrrolines from Propargylamines

Alkene–alkyne (enyne) metathesis is considered as a popular way to synthesize pyrroline derivatives from propargylamines. A ring-opening/ring-closing enyne metathesis cascade approach to access pyrrolines **133** from 1,6-cyclopropene-ynes **132** has been described by Zhu and Shi. The 1,6-cyclopropene-yne **132** reacts with the first-generation Grubbs' catalysts to afford the intermediate **134**, which evolves to **136** via intramolecular ring-closing enyne metathesis. Compound **136** then reacts with an external alkene to afford pyrroline **133** and regenerate the ruthenium catalyst (Scheme 4.47a) [83].

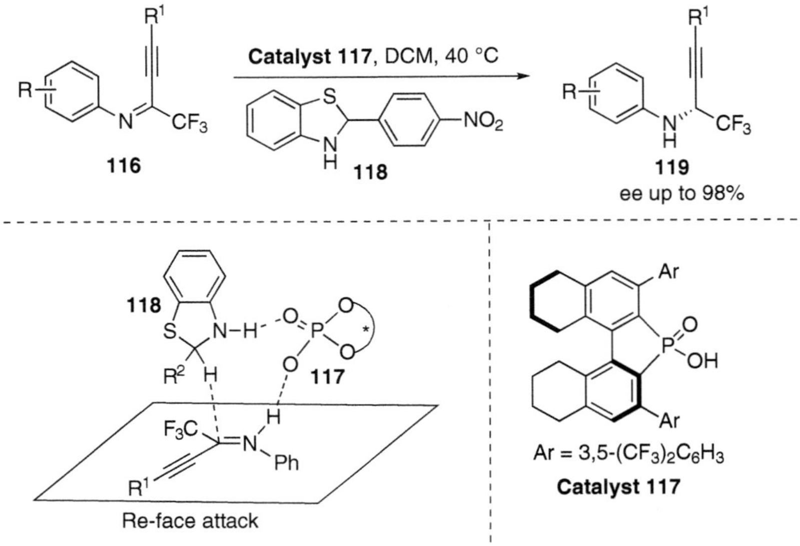

Scheme 4.45 Organocatalytic reduction of alkynyl ketimines. Source: Modified from Chen et al. [78].

In addition to enyne metathesis, other approaches were reported to synthesize pyrrolines from propargylamines. Polindara-García and Miranda described a base-mediated synthesis of 2-pyrroline **138** using propargylamine (Scheme 4.47b). The method involves an initial indium-catalyzed Ugi four-component reaction with commercial propargylamine leading to **137**, which is then treated with *t*-BuOK to afford 2-pyrroline **138** [84]. Finally, Yamada and coworkers developed a novel approach to obtain pyrrolinone **140** from propargylamine **139** via a silver-catalyzed CO_2 incorporation reaction, followed by intramolecular cyclization rearrangement. The propargylamine was converted into the corresponding pyrroline **140** using $AgNO_3$ as a catalyst, and DBU as the base, under an atmospheric pressure of CO_2 (Scheme 4.47c) [85].

4.6.3 Synthesis of Pyridines from Propargylamines

Cacchi et al. reported an early example of pyridine synthesis from *N*-propargylic β-enaminones **141**. The reaction mechanism involves the formation of an alkyne–copper complex followed by a 6-endo-dig cyclization to afford a dihydropyridine intermediate. The latter is then oxidized into pyridine **142** (Scheme 4.48a) [86].

Abbiati et al. described the use of commercially available propargylamine to synthesize a series of pyridine derivatives **143**. Propargylamine was reacted with various ketones in the presence of $NaAuCl_4 \cdot 2H_2O$ to yield pyridines **143** with good yields. It was suggested that the pyridine formation occurs through the sequential amination of the carbonyl substrates, followed by a regioselective 6-endo-dig cyclization of the *N*-propargylenamine intermediate and subsequent aromatization. It is worth mentioning that copper catalysts were also found to be efficient in this reaction with reactive ketones (Scheme 4.48b) [87].

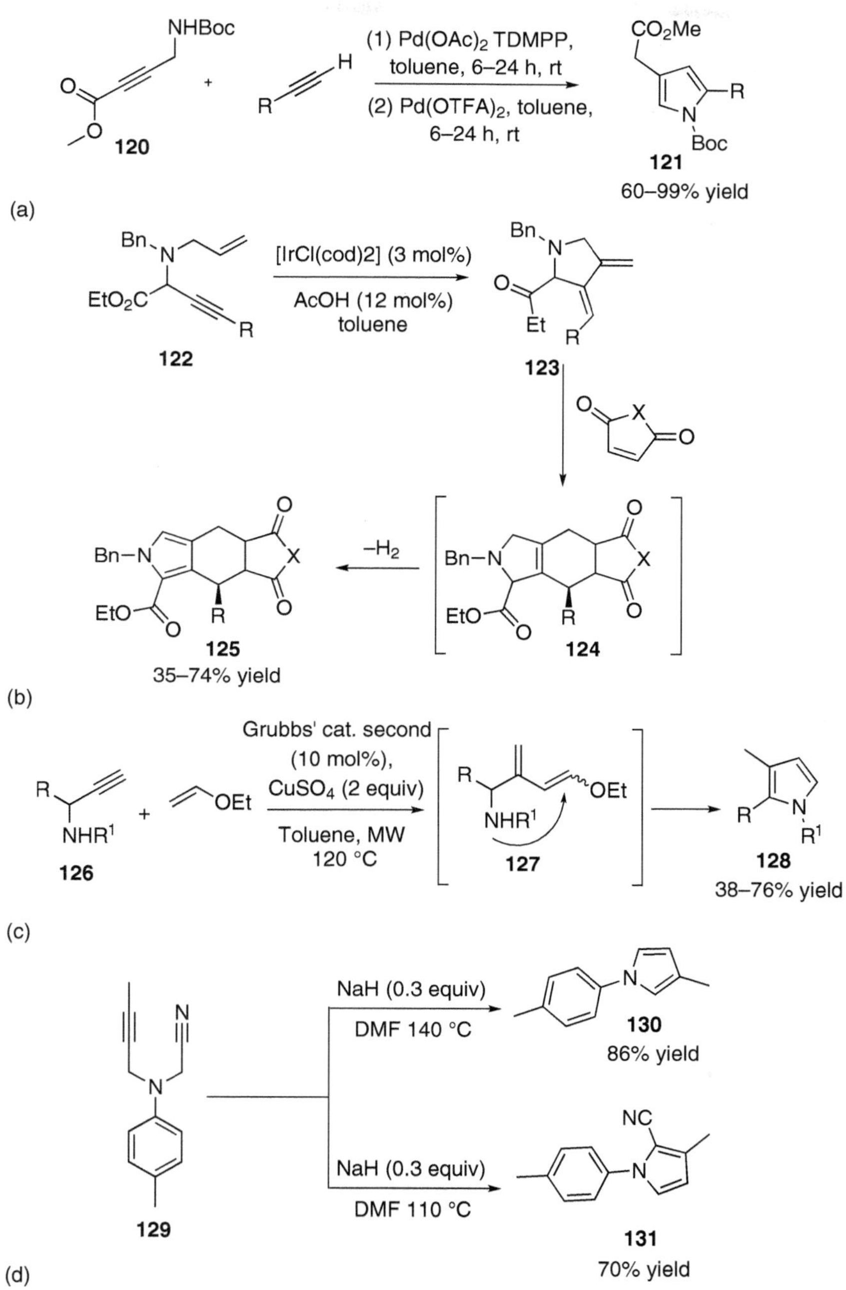

Scheme 4.46 Synthesis of pyrroles from propargylamines. Source: (a) Modified from Trost et al. [79], (b) modified from Yamamoto et al. [80], and (c) modified from Chachignon et al. [81].

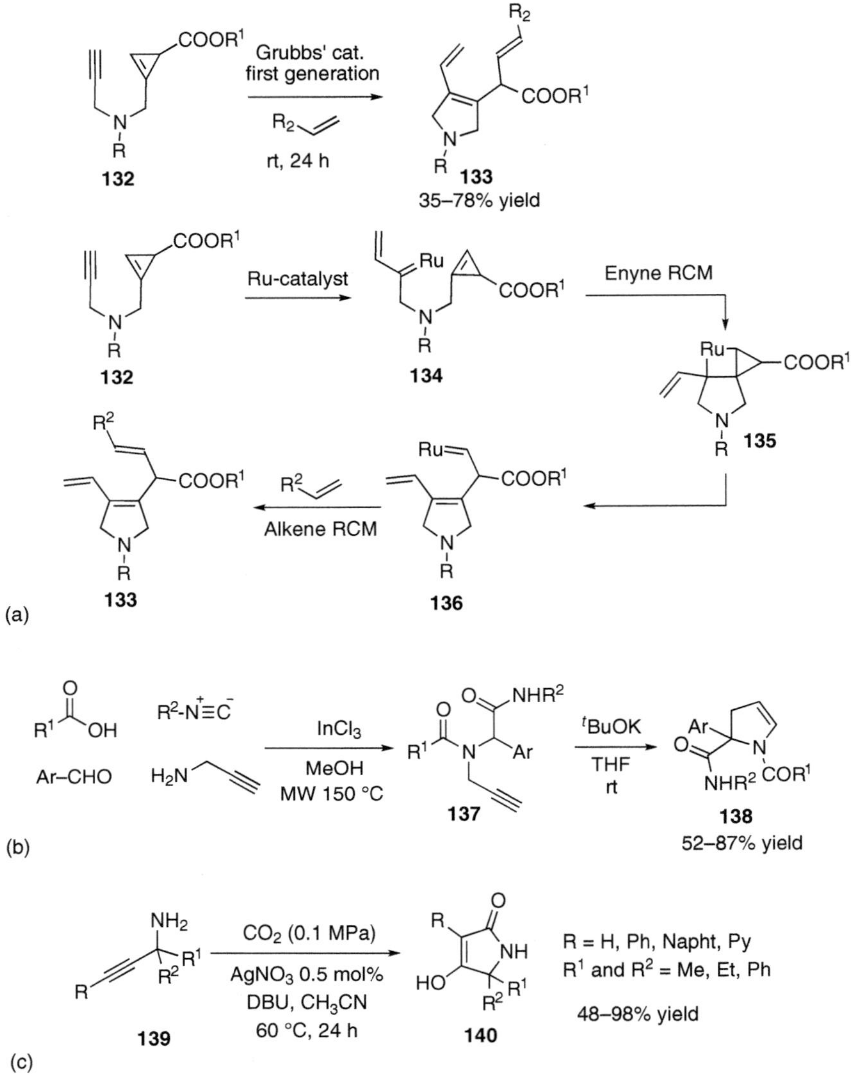

Scheme 4.47 Synthesis of pyrrolines from propargylamines. Source: (a) Modified from Zhu et al. [83], (b) modified from Polindara-García et al. [84], and (c) modified from Ishida et al. [85].

Finally, dihydropyridines **145** can be synthesized from *N*-propargyl-enamines **144** via a rhodium-catalyzed cyclodimerization in the presence of DABCO (Scheme 4.48c) [88].

4.6.4 Synthesis of Quinolines from Propargylamines

Huma et al. reported the first example of quinoline synthesis from *N*-aryl-propargylamine **146** via propargyl-allenyl isomerization, followed by copper(I)-catalyzed cyclization to give quinoline **147** (Scheme 4.49a) [89].

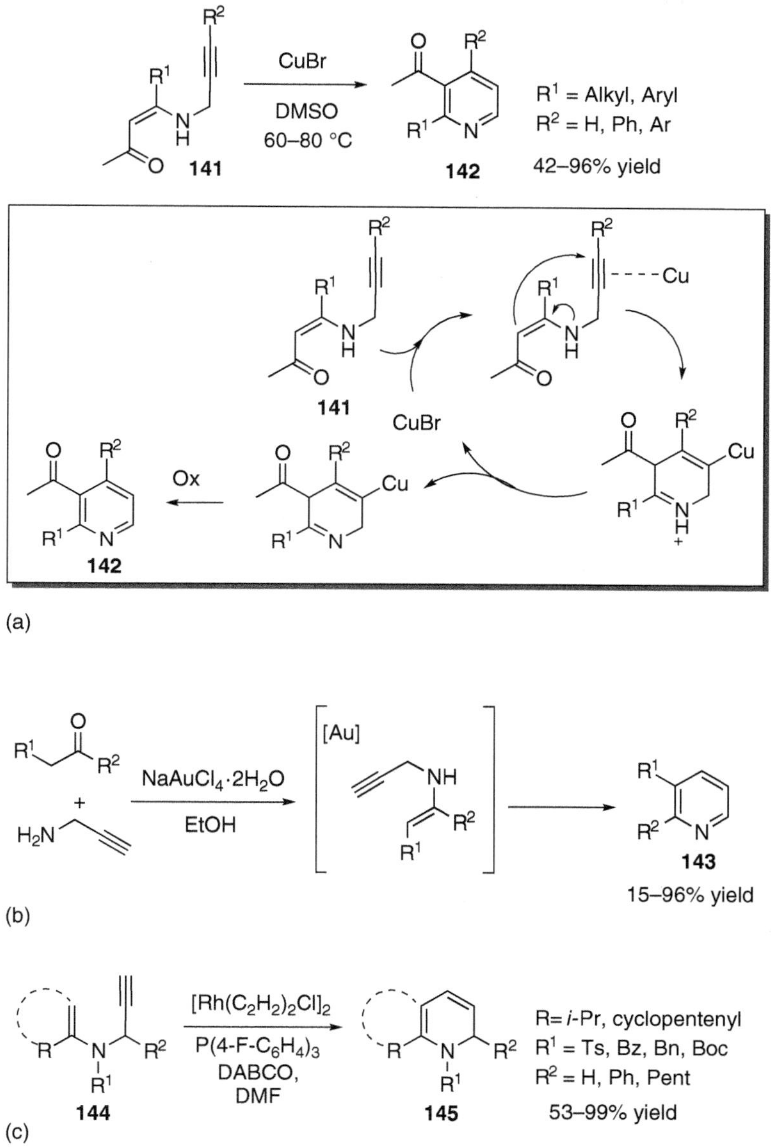

Scheme 4.48 Synthesis of pyridines from propargylamines. Source: (a) Modified from Cacchi et al. [86], (b) modified from Abbiati et al. [87], and (c) modified from Kim et al. [88].

A similar approach has been later reported by several research groups using gold catalysts. The *N*-aryl-propargylamines **148** can be transformed into quinolines **149** under the catalysis of $AuCl_3$ [90], $Ph_3PAuCl/AgOTf$ [91], or most recently Au/TiO_2 [92]. The cyclization reaction proceeds via the formation of a gold-π-alkyne complex which in turn favors a 6-endo-dig cyclization to afford a dihydroquinoline intermediate. The latter is finally converted into quinoline by spontaneous air oxidation (Scheme 4.49b). More

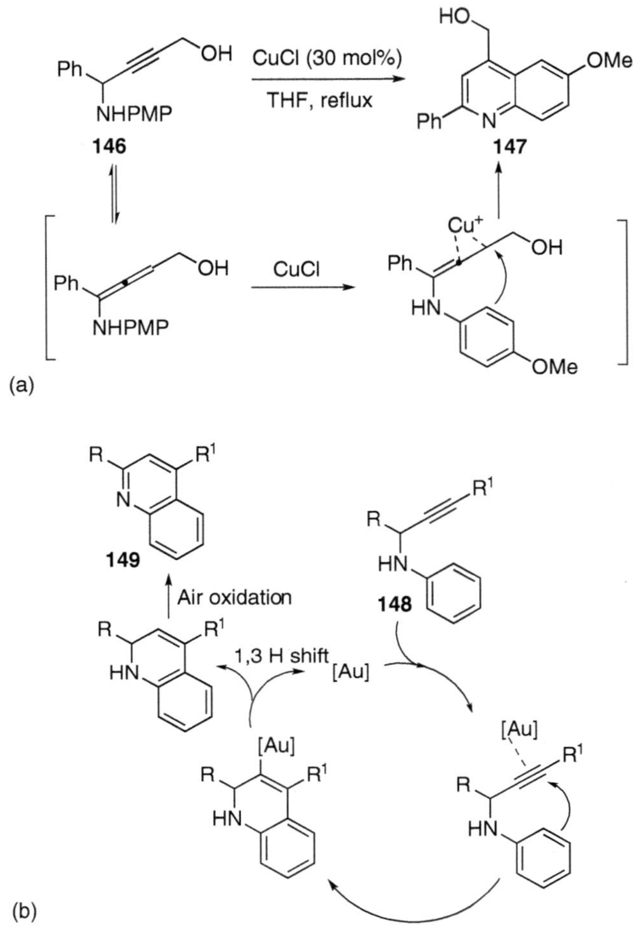

Scheme 4.49 Copper- and gold-catalyzed synthesis of quinolines from propargylamines. Source: (a) Modified from Huma et al. [89].

recently, the synthesis of quinoline derivatives via a palladium-catalyzed dimerization of *N*-aryl propargylamines has been described by Li et al. [93].

An interesting approach to synthesize quinoline **152** was reported by Kulkarni and Török. The authors synthesized the propargylamine **150** via the in situ Montmorillonite **K-10** catalyzed coupling of benzaldehyde, aniline, and phenylacetylene. Then, **150** cyclizes to give the cyclic intermediate **151** by intramolecular nucleophilic attack. Finally, the quinoline **152** is obtained through oxidative aromatization of **151** (Scheme 4.50a) [94].

Finally, Tang and coworkers reported the synthesis of 3-arylsulfonylquinolines **154** via TBHP-mediated sulfonylation–cyclization–aromatization of propargyl anilines **153** with arylsulfonylhydrazide (Scheme 4.50b). The author suggested that a sulfonyl radical is initially formed from the reaction of sulfonylhydrazide with TBHP. This radical then reacts with the triple bond of **153** to provide an alkenyl radical intermediate, which attacks

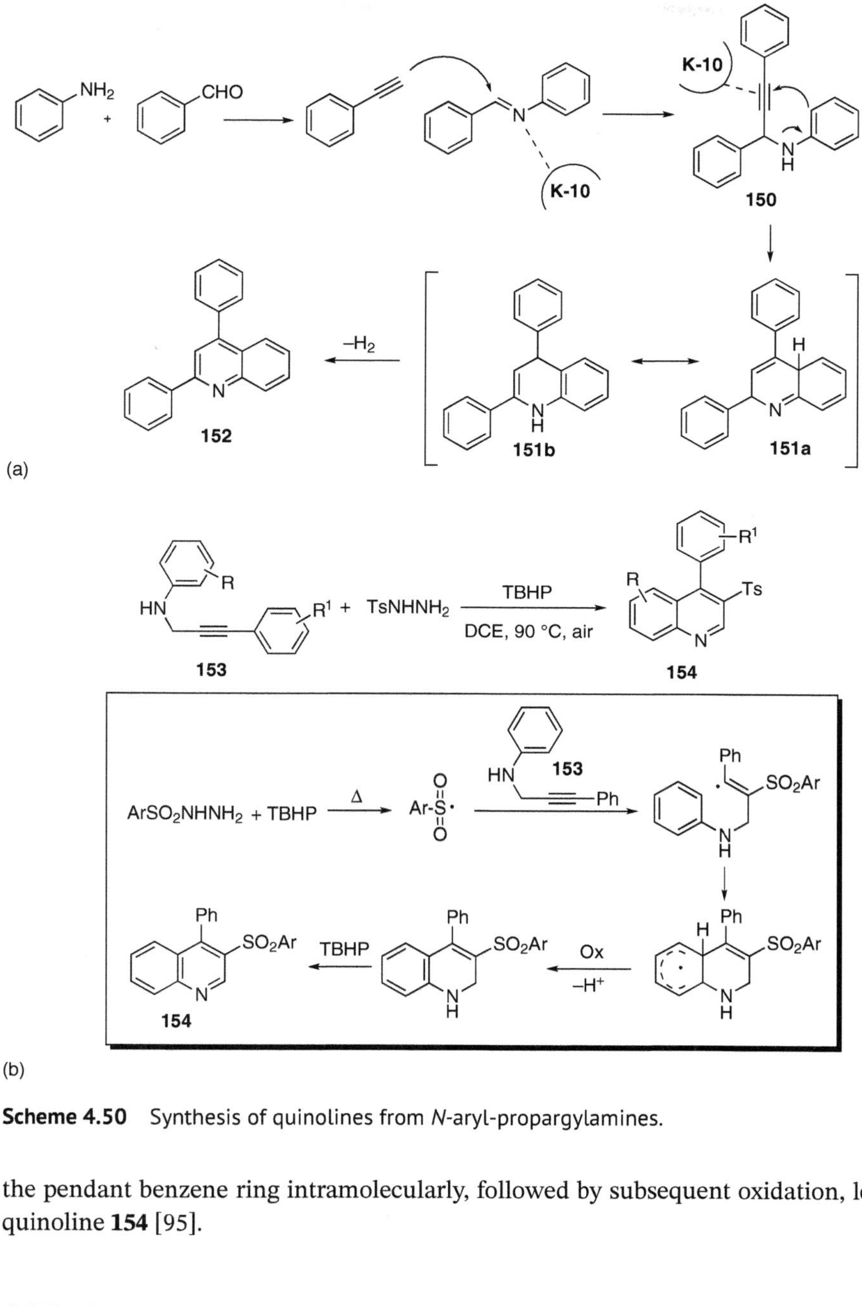

Scheme 4.50 Synthesis of quinolines from *N*-aryl-propargylamines.

the pendant benzene ring intramolecularly, followed by subsequent oxidation, leading to quinoline **154** [95].

4.6.5 Synthesis of Oxazoles from Propargylamines

N-Acyl-propargylamines represent an ideal class of substrates for the synthesis of oxazole derivatives because of the nucleophilic ability of the oxygen of the acyl group to attack the carbon–carbon triple bond of the propargylamine backbone.

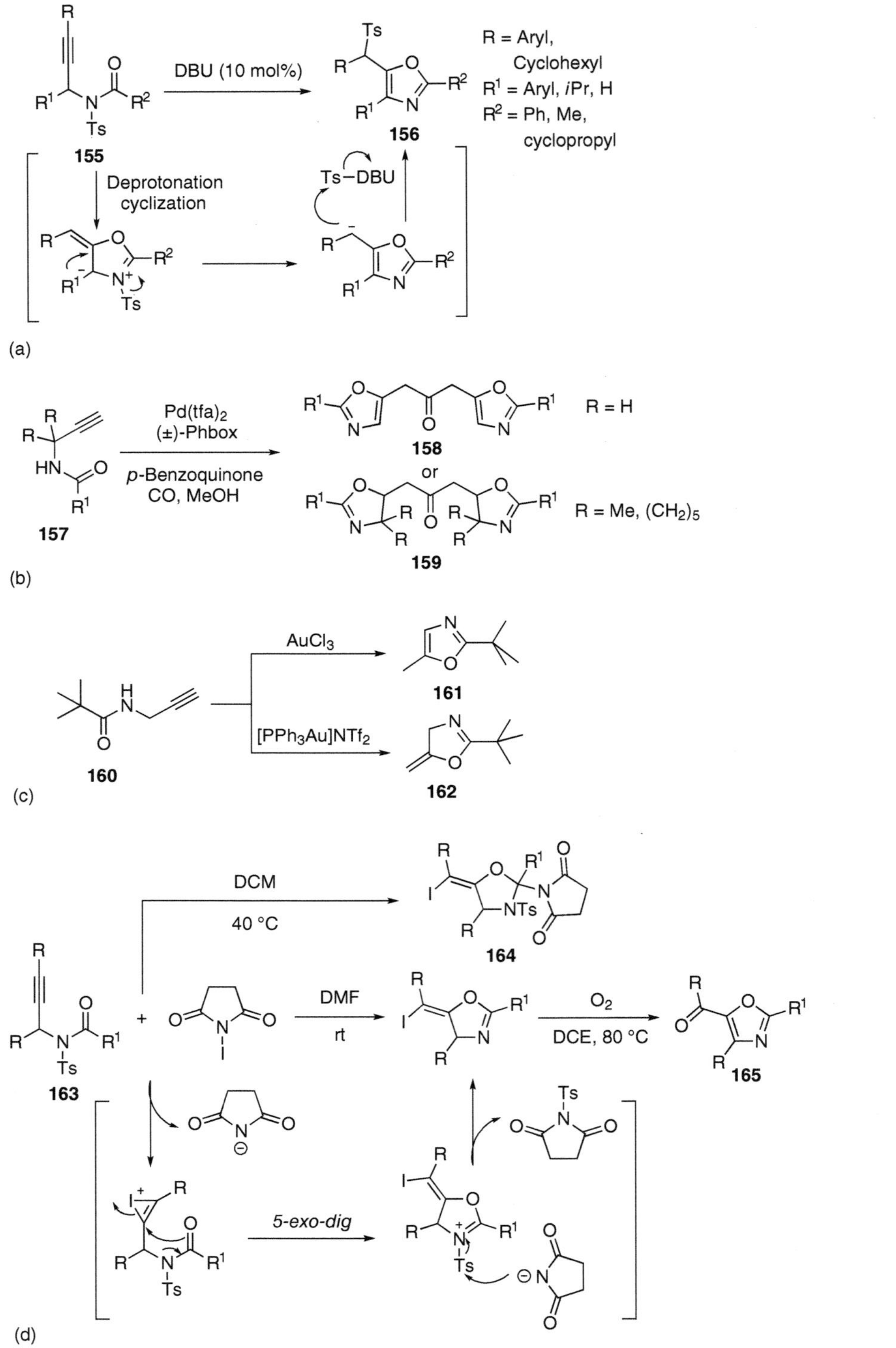

Scheme 4.51 Synthesis of oxazoles from propargylamines. Source: (a) Modified from Yu et al. [99], (b) modified from Yasuhara et al. [96], (c) modified from Weyrauch et al. [97], and (d) modified from Hu et al. [98].

Yu et al. reported the synthesis of the oxazole **156** from the *N*-tosyl-propargylamine **155** through DBU-catalyzed cyclization. Interestingly, during the reaction, a 1,4-sulfonyl migration occurs as shown in Scheme 4.51a [99].

Oxazoles have also been synthesized from propargylamines using transition metal catalysts. Yasuhara et al. described the synthesis of oxazole **158** via a cyclization–carbonylation–cyclization coupling reaction using Pd(II)-bisoxazoline catalysts (Scheme 4.51b). It has been hypothesized that the Pd(II)–bisoxazoline complex promotes the cyclization of the propargylamide **157**, which is carbonylated and coupled to a second oxazole molecule affording **158** [96]. Interestingly, depending on the substituents, the oxazolidine **159** can be obtained. Weyrauch et al. reported the synthesis of oxazoles from propargylamide **160** using the gold catalysts $AuCl_3$, $[PPh_3Au]NTf_2$, or $[PPh_3Au]OTs$. Depending on the oxidation state of the gold catalysts, either oxazole **161** or methylenedihydrooxazole **162** were obtained (Scheme 4.51c) [97].

Finally, it has been shown that *N*-iodosuccinimide (NIS) can promote the cyclization of *N*-sulfonyl propargylamides **163** via a 5-exo-dig intramolecular process to obtain oxazole **165** (Scheme 4.51d) [98].

More recently, Liu and coworkers reported an interesting cyclization of propargylamides **166** promoted by $PhI(OAc)_2$. The reaction occurs in the presence of oxygen and light and proceeds through the formation of the iodo-intermediate **168** as shown in Scheme 4.52a [100]. A similar approach has also been reported by An et al. who described the intramolecular cyclization of internal propargylamides **169** into oxazoles **170** using the gold catalyst PPh_3AuCl with 4-MeO-TEMPO as an oxidant (Scheme 4.52b) [101].

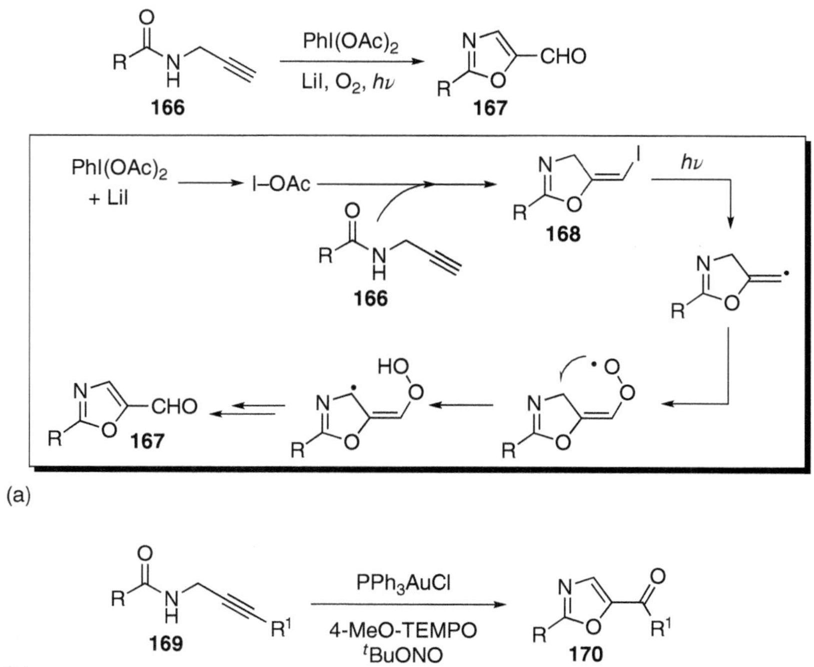

Scheme 4.52 Recent syntheses of oxazoles from propargylamides. Source: (a) Modified from Yi et al. [100] and (b) modified from An et al [101].

4.6.6 Synthesis of Thiazoles from Propargylamines

Similar to oxazoles, propargylamines can also be exploited as building blocks in the synthesis of thiazoles. The first cyclization of a propargylamine with CS_2 into a thiazole derivative **174** was reported by Batty and Weedonin 1949 (Scheme 4.53a) [102]. More recently, CS_2 has been exploited by various groups in combination with metal catalysts to synthesize thiazole derivatives from propargylamines. Stevens and coworkers described a gold-catalyzed, two-step synthesis of 5-alkylidene dihydrothiazoles **178** (Scheme 4.53b). The reaction occurs via the formation of *N*-propargylic dithiocarboimidate **175** from commercial propargylamine. Then, **175** is converted into dihydrothiazole **178** through a $AuCl_3$-catalyzed 5-exo-dig cyclization, followed by a 1,3-alkyl migration [103]. Van der Eycken also used CS_2 in the synthesis of thiazole derivatives **180** (Scheme 4.53c). The thiazolidine-2-thiones **180** were obtained through a one-pot process involving the

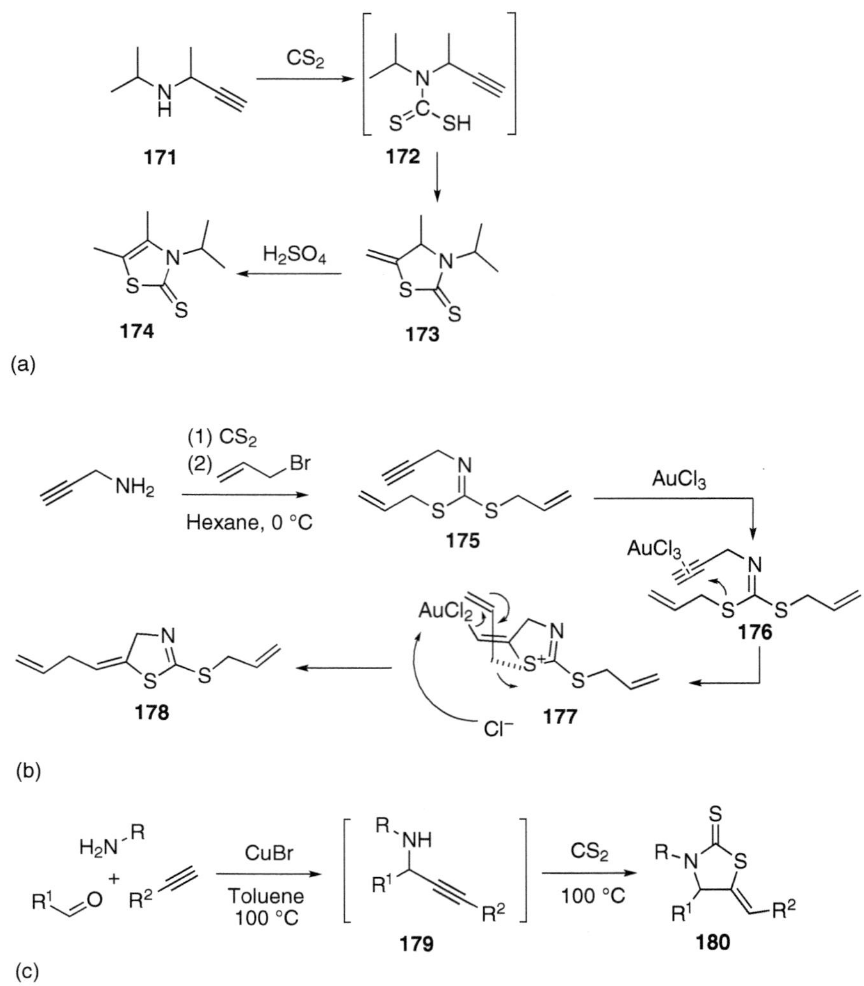

Scheme 4.53 Reactions of propargylamine with CS_2. Source: (a) Modified from Batty and Weedon [102].

formation of the propargylamine intermediate **179** via a Cu-catalyzed A^3 coupling of an alkyne, aldehyde, and amine, followed by cyclization in the presence of CS_2 [104].

Isothiocyanates can be used in place of CS_2 for the synthesis of thiazoles from propargylamines. In 1964, Easton et al. firstly described the synthesis of 2-iminothiazolidines **181** through the reaction of substituted propargylamines with various isothiocyanates (Scheme 4.54a) [105].

Since then, the synthesis of thiazoles using isothiocyanate reagents has been described in many studies. Among them, Pieters and coworkers reported the synthesis of lactose-2-aminothiazoline conjugate **183**, which showed high affinity for cholera toxin B, from a *N*-Boc-protected propargylamine through deprotection and functionalization with lactose β-isothiocyanate (Scheme 4.54b) [106].

More recently, our laboratory developed a microwave-assisted method to selectively synthesize 2-aminothiazoles **184** or 5-methylthiazol-2-amines **185** using isothiocyanate and propargylamines. When the reaction was carried out in DCE under microwave irradiation at 130 °C, thiazole **184** was formed as the only product in good yields, while in CH_3CN at 100 °C, the thiazoline **185** was obtained as the main reaction product (Scheme 4.54c) [107].

Finally, Alhalib and Moran described the synthesis of dihydrothiazole by treatment of *N*-propargylamides with Lawesson's reagent. The reaction proceeds through the formation of a thioamide intermediate **186**, which spontaneously cyclizes to afford thiazole derivative **187** (Scheme 4.55) [108].

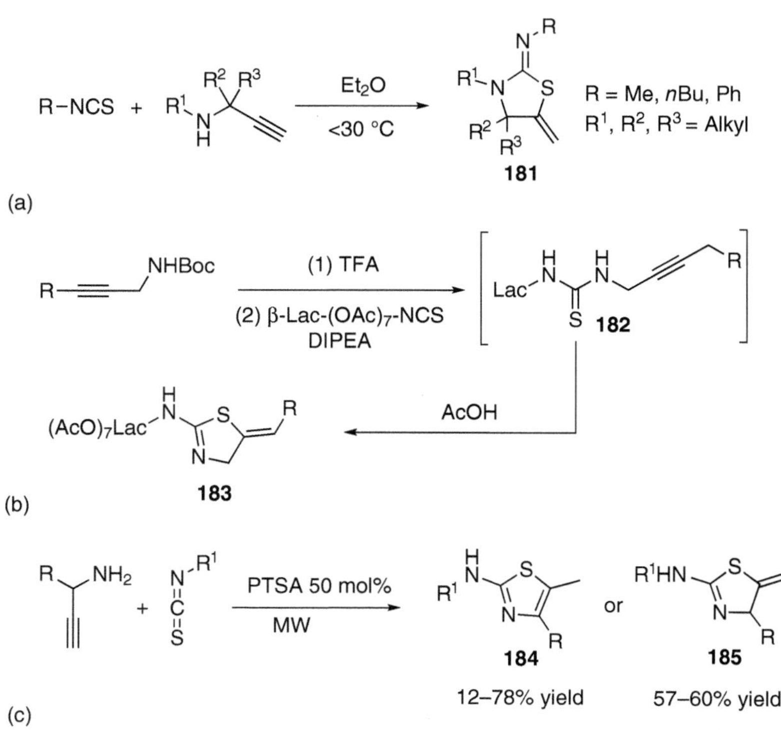

Scheme 4.54 Reactions of propargylamine with isothiocyanates. Source: (a) Modified from Easton et al. [105], (b) modified from Vrasidas et al. [106], and (c) modified from Scalacci et al. [107].

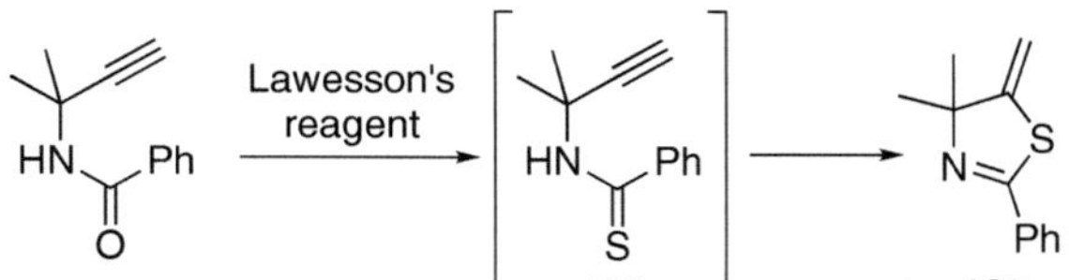

186 187

Scheme 4.55 Synthesis of thiazoline derivatives from *N*-propargylamines using the Lawesson's reagent.

4.7 Conclusions

The present chapter has highlighted the main methodologies currently used in organic chemistry to access propargylamines, a class of amine compounds possessing unique properties and reactivities. The high interest of the scientific community toward this class of molecules becomes evident by the high number of new synthetic methodologies published every year in literature, also exploiting new approaches such as biocatalysis, organocatalysis, and photoredox catalysis. This chapter has highlighted the recent progresses made in the past few years on the synthesis of chiral propargylamines, showing how chiral metal catalysts, enzymes, or organocatalysts can be used for the construction of enantiomerically pure stereocenters on a propargylamine backbone. Not least, photoredox catalysis has contributed to expand the scope of methodologies to synthesize functionalized propargylamines under mild and more sustainable conditions. Finally, the use of propargylamines as building blocks in the synthesis of a variety of organic compounds, especially heterocycles, is presented in the last part of this chapter. The electrophilic/nucleophilic properties of propargylamines combined in a single molecule make them unique in the amine compounds panorama and ideal substrates for the synthesis of pyrroles, pyridines, quinolines, thia/oxazoles, and other heterocycles. Since the publication of our review few years ago [10], a number of new methodologies for the synthesis and derivatization of propargylamines has been reported in the literature, making clear once more the importance and the role of these molecules in chemistry and related disciplines. It is certain that new synthetic approaches, exploiting for instance electrochemistry or flow chemistry, and novel transformations of propargylamines, will appear in the near future.

References

1 Trost, B. and Li, C.-J. (2014). *Modern Alkyne Chemistry*. Wiley-VCH.

2 Shachat, N. and Bagnell, J.J. (1963). *The Journal of Organic Chemistry* 28: 991–995.

3 Easton, N.R., Cassady, D.R., and Dillard, R.D. (1963). *The Journal of Organic Chemistry* 28: 448–453.

4 Dillard, R.D. and Easton, N.R. (1964). *The Journal of Organic Chemistry* 29: 2464–2467.

5 (a) Baranyi, M., Porceddu, P.F., Gölöncsér, F. et al. (2016). *Molecular Neurodegeneration* 11: 1–21. (b) Huleatt, P.B., Khoo, M.L., Chua, Y.Y. et al. (2015). *Journal of Medicinal Chemistry* 58: 1400–1419.

6 Bolea, I., Gella, A., and Unzeta, M. (2013). *Journal of Neural Transmission* 120: 893–902.

7 Mao, F., Li, J., Wei, H. et al. (2015). *Journal of Enzyme Inhibition and Medicinal Chemistry* 30: 995–1001.

8 Kopka, I.E., Fataftah, Z.A., and Rathke, M.W. (1980). *The Journal of Organic Chemistry* 45: 4616–4622.

9 Hennion, G.F. and Hanzel, R.S. (1960). *Journal of the American Chemical Society* 82: 4908–4912.

10 Lauder, K., Toscani, A., Scalacci, N., and Castagnolo, D. (2017). *Chemical Reviews* 117: 14091–14200.

11 Rokade, B.V., Barker, J., and Guiry, P.J. (2019). *Chemical Society Reviews* 48: 4766–4790.

12 Wei, C. and Li, C.-J. (2002). *Journal of the American Chemical Society* 124: 5638–5639.

13 Wei, C., Mague, J.T., and Li, C.-J. (2004). *Proceedings of the National Academy of Sciences of the United States of America* 101: 5749–5754.

14 Nakamura, S., Ohara, M., Nakamura, Y. et al. (2010). *Chemistry - A European Journal* 16: 2360–2362.

15 Benaglia, M., Negri, D., and Dell'Anna, G. (2004). *Tetrahedron Letters* 45: 8705–8708.

16 Liu, B., Huang, L., Liu, J. et al. (2007). *Tetrahedron: Asymmetry* 18: 2901–2904.

17 Hatano, M., Asai, T., and Ishihara, K. (2008). *Tetrahedron Letters* 49: 379–382.

18 Gommermann, N., Koradin, C., Polborn, K., and Knochel, P. (2003). *Angewandte Chemie, International Edition* 42: 5763–5766.

19 Knöpfel, T.F., Aschwanden, P., Ichikawa, T. et al. (2004). *Angewandte Chemie, International Edition* 43: 5971–5973.

20 Cardoso, F.S.P., Abboud, K.A., and Aponick, A. (2013). *Journal of the American Chemical Society* 135: 14548–14551.

21 Paioti, P.H.S., Abboud, K.A., and Aponick, A. (2016). *Journal of the American Chemical Society* 138: 2150–2153.

22 Paioti, P.H.S., Abboud, K.A., and Aponick, A. (2017). *ACS Catalysis* 7: 2133–2138.

23 Rokade, B.V. and Guiry, P.J. (2017). *ACS Catalysis* 7: 2334–2338.

24 Detz, R.J., Delville, M.M.E., Hiemstra, H., and van Maarseveen, J.H. (2008). *Angewandte Chemie, International Edition* 47: 3777–3780.

25 Detz, R.J., Abiri, Z., le Griel, R. et al. (2011). *Chemistry - A European Journal* 17: 5921–5930.

26 Hattori, G., Sakata, K., Matsuzawa, H. et al. (2010). *Journal of the American Chemical Society* 132: 10592–10608.

27 Yoshida, A., Hattori, G., Miyake, Y., and Nishibayashi, Y. (2011). *Organic Letters* 13: 2460–2463.

28 Shibata, M., Nakajima, K., and Nishibayashi, Y. (2014). *Chemical Communications* 50: 7874–7877.

29 Hattori, G., Matsuzawa, H., Miyake, Y., and Nishibayashi, Y. (2008). *Angewandte Chemie, International Edition* 47: 3781–3783.

30 Zhang, C., Wang, Y.-H., Hu, X.-H. et al. (2012). *Advanced Synthesis and Catalysis* 354: 2854–2858.

31 Watanabe, K., Miyazaki, Y., Okubo, M. et al. (2018). *Organic Letters* 20: 5448–5451.

32 Hattori, G., Yoshida, A., Miyake, Y., and Nishibayashi, Y. (2009). *The Journal of Organic Chemistry* 74: 7603–7607.
33 Gómez, J.E., Guo, W., Gaspa, S., and Kleij, A.W. (2017). *Angewandte Chemie, International Edition* 56: 15035–15038.
34 Tian, L., Gong, L., and Zhang, X. (2018). *Advanced Synthesis and Catalysis* 360: 2055–2059.
35 Koradin, C., Polborn, K., and Knochel, P. (2002). *Angewandte Chemie, International Edition* 41: 2535–2538.
36 Bai, X.-Y., Zhang, W.-W., Li, Q., and Li, B.-J. (2018). *Journal of the American Chemical Society* 140: 506–514.
37 Liu, T.-L., Wang, C.-J., and Zhang, X. (2013). *Angewandte Chemie, International Edition* 52: 8416–8419.
38 Chen, M.-W., Wu, B., Chen, Z.-P. et al. (2016). *Organic Letters* 18: 4650–4653.
39 Fan, D., Hu, Y., Jiang, F. et al. (2018). *Advanced Synthesis and Catalysis* 360: 2228–2232.
40 Wang, Y., Wen, X., Cui, X., and Zhang, X.P. (2018). *Journal of the American Chemical Society* 140: 4792–4796.
41 Wen, X., Wang, Y., and Zhang, X.P. (2018). *Chemical Science* 9: 5082–5086.
42 Li, C., Lang, K., Lu, H. et al. (2018). *Angewandte Chemie, International Edition* 57: 16837–16841.
43 Xing, Q., Chan, C.-M., Yeung, Y.-W., and Yu, W.-Y. (2019). *Journal of the American Chemical Society* 141: 3849–3853.
44 Wang, H., Park, Y., Bai, Z. et al. (2019). *Journal of the American Chemical Society* 141: 7194–7201.
45 Patel, R.N. (2011). *ACS Catalysis* 1: 1056–1074.
46 Devine, P.N., Howard, R.M., Kumar, R. et al. (2018). *Nature Reviews Chemistry* 2: 409–421.
47 Sheldon, R.A. and Brady, D. (2019). *ChemSusChem* 12: 2859–2881.
48 Messina, F., Botta, M., Corelli, F. et al. (1999). *The Journal of Organic Chemistry* 64: 3767–3769.
49 Castagnolo, D., Armaroli, S., Corelli, F., and Botta, M. (2004). *Tetrahedron: Asymmetry* 15: 941–949.
50 Botta, M., Corelli, F., Gasparrini, F. et al. (2000). *The Journal of Organic Chemistry* 65: 4736–4739.
51 Castagnolo, D., Giorgi, G., Spinosa, R. et al. (2007). *European Journal of Organic Chemistry* 22: 3676–3686.
52 (a) Topgi, R.S., Ng, J.S., Landis, B. et al. (1999). *Bioorganic & Medicinal Chemistry* 7: 2221–2229. (b) Landis, B.H., Mullins, P.B., Mullins, K.E., and Wang, P.T. (2002). *Organic Process Research and Development* 6: 539–546.
53 Schmidt, N.G., Simon, R.C., and Kroutil, W. (2015). *Advanced Synthesis and Catalysis* 357: 1815–1821.
54 Aleku, G.A., France, S.P., Man, H. et al. (2017). *Nature Chemistry* 9: 961–969.
55 (a) Tauber, K., Fuchs, M., Sattler, J.H. et al. (2013). *Chemistry - A European Journal* 19: 4030–4035. (b) Mutti, F.G., Knaus, T., Scrutton, N.S. et al. (2015). *Science* 349: 1525–1529.

56 Montgomery, S.L., Mangas-Sanchez, J., Thompson, M.P. et al. (2017). *Angewandte Chemie, International Edition* 56: 10491–10494.

57 Hussain, S., Leipold, F., Man, H. et al. (2015). *ChemCatChem* 7: 579–583.

58 (a) Wetzl, D., Gand, M., Ross, A. et al. (2016). *ChemCatChem* 8: 2023–2026. (b) Scheller, P.N., Lenz, M., Hammer, S.C. et al. (2015). *ChemCatChem* 7: 3239–3242. (c) Huber, T., Schneider, L., Präg, A. et al. (2014). *ChemCatChem* 6: 2248–2252.

59 Matzel, P., Gand, M., and Höhne, M. (2017). *Green Chemistry* 19: 385–389.

60 Wetzl, D., Berrera, M., Sandon, N. et al. (2015). *ChemBioChem* 16: 1749–1756.

61 Hyslop, J.F., Lovelock, S.L., Sutton, P.W. et al. (2018). *Angewandte Chemie, International Edition* 57: 13821–13824.

62 Odachowski, M., Greaney, M.F., and Turner, N.J. (2018). *ACS Catalysis* 8: 10032–10035.

63 (a) Prier, C.K., Rankic, D.A., and MacMillan, D.W.C. (2013). *Chemical Reviews* 113: 5322–5363. (b) Parasram, M. and Gevorgyan, V. (2017). *Chemical Society Reviews* 46: 6227–6240.

64 Rueping, M., Koenigs, R.M., Poscharny, K. et al. (2012). *Chemistry - A European Journal* 18: 5170–5174.

65 Zhou, Q.-Q., Guo, W., Ding, W. et al. (2015). *Angewandte Chemie, International Edition* 54: 11196–11199.

66 Zhang, H., Zhang, P., Jiang, M. et al. (2017). *Organic Letters* 19: 1016–1019.

67 Le Vaillant, F., Courant, T., and Waser, J. (2015). *Angewandte Chemie, International Edition* 54: 11200–11204.

68 Xie, J., Shi, S., Zhang, T. et al. (2015). *Angewandte Chemie, International Edition* 54: 6046–6050.

69 Sagadevan, A., Pampana, V.K.K., and Hwang, K.C. (2019). *Angewandte Chemie, International Edition* 58: 3838–3842.

70 (a) Kano, T., Yurino, T., Asakawa, D., and Maruoka, K. (2013). *Angewandte Chemie, International Edition* 52: 5532–5534. (b) Kano, T., Yurino, T., and Maruoka, K. (2013). *Angewandte Chemie, International Edition* 52: 11509–11512. (c) Kano, T., Aota, Y., Asakawa, D., and Maruoka, K. (2015). *Chemical Communications* 51: 16472–16474.

71 Gómez-Bengoa, E., Jiménez, J., Lapuerta, I. et al. (2012). *Chemical Science* 3: 2949–2957.

72 Lapuerta, I., Vera, S., Oiarbide, M., and Palomo, C. (2016). *Chemistry - A European Journal* 22: 7229–7237.

73 Wang, Y., Mo, M., Zhu, K. et al. (2015). *Nature Communications* 6: 2–10.

74 Wang, Y., Jiang, L., Li, L. et al. (2016). *Angewandte Chemie, International Edition* 55: 15142–15146.

75 Feng, F.-F., Li, S., Cheung, C.W., and Ma, J.-A. (2019). *Organic Letters* 21: 8419–8423.

76 Klauber, E.G., De, C.K., Shah, T.K., and Seidel, D. (2010). *Journal of the American Chemical Society* 132: 13624–13626.

77 Kolleth, A., Christoph, S., Arseniyadis, S., and Cossy, J. (2012). *Chemical Communications* 48: 10511–10513.

78 Chen, M.-W., Yang, Q., Deng, Z. et al. (2018). *The Journal of Organic Chemistry* 83: 8688–8694.

79 Trost, B.M., Lumb, J.P., and Azzarelli, J.M. (2011). *Journal of the American Chemical Society* 133: 740–743.
80 Yamamoto, Y., Hayashi, H., Saigoku, T., and Nishiyama, H. (2005). *Journal of the American Chemical Society* 127: 10804–10805.
81 Chachignon, H., Scalacci, N., Petricci, E., and Castagnolo, D. (2015). *The Journal of Organic Chemistry* 8: 5287–5295.
82 Meng, J., Li, Y.J., Zhao, Y.L. et al. (2014). *Chemical Communications* 50: 12490–12492.
83 Zhu, Z.-B. and Shi, M. (2010). *Organic Letters* 12: 4462–4465.
84 Polindara-García, L.A. and Miranda, L.D. (2012). *Organic Letters* 14: 5408–5411.
85 Ishida, T., Kobayashi, R., and Yamada, T. (2014). *Organic Letters* 16: 2430–2433.
86 Cacchi, S., Fabrizi, G., and Filisti, E. (2008). *Organic Letters* 10: 2629–2632.
87 Abbiati, G., Arcadi, A., Bianchi, G. et al. (2003). *The Journal of Organic Chemistry* 68: 6959–6966.
88 Kim, H. and Lee, C. (2006). *Journal of the American Chemical Society* 128: 6336–6337.
89 Huma, H.Z.S., Halder, R., Kalra, S.S. et al. (2002). *Tetrahedron Letters* 43: 6485–6488.
90 Xiao, F., Chen, Y., Liu, Y., and Wang, J. (2008). *Tetrahedron* 64: 2755–2761.
91 Zhu, M., Fu, W., Zou, G. et al. (2012). *Journal of Fluorine Chemistry* 135: 195–199.
92 Symeonidis, T.S., Lykakis, I.N., and Litinas, K.E. (2013). *Tetrahedron* 69: 4612–4616.
93 Li, X.-F., Zhang, X.-G., Hua, B.L., and Zhang, X.H. (2018). *Organic & Biomolecular Chemistry* 16: 1736–1744.
94 Kulkarni, A. and Török, B. (2010). *Green Chemistry* 12: 875–878.
95 Zhang, L., Chen, S., Gao, Y. et al. (2016). *Organic Letters* 18: 1286–1289.
96 Yasuhara, S., Sasa, M., Kusakabe, T. et al. (2011). *Angewandte Chemie, International Edition* 50: 3912–3915.
97 Weyrauch, J.P., Hashmi, A.S.K., Schuster, A. et al. (2010). *Chemistry - A European Journal* 16: 956–963.
98 Hu, Y., Yi, R., Wang, C. et al. (2014). *The Journal of Organic Chemistry* 79: 3052–3059.
99 Yu, X., Xin, X., Wan, B., and Li, X. (2013). *The Journal of Organic Chemistry* 78: 4895–4904.
100 Yi, W., Liu, Q.-Y., Fang, X.-X. et al. (2018). *Organic & Biomolecular Chemistry* 16: 7012–7018.
101 An, H., Mai, S., Xuan, Q. et al. (2019). *The Journal of Organic Chemistry* 84: 401–408.
102 Batty, J.W. and Weedon, B.C.L. (1949). *Journal of the Chemical Society*: 786–789.
103 Heugebaert, T.S.A., Vervaecke, L.P.D., and Stevens, C.V. (2011). *Organic & Biomolecular Chemistry* 9: 4791–4794.
104 Nechaev, A.A., Peshkov, A.A., Van Hecke, K. et al. (2017). *European Journal of Organic Chemistry* 2017: 1063–1069.
105 Easton, N.R., Cassady, D.R., and Dillard, R.D. (1964). *The Journal of Organic Chemistry* 29: 1851–1855.
106 Vrasidas, I., Kemmink, J., Liskamp, R.M.J., and Pieters, R.J. (2002). *Organic Letters* 4: 1807–1808.
107 Scalacci, N., Pelloja, C., Radi, M., and Castagnolo, D. (2016). *Synlett* 27: 1883–1887.
108 Alhalib, A. and Moran, W.J. (2014). *Organic & Biomolecular Chemistry* 12: 795–800.

5

Transition-Metal-Catalyzed Chiral Amines Synthesis

Anton Cunillera[1], Carmen Claver[1], Cyril Godard[1], Martine Urrutigoïty[2] and Philippe Kalck[2]

[1] *Universitat Rovira i Virgili, Departament de Química Física i Inorgánica, Marcel.lí Domingo s/n, Campus Sescelades, Tarragona 43007, Spain*
[2] *University of Toulouse UPS-INP, Laboratoire de Chimie de Coordination du CNRS UPR 8241, Composante ENSIACET de l'Institut National Polytechnique de Toulouse, 4 allée Emile Monso, 31030 Toulouse Cedex 4, France*

5.1 Introduction

Chiral amine frameworks are present in around 40% of the pharmaceutical products. Therefore, in a multistep synthesis of drugs, it is important to include such building chiral components to perform efficient processes [1, 2]. Indeed, one enantiomer develops the expected pharmacological activity, whereas the other enantiomer is inactive or has a different biological response, even a poisonous effect. The active pharmaceutical ingredient market is in a constant growth rate of around 6% per year and is estimated to be US$ 182.2×10^9 in 2019 and to reach US$ 245.2×10^9 by 2024 [3]. As enantiomerically pure amines, amino acids and aminoalcohols (with ee >99%) are required in large quantities as building blocks in the pharmaceutical and agrochemical industry, the development of efficient methodologies for the preparation of chiral amines in high optical purity is of great significance and highly desirable. Since the "chiral switch" of 1997 in pharmaceutical industry, many investigations have been done to develop biocatalytic routes in parallel to the established chemical methods [4]. In order to avoid processes of low efficiency, poor selectivity, to reduce the wastes, and promote green chemistry, the pharmaceutical research explored biocatalysis, especially amine dehydrogenase, monoamine oxidases, imine reductases, ammonia lyases, and transaminases enzymes [5]. As reported elsewhere in this book by Turner and coworkers, genetic engineering techniques combined with powerful predictive models have significantly increased the optimization of catalytic efficiency and stereoselectivity of enzymes, especially now that their immobilization allows their recovery and recycling [6–10]. Even if these synthetic routes avoid the use of protecting groups, resolutions, and by-product formation, the limited substrate scope of biocatalysts as well as their possible fragility in the context of industrial processing conditions still require intense research [11]. Thus, transition metal catalysts and enzymes will be complementary in the future [12–14]. A recent review describes some tandem reactions combining transition metal catalysts and biocatalysts, which allow to reach chemo-, regio-, and stereoselectivities [15].

Methodologies in Amine Synthesis: Challenges and Applications, First Edition.
Edited by Alfredo Ricci and Luca Bernardi.

In the context of green chemistry, research on organometallic catalysis is still progressing to build the best tool to access and produce optically pure chiral amines, especially concerning the recycling of the metal complex as pointed out by Scholz and coauthors in another chapter of this book. Of course, significant efforts to reach the appropriate coordination sphere around the transition metal center are still necessary, in order to obtain full selectivity and to separate efficiently the catalyst from the products.

The goal of this chapter is to focus on the recent advances in enantioselective metal-catalyzed reactions for the synthesis of chiral amines, i.e. reductive amination, hydroamination, hydroaminomethylation, and hydroaminoalkylation. In addition, we will analyze the role of a chiral metal center in C–H and C–C couplings. Among the most explored methods, asymmetric hydrogenation (AH) of imines or enamines using hydrogen is the most efficient and elegant method. This reaction is still the subject of studies in view of the very recent publications that can be found in the literature [16–18]. The analysis of future opportunities for the hydrogenation of imines and enamines points to the fact that not only noble metal catalysts but also Mn-, Fe-, Co-, Ni-, or Cu-complexes for having access to pharmaceutical and biologically active agents are actively pursued [19]. Since a recent report written by some of us deals with the enantioselective hydrogenation of imines and enamines [20], this topic will not be treated in the present chapter.

5.2 Asymmetric Reductive Amination

The reductive amination reaction is a one-pot process, in which aldehydes or ketones react with primary or secondary amines to form an intermediate carbinolamine that gives after dehydration an imine, which is then hydrogenated [20, 21]. This reaction proceeds in two consecutive steps and can be performed with high atom efficiency and very limited waste production. In the global reaction drawn in Scheme 5.1, an aldehyde (R^1 = H) reacts with NH_3 or a primary (R^3 = H) or a secondary amine (R^3 and R^4 being alkyl or aryl groups) to provide after hydrogenation of the corresponding amine.

The choice of the reducing agents plays a crucial role in the success of the reaction, as the imine/enamine reduction needs to be selective over the carbonyl reduction under the same reaction conditions. Indeed, the formation of alcohol resulting from the direct reduction of the carbonyl group should be largely minimized.

An important breakthrough in enantioselective reductive amination was presented by Blaser et al. for the synthesis of (*S*)-metolachlor, resulting in the condensation of 2-methyl-6-ethyl-aniline with methoxyacetone and the in situ hydrogenation of the intermediate imine catalyzed by $[Ir_2(\mu\text{-}Cl)_2(C_8H_{12})_2]$/xyliphos chiral complex [22]. It is necessary to operate in the presence of $[NBu_4]I$ and CF_3COOH or CH_3SO_3H at 50 °C and 80 bar H_2 to obtain the (*S*)-*N*-(2-ethyl-6-methylphenyl)-*N*-(1′-methoxymethyl)-ethyl-amine

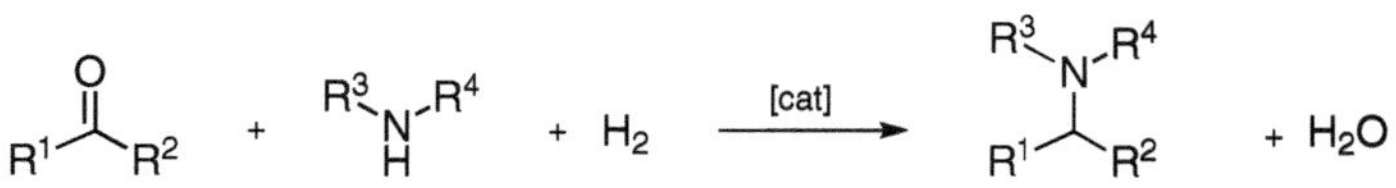

Scheme 5.1 General equation for the reductive amination of an aldehyde or a ketone.

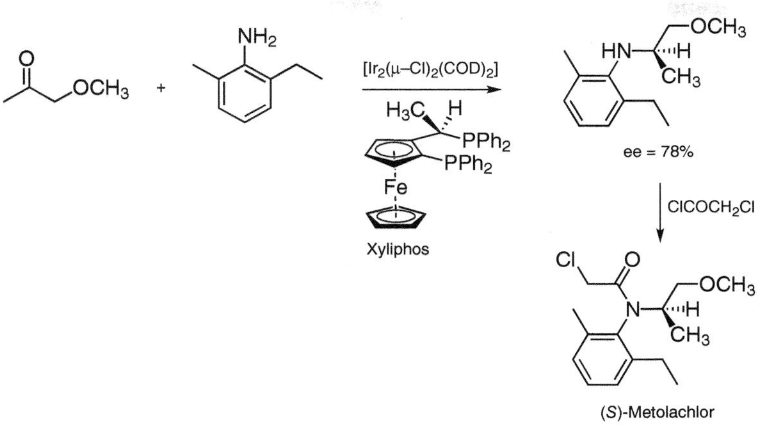

Scheme 5.2 Reductive amination of 2-methyl-6-ethyl-aniline and methoxyacetone, catalyzed by $[Ir_2(\mu\text{-}Cl)_2(C_8H_{12})_2]$/xyliphos in the presence of $[NBu_4]I$ and CF_3COOH. Source: Modified from Blaser et al. [22].

in 99% conversion and 78% ee. Chloroacetylation of this precursor provided the active (*S*)-enantiomer of metolachlor (Scheme 5.2).

A short time later, Börner and coworkers have shown that the reaction of keto-acids with benzylamine gave chiral amino acids in the presence of the $[Rh(COD)L_2^*]$ catalyst, bearing (*R*,*R*)-norphos or (*R*,*R*)-deguphos diphosphines. Maximum yields and ee up to 98% were obtained using methanol as a solvent, operating at room temperature, and 60 bar H_2 pressure for 24 hours [23, 24].

The first direct asymmetric intramolecular reductive amination of a dialkyl ketone with an alkylamine was developed using the [(*S*,*S*)-RuCl(*p*-cymene)($ArSO_2$DPEN)] catalyst. This reaction proceeds with high yield (>97%) and high enantioselectivity (>94%) to form a diazepane ring in the key step of the synthesis of a novel dual orexin inhibitor (Suvorexant) and has been run on >100 kg scale (Scheme 5.3) [25].

During this work, the deep impact of CO_2 on the balance between a ruthenium hydride and the ruthenium formate catalyst as well as its negative effects on the reaction rate were evidenced. The formation of carbamates by reaction of the products with CO_2 occurred,

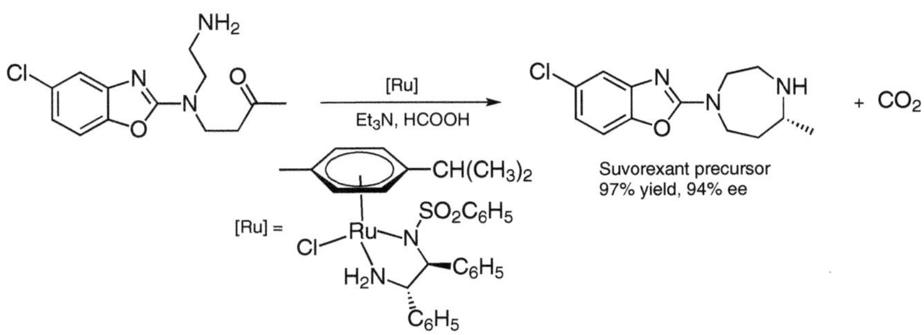

Scheme 5.3 Reductive amination of the ketone function to produce the chiral diazepane ring of a Suvorexant precursor. Source: Modified from Strotman et al. [25].

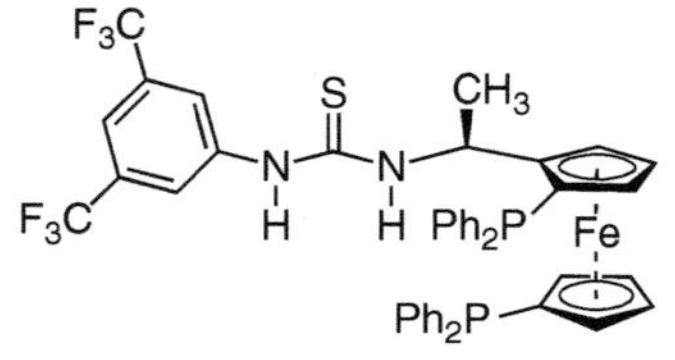

Scheme 5.4 Chiral bisphosphine-thiourea ligand (ZhaoPhos ligand). Source: Modified from Zhao et al. [27].

reducing the overall isolated yield. Incorporating a purge of CO_2 into the process led to an increase in the rate of the reaction and allowed for product isolation with a nearly quantitative yield. As other intramolecular versions of the reaction, chiral tetrahydroquinolines and tetrahydroisoquinolines were developed with the iridium/ZhaoPhos ligand catalytic system in the presence of HCl, which plays a critical role in the synthesis [26]. The ZhaoPhos ligand [27] (Scheme 5.4) is based on a ferrocenyl bisphosphine backbone and a thiourea subunit, which strongly interacts with the chloride ions present in the medium offering excellent enantiocontrol of the reaction as ee up to 98% was attained together with a large substrate scope.

Similarly, iridium catalysts allow the synthesis of useful dibenz[*c*,*e*]azepines containing both central and axial chiralities, produced with excellent enantiocontrol with ee up to 97% [28]. The method is described through a one pot *N*-Boc deprotection/intramolecular asymmetric reductive amination sequence of bridged biaryl derivatives. The catalyst system is based on $[Ir(COD)Cl]_2$ dimer, chiral diphosphines, particularly (*S*)-difluorophos, and $Ti(OiPr)_4$ as an additive, presumably for accelerating the reduction process (Scheme 5.5).

Because of the successful synthesis of the dual orexin inhibitor Suvorexant, the direct asymmetric reductive amination reaction experienced significant developments aimed at improving substrate scope and catalyst reactivity and to make this reaction a general tool in the synthesis of chiral amines [29]. Recently, (*S*)-rivastigmine, an active pharmaceutical ingredient (API) for the treatment of Alzheimer-type disease, was synthesized in two steps with 92% yield and 91% ee starting from *meta*-hydroxyacetophenone and dimethylamine

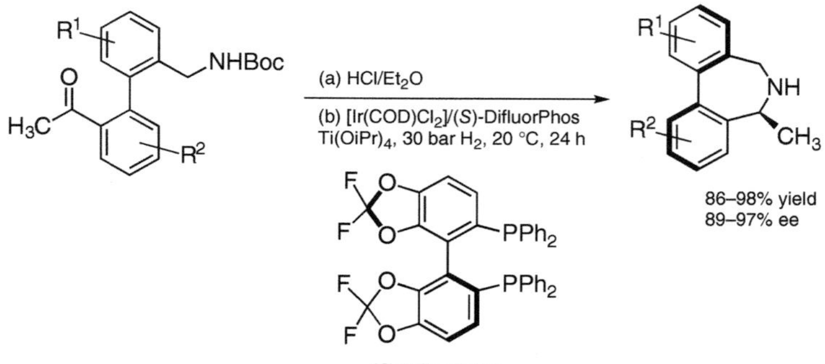

Scheme 5.5 Intramolecular asymmetric reductive amination using $[Ir(COD)Cl]_2$/(*S*)-difluorPhos/$Ti(OiPr)_4$ to synthesize seven-membered dibenz[*c*,*e*]azepines. Source: Modified from Yang et al. [28].

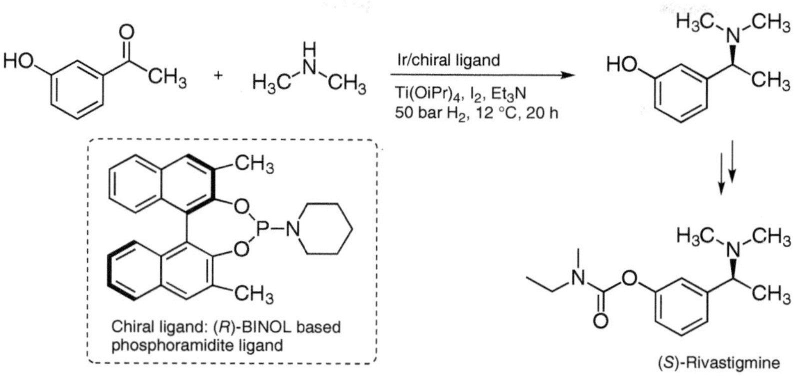

Scheme 5.6 Synthesis of (*S*)-rivastigmine from *meta*-hydroxyacetophenone and dimethylamine, catalyzed by a chiral iridium–phosphoramidite complex. Source: Modified from Wu et al. [30].

[30]. The optimized catalytic system involved is composed by iridium-chiral monodentate phosphoramidite ligand and additives such as NEt_3 (10 mol%) and $Ti(OiPr)_4$ (1.2 equiv), the reaction being performed at 50 bar H_2 pressure and 12 °C temperature for 20 hours (Scheme 5.6).

This catalytic system has proven to be effective (87–98% yields and 90–97% ee) for a wide range of ketones bearing electron-withdrawing or electron-donating substituents at the para-, meta-, or ortho-positions of the aryl ring. This study also shows that secondary amines can be used as N-sources and that the ligand involved can be easily modulated enough to accommodate different electronic or steric features.

The synthesis of a key intermediate for tofacitinib, a drug for the treatment of rheumatoid arthritis, has been successfully completed by applying an efficient protocol based on asymmetric reductive amination under dynamic kinetic resolution conditions. High-throughput screening was the key to identify the chiral iridium/BiPheP catalyst able to perform the reaction under acidic conditions. The starting ketone was fully converted into the desired (*R*,*R*)-amine with 72% ee and 95% de. The final product was isolated in 86% yield with a chemical purity of 97% (Scheme 5.7) [31].

Using $[Ir(COD)Cl]_2$/f-binaphane complex, the reductive amination of acetophenone and derivatives and phenylhydrazide as nitrogen source was explored. Additives such

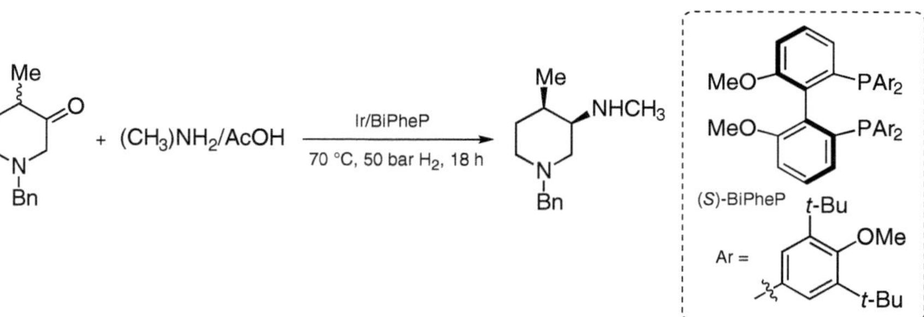

Scheme 5.7 Asymmetric reductive amination under dynamic kinetic resolution conditions. Source: Modified from Verzijl et al. [31].

as molecular sieves, iodine, and *p*-toluene sulfonic acid are key to the success of the synthesis of various chiral hydrazides in excellent yields and enantioselectivities. Indeed, the presence of protons facilitates the imine formation, and molecular sieves help to remove water formed during the condensation between the aldehyde and the amine, as well as to promote the reduction of imine; iodine allows the formation of an effective [$Ir^{(III)}I_2Cl$(f-Binaphane)] that is able to reduce the imine [32].

Iridium has also been investigated with a wide spectrum of ketones and *para*-methoxyaniline giving 90% yields and 80% ee for a large variety of amines [33]. The reaction is carried out by cooperative catalysis of the metal complex and a chiral Brønsted acid BH with the former reducing an in situ generated iminium cation via ionic hydrogenation and the latter promoting enantioselective hydride transfer via ion pairing of its conjugate base X^- with the iminium ion, in the presence of molecular sieves.

Diphenylmethanamine is relevant as an effective coupling partner for ketones to provide chiral amines in high yields up to 95% and enantioselectivities up to 99% [34]. The catalytic system based on the [$Ir_2(\mu\text{-}Cl)_2(COD)_2$] precursor and monodentate phosphoramidite ligands is performant for a wide range of amines, also in the presence of additives such as an acid, molecular sieves, and $Ti(OiPr)_4$.

Recently, reaction of alkyl aryl ketones with ammonium acetate (NH_4OAc) as the amine source using the highly chemo- and enantioselective ruthenium/diphosphineC3-TunePhos catalytic system affords a wide range of primary amines with 90% ee in most cases (Scheme 5.8). Both electron-donating and electron-withdrawing substituents on the benzene ring were tolerated. The potential of this procedure is illustrated in the scale-up syntheses of key intermediates of drug molecules from commercially available ketones, such as tecalcet hydrochloride, cinacalcet, and (*R*)-rivastigmine [35] (shown in Scheme 5.6).

Iron, as a more economical and environmentally friendly catalyst, has been investigated as an alternative metal. Beller and his group used a system based on combination of the nonchiral iron Knölker catalyst [36] and TRIP as a chiral Brønsted acid (Scheme 5.9) [37].

Various ketones including aromatic, heteroaromatic, and aliphatic ones were transformed by reaction with aniline or meta- and para-substituted anilines to the corresponding chiral amines. Moderate to good yields were reached (60–80%), and excellent ee's up to 94% were achieved in the following catalytic conditions: 5 mol% iron catalyst/3 mol% TRIP,

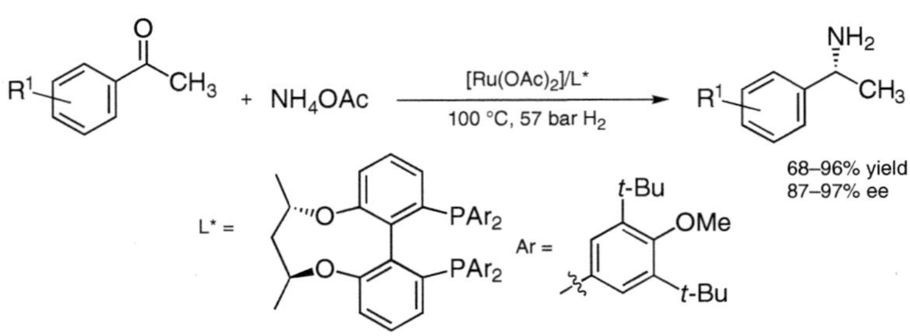

Scheme 5.8 Reductive amination of alkyl aryl ketones catalyzed by [$Ru(OAc)_2$]/C3-TunePhos ligand. Source: Modified from Tan et al. [35].

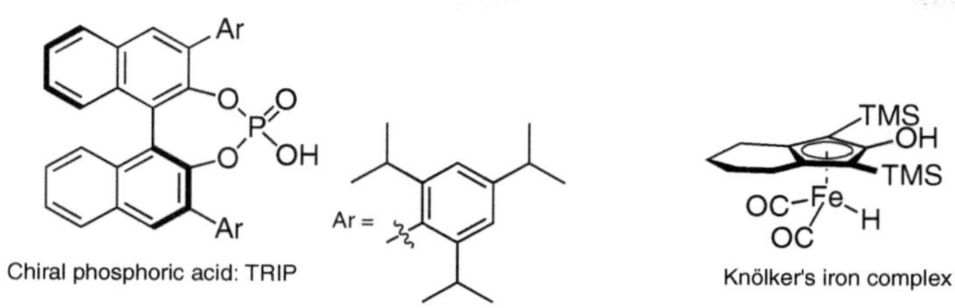

Scheme 5.9 The Knölker iron catalyst and the TRIP Brønsted acid. Source: Modified from Zhou et al. [37].

50 bar H_2, 65 °C, and 24 hours in toluene. The best result (80% yield and 91% ee) was obtained in the case of the *para*-trifluoroacetophenone bearing an electron-withdrawing substituent and aniline. In contrast, no reaction occurs with ortho-substituted aromatic ketones as well as with ortho-substituted aniline.

Another study shows that *cis*-*N*-(2-methylcyclopentyl)- and *cis*-*N*-(2-methylcyclohexyl) amines can be obtained in 70–91% yields and good to excellent ee's of 62–99% starting from 2-methylcyclopentanone, or 2-methylcyclohexanone, and functionalized anilines bearing ortho-, meta-, and para-substitutions in the aryl ring [38]. The reaction is influenced by the bulkiness of the substituents in the aniline as no reaction is observed when using 2,4,6-trimethylaniline or 2,3,4,5,6-pentafluoroaniline. The reaction is catalyzed by the stable chiral preformed [Pd((*R*)-BINAP)Br_2] complex, providing a direct and efficient route toward enantioselective synthesis of *N*-cycloalkylamines from 2-methyl-cyclopentanone or 2-methyl-cyclohexanone (Scheme 5.10).

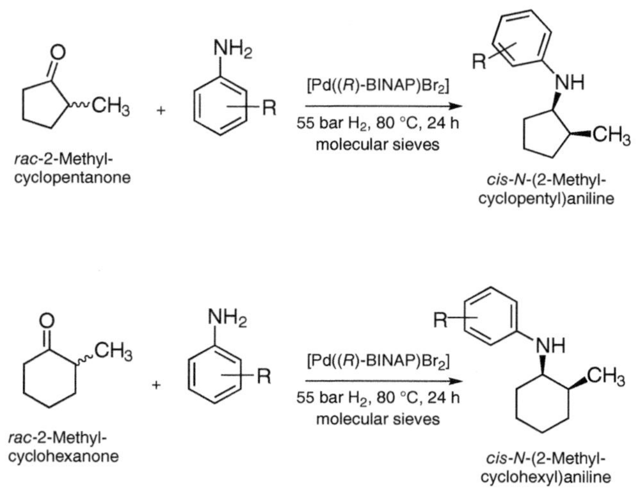

R	Yield (%)	ee (%)
m-CH_3	90	93
p-CH_3	97	93
p-C_2H_5	86	83
p-CH_3O	78	96
m-CF_3	80	98
m-Cl	86	91
p-Cl	88	84
o-Br	75	18
o-CF_3	69	95

R	Yield (%)	ee (%)
m-CH_3	90	93
p-CH_3	97	93
p-C_2H_5	86	83
p-CH_3O	78	96
m-CF_3	80	98
m-Cl	86	91
p-Cl	88	84
o-Br	75	18
o-CF_3	69	95

Scheme 5.10 Asymmetric reductive amination of 2-methyl cyclopentanone and cyclohexanone with anilines catalyzed by [Pd((*R*)-BINAP)Br_2]. Source: Modified from Cabrera et al. [38].

5.3 Asymmetric Hydroamination

Various examples can be found in the literature on the hydroamination reaction that allows the direct formation of a new carbon–nitrogen bond by addition of an amine to the unsaturated carbon–carbon bond of an unactivated alkene. This reaction is of some interest because it presents 100% atom economy and uses simple starting materials such as amines and alkynes or alkenes without any further stoichiometric additives. Numerous research efforts have been made to develop efficient catalytic systems for the inter- or intramolecular reaction. Catalysts based on transition metals of groups 3–5 and 8–10 and rare earth metals have been developed over the past decade [39–43]. Because of the relevance of the reaction for the synthesis of chiral amines for pharmaceuticals and biological active molecules, the development of the asymmetric version has raised an increasing interest (Scheme 5.11). To this goal, it is thus necessary to obtain preferentially, or even exclusively, the branched isomer and to achieve asymmetric induction in the reaction.

Recent reviews describe the most significant advances in the field [44–49], from which general trends can be extracted.

In the intermolecular version of the hydroamination, the first enantioselective example was reported by Widenhoefer and coworkers with a gold(I) complex containing the chiral diphosphine (*S*)-3,5-*t*-Bu-4-MeO-MeOBiPheP ligand in the presence of AgOTf (the trifluoromethanesulfonate salt of Ag^+) [50]. The branched regioisomer was formed in 89% yield and 78% ee in the addition of the cyclic *t*-Bu-urea to 1-octene (Scheme 5.12).

The same chiral catalytic system allows to perform the stereoconvergent, enantioselective addition of a benzylcarbamate to 1,3-disubstituted allenes, giving *N*-allylic carbamates with ee reaching 92% [51].

The mixture of [AuCl(P(*t*-Bu)$_2$-*o*-biphenyl)]/$AgSbF_6$ catalyzes the amination of enantiomerically pure allyl alcohols with 1-methylimidazolidin-2-one with ee as high as 92% [52]. Interestingly, with γ-unsubstituted or γ-methyl-substituted allylic alcohols, a high

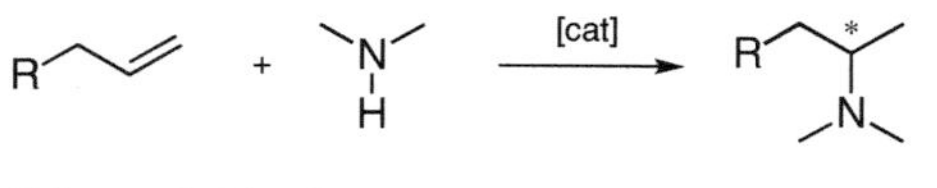

Scheme 5.11 Asymmetric hydroamination of an unactivated alkene with a secondary amine.

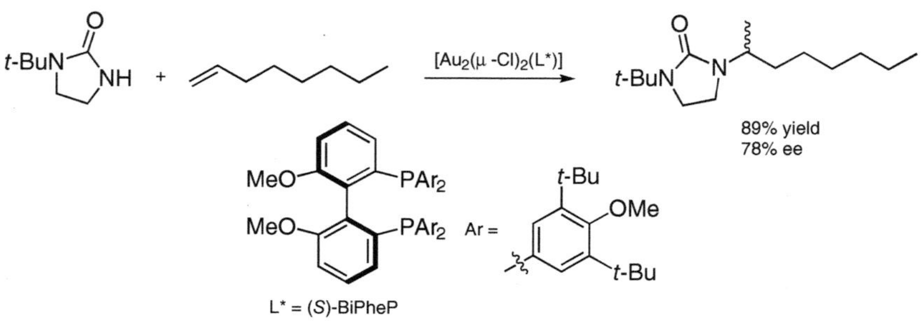

Scheme 5.12 Enantioselective hydroamination of 1-octene with 1-*tert*-butylimidazolidin-2-one catalyzed by [Au$_2$(μ-Cl)$_2$((*S*)-L*)] in the presence of AgOTf. Source: Modified from Zhang et al. [50].

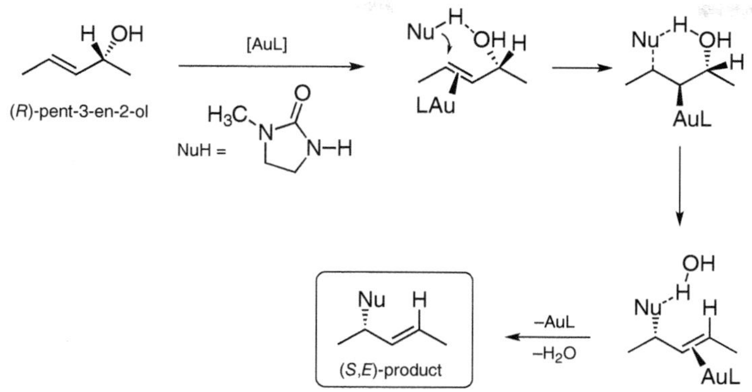

Scheme 5.13 Proposed mechanism for the asymmetric hydroamination of (*R*)-3-penten-2-ol catalyzed by a gold complex. Source: Modified from Mukherjee and Widenhoefer [52].

γ-regioselectivity and *syn*-stereoselectivity are observed, consistent with the coordination of the C=C double bond, followed by the nucleophilic attack of the cyclic urea and departure of the hydroxyl group [52]. Scheme 5.13 shows the mechanistic pathway proposed by the authors for the allylic amination of (*R*)-pent-3-en-2-ol.

Other metals have been involved in asymmetric intermolecular hydroamination. The chiral palladium complex [Pd(OTf)$_2$((*R*)-BINAP)] catalyzes the addition at 25 °C of aniline to (4-trifluoromethyl)styrene to generate (*S*)-*N*-(1-(4-(trifluoromethyl)phenyl)ethyl)aniline with 80% yield and 81% ee. The reaction seems to proceed through a nucleophilic attack on the C=C coordinated bond of styrene, in a similar manner to the mechanism sketched in Scheme 5.13 [53].

The hydrido-copper(I) complex bearing a chiral diphosphine ligand catalyzes the enantioselective intermolecular hydroamination of electrophilic alkenes [54, 55]. Styrene and various substituted analogs are exclusively transformed into the α-branched amines at 40 °C after 36 hours, using *O*-benzoylhydroxylamines as reactants in the presence of SiH(Me)(OEt)$_2$ as hydride source (Scheme 5.14). The styrene substrate coordinates to the CuH species, and the nucleophilic attack of the hydrido ligand occurs on the CH_2 moiety to produce a chiral benzyl copper species. Then, oxidative addition of the N-protected-*O*-benzoylhydroxylamine generates the [Cu(PhCH$_2$CH$_3$)(OBz)(N(CH$_2$Ph)$_2$)L*] intermediate from which reductive elimination produces the chiral amine, with the silane regenerating the Cu–H-active species by elimination of benzoic acid.

This reaction was extended by Buchwald and coworkers to the synthesis of secondary chiral amines starting from aminobenzoate reactants with performances reaching 93% yield and 97% ee [56]. In addition, γ-substituted allylic amines can be functionalized, provided that these amines bear a pivaloyl protecting group. With the same catalytic system shown in Scheme 5.14, the resulting 1,2-diamines are obtained with regioisomeric ratios from 12 to 20/1 and ee's as high as 98–99% [57]. In a similar approach, γ-substituted 1,3-aminoalcohols are prepared by hydroamination of unprotected allylic alcohols with 99% ee [58]. On the other hand, the hydroamination of 1-substituted cyclobutenes and cyclopropenes affords exclusively the Markovnikov product, and the

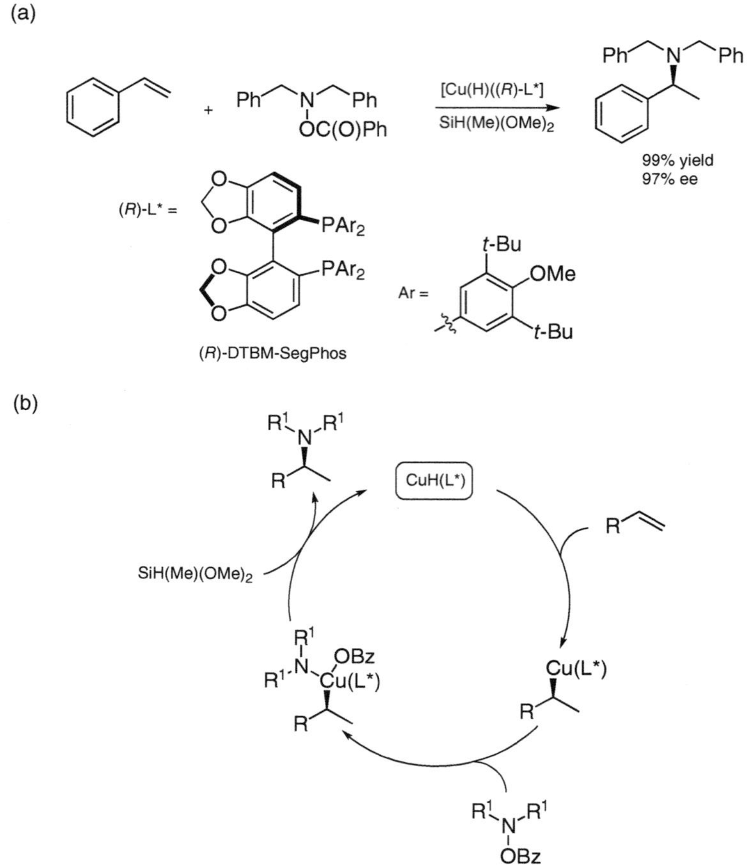

Scheme 5.14 (a) Enantioselective hydroamination of styrene catalyzed by the chiral Cu–H complex containing the (*R*)-DTBM-SegPhos bidentate ligand. Source: Modified from Miki et al. [54]. (b) The proposed catalytic cycle. Source: Modified from Zhu et al. [55].

diastereo- and enantioselective synthesis of various polysubstituted aminocyclobutanes and aminocyclopropanes can thus be realized [59].

The [Cu(H)((*R*)-DTBM-SegPhos)] catalytic system in conjunction with *O*-acylhydroxylamines has also been used by Somfai and cowoker to perform the hydroamination reaction of enamines [60], leading to the synthesis of 1,2-diamines with high efficiency and excellent enantioselectivities (75–99% ee). Conversely, the same SegPhos ligand was used in combination with iridium in the reaction between norbornene and 4-bromoaniline, affording the final amine with 96% ee (Scheme 5.15) [61].

This reaction has been extended to the hydroamination of unactivated alkenes such as hex-1-ene and oct-1-ene with indoles, and the reaction provides exclusively *N*-alkylindoles with good yields and 63–88% ee's. Concerning the mechanism, Hartwig and coworkers proposed that the iridium complex is mononuclear, and that the indole reacts through an oxidative addition of the N—H bond, followed by the nucleophilic attack of the amido ligand to the coordinated alkene on the 2-position, giving an alkyl species [62].

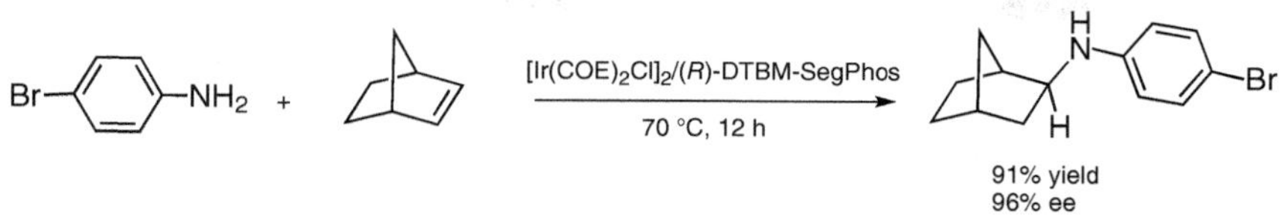

Scheme 5.15 Iridium-catalyzed hydroamination of norbornene by 4-bromoaniline. Source: Modified from Zhou and Hartwig [61].

Recent works report the chiral functionalization of conjugated dienes catalyzed by $[Ni(C_8H_{12})_2]$ complexes in the presence of the diphosphine [(2*S*,2′*S*,5*S*,5′*S*)-2,2′,5,5′-tetramethyl-1,1′-(*o*-phenylene)diphospholane] (or (*S*,*S*)-Me-DuPhos) and phthalic acid as a cocatalyst [63]. Phenyl-1-buta-1,3-diene reacts with various primary and secondary amines to provide the corresponding secondary and tertiary allylic amines with high yields and excellent enantioselectivities (90–99% ee). Aryl- as well as alkyl-substituted butadienes are converted into the Markovnikov hydroamination products. More complex amines, including aminoethanol, also react to give exclusively the 1,2-hydroamination products. The role of the Brønsted acid (BH) cocatalyst is to generate the $[Ni^{II}(H)(B)(L^*)]$ complex through the oxidative addition to the $[Ni^0(COD)(L^*)]$ species. This Ni–H complex coordinates the diene and forms a $[Ni^{II}(B)(allyl)(L^*)]$ species on which the amine reactant produces the chiral allylic amine and regenerates the active Ni^0 species.

In the intramolecular reaction, pyrrolidine and piperidine derivatives are synthesized from unactivated alkenes tethered by primary or secondary amines in high enantioselectivities promoted by rare earth and group 4 elements. Generally, the cyclization is accelerated by the presence of substituents on the tether, according to the Thorpe–Ingold effect. As for the intermolecular reaction, it is performed by the use of late transition metals. The alkenes involved in the reaction are often activated alkenes such as vinylarenes, norbornene, or 1,3-dienes, and the amines should possess a low basicity to prevent a tight coordination on the metal center.

A highly active oxazolinylborato yttrium catalyst provides 2-methyl pyrrolidines in high yields (95–100%) and excellent enantioselectivities (89–96%) (Scheme 5.16) [64].

Even if the catalyst provides various disubstituted 2-methyl pyrrolidines with high ee's, it is less effective for the synthesis of piperidines and unsubstituted pyrrolidines.

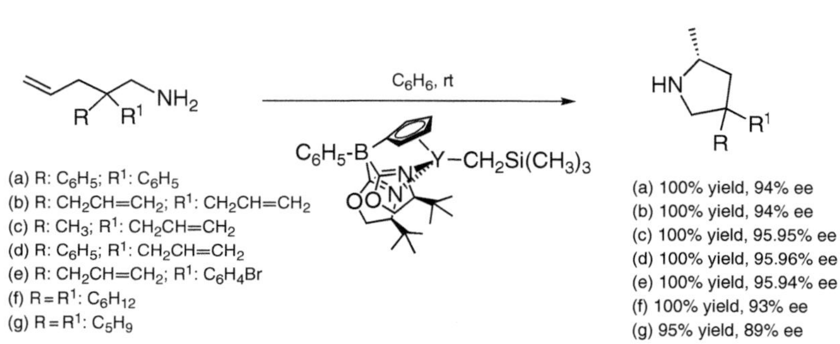

Scheme 5.16 Intramolecular hydroaminomethylation of 5-hex-1-ene amines catalyzed by an oxazolinylborato yttrium complex. Source: Modified from Manna et al. [64].

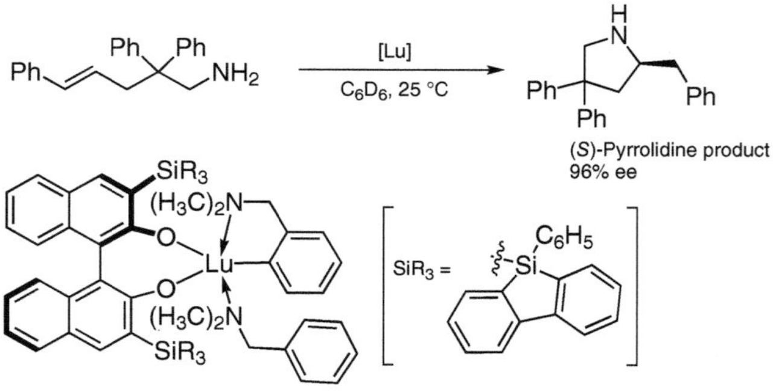

Scheme 5.17 Asymmetric intramolecular hydroamination of 2,2,5-triphenylpent-4-en-1-amine catalyzed by a lutetium complex. Source: Modified from Nguyen et al. [65].

Recently, another study described the use of a variety of novel 3,3′-bis(trisarylsilyl)- and 3,3′-bis(trisalkylsilyl)-substituted binaphtolate yttrium and lutetium complexes as effective catalysts in the cyclization of different aminoalkenes [65]. Moreover, the cyclization of an internal alkene has also been successfully carried out, although the reaction is slower than that with a terminal alkene. The (*R*)-binaphtolate yttrium and lutetium complexes furnish the corresponding 2-substituted pyrrolidines with good to excellent enantioselectivities (75–96%), the lutetium complexes providing the highest 96% selectivity (Scheme 5.17).

Among metal catalysts, gold and copper complexes have been applied to this reaction. Few works have been published, but recently, an interesting study reports the first silver-dependent enantiodivergent gold-catalyzed reaction [66]. The binuclear gold(I)chloride complex coordinated to (*S*)-DTBM-SegPhos efficiently catalyzes the intramolecular hydroamination of 2,2-diphenylpent-4-en-1-amine at mild temperature in the presence of water with high yield and good ee (75%). The combination of a single chiral binuclear Au(I) chloride complex and $AgClO_4$ affords both enantiomers of the products by simply changing the solvent from toluene to methanol. The authors show through various catalyst analyses and DFT calculations that the enantioinversion is controlled by the bimetallic activation of gold–silver adducts, which occurs in methanol, affording the (*S*)-isomer instead of the (*R*)-one obtained in toluene.

A series of gold-catalyzed anti-Markovnikov hydroamination-initiated azidation, allylation, and indolation has been developed to build pyrrolidines bearing functional moieties in the 2-position, starting from readily available chiral homopropargyl sulfonamides [67]. Using azidosilane as a nucleophile, the desired 2-azido-substituted pyrrolidines were achieved in relatively high yields (65–85%) with high cis-diastereoselectivity up to 10 : 1. Using allylsilane, the 2-allyl-substituted products were obtained in 80–94% yield with a lower diastereoselectivity (up to 5 : 1). Other heterocycles were also involved as carbon nucleophiles, giving the corresponding heterocycle-substituted pyrrolidines in good to excellent yields and diastereoselectivities. The catalyst can be $Ph_3PAuNTf_2$, instead of $IPrAuNTf_2$ (IPr = 1,3-bis(2,6-diisopropylphenyl)imidazol-2-ylidene) and Et_3N is used as an additive (Scheme 5.18).

The synthesis of 2-aryl- and 2-heteroarylpyrrolidines including pyrazole, pyrimidine, pyridine, and azaindole has been carried out with a chiral CuH catalyst generated from

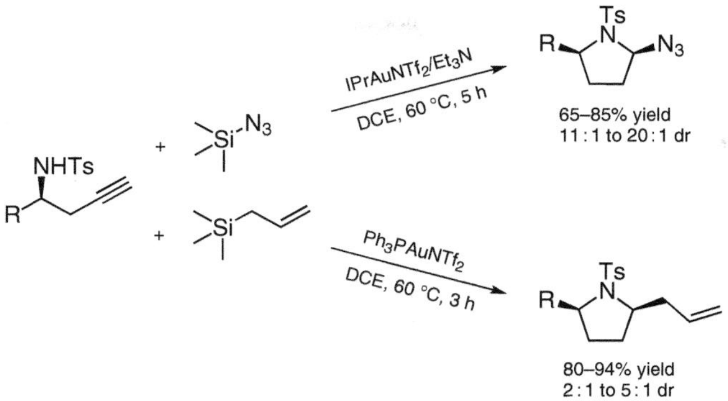

Scheme 5.18 Stereoselective synthesis of 2,5-substituted pyrrolidines from chiral homopropargyl sulfonamides in the presence of trimethylsilyl azide or allyltrimethylsilane. Source: Modified from Tan et al. [67].

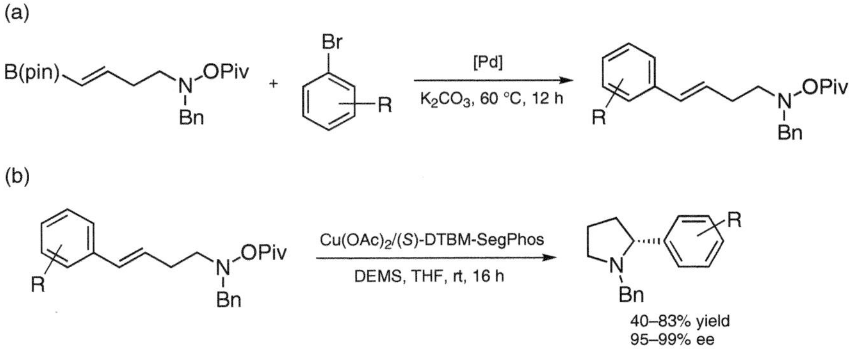

Scheme 5.19 Synthesis of the enantiopure α-arylpyrrolidines by intramolecular hydroamination (b) of substrates first prepared by Suzuki–Miyaura reaction (a). Source: Modified from Dai et al. [68].

$Cu(OAc)_2$ and stoichiometric amounts of diethoxy(methyl)silane (DEMS) as the hydride source [68].The chiral ligand identified as optimal for the synthesis of these enantiopure substituted-pyrrolidines was (*S*)-DTBM-SegPhos. To avoid competing N—O bond reduction of the electrophilic hydroxylamine source by the CuH catalyst accounting for the side product, the authors first prepared the hydroxylamine pivalate substrates by using the Suzuki–Miyaura cross-coupling reaction of a vinylboronate ester precursor with a wide range of aryl bromides. The excellent stereoselectivity obtained in the intramolecular hydroamination and the transformation of various substrates to pharmaceutically relevant heteroarenes make this synthetic method practical and versatile for the synthesis of pyrrolidine derivatives (Scheme 5.19).

A quite similar catalytic system consisting of $Cu(OAc)_2$/(*S*)-DTBM-SegPhos/dimethoxy (methyl)silane (DMMS) with added PPh_3 has been involved in the synthesis of chiral aziridines [69], well known as being key functional groups in natural product and pharmaceutical chemistry (Scheme 5.20).

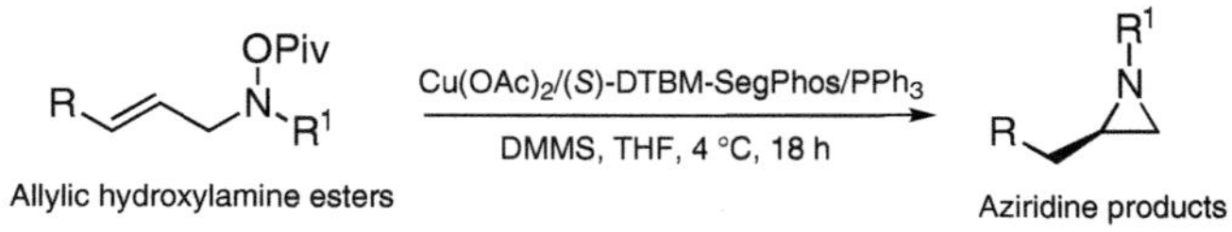

Scheme 5.20 Chiral aziridine synthesis catalyzed by a Cu–H/(*S*)-DTBM-SegPhos catalyst. Source: Modified from Wang et al. [69].

After optimization of the operating conditions, the substrate scope was examined by varying electronic and steric properties of the N-substituent or of the substituents appended to the olefin. In both cases, good to excellent yields and enantioselectivities were obtained, demonstrating that the catalytic system was highly efficient in this aziridination reaction.

5.4 Asymmetric Hydroaminoalkylation

In a similar way to the hydroamination protocol, the catalytic hydroaminoalkylation reaction [49, 70–72], that is the addition of a C–H α to an amine to alkenes, also called α-alkylation of amines, is a 100% atom economical synthesis because the reaction takes place without the formation of any coproduct. The exclusive C–H activation by the metal center and the insertion of the alkene result in the production of a chiral-branched alkylamine and of its linear isomer, whose formation must be avoided in the asymmetric synthesis (Scheme 5.21).

The first C-alkylation of secondary amines with alkenes was described by Clerici and Maspero [70] and Nugent et al. [73] using at high temperatures early transition d^0 metal complexes, such as $[Nb(NMe_2)_5]$, $[Ta(NMe_2)_5]$, or $[Zr(NMe_2)_4]$. The coupling of *N*-methylaniline with terminal alkenes was shown by Hartwig and Herzon to be catalyzed by $[Ta(NMe_2)_5]$ at 160–165 °C in toluene, giving in 24 hours yields as high as 96% for norbornene, with the branched alkylation product being the sole regioisomer [74]. Using $[TaCl_3(NMe_2)_2]$ at lower temperatures, the same authors showed that the addition occurs on the methyl group of various linear alkyl- and branched alkyl methylamines in high yields [75]. With these two tantalum precursors, it was shown that a large variety of alkenes and amines could be used to obtain almost exclusively the branched regioisomer [76, 77]. The enantioselective reaction has been reported recently by Schafer and her group, using tantalum–amidate complexes [78]. Indeed, addition of the axially chiral (*S*)-N^2,$N^{2\prime}$-bis[2,6-diisopropylphenyl]-6,6′-dimethylbiphenyl-2,2′-dicarboxamide to the $[Ta(NMe_2)_5]$ complex provides the dicarboxamidate precatalyst (Scheme 5.22), where the chelate ring strain induces exclusively a κO, κO′ coordination mode as shown by the X-ray

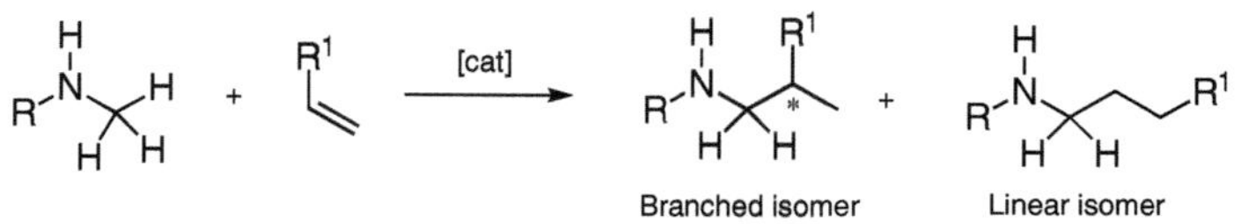

Scheme 5.21 Hydroaminoalkylation of a secondary amine through C=C bond formation by insertion of the alkene into a C–H methyl bond.

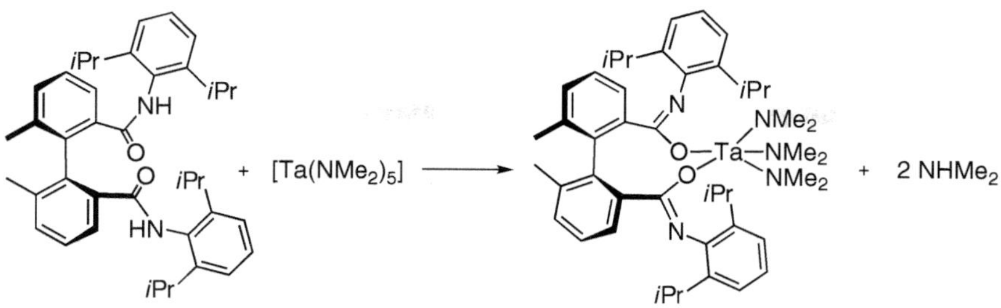

Scheme 5.22 Synthesis of the tantalum catalyst precursor in the hydroaminoalkylation of secondary amines. Source: Modified from Eisenberger et al. [78].

crystal structure. Operating at 130 °C for 46 hours, the best 61% ee was observed for the reaction of norbornene with *N*-methylaniline with a yield of 80%.

Few modifications to the ligand, particularly by replacing the two isopropyl- by three methyl-substituents in 2,4,6, and 2′,4′,6′ positions, lead to improved performances [79]. Indeed, the reaction of *N*-methylaniline with norbornene provides at 130 °C and for 120 hours a conversion of 89% and an enantiomeric excess of 93% in the exo-(*R*)-isomer of *N*-(bicyclo[2.2.1]heptan-2-ylmethyl)aniline. Various chiral biarylamidate ligands have been coordinated to the Ta metal center, and the corresponding X-ray crystal structures show clearly that from the two N͡O functionalities, one is only κO coordinated, whereas the second is κO and κN coordinated in a bidentate mode [79, 80]. The Ta-catalyzed reaction is also operating in the condensation of *N*-methylaniline and vinylcyclohexane or oct-1-ene. At 160 °C for 120 hours, the reaction of 1,2,3,4-tetrahydroquinoline and norbornene still occurs but with significantly lower yields (20–40%) and ee's (31–43%). In addition, the authors observed that the niobium corresponding complexes are not active in this hydroaminoalkylation reaction.

In the early observations of the catalytic activity of the tantalum complex [Ta(NMe$_2$)$_5$], including the H/D exchange in the methylamine, a metallaaziridine intermediate was proposed [73]. By reacting *t*-Bu(CO)NH(2,6Me$_2$Ph) with [Ta(NMe$_2$)$_5$], Schafer and her group isolated the bis-(*N*-[2,6-dimethylphenyl]pivalamidate)(dimethylamido)*N*-methyltantalaaziridine complex, whose X-ray crystal structure shows clearly the presence of an aziridine ligand [78]. Recent DFT calculations evidenced that the mechanism of the reaction involves the coordination of the alkene to the tantalum complex containing the aziridine framework, its insertion, presumably through a nucleophilic intrasphere attack of the CH$_2$ group (in red on Scheme 5.23), giving exclusively the branched product by 1,2 insertion. Then, the coordination of *N*-methylaniline to the five-membered metallacycle occurs, and a H-atom is transferred from the amine to the CH$_2$ group of the metallacycle, giving rise to the two σ-NMePh and σ-NPh(CH$_2$)(CHMe$_2$) ligands. Hydrogen abstraction from the methyl group of the σ-NMePh ligand to the *N*-amido ligand affords the tantalaaziridine structure, from which the amine is generated upon leaving the coordination sphere [81].

Hultzsch and his group reported the catalytic activity of 3,3′-silyl-substituted binaphtholate tris(dimethylamido) niobium or tantalum complexes in the asymmetric intermolecular hydroaminoalkylation of various unactivated alkenes and vinyl arenes with secondary

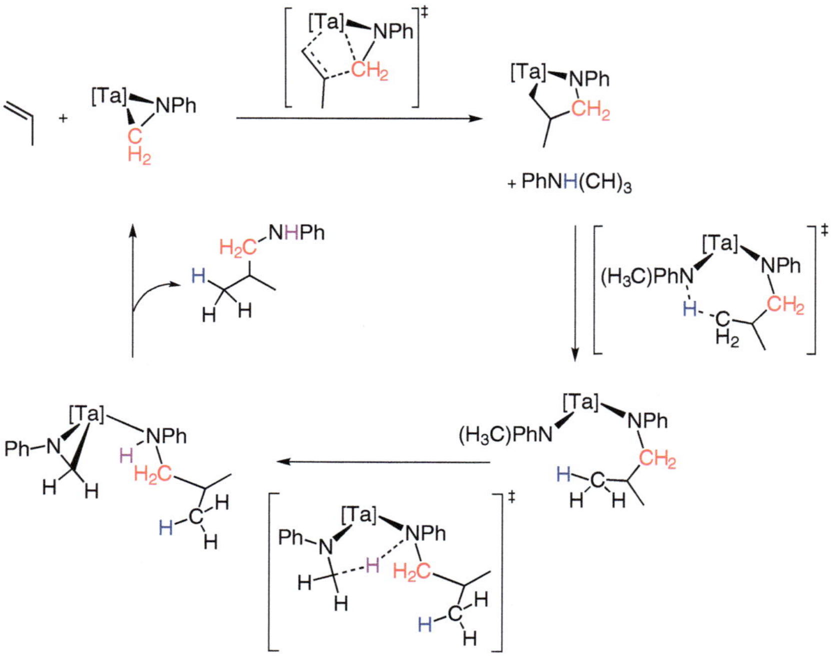

Scheme 5.23 Main steps of the catalytic cycle calculated by DFT, showing the main role played by the tantalaaziridine species. Source: Modified from Gilmour et al. [81].

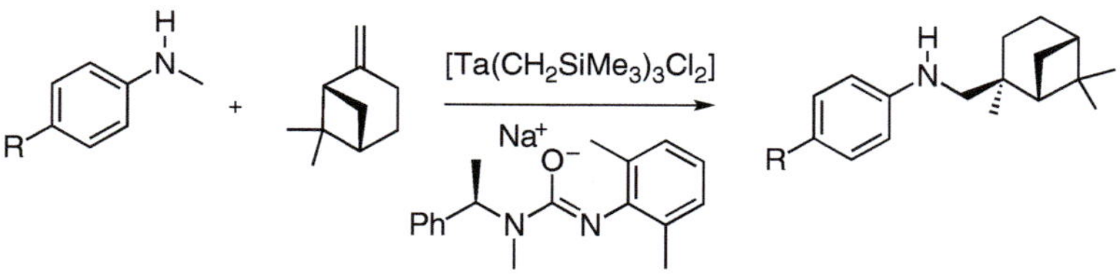

Scheme 5.24 Intermolecular coupling of *N*-methylaniline with (1*S*)-(−)-β-pinene catalyzed by a 1 : 1 mixture of [Ta(CH_2SiMe_3)$_3$$Cl_2$] and a sodium ureate salt. Source: Modified from DiPucchio et al. [84].

amines. The niobium complex with the silyl-substituent $SiPhMe_2$ allowed to reach 98% ee for the reaction of *N*-methylaniline with trimethyl(vinyl)silane [82].

Improvement of the tantalum catalytic system has been developed by Schafer and coworkers using the [Ta(CH_2SiMe_3)$_3$$Cl_2$] precursor in the presence of equimolecular amounts of sodium ureate so that higher turnover frequencies and turnover numbers can be obtained [83]. Even in the absence of amine protecting groups, directing groups, or additives, an excess of alkene is no more required. This catalyst allowed to perform the hydroaminoalkylation of limonene and β-pinene with *N*-methylaniline-derived substrates giving rise to the branched product with an excellent diastereoselectivity (Scheme 5.24) [84].

5.5 Asymmetric Hydroaminomethylation

The rhodium catalyzed hydroaminomethylation of alkenes is a promising strategy for the direct obtainment of chiral amines from alkenes in a single reaction. This one-pot, atom-efficient reaction for the synthesis of amines consists of an initial hydroformylation of olefin to an aldehyde, subsequent formation of an enamine or imine, followed by hydrogenation (Scheme 5.25) [85].

Nonetheless, to the best of our knowledge, only one report was published on the asymmetric version of this reaction using a single catalyst.

Kalck and coworkers performed a thorough study on the asymmetric hydroaminomethylation of styrene with piperidine, using a variety of chiral diphosphine ligands, which had been previously reported as very efficient in asymmetric hydroformylation and asymmetric hydrogenation reactions (Scheme 5.26) [86]. The study also included ^{1}H and ^{31}P NMR investigations and DFT calculations.

In this case, it was observed that the hydroformylation and the condensation proceeded rapidly, with the hydrogenation step being slower and therefore rate determining. In this reaction, the enantioselectivity should be induced in the hydrogenation step because the

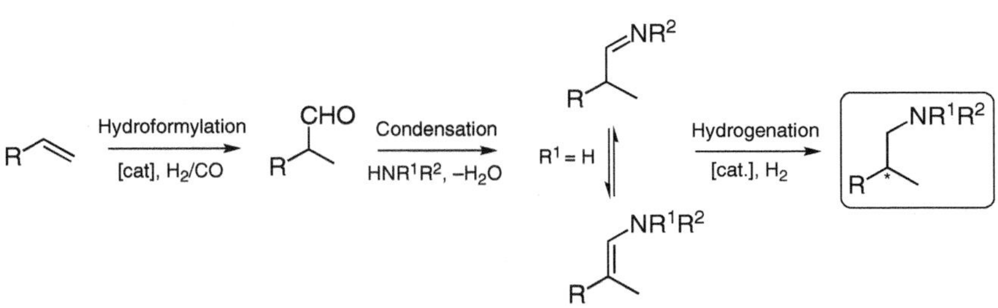

Scheme 5.25 Hydroaminomethylation reaction of olefins: synthesis of amines. Source: Modified from Kalck and Urrutigoïty [85].

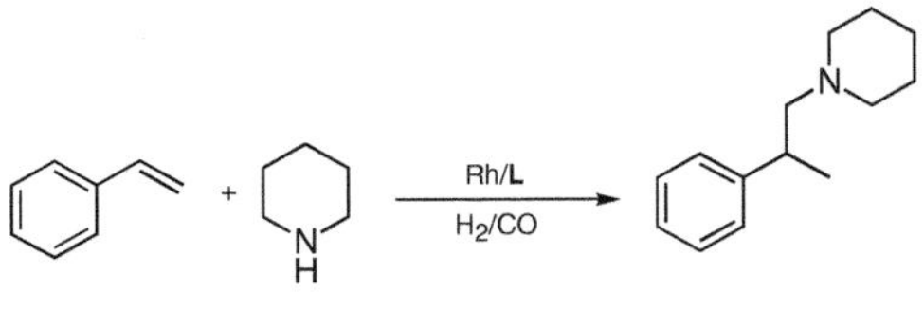

L	Conversion	Aldehyde (%)	Enamine (%)	Amine % (iso/n)	ee (%)
L1	65	4	85	5 (41 : 59)	0
L2	>99	8	40	48 (64 : 36)	0
L3	91	8	14	74 (75 : 25)	0
L4	>99	10	47	42 (48 : 52)	0

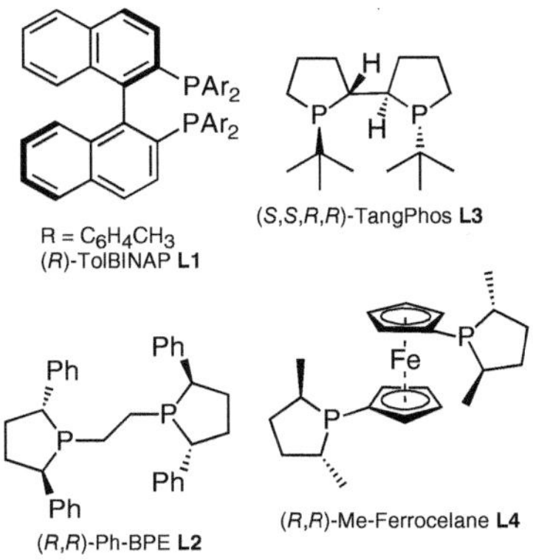

Scheme 5.26 Rh-catalyzed hydroaminomethylation of styrene using **L1–L4** chiral ligands. Source: Modified from Crozet et al. [86].

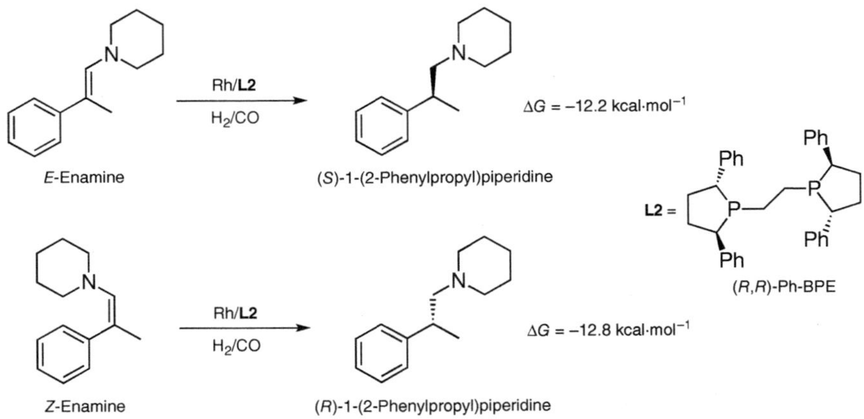

Scheme 5.27 Energy values for the hydrogenation of *E*- and *Z*-enamines. Source: Modified from Crozet et al. [86].

chiral information obtained in the hydroformylation is lost after condensation of the amine with the chiral aldehyde.

High to full conversions were afforded, and good chemo- and regioselectivity toward the branched intermediates was also obtained, even if only diphospholane ligand **L3** provided the amine with high selectivity (Scheme 5.26). However, in all the cases, no enantiomeric excess was afforded.

As observed by NMR spectroscopy, upon condensation of the amine, two isomeric (*E*)- and (*Z*)-enamines are generated. This is a crucial fact because the asymmetric hydrogenation of these enamines will lead to the (*S*)- or (*R*)-products, respectively. DFT calculations demonstrated that the hydrogenation of both enamines is very close in energy (Scheme 5.27), thus explaining why no enantiomeric excess could be obtained.

In view of this issue, several groups attempted to overcome this problem using alternative approaches. List and coworkers reported the catalytic asymmetric amination of aldehydes via dynamic kinetic resolution [87]. Such strategy takes advantage of the imine–enamine equilibrium, via the use of a chiral phosphoric acid that promotes the formation of an iminium of one enantiomer, which is later reduced with a Hantzsch ester (HEH) to afford the desired chiral amine (Scheme 5.28).

Xiao and coworker adapted this methodology to the rhodium-catalyzed HAM in a metal and organocatalyzed asymmetric hydroaminomethylation [88]. Styrene derivatives are first converted into the corresponding branched aldehyde via rhodium catalyzed hydroformylation using a triphenylphosphine derivative as a ligand. Subsequently, the condensation takes place in the presence of anilines to produce the corresponding imines, which are reduced via the same strategy depicted in Scheme 5.28 to produce the final amines in moderate to good yields and moderate to excellent enantioselectivities (Scheme 5.29). The system required stoichiometric amounts of Et-HEH, the use of a chiral phosphoric acid in catalytic amounts, long reaction times (up to 72 hours), and the system is limited to aniline derivatives.

Recently, Han and coworkers improved the system using a bulky phosphoric acid, (*R*,*R*)-Ph-BPE ligand in the hydroformylation, and *t*Bu-HEH as the reducing agent

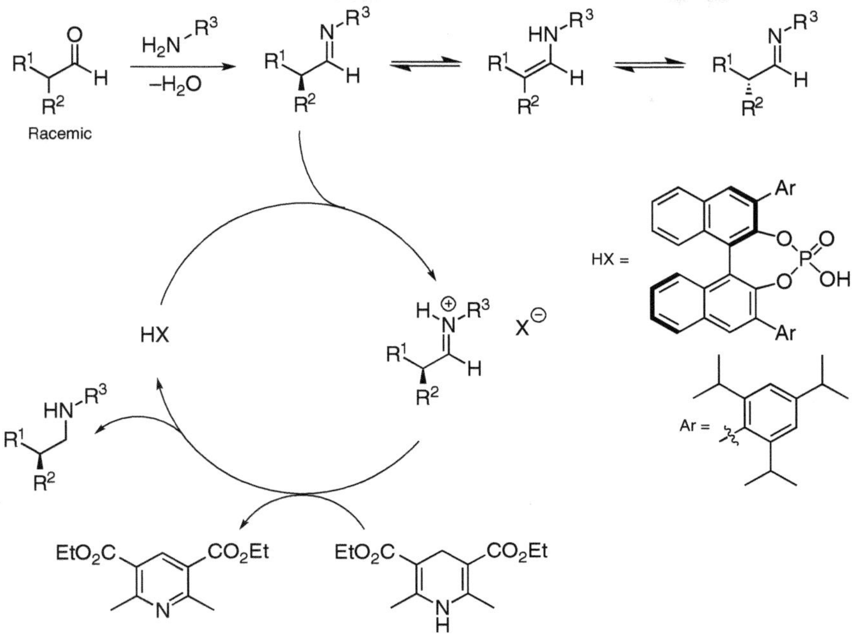

Scheme 5.28 Dynamic kinetic reductive amination of aldehydes developed by List and coworkers. Source: Modified from Hoffmann et al. [87]

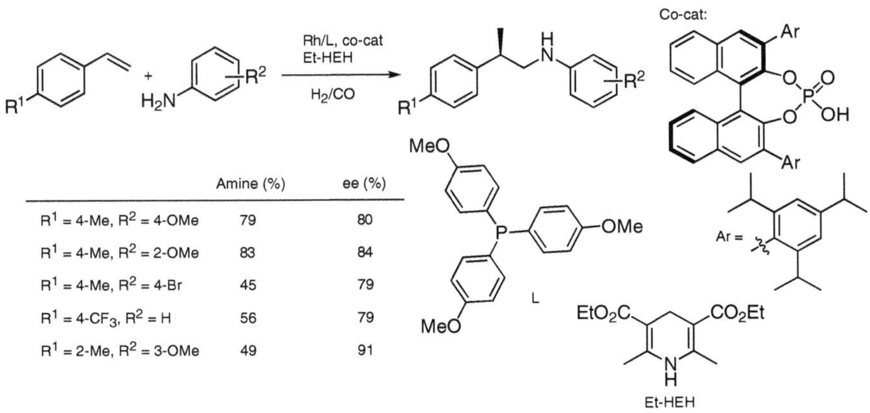

	Amine (%)	ee (%)
R^1 = 4-Me, R^2 = 4-OMe	79	80
R^1 = 4-Me, R^2 = 2-OMe	83	84
R^1 = 4-Me, R^2 = 4-Br	45	79
R^1 = 4-CF_3, R^2 = H	56	79
R^1 = 2-Me, R^2 = 3-OMe	49	91

Scheme 5.29 Rh- and organocatalyzed asymmetric hydroaminomethylation of styrene derivatives with anilines. Source: Modified from Villa-Marcos et al. [88].

(Scheme 5.30) [89]. These changes allowed reducing the total pressure to 1 bar, using lower temperatures, while provided chiral amines in higher yields and enantioselectivities compared with the system used by Xiao and coworker. Moreover, the scope of alkenes was expanded to other aromatic groups such as thiophene, and other olefins such as acrylamides and acrylates. However, the reaction still requires long reaction times (up to 72 hours) and additional reagents, and the system was still limited to aniline derivatives.

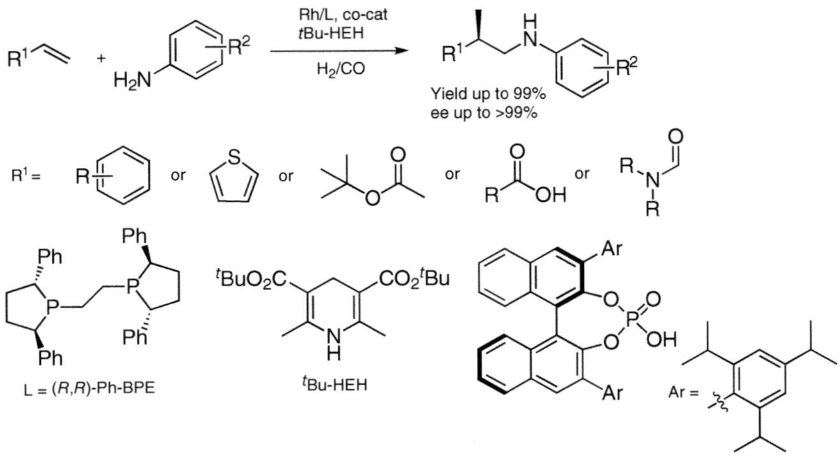

Scheme 5.30 Improved system in the metal and organocatalyzed asymmetric HAM of alkenes with anilines. Source: Modified from Meng et al. [89].

In contrast to the previous strategy, Zhang and coworkers reported the so-called interrupted hydroaminomethylation of trans-1,2-disubstituted olefins (Scheme 5.31) [90]. In order to avoid the problems produced in the equilibrium between imines and enamines (Scheme 5.27), Zhang and coworkers interrupted the formation of these species by the use of external oxidants or reducing agents. The strategy proceeds as follows: first, the Rh-catalyzed asymmetric hydroformylation of three-substituted allylamines was carried out with a system bearing the (*S,R*)-YanPhos ligand, then intramolecular condensation provided a stable hemiacetal that was subsequently either oxidized using pyridinium chlorochromate (PCC) to obtain chiral pyrrolidinones or reduced using triethylsilane and boron trifluoride diethyl etherate to access chiral pyrrolidines.

Both products were afforded in excellent yields (up to 99%) and excellent enantioselectivities (up to 95%). In this case, the enantioselectivity was established in the hydroformylation step in contrast to the classical approach, where the enantioselectivity is induced in the hydrogenation step.

Moreover, the system was also tolerant to other heteroaryl groups such as thiophenes and furans, and the new ring could be expanded from five to six members. However, the system suffered from drawbacks such as long reaction times (up to 70 hours), the requirement of additional reagents in stoichiometric amount, some of them very toxic such as PCC, and the restriction to tosyl group in the amine moiety. Despite these disadvantages, the afforded molecules were envisioned as possible intermediates for the synthesis of Vernakalant [91] and Enablex [92].

Recently, Godard and coworkers reported the first successful example of rhodium-catalyzed intramolecular asymmetric hydroaminomethylation using one single catalyst (Scheme 5.32) [93]. In this work, the authors focused on the asymmetric hydroaminomethylation of α-alkyl acrylates using $[Rh(COD)_2]BF_4$/(*R,R*)-QuinoxP* as the catalytic system to access chiral γ-aminobutyric esters in poor to high yields (up to 98%) and high enantioselectivities (up to 84%). A variety of acrylates were used. It was observed that the steric

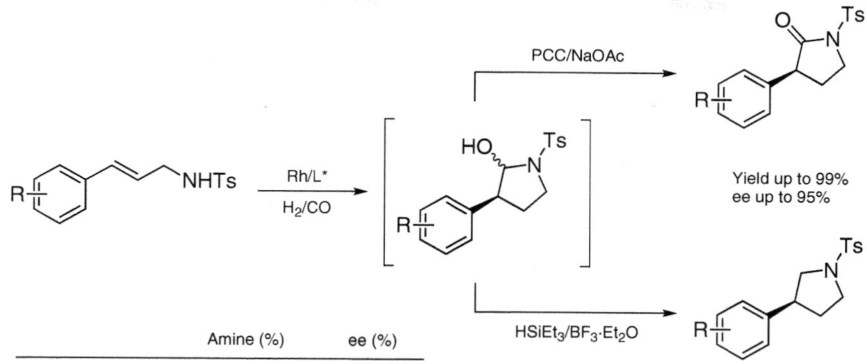

	Amine (%)	ee (%)
R = H	93	95
R = 4-Me	90	93
R = 4-OMe	85	94
R = 4-CF_3	96	90
R = 4-F	95	94

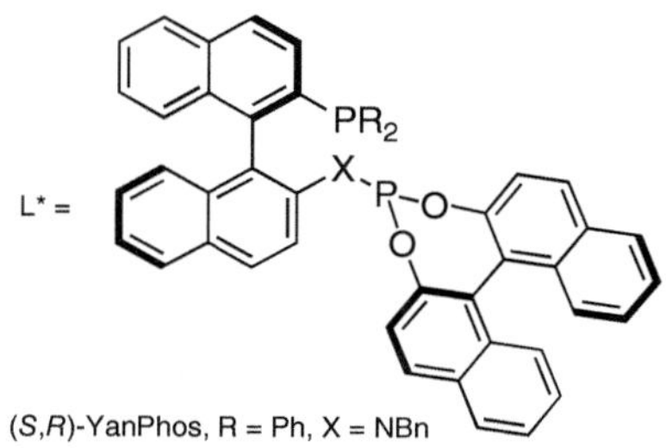

Scheme 5.31 Rh-catalyzed asymmetric interrupted hydroaminomethylation of styrene derivatives using (*S*,*R*)-YanPhosas ligand. Source: Modified from Chen et al. [90].

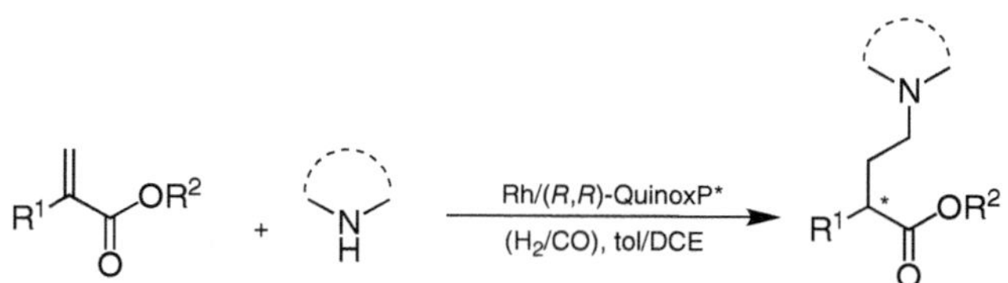

	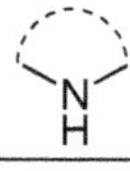	yield (%)	ee (%)
R^1 = Me, R^2 = Me	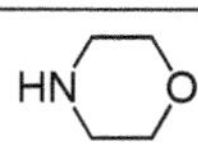	60	73
R^1 = *i*Pr, R^2 = Et		98	84
R^1 = Bn, R^2 = Et	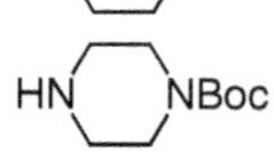	50	71
R^1 = Cp, R^2 = Et		74	84
R^1 = Me, R^2 = Me	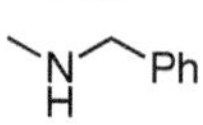	50	62
R^1 = Bn, R^2 = Et		37	76

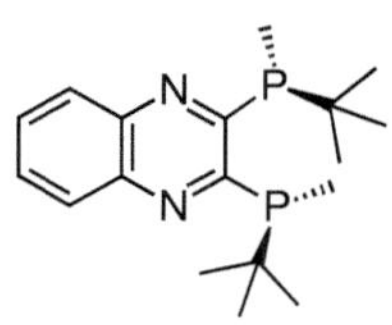

(*R*,*R*)-QuinoxP*

Scheme 5.32 Rh-catalyzed asymmetric hydroaminomethylation of α-alkyl acrylates. Source: Modified from Cunillera et al. [93].

hindrance in α-position was crucial to achieve high enantioselectivities. Furthermore, various amines were well accommodated, even those containing protecting groups such as Boc or benzyl. These products are derivatives of γ-amino acids, which act as major inhibitory transmitters in the mammalian central nervous system [94]. Indeed, drugs containing γ-amino acids are currently commercialized as pharmaceutical agents [95].

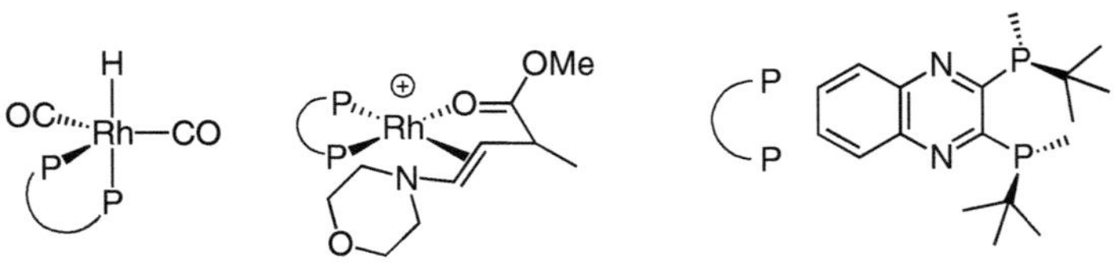

Scheme 5.33 Rhodium species observed in the Rh-catalyzed asymmetric hydroaminomethylation of α-alkyl acrylates. Source: Modified from Cunillera et al. [93].

Following this strategy, the authors avoided the previously mentioned isomerization issue in the hydrogenation step (Scheme 5.27) because the chiral position generated during the hydroformylation was not affected in the subsequent condensation and hydrogenation steps. The challenges in this approach relied on the use of the appropriate ligand to induce high chirality in the hydroformylation step and to find the appropriate combination of reaction conditions. Mechanistic studies revealed that the coexistence of a neutral rhodium hydride involved the hydroformylation, and a cationic species involved in the hydrogenation is crucial (Scheme 5.33). It was also evidenced that a mixture of toluene/DCE (dichloroethane) was necessary to achieve the coexistence of both species. In this case, no extra reducing agents or cocatalysts are required, highlighting the efficiency that characterizes this sort of reaction.

The same group reported the regioselective hydroaminomethylation of α-substituted acrylates to access $\beta^{2,2}$-amino esters [96]. In this specific case, the products were not chiral but provided $\beta^{2,2}$-amino acid derivatives that are scaffolds present in various medicinal drugs [13], via the generation a quaternary center [97].

5.6 Coupling on a Chiral Metal Center

Photoredox catalysts have the capacity to efficiently convert visible light into chemical energy. They can offer the access to highly reactive radical intermediates from bench-stable precursors. Particularly interesting are ruthenium(II) and iridium(III) complexes bearing bipyridine or phenylpyridine ligands that are able, under mild conditions, to promote, in a single-electron transfer mode, the access to open-shell reactive species. Indeed, the $(t_2g)^6$ ground state of [Ir(ppy)$_3$] is easily transformed into the $(t_2g)^5(\pi^*)^1$ [*Ir(ppy)$_3$] excited state, which has the capacity to act as a reductant, by giving its π* electron, or to be an oxidant, by completing to six electrons the $(t_2g)^5$ level [98, 99]. These ruthenium(II) and iridium(III) complexes also allow an energy transfer photosensitivitization catalysis [98, 100]. They are particularly useful to cleave an E—H σ-bond and to form E· and H· radicals. They can be applied in the so-called "proton-coupled electron transfer" and Knowles and coworker generated by this way nitrogen-centered highly reactive radicals for enantioselective reactions [101]. Meggers and his group described that the asymmetric β-amination of α,β-unsaturated 2-acylimidazoles can be performed with 85% yield and 97% ee (when R is 4-methoxyphenyl) by using the *chiral-only-at-metal*-Rh cationic precursor, bearing two 5-*tert*-butyl-2-phenylbenzooxazole ligands (Scheme 5.34) [102–105].

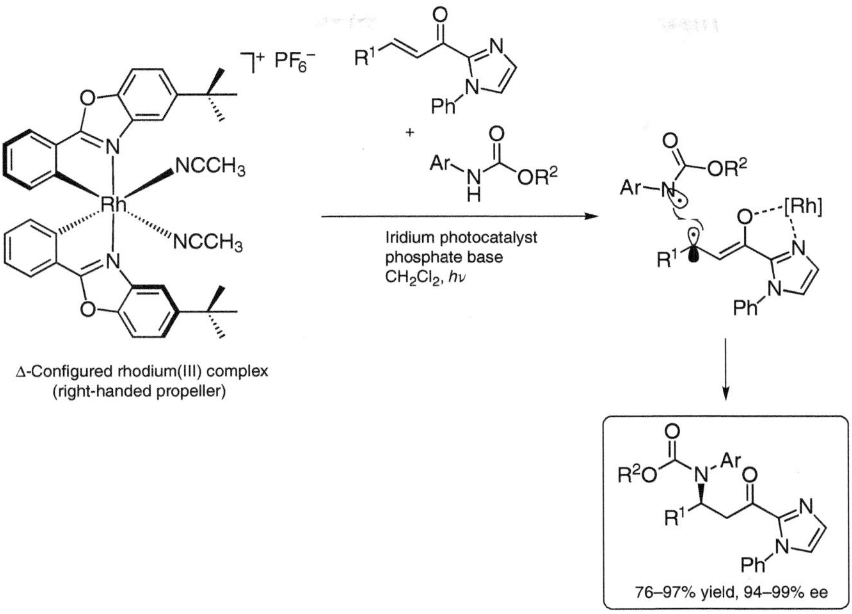

Scheme 5.34 Asymmetric photoredox β-amination of α,β-unsaturated 2-acylimidazoles. Source: Modified from Zhou et al. [102].

A significant number of reactions catalyzed by similar bis-cyclometalated rhodium(III) complexes demonstrated the control of asymmetric radical reactions in α- and β-amination reactions [106] and 1,3-dipolar cycloadditions or alkylation of indoles to obtain high yields (up to 99%) and ee's up to 99% [107, 108].

Moreover, the Δ-Ir cationic complex bearing the same ligands as the rhodium complex of Scheme 5.34 is also efficient in the asymmetric transfer hydrogenation of imines [109]. On the other hand, chiral-at-cobalt and -iron or -ruthenium complexes have recently been designed to perform highly enantioselective reactions [110–112].

5.7 Conclusion

As chiral amines constitute important building blocks in the pharmaceutical and agrochemical industry, it appears that the development of efficient methodologies for their preparation in high optical purity is ensured through metal catalysts. These processes use simple and abundant feedstock and most of the time do not generate waste. Reductive amination, hydroamination, hydroaminomethylation, and hydroaminoalkylation are among the most versatile and useful synthetic tools for the preparation of amines in chemical systems.

The most modern trends are based on the use of first-row transition metals, and for the corresponding precatalysts, it is necessary to adjust their coordination sphere in order to have high chemo-, regio-, and enantioselectivities.

The *chiral-only-at-metal* approach seems to be a recent and pertinent strategy using visible-light-induced asymmetric photocatalysis to obtain chiral amines under mild conditions. Presumably, this methodology is bound to develop for new efficient asymmetric reactions in the near future.

References

1 Ghislieri, D. and Turner, N.J. (2014). Biocatalytic approaches to the synthesis of enantiomerically pure chiral amines. *Topics in Catalysis* 57: 284–300.

2 Nugent, T.C. (2010). *Chiral Amines Synthesis: Methods, Developments and Applications* (ed. T.C. Nugent). Weinheim: Wiley-VCH.

3 Active Pharmaceutical Ingredient (API Market) (2019). marketsandmarkets.com (accessed 18 June 2020).

4 Breuer, M., Ditrich, K., Habicher, T. et al. (2004). Industrial methods for the production of optically active intermediates. *Angewandte Chemie International Edition* 43: 788–824.

5 Constable, D.J.C., Dunn, P.J., Hayler, J.D. et al. (2007). Key green chemistry research areas – perspective from pharmaceutical manufacturers. *Green Chemistry* 9: 411–420.

6 Fasan, R., Kan, S.B.J., and Zhao, H. (2019). A continuing career in biocatalysis: F.H. Arnold. *ACS Catalysis* 9: 9775–9788.

7 Nestl, B.M., Nebel, B.A., and Hauer, B. (2011). Recent progress in industrial biocatalysis. *Current Opinion in Chemical Biology* 15: 187–193.

8 Sheldon, R.A. and Woodley, J.M. (2018). Role of biocatalysis in sustainable chemistry. *Chemical Reviews* 118: 801–838.

9 Sheldon, R.A. and Brady, D. (2018). The limits to biocatalysis: pushing the envelope. *Chemical Communications* 54: 6088–6104.

10 Sheldon, R.A. and Brady, D. (2019). Broadening the scope of biocatalysis in sustainable organic synthesis. *ChemSusChem* 12: 2859–2881.

11 Adams, J.P., Brown, M.J.B., Diaz-Rodriguez, A. et al. (2019). Biocatalysis: a pharma perspective. *Advanced Synthesis and Catalysis* 361: 2421–2432.

12 Reetz, M.T. (2013). Biocatalysis in organic chemistry and biotechnology: past, present, and future. *Journal of the American Chemical Society* 135: 12480–12496.

13 Hönig, M., Sondermann, P., Turner, N.J., and Carreira, E.M. (2017). Enantioselective chemo- and biocatalysis: partners in retrosynthesis. *Angewandte Chemie International Edition* 56: 8942–8973.

14 Reetz, M.T. (2019). Directed evolution of artificial metalloenzymes: a universal means to tune the selectivity of transition metal catalysts? *Accounts of Chemical Research* 52: 336–344.

15 Wang, Y. and Zhao, H. (2016). Tandem reactions combining biocatalysts and chemical catalysts for asymmetric synthesis. *Catalysts* 6: 1–21.

16 De Vries, A., Lefort, L., and Verzijl, G. (2017). Process for preparing chiral amines. Patent WO 2018/060512 A1 (October 02 2017) granted on April 05 2018.

17 Gao, Y., Yang, F., Pu, D. et al. (2018). Pd and $Zn(OTf)_2$ cocatalyzed asymmetric hydrogenation of imines under normal pressure of hydrogen. *European Journal of Organic Chemistry* 45: 6274–6279.

18 Feng, G.-S., Shi, L., Meng, F.-J. et al. (2018). Iridium-catalyzed asymmetric hydrogenation of 4,6-disubstituted 2-hydroxypyrimidines. *Organic Letters* 20: 6415–6419.

19 Seo, C.S.G. and Morris, R.H. (2019). Catalytic homogeneous asymmetric hydrogenation: successes and opportunities. *Organometallics* 37: 48–65.

20 Claver, C., Peñafiel, I., Urrutigoïty, M., and Kalck, P. (2018). Reduction of imines and reductive amination of aldehydes and ketones. In: *Science of Synthesis: Catalytic Reduction in Organic Synthesis*, vol. 2 (ed. J.C. de Vries), 287–330. Thieme.

21 Raoufmoghaddam, S. (2014). Recent advances in catalytic C—N bond formation: a comparison of cascade hydroaminomethylation and reductive amination reactions with the corresponding hydroamidomethylation and reductive amidation reactions. *Organic & Biomolecular Chemistry* 12: 7179–7193.

22 Blaser, H.U., Buser, H.-P., Jalett, H.-P. et al. (1999). Iridium ferrocenyl diphosphine catalyzed enantioselective reductive alkylation of a hindered aniline. *Synlett*: 867–868.

23 Tararov, V.I., Kadyrov, R., Riermeier, T.H. et al. (1999). Phosphines versus phosphinites as ligands in the rhodium catalyzed asymmetric hydrogenation of imines: a systematic study. *Tetrahedron: Asymmetry* 10: 4009–4015.

24 Tararov, V.I. and Börner, A. (2005). Approaching highly enantioselective reductive amination. *Synlett*: 203–211.

25 Strotman, N.A., Baxter, C.A., Brands, K.M.J. et al. (2011). Reaction development and mechanistic study of a ruthenium catalyzed intramolecular asymmetric reductive amination en route to the dual Orexin inhibitor Suvorexant (MK-4305). *Journal of the American Chemical Society* 133: 8362–8371.

26 Yang, T., Yin, Q., Gu, G., and Zhang, X. (2018). A one-pot process for the enantioselective synthesis of tetrahydroquinolines and tetrahydroisoquinolines via asymmetric reductive amination (ARA). *Chemical Communications* 94: 7247–7250.

27 Zhao, Q., Li, S., Huang, K. et al. (2013). A novel chiral bisphosphine-thiourea ligand for asymmetric hydrogenation of β,β-disubstituted nitroalkenes. *Organic Letters* 15: 4014–4017.

28 Yang, T., Guo, X., Yin, Q., and Zhang, X. (2019). Intramolecular asymmetric reductive amination: synthesis of enantioenriched dibenz[c,e]azepines. *Chemical Science* 10: 2473–2477.

29 Afanasyev, O., Kuchuk, E., Usanov, D., and Chusov, D. (2019). Reductive amination in the synthesis of pharmaceuticals. *Chemical Reviews* 119: 11857–11911.

30 Wu, Z., Du, S., Gao, G. et al. (2019). Secondary amines as coupling partners in direct catalytic asymmetric reductive amination. *Chemical Science* 10: 4509–4514.

31 Verzijl, G.K.M., Schuster, C., Dax, T. et al. (2018). Asymmetric synthesis of a key intermediate for Tofacitinib via a dynamic kinetic resolution-reductive amination protocol. *Organic Process Research and Development* 22: 1817–1822.

32 Chang, M., Liu, S., Huang, K., and Zhang, X. (2013). Direct catalytic asymmetric reductive amination of simple aromatic ketones. *Organic Letters* 15: 4354–4357.

33 Li, C., Villa-Marcos, B., and Xiao, J. (2009). Metal-Brønsted acid cooperative catalysis for asymmetric reductive amination. *Journal of the American Chemical Society* 131: 6967–6969.

34 Huang, H., Zhao, H., Yang, Y. et al. (2017). Direct catalytic asymmetric reductive amination of aliphatic ketones utilizing diphenylmethanamine as coupling partner. *Organic Letters* 19: 1942–1945.

35 Tan, X., Gao, S., Zeng, W. et al. (2018). Asymmetric synthesis of chiral primary amines by ruthenium-catalyzed direct reductive amination of alkyl aryl ketones with ammonium salts and molecular H_2. *Journal of the American Chemical Society* 140: 2024–2027.

36 Pagnoux-Ozherelyeva, A., Pannetier, N., Mbaye, M.D. et al. (2012). Knolker's iron complex: an efficient in situ generated catalyst for reductive amination of alkyl aldehydes and amines. *Angewandte Chemie International Edition* 51: 4976–4980.

37 Zhou, S., Fleischer, S., Jiao, H. et al. (2014). Cooperative catalysis with iron and a chiral Brønsted acid for asymmetric reductive amination of ketones. *Advanced Synthesis and Catalysis* 356: 3451–3455.

38 Cabrera, A., Sharma, P., Perez-Flores, J. et al. (2014). Diastereo- and enantioselective reductive amination of cycloaliphatic ketones by preformed chiral palladium complexes. *Catalysis Science & Technology* 4: 2626–2630.

39 Müller, T.E. and Beller, M. (1998). Metal-initiated amination of alkenes and alkynes. *Chemical Reviews* 98: 675–703.

40 Müller, T.E., Hultzsch, K.C., Yus, M. et al. (2008). Hydroamination: direct addition of amines to alkenes and alkynes. *Chemical Reviews* 108: 3795–3892.

41 Huang, L., Arndt, M., Goossen, K. et al. (2015). Late transition metal-catalyzed hydroamination and hydroamidation. *Chemical Reviews* 115: 2596–2697.

42 Lepori, C. and Hannedouche, J. (2017). First-row late transition metals for catalytic (formal) hydroamination of unactivated alkenes. *Synthesis* 49: 1158–1167.

43 Blieck, R., Perego, L.A., Ciofini, I. et al. (2019). Copper-catalysed hydroamination of N-allenyl sulfonamides: the key role of ancillary coordinating groups. *Synthesis* 51: 1225–1234.

44 Reznichenko, A.L. and Hultzsch, K.C. (2010). Asymmetric hydroamination. In: *Chiral Amine Synthesis: Methods, Developments and Applications* (ed. T.C. Nugent), 341–375. Weinheim: Wiley-VCH.

45 Hannedouche, J. and Schulz, E. (2013). Asymmetric hydroamination: a survey of the most recent developments. *Chemistry A European Journal* 19: 4972–4985.

46 Hannedouche, J. (2018). Mechanistic insights into first-row late transition metal-catalysed (formal) hydroamination of unactivated alkenes. *Chimia* 72: 635–641.

47 Pellissier, H. (2017). Recent developments in enantioselective lanthanide-catalyzed transformations. *Coordination Chemistry Reviews* 336: 96–151.

48 Michon, C., Abadie, M.-A., Medina, F., and Agbossou-Niedercorn, F. (2017). Recent metal-catalysed asymmetric hydroaminations of alkenes. *Journal of Organometallic Chemistry* 847: 13–27.

49 Hannedouche, J. and Schulz, E. (2018). Hydroamination and hydroaminoalkylation of alkenes by group 3-5 elements: recent developments and comparison with late transition metals. *Organometallics* 37: 4313–4326.

50 Zhang, Z., Lee, S.D., and Widenhoefer, R.A. (2009). Intermolecular hydroamination of ethylene and 1-alkenes with cyclic ureas catalyzed by achiral and chiral gold(I) complexes. *Journal of the American Chemical Society* 131: 5372–5373.

51 Butler, K.L., Tragnis, M., and Widenhoefer, R.A. (2012). Gold(I)-catalyzed stereoconvergent, intermolecular enantioselective hydroamination of allenes. *Angewandte Chemie International Edition* 51: 5175–5178.

52 Mukherjee, P. and Widenhoefer, R.A. (2010). Gold(I)-catalyzed amination of allylic alcohols with cyclic ureas and related nucleophiles. *Organic Letters* 12: 1184–1187.

53 Kawatsura, M. and Hartwig, J.F. (2000). Palladium-catalyzed intermolecular hydroamination of vinylarenes using arylamines. *Journal of the American Chemical Society* 122: 9546–9547.

54 Miki, Y., Hirano, K., Satoh, T., and Miura, M. (2013). Copper-catalyzed intermolecular regioselective hydroamination of styrenes with polymethylhydrosiloxane and hydroxylamines. *Angewandte Chemie International Edition* 52: 10830–10834.

55 Zhu, S., Niljianskul, N., and Buchwald, S.L. (2013). Enantio- and regioselective CuH-catalyzed hydroamination of alkenes. *Journal of the American Chemical Society* 135 (135): 15746–15749.

56 Pirnot, M.T., Wang, Y.-M., and Buchwald, S.L. (2016). Copper hydride catalyzed hydroamination of alkenes and alkynes. *Angewandte Chemie International Edition* 55: 48–57.

57 Ichikawa, S., Dai, X.-J., and Buchwald, S.L. (2019). Regio- and enantioselective synthesis of 1,2-diamine derivatives by copper-catalyzed hydroamination. *Organic Letters* 21: 4370–4373.

58 Ichikawa, S. and Buchwald, S.L. (2019). Asymmetric synthesis of γ-amino alcohols by copper-catalyzed hydroamination. *Organic Letters* 21: 8736–8739.

59 Feng, S., Hao, H., Liu, P., and Buchwald, S.L. (2020). Diastereo- and enantioselective CuH-catalyzed hydroamination of strained trisubstituted alkenes. *ACS Catalysis* 10: 282–291.

60 Yu, L. and Somfai, P. (2019). Regio- and enantioselective formal hydroamination of enamines for the synthesis of 1,2-diamines. *Angewandte Chemie International Edition* 58: 8551–8555.

61 Zhou, J. and Hartwig, J.F. (2008). Intermolecular, catalytic asymmetric hydroamination of bicyclic alkenes and dienes in high yield and enantioselectivity. *Journal of the American Chemical Society* 130: 12220–12221.

62 Sevov, C.S., Zhou, J., and Hartwig, J.F. (2014). Iridium-catalyzed, intermolecular hydroamination of unactivated alkenes with indoles. *Journal of the American Chemical Society* 136: 3200–3207.

63 Long, J., Wang, P., Wang, W. et al. (2019). Nickel/Bronsted acid-catalyzed chemo- and enantioselective intermolecular hydroamination of conjugated dienes. *iScience* 22: 369–372.

64 Manna, K., Kruse, M., and Aaron, D. (2011). Concerted C—N/C—H bond formation in highly enantioselective Yttrium(III)-catalyzed hydroamination. *ACS Catalysis* 1: 1637–1642.

65 Nguyen, H.N., Lee, H., Audorsch, S. et al. (2018). Asymmetric intra- and intermolecular hydroamination catalyzed by 3,3′-bis(trisarylsilyl)- and

3,3′-bis(arylalkylsilyl)-substituted binaphtholate rare-earth-metal complexes. *Organometallics* 37: 4358–4379.

66 Abadie, M.-A., Trivelli, X., Medina, F. et al. (2017). Gold(I)-catalysed asymmetric hydroamination of alkenes: a silver- and solvent-dependent enantiodivergent reaction. *Chemistry A European Journal* 23: 10777–10788.

67 Tan, T.-D., Chen, Y.-B., Yang, M.-Y. et al. (2019). Stereoselective synthesis of 2,5-disubstituted pyrrolidines via gold-catalysed anti-Markovnikov hydroamination-initiated tandem reactions. *Chemical Communications* 55: 9923–9926.

68 Dai, X.-J., Engl, O.D., León, T., and Buchwald, S.L. (2019). Catalytic asymmetric synthesis of a-arylpyrrolidines and benzo-fused nitrogen heterocycles. *Angewandte Chemie International Edition* 58: 3407–3411.

69 Wang, H., Yang, J., and Buchwald, S.L. (2017). CuH-catalyzed regioselective intramolecular hydroamination for the synthesis of alkyl-substituted chiral aziridines. *Journal of the American Chemical Society* 139: 8428–8431.

70 Clerici, M.G. and Maspero, F. (1980). Catalytic C-alkylation of secondary amines with alkenes. *Synthesis*: 305–306.

71 Roesky, P.W. (2009). Catalytic hydroaminoalkylation. *Angewandte Chemie International Edition* 48: 4892–4894.

72 Edwards, P.M. and Schafer, L.L. (2018). Early transition metal-catalyzed C—H alkylation: hydroaminoalkylation for C_{sp3}—C_{sp3} bond formation in the synthesis of selectively substituted amines. *Chemical Communications* 54: 12543–12560.

73 Nugent, W.A., Ovenall, D.W., and Holmes, S.J. (1983). Catalytic C—H activation in early transition-metal dialkylamides and alkoxides. *Organometallics* 2: 161–162.

74 Herzon, S.B. and Hartwig, J.F. (2007). Direct, catalytic hydroaminoalkylation of unactivated olefins with N-alkyl arylamines. *Journal of the American Chemical Society* 129: 6690–6691.

75 Herzon, S.B. and Hartwig, J.F. (2008). Hydroaminoalkylation of unactivated olefins with dialkylamines. *Journal of the American Chemical Society* 130: 14940–14941.

76 Zhang, Z., Hamel, J.-D., and Schafer, L.L. (2013). $TaMe_3Cl_2$-catalyzed intermolecular hydroaminoalkylation: a simple complex for enhanced reactivity and expanded substrate scope. *Chemistry A European Journal* 19: 8751–8754.

77 Dörfler, J. and Doye, S. (2014). A commercially available tantalum catalyst for the highly regioselective intermolecular hydroaminoalkylation of styrenes. *European Journal of Organic Chemistry*: 2790–2797.

78 Eisenberger, P., Ayinla, R.O., Lauzon, J.M.P., and Schafer, L.L. (2009). Tantalum–amidate complexes for the hydroaminoalkylation of secondary amines: enhanced substrate scope and enantioselective chiral amine synthesis. *Angewandte Chemie International Edition* 48: 8361–8365.

79 Zi, G., Zhang, F., and Song, H. (2010). Highly enantioselective hydroaminoalkylation of secondary amines catalyzed by group 5 metal amides with chiral biarylamidate ligands. *Chemical Communications* 46: 6296–6298.

80 Zhang, F., Song, H., and Zi, G. (2011). Synthesis and catalytic activity of group 5 metal amides with chiral biaryldiamine-based ligands. *Dalton Transactions* 40: 1547–1566.

81 Gilmour, D.J., Lauzon, J.M.P., Clot, E., and Schafer, L.L. (2018). Ta-catalyzed hydroaminoalkylation of alkenes: insights into ligand-modified reactivity using DFT. *Organometallics* 37: 4387–4394.

82 Reznichenko, A.L. and Hultzsch, K.C. (2012). The mechanism of hydroaminoalkylation catalyzed by group 5 metal binaphtholate complexes. *Journal of the American Chemical Society* 134: 3300–3311.

83 DiPucchio, R.C., Rosca, S.-C., and Schafer, L.L. (2018). Catalytic and atom-economic C_{sp3}—C_{sp3} bond formation: alkyl tantalum ureates for hydroaminoalkylation. *Angewandte Chemie International Edition* 57: 3469–3472.

84 DiPucchio, R.C., Rosca, S.-C., Athavan, G., and Schafer, L.L. (2019). Exploiting natural complexity: synthetic terpenoid-alkaloids by regioselective and diastereoselective hydroaminoalkylation catalysis. *ChemCatChem* 11: 3871–3876.

85 Kalck, P. and Urrutigoïty, M. (2018). Tandem hydroaminomethylation to synthesize amines from alkenes. *Chemical Reviews* 118: 3833–3861.

86 Crozet, D., Kefalidis, C.E., Urrutigoïty, M. et al. (2014). Hydroaminomethylation of styrene catalyzed by rhodium complexes containing chiral diphosphine ligands and mechanistic studies: Why is there a lack of asymmetric induction? *ACS Catalysis* 4: 435–447.

87 Hoffmann, S., Nicoletti, M., and List, B. (2006). Catalytic asymmetric reductive amination of aldehydes via dynamic kinetic resolution. *Journal of the American Chemical Society* 128: 13074–13075.

88 Villa-Marcos, B. and Xiao, J. (2015). Metal and organo-catalysed asymmetric hydroaminomethylation of styrenes. *Chinese Journal of Catalysis* 36: 106–112.

89 Meng, J., Li, X.-H., and Han, Z.-Y. (2017). Enantioselective hydroaminomethylation of olefins enabled by Rh/Brønsted acid relay catalysis. *Organic Letters* 19: 1076–1079.

90 Chen, C., Jin, S., Zhang, Z. et al. (2016). Rhodium/yanphos-catalyzed asymmetric interrupted intramolecular hydroaminomethylation of trans-1,2-disubstituted alkenes. *Journal of the American Chemical Society* 138: 9017–9020.

91 Limanto, J., Ashley, E.R., Yin, J. et al. (2014). A highly efficient asymmetric synthesis of Vernakalant. *Organic Letters* 16: 2716–2719.

92 Cross, P.E. and MacKenzie, A.R. (1992). Pyrrolidine derivatives US Patent 005096890A.

93 Cunillera, A., de los Bernardos, M.D., Urrutigoïty, M. et al. (2020). Efficient synthesis of chiral γ-aminobutyric esters via direct rhodium-catalyzed enantioselective hydroaminomethylation of acrylates. *Catalysis Science & Technology* 10: 630–634.

94 (a) Brown, K.M., Roy, K.K., Hockerman, G.H. et al. (2015). Activation of the γ-aminobutyric acid type B (GABAB) receptor by agonists and positive allosteric modulators. *Medicinal Chemistry* 58: 6336–6347. (b) Abdel-Halim, H., Hanrahan, J.R., Hibbs, D.E. et al. (2008). A molecular basis for agonist and antagonist actions at GABAC receptors. *Chemical Biology & Drug Design* 71: 306–327. (c) Lauria-Homer, B.A. and Phol, R.B. (2003). Pregabalin: a new anxiolytic. *Expert Opinion on Investigational Drugs* 12: 663–672. (d) Lapin, I. (2001). Phenibut(β-phenyl-GABA): A tranquilizer and nootropic drug. *CNS Drug Reviews* 7: 471–481. (e) Tomic, M., Pecikoza, U., Micov, A. et al. (2018). Antiepileptic drugs as analgesics/adjuvants in inflammatory pain: current preclinical evidence. *Pharmacology & Therapeutics* 192: 42–64.

95 (a) Kumar, K., Sharma, S., Kumar, P., and Deshmukh, R. (2013). Therapeutic potential of GABAB receptor ligands in drug addiction, anxiety, depression and other CNS disorders. *Pharmacology, Biochemistry, and Behavior* 110: 174–184. (b) Tyacke, R.J., Lingford-Hughes, A., Reed, L.J., and Nutt, D.J. (2010). GABAB receptors in addiction and its treatment. *Advances in Pharmacology* 58: 373–396.

96 Cunillera, A., Ruiz, A., and Godard, C. (2019). Synthesis of β2,2-amino esters via Rh-catalysed regioselective hydroaminomethylation. *Advanced Synthesis and Catalysis* 361: 4201–4207.

97 (a) Corey, E.J. and Guzman-Perez, A. (1998). The catalytic enantioselective construction of molecules with quaternary carbon stereocenters. *Angewandte Chemie International Edition* 37: 388–401. (b) Fuji, K. (1993). Asymmetric creation of quaternary carbon centers. *Chemical Reviews* 93: 2037–2066.

98 Shaw, M.H., Twilton, J., and MacMillan, D.W.C. (2016). Photoredox catalysis in organic chemistry. *The Journal of Organic Chemistry* 81: 6898–6926.

99 Staveness, D., Bosque, I., and Stephenson, C.R.J. (2016). Free radical chemistry enabled by visible light-induced electron transfer. *Accounts of Chemical Research* 49: 2295–2306.

100 Strieth-Kaltoff, F., James, M.-J., Teders, M. et al. (2018). Energy transfer catalysis mediated by visible light: principles, applications, directions. *Chemical Society Reviews* 47: 7190–7202.

101 Gentry, E. and Knowles, R.R. (2016). Synthetic applications of proton-coupled electron transfer. *Accounts of Chemical Research* 49: 1546–1556.

102 Zhou, Z., Li, Y., Han, B. et al. (2017). Enantioselective catalytic γ-amination through proton-coupled electron transfer followed by stereocontrolled radical-radical coupling. *Chemical Science* 8: 5757–5763.

103 Wang, C., Chen, L.-A., Huo, H. et al. (2015). Asymmetric Lewis acid catalysis directed by octahedral rhodium centrochirality. *Chemical Science* 6: 1094–1100.

104 Ma, J., Shen, X., Harmsa, K., and Meggers, E. (2016). Expanding the family of bis-cyclometalated chiral-at-metal rhodium(III) catalysts with a benzothiazole derivative. *Dalton Transactions* 45: 8320–8323.

105 Ma, X., Zhang, X., Huang, X. et al. (2018). Preparation of chiral-at-metal catalysts and their use in asymmetric photoredox chemistry. *Nature Protocols* 13: 605–632.

106 Huang, X. and Meggers, E. (2019). Asymmetric photocatalysis with bis-cyclometalated rhodium complexes. *Accounts of Chemical Research* 52: 833–847.

107 Gong, J., Wan, Q., and Kang, Q. (2018). Asymmetric [3+2] cycloaddition employing N,N′-cyclic azomethine imines catalyzed by chiral-at-metal rhodium complex. *Organic Letters* 20: 3354–3357.

108 Carmona, M., Rodriguez, R., Passarelli, V., and Carmona, D. (2019). Mechanism of the alkylation of indoles with nitrostyrenes catalyzed by chiral-at-metal complexes. *Organometallics* 38: 988–995.

109 Li, Y., Lei, M., Yuan, W. et al. (2017). An N-heterocyclic carbene iridium catalyst with metal-centered chirality for enantioselective transfer hydrogenation of imines. *Chemical Communications* 53: 8089–8092.

110 Cruchter, T. and Larionov, V.A. (2018). Asymmetric catalysis with octahedral stereogenic-at-metal complexes featuring chiral ligands. *Coordination Chemistry Reviews* 376: 95–113.

111 Hong, Y., Jarrige, L., Harms, K., and Meggers, E. (2019). Chiral-at-iron catalyst: expanding the chemical space for asymmetric earth-abundant metal catalysis. *Journal of the American Chemical Society* 141: 4569–4572.

112 Zhou, Z., Chen, S., Qin, J. et al. (2019). Catalytic enantioselective intramolecular $C(sp^3)$-H amination of 2-azidoacetamides. *Angewandte Chemie International Edition* 58: 1088–1093.

6

Industrial Relevance of Asymmetric Organocatalysis in the Preparation of Chiral Amine Derivatives

Luca Bernardi[1], *Armando Carlone*[2] *and Francesco Fini*[3]

[1] *University of Bologna, Department of Industrial Chemistry "Toso Montanari" & INSTM RU Bologna, Alma Mater Studiorum, V. Risorgimento 4, 40136 Bologna, Italy*
[2] *Università degli Studi dell'Aquila, Department of Physical and Chemical Sciences, via Vetoio, 67100 L'Aquila, Italy*
[3] *Università degli studi di Modena e Reggio Emilia, Department of Life Sciences, via Giuseppe Campi 103, 41125 Modena, Italy*

6.1 Introduction

Organocatalysis has generated a great deal of attention since 2000, when seminal contributions that ultimately defined organocatalysis as the third pillar of asymmetric catalysis, beside metal and enzymatic catalysis, started to appear in the literature. Since then, the number of disclosures has dramatically increased, reaching thousands of articles every year [1].[1] Interestingly, industrialists introduced the use of organic molecules as chiral catalysts much earlier. In the early 1970s, Hajos and Parrish (at Hoffmann-La Roche) and Eder, Sauer, and Wiechert (at Schering AG) reported enantioselective L-proline **1** catalyzed intramolecular aldol reactions of triketone **2** for the synthesis of bicyclic intermediates (**3**, **4**) for steroid synthesis (Scheme 6.1) [2]. The steroid core **4** was synthesized with excellent yields and moderate to high enantioselectivities, with functional groups suitable for subsequent synthetic manipulations.

In the 1980s, Dolling and Grabowski, at Merck Sharp & Dohme, reported a straightforward asymmetric phase transfer catalyzed (PTC) alkylation of a substituted indanone **6** toward the synthesis of the pharmaceutical candidate (+)-indacrinone (MK-0197), with good results in terms of yield and enantioselectivity (Scheme 6.2) [3]. The alkylation, delivering intermediate **7**, was mediated by the *Cinchona*-derived quaternary ammonium salt **5** under biphasic reaction conditions (toluene/NaOH mixture). The simple operational procedure enabled a multikilogram-scale process.

These two contributions showed groundbreaking results, but only for very narrow reaction scope and conditions. They had been, in fact, designed for specific aims. After several disclosures in the 1990s delineating different modes of activation [1], in 2000, two seminal

1 Searching on Sci Finder scifinder.cas.org the term "Organocatalysis" as word or concept for the year 2019, resulted in 1559 entries.

Methodologies in Amine Synthesis: Challenges and Applications, First Edition.
Edited by Alfredo Ricci and Luca Bernardi.

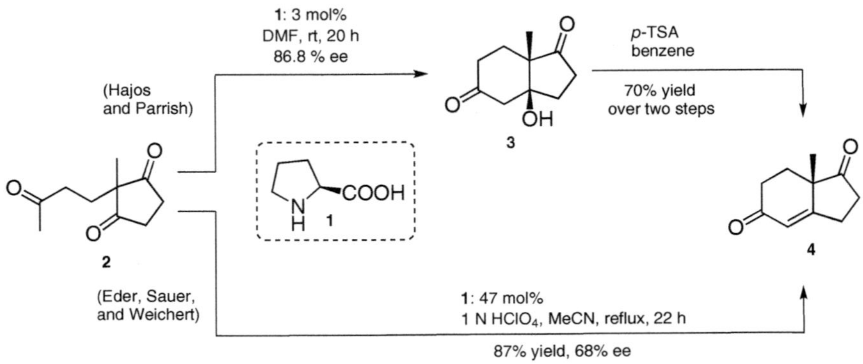

Scheme 6.1 Hajos–Parrish–Eder–Sauer–Wiechert reactions for the synthesis of bicyclic core **4**, a chiral building block for steroids.

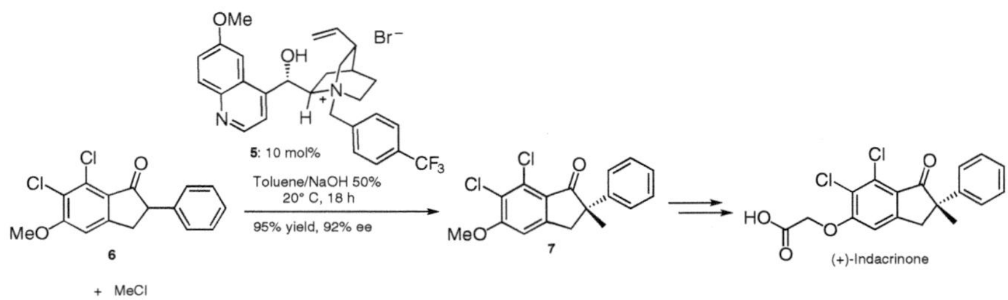

Scheme 6.2 Alkylation of a substituted indanone **6** catalyzed by *Cinchona* alkaloid-derived quaternary ammonium salt **5** under biphasic conditions.

contributions appeared in the literature disclosing enamine catalysis by Barbas, Lerner, and List and coworkers and iminium ion catalysis by MacMillan [4]. Both articles showed broad reaction conditions and scope. In addition, the latter one defined the field, introducing the word *organocatalysis*. Academicians immediately swam in this virgin grassland adding several (biomimetic) modes of activation aided by the premise of the field: organocatalysts are generally cheap, relatively nontoxic, and derive from the chiral pool. Furthermore, reactions do not require strictly inert conditions, and the scalability of the process is usually quite straightforward [5].

Despite academic recognition, industry seemed to embrace organocatalysis at a much slower pace perhaps due to the initial tendency to evaluate catalysts in terms of pure prowess (TON and TOF) [6]. However, notwithstanding the well-known veil of secrecy around industrial preparation of active pharmaceutical ingredients (APIs) [7], nowadays, several prospective processes employing organocatalysis have been divulged, showing its potential in manufacture [8]. Some of these processes, targeting nitrogen-containing compounds, are summarized in Section 6.2. Because the development of these processes has been exhaustively detailed elsewhere, discussion is kept to a minimum. In Section 6.3, three case studies are discussed, wherein organocatalysis proved to be a versatile synthetic platform to approach targets in a scale-up perspective.

6.2 Organocatalysis in Manufacture: Representative Examples

Two landmark industrial applications of organocatalysis were disclosed and divulged by Xu at Merck, already in the first decade of 2000 [9]. One example involved iminium ion catalysis used for the enantioselective addition of nitromethane to an α,β-unsaturated aldehyde, as the key enantiodetermining step of the synthesis of telcagepant, an antimigraine drug candidate featuring a chiral α-amidocaprolactam ring (Scheme 6.3) [10]. The retrosynthetic approach envisioned urea formation from **8** and **9**. Caprolactam **9** derives from intermediate **10**, assembled through a Doebner–Knoevenagel reaction between **11** and **12**. Compound **12** results from the key organocatalytic step between enal **13** and nitromethane [11].

Following Hayashi's disclosure [12] regarding the addition of nitromethane to α,β-unsaturated aldehydes using Jørgensen–Hayashi catalyst **14** [13], the same catalytic strategy was planned and applied to aldehyde **13**, for the synthesis of key intermediate **12** (Scheme 6.4). Literature-reported conditions (**14** and benzoic acid at 20 mol% loadings and alcoholic solvents) led to very promising initial results. Stereoinduction was very good (95% ee), accompanied by moderate (50–75%) yields. Unfortunately, several side products, such as multiple condensation adducts and acetals, were also produced at unacceptable levels during the catalytic reaction. An extensive screening of reaction conditions was thus undertaken, and information was gathered: among others, alcoholic solvents performed better under dry conditions, and a carboxylic acid had to be used as a cocatalyst. However, ultimately, a perfect balance between reaction rate/conversion and suppression of impurities was achieved by switching the alcoholic medium to aqueous THF and using a unique mixture of pivalic acid (5 mol%) and boric acid (50 mol%) as cocatalysts.

Under these optimized reaction conditions, diphenylprolinol trimethylsilyl ether **14** was able to promote the reaction at just 5 mol% loading, affording **12** with 73% in-process yield and 95% ee. Interestingly, the process resulted to be quite robust. Moving to using

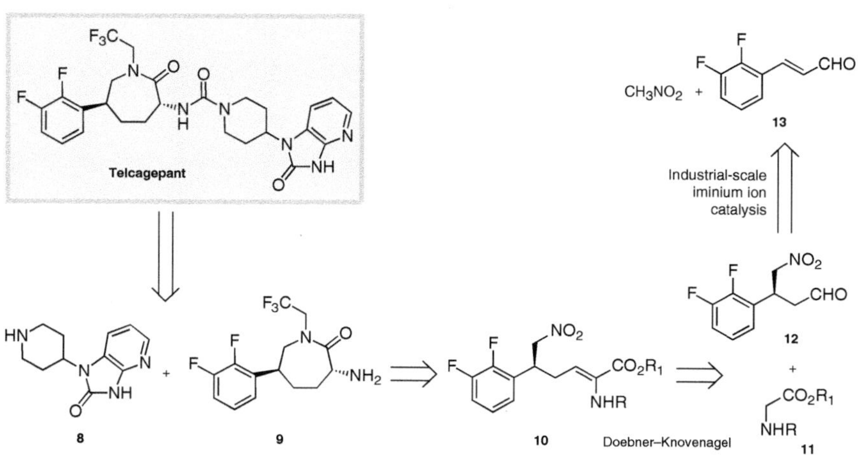

Scheme 6.3 Retrosynthetic approach for the synthesis of telcagepant.

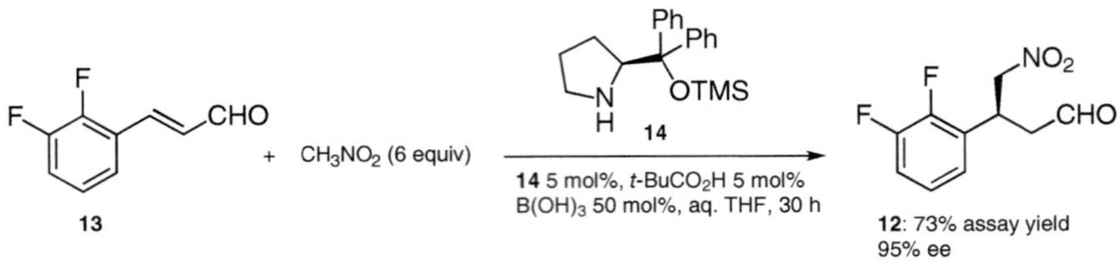

Scheme 6.4 Optimized organocatalytic asymmetric addition of nitromethane to aldehyde **13**.

unpurified catalyst **14** and aldehyde **13** did not affect the reaction outcome, with results very similar to the one with purified materials. Furthermore, the catalytic transformation remained reproducible even on a 250 kg scale, demonstrating the ability of organocatalysis to make an overall competitive process. The synthesis was taken forward with the planned Doebner–Knoevenagel reaction, followed by a reduction/lactamization sequence to afford **9**. Finally, 1,1′-carbonyldiimidazole (CDI) was the reagent of choice to bring a cost-effective synthesis of telcagepant to completion, with an overall yield of 27%, >99.8% purity, and >99.9% ee.

The second remarkable example developed at Merck in the early 2000s regards the synthesis of MK-8613, a selective protease inhibitor and a drug candidate for the treatment of type 2 diabetes mellitus. The molecule is constituted by a piperidine moiety, fused in a tricyclic system, containing two adjacent stereogenic centers (Scheme 6.5). The initial synthesis was racemic and contemplated a quite inefficient late-stage chromatographic resolution [15]. To increase the effectiveness, Xu and coworkers, at Merck, worked on a retrosynthetic plan involving an early organocatalytic enantioselective Michael reaction between a malonate and nitroalkene **18**, followed by a novel three-component cascade coupling reaction

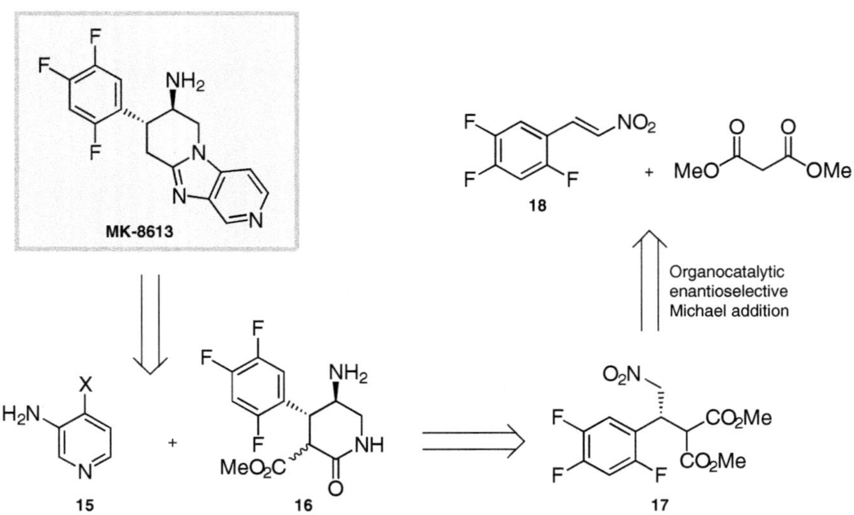

Scheme 6.5 Retrosynthetic plan to obtain MK-8613, by using asymmetric organocatalysis. Source: Xu [9]; and Xu et al. [14c].

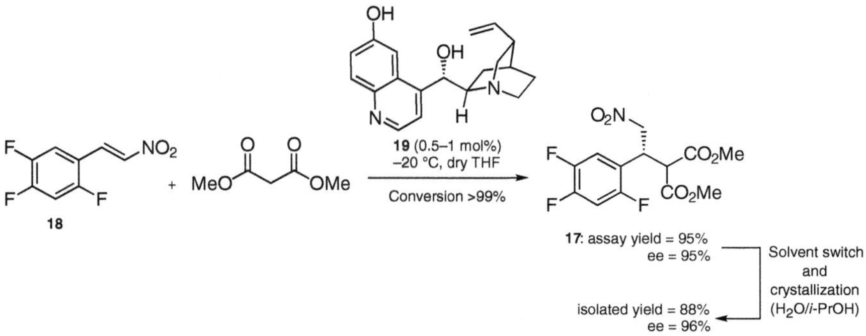

Scheme 6.6 Enantioselective malonate addition to nitrostyrene **18** promoted by desmethyl quinidine **19**.

on the Michael adduct **17**. The resulting product **16**, coupled with pyridine derivative **15**, delivers the target MK-8613 (Scheme 6.5) [14].

The protocol for the addition of malonates to nitroalkenes developed by Deng and coworkers [16], exploiting bifunctional desmethyl quinidine **19** as organocatalyst (Scheme 6.6), was taken as the starting point, although other procedures had probably been already reported at that time [17]. The key enantioselective step was then custom-optimized for the addition of dimethyl malonate to **18**. Relevant results are listed below:

- Running the reaction at temperatures lower than −25 °C gave better results in terms of stereocontrol, but the addition was sluggish. In contrast, at −20 to −15 °C, an optimal compromise between conversion and enantiocontrol was achieved.
- Control of water content was found to be essential for a highly enantioselective process. As a matter of fact, by carrying out the reaction in dry THF at −20 °C, adduct **17** was obtained with an enantiomeric excess of 95%, with high reproducibility.
- A further optimization was performed to lower the amount of dimethyl malonate down to 1.1 equiv; reaction rate and stereocontrol were not affected, making the process more cost efficient.

Thus, the organocatalytic Michael addition of dimethyl malonate to nitroalkene **18** was significantly remodeled, including product isolation. Using just a 0.5–1 mol% of catalyst **19** loading, an isolated yield of 88% and an ee of 96% for product **17** were attained, by switching solvent to isopropanol after the catalytic reaction, and performing a recrystallization with isopropanol/water mixture (Scheme 6.6). The process could be run on a >100 kg scale. Furthermore, the synthesis of catalyst **19** was improved, compared with the original Deng procedure, by developing a crystallization from *i*-PrOAc/cyclohexane as a purification protocol.

The synthesis of MK-8613 was then completed through the planned one-pot cascade coupling reaction, followed by a copper-catalyzed heterocyclization (Scheme 6.5).

A more recent example of the application of another bifunctional catalyst derived from a *Cinchona* alkaloid to a large-scale reaction was reported by researchers at TEVA, targeting funapide. Funapide is a small-molecule oxindole with a chiral spiro-ether, which was under

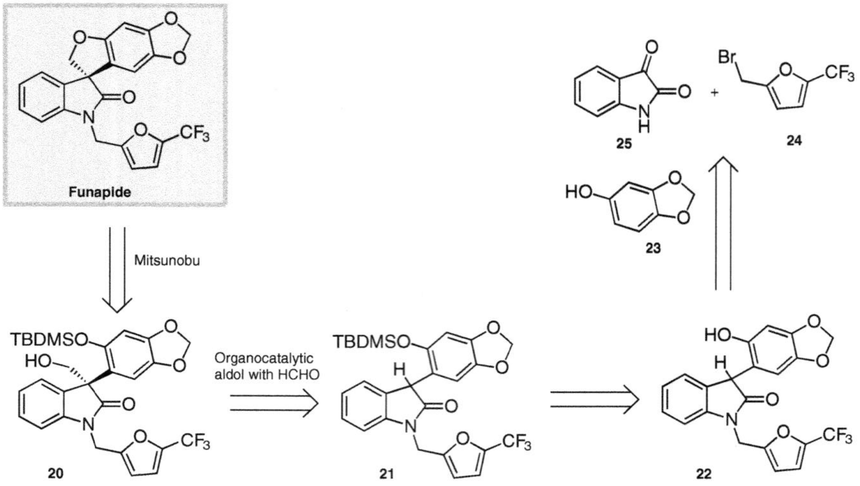

Scheme 6.7 Retrosynthetic approach to funapide encompassing the organocatalytic aldol reaction as the stereodetermining step (TBDMS = *tert*-butyldimethylsilyl). Source: Sclafani et al [18a] and Sun et al. [18b].

development for postherpetic neuralgia (Scheme 6.7). Early synthetic approach to funapide [19], able to sustain early phase-I clinical trials, suffered few drawbacks: the use of neat trifluoroacetic acid, late-stage chiral stationary-phase chromatography, extensive usage of protecting groups, and the employment of a genotoxic alkylating agent in the last step of the sequence, resulting in the risk of contamination in the final API.

With the information gained during the exploration of different enantioselective routes, including a PTC one, a novel and more straightforward retrosynthetic approach was introduced (Scheme 6.7) [18]. This approach entailed cyclizing the spiro-ether of funapide via a Mitsunobu-like reaction on deprotected **20** as the last step of the synthesis. Such precursor **20** could be obtained via the key organocatalytic asymmetric aldol reaction between formaldehyde and **21**. Decorated oxindole **21** could be assembled from starting materials **23–25** through the initial N-alkylation of **25** with genotoxic **24**, addition of sesamol **23**, followed by tertiary alcohol reduction and phenol protection with TBDMSCl.

Focusing on the enantioselective aldol reaction between **21** and formaldehyde, this reaction could be promoted by a variety of chiral tertiary bases of the organocatalytic pool. After an extensive optimization, it was found that the dihydroquinine-derived thiourea **26** could catalyze the reaction efficiently at 1.1 mol% loading, converting 95% of the starting material **21** to the desired aldol adduct **20** in 70.5% ee (Scheme 6.8). The reaction was better performed under slurry conditions by using a mixture of formaldehyde/paraformaldehyde in heptane as the dispersing medium. The utilization of such slurry conditions, inducing precipitation of adduct **20** as soon as formed during the reaction, was important to increase catalyst throughput. In fact, product **20** could inhibit catalyst **26** by binding it via multiple hydrogen bond interactions.

Despite the modest enantioinduction, this catalytic reaction was still useful for scale-up. In fact, by adding methanol to the mixture, a preferential crystallization of nearly racemic **20** occurred, enabling its removal from the mixture. Enantiomerically pure **20** could then

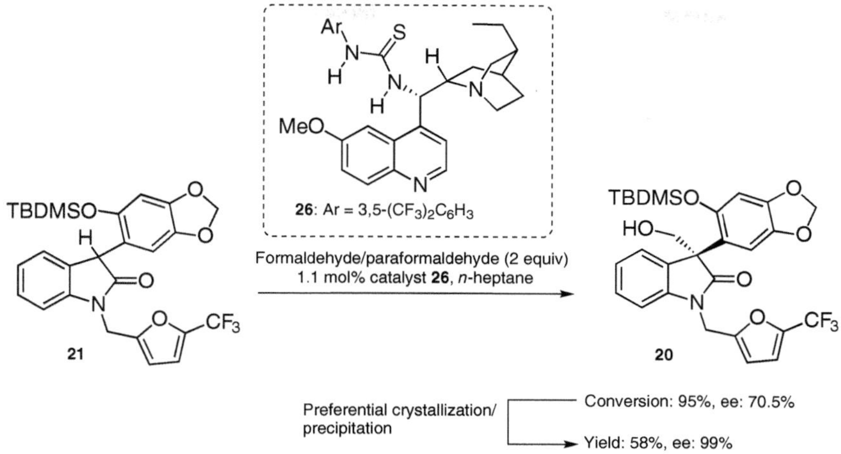

Scheme 6.8 Optimized enantioselective aldol reaction between **21** and formaldehyde catalyzed by hydroquinine derivative **26**.

be obtained (98–99% ee) from the reaction mixture by precipitation from water. The final steps of the synthesis were brought to completion via silyl group removal with HBr 37% in water, followed by a Mitsunobu-like cyclization based on chlorodiphenylphosphine as a condensating reagent (Scheme 6.7).

With the novel process, implemented to the pilot plant, all the major drawbacks of the early synthesis were solved. The genotoxic alkylating agent **24** was used at the beginning of the synthetic sequence; purifications were carried out by crystallization or precipitation, without chromatography; only one protecting group was used. Funapide was obtained with an overall 35% yield from **25** in a short five-step reaction sequence, by using an organocatalytic aldol reaction with low catalyst **26** loading (1.1 mol%) as the key stereodetermining step. The authors reported the recycle of the nearly racemic **20** by using a retroaldol reaction to give back the reagent **21** but did not comment about the recycle of the catalyst **26**. Nevertheless, they reported an interesting addendum regarding the development of a plant-scale access to catalyst **26** and its quinine derivative [20], modifying the original procedure reported by Soós and subsequent academic versions [21]. The new process consists of three reactions and one recrystallization. It was performed in a single reactor, providing the quinine derived thiourea catalyst, as efficient as **26** in funapide synthesis, at an estimated cost of USD2750 per kg. Useful insights into cost contribution of a nontraditional industrial catalyst for manufacturing can be, therefore, inferred.

An impressive demonstration of the potential of organocatalysis in API manufacture has been given by recent disclosures regarding the active principle of the antiviral drug Prevymis™:letermovir. Letermovir is constituted by an 8-fluorodihydroquinazoline core, bearing a piperazine ring at C-2, an aryl group at N-3, and an acetyl stereogenic center at C-4 (Scheme 6.9). The early synthetic route to letermovir exploited a late-stage classical resolution [23], an undesirable strategy from an industrial point of view. Scientists at Merck & Co. aimed to replace the resolution with asymmetric catalysis. Although exploration of several metal-catalyzed routes showed severe limitations (insufficient catalyst activity and

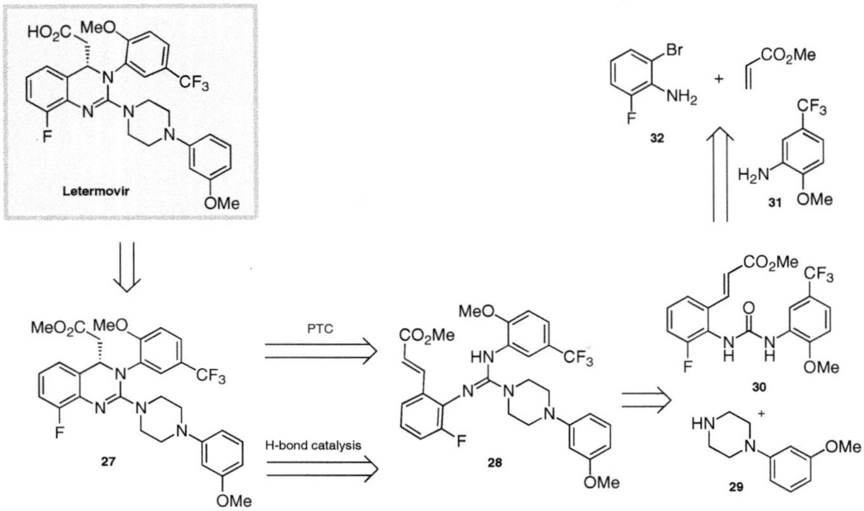

Scheme 6.9 Retrosynthetic analysis of letermovir encompassing the key organocatalytic enantioselective intramolecular aza-Michael reaction. Source: Humphrey et al. [22a], Luzung et al [22b], and Chung et al. [24a].

chemoselectivity), the use of organocatalysis successfully delivered letermovir with high yield and enantiopurity, through the retrosynthetic approach outlined in Scheme 6.9 [22]. In this approach, letermovir derives from the corresponding stereodefined ester **27**. The synthesis of such ester **27**, through the intramolecular aza-Michael addition of the guanidine function of **28** to the α,β-unsaturated ester, is the stereodetermining step of the sequence. Remarkably, two distinct organocatalytic processes were developed and implemented on scale for this reaction, one based on chiral phase transfer catalysis [22] and the another one on H-bond donor catalysts [24]. Substrate **28** for the catalytic process results from the condensation of piperazine **29** with urea **30**. This latter compound derives from the functionalization of aniline **32** with methyl acrylate via Heck reaction, followed by urea formation with a second aniline **31**.

The first approach to the enantioselective aza-Michael reaction **28** → **27** was based on PTC. An extensive screening of phase transfer catalysts and reaction conditions (solvents, inorganic bases, etc.) was carried out, resulting in the optimal protocol reported in Scheme 6.10. In brief, this screening indicated that only solvents of low polarity gave good results in term of enantioselectivity, with toluene further helping in preventing a background – uncatalyzed – cyclization. Many *Cinchona* alkaloid-derived quaternary ammonium salts were tested. Traditional mono-quaternized salts provided modest enantioselectivities, in the range of 10–40% ee, while, surprisingly, bis-quaternized ammonium salts [25], in which both quinolinic and quinuclidinic nitrogens are alkylated, gave remarkably better results. By using catalyst **33** bearing trifluoromethyl-decorated benzyl substituents at 3–5 mol% loadings, compound **27** was obtained with 98% yield and 76% ee. A lowering of the catalyst loading was not feasible, probably because of its Hofmann degradation. Furthermore, the optimization process pointed toward aqueous K_3PO_4 as the most suitable inorganic base and to bromides as best catalyst counterions.

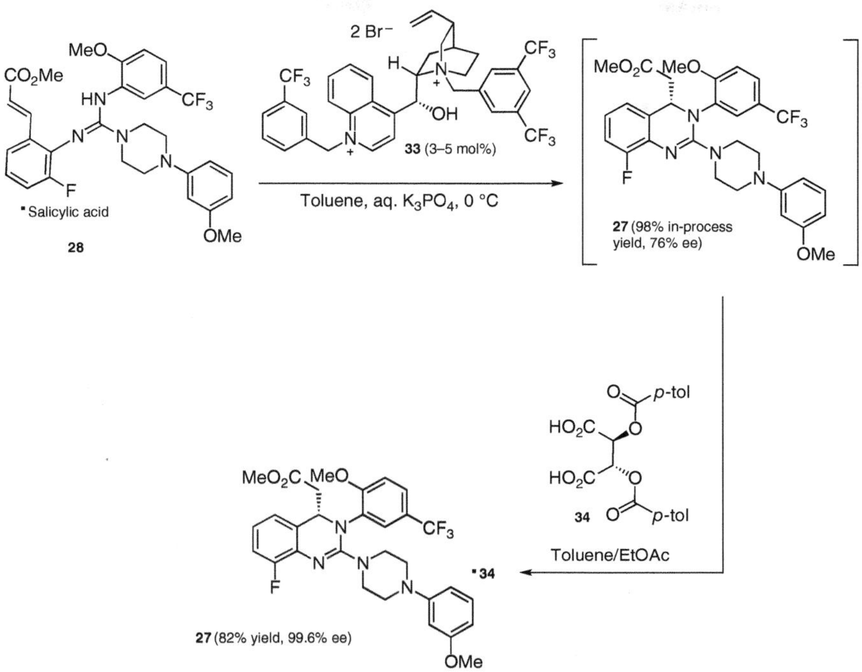

Scheme 6.10 First-generation enantioselective aza-Michael cyclization based on PTC.

The enantioselectivity was in fact quite sensitive to catalyst counterions. Although the in-process enantiomeric excess of product **27** was only moderate (76% ee), crystallization with the tartaric acid derivative **34** upgraded the enantiopurity to 99.6% ee, while keeping a respectable 82% yield for the overall process. Indeed, this first-generation process based on PTC, enclosed in the synthetic plan sketched in Scheme 6.9, was used to produce over 1 tonne of letermovir to sustain early and advanced clinical trials.

Nevertheless, this PTC-based process showed some nonoptimal features, such as a relatively high catalyst loading, and the requirement of an efficient agitation, typical of PT processes. Therefore, studies were continued aiming at switching from a biphasic PTC to a homogeneous catalytic protocol. Observing that by using protic solvents, such as 2,2,2-trifluoroethanol, compound **28** was fully converted to **27** in 15 minutes, an asymmetric H-bond catalysis approach was considered. A wide range of screening of H-bond donor catalysts indicated that strong donors such as chiral phosphoric acids were not suitable, promoting the aza-Michael **28** → **27** reaction with only moderate enantioselectivity (<58% ee) and conversion (<60%). Conversely, C_2-symmetric, chiral bis-triflamides provided promising results in terms of both conversion and enantioinduction. After an extensive optimization, organocatalyst **35** was identified, promoting the cyclization with 95% yield and 94% ee on a small scale, with a loading of 5 mol% (Scheme 6.11). Cyclopentyl methyl ether (CPME), a relatively harmless solvent, could be used as reaction medium. Even if no details on large-scale implementations have been published so far, the overall process to letermovir, encompassing the second-generation H-bond catalysis, was awarded the 2017 Presidential

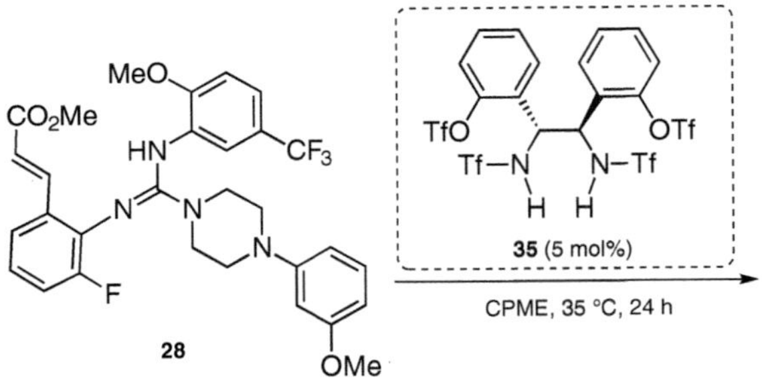

Scheme 6.11 Second-generation enantioselective aza-Michael cyclization based on H-bond donor catalysis. Tf = SO_2CF_3.

green chemistry challenge award from the US Environmental Protection Agency. Furthermore, these two generations of organocatalytic processes have been taken to exemplify the necessity of innovating synthetic chemistry for advancing new medicines [26].

Interestingly, the mechanism of this weak H-bond catalysis was studied by DFT calculations, kinetic isotopic effect, and control experiments, hinting toward the coordination of **35** with the α,β-unsaturated ester, thus activating the ester moiety for the intramolecular aza-Michael reaction, via transition state **A** (Scheme 6.12). Such transition state was considered to be favored over an alternative electrocyclic transition state **B**, triggered by the tautomerization of **28** to an aza-*ortho*-quinone methide.

Lewis bases, another class of organic catalysts, were also successfully employed in industry, and herein, two examples are illustrated. Censavudine, a nucleoside reverse-transcriptase inhibitor, formerly known as festinavir, was disclosed at Yale University and developed by Bristol-Myers Squibb (BMS). Structurally speaking, censavudine is constituted by a pyrimidine derivative connected to a furanose-derived five-membered ring through a β-*N*-glycosidic bond (Scheme 6.13). Among the different synthetic approaches considered, the challenging retrosynthetic tactic sketched in Scheme 6.13 ultimately led to the most efficient process. This approach contemplated several "risky" synthetic steps, which did not have solid literature precedents at the development set off: the Achmatowicz rearrangement on scale (**40** → **39**), the resolution of pyranone **39** to give protected, enantioenriched, counterparts **38**, and the furanose trapping step (**37** → **36**) [27]. On the

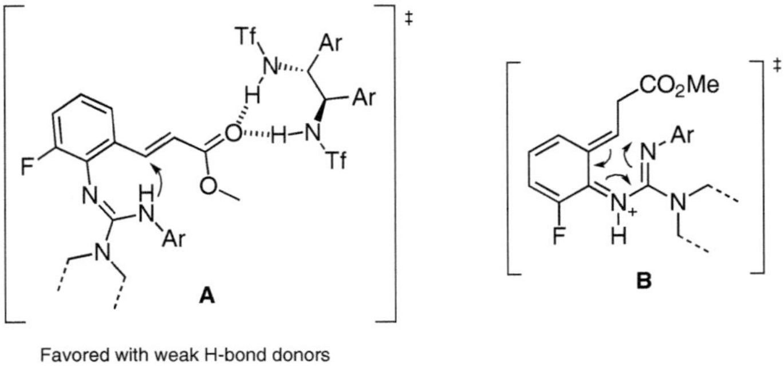

Scheme 6.12 Possible transition states for the aza-Michael reaction with H-bond donor catalysts.

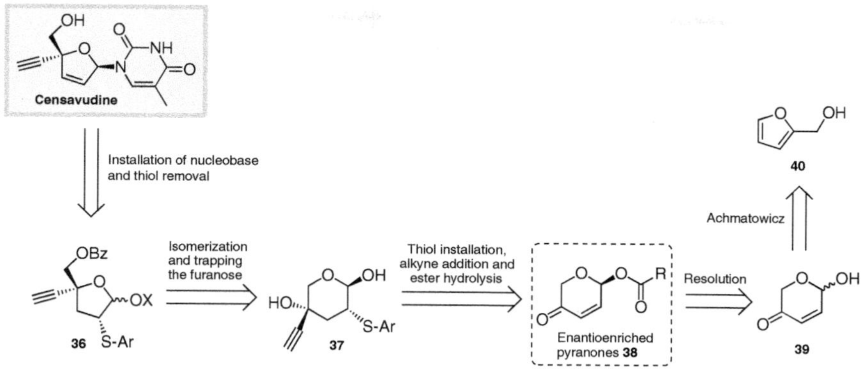

Scheme 6.13 Retrosynthetic approach for the synthesis of censavudine.

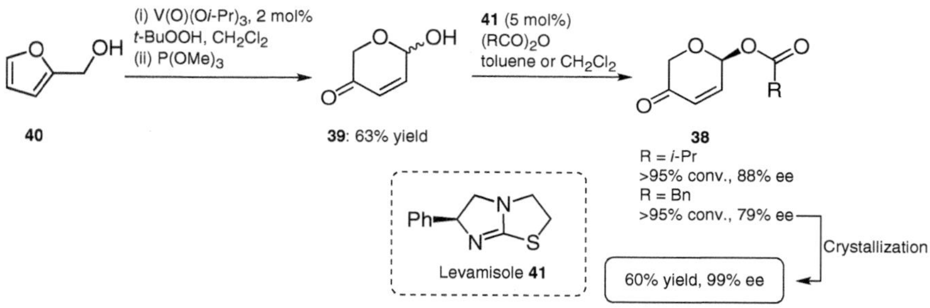

Scheme 6.14 Synthesis of enantiopure pyranones **38** via a dynamic kinetic acylative resolution catalyzed by levamisole **41**.

other hand, this approach could guarantee fewer steps and easier purifications compared to a more classical synthesis starting from the nucleoside 5-methyluridine [28].

The synthesis of **39** was brilliantly achieved by using a vanadium (V) catalyst and *tert*-butyl hydroperoxide as a stoichiometric oxidant, novel and scalable reaction conditions for the Achmatowicz rearrangement of furfurol **40** (Scheme 6.14) [29]. An initially developed approach to enantioenriched pyranones **38** involved kinetic resolutions by a lipase [30]. Searching for more efficient entries to these compounds, it was surprisingly found that a very cheap enantiopure isothiourea, levamisole **41**, a commercial anthelmintic pharmaceutical, could catalyze dynamic kinetic acylative resolutions of alcohol **39**. Using a hindered, aliphatic anhydride for acylation resulted in the best enantioselectivity (88% ee). However, a phenylacetic acid derivative, although providing lower stereoselectivity (79% ee), gave a crystalline compound allowing enantiopurity upgrade by crystallization, and was thus ultimately preferred. Overall, the key pyranone building block **38** was obtained from Achmatowicz product **39** with 60% isolated yield and in 99% ee.

The subsequent synthetic steps were preparatory for compound **37**. Starting from **38**, thiocresol, as a directing group, was anchored via a Michael addition, then a terminal alkyne was added to the ketone with complete diastereoselectivity, thanks to the previously installed thiol directing group. Eventually, the phenylacetic acid ester was hydrolyzed obtaining **37** as an equilibrating mixture of pyranose **37** and furanose **37′** isomers, both

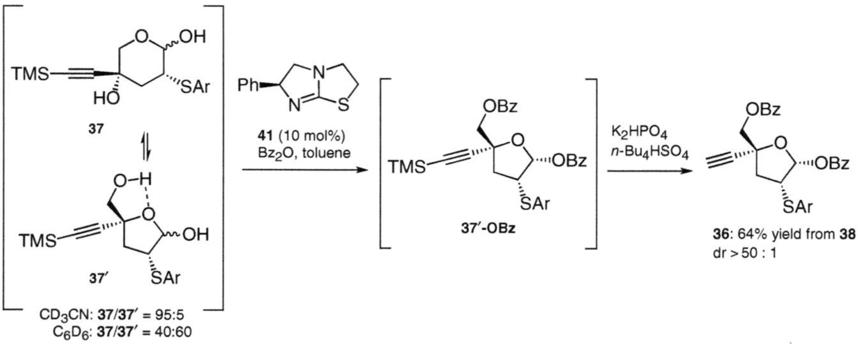

Scheme 6.15 Trapping of the furanose form **37′** with benzoic anhydride using levamisole **41** as the catalyst, followed by TMS removal. Ar = 4-MeOC_6H_4.

present as diastereomeric mixture. The equilibrium between **37** and **37′** was studied by NMR in different solvents (Scheme 6.15). While polar solvents, such as acetonitrile, favored pyranose **37** with a 95 : 5 ratio, in less polar media, such as benzene, a more balanced 40 : 60 ratio toward **37′** was observed. With this information in hand, the furanose trapping event was performed in toluene. Remarkably, levamisole **41** was used to guide the acylative trapping reaction, although 4-(*N*,*N*-dimethylamino)pyridine(DMAP) could also be employed [27b]. As a matter of fact, **37′-OBz** was obtained in a diastereomeric pure form and was subsequently treated with a base, removing the TMS group and affording **36**. Overall, **36** was obtained in five steps and 64% yield from pyranone **38**.

The final part of the synthesis involved Vorbrüggen installation of the nucleobase, removal of the thiocresol restoring the olefin under mild oxidative (chloramine-T), thermolytic conditions, and a final transesterification leading to censavudine with 69% yield from **36**. The overall synthesis, possibly executed in plant, features a remarkable 44% overall yield from chiral building block **38**, without any chromatographic purification.

The significance of this process goes beyond the novel transformations that were developed, pointing to the importance of unconventional decision-making in the accomplishment of an efficient industrial route. To quote one of the research coordinators [31]: "the multiple innovations of this route highlight the need for vision in selecting a commercial process: the focus on the pyranone strategy came at a time when the furan oxidation catalyzed by vanadium, the levamisole chemistries for enantioselection, and the chloramine sulfur elimination were yet to be invented." Remarkably, organocatalysis enabled the innovation required in two of the key steps of the synthesis.

Another antiviral nucleoside analog for which an innovative organocatalytic transformation was the key for a successful process is uprifosbuvir [32]. Developed by Merck & Co and currently halted during phase II clinical trials, uprifosbuvir is constituted by a 2′-chloro nucleoside core linked at 5′ with a phosphoramidate moiety of defined R configuration at phosphorous (Scheme 6.16). One of the most challenging parts of the synthesis of this, and of related prodrug nucleotides, is the regio- and stereocontrolled introduction of the phosphoramidate moiety on the nucleoside core. The usual approach involves a Grignard reagent as a base to deprotonate the nucleoside, which then stereospecifically attacks a diastereo- and enantiopure phosphoramidating agent [33]. Although good results in terms

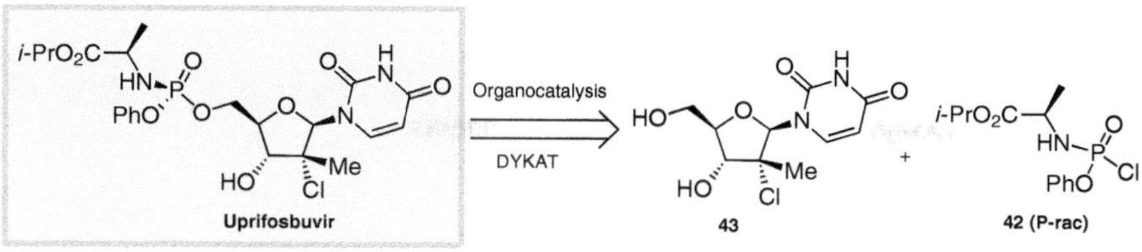

Scheme 6.16 Regio- and stereoselective (at P) introduction of the phosphoramidate moiety on nucleoside **43** using P-racemic **42**.

of stereoselectivity and yield are achieved, the synthesis of the stereodefined phosphoramidating reagent is usually accomplished by selective crystallization of the stereoisomeric mixture, implying low efficiency. In addition, safety concerns regarding the use of the Grignard (usually *tert*-butylmagnesium bromide) as the strong base need to be considered. The process was thus rethought by scientists at Merck, planning to introduce organocatalysis and in more detail Lewis base catalysis, as the stereocontrol element in the phosphoramidating key step (Scheme 6.16) [34]. Even if solid literature examples were lacking, a dynamic stereoselective process (dynamic kinetic asymmetric transformation, DYKAT) was envisioned. In this process, a chiral Lewis base activates the correct epimer at P of the racemic phosphoramidating agent **42**, which is attacked regioselectively by the 5′-OH of the nucleoside **43**. The off-epimer of **42** racemizes in situ in a dynamic manner [35].

After having confirmed the lack of strong substrate bias on the stereoselectivity of the reaction using achiral *N*-methylimidazole **44** as a catalyst, an initial screening was performed with various bicyclic imidazole acylating agents (e.g.**45**, **46**), which gave modest but promising results (Scheme 6.17). For example, 62% yield and 89 : 11 diastereomeric ratio favoring the right configuration at phosphorous were achieved by using 20 mol% loading of catalyst **46**. In order to improve the catalyst performance, a systematic mechanistic study was undertaken, discovering a fast epimerization and a constant P-epimeric ratio in the phosphoramidating reagent **42**. Most importantly, kinetics was found to be second order in organocatalyst. Thus, two imidazole groups are involved in transition state **C**, justifying the second-order kinetics. DFT calculations showed additional interactions between the strongly Lewis basic P=O and some aromatic C–H of the catalyst, and the phenolic substituent. These interactions assist the reaction by oxyanion stabilization. The origin of diastereoselectivity was found to be due to the lower energy of transition state **C** compared to its counterpart giving the opposite configuration at phosphorus and not to the equilibrium between **A** and **B** (Curtin–Hammett control). With this information in hand, dimeric catalysts containing two imidazole units were introduced. Carbamate catalyst **47** with a semiflexible chain connecting the two imidazoles gave excellent results in terms of yield and stereoselectivity (92%, 99 : 1 d.r. at P). The catalyst loading could be even lowered down to 2 mol%, while the reaction proceeded with full regioselectivity.

The optimized process with **47** as an organocatalyst could be successfully applied to several other nucleoside analogs, demonstrating the robustness of the methodology and its generality. It must be stressed that other commercial nucleoside analogs (e.g. sofosbuvir) feature the same phosphoramidate group, which can be thus easily and elegantly introduced with this Lewis base organocatalyzed reaction.

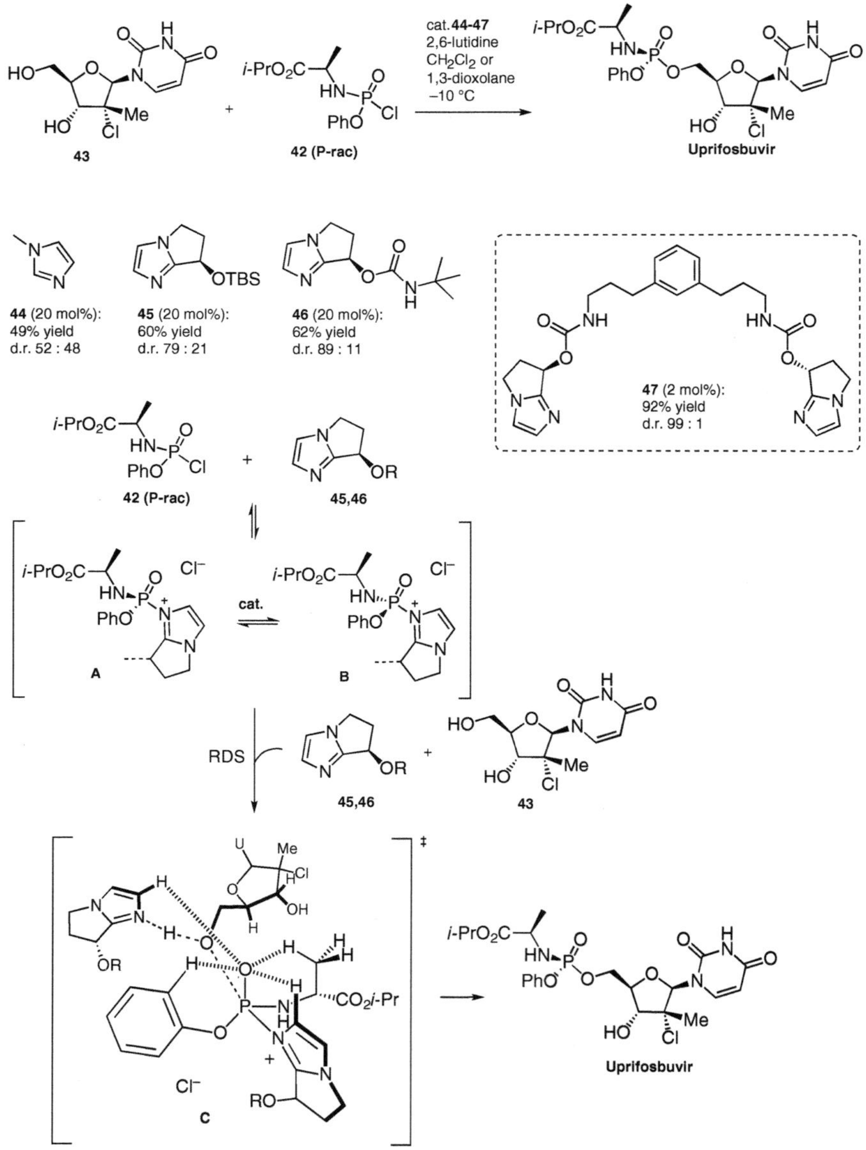

Scheme 6.17 Stereoselective phosphoramidate formation in uprifosbuvir synthesis catalyzed by imidazoles **44–47**.

6.3 Case Studies

6.3.1 Pregabalin

Pfizer's Lyrica™ and associated brand name drugs are useful treatments for many conditions related to the central nervous system, such as seizure and anxiety disorders, neuropathic pain, fibromyalgia, and epilepsy. In commercial terms, these drugs have been among the most successful of the past decades [36]. Their API is a relatively simple γ-amino acid in enantiopure (*S*)-form: pregabalin, that is (*S*)-3-aminomethyl-5-methyl-hexanoic acid [37]. In the late 1990s, a first assessment of several synthetic approaches to this γ-amino acid was carried out by the originator (Pfizer), leading to a useful manufacturing process [38] based on a classical resolution with (*S*)-mandelic acid in the last step of the synthesis, i.e. of the racemic γ-amino acid. The demand of large volumes of pregabalin, due to outstanding commercial success, prompted further efforts toward improving its preparation. To switch from a resolution-based to potentially higher yielding enantioselective syntheses, catalytic hydrogenation [39] and ene-reductase [40] approaches were evaluated. However, ultimately, the linchpin to advance a more efficient process for pregabalin has been a lipolase-catalyzed kinetic resolution of intermediate **49**, prepared from alkylidene malonate **48** (Scheme 6.18) [41]. Remarkable assets of this manufacturing route – in all respects a text book example in green chemistry – are the streamlined and simple chemistry, the high concentration and the low enzyme loading of the lipolase reaction, the early-stage resolution strategy accompanied by easy recycling of the off-enantiomer (*R*)-**49**, use of water as the reaction medium in the last steps (*rac*-**49** →(*S*)-**49** → **50** → **51** → pregabalin), and very high throughput [42]. Overall, the efficiency of the process is reflected in a remarkably low *E*-factor of 17. For comparison, the *E*-factor of the first-generation route, based on the classical resolution, was 86.

The commercial success of pregabalin has prompted the search for alternative, non-infringing routes directed at its production for the generic market. Its relatively simple and nonfunctionalized structure points to resolution approaches as first choice. Indeed, classical and kinetic resolutions, mostly enzymatic, of pregabalin and its intermediates are widespread in the patent literature [43]. However, the simple γ-amino acid framework [44] suggests several simple retrosynthetic disconnections very well posited for catalytic enantioselective transformations (Scheme 6.19). For example, besides the reduction approaches explored by Pfizer and aforementioned, pregabalin can result from a nitrene-based rearrangement (i.e. Curtius or Hofmann) on an intermediate **52** derived from the desymmetrization of glutaric anhydride **53** (pathway A). Moving to C—C bond forming reactions, the γ-amino acid framework can be built by the addition of an amino

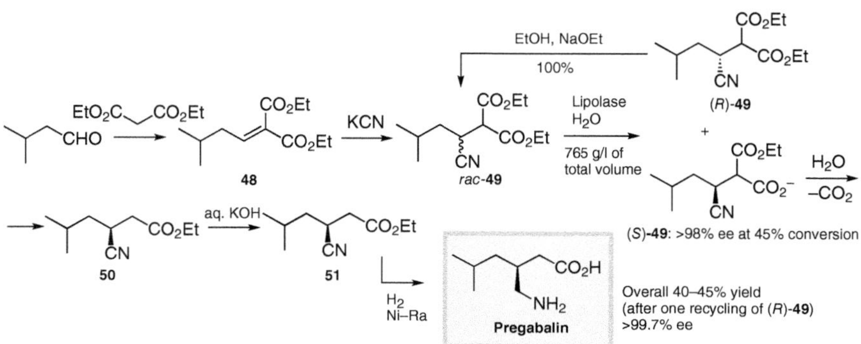

Scheme 6.18 Pfizer manufacturing route to pregabalin encompassing an early-stage lipolase kinetic resolution.

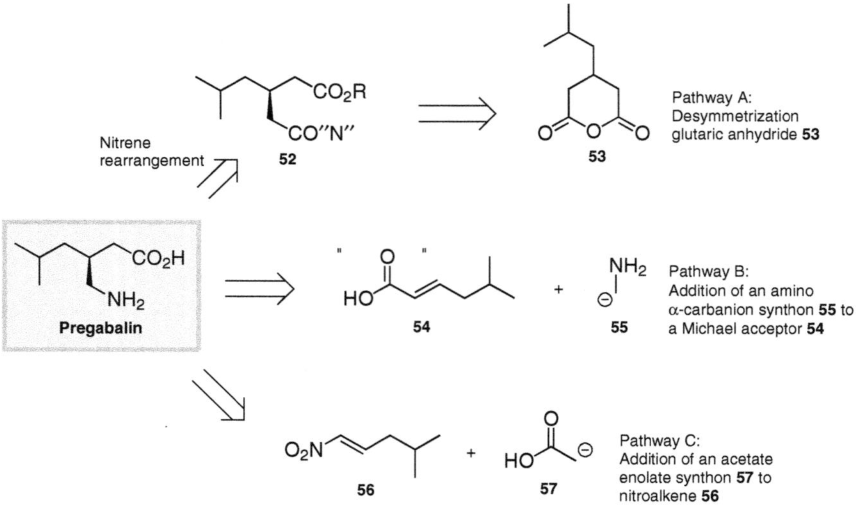

Scheme 6.19 Retrosynthetic disconnections of the γ-amino acid pregabalin.

α-carbanion synthon **55** to a Michael acceptor **54** (pathway B), or, in a complementary manner, by addition of an acetic acid α-anion synthon **57** to nitroalkene **56** (pathway C). In the next Sections 6.3.1.1–6.3.1.3, we will focus on some of the most relevant organocatalytic transformations that, following these strategies, introduced several appealing and, in some cases, scalable protocols. In fact, the preparation of enantioenriched pregabalin has been considered in several instances a platform to validate the synthetic utility of innovative enantioselective methodologies.

6.3.1.1 Pathway A: Desymmetrization of Glutaric Anhydride 53

A convincing demonstration of the glutaric anhydride pathway A for the synthesis of pregabalin was reported by Hameršak and coworkers in 2007 (Scheme 6.20) [45]. Bolm's protocol for the desymmetrization of anhydrides, based on stoichiometric amounts of quinine **58** as chiral promoter [46], was adapted targeting the desired compound. In more detail, anhydride **53**, prepared from cyanoacetamide and isovaleraldehyde, reacted with cinnamyl alcohol, delivering the ring-opened product **59** with moderate enantioselectivity (c. 70% ee). Crystallization with (*S*)-1-phenylethylamine improved the enantiomeric excess to 97%. Cinnamyl alcohol, instead of simpler alcohols, was chosen with the purpose of performing a catalytic deprotection later in the sequence. Conversion of the acid to an acyl azide followed by Curtius rearrangement and trapping of the isocyanate with benzyl alcohol afforded the bis-protected pregabalin **60**, which did not require purification. Two-step removal of the protecting groups, via **61**, led to the target γ-amino acid in an essentially enantiopure form. An alternative one-step bis-deprotection approach was investigated on a related derivative but gave less satisfactory results. The overall sequence was carried out on a multigram scale and avoided chromatographic purifications. Even if quinine **58** was used in stoichiometric amounts, it is readily available; it must be taken into account, however, the worldwide availability of quinine **58** vs. the quantity needed in a manufacture scenario. Furthermore, the acid functionality is unveiled in **61** before the free amine one, avoiding formation of

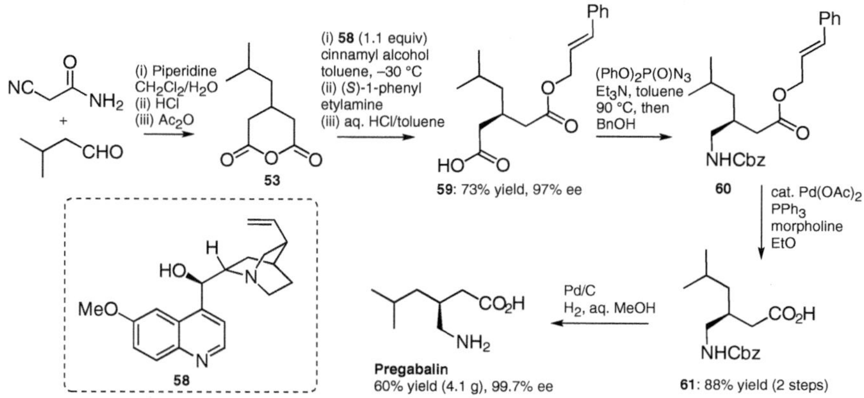

Scheme 6.20 Desymmetrization of glutaric anhydride **53** promoted by quinine **58** applied to the synthesis of pregabalin.

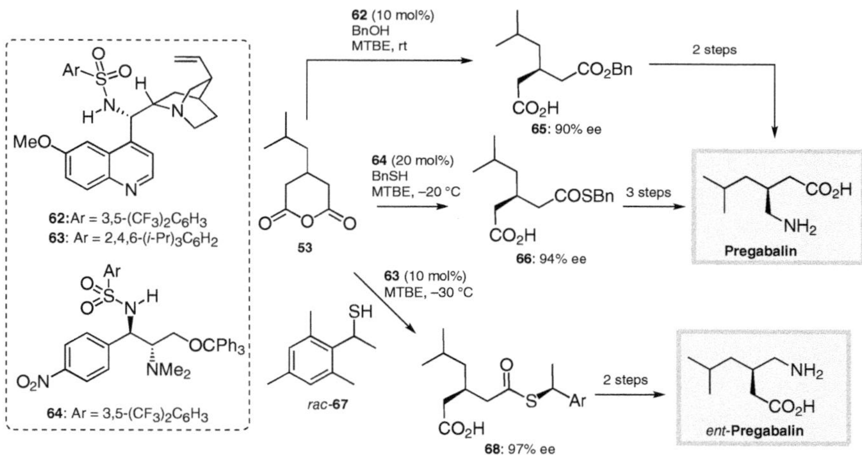

Scheme 6.21 Synthesis of pregabalin and *ent*-pregabalin by catalytic desymmetrization of glutaric anhydride **53**.

the corresponding γ-lactam. Thus, the overall approach appears well designed for further scale-up.

In general terms, the synthetic utility of the desymmetrization of anhydrides has ignited considerable interest, resulting in several highly efficient organocatalytic protocols. Some of these were applied to milligram-scale pregabalin syntheses (Scheme 6.21). For example, Song and coworkers described the alcoholytic desymmetrization of glutaric anhydrides catalyzed by the bifunctional *epi*-9-sulfonamido-9-deoxy quinine catalyst **62** [47]. Targeting pregabalin, the desymmetrized benzyl ester product **65** was obtained with good enantioselectivity and then converted to the γ-amino acid in two steps (Curtius rearrangement and trapping of the isocyanate with benzyl alcohol, double deprotection). An additional asset of this protocol is the subsequent development of the Nylon-supported version of catalyst **62**, enabling its recovery and reuse [48]. When applied to related desymmetrization reactions, such immobilization strategy has resulted in outstanding turnover numbers. The thiolytic desymmetrization of anhydride **53** with benzyl thiol was instead reported by Chen and coworkers with a structurally distinct bifunctional catalyst **64** [49]. Using a three-step sequence (Curtius rearrangement and trapping of the isocyanate with benzyl alcohol, hydrolysis of the thioester and deprotection), the catalytic product **66** was converted to pregabalin. Conversely, the kinetic resolution of chiral racemic thiols such as **67** developed by Connon and coworkers allowed to simultaneously desymmetrize the same glutaric anhydride **53**, using another *Cinchona*-derived sulfonamide catalyst **63** [50]. Ammonolysis of the thioester of **68** gave a primary amide, resulting in a formal synthesis of the antipode of pregabalin (via Hofmann rearrangement).

6.3.1.2 Pathway B: Addition of an Amino α-Carbanion 55 to Michael Acceptors

Nitromethane and cyanide are typical synthetic equivalents of the amino α-carbanion synthon **55**. They can undergo nucleophilic additions and then unveil the amino functionality upon reduction. However, conjugate additions to α,β-unsaturated esters and acids are often hampered by their low reactivity. Performing these reactions under the mild conditions usually required for stereocontrol mandates the use of surrogates conferring a more pronounced electrophilic character to the double bond [51].

On the other hand, the Pfizer route to the racemic pregabalin intermediate **49** for the key lipolase kinetic resolution exploited the addition of an amino α-carbanion equivalent – cyanide – to a masked α,β-unsaturated ester acceptor – the alkylidene malonate **48** (Scheme 6.18). Maruoka and coworkers reinvestigated the same reaction by using chiral phase transfer catalysts to render it enantioselective [52], thus potentially leading to a pregabalin synthesis mirroring Pfizer's route, but not requiring the resolution step. Optimal conditions for the enantioselective reaction entailed the use of the quaternary ammonium salt **69** as a catalyst, in a biphasic cyclopentane/water mixture, and using sodium or potassium cyanide as cyanide sources (Scheme 6.22). The use of protic additives, such as ammonium or hydrogen chloride, was found to be essential to achieve good reactivity. It was hypothesized that these additives serve to quench the enolate, resulting from cyanide addition, which is tightly coordinated to the cationic catalyst, thus assisting catalyst turnover. As shown in Scheme 6.22, bulky alkylidene malonates **70** and **71** gave better enantioselectivities than **48** in the reaction, which delivered, with the catalyst enantiomer shown, intermediates **49**, **72**, and **73** useful for the synthesis of *ent*-pregabalin.

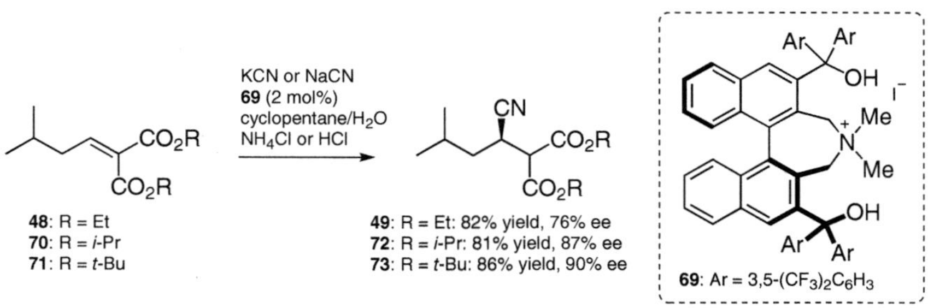

Scheme 6.22 Enantioselective conjugate addition of cyanide to alkylidene malonates **48**, **70**, **71** toward *ent*-pregabalin.

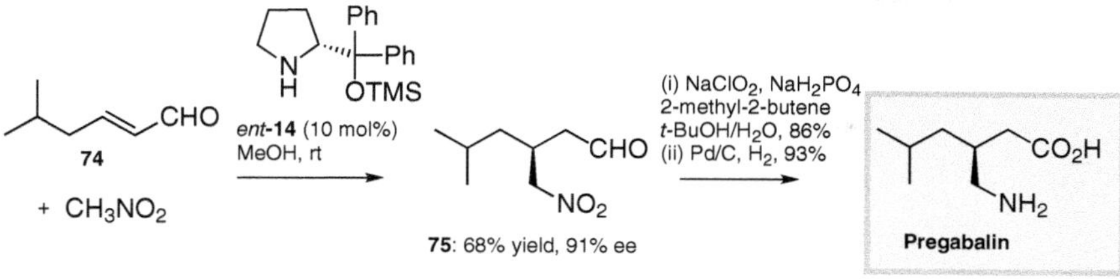

Scheme 6.23 Enantioselective addition of nitromethane to **74** catalyzed by *ent*-**14**.

Another α,β-unsaturated ester surrogate that has been employed in pregabalin synthesis has been 5-methylhex-2-enal **74**, in combination with nitromethane as the nucleophilic component under aminocatalytic conditions. The first example of this reaction was reported by Hayashi and coworkers [12] and entailed the use of secondary amine *ent*-**14** as catalyst, in MeOH without additives (Scheme 6.23). The catalytic product **75**, obtained in moderate yield and good enantiomeric excess, was converted to pregabalin in two steps (Pinnick oxidation and nitro group reduction). It is worth to note that the addition of nitromethane to enals via aminocatalysis has received considerable attention in manufacturing settings, as highlighted in Section 6.2. However, the specific reaction useful for pregabalin synthesis does not appear to have received equal consideration, with the exception of a recent patent describing the reaction on a gram scale, under exactly the same conditions reported in Scheme 6.23 [53]. Besides Hayashi's example, supposedly serving to demonstrate the synthetic usefulness of the developed methodology, only two additional works describe this reaction. In one case, the reaction was studied in the context of the disclosure of a new aminocatalyst [54]. In the second instance, an innovative merger of organocatalytic activation concepts and enzymatic catalysis, resulting in a modified enzyme able to catalyze reactions via iminium ion activation, was applied to this transformation [55]. Neither of these two studies seems to consider the reaction from a manufacturing perspective.

In contrast, the development of a new and convenient manufacturing route to pregabalin was the acknowledged aim of the investigation carried out by Moccia, Adamo, and coworkers with their brand α,β-unsaturated acid acceptor surrogates derived from 3,5-dimethyl-4-nitroisoxazole **78** [56]. The synthesis, reported in Scheme 6.24, was demonstrated on kilogram scale all through the sequence. The 3,5-dimethyl-4-nitroisoxazole **78** heterocycle was first assembled by standard chemistry (condensation of hydroxylamine with acetylacetone, giving **77**, followed by nitration). Intermediate **78** was reacted with isovaleraldehyde in a two-step sequence, resulting in the unsaturated acceptor **79** for the catalytic reaction. Phase transfer catalysis was used to achieve stereocontrol during the addition of nitromethane to **79**. The quinidine-derived quaternary ammonium salt **76** was able to induce moderately high levels of enantioselectivity (72% ee) in this reaction, even at just 3 mol% loadings. A relatively high 0.5 M concentration was preferred for practical reasons, even if enantioselectivity could be higher (up to 86% ee) at higher dilutions. Anyhow, enantioenrichment could be enhanced to >99% ee by crystallization of the product **80**, with an overall yield (catalytic reaction followed by crystallization) of 65%. The carboxylic acid function was unveiled by basic hydrolysis of the 4-nitroisoxazole ring, and the resulting sodium carboxylate **81** telescoped to the last hydrogenation step. Precipitation by pH adjustment gave pregabalin in 95% yield and >99% ee. Assets of this work, which

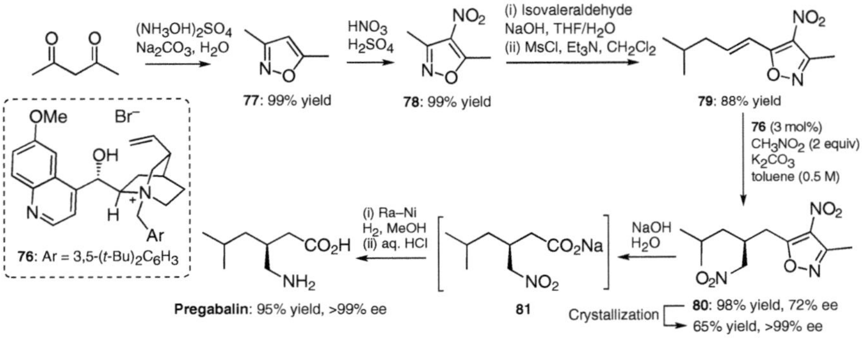

Scheme 6.24 Synthesis of pregabalin on kilogram scale via enantioselective conjugate addition of nitromethane to Michael acceptor **79** as an α,β-unsaturated acid surrogate.

might make it competitive with resolution approaches, are (i) the very good overall yield (54% over six steps); (ii) the cheapness of all carbon building blocks employed (acetylacetone, nitromethane, and isovaleraldehyde), and of the chirality source (quinidine); (iii) the possibility to recover and reuse the phase transfer catalyst **76**: following an approach pioneered by Itsuno and coworkers [57], catalyst **76** was immobilized by electrostatic interactions with a sulfonate polystyrene resin; the resulting catalyst could be recovered by filtration and reused for at least 10 runs with similar results.

6.3.1.3 Pathway C: Addition of Acetate Enolate Equivalents to Nitroalkene 56

Similar to α,β-unsaturated esters and acids, whose poor reactivity mandates the use of surrogates, catalytic asymmetric transformations requiring the acetate enolate synthon **57** are usually not performed by deprotonating directly the acetate because of its low acidity. Various synthetic equivalents, posited to catalytic processes, are available [58]. The desired acetate moiety can then be unveiled once the catalytic reaction has taken place. In the frame of organocatalytic synthesis of pregabalin, mainly three synthetic equivalents of the acetate enolate **57** have been used in combination with nitroalkene **56**: acetaldehyde, malonates, and Meldrum's acid.

Acetaldehyde suggests the use of aminocatalysis for its activation. However, efficient enamine-mediated reactions of acetaldehyde are much more challenging than reactions with longer chain aldehydes. Acetaldehyde and the resulting products are very reactive aldol acceptors and can readily give (self-)aldol additions/condensations. Furthermore, the product of an enamine-mediated reaction of acetaldehyde is an aldehyde carrying a methylene group, which still displays considerable reactivity toward an aminocatalyst. It is, thus, highly prone to further functionalization giving an α,α-disubstituted adduct. Nevertheless, these challenges could be overcome by a careful tuning of reaction conditions, and efficient aminocatalytic enantioselective reactions of acetaldehyde were ultimately developed [59]. Regarding conjugate additions to nitroalkenes, a handful of examples, which include in their scope nitroalkene **56**, resulting in the same pregabalin precursor **75** prepared via the addition of nitromethane to enal **74** (Scheme 6.23), are available [60]. Two closely related protocols were proposed nearly simultaneously by List, Hayashi, and their coworkers in 2008 (Scheme 6.25). In one case [60a], DMF was used as the reaction

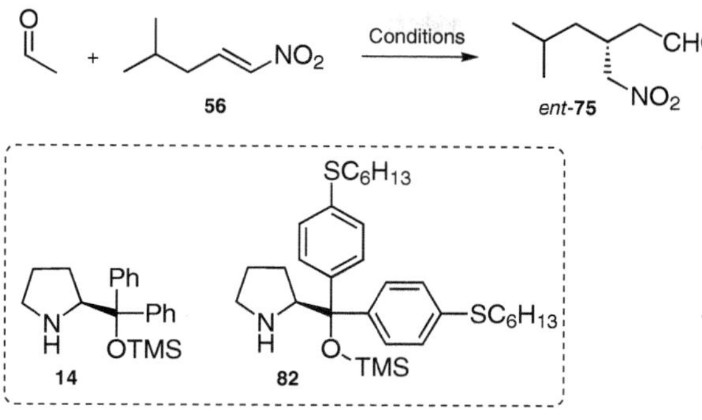

Reaction conditions	Results
14 (20 mol%) 10 equiv CH_3CHO DMF/*i*-PrOH, 4 °C	52% yield 94% ee
14 (10 mol%) 5 equiv CH_3CHO 1,4-dioxane, rt	61% yield >99% ee
82 (5 mol%) 2 equiv CH_3CHO BzOH (5 mol%) PEG-400, 4 °C	65% yield 98% ee

Scheme 6.25 Enantioselective additions of acetaldehyde to nitroalkene **56** under aminocatalytic conditions.

medium in the presence of small amounts of a protic solvent (*iso*-propanol). Acetaldehyde (5 equiv) was added slowly via a syringe pump to limit its self-aldol consumption. In the second example [60b], the reaction was performed in 1,4-dioxane, and acetaldehyde (10 equiv) was added all at once. Both protocols employed the same catalyst **14** at relatively high loadings and afforded compound *ent*-**75**, precursor of the pregabalin antipode, in moderate yields and very good enantioselectivities. Later, the same reaction was reported in the frame of a study demonstrating the activity of catalyst **82** in additions of aldehydes to nitroalkenes, performed in an environmentally friendly and recyclable reaction medium such as PEG-400 [60c]. In this medium, this tailored catalyst appeared more active than standard structures (like **14**), and even for the challenging reaction with acetaldehyde, the results were slightly better. Compared with the previous protocols, comparable outcomes were obtained even by using lower excess of acetaldehyde (2 instead of 5–10 equiv) and catalyst loading (5 instead of 10–20 mol%).

In analogy with the addition of nitromethane to enals [55], this transformation was also developed by means of an artificial enzyme, a "Michaelase," whose activity is based on organocatalytic concepts (enamine activation of the aldehyde) [60d]. Interestingly, this enzymatic process could be followed by an aldehyde dehydrogenase oxidative reaction, implemented with enzymatic cofactor recycling, and a subsequent nickel boride reduction of the nitro group of nitro acid **83** (Scheme 6.26). The three steps could be executed in a one-pot manner, affording the target pregabalin directly from acetaldehyde and nitroalkene **56** in an impressive 55% isolated yield and 96% ee.

In these disclosures, the synthesis of pregabalin, or of its intermediates, was carried out to demonstrate the synthetic usefulness of these highly innovative protocols, without discussing potential industrialization issues. In fact, the utilization of acetaldehyde on scale might be problematic because of poor stability, high volatility, difficult handling, and toxicity. In this context, Pericàs and coworkers reported an interesting approach to the use of acetaldehyde in aminocatalytic additions to nitroalkenes [61]. They exploited a catalyst compartmentalization strategy, enabling the coexistence in the reaction mixture of an aminocatalyst, promoting the enantioselective reaction, with a second, relatively highly acidic, cocatalyst (a sulfonic acid resin). The acidic catalyst generates acetaldehyde in situ by depolymerization of its much safer precursor paraldehyde, thus avoiding its direct

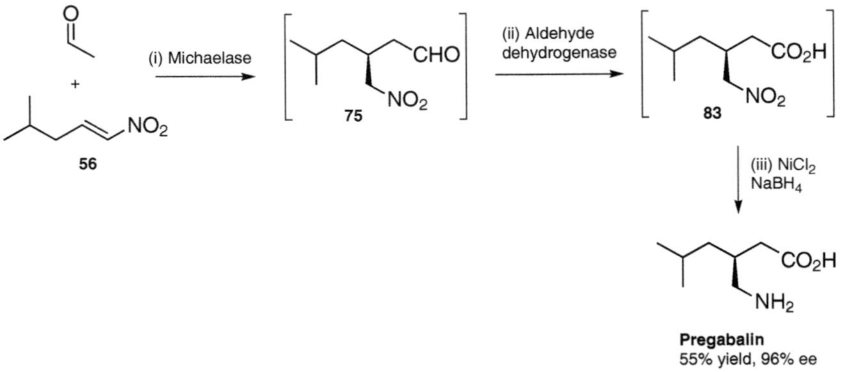

Scheme 6.26 One-pot synthesis of pregabalin by a "Michaelase" enantioselective enzyme.

manipulation. Another interesting strategy to generate acetaldehyde in situ, which was combined successfully with organocatalytic enantioselective cross-aldol reactions [62], is through the lipase-promoted hydrolysis of vinyl acetate. None of these strategies was however applied to the synthesis of pregabalin or of its intermediates.

Moving to another equivalent of the acetate enolate synthon **57**, the addition of malonates to nitroalkene **56** is without doubt the catalytic asymmetric approach to pregabalin that presents the largest number of examples, including scale-up studies. On the other hand, more generally, conjugate additions of malonates to nitroalkenes are among the most direct entries to β-substituted-γ-amino acids and derivatives thereof (γ-lactams). These transformations, in their catalytic asymmetric versions, have been the subject of countless investigations. Regarding industrialization, it is worth to first mention the chiral Lewis acid promoted reaction – state of the art at that time – of the related addition of β-keto esters to aromatic nitroalkenes, studied at Merck in the late 1990s toward the preparation of pyrrolidine drug clinical candidates [63]. In fact, chiral Lewis acids have also been used to approach pregabalin via addition of malonates to **56** [64]. However, the major impulse to the development of this reaction has been given by the use of bifunctional organic catalysts, i.e. structures bearing a tertiary amine flanked by hydrogen-bond donors, as also shown for the synthesis of the drug candidate MK-8613 mentioned in Section 6.2. Actually, some of the most efficient, and nowadays considered very general, catalysts of this class were disclosed by using the addition of malonates to nitroalkenes as a benchmark. At those times, such a reaction still called for innovative solutions. Representative examples are Takemoto's catalyst [65], des-methyl quinine and quinidine derivatives [16], (thio)ureas derived from 9-amino-9-deoxy *epi-Cinchona* alkaloids [21, 66], and their squaramide [67] counterparts.

The efficiency of these catalysts in imparting enantioselectivity to these and many other C—C bond forming reactions can be understood considering the simplified pathway sketched in Scheme 6.27 for Takemoto's catalyst **84**. This pathway involves first the soft-enolization of the malonate, resulting in the tri-coordinated intermediate **A**. Thereafter, addition to the nitroalkene occurs within the catalyst coordination sphere. Several transition states, featuring different coordination modes, are possible and have been validated computationally [68]. The two representative ones **B** and **C** reported in Scheme 6.27 involve either coordination of the nitroalkene to the thiourea moiety (**B**) or

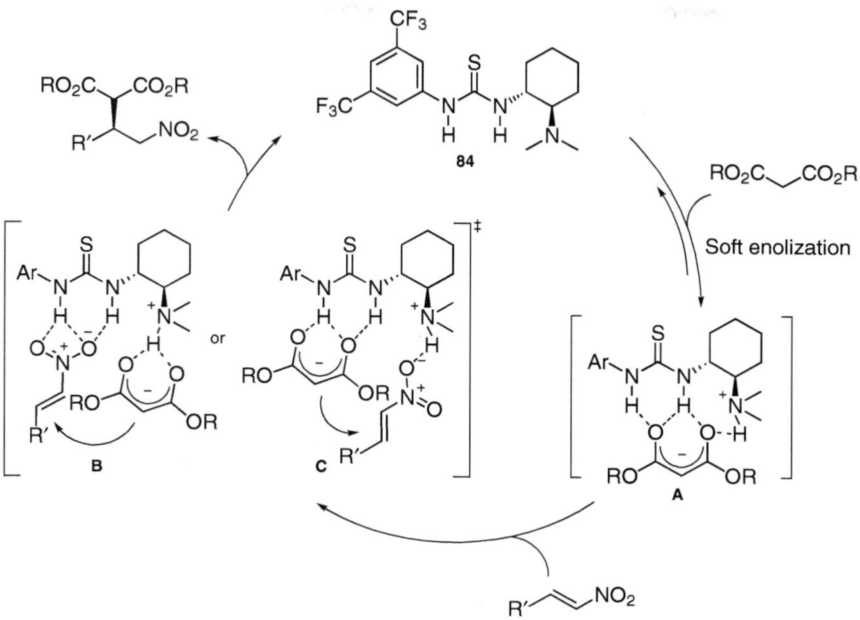

Scheme 6.27 Simplified reaction pathway, intermediates, and transition states in the addition of malonates to nitroalkenes catalyzed by Takemoto's catalyst **84**.

to the ammonium group (**C**). Ensuing proton transfer processes deliver the product while restoring the catalyst **84**.

While the relevance of these and related transition states is highly dependent on the reaction under study and on the specific catalyst structure, few general conclusions can be drawn from this simplified model pathway: (i) these catalysts are able to promote efficient soft-enolization of pronucleophiles, thanks to the presence of both hydrogen-donating and hydrogen-accepting groups; (ii) the stereodetermining C—C bond forming step occurs with the substrates coordinated to a triple hydrogen bond donor network, suggesting a highly ordered transition state. Such network has been defined as a "synthetic oxyanion hole," in analogy with the active sites of some enzymes.

This class of catalysts has been applied to the specific transformation required for pregabalin synthesis, i.e. the conjugate addition of malonates to nitroalkene **56**. A small selection of catalysts is reported in Scheme 6.28, which represents a small portion out of the very large number of examples available (>30). This selection has been done with an eye on the scalability of the proposed protocols, according to the results reported with these catalysts, displayed in Table 6.1. It is worth to mention that, compared to the reaction used to assemble MK-8613 (see Section 6.2), the addition of malonates to nitroalkene **56** brings additional challenges. Aliphatic nitroalkenes such as **56** are more difficult to prepare and purify (vide infra), can give rise to secondary reactions such as isomerization of the double bond, and are less electrophilic than aromatic ones. Their lower reactivity might call for higher catalyst loading. The amount of catalyst used might have a strong influence on the cost of the process, especially for a relatively simple molecule such as pregabalin.

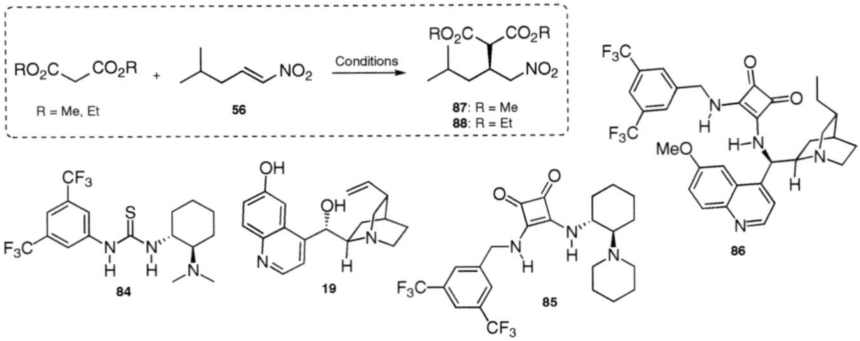

Scheme 6.28 Addition of malonates to nitroalkene **56**. Selected bifunctional catalysts.

Table 6.1 Addition of malonates to nitroalkene **56**.

Entry [references]	R	Catalyst (mol%)	Reaction conditions	Results
1 [69]	Et	**84** (10)	Neat, −20 °C, 24 h, >2 g scale	**88** : 73% Yield, 88% ee
2 [70]	Me	**19** (9)	MTBE, −10 °C, 40 h, >10 g scale	**87**: Full Conversion, 86% ee, 91% catalyst **19** recovery by aq. extraction
3 [71]	Me	**85** (5)	CH_2Cl_2, rt, 96 h, 1 mmol scale	**87** : 84% Yield, 88% ee
4 [72]	Me	**86** (5)	Brine, 0 °C, 24 h, >30 g scale	**87**: >95% Conversion, 91% ee, 99% catalyst **86** recovery by filtration
5 [73]	Me	**86** (0.5)	Neat, rt, 24 h, >2 kg scale	**87** : 90% Conversion, 91% ee

Selected results with catalysts **19** and **84–86**.

The first entry of Table 6.1 reports the result of an elaboration of the original Takemoto's protocol [65]. This more recent process, developed by Li and coworkers [69], runs under neat conditions at −20 °C, leading to the product **88** in moderate yield and good enantioselectivity (88% ee). The reaction, performed on a multigram scale, gave, in fact, better results compared to the previous protocol (toluene, rt, 81% ee), although the same rather large catalyst **84** loading (10 mol%) is required. The results of a reaction performed with des-methyl quinidine catalyst **19** are instead reported in entry 2 [70]. The reaction conditions and the catalyst are derived from the previous disclosure by Deng and coworkers [16] and are related to MK-8613 synthesis (see Section 6.2). It is worth to note the multigram scale of the reaction, as well as the possibility to recover >90% of the catalyst **19** by a simple aqueous extraction, after the catalytic process.

Furthermore, the main purpose of this disclosure by Siegfried Ltd. is to improve the downstream chemistry from **87** to pregabalin. A typical approach to convert **87** to pregabalin involves reduction of the nitro group, with concomitant lactamization to **89**. Hydrolysis and decarboxylation, followed by neutralization, give pregabalin (Scheme 6.29, top). Even though carrying a nitro group until the latest stages of a synthesis is not ideal because it

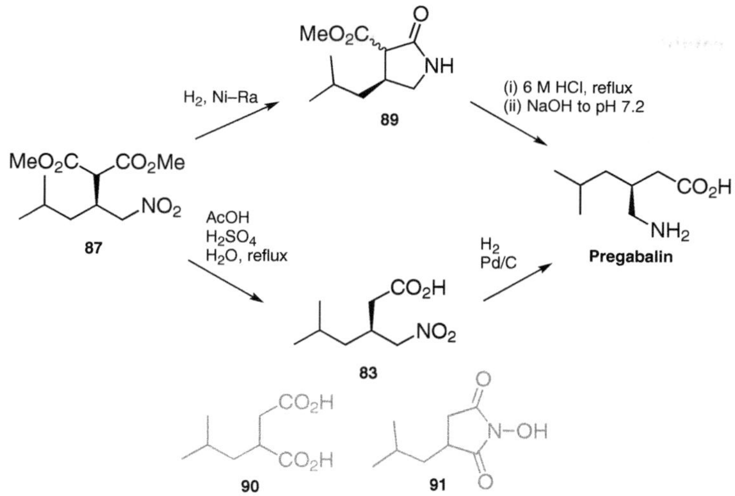

Scheme 6.29 Downstream routes to pregabalin from catalytic product **87**.

might give rise to concerns about the presence of potential genotoxic impurities (PGIs) in the final product, synthetically speaking it might be more convenient to postpone the reduction of the nitro group after decarboxylation, avoiding the formation of the γ-lactam **89** and the requirement of the neutralization step. However, acidic treatment of **87**, to effect decarboxylation, may lead to the formation of impurities **90** and **91** next to the desired nitroacid **83**. Researchers at Siegfried Ltd. found that the treatment of **87** under specific conditions (AcOH and H_2SO_4) could promote decarboxylation while keeping impurities **90** and **91** at minimum levels (Scheme 6.29, bottom). Catalytic hydrogenation gave access to pregabalin in just two steps from **87**.

Moving back to the catalytic reaction (Table 6.1), a general analysis of the literature reveals that the most successful organic catalysts for these reactions embed squaramides as hydrogen bond donor motives. Out of the plethora of examples available, a reaction developed by Šebesta and coworkers is reported in entry 3 [71]. In this work, catalyst **85** was identified as a competent structure for the reaction between dimethyl malonate and **56** based on the screening of few other squaramide-based catalysts. With catalyst **85**, product **87** was obtained with good results (88% ee), even if catalyst turnover numbers and frequencies are not exceptional (5 mol%, 96 hours), at least under the reported conditions.

The 9-amino-9-deoxy-*epi*-dihydroquinidine derivative **86** is another squaramide-based bifunctional catalyst that has been employed with great success in this reaction. Song and coworker have reported an application of this catalyst to the reaction between dimethyl malonate and **56** using brine as the bulk reaction medium (entry 4) [72]. Curiously, this study established that, in these reactions performed in brine, dihydro-*Cinchona* catalysts such as **86** are more active than their unsaturated counterparts (i.e. with a vinyl instead of ethyl group at the quinuclidine). It was hypothesized that the higher lipophilicity of dihydro-*Cinchona* structures could assist the on-water catalytic process via hydrophobic effects. The reaction appears indeed efficient, delivering the product **87** with full conversion and good enantioselectivity (91% ee), although catalyst loading is not low (5 mol%).

This multigram-scale reaction included a remarkable implementation to compensate the high loading: after the catalysis, addition of methylcyclohexane allowed to recover quantitatively the valuable catalyst **86** by filtration. Crude **87** obtained by evaporation of the organic phase was subjected to the three-step protocol summarized on the top of Scheme 6.29 to convert it to pregabalin. Although a chromatographic purification was performed on the intermediate lactam **89**, the proposed protocol afforded pregabalin (91% ee) in an overall 73% yield from nitroalkene **56**; enantiopurity of the pregabalin product could be upgraded by crystallization.

From a manufacturing perspective, utilization of the rather expensive 3,5-(trifluoromethyl)benzyl amino appendage on catalyst **86** is not desirable because of its cost. However, this moiety was found to be necessary from a large screening of different squaramide structures in an investigation performed at Dr. Reddy's laboratories [73]. Nevertheless, the catalyst loading could be lowered considerably, down to 0.5 mol%, by running the reaction under neat conditions (entry 5). The reaction was demonstrated on a >2 kg scale, highlighting the potential of this methodology in a scale-up perspective. It is also interesting to note that the dihydro catalyst **86** was also preferred in this case, but because of its easier preparation on scale and slightly higher activity, compared to its unsaturated counterpart. Furthermore, it was noted that mixing catalyst **86** with nitroalkene **56** in the absence of dimethyl malonate had to be avoided, as it resulted in a dramatic catalyst deactivation/degradation. An additional point raised by this investigation was the synthesis and availability of nitroalkene **56**, a neglected aspect in most other contributions. Typical preparation of **56** involves a two-step synthesis via dehydration of the Henry adduct **92**, followed by chromatographic purification (Scheme 6.30, top). Crystallization is not viable, being **56** a liquid; distillation must be avoided because of poor thermal stability. Implementing a low-yielding batch precedent [74], a flow set up was developed (Scheme 6.30, bottom), which furnished **56** of sufficient quality for the catalytic reaction, without requiring purification. Apparently, the fast mixing typical of a flow process allowed to control the selectivity as desired during the acidic treatment of

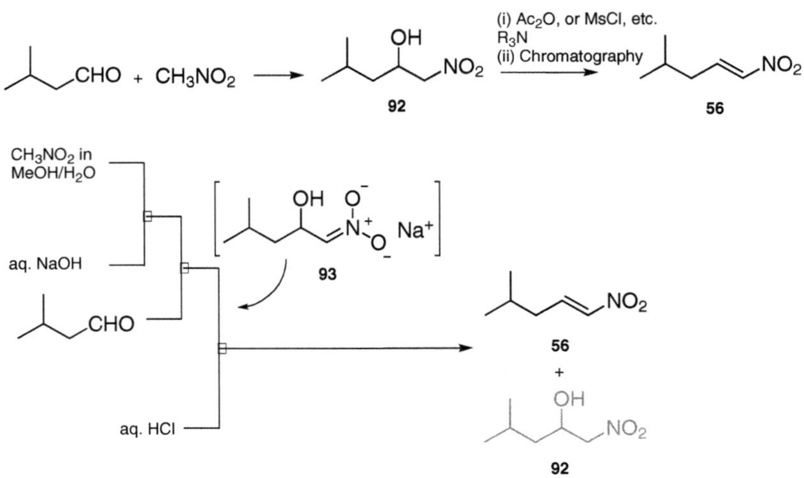

Scheme 6.30 Typical preparation of nitroalkene **56** vs. flow chemistry approach.

intermediate **93**, by ensuring a highly acidic local environment at all stages, thus leading to **56** with good selectivity vs. Henry adduct **92**. Furthermore, the flow set up addressed safety issues related to unstable intermediate species, such as nitronate salts.

Alternatively, the issue of nitroalkene **56** preparation was tackled by Corma and coworkers in a very elegant manner [75]. They proposed a multisite organic/inorganic hybrid catalyst **94**, able to promote both nitroalkene generation, possibly through a nitro amine intermediate, instead of a nitroalcohol, and subsequent enantioselective addition of dimethyl malonate (Scheme 6.31). Unfortunately, results of the application of this approach to pregabalin precursor **87**, which has a precedent in some work by McQuade and coworkers based on chiral Lewis acids [76], are not satisfactory.

The last transformation highlighted in this subsection is the addition of Meldrum's acid to nitroalkene **56**. Two versions of this reaction, which is closely related to the addition of malonates, have been reported (Scheme 6.32). Koskinen and coworkers applied thiourea catalyst **95**, obtaining the product **97** with moderate results [77]. Šebesta and coworkers found instead catalyst **96** as a more efficient structure for the same transformation, providing product **97** in highly enantioenriched form (94% ee) [71]. This adduct **97** can be converted to pregabalin hydrochloride analogously to the products **87** and **88** resulting from the addition malonates.

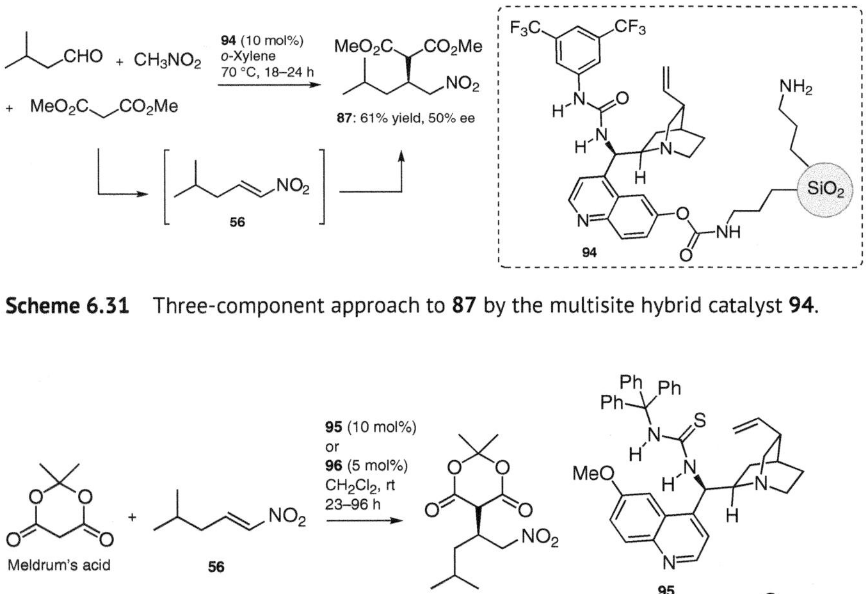

Scheme 6.31 Three-component approach to **87** by the multisite hybrid catalyst **94**.

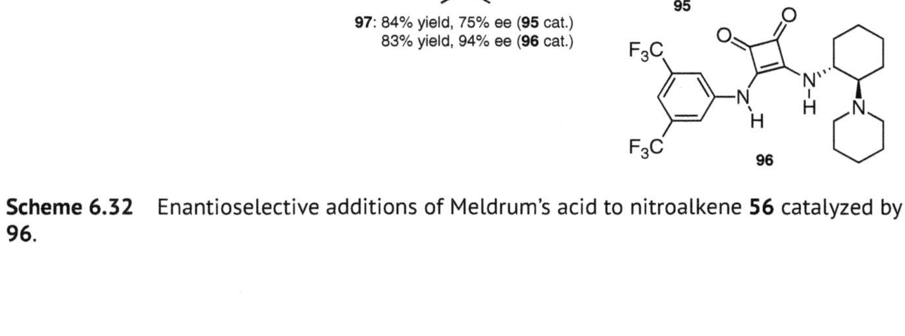

Scheme 6.32 Enantioselective additions of Meldrum's acid to nitroalkene **56** catalyzed by **95** and **96**.

Thus, pregabalin has been successfully accessed through a large number of organocatalytic methods. In most cases, these innovative methodologies made the most of pregabalin notoriety to show their synthetic utility. However, industry-oriented protocols have been developed too. Several gram-scale and at least two kilogram-scale processes have been disclosed.

6.3.2 Bicyclic α-Amino Acid Core of Telaprevir

Telaprevir [78], an antiviral polyamide developed jointly by Vertex Pharmaceuticals and Johnson & Johnson, has been a key ingredient for the treatment of patients affected by HCV, before the rise of Gilead's pro-Tide anti-HCV agent sofosbuvir. The logical synthetic assembly of telaprevir combines the single amino acid components, among which the peculiar central unit **98** stands out as being synthetically challenging (Scheme 6.33). It presents a bicyclic structure with defined absolute and relative configuration in three contiguous chirality centers. A manufacture route to this amino acid, developed at Vertex Pharmaceuticals, involves exo-diastereoselective introduction of the carboxylic group by cryogenic lithiation and carboxylation of *N*-Boc protected 3-azabicyclo[3.3.0]octane, via **100**, followed by classical resolution of *rac*-**99** by diastereomeric salt [79]. An alternative appealing approach relies on enzymatic desymmetrization of the same starting material **101** [80, 81]. A monoamine oxidase N (MAO-N) enzyme from *Aspergillus Niger*, optimized by directed evolution, affords the Δ^1-pyrroline **102** in a highly enantioenriched form. Then, the carboxylic function is installed via a challenging exo-diastereoselective cyanide addition and hydrolysis.

Although ultimately unfulfilled, telaprevir did hold great promise as a blockbuster drug; it was projected to be a large volume product for the generic market, prompting a variety of studies on the synthesis of its most challenging part, the bicyclic amino acid core **98**, and on its straightforward assembly in the polyamide sequence. Classical resolutions of the amino

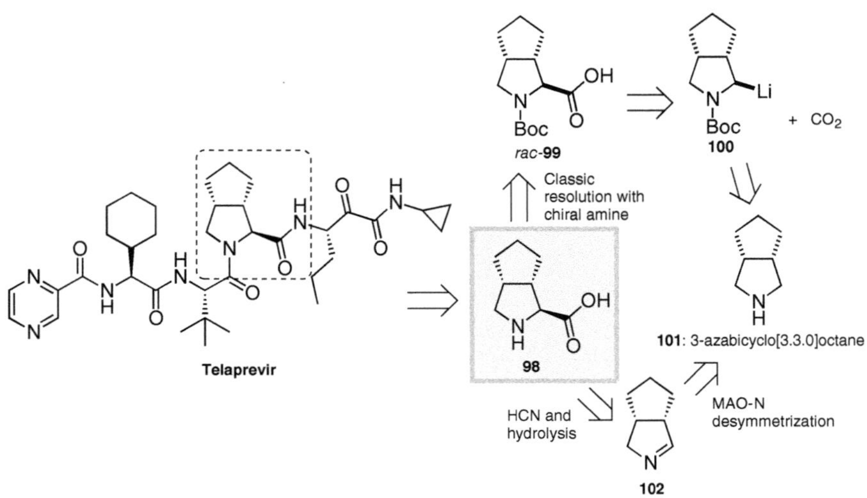

Scheme 6.33 Telaprevir and synthetic approaches to its bicyclic amino acid core **98**.

ester encompassing recycling of the undesired enantiomer [82], improvements in cyanide addition step by pyrroline **102** trapping as bisulfite adduct [81, 82], cyanide replacement with nitromethane [83], enzymatic desymmetrization of a mono-cyclic diol intermediate [84], and Ugi chemistry for the insertion of the enantioenriched Δ^1-pyrroline **102** in the polyamide sequence [85] are all relevant references.

In addition to these approaches based on enantiomers separation and biocatalysis, Xu, Sun, and coworkers applied Bolm's desymmetrization of *meso*-anhydrides mediated by natural *Cinchona* alkaloids [46] to the synthesis of telaprevir core in its *N*-benzyl protected form (**111**) [86]. As shown in Scheme 6.34, methanolysis of *meso*-anhydride **104** in the presence of quinidine **103** (1.1 equiv) affords the corresponding acid ester **105** in 98.7% ee. Amidation with *N*,*N*-dibenzylamine (**105** → **106**), followed by borane reduction of both amide and ester, leads to the amino alcohol **107**, which, subjected to additional manipulations (partial debenzylation and Boc protection (**107** → **108**), PCC oxidation (**108** → **109**), Boc deprotection, iminium ion formation, and addition of TMSCN), leads to the formation of compound **110** featuring the full carbon skeleton of the target. Upon acidic hydrolysis, the equilibrating mixture of exo and endo amino nitriles **110** gives nearly exclusively the amino acid **111** with the desired exo relative configuration, presumably via a dynamic resolution process. Thereafter, this multistep protocol, performed on a gram scale, was also applied successfully to several 3,4-cis-disubstituted proline analogs.

From a retrosynthetic point of view, this and the aforementioned protocols employ meso-starting materials, such as bicyclic amine **101**, anhydride **104**, or diol **112**, featuring the carbon framework of the bicyclic amino acid **98**, except the carboxylic function, which has to be installed, via, e.g. cyanide chemistry, on the "de-symmetrized" Δ^1-pyrroline**102** (Scheme 6.35). Alternative strategies, avoiding the use of possibly nonreadily available bicyclic meso-structures, can also be envisioned. As shown in Scheme 6.35, the "de-symmetrized" pyrroline **102** can result from the cyclization of compound **113**, derived from the conjugate addition of the α-amino carbanion synthon **55** to readily available cyclopent-1-enecarbaldehyde **114**. In a more convergent manner, the target compound can be assembled by adding a glycine α-anion synthon **116** to the same aldehyde **114**,

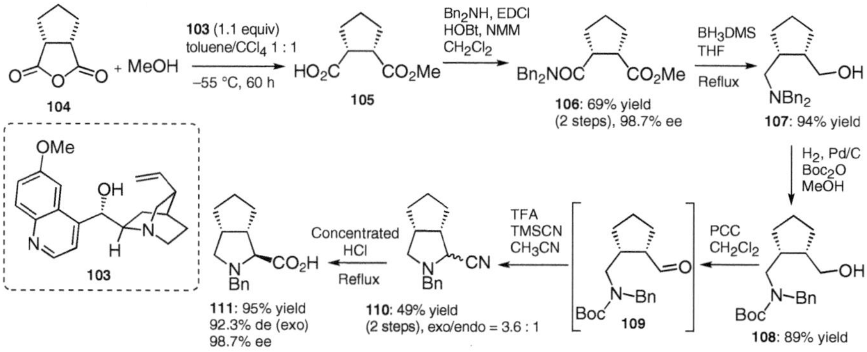

Scheme 6.34 Synthesis of *N*-benzyl telaprevir core **111** based on quinidine **103**-mediated desymmetrization of *meso*-anhydride **104**.
EDCI = *N*-(3-dimethylaminopropyl)-*N*′-ethylcarbodiimide hydrochloride;
HOBt = 1-hydroxybenzotriazole; NMM = *N*-methylmorpholine; PCC = pyridinium chlorochromate.

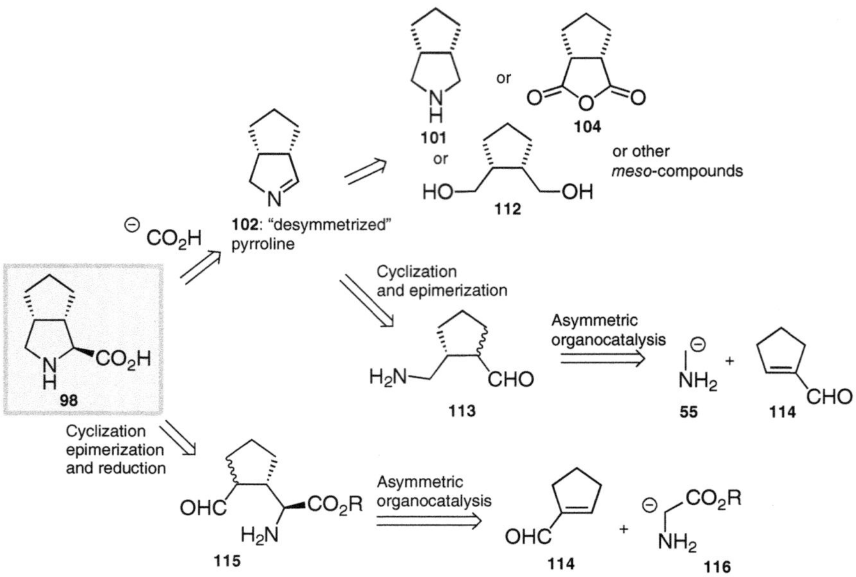

Scheme 6.35 Retrosynthetic approaches to the telaprevir core **98**: desymmetrization of *meso*-compounds **101, 104, 112** vs. cyclopent-1-enecarbaldehyde **114**-based routes.

followed by reductive cyclization of **115**. Assumptions behind these approaches are that (i) regioselective 1,4- vs. 1,2-addition to enal **114** is possible and that (ii) cyclization events on **113** and **115** forming the fused bicyclic [3.3.0] pyrroline can force the desired cis-configuration at the fusion between the two rings, by epimerization at the α position of the aldehyde function.

In this Section 6.3.2, we will summarize how these less conventional, but appealing, approaches to the telaprevir core were realized with asymmetric organocatalysis. However, it is important to recognize that the application of one of these strategies – the addition of a glycine α-anion equivalent to **114** – was first reported at Ajinomoto Co. by using Belokon Ni(II) Schiff bases complexes **117** and **118** as chiral masked glycine equivalents (Scheme 6.36) [87]. The addition products **119** and **120** can be converted to **98** through a two-step sequence via an intermediate imine.

Asymmetric catalysis in such type of transformation was instead first described by Maruoka and coworkers using glycinate imines as glycine α-anion equivalents [88]. Interestingly, this study was not restricted to cyclopentene-1-carbaldehyde **114**. The phase-transfer-catalyzed addition of benzophenone *tert*-butyl glycinate imine **122** to enals was shown with several β-substituted α,β-unsaturated aldehydes **123** (Scheme 6.37). The regioselectivities (1,4- vs. 1,2-addition) in these reactions appear to be very good; the crude adducts **124** can be converted to highly valuable three-substituted pyrrolidine derivatives (**126**) via a transimination reaction promoted by simple silica gel purification giving pyrrolines **125**, followed by hydrogenation and additional chromatography. By using one of their brand binaphthyl-derived second-generation quaternary ammonium salts (**121**) as a catalyst, the desired pyrrolidines (**126**) were obtained in good yields (over three steps)

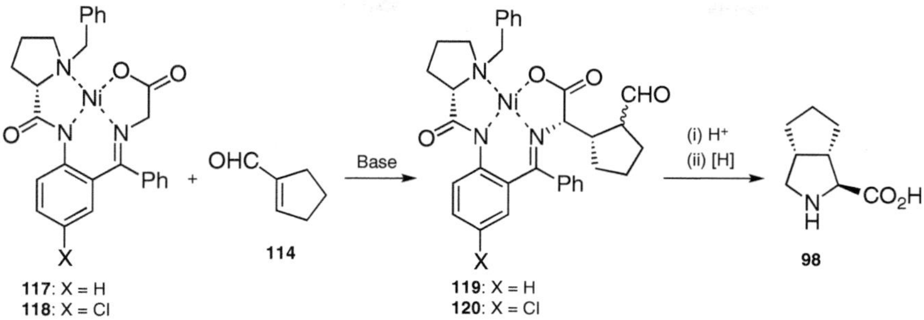

Scheme 6.36 Enantioselective approach to telaprevir core **98** by chiral masked glycine equivalents.

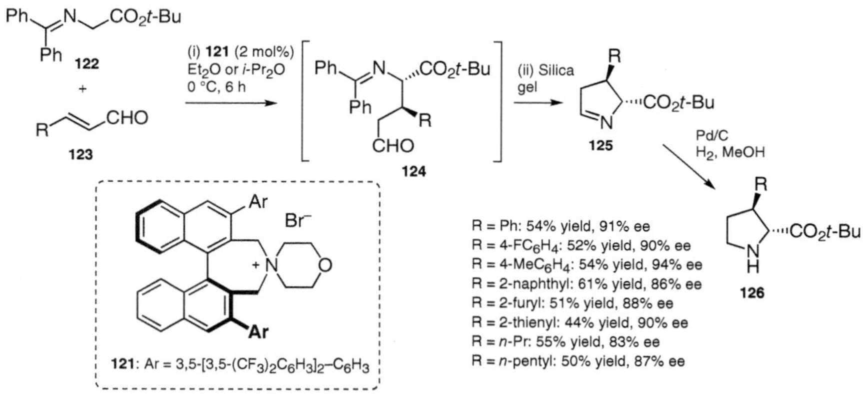

Scheme 6.37 PTC in enantioselective conjugate additions of benzophenone glycinate imine **122** to β-substituted enals **123**.

and enantioselectivities. The perfect trans-stereochemistry suggests very good anti simple diastereoselectivity in the catalytic step giving adducts **124**.

Curiously, the reaction of **122** with cyclopentene-1-carbaldehyde **114** targeting telaprevir core required a different catalyst (**127**) to achieve good enantioselectivity (Scheme 6.38), with previously employed catalyst **121** providing only 6% ee with this substrate. Catalyst **127**, which bears the same extended aromatic groups ($3,5\text{-}[3,5\text{-}(CF_3)_2C_6H_3]_2C_6H_3$) at 3,3′ positions of **121**, but entails a second unsubstituted binaphthyl unit ("first generation" catalyst), promotes the conjugate addition smoothly and with high enantioselectivity. Interestingly, the catalytic reaction gives good (13 : 1) cis-diastereoselectivity at the cyclopentane ring of adduct **128**: the catalytic reaction defines all three chiral centers featuring the configuration of the target bicyclic compound **130**. Thus, epimerization is not required during the cyclization to afford the desired cis-fused bicyclic system of pyrroline **129**. The overall sequence was demonstrated on gram scale, leading to the target bicyclic amino ester **130** in 53% yield (three steps) and 93% ee.

In a joint work by the University of Bologna and Dr. Reddy's Laboratories [89], the PTC reaction between glycinate imines, such as **122**, and cyclopent-1-enecarbaldehyde **114** was reinvestigated, with the goal of employing a different class of catalysts, namely, quaternary

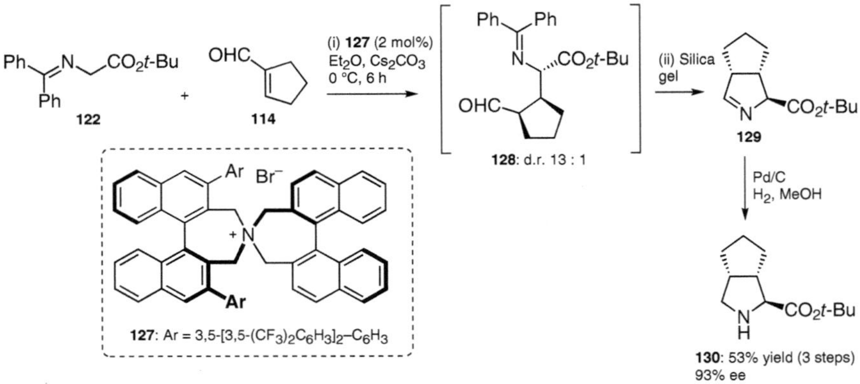

Scheme 6.38 PTC addition of benzophenone glycinate imine **122** with binaphthyl catalyst **127**, applied to the telaprevir bicyclic core.

ammonium salts derived from *Cinchona* alkaloids. In fact, although catalyst **127** shows excellent performances at relatively low loadings (2 mol%), and is at least partially (80%) recoverable after the process, its availability and accessibility, accompanied by the need of freedom to operate, prompted investigation for alternative structures. A large screening of catalysts and reaction parameters (inorganic base, solvent, and temperature) did identify conditions enabling to perform the target reaction with moderately good results even with *Cinchona* alkaloid-derived catalysts (Scheme 6.39). More in detail, reaction of the typical *tert*-butyl glycinate benzophenone imine **122** with cyclopent-1-enecarbaldehyde **114**, in the presence of just 1 mol% Corey–Lygo cinchonidine catalyst **131**, with an excess of the inexpensive aqueous KOH as a base and dichloromethane as a solvent at −20 °C, gives a 3 : 1 cis/trans mixture of adducts **128**; conversion of the cis isomer to the bicyclic structure **129** was performed by modifying Maruoka's conditions, i.e. by employing aqueous citric acid, instead of silica gel treatment, for the transimination step. Ultimately, after reduction of **129**, the desired product **130** was obtained in 60% overall yield (three steps) and 81% enantiomeric excess. Interestingly, catalyst **132** bearing the free 9-hydroxy function gave comparable enantioinduction (75% ee) but induced a very rapid epimerization at the stereogenic center in α-position to the aldehyde function, tending to lead to the prevalence of the more stable trans-isomer of **128** (cis/trans ratio after 0.5, 1, and 2.5 hours: 1 : 1, 0.5 : 1, and 1 : 9, respectively). Such epimerization is a much slower process with 9-hydroxyl-alkylated catalysts such as optimal **131**.

Substrates possibly cheaper than the standard *tert*-butyl glycinate benzophenone imine **122** were also tested in the reaction (Scheme 6.40). The less-hindered *iso*-propyl ester substrate **133** gave the corresponding product **135**, via **134**, with lower enantioselectivity (72% vs. 81% ee), in line with Maruoka's results [88c]. An aldimine, instead of ketimine, substrate (**136**) also gave lower enantioselectivity (58% ee) and furnished the trans-adduct **137**. Interestingly, the addition reaction appears to provide the product with simple anti diastereoselectivity required for the target amino acid. This opens possibilities for applying cheaper aldimine substrates [90] to this reaction, if conditions for simultaneous cyclization and epimerization are developed.

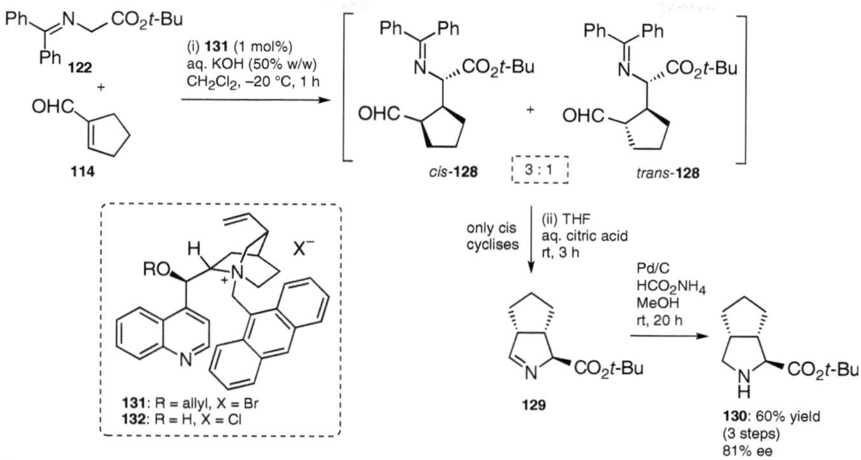

Scheme 6.39 PTC addition of benzophenone imine glycinate **122** to **114** catalyzed by *Cinchona* alkaloid derivative **131**.

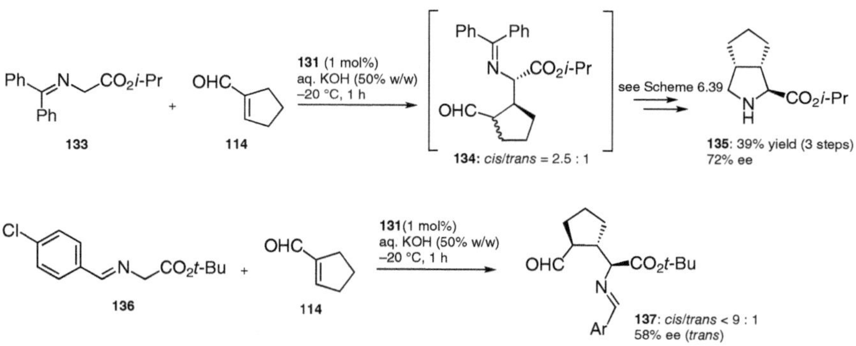

Scheme 6.40 Alternative glycinate imine substrates **133** and **136** applied to the conjugate addition to **114**.

These studies were part of a wide-range exploration of several α-amino carbanion (nitromethane and cyanide) and glycine α-anion (2-amidomalonates, nitroacetates, isocyanoacetates, azlactones, and glycinate imines) equivalents in the reaction with cyclopent-1-enecarbaldehyde **114**, under organocatalytic conditions, along the lines of the retrosynthetic logic reported in Scheme 6.35 [89]. In this context, two additional synthetic equivalents of the glycine α-anion synthon **116** – 2-amidomalonates and nitroacetates – gave promising results. These reactions, which proceeded with most success under amino-catalytic conditions, that is by activating the enal through an iminium ion intermediate, were extensively optimized, with a special focus on reducing the catalyst loadings.

Considering the reaction with 2-amidomalonates, the use of additives was essential in achieving an efficient protocol. Using 5 mol% of Jørgensen–Hayashi catalyst *ent*-**14**, in EtOH as a solvent and with a mixture of AcOH and LiOAc (5 mol% each) as additives, the trans-product **140** of the reaction between 2-acetamidomalonate **138** and

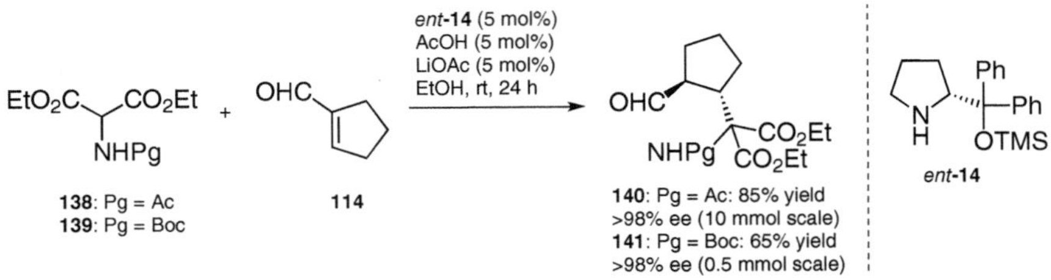

Scheme 6.41 Aminocatalytic additions of 2-amidomalonates **138** and **139** to cyclopent-1-enecarbaldehyde **114**, en route to telaprevir core.

cyclopent-1-enecarbaldehyde **114** was obtained in 85% yield and in essentially enantiopure form, even on 10 mmol scale (Scheme 6.41). It was remarkable that an increase of 100-fold in scale, from 0.1 mmol of the screening conditions to the preparative 10 mmol results, did not have any influence on the results obtained, suggesting that further scale-up would be straightforward. A variation in the amido portion, using *N*-Boc 2-aminomalonate **139** instead of **138**, decreases the reactivity but does not affect the stereoselectivity of the process. Unfortunately, the downstream chemistry to convert the catalytic products **140** and **141** to the target bicyclic structure **98** (reductive cyclization and epimerization, decarboxylation, and equilibration to more stable exo-adduct) was proposed but not investigated. It is instructive to compare the efficiency of this optimized protocol with published additions of 2-acetamidomalonates to enals [91]. The latter reactions use 20 mol% loading of catalyst **14** and long reaction times (144 hours), at low temperatures, to achieve satisfactory enantioselectivities. This comparison highlights the importance of the careful choice of additives and reaction conditions in aminocatalytic processes.

The other glycine α-anion equivalent studied, which also performed best with aminocatalysis as activation strategy, was ethyl nitroacetate **142**. Considering the different acidity of this pronucleophile compared to 2-amidomalonates, it is not surprising that drastically different reaction conditions were required, in accordance with literature precedents [92]. As shown in Scheme 6.42, optimal conditions for the reaction between cyclopent-1-enecarbaldehyde **114** and ethyl nitroacetate **142** entail the use of toluene as a solvent, with an acidic additive (benzoic acid). Under these conditions, the loading of catalyst **14** (the enantiomer of the catalyst employed with 2-amidomalonates) could be lowered down to 0.5 mol%, without compromising neither on reactivity nor on enantioselectivity, determined after Wittig olefination of adduct **143** to obtain the more easily analyzable product **144**. On the other hand, a previous optimization of the reaction of nitroacetates with crotonaldehyde, directed at the synthesis of a natural product [92b], allowed a similarly low loading (of a closely related catalyst). The catalytic product **143** was obtained in good yield, as a trans-isomer at the cyclopentane ring, and as an equilibrating diastereomeric mixture at the α-nitro chiral center. In this case as well, the downstream conversion of **143** to the telaprevir bicyclic core was not demonstrated.

Among the two α-amino carbanion equivalents tested – nitromethane and cyanide – only nitromethane reacted with cyclopent-1-enecarbaldehyde **114** in the desired manner, once again under aminocatalytic conditions. On the other hand, the conjugate addition of nitromethane to enals is known to proceed very well using Jørgensen–Hayashi-type

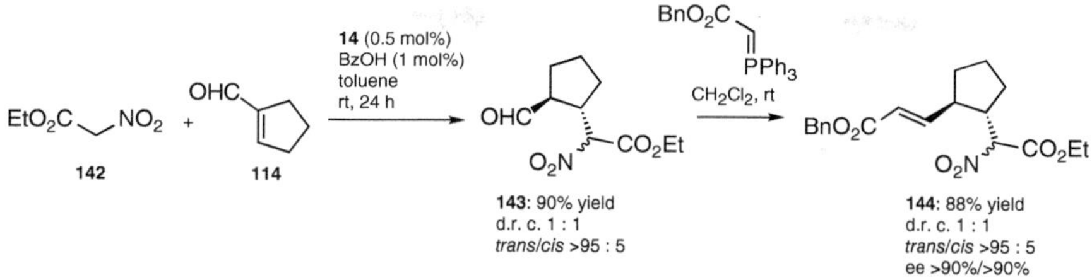

Scheme 6.42 Aminocatalytic addition of ethyl nitroacetate **142** to cyclopent-1-enecarbaldehyde **114**.

catalysts. As discussed in Sections 6.2 and 6.3.1.2, this reaction, first developed by several research groups around 2007 [12, 93], has been the subject of intense research, being a convenient entry to valuable enantioenriched γ-amino acids. It has been even implemented on a large scale to manufacture the pharmaceutical candidate telcagepant (see Section 6.2) [9]. This amino-catalytic reaction is usually performed in protic solvents. Acidic (e.g. benzoic or acetic acid) or basic (e.g. lithium acetate) cocatalysts are also required for optimal results. Curiously, the very same reaction targeted for telaprevir core synthesis – the addition of nitromethane to cyclopent-1-enecarbaldehyde **114** – had been already reported by Gellman et al. but was used for the synthesis of a monocyclic cyclopentane-based γ-amino acid [94]. The reaction proceeds fast and very efficiently, using a relatively large (20 mol%) catalyst **14** loading, in ethanol as a solvent (Scheme 6.43, top). A "buffered" cocatalyst system of benzoic acid and 2,4,6-collidine, in a 3 : 1 ratio, was found to improve reaction rate, compared to the use of acids alone. The product **145**, more conveniently isolated as alcohol **146** after reduction of the aldehyde function, is obtained in nearly quantitative yield and as a single stereoisomer.

Aiming at devising a new protocol for the preparation of the bicyclic core of telaprevir, possibly amenable to industrial synthesis, attempts were made to lower the less-desirable high catalyst loading of Gellman's procedure [89]. Keeping the same "buffering" strategy, a screening of different cocatalyst combinations was performed. An equimolar mixture of acetic acid and lithium acetate provided the cleanest reaction profile and enabled to lower the catalyst **14** loading down to 1 mol% without compromising the outstanding enantioselectivity of the reaction (Scheme 6.43, middle), increasing the reaction time. It is interesting to note the similarity of these optimal conditions to the ones devised for 2-amidomalonates **138** and **139** (Scheme 6.41) and conversely the differences with the conditions used for the more acidic nitroacetate pronucleophile **142** (Scheme 6.42). Such observation suggests a possible rationalization of the choice of reaction conditions (solvents, additives, etc.) based on the (pro)nucleophile used in aminocatalytic reactions [95].

Going back to the reaction between nitromethane and cyclopent-1-enecarbaldehyde **114**, this reaction, directed at the preparation of telaprevir core, has also been under study at Sumitomo Chemical Company [96]. Discarding "buffering" additives, several useful protocols were disclosed, wherein the addition of a small amount of water to the alcoholic reaction mixture improved catalytic performances. One of the protocols, reported at the bottom of Scheme 6.43 and demonstrated on a multigram scale, highlights the very good results obtainable with this approach, based on a more conventional acid cocatalysis, as well

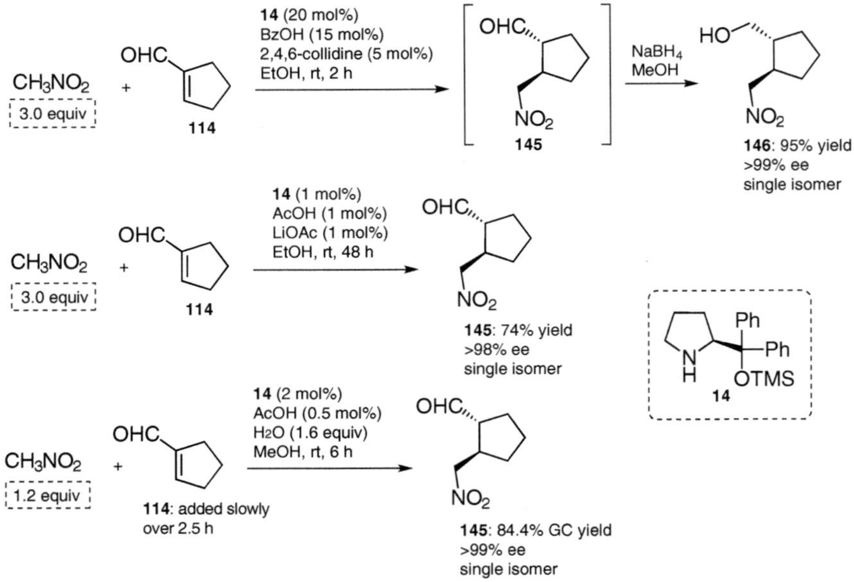

Scheme 6.43 Aminocatalytic conjugate additions of nitromethane to cyclopent-1-enecarbaldehyde **114**. Top: Gellman's protocol; middle: Dr. Reddy's low loading protocol; bottom: an example by Sumitomo Chemical Company.

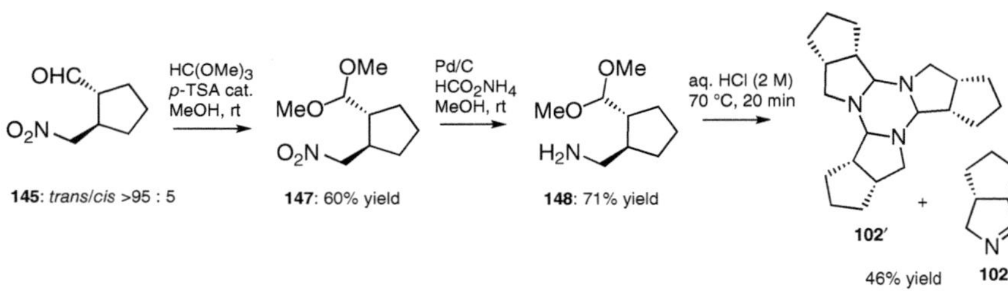

Scheme 6.44 Conversion of the γ-nitroaldehyde **145** to the "desymmetrized" pyrroline **102/102′** based on direct epimerization/cyclization under acidic conditions.

as the important possibility of limiting the amount of nitromethane used in the reaction. In contrast, the other works employ an excess of this dangerous chemical (3 equiv).

Downstream chemistry delivering the telaprevir bicyclic core from the γ-nitroaldehyde **145** was demonstrated in this case. Three main approaches have been reported. Work at the University of Bologna and Dr. Reddy's proposed a three-step conversion from **145** to the bicyclic "desymmetrized" pyrroline **102**, in mixture with its equally useful, equilibrating under appropriate conditions, trimer **102′** [89]. Aldehyde protection by acetalization gave **147**, whose nitro group was reduced to amine rendering compound **148**. Heating **148** under acidic conditions induced its epimerization–cyclization, affording **102** and **102′** (Scheme 6.44).

Concurrently, several alternative approaches were disclosed by the Sumitomo Chemical Company [96], of which a representative streamlined one is detailed in Scheme 6.45.

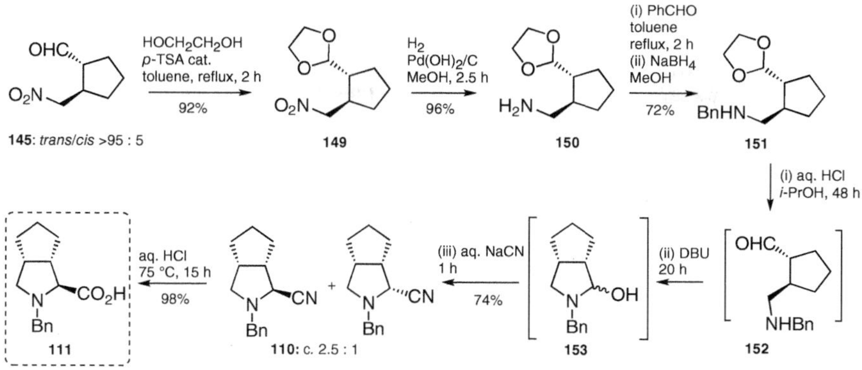

Scheme 6.45 Conversion of the γ-nitroaldehyde adduct **145** to the *N*-benzyl-protected telaprevir core **111**: a representative example.

The catalytic product **145** is first converted to the γ-amino aldehyde **150**, with the aldehyde function protected in this case as a 1,3-dioxolane, through a two-step sequence proceeding via compound **149**. Then, instead of performing a direct cyclization, the amino function of **150** is benzylated. The resulting product **151** is treated with aqueous hydrogen chloride in *iso*-propanol, unveiling the aldehyde function, and then with the relatively strong base DBU, which induces epimerization of this intermediate **152** and its subsequent cyclization to the cis-fused aminal **153**. Addition of aqueous sodium cyanide delivers the *N*-benzyl amino nitrile **110** in very good yield (over three steps). Even if not specified, the moderate control of the exo/endo diastereoselectivity in the cyanide addition step might be unconsequential in the subsequent hydrolysis delivering the *N*-benzyl-protected telaprevir core **111**, according to the dynamic process reported in Scheme 6.34. Other examples, which include an alternative nitrogen-protecting group such as CBz, show alternative cyanide insertion and base-promoted epimerization procedures, and the possibility of obtaining the deprotected amino acid of telaprevir (**98**) and/or its *tert*-butyl ester (**130**).

In summary, organocatalysis allowed to validate the appealing retrosynthetic logic reported in Scheme 6.35 based on the enantioselective conjugate additions of synthetic equivalents of the α-amino and glycine α-anion synthons **55** and **116** to cyclopent-1-enecarbaldehyde **114**. The possibility to force the system to the cis-fused configuration during the cyclization step building the bicyclic structure, under both acidic and basic conditions, has been demonstrated too.

6.3.3 5-(Trifluoromethyl)-2-isoxazolines as Antipest Agents

Several recently developed veterinary drugs are based on a 3,5-diaryl-5-(trifluoromethyl)-2-isoxazoline scaffold **154** (Scheme 6.46). Indeed, following the disclosure at Nissan Chemical Industry in 2004–2005 that benzamides presenting such a scaffold show interesting antipest properties [97], a very large number of structurally varied 2,5-diaryl-5-(trifluoromethyl)-2-isoxazolines have been screened for activity by several agrochemical companies. Such screenings ultimately resulted in the commercialization of some of these compounds as oral antiflea and antitick treatments for dogs and cats,

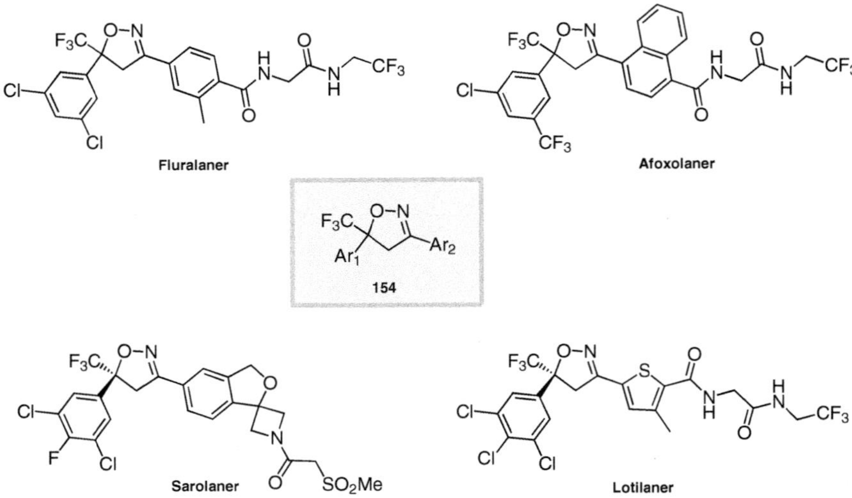

Scheme 6.46 5-(Trifluoromethyl)-2-isoxazoline veterinary drugs.

such as fluralaner [98] (Brevecto™), developed by Nissan and Merck Co., and afoxolaner [99] (NextGard™), by Merial and DuPont. These heterocyclic molecules, administered in racemic form, exert their activity by blocking the insect GABA- and glutamate-gated chloride channel [100]. Given that (*S*)-enantiomers are more active than their (*R*)-counterparts, it is not surprising that some newly disclosed commercialized examples of this class of compounds entail, instead racemic, enantiopure (*S*)-ingredients, such as Pfizer-Zoetis sarolaner [101] (Simparica™) and Elanco-Eli Lilly's lotilaner [102] (Credelio™).

A straightforward synthetic approach to the 3,5-diaryl-5-(trifluoromethyl)-2-isoxazoline scaffold **154** was reported by Nissan Chemical Industries in 2008 [103]. In this approach (Scheme 6.47), the 2-isoxazoline heterocycle is built by the reaction of a β-trifluoromethyl enone **155** with hydroxylamine. Enone **155** is itself obtained by aldol condensation of a 2,2,2-trifluoromethylacetophenone **156** with an acetophenone **157** or via a Wittig reaction with phosphorous ylide **158**.

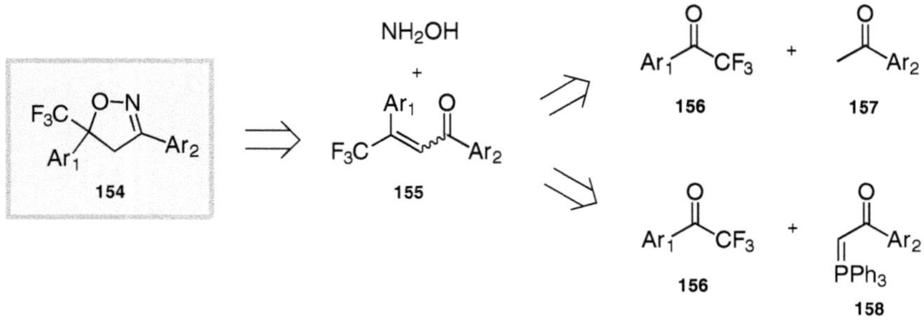

Scheme 6.47 Nissan Chemical Industry approach to the 5-(trifluoromethyl)-2-isoxazoline scaffold **154**.

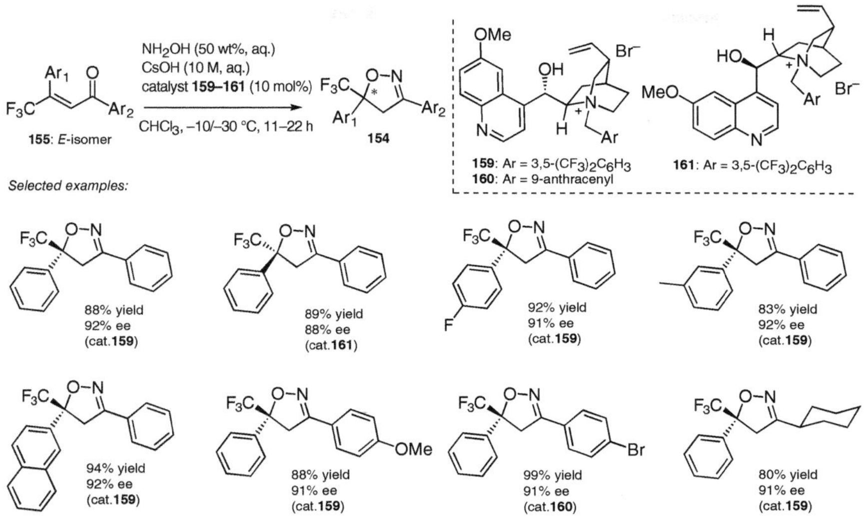

Scheme 6.48 Enantioselective preparation of 5-(trifluoromethyl)-2-isoxazolines **154** by PTC: selected examples.

Such an approach, amenable to phase transfer catalytic reaction conditions in the reaction between hydroxylamine and the enone **155**, lent itself to the development of an enantioselective version by using chiral catalysts. Work at Nissan Chemical Industry, complemented by Shibata and coworkers at the University of Nagoya, showed indeed that *Cinchona* alkaloid-derived quaternary ammonium salts can be used to promote this reaction in an enantioselective manner [104]. As sketched in Scheme 6.48, several β-trifluoromethyl enones **155** can be engaged in quinidinium **159–160** catalyzed reactions, delivering the corresponding (*R*)-configured 2-isoxazolines **154** with very good results in terms of yields and enantioselectivities [104b]. While most substrates require the 3,5-bis(trifluoromethyl)benzyl catalyst **159** for optimal results, in some cases, the 9-anthracenylmethyl derivative **160** gives better selectivities. Furthermore, the reaction tolerates an aliphatic ketone, and selective access to the (*S*)-enantiomers of the 5-trifluoromethyl-2-isoxazoline **154** products – i.e. the biologically active enantiomeric form of the 2-isoxazolines – is possible by using the pseudo-enantiomeric catalyst **161** derived from quinine.

The proposed reaction pathway of 2-isoxazoline formation involves the oxa-Michael addition of the deprotonated hydroxylamine to the enone **155**, forming intermediate **B**, which then undergoes ring closure and dehydration to product **154** (Scheme 6.49). This pathway, entailing a 5-exo-trig cyclization, was thought to be favored over an alternative sequence involving oxime formation, followed by an anti-Baldwin 5-endo-trig cyclization. Indeed, a control experiment demonstrated the correctness of this oxa-Michael hypothesis. Furthermore, results obtained by modifying the catalyst structure (alkylating the hydroxyl group and using cinchonine instead of quinidine), combined with a conformational analysis (MOPAC PM3 calculation for the enone, X-ray crystallography for the cinchonine catalyst), prompted the proposal of the transition state model **A**, in which the two aryl

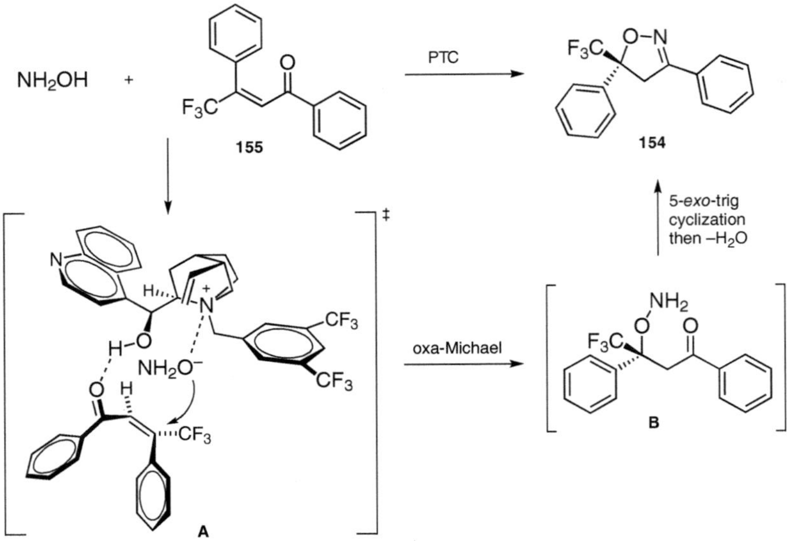

Scheme 6.49 Proposed reaction pathway and transition-state model **A** with a cinchonine catalyst.

groups of the enone are twisted, the ketone interacts with the hydroxyl group, and the deprotonated hydroxylamine, associated with the cationic nitrogen, is sandwiched between the catalyst and the enone.

Although fluralaner is currently commercialized as a racemate, the (*S*)-enantiomer is more active than its (*R*) counterpart. Accordingly, this methodology was applied to the preparation of the potential (*S*)-fluralaner precursor **165** from enone **164** (Scheme 6.50) [105]. Using quininium catalyst **162**, the desired fluralaner precursor **165** was obtained in quantitative yield and good enantioselectivity by applying the previously established reaction conditions. Structure **165** can be potentially converted to fluralaner by a palladium-catalyzed aminocarbonylation using amine **166**. As summarized in previous highlights [8b, c], similar results could be obtained even by using the enone as an *E*/*Z* 3 : 1 mixture, replacing $CHCl_3$ with 1,2-dichloroethane as a solvent, and with cheaper NaOH as an inorganic base instead of CsOH [104a]. Most remarkably, later investigation at Nissan Chemical Industry disclosed that the synthesis of (*S*)-fluralaner can be accomplished by using the enantioselective PTC 2-isoxazoline formation step as the last reaction of the sequence [104c]. As shown in Scheme 6.50, (*S*)-fluralaner was obtained with very satisfactory results in terms of both yield and enantioselectivity, by using the quininium phase transfer catalyst **163** in the reaction between the benzamide enone **167** and hydroxylamine. Less-expensive sodium hydroxide was employed as an inorganic base.

The application of the phase-transfer-catalyzed enantioselective 2-isoxazoline forming reaction as the last step of the synthesis of this class of insecticides, or for the preparation of late-stage intermediates, is a relatively common theme in the relevant patent literature. For example, researchers at Pfizer/Zoetis [106] have reported protocols closely related to the one described for (*S*)-fluralaner, but directed at the preparation of **169**, a precursor of an isoxazoline azetidine drug candidate, and of **171**, an intermediate for (*S*)-sarolaner synthesis (Scheme 6.51).

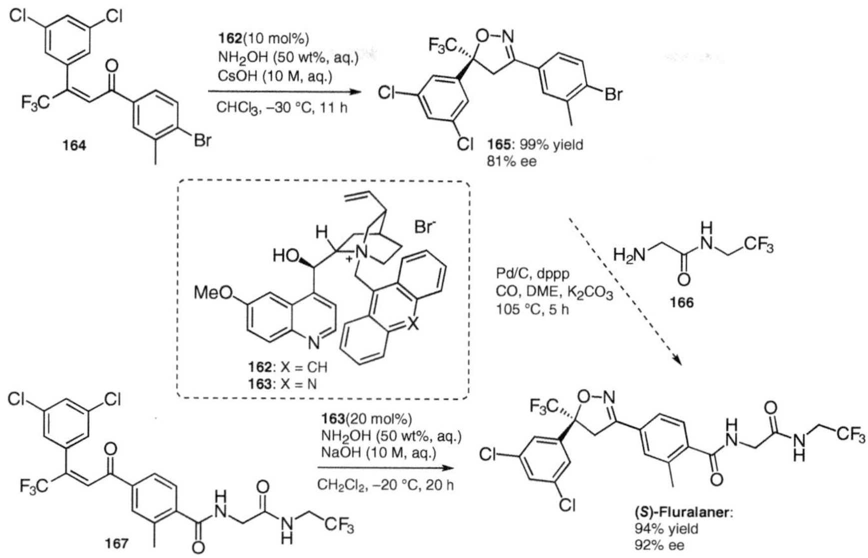

Scheme 6.50 Application of the PTC 5-(trifluoromethyl)-2-isoxazoline formation to (*S*)-fluralaner. dppp = 1,3-(diphenylphosphino)propane, DME = 1,2-dimethoxyethane.

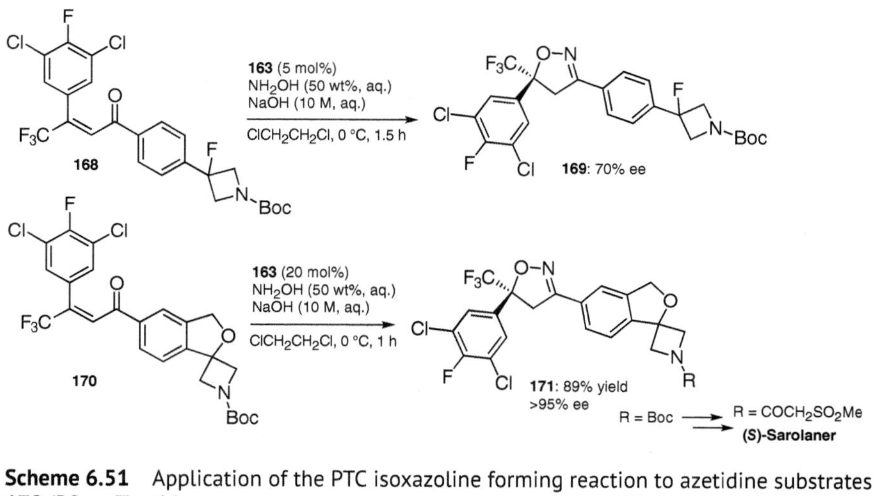

Scheme 6.51 Application of the PTC isoxazoline forming reaction to azetidine substrates **168** and **170** (Pfizer/Zoetis).

For such isoxazoline forming reaction, work by Merial disclosed the superior selectivity of quininium salts bearing electron-donating groups at appropriate positions of the *N*-benzyl substituent, by covering the preparation of (*S*)-fluralaner, (*S*)-afoxolaner, (*S*)-lotilaner, (*S*)-sarolaner, and an additional (*S*)-2,5-diaryl-5-(trifluoromethyl)-2-isoxazoline compound [107]. These disclosures were exemplified for the preparation of (*S*)-afoxolaner from enone **173**: two multigram–kilogram-scale procedures, based on the use of catalyst **172** bearing a 3,4,5-tris(benzyloxy)benzyl substituent in 3–5 mol% charge, were described (Scheme 6.52).

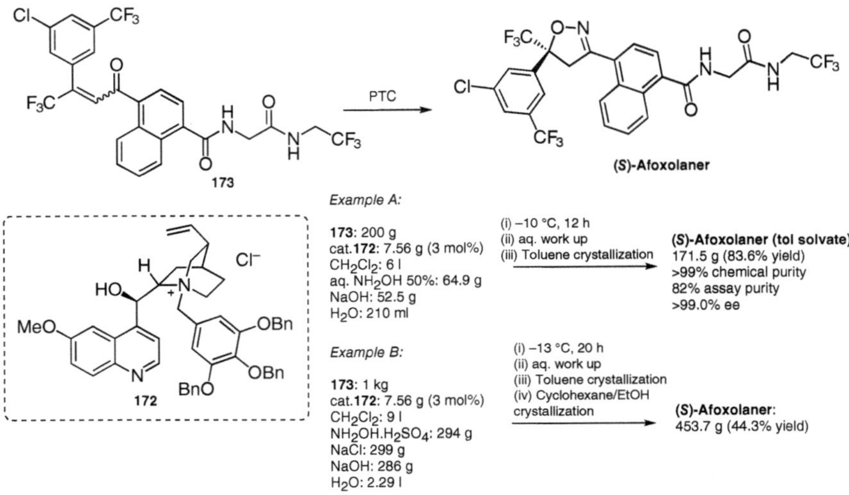

Scheme 6.52 Large-scale examples of the utilization of catalyst **172** in (*S*)-afoxolaner preparation, reported by Merial.

The optical enrichment of the target (*S*)-afoxolaner, obtained in about 84% ee with the catalytic reaction, can be upgraded to >99% by simple crystallization from toluene, which however affords a toluene solvate. Further crystallization from cyclohexane/EtOH delivers pure (*S*)-afoxolaner.

As summarized in a previous highlight [8b], this contribution provides further information about catalyst requirements. Besides optimal catalyst **172**, other 3,4,5-trialkoxy substituted derivatives gave similar, although slightly worse, results; other substitution patterns proved to be unsuitable. In line with the results reported by Nissan Co. and Shibata and coworkers, the presence of the free hydroxyl group is mandatory to achieve enantioselectivity in the reaction; however, the 9-anthracenylmethyl quininium catalyst **162**, which gave good results for the (*S*)-fluralaner precursor (Scheme 6.50), was much less enantioselective in the case of afoxolaner.

Several studies on quininium catalyst variations in this reaction, but applied to their isoxazoline candidates, have been performed and disclosed by Syngenta [108]. For example, a large screening of different *N*-benzyl quininium catalyst substituents was carried out in the reaction with a thietane derivative **180**, its corresponding sulfoxide **181**, and the sulfone **182**, delivering the corresponding products **183–185** (Scheme 6.53) [108a]. The selected results for the thietane substrate **180** reported in Scheme 6.53 show that the 9-anthracenylmethyl-derived catalyst **162**, selective for fluralaner, gives poor results in this reaction. Among the catalysts **174–179** shown, two stand out for their superior enantioselectivities: the electron-rich 3,4,5-trimethoxybenzyl derivative **179**, in line with the results later reported by Merial, and the electron poor pentafluorobenzyl catalyst **175**. Thus, even electron-poor substituents can be effective for this reaction. Furthermore, the influence of the *E*/*Z* ratio in the enone substrate was shortly investigated, indicating a measurable effect of the *E*/*Z* ratio on both yield and enantioselectivity of the resulting isoxazoline.

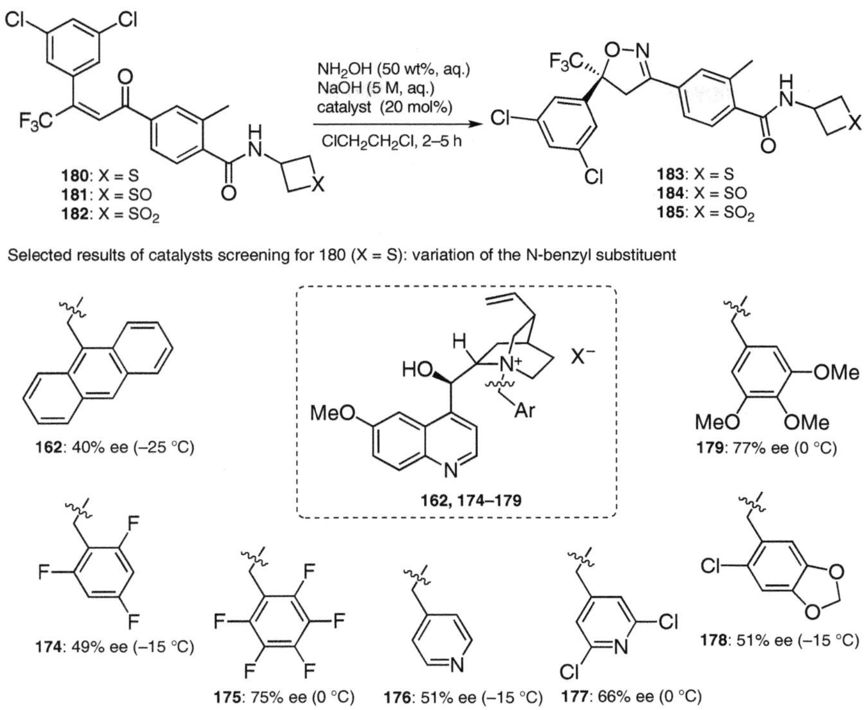

Scheme 6.53 PTC enantioselective reaction with Syngenta's candidates: selected results of catalysts screening.

A subsequent contribution by Syngenta disclosed the utility of another class of quininium catalysts, in the application to another 5-(trifluoromethyl)-2-isoxazoline candidate (**189**) (Scheme 6.54). In this case, being the enone substrate **188** chiral and enantiopure, the process was targeted to favor the formation of the (*S*,*R*)-product **189** over its (*R*,*R*)-diastereoisomer. However, four products **189** were observed: next to the (*S*,*R*) and (*R*,*R*) isoxazolines, their isomers (*S*,*S*) and (*R*,*S*), deriving from epimerization of the already existing chirality center, formed at different extents under the basic reaction conditions. The dimeric, electron-poor tetrafluoro catalyst **186** was found to be the most promising structure; both halide substituents and dimeric features are essential because previously optimal pentafluoro catalyst **175** and a dimeric, but nonhalogenated, catalyst give poorer results. On the other hand, a tetrachloro derivative performs rather similarly to **186**. Few multigram (1–10 g) examples showed that by tuning the reaction conditions in terms of basicity (i.e. using a substoichiometric amount of hydroxides), catalyst loading, solvent, temperature, and reaction time, it is possible to limit the extent of epimerization observed in the reaction. Thus, careful control of reaction parameters enables the obtainment of the desired (*S*,*R*)-stereoisomer with excellent selectivities, using dimeric quininium bromide salt **186** or its tetrafluoroborate counterpart **187** as phase transfer catalysts.

To conclude, the phase-transfer-catalyzed isoxazoline ring forming reaction appears to be an appealing asymmetric entry to this class of trifluoromethylated veterinary

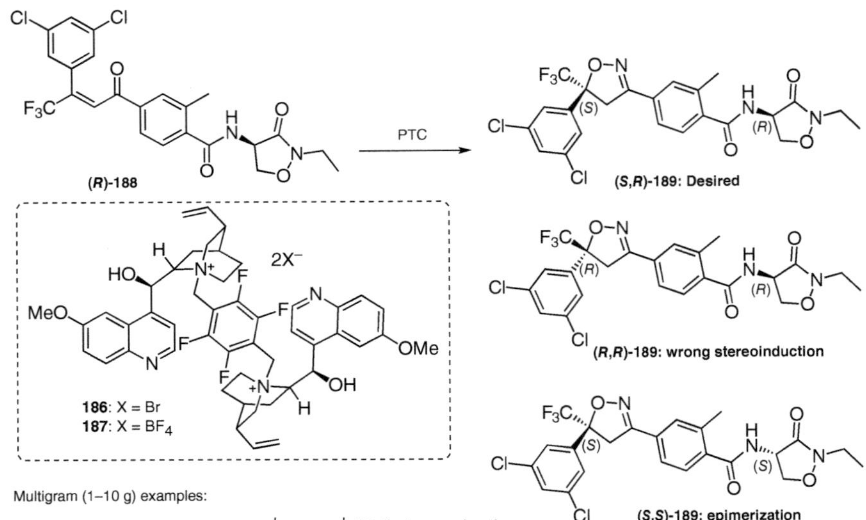

Multigram (1–10 g) examples:

Conditions	Yield (%)	189 diastereomeric ratio (S,R):(R,R):(S,S):(R,S)
186 (2.5 mol%), NH_2OH (1.1 equiv) KOH (0.16 equiv) CH_2Cl_2, −10 °C, 1.5 h	86	**89** : 9 : 1 : 0
186 (5 mol%), NH_2OH (1.1 equiv) NaOH (0.16 equiv), $ClCH_2CH_2Cl$ rt, 20 min	83	**85** : 7 : 7 : 1
186 (10 mol%), NH_2OH (2.0 equiv) KOH (0.1 equiv), *i*-PrOH, 0 °C, 3 h	60	**93** : 7 : 0 : 0
187 (5 mol%), NH_2OH (1.1 equiv) NaOH (0.16 equiv), $ClCH_2CH_2Cl$ −10 °C, 4.5 h	79	**92** : 5 : 3 : 0

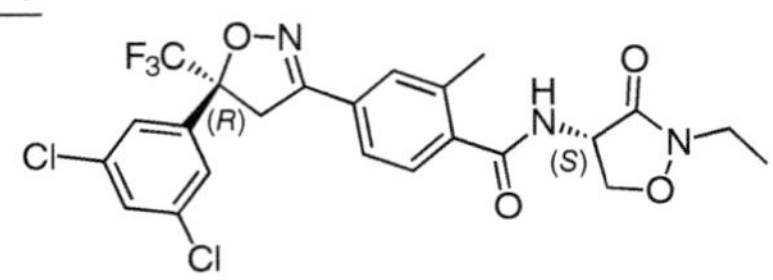

Scheme 6.54 Dimeric quininium catalysts **186**, **187** in the diastereoselective isoxazoline forming reaction from (*R*)-**188** reported by Syngenta.

drugs, with quinine derivatives being the catalysts of choice. Available reports describe small/preparative scale protocols, with the exception of the kilogram-scale reaction reported by Merial. In this context, a disclosure by Novartis has recently described a protocol on multikilogram scale for (*S*)-lotilaner based on classical resolution and off-enantiomer recycling (base and heat promoted racemization) [109]. Nevertheless, it is clear that several agrochemical companies have put considerable investments and considerable time in the development and coverage of such phase transfer catalytic enantioselective isoxazoline formation. Such investments are convincing evidence of the interest and potential application of this catalytic approach.

6.4 Summary and Conclusions

The examples highlighted in this chapter provide convincing evidence that organocatalysis is an exceedingly useful synthetic tool for the preparation of APIs in industrial settings.

An overview of Section 6.2, which highlights divulged manufacturing processes, might lead to the following deductions:

- In light of the Anderson adage "With few exceptions, manufacturing processes are not divulged" [7], it is remarkable that several processes that are based on organocatalysis as stereoestablishing element have been disclosed. Interestingly, most targets highlighted in Section 6.2 were abandoned or stalled during late-stage clinical development (e.g. uprifosbuvir, telcagepant, MK-8613, and funapide), thus enforcing the impression that other commercial processes, not divulged for competitive reasons, exist.
- The cheapness and availability of simple organic catalysts make processes utilizing relatively high catalyst loadings economically feasible. Thus, in a manufacturing setting, the poor activity of organic catalysts is alleviated, at least partially, by their low cost.
- Applications of organocatalysis are not restricted to a single catalyst class or activation mode. Phase transfer catalysis with quaternary ammonium salts has been the most widely explored, perhaps for both historical reasons and its high activation, thanks to strongly basic conditions. However, examples in Chapter 2 also include catalyses based on secondary amines (aminocatalysis), hydrogen bond donors, bifunctional catalysts, and Lewis bases. This indicates the wide-range applicability of this relatively new platform to asymmetric synthesis.

Section 6.3, which describes the variety of organocatalytic approaches toward some simple targets, mainly for the generic market, well outlines the strengths and weaknesses of this methodology. The many examples reported demonstrate its versatility. Pregabalin and the telaprevir core have been tackled by using a number of different synthetic strategies, and all have led to very good results in terms of yields and enantioselectivities. By tuning the catalyst structure, all targeted 5-(trifluoromethyl)-2-isoxazolines could be accessed with acceptable stereoinductions. On the other hand, looking at these innovative procedures from an industrial point of view, several issues can be pointed out; many of these protocols employ difficult-to-handle, highly reactive chemicals, such as acetaldehyde, nitromethane, nitroalkenes, and the activation offered by these catalyzes is not exceptional. As a result, turnover numbers and frequencies are generally low. Not all organic catalysts are sufficiently cheap to offset this issue. For simple (i.e. low value) compounds, the cost of the catalyst can be, in fact, sometimes prohibitive and might hinder its utilization for commercial purposes. Furthermore, substrate scope can be limited: for example, the enantioselective addition of nitromethane to alkylidene malonates targeting pregabalin has not been reported, despite being an appealing alternative to the addition of malonates to nitroalkene **56**. Nevertheless, it must be recognized that even for the simple γ-amino acid pregabalin at least two kilogram-scale processes based on organocatalysis have been demonstrated. Considering also the possibility of catalyst recovery and reuse, this leaves optimism for growing commercial applications of organocatalysis in the near future.

References

1 MacMillan, D.W.C. (2008). The advent and development of organocatalysis. *Nature* 455: 304–308.

2 (a) Hajos, Z.G. and Parrish, D.R. (1974). Asymmetric synthesis of bicyclic intermediates of natural product chemistry. *The Journal of Organic Chemistry* 39: 1615–1621. (b) Eder, U., Sauer, G., and Wiechert, R. (1971). New type of asymmetric cyclization to

optically active steroid CD partial structures. *Angewandte Chemie, International Edition* 10: 496–497.

3 Dolling, U.-H., Davis, P., and Grabowski, E.J.J. (1984). Efficient catalytic asymmetric alkylations. 1. Enantioselective synthesis of (+)-indacrinone via chiral phase-transfer catalysis. *Journal of the American Chemical Society* 106: 446–447.

4 (a) List, B., Lerner, R.A., and Barbas, C.F. III (2000). Proline-catalyzed direct asymmetric aldol reactions. *Journal of the American Chemical Society* 122: 2395–2396. (b) Ahrendt, K.A., Borths, C.J., and MacMillan, D.W.C. (2000). New strategies for organic catalysis: the first highly enantioselective organocatalytic Diels–Alder reaction. *Journal of the American Chemical Society* 122: 4243–4424.

5 (a) Dalko, P. (ed.) (2013). *Comprehensive Enantioselective Organocatalysis*. Weinheim, Germany: Wiley-VCH. (b) List, B. (ed.) (2010). Asymmetric organocatalysis. In: *Topics in Current Chemistry*. Springer-Verlag: Berlin and New York.

6 Blaser, H.-U. and Schmidt, E. (eds.) (2004). *Asymmetric Catalysis on Industrial Scale: Challenges, Approaches and Solutions*. Weinheim, Germany: Wiley-VCH.

7 Anderson, N.G. (2012). *Practical Process Research & Development*, 2e. Academic Press.

8 (a) Carlone, A. and Bernardi, L. (2019). Enantioselective organocatalytic approaches to active pharmaceutical ingredients – selected industrial examples. *Physical Sciences Reviews* 4 https://doi.org/10.1515/psr-2018-0097. (b) Hughes, D.L. (2018). Asymmetric organocatalysis in drug development – highlights of recent patent literature. *Organic Process Research and Development* 22: 574–584. (c) Tan, J. and Yasuda, N. (2015). Contemporary asymmetric phase-transfer catalysis: large scale industrial applications. *Organic Process Research and Development* 19: 1731–1746. (d) Blaser, H.-U. and Federsel, H.-J. (eds.) (2010). *Asymmetric Catalysis on Industrial Scale: Challenges, Approaches and Solutions*, 2e. Weinheim, Germany: Wiley-VCH.

9 Xu, F. (2013). Organocatalysis for asymmetric synthesis: from lab to factory. In: *Sustainable Catalysis* (eds. P.J. Dunn, K.K. Hii, M.J. Krische and M.T. Williams), 317–337. Hoboken: Wiley.

10 Paone, D.V., Shaw, A.W., Nguyen, D.N. et al. (2007). Potent, orally bioavailable calcitonin gene-related peptide receptor antagonists for the treatment of migraine: discovery of *N*-[(3*R*,6*S*)-6-(2,3-difluorophenyl)-2-oxo-1-(2,2,2-trifluoroethyl)azepan-3-yl]-4-(2-oxo-2,3-dihydro-1*H*-imidazo[4,5-*b*]pyridin-1-yl) piperidine-1-carboxamide (MK-0974). *Journal of Medicinal Chemistry* 50: 5564–5567.

11 Xu, F., Zacuto, M., Yoshikawa, N. et al. (2010). Asymmetric synthesis of telcagepant, a CGRP receptor antagonist for the treatment of migraine. *The Journal of Organic Chemistry* 75: 7829–7841.

12 Gotoh, H., Ishikawa, H., and Hayashi, Y. (2007). Diphenylprolinol silyl ether as catalyst of an asymmetric, catalytic, and direct Michael reaction of nitroalkanes with α,β-unsaturated aldehydes. *Organic Letters* 9: 5307–5309.

13 (a) Marigo, M., Wabnitz, T.C., Fielenbach, D. et al. (2005). Enantioselective organocatalyzed α sulfenylation of aldehydes. *Angewandte Chemie, International Edition* 44: 794–797. (b) Hayashi, Y., Gotoh, H., Hayashi, T. et al. (2005). Diphenylprolinol silyl ethers as efficient organocatalysts for the asymmetric Michael reaction of aldehydes and nitroalkenes. *Angewandte Chemie, International Edition* 44: 4212–4215. (c) Reyes-Rodríguez, G.J., Rezayee, N.M., Vidal-Albalat, A. et al. (2019). Prevalence of diarylprolinol silyl ethers as catalysts in total synthesis and patents. *Chemical Reviews* 119: 4221–4260.

14 (a) Jacobo, R., Cota, A., Rogel, E. et al. (1994). Synthesis of some fluorinated nitro-olefins. *Journal of Fluorine Chemistry* 67: 253–255. (b) Xu, F., Corley, E., Murry, J.A. et al. (2007). Highly selective cascade couplings for the syntheses of functionalized piperidinones and bispidines. *Organic Letters* 9: 2669–2672. (c) Xu, F., Corley, E., Zacuto, M. et al. (2010). Asymmetric synthesis of a potent, aminopiperidine-fused imidazopyridine dipeptidyl peptidase IV inhibitor. *The Journal of Organic Chemistry* 75: 1343–1353.

15 (a) Edmondson, S.D., Mastracchio, A., Cox, J.M. et al. (2009). Aminopiperidine-fused imidazoles as dipeptidyl peptidase-IV inhibitors. *Bioorganic & Medicinal Chemistry Letters* 19: 4097–4101. (b) Cox, J.M., Edmondson, S.D., Harper, B., et al. (2005), Aminopiperidines as dipeptidyl peptidase-IV inhibitors for the treatment or prevention of diabetes. World Patent WO/2006/039325 A3, filed 27 September 2005 and issued 13 April 2006.

16 Li, H., Wang, Y., Tang, L., and Deng, L. (2004). Highly enantioselective conjugate addition of malonate and β-ketoester to nitroalkenes: asymmetric C–C bond formation with new bifunctional organic catalysts based on Cinchona alkaloids. *Journal of the American Chemical Society* 127: 9906–9907.

17 Palomo, C., Oiarbide, M., and López, R. (2009). Asymmetric organocatalysis by chiral Brønsted bases: implications and applications. *Chemical Society Reviews* 38: 632–653.

18 (a) Sclafani, J.A., Chen, J., Levy, D.V. et al. (2017). The first asymmetric pilot-scale synthesis of TV-45070. *Organic Process Research and Development* 21: 1616–1624. (b) Sun, S., Fu, J., Chowdhury, S., et al. (2013). Asymmetric syntheses of spiro-oxindole compounds useful as therapeutic agents. US Patent 9487535 B2, filed 11 March 2013 and issued 17 October 2013.

19 Cadieux, J.-J., Chafeev, M., Chowdhury, S., et al. (2010). Synthetic methods for spiro-oxindole compounds. US Patent 8,445,696 B2, filed 14 October 2010 and issued 13 April 2011.

20 Wang, Y., Milkiewicz, K.L., Kaufman, M.L. et al. (2017). Plant process for the preparation of Cinchona alkaloid-based thiourea catalysts. *Organic Process Research and Development* 21: 408–413.

21 (a) Vakulya, B., Varga, S., Csámpai, A. et al. (2005). Highly enantioselective conjugate addition of nitromethane to chalcones using bifunctional cinchona organocatalysts. *Organic Letters* 7: 1967–1969. (b) McCooey, S.H. and Connon, S.J. (2007). Accessible 9-epi-amino Cinchona alkaloid derivatives promote efficient, highly enantioselective additions of aldehydes and ketones to nitroolefins. *Organic Letters* 9: 599–602; (c) Cassani, C., Martín-Rapún, R., Arceo, E. et al. (2013). Synthesis of 9-amino(9-deoxy)epi Cinchona alkaloids, general chiral organocatalysts for the stereoselective functionalization of carbonyl compounds. *Nature Protocols* 8: 325–344.

22 (a) Humphrey, G.R., Dalby, S.M., Andreani, T. et al. (2016). Asymmetric synthesis of letermovir using a novel phase-transfer-catalyzed aza-Michael reaction. *Organic Process Research and Development* 20: 1097–1103. (b) Luzung M., Humphrey G.R., Xiang B., et al. (2014). Process for the preparation of substituted quinazoline compounds. World Patent WO2015088931 A1, filed 8 December 2014 and issued 18 June 2015.

23 Goossen, K., Kuhn, O., Berwe, M., et al. (2006). Process for the preparation of dihydroquinazolines. US Patent 20090221822 A1, filed 2 June 2006 and issued 3 September 2009.

24 (a) Chung, C.K., Liu, Z., Lexa, K.W. et al. (2017). Asymmetric hydrogen bonding catalysis for the synthesis of dihydroquinazoline-containing antiviral, letermovir. *Journal of the American Chemical Society* 139: 10637–10640. (b) Metsänen, T.T., Lexa, K.W., Santiago, C.B. et al. (2018). Combining traditional 2D and modern physical organic-derived descriptors to predict enhanced enantioselectivity for the key aza-Michael conjugate addition in the synthesis of Prevymis™ (letermovir). *Chemical Science* 9: 6922–6927.

25 Xiang, B., Belyk, K.M., Reamer, R.A. et al. (2014). Discovery and application of doubly quaternized Cinchona-alkaloid-based phase-transfer catalysts. *Angewandte Chemie, International Edition* 53: 8375–8378.

26 Campos, K.R., Coleman, P.J., Alvarez, J.C. et al. (2019). The importance of synthetic chemistry in the pharmaceutical industry. *Science* 363 (6424) https://doi.org/10.1126/science.aat0805.

27 (a) Ortiz, A., Benkovics, T., Beutner, G.L. et al. (2015). Scalable synthesis of the potent HIV inhibitor BMS-986001 by non-enzymatic dynamic kinetic asymmetric transformation (DYKAT). *Angewandte Chemie, International Edition* 54: 7185–7188. (b) Ortiz A., Benkovics T., Shi Z., et al.(2013). Sulfilimine and sulfoxide methods for the preparation of festinavir. World Patent WO2013177243 A1, filed 22 May 2013 and issued 28 November 2013.

28 (a) Gallagher, W.P., Deshpande, P.P., Li, J. et al. (2015). Claisen approach to 4′-Ed4T. *Organic Letters* 17: 14–17. (b) Gallagher, W.P., Deshpande, P.P., Li, J., et al. (2014). Method for producing festinavir using 5-methyluridine as starting material. World Patent WO 2014172264 A1, filed 14 April 2014 and issued 23 October 2014.

29 Ji, Y., Benkovics, T., Beutner, G.L. et al. (2015). Mechanistic insights into the vanadium-catalyzed Achmatowicz rearrangement of furfurol. *The Journal of Organic Chemistry* 80: 1696–1702.

30 Benkovics, T., Ortiz, A., Guo, Z. et al. (2014). Enantioselective preparation of (*S*)-5-oxo-5,6-dihydro-2*H*-pyran-2-yl benzoate. *Organic Syntheses* 91: 293–306.

31 Eastgate, M.D., Schmidt, M.A., and Fandrick, K.R. (2017). On the design of complex drug candidate syntheses in the pharmaceutical industry. *Nature Reviews* 1: 1–16.

32 (a) Mayes B.A., Stewart A.J., and Moussa A.M. (2013). D-Amino acid compounds for liver disease. World Patent WO 2013177219 A1, filed 21 May 2013 and issued 28 November 2013; (b) Gosselin G., Parsy C.C., Alexandre F.-R., et al. 2′-Chloro nucleoside analogs for HCV infection. World Patent WO2014058801 A1, filed 7 October 2013 and issued 17 April 2014.

33 Ross, B.S., Reddy, P.G., Zhang, H.-R. et al. (2011). Synthesis of diastereomerically pure nucleotide phosphoramidates. *The Journal of Organic Chemistry* 76: 8311–8319.

34 Simmons B.L., Campos K.R., Klapars A., et al. (2015). Process for making nucleoside phosphoramidate compounds. World patent WO2016064797 A1, filed 20 October 2015 and issued 28 April 2016.

35 DiRocco, D.A., Ji, Y., Sherer, E.C. et al. (2017). A multifunctional catalyst that stereoselectively assembles prodrugs. *Science* 356: 426–430.

36 McGrath, N.A., Brichacek, M., and Njardarson, J.T. (2010). A graphical journey of innovative organic architectures that have improved our lives. *Journal of Chemical Education* 87: 1348–1349.

37 Silverman, B. (2008). From basic science to blockbuster drug: the discovery of Lyrica. *Angewandte Chemie, International Edition* 47: 3500–3504.

38 Hoekstra, M.S., Sobieray, D.M., Schwindt, M.A. et al. (1997). Chemical development of CI-1008, an enantiomerically pure anticonvulsant. *Organic Process Research and Development* 1: 26–38.

39 (a) Burk, M.J., de Koning, P.D., Grote, T.M. et al. (2003). An enantioselective synthesis of(*S*)-(+)-3-aminomethyl-5-methylhexanoicacid via asymmetric hydrogenation. *The Journal of Organic Chemistry* 68: 5731–5734. (b) Hoge, G., Wu, H.-P., Kissel, W.S. et al. (2004). Highly selective asymmetric hydrogenation using a three hindered quadrant bisphosphine rhodium catalyst. *Journal of the American Chemical Society* 126: 5966–5967.

40 (a) Burrell, A., Martinez, C., McDaid, P.,et al. (2011). Process for the preparation of (*S*)-3-cyano-5-methyl-hexanoic acid derivatives and of pregabalin. World Patent WO2012025861 A1, filed 19 August 2011 and issued 1 March 2012; (b) Winkler, C.K., Clay, D., Davies, S. et al. (2013). Chemoenzymatic asymmetric synthesis of pregabalin precursors via asymmetric bioreduction of β-cyanoacrylate esters using ene-reductases. *The Journal of Organic Chemistry* 78: 1525–1533. (c) Debarge, S., McDaid, P., O'Neill, P. et al. (2014). Evaluation of several routes to advanced pregabalin intermediates: synthesis and enantioselective enzymatic reduction using ene-reductases. *Organic Process Research and Development* 18: 109–121.

41 Martinez, C.A., Hu, S., Dumond, Y. et al. (2008). Development of a chemoenzymatic manufacturing process for pregabalin. *Organic Process Research and Development* 12: 392–398.

42 Dunn, P.J., Hettenbach, K., Kelleher, P. et al. (2010). The development of a green, energy efficient, chemoenzymatic manufacturing process for pregabalin. In: *Green Chemistry in the Pharmaceutical Industry* (eds. P.J. Dunn, A.S. Wells and M.T. Williams), 161–177. Weinheim, Germany: Wiley-VCH.

43 Adamo, M.F.A., Kelly, B.G., and Moccia, M. (2016). Recent advances in the preparation of active pharmaceutical ingredient (*S*)-pregabalin. *Chimica Oggi/Chemistry Today* 34: 54–57.

44 (a) Ordóñez, M. and Cativiela, C. (2007). Stereoselective synthesis of γ-amino acids. *Tetrahedron: Asymmetry* 18: 3–99. (b) Ordóñez, M., Cativiela, C., and Romero-Estudillo, I. (2016). An update on the stereoselective synthesis of γ-amino acids. *Tetrahedron: Asymmetry* 27: 999–1055.

45 Hameršak, Z., Stipetić, I., and Avdagić, A. (2007). An efficient synthesis of (*S*)-3-aminomethyl-5-methylhexanoicacid (pregabalin) via quinine-mediated desymmetrizationof cyclic anhydride. *Tetrahedron: Asymmetry* 18: 1481–1485.

46 Bolm, C., Schiffers, I., Dinter, C.L. et al. (2000). Practical and highly enantioselective ring opening of cyclic *meso*-anhydrides mediated by *Cinchona* alkaloids. *The Journal of Organic Chemistry* 65: 6984–6991.

47 Park, S.E., Nam, E.H., Jang, H.B. et al. (2010). Enantioselective alcoholysis of *meso*-glutaric anhydrides catalyzed by *Cinchona*-based sulfonamide catalysts. *Advanced Synthesis and Catalysis* 352: 2211–2217.

48 Lee, J.-W., Mayer-Gall, T., Opwis, K. et al. (2013). Organotextile catalysis. *Science* 341: 1225–1229.

49 Yang, H.-J., Xiong, F.-J., Chen, X.-F. et al. (2013). Highly enantioselective thiolysis of prochiral cyclic anhydrides catalyzed by amino alcohol bifunctional organocatalysts and its application to the synthesis of pregabalin. *European Journal of Organic Chemistry* 2013: 4495–4498.

50 Peschiulli, A., Procuranti, B., O'Connor, C.J. et al. (2010). Synergistic organocatalysis in the kinetic resolution of secondary thiols with concomitant desymmetrization of an anhydride. *Nature Chemistry* 2: 380–384.

51 Monge, D., Jiang, H., and Alvarez-Casao, Y. (2015). Masked unsaturated esters/amides in asymmetric organocatalysis. *Chemistry - A European Journal* 21: 4494–4504.

52 Liu, Y., Shirakawa, S., and Maruoka, K. (2013). Phase-transfer-catalyzed asymmetric conjugate cyanation of alkylidenemalonates with KCN in the presence of a Brønsted acid additive. *Organic Letters* 15: 1230–1233.

53 Zhang, Y., Zhang, K., Xie, S., et al. (2017). Synthesis methods of pregabalin and intermediate thereof. China patent CN108218649 A, filed 29 December 2017 and issued 29 June 2018.

54 Liu, W.-X., Chen, S.-K., Tian, J.-M. et al. (2015). Atriazole organocatalyst with spiropyrrolidine framework and its application to the catalytic asymmetric addition of nitromethane to α,β-unsaturated aldehydes. *Advanced Synthesis and Catalysis* 357: 3831–3835.

55 Guo, C., Saifuddin, M., Saravanan, T. et al. (2019). Biocatalytic asymmetric Michael additions of nitromethane to α,β-unsaturated aldehydes via enzyme-bound iminium ion intermediates. *ACS Catalysis* 9: 4369–4373.

56 (a) Adamo, M. (2012). Process for the preparation of gamma amino acids and intermediates used in said process. World patent WO2013076225 A1, filed 22 November 2012 and issued 30 May 2013; (b) Moccia, M., Cortigiani, M., Monasterolo, C. et al. (2015). Development and scale-up of an organocatalytic enantioselective process to manufacture (*S*)-pregabalin. *Organic Process Research and Development* 19: 1274–1281.

57 Arakawa, Y., Haraguchi, N., and Itsuno, S. (2008). An immobilization method of chiral quaternary ammonium salts onto polymer supports. *Angewandte Chemie, International Edition* 47: 8232–8235.

58 Bernardi, L., Ricci, A., and Comes Franchini, M. (2011). Organocatalytic asymmetric Mannich reactions in the preparation of enantioenriched β^3-amino acid derivatives. *Current Organic Chemistry* 15: 2210–2226.

59 (a) Yang, J.W, Chandler, C., Stadler, M., et al. (2008). Proline-catalysed Mannich reactions of acetaldehyde. *Nature* 452, 453–455; (b) Hayashi, Y., Itoh, T., Aratake, S. et al. (2008). A diarylprolinol in an asymmetric, catalytic, and direct crossed-aldol reaction of acetaldehyde. *Angewandte Chemie, International Edition* 47: 2082–2084.

60 (a) García-García, P., Ladépêche, A., Halder, R. et al. (2008). Catalytic asymmetric Michael reactions of acetaldehyde. *Angewandte Chemie, International Edition* 47: 4719–4721. (b) Hayashi, Y., Itoh, T., Ohkubo, M. et al. (2008). Asymmetric Michael reaction of acetaldehyde catalyzed by diphenylprolinol silyl ether. *Angewandte Chemie, International Edition* 47: 4722–4724. (c) Feu, K.S., de la Torre, A.F., Silva, S. et al. (2014). Polyethylene glycol (PEG) as a reusable solvent medium for an asymmetric organocatalytic Michael addition Application to the synthesis of bioactive compounds. *Green Chemistry* 16: 3169–3174. (d) Biewenga, L., Saravanan, T., Kunzendorf, A. et al. (2019). Enantioselective synthesis of pharmaceutically active γ-aminobutyric acids using a tailor-made artificial Michaelase inone-pot cascade reactions. *ACS Catalysis* 9: 1503–1513.

61 Fan, X., Rodríguez-Escrich, C., Sayalero, S., and Pericàs, M.A. (2013). Paraldehyde as an acetaldehyde precursor in asymmetric Michael reactions promoted by site-isolated incompatible catalysts. *Chemistry - A European Journal* 19: 10814–10817.

62 Kumar, M., Kumar, A., Rizvi, M. et al. (2014). Synthesis of α,β-unsaturated δ-lactones by vinyl acetate mediated asymmetric cross-aldol reaction of acetaldehyde: mechanistic insights. *European Journal of Organic Chemistry* 2014: 5247–5255.

63 Barnes, D.M. (2010). Discovery and development of a catalytic asymmetric conjugate addition of ketoesters to nitroalkenes and its use in the large-scale preparation of ABT-546. In: *Asymmetric Catalysis on Industrial Scale*, 2e (eds. H.-U. Blaser and H.-J. Federsel), 439–456. Weinheim: Wiley-VCH.

64 Evans, D.A., Mito, S., and Seidel, D. (2007). Scope and mechanism of enantioselective Michael additions of 1,3-dicarbonyl compounds to nitroalkenes catalyzed by Nickel(ii)–diamine complexes. *Journal of the American Chemical Society* 129: 11583–11592.

65 Okino, T., Hoashi, Y., and Takemoto, Y. (2003). Enantioselective Michael reaction of malonates to nitroolefins catalyzed by bifunctional organocatalysts. *Journal of the American Chemical Society* 125: 12672–12673.

66 (a) McCooey, S.H. and Connon, S.J. (2005). Urea- and thiourea-substituted Cinchonaalkaloid derivatives as highly efficient bifunctional organocatalysts for the asymmetric addition of malonate to nitroalkenes: inversion of configuration at C9 dramatically improves catalyst performance. *Angewandte Chemie, International Edition* 44: 6367–6370. (b) Ye, J., Dixon, D.J., and Hynes, P.S. (2005). Enantioselective organocatalytic Michael addition of malonate esters to nitro olefins using bifunctional cinchonine derivatives. *Chemical Communications*: 4481–4483.

67 Malerich, J.P., Hagihara, K., and Rawal, V.H. (2008). Chiral squaramide derivatives are excellent hydrogen bond donor catalysts. *Journal of the American Chemical Society* 130: 14416–14417.

68 (a) Hamza, A., Schubert, G., Soós, T. et al. (2006). Theoretical studies on the bifunctionality of chiral thiourea-based organocatalysts: competing routes to C–C bond formation. *Journal of the American Chemical Society* 128: 13151–13160. (b) Zhu, J.-L., Zhang, Y., Liu, C. et al. (2012). Insights into the dual activation mechanism involving bifunctional Cinchona alkaloid thiourea organocatalysts: an NMR and DFT study. *The Journal of Organic Chemistry* 77: 9813–9825. (c) Kótai, B., Kardos, G., Hamza, A. et al. (2014). On the mechanism of bifunctional squaramide-catalyzed organocatalytic Michael addition: a protonated catalyst as an oxyanion hole. *Chemistry - A European Journal* 20: 5631–5639.

69 Liu, J.-M., Wang, X., Ge, Z.-M. et al. (2011). Solvent-free organocatalytic Michael addition of diethyl malonate to nitroalkenes: the practical synthesis of Pregabalin and γ-nitrobutyric acid derivatives. *Tetrahedron* 67: 636–640.

70 Suzuki, Y., Lehmann, I., Bichsel, H.U., et al. (2015). Method for the preparation of beta-substituted gamma-amino carboxylic acids. World patent WO2015289068 A1, filed 2 June 2015 and issued 17 December 2015.

71 Baran, R., Veverková, E., Škvorcová, A. et al. (2013). Enantioselective Michael addition of 1,3-dicarbonylcompounds to a nitroalkene catalyzed by chiral squaramides – a key step in the synthesis of pregabalin. *Organic & Biomolecular Chemistry* 11: 7705–7711.

72 Bae, H.Y. and Song, C.E. (2015). Unprecedented hydrophobic amplification in noncovalent organocatalysis "on water": hydrophobic chiral squaramide catalyzed Michael addition of malonates to nitroalkenes. *ACS Catalysis* 5: 3613–3619.

73 Carlone, A., Bernardi, L., de Koning, P., et al. (2014). Process for preparation of pregabalin. India patent IN2013CH04788A20160617, filed 21 October 2014.

74 Zhang, G. (2012). A heterotrimetallic Pd–Sm–Pd complex for asymmetric Friedel–Crafts alkylations of pyrroles with nitroalkenes. *Organic & Biomolecular Chemistry* 10: 2534–2536.

75 Leyva-Pérez, A., García-García, P., and Corma, A. (2014). Multisite organic–inorganic hybrid catalysts for the direct sustainable synthesis of GABAergic drugs. *Angewandte Chemie, International Edition* 53: 8687–8690.

76 Poe, S.L., Kobašlija, M., and McQuade, D.T. (2007). Mechanism and application of a microcapsule enabled multicatalyst reaction. *Journal of the American Chemical Society* 129: 9216–9221.

77 Bassas, O., Huuskonen, J., Rissanen, K. et al. (2009). A Simple organocatalytic enantioselective synthesis of pregabalin. *European Journal of Organic Chemistry*: 1340–1351.

78 (a) Lin, C., Kwong, A.D., and Perni, R.B. (2006). Discovery and development of VX-950, a novel, covalent, and reversible inhibitor of Hepatitis C Virus NS3•4A serine protease. *Infectious Disorders Drug Targets* 6: 3–16. (b) Fowell, A.J. and Nash, K.L. (2010). Telaprevir: a new hope in the treatment of chronic hepatitis C? *Advances in Therapy* 27: 512–522.

79 Tanoury, G.J., Chen, M., Dong, Y. et al. (2014). Stereoselective lithiation and carboxylation of Boc-protected bicyclopyrrolidine: synthesis of a key building block for HCV protease inhibitor telaprevir. *Organic Process Research and Development* 18: 1234–1244.

80 Köhler, V., Bailey, K.R., Znabett, A. et al. (2010). Enantioselective biocatalytic oxidative desymmetrization of substituted pyrrolidines. *Angewandte Chemie, International Edition* 49: 2182–2184.

81 Mijts, B., Muley, S., Liang, J., et al. (2009). Biocatalytic processes for the preparation of substantially stereochemically pure fused bicyclic proline compounds. World patent WO2010008828, filed 23 June 2009 and issued 21 January 2010.

82 Liu, L.-W., Wang, F.-Y., Tian, F. et al. (2016). An improved and enantioselective preparation of the telaprevir bicyclic [3.3.0] proline intermediate and reuse of unwanted enantiomer. *Organic Process Research and Development* 20: 320–324.

83 Iuliano, A., Jumde, V.R., Attolino, E., et al. (2014). Process for the preparation of a viral protease inhibitor and intermediates thereof. US Patent US20140323689, filed 21 April 2014 and issued 30 October 2014.

84 Moni, L., Banfi, L., Basso, A. et al. (2015). Ugi and Passerini reactions of biocatalytically derived chiral aldehydes: application to the synthesis of bicylic pyrrolidines and of antiviral agent telaprevir. *The Journal of Organic Chemistry* 80: 3411–3428.

85 Znabett, A., Polak, M.M., Janssen, E. et al. (2010). A highly efficient synthesis of telaprevir by strategic use of biocatalysis and multicomponent reactions. *Chemical Communications* 46: 7918–7920.

86 Zhang, F., Wen, X., Xu, Q.-L. et al. (2014). Asymmetric synthesis of 3,4-disubstituted proline derivatives: application in synthesis of Hepatitis C virus protease inhibitor telaprevir. *European Journal of Organic Chemistry* 2014: 8101–8109.

87 Hamada, T. (2008). Process for production of bicyclic proline compound. World patent WO2008090819 A1, filed 18 January 2008 and issued 31 July 2008.

88 (a) Shirakawa, S., Liu, Y., Usui, A. et al. (2012). Efficient asymmetric synthesis of a bicyclic amino acid as a core structure of telaprevir. *ChemCatChem* 4: 980–982. (b) Liu, Y., Usui, A., Shirakawa, S. et al. (2012). Catalytic asymmetric synthesis of 3-substituted proline derivatives by using phase-transfer-catalyzed conjugate addition. *Chemistry, an Asian Journal* 1: 180–186. (c) Maruoka, K., Shirakawa, S., Daisuke, S., et al. (2012). Optically active quaternary ammonium salt, and method for producing optically active compound. Japan patent JP 2013063942 A, filed 10 January 2012 and issued 11 April 2013.

89 Bernardi, L., Fochi, M., Carbone, R. et al. (2015). Organocatalytic asymmetric conjugate additions to cyclopent-1-enecarbaldehyde: a critical assessment of organocatalytic approaches towards the telaprevir bicyclic core. *Chemistry - A European Journal* 21: 19208–19222.

90 Ooi, T., Arimura, Y., Hiraiwa, Y. et al. (2006). Highly enantioselective monoalkylation of *p*-chlorobenzaldehyde imine of glycine *tert*-butyl ester under mild phase-transfer conditions. *Tetrahedron* 17: 603–606.

91 Rios, R., Ibrahem, I., Vesely, J. et al. (2007). Organocatalytic asymmetric 5-hydroxypyrrolidine synthesis: a highly enantioselective route to 3-substituted proline derivatives. *Tetrahedron Letters* 48: 8695–8699.

92 (a) Han, M.-Y., Zhang, Y., Wang, H.-Z. et al. (2012). Organocatalytic Michael addition of nitro esters to α,β-unsaturated aldehydes: towards the enantioselective synthesis of trans-3-substituted proline derivatives. *Advanced Synthesis and Catalysis* 354: 2635–2640. (b) Han, M.-Y., Wang, H.-Z., An, W.-K. et al. (2013). A concise synthesis of L-pyrrolysine. *Chemistry - A European Journal* 19: 8078–8081.

93 (a) Zu, L., Xie, H., Li, H. et al. (2007). Highly enantioselective organocatalytic conjugate addition of nitromethane to α,β-unsaturated aldehydes: three-step synthesis of optically active baclofen. *Advanced Synthesis and Catalysis* 349: 2660–2664. (b) Palomo, C., Landa, A., Mielgo, A. et al. (2007). Water-compatible iminium activation: organocatalytic Michael reactions of carbon-centered nucleophiles with enals. *Angewandte Chemie, International Edition* 46: 84341–88435. (c) Wang, Y., Li, P., Liang, X. et al. (2008). An efficient enantioselective method for asymmetric Michael addition of nitroalkanes to α,β-unsaturated aldehydes. *Chemical Communications*: 1232–1234.

94 (a) Gellman, S.H., Guo, L., and Giuliano, M. (2010). Gamma amino acid building blocks. World patent WO 2011047190A1, filed 14 October 2010 and issued 21 April 2011; (b) Giuliano, M.W., Maynard, S.J., Almeida, A.M. et al. (2013). Evaluation of a cyclopentane-based γ-amino acid for the ability to promote α/γ-peptide secondary structure. *The Journal of Organic Chemistry* 78: 12351–12361.

95 Duce, S., Alonso, I., Lamsabhi, A.M. et al. (2018). The acidity of a carbon nucleophile dictates enantioselectivity and reactivity in Michael additions to aromatic and aliphatic enals via iminium activation. *ACS Catalysis* 8: 22–34.

96 Ikemoto, T., Lutete, L.M., and Aikawa, T. (2012). Process for producing ester compound. US Patent US20140094616A1, filed 17 May 2012 and issued 3 April 2014.

97 (a) Mita, T., Kudo, Y., Mizukoshi, T., et al. (2005). Substituted benzanilide compound and noxious organism controlling agent. US Patent US20050250822 A1, filed 25 February 2005 and issued 10 November 2005; (b) Mita, T., Kikuchi, T., Mizukoshi, T., et al. (2005). Isoxazoline-substituted benzamide compound and noxious organism control agent. World patent WO2005085216 A1, filed 4 March 2005 and issued 15 September 2005.

98 Ozoe, Y., Asahi, M., Ozoe, F. et al. (2010). The antiparasitic isoxazoline A1443 is a potent blocker of insect ligand-gated chloride channels. *Biochemical and Biophysical Research Communications* 391: 744–749.

99 Shoop, W.L., Hartline, E.J., Gould, B.R. et al. (2014). Discovery and mode of action of afoxolaner, a new isoxazoline parasiticide for dogs. *Veterinary Parasitology* 201: 179–189.

100 Casida, J.E. (2015). Golden age of RyR and GABA-R diamide and isoxazoline insecticides: common genesis, serendipity, surprises, selectivity, and safety. *Chemical Research in Toxicology* 28: 560–566.

101 McTier, T.L., Chubb, N., Curtis, M.P. et al. (2016). Discovery of sarolaner: a novel, orally administered, broad-spectrum, isoxazoline ectoparasiticide for dogs. *Veterinary Parasitology* 222: 3–11.

102 Rufener, L., Danelli, V., Bertrand, D. et al. (2017). The novel isoxazoline ectoparasiticide lotilaner (Credelio™): a non-competitive antagonist specific to invertebrates γ-aminobutyric acid-gated chloride channels (GABACls). *Parasites & Vectors* 10: 530.

103 Yosaka, M., Utsunomiya, T., Moriyama, Y., et al. (2008). Method for production of 3-hydroxypropan-1-one compound, method for production of 2-propen-1-one compound, and method for production of isoxazoline compound. World patent WO 2009001942 A1, filed 27 June 2008 and issued 31 December 2008.

104 (a) Matoba, K. (2008). Method for catalytic asymmetric synthesis of optically active isoxazoline compound. World patent WO2009063910 A1, filed 12 November 2008 and issued 22 May 2009; (b) Matoba, K., Kawai, H., Furukawa, T. et al. (2010). Enantioselective synthesis of trifluoromethyl-substituted 2-isoxazolines: asymmetric hydroxylamine/enone cascade reaction. *Angewandte Chemie, International Edition* 49: 5762–5766. (c) Toyama, K.-I., Moriyama, Y., Matoba, K., et al. (2012). Method for catalytic asymmetric synthesis of optically active isoxazoline compound, and optically active isoxazoline compound. World patent WO2013069731, filed 8 November 2012 and issued 16 May 2013.

105 Kagami, T. (2015). Insecticidal, miticidal, nematicidal, molluscicidal, microbicidal, or bactericidal composition and method for controlling pest. US Patent US20170135347 A1, filed 30 June 2015 and issued 18 May 2017.

106 (a) Curtis, M., Menon, S., Vaillancourt, V.A., et al. (2012). Spirocyclic isoxazoline derivatives as antiparasitic agents. US Patent US20120232026, filed 13 September 2012 and issued 13 September 2012; (b) Billen, D., Greenwood, S.D.W., and Stuk, T.L., (2013). Process for the preparation of chiral isoxazoline azetidine derivatives as antiparasitic agents. WO Patent WO2013116236 A1, filed 29 January 2013 and issued 8 August 2013.

107 Yang, C., Le Hir de Fallois, L.P., Meng, C.Q., et al. (2017). Process for the preparation of isoxazoline compounds. US Patent US20170311601 A1, filed 5 April 2017 and issued 2 November 2017.

108 (a) Mulholland, N.P., Godineau, E., Cassayre, J.Y., et al. (2011). Process for the preparation of isoxazoline derivatives. World patent WO2011104089 A1, filed on 3 February 2011 and issued 1 September 2011; (b) Smejkal, T., and Smits, H. (2015). Process for the preparation of optically active isoxazoline compounds. US Patent US9815829 B2, filed 4 August 2015 and issued 14 November 2017.

109 Nanchen, S. (2013). Process for the enantiomeric enrichment of diaryloxazoline derivatives. World patent WO 201490918 A1, filed 12 December 2013 and issued 14 June 2014.

7

Biocatalytic Synthesis of Chiral Amines Using Oxidoreductases

Sebastian C. Cosgrove[1], Jeremy I. Ramsden[1], Juan Mangas-Sanchez[1,2] and Nicholas J. Turner[1]

[1] *University of Manchester, Department of Chemistry, Manchester Institute of Biotechnology, 131 Princess Street, Manchester M1 7DN, UK*
[2] *ARAID Foundation, Institute of Chemical Synthesis and Homogeneous Catalysis, C.S.I.C - University of Zaragoza, Pedro Cerbuna, 12, Zaragoza 50009, Spain*

7.1 Introduction

In the past two decades, the development of more effective tools for enzyme discovery and engineering has greatly expanded the biocatalytic toolbox for chiral amine synthesis (Scheme 7.1). Currently, several classes of enzymes are available for this task, and some of them, such as transaminases (TAs), are now widely used in process chemistry for the manufacture of a broad range of chiral amines [1]. Lipases are also widely used and, similarly, they have been thoroughly reviewed [2]. The focus of this chapter is oxidoreductases, enzymes that use nicotinamide or flavin cofactors to perform redox chemistry, which represent an exciting and a promising field of research in the coming years [3].

7.2 Amine Oxidases

7.2.1 Introduction

Amine oxidases (AOs) catalyze the oxidation of amines to the corresponding imines. AOs can be divided into two classes depending on the cofactor they use. On the one hand, copper-dependent oxidases (EC 1.4.3.21 and 1.4.3.22) use copper and different quinone species as cofactors. One of them is topaquinone (TPQ), which is formed from tyrosine via Cu(II) oxidation. The amine binds to form a Schiff base that is subsequently released after hydrolysis to yield the aldehyde and the reduced TPQ [4]. The catalytic cycle is closed by reoxidation of TPQ by O_2. To date, these enzymes have shown limited application in amine synthesis and therefore are not the focus of this chapter. On the other hand, flavin-dependent AOs belong to a large family of enzymes that use the vitamin B_2-derived cofactors flavin adenine dinucleotide (FAD) or flavin mononucleotide (FMN) to perform chemical reactions. Around 90% of flavin-dependent enzymes are oxidoreductases [5], and

Methodologies in Amine Synthesis: Challenges and Applications, First Edition.
Edited by Alfredo Ricci and Luca Bernardi.

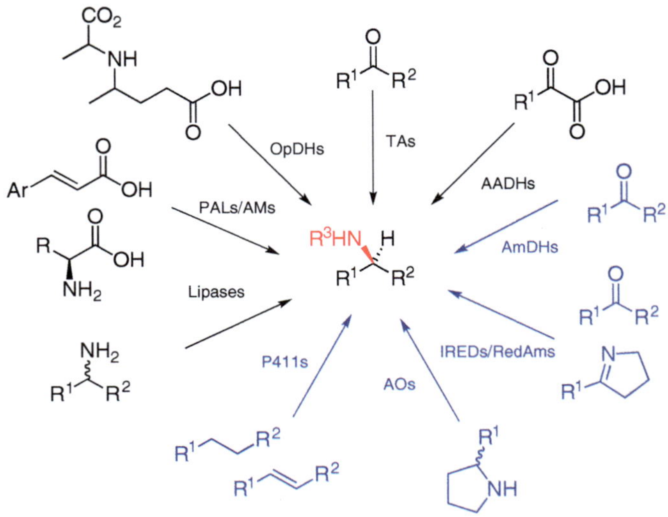

Scheme 7.1 Biocatalytic strategies for chiral amine synthesis. Opine dehydrogenases (OpDHs), phenyl ammonia lyases (PALs), aminomutases (AMs), lipases, amine oxidases (AOs), imine reductases (IREDs), reductive aminases (RedAms), amino dehydrogenases (AmDHs), engineered cytochrome P450s (P411s), amino acid dehydrogenases (AADHs), and transaminases (TAs). Oxidoreductases discussed in this chapter are highlighted in blue.

in particular, AOs have found broad applications in synthetic chemistry. Therefore, this family of enzymes is the focus of this section.

7.2.2 (*S*)-Selective Amine Oxidases

7.2.2.1 Monoamine Oxidase from *Aspergillus niger*

The monoamine oxidase from *Aspergillus niger* (MAO-N) is the most studied AO for synthetic applications. It was originally identified, cloned, and heterologously expressed in *Escherichia coli* by Schilling and Lerch [6, 7]. It is a 55 kDa tetrameric enzyme, which showed many characteristic properties of the mammalian monoamine oxidases MAO-A and MAO-B. This enzyme was able to oxidize small aliphatic amines such as butylamine but also α-methyl benzylamine displaying a preference for the (*S*)-enantiomer. Despite this promising activity, the wild-type enzyme is very limited in both substrate scope and catalytic efficiency. For the past 20 years, directed evolution has proven to be an exceptional tool to overcome these limitations [8]. It comprises different approaches for introducing genetic diversity in enzymes, followed by screening for improved characteristics to generate fit-to-purpose biocatalysts; MAO-N is one of the best enzymes to show the power of this tool. A summary of the evolution of the substrate scope accessible by engineered variants of MAO-N is displayed in Scheme 7.2.

7.2.2.2 Directed Evolution of MAO-N

Even though MAO-N displayed poor activity toward α-methyl benzylamine, it represented a starting point for the Turner group to engineer the enzyme to improve the catalytic

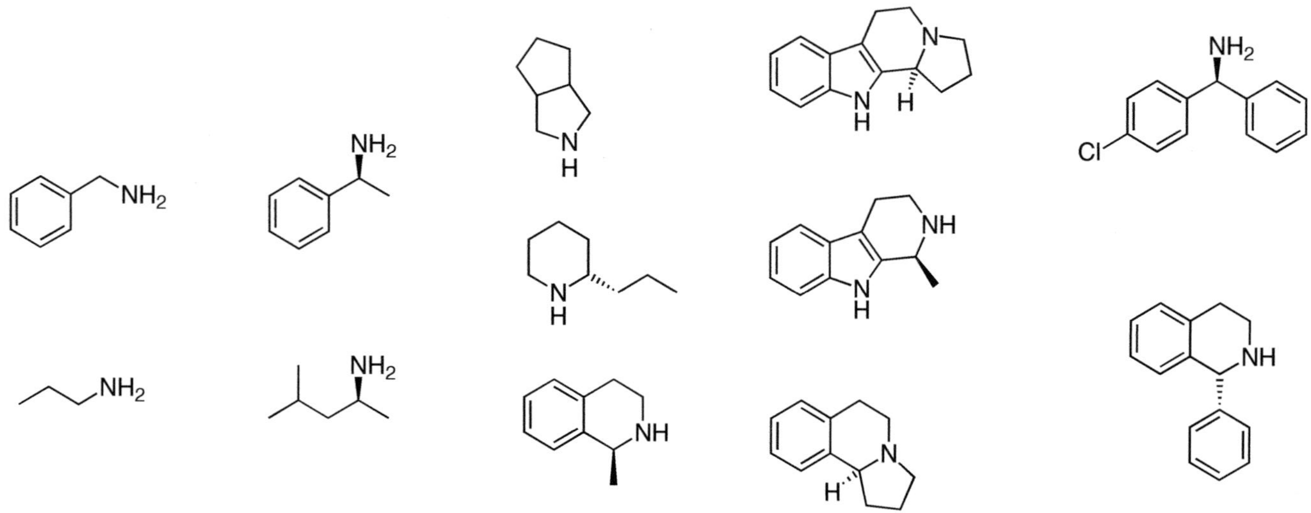

Wild-type D3 D5 D9 D11

Scheme 7.2 Evolution of the substrate scope of the monoamine oxidase from *Aspergillus niger*.

efficiency [9]. Using the *E. coli* XL1-Redmutator strain, they screened around 150 000 clones that led to the identification of 35 variants with enhanced activity toward α-methyl benzylamine. Most of the variants obtained were expression variants, i.e. bearing mutations that led to a higher protein production. Nevertheless, one of the variants (N336S) displayed a 47-fold improvement in activity toward (*S*)-α-methyl benzylamine with similar expression levels. The substrate scope of this variant was subsequently evaluated toward a large panel of structurally diverse primary and secondary amines [10]. Besides the increase in the activity found, the access to a variety of aliphatic, aromatic, and cyclic amines was possible via either kinetic resolution or deracemization in combination with a nonselective hydride source such as ammonia borane. Further engineering was conducted following a similar strategy, and two new variants (MAO-N D3 and D5) showing increased activity and substrate scope were identified as a result [11, 12]. Remarkably, MAO-N D5 showed activity toward tertiary amines, *O*-methyl-*N*-hydroxylamines [13], and it was used to synthesize a precursor of telaprevir, a protease inhibitor of hepatitis C virus [14]. In 2008, Grogan and coworkers solved the structures of both MAO-N D3 and D5, providing a platform to carry out structure-guided engineering [15]. In this manner, Rowles et al. modelled the alkaloid (*S*)-crispine A into the active site of MAO-N [16]. They identified four residues at the entrance of the active site channel (F210, L213, M242, and M246) that could be responsible for steric interactions with the substrate. Two libraries targeting those residues were screened separately and the hits were combined to yield a variant with four additional mutations (F210L/L213T/M242Q/M246T), now termed MAO-N D9, displaying a 990-fold increase in activity toward the target substrate. Further engineering was carried out to increase the volume of the active site and, therefore, expand the scope toward bulky amines. In this manner, Ghislieri et al. identified two amino acids (A429 and W430) as potential targets for site-directed mutagenesis [17]. Using MAO-N D5 as a template, a variant W430G with enhanced activity toward aminodiphenylmethane was identified and a homology model was built, finding an increase in the active site volume from 140 to 195 Å^3. The authors then introduced the mutations previously described in the active site channel that led to an increase in activity toward bulky substrates obtaining the MAO-N D11 variant presenting a larger active site volume (464 Å^3). Using this variant, building blocks for the active pharmaceutical ingredients (APIs) levocetirizine and solifenacin were synthesized in good to excellent yields (45–90%) and enantiomeric excess (>97%).

Aside from the Turner laboratory, other groups have shown interest in MAO-N because of its potential in synthetic chemistry. A semirational approach was employed by Chen et al. to engineer MAO-N D5 for the preparation of (*R*)-mexiletine, an antiarrhythmic, antimyotonic, and analgesic drug that is marketed as a racemate [18]. After modeling the substrate into the active site, three saturation libraries targeting two residues simultaneously (F210-L213, M242-L245, and Y365-I367) were constructed and screened. From the first round, two variants (F210V and F210V/L213C) showed activity toward the racemic amine. A second round combining the active hits with the other two libraries afforded a new active variant F210V/I367T displaying a similar catalytic efficiency. Employing the D5 variants F210V/L213C and F210V/I367T, the kinetic resolution of the racemic amine was possible, obtaining (*R*)-mexiletine in 92% and 96% ee, respectively. Recently, Reetz and coworkers engineered MAO-N D3 by targeting simultaneously residues in the entrance tunnel and the active site via iterative saturation mutagenesis (ISM) [19] to improve the activity toward compounds bearing the benzo-piperidine scaffold to which the wild-type presents no activity. There were 23 residues selected and combined in six different libraries, targeting either four or three residues simultaneously in each of them. The best variants were found in the library targeting the residues W230, T354, and W430, and the two best-performing biocatalysts (W230R/W430C/C214L and W230I/T354S/W430R/M242R/Y365V) were used for the deracemization of 1,2,3,4-tetrahydro-1-methylisoquinoline and 1,2,3,4-tetrahydro-1-phenylisoquinoline, obtaining the products in 73% and 86% isolated yields, respectively, and high enantiomeric excesses. Remarkably, the (*S*)-amines were recovered, indicating that a switch in the enantiopreference of MAO-N was achieved with these two variants. In 2012, Merck and Codexis partnered to develop an oxidative Strecker reaction using MAO-N for the preparation of a key chiral intermediate in the synthesis of boceprevir, a drug developed for treating chronic hepatitis C infection [20]. They initially identified wild-type MAO-N from *Aspergillus niger* as the best candidate for directed evolution because of its robustness. Initial rounds of evolution with the focus on activity yielded a MAON156 variant bearing two mutations, which showed a 2.4-fold increase in activity. Further rounds combining recombination, shuffling, and active site engineering focusing on activity, solubility, and thermal stability yielded a MAO401 variant, which was used to prepare a precursor of boceprevir in >99% ee.

Recently, Currin et al. performed an extensive scanning in MAO-N sequence by mutating every residue in the secondary structure-containing regions (276 out of a total 495 amino acids) to screen for improvements in catalytic activity [21]. Codon degeneracies were carefully selected depending on the position to ensure that the searched sequence space encodes enzymes with a similar secondary structure pattern and, therefore, increasing the chances of finding functional variants. Using α-methyl benzylamine as a model substrate, a variant A289V was found displaying a 1.6-fold increase in k_{cat} compared to that of MAO-ND5.

7.2.2.3 Synthetic Applications and Cascades

The D5 variant was also employed with alternative systems for imine reduction to obtain the enantiopure amine after iterative cycles of selective oxidation followed by selective/nonselective reduction. In this manner, Foulkes et al. used Pd(0) supported on *E. coli* cells expressing MAO-N D5 as a catalyst to hydrogenate the corresponding imine employing either ammonium formate or H_2 as a hydrogen source [22]. An elegant approach involving the combination of an artificial metalloenzyme developed by Ward and coworkers and MAO-N D5 allowed for the access to a set of enantiomerically pure tetrahydroisoquinolines (THIQs) and pyrrolidines [23]. In this process, the (*S*)-enantiomer was oxidized by the amine oxidase to form the imine that was reduced by a (*R*)-selective artificial imine reductase (ATHase). The addition of catalase was necessary to deplete the hydrogen peroxide produced in the oxidation step that inhibited the ATHase.

Besides the extensive use for the preparation of chiral amines, MAO-N was also employed to generate imines and iminium ions and combine them with either further enzymatic or chemical steps to access a variety of compounds. For instance, Ghislieri et al. developed the synthesis of (*R*)-harmacine via a Pictet–Spengler reaction upon imine formation followed by a deracemization of the corresponding amine [17]. Different MAO-N variants

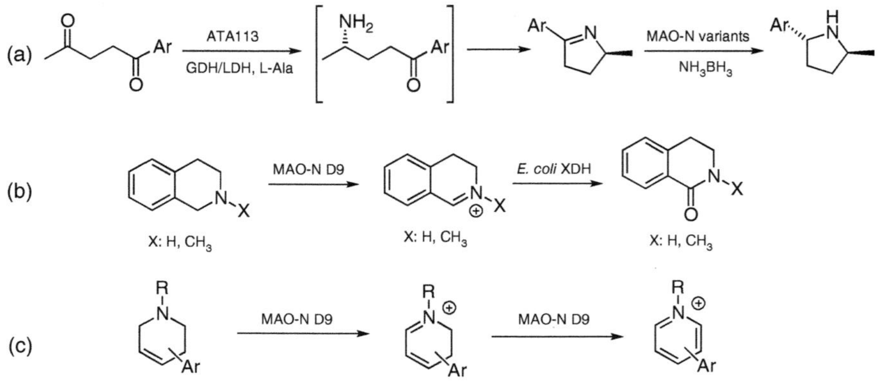

Scheme 7.3 Enzymatic cascades involving MAO-N variants for the preparation of chiral amines and lactams.

were also combined with other enzymes to build synthetic complexity in a one-pot manner. For instance, MAO-N D9 was combined with a transaminase (ATA113) for the preparation of chiral pyrrolidines [24] (Scheme 7.3a). Starting from diketones, a transamination step to regioselectively aminate one of the ketone groups followed by spontaneous cyclization to form the corresponding cyclic imine was conducted on the first place. Subsequent deracemization catalyzed by MAO-N D9 in the presence of ammonia borane afforded the disubstituted chiral pyrrolidines in high yields as well as excellent enantio- and diastereoselectivity. This MAO-N variant was also combined with xanthine dehydrogenase for the synthesis of lactams [25] (Scheme 7.3b). More recently, MAO-N D9 was used to oxidase a series of tetrahydropyridines to access pyridines or pyridinium ions [26] (Scheme 7.3c). In this case, MAO-N catalyzed a first oxidation step to form the 2,3-dihydropyridines that, after tautomerization to the corresponding 1,2-dihydropyridines, were further oxidized to access the final pyridines and pyridinium ions.

Besides the combination with other enzymes, MAO-N has also been used in sequential processes with other catalytic steps. For instance, Scalacci et al. reported a ring-closing metathesis of diallyl amines using Grubbs catalyst followed by oxidation of the resulting cyclic amine to aromatize and access the corresponding pyrroles [27] (Scheme 7.4a). MAO-N has also been used as a tool to generate and develop imine chemistry in aqueous media. In 2010, the Orru group employed MAO-N to perform the desymmetrization of 3,4-substituted *meso*-pyrrolidines to access 1-pyrrolines in high enantiomeric excess to subsequently carry out an Ugi multicomponent reaction to prepare highly functionalized and enantiopure 3,4-substituted prolyl peptides [28] (Scheme 7.4b). The same group in 2016 expanded this approach to access chiral 2-substituted pyrrolidines via aza-Friedel–Crafts reaction [29]. More recently, in a collaboration between the groups of Turner and Greaney, MAO-N D5 was used to generate different iminium ions starting from *N*-alkyl tetrahydroisoquinolines followed by gold-catalyzed C—C bond formation to access a set of C-(1) functionalized *N*-alkyl THIQs in high yields, under mild conditions and with no compartmentalization required (Scheme 7.4c) [30].

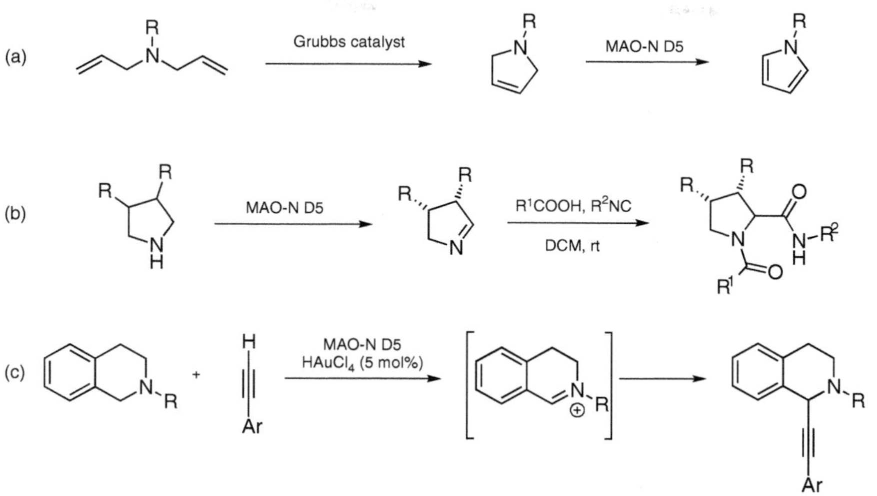

Scheme 7.4 Chemoenzymatic cascades involving MAO-N variants for the preparation of nitrogenated heterocycles.

7.2.2.4 Monoamine Oxidase from *Pseudomonas monteilii* ZMU-T01

In 2016, Chen and coworkers described the oxidation of a series of 2-substituted 1,2,3,4-tetrahydroquinolines (THQs) employing whole cells from *Pseudomonas monteilii* ZMU-T01 [31], obtaining the corresponding 3,4-dihyroquinolines and the chiral amines in up to 50% conversion and >99% ee, showing a preference for the (*S*)-enantiomer. Subsequently, the same group identified five MAO sequences in *Pseudomonas monteilii* ZMU-T01 that were amplified, cloned, and expressed in *E. coli* [32]. After initial screening, they identified an active enzyme sharing 14.5% and 25.2% sequence identity with MAO-N and cyclohexylamine oxidase (CHAO), respectively. The synthetic potential of this enzyme, now termed MAO5-ZMU6, was investigated in its ability to perform the kinetic resolution of a series of THQs. It was found that the enzyme can tolerate a good range of different substituents in the aromatic ring, although both conversion and enantiomeric excess are restricted to small aliphatic groups in the piperidine ring. Only high enantiomeric excesses were obtained for 2-methyl-substituted THQs.

7.2.2.5 Cyclohexylamine Oxidase from *Brevibacterium oxydans* (CHAO)

The CHAO from *Brevibacterium oxydans* is a 50 kD flavin-dependent oxidase that is responsible for the oxidation of cyclohexylamine to cyclohexanone. It was isolated and characterized by Iwaki et al. in 1999, but its potential for synthetic chemistry was not assessed until the 2010s [33, 34]. Lau and coworkers investigated the substrate scope of the wild-type enzyme and found that it presented a strong preference for aliphatic and cyclic primary amines such as cyclohexylamine derivatives or cycloheptylamine and also for aromatic amines such as α-methyl benzylamine and derivatives thereof, where it shows a strong preference for the (*S*)-enantiomer [35]. In a collaboration between the groups of Zhu and Lau, 11 amino acid residues in the active site within a 5 Å distance of cyclohexanone were targeted via saturation mutagenesis [36]. Four variants were found to bear beneficial mutations toward the oxidation of 2-methyl-1,2,3,4-THQ, i.e. T198F, L199T, M226F, and Y459T. ISM was then performed in those positions finding a triple variant (T198F/L199S/M226F) showing a 100-fold increase in k_{cat} toward the oxidation of 2-methyl-1,2,3,4-THQ. Subsequently, Li et al. further explored the synthetic potential of CHAO by creating libraries in certain positions after superimposing the CHAO structure to that of a MAO B homolog [37]. In the substrate-binding domain, they identified Y321, F368, and Y459 as responsible for creating a pocket where the amine binds. Moreover, they selected T198, L199, M226, and F351 as residues likely to interact with the substrate in the entrance channel. Seven positions were created generating 51 variants, and their activity was evaluated toward a series of chiral amines, finding a preference for primary amines and the (*S*)-enantiomer. Among all the variants generated, Y321I and the double variant M226T/Y321I were used to prepare the API (*R*)-Rasagiline and an intermediate of the drug dextromethorphan in 51% and 78% yield, respectively. Further research into the ability to synthesize enantioenriched THQs was recently performed by the same research group. Insight into further residues to target was obtained from the crystal structure, and a new set of 71 variants was created and screened against a panel of structurally diverse THQs [38]. Among them, variants L225A and Q233A were found to be the best-performing catalysts, and they were employed to prepare both enantiomers of several THQs (Scheme 7.5). Recently, Cosgrove et al. combined the Rh(I)-catalyzed conjugate addition/condensation of alkyl vinyl ketones and aminophenylboronic acids with the triple variant CHAO T198F/L199S/M226F developed by Li et al. [36] to prepare a series of 2-substituted THQs, including the des-methyl variants of the alkaloids Cuspareine and Galipeine [39].

Despite the fact that directed evolution is a well-established methodology and is routinely applied in academic and industrial research, it still comprises laborious steps in diversity generation and screening of a large number of transformants. Recently, Hilvert and coworkers developed a high-throughput screening that enables the rapid and effective test of a large number of variants of CHAO [40]. In this system, called fluorescence detection and fluorescence-activated droplet sorting (FADS), single cells are coencapsulated with lysis agents, the substrate of interest, and the detection cascade, which contains a fluorogenic substrate and horseradish peroxidase (HRP). If the variant is active, H_2O_2 is released, and the fluorogenic substrate is oxidized by HRP. Upon cell lysis, the droplet fluorescence is analyzed and droplets exceeding a certain threshold are sorted and collected. They initially identified eight positions either in close contact with the model substrate, 1-phenyl-1,2,3,4-tetrahydroisoquinoline for which the wild type shows no activity, or in the diffusion channel. All residues were randomized simultaneously using different codon degeneracies to build a library containing 1.7×10^6 possible variants. Three rounds of evolution collecting 0.1% of most active variants of each round were performed finding a variant containing five mutations compared to the wild type (L199V, M226T, Y321S, L353I, and P422S). The variant was characterized and found to have a 960-fold improvement in catalytic efficiency (k_{cat}/K_M) compared to that of the wild type.

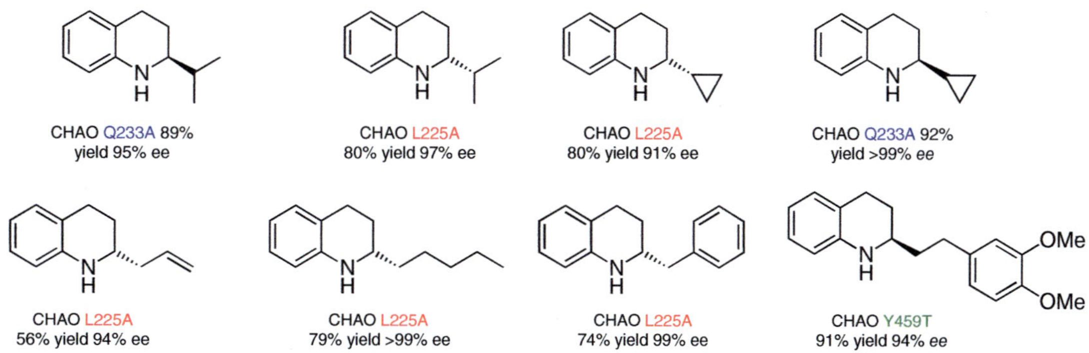

Scheme 7.5 CHAO variants in the biocatalytic deracemization of THQs.

7.2.3 (*R*)-Selective Amine Oxidases

Despite all directed evolution undertaken on MAO-N and CHAO, only some variants developed in the Reetz group showed a preference for the (*R*)-enantiomer. In 2014, the groups of Asano and Turner independently developed two different (*R*)-selective amine oxidases, i.e. a porcine kidney D-amino acid oxidase (pkDAO) and 6-hydroxy-D-nicotine oxidase (6-HDNO), respectively.

7.2.3.1 D-Amino Acid Oxidase (pkDAO)

In 2014, Asano and coworkers engineered pkDAO to develop a (*R*)-selective amine oxidase [41]. The wild-type enzyme catalyzes the oxidative deamination of α-amino acids to the corresponding α-keto acids, but it showed no activity toward simple amines. The crystal structure of this enzyme suggested an interaction between R283 and Y228 in the active site, so the authors hypothesized that by replacing those two residues, the enzyme would be able to recognize substrates bearing no carboxylate moiety. Using (*R*)-α-methyl benzylamine as model substrate, they found that a double variant Y228L/R283G showed no activity toward (*R*)-phenyl alanine but was able to oxidize the substrate, displaying a remarkable specific activity of 21.5 U/mg. Nevertheless, the substrate scope was still very narrow, so further engineering was required. Saturation libraries were designed at three different residues in the active site, which were suggested to create steric constrains with bulky substrates [42]. A variant I230A/R283G was found to be very active toward benzhydrylamines and the deracemization of rac-(4-chlorophenyl)(phenyl)methanamine was achieved in 1 hour.

7.2.3.2 6-Hydroxy-D-nicotine Oxidase (6-HDNO) from *Arthrobacter nicotinovorans*

6-HDNO is involved in the degradation of nicotine, and it is reported to catalyze the oxidation of (*R*)-6-hydroxy-D-nicotine to myosmine [43]. The crystal structure was solved in 2005 [44] and shares similar structural features to other oxidases, although the sequence similarity to MAO-N is very low. The wild-type enzyme presents a very narrow substrate scope that mainly consists of nicotine and its derivatives. In 2014, Heath et al. reported the directed evolution of 6-HDNO by modifying the residues involved in substrate binding and active site volume via CASTing [45]. By targeting three pairs of residues in the active site by site-directed mutagenesis, they found a double variant E350L/E352D with a much larger substrate scope and activity toward a panel of structurally diverse and synthetically relevant cyclic amines. Deracemization using ammonia borane as a reductant and scale up of selected substrates led to the access of a panel of chiral amines in (*S*)-configuration in good to excellent yields and excellent optical purity (Scheme 7.6).

7.3 Amine Dehydrogenases

7.3.1 Introduction

Amine dehydrogenase (AmDH) biocatalysts catalyze the stereoselective reductive amination of simple carbonyl compounds using an ammonium salt as a source of ammonia to afford chiral primary amines. They were first evolved from amino acid dehydrogenase (AADH) enzymes with two single-point mutations altering the specificity, so they did

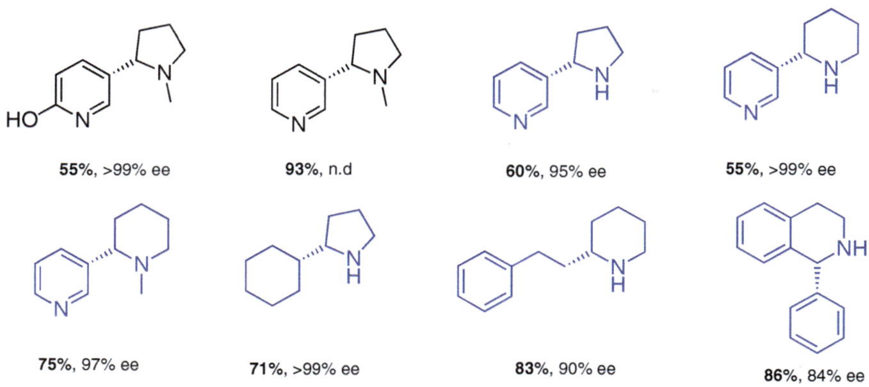

Scheme 7.6 6-HDNO wild-type (black) and D350L/E352D (blue) cyclic amine substrate scope. Isolated yields and enantiomeric excess after deracemization by combination with ammonia borane.

not only produce amino acids. They are nicotinamide dependent and are often coupled with formate dehydrogenase (FDH) recycling systems, permitting the use of ammonium formate as buffer for both amine and sacrificial reductant source. Much work has looked into expanding the substrate scope of AmDH as initially this was quite limited; however, the appeal of cheap and readily available ammonium sources makes this biocatalyst attractive for industrial applications.

7.3.2 Discovery and Engineering of AmDH

7.3.2.1 Leucine Dehydrogenase

The first AmDH variant was evolved from a leucine dehydrogenase (LeuDH) by Bommarius and coworkers [46]. The LeuDH from *Bacillus stearothermophilus* is responsible for the reductive amination of an α-keto acid to afford amino acid leucine. The authors reasoned that through semirational engineering of the LeuDH, they could engineer away from AADH activity and generate an AmDH.

As no crystal structure was available, the crystal structure of phenylalanine dehydrogenase (PheDH) from *Rhodococcus* sp. M4 with high sequence similarity to the LeuDH was used to design libraries at key positions. Two amino acids in particular, namely, K68/67 (LeuDH/PheDH) and N261/263 (LeuDH/PheDH), were shown to interact with the carbonyl moiety in the acid: the lysine residue interacted directly with it, and the asparagine residue, despite interacting with the cofactor via the backbone amino group, also interacted through the tethered amide group. After their initial library screening, the authors found that a synergistic relationship between K68 and N261 was key to enabling high levels of AmDH activity. Further studies have also demonstrated that mutation of the preserved K68 and N261 amino acids alters the specificity of multiple AADHs, and this is the first target when altering the specificity of AADHs to AmDHs.

The initial work reported a four-point variant of the LeuDH (K68S/E114V/N261L/V291C), with activity engineered toward methyl isobutyl ketone (MIBK), the reduced ketone form of the keto-acid precursor of leucine. While the wild-type AADH had no detectable

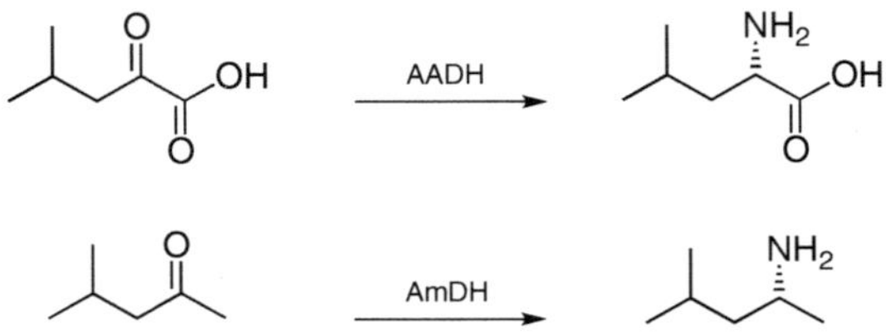

Scheme 7.7 Difference between reactions catalyzed by AADH and AmDH.

activity toward MIBK, the AmDH had a specific activity of 0.7 U/mg and a K_M of 15.1 mM (Scheme 7.7).

Not only did the new AmDH variant demonstrate good activity for reductive amination of ketones, it also retained the high levels of stereoselectivity associated with the wild-type AADH. Using α-methyl benzylamine as a model substrate, activity for the (*R*)-enantiomer (in the deamination direction) was recorded as being around 500 times higher than for the (*S*)-enantiomer, and a preparative scale synthesis of 1,3-dimethylbutylamine also delivered the product in high ee (99.8%).

7.3.2.2 Phenylalanine Dehydrogenase and Chimeric Amine Dehydrogenase

Using the LeuDH as a model study, the Bommarius group applied a similar strategy to a PheDH from *Bacillus badius*, which had a 48% sequence identity to the LeuDH [47]. After sequence alignment, it was found that residues K77 and N276 were the carboxyl-binding amino acids, and saturation mutagenesis at these positions using the reduced amino acid DDK libraries unveiled K77S/N276L as the most active variant toward 4-fluorophenylacetone. This additional AmDH mutant, as highlighted by the authors, was thermostable at 50 °C and thus could provide a good point for evolution of AmDH variants with activity toward more complex organic molecules.

This study was followed up with the development of a chimeric biocatalyst, produced by combining the LeuAmDH and PheAmDH [48]. The primary aim of this study was to increase the substrate scope of AmDH enzymes, with the resulting chimera successfully able to catalyze the reductive amination of new ketones, such as the benzylic ketone, acetophenone, and adamantyl methyl ketone, while also retaining the activity and high levels of stereoselectivity of the two parent enzymes (Scheme 7.8).

7.3.2.3 Native Amine Dehydrogenase

Before the significant engineering by the Bommarius group, only one natural AmDH had been reported in the literature. This involved a report by Itoh et al. [49] of an AmDH from *Streptomyces virginiae* 2827 being shown to catalyze the reductive amination of a large range of carbonyl substrates. However, the stereoselectivities were not very high, the report did not reveal any sequence information, and no subsequent studies have been published. Until 2019, this was the only natural AmDH report; then Grogan, Vergne-Vaxelaire and coworkers discovered a family of what they termed native AmDHs (natAmDHs) based on sequence data of a previously published homolog [50, 51]. The previously described "ω-AmDH" from *Petrotoga mobilis* (termed AmDH4) was found to catalyze the reductive amination of 4-oxopentanoic acid to afford (*S*)-4-aminopentanoic acid, although the presence of the γ-carboxylate was essential for reductive amination activity [50] (Scheme 7.9a).

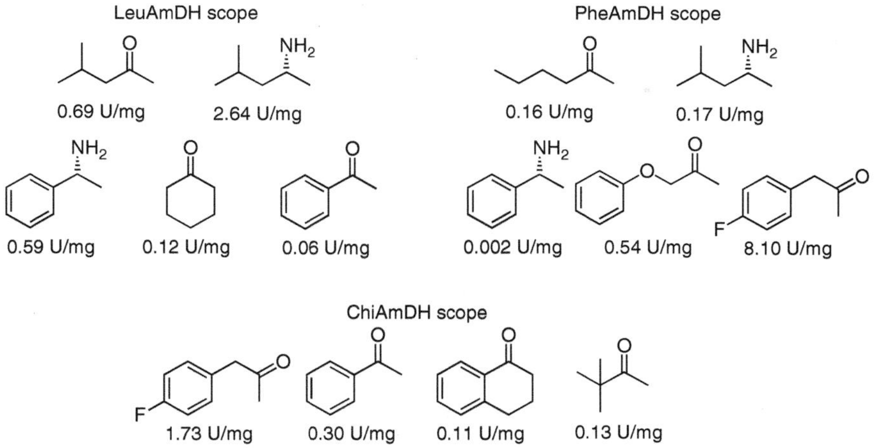

Scheme 7.8 Selected substrate scope of the original AmDH variants developed by the Bommarius group.

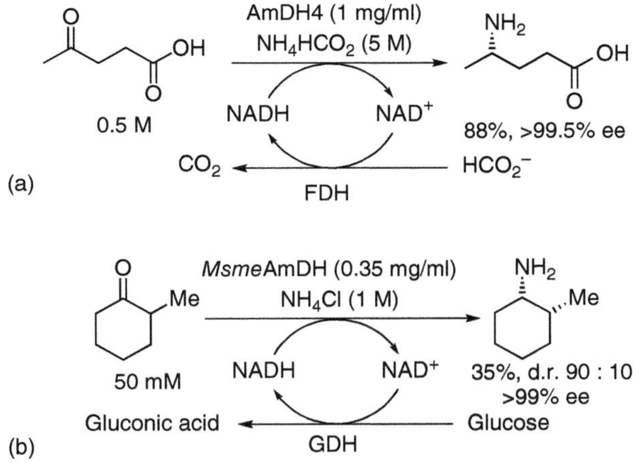

Scheme 7.9 Examples of preparative scale reactions using natural AmDH.

The AmDH4 protein was used as a basis to screen for natAmDH sequences, as it shared little sequence similarity with engineered AmDH, imine reductases (IREDs), or reductive aminases (RedAms) [51–53]. The report disclosed several proteins that were capable of performing reductive amination on a considerable substrate set and demonstrated preparative scale examples on some cyclohexanone derivatives [51] (Scheme 7.9b). The study demonstrated not only that naturally occurring AmDH exist, but that they also provide a platform for further engineering that could increase the substrate scope of this class of biocatalysts.

The combined scope of all reported AmDH enzymes demonstrates a useful set of biocatalysts that can be applied to the synthesis of a large set of chiral amines [54–58]. There have been other engineering efforts toward further expanding the application of AmDH, and recent reports have even demonstrated reductive amination activity with small alkylamines for some AmDHs [51, 59]. These different applications will be discussed in Section 7.3.3.

7.3.3 Synthetic Applications of AmDH

7.3.3.1 Primary Amine Synthesis with Engineered AmDH

The primary application of AmDH has been the reductive amination of prochiral ketones to afford chiral primary amines, with the use of inorganic ammonium sources as an amine donor. These are commonly used with cofactor recycling systems that permit the use of catalytic amounts of the expensive nicotinamide cofactors $NAD(P)^+$ [60]. Often coupled to AmDH reactions are glucose dehydrogenase (GDH) and FDH. FDH allows ammonium formate to be used as a reaction buffer, providing both the amine donor and sacrificial substrate to enable cofactor turnover.

The variants generated by Bommarius and coworkers significantly expanded the chiral amine chemical space accessible by biocatalysts [46–48]. The LeuAmDH variants from the initial work displayed limited amination activity [46], which was complemented by the PheAmDH variant [47], and then, the activity of both was retained in the chimera of the two [48].

As stated, there have been numerous further efforts to increase the scope and synthetic utility of AmDH since the first reports. The Li group, in collaboration with GlaxoSmithKline, expanded the substrate scope of PheDH from *Rhodococcus* sp. M4, as studies had uncovered the structure and mechanism of action, providing a basis for evolution [61]. The authors also pointed out that this PheDH only shared 32% sequence ID with the PheDH from *B. badius* and thus could potentially broaden the AmDH scope. Libraries at the K66/N262 carboxylate binding residues employing NNK degeneracy found that the K66Q/N262C variant was the most active toward phenylacetone and 4-phenyl-2-butanone. This was different from the prior "best" variant from Bommarius and colleagues, who frequently found that K68S/N261L were the most active mutants [46, 47]. An additional library at S149 found that the triple mutant K66Q/S149G/N262C was considerably more active. The authors rationalized that this expanded the binding pocket by around 2 Å, permitting easier access to the active site. Preparative scale reductive amination of 4-phenyl-2-butanone was achieved, using GDH and glucose to recycle NAD^+ and affording >95% conversion at 15 mM concentration, and total turnover numbers of 2800 with respect to NAD^+ [61]. While not engineering the biocatalyst, the same group published a biphasic organic/aqueous system that could be used to expand the scope of the PheAmDH as well; however, only two additional substrates were tested [62].

An approach to increase the binding pocket was taken by Chen et al. using LeuDHs from *Exiguobacterium sibiricum*, *Lysinibacillus fusiformis*, and *Bacillus sphaericus* as templates for evolution [54]. After application of the standard mutations to alter the binding pocket (K68S/N261L), these three variants displayed high thermostability and good activity against some substrates but poor activity toward bulkier ketones, namely, 2-heptanone and 2-octanone. Using a homology model, the authors rationalized that several key amino acids could be targeted for further manipulation to enlarge the active site. The first mutation was A113G, with the triple mutant demonstrating improved activity against 2-hexanone. To enhance the activity further, alanine-scanning mutagenesis (replacing

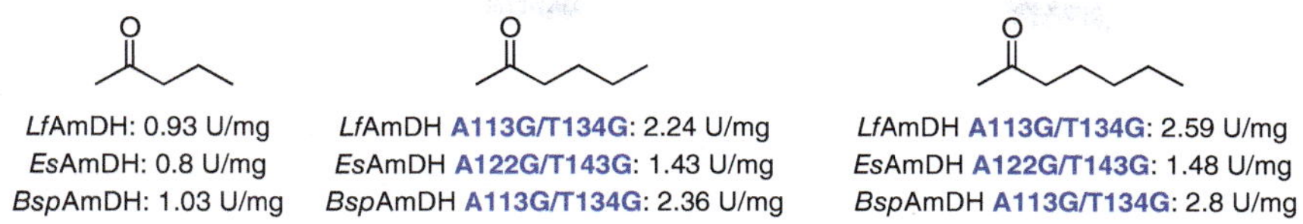

Scheme 7.10 Longer chain ketones accepted by the pocket-expanded AmDH described by Chen et al. [54].

selected amino acids with alanine) at other spots in the active site delivered the variant K68S/A113G/T134A/N261L, which displayed good activity (>1 U/mg) toward 2-hexanone and 2-heptanone. Altering the T134A residue to T134G afforded the final variant, termed *Lf*AmDH-M3. The volume of the active site of the variant with activity toward longer chain ketones was calculated to be 43 Å^3 higher than the wild type. Interestingly, by applying the mutations to the other two LeuAmDH in this study (*Es*AmDH and *Bsp*AmDH), both showed similar improvements in activity to bulkier ketones, thereby offering a new synergistic pair of mutations that could potentially be applied to other AmDH (Scheme 7.10).

In 2017, Mutti and coworkers disclosed one of the most comprehensive studies to date of AmDH [56]. The study compared PheAmDH from *B. badius* generated by the Bommarius group [47], PheAmDH from *Rhodococcus* sp. M4 by the Li group [61], and ChiAmDH [48]. The study initially focused on recycling system selection and looked at the varying stabilities of cofactors and the recycling enzymes in a range of buffers [56]. For all the three AmDHs, an FDH/ammonium formate combination was found to be the most efficient because of the stability of the recycling enzyme and overall efficiency of the transformation. Thermostability studies showed that the enzymes all performed well at a range of temperatures, and importantly, no detrimental effect on the stereoselectivity with >99% ee was obtained. Perhaps the most interesting synthetic case study was the substrate scope determination. Four categories of substrates, with a total of 25 distinct molecules, were all subjected to reductive amination at 50 mM concentration with each of the biocatalysts. Aside from the phenylacetone category, for which *Rs*AmDH proved to be the most effective, ChiAmDH had the broadest scope against all the other substrates, which was perhaps to be expected because it is a combination of two AmDHs. Interestingly, poor chemoselectivity was observed with hydrocinnamaldehyde, where when using ChiAmDH, 30% of the aldehyde substrate was reduced to hydrocinnamyl alcohol. The authors stated that they believed this to be the first instance of carbonyl reductase activity by AmDHs.

Despite the increase in the substrate scope of the engineered AmDHs, the highest k_{cat} observed for these enzymes is 6.85 s^{-1} for PheAmDH from *B. badius* and 4-fluorophenylacetone, which is around 12 times lower than the wild-type AADH for the natural substrate [47, 63]. Researchers from Johnson Matthey tried to address this issue through engineering of a thermostable AmDH from *Caldalkalibacillus thermarum* [63]. They also noted that no previous AmDH had operated with a substrate concentration greater than 80 mM, highlighting their limitations. Their rational was a more thermotolerant wild-type enzyme that could offer a better basis for engineering, allowing more mutations to be applied to increase the substrate tolerance beyond previous AmDHs. The resulting *Cal*AmDH was highly thermostable, however displayed similar activities to PheAmDH across a panel of 18 substrates, and was not further engineered in this

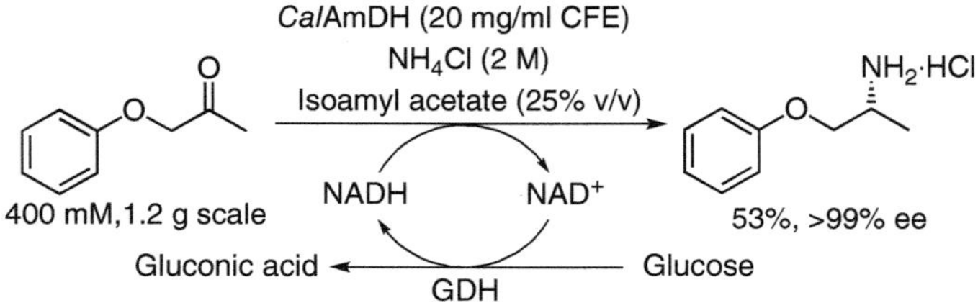

Scheme 7.11 Scale-up reductive amination of 2-phenoxypropanone. CFE, cell-free extract.

study. Careful inspection and optimization of the reaction conditions did allow conversion of phenoxy-2-propanone at 400 mM to the desired amine in 96% conversion and a 53% isolated yield within 48 hours, by applying a biphasic system and lyophilized whole cells (Scheme 7.11).

Although this approach did achieve higher substrate concentrations and conversions, it fell short of solving the issue of low substrate tolerance and low k_{cat}. The authors did consider the binding affinities of both *Cal*AADH and *Cal*AmDH for their respective substrates to try and explain the low k_{cat} values. The K_M values for binding each of the reactants were substantially different and were attributed in part to the lack of carboxylate binding residues that stabilize the amino acid precursors. Furthermore, computational analysis predicted that the binding energies for the imines were always higher than alkyl ketones, adding weight to the argument that AmDHs were not as well set up to perform the reduction as their AADH counterparts.

7.3.3.2 Primary Amine Synthesis with Natural AmDH

As discussed previously, some AmDHs have recently been discovered that naturally catalyze the reductive amination of isolated ketones [50, 51]. The first report from Mayol et al. used metabolic data mining to find a series of what they termed the first ω-AmDH, as no α- or β-carboxylate was necessary for reductive amination activity in the wild type [50]. The authors described finding nine sequences based on similarity to the 2,4-diaminopentanoate dehydrogenase from *Clostridium sticklandii*. The nine enzymes were expressed and screened against a panel of substrates derived from 2,4-diaminopentanoate to reveal the ω-AmDH, which they referred to as AmDH4. The AmDH4 biocatalyst had high activity against 4-oxo-pentanoic acid, revealing the lack of requirement for an α- or β-carboxylate; however, the substrate scope was limited to only a few compounds. The thermostability of the enzyme was revealed to be very high, with a half-life of 65 hours at 60 °C. This proved to be a key point for the authors, who stressed that the enzyme may provide a platform to engineer a new AmDH that would be distinct from the original AADH-derived families.

This vision was later realized by Mutti and coworkers [58]. Taking a similar approach to the previous report, a lysine dehydrogenase from *Geobacillus stearothermophilus*, which is responsible for the deamination of the remote amino group in L-lysine, was tested for activity in the reverse direction toward 6-oxo-hexanoic acid. At a 10 mM substrate concentration, 44% conversion was measured. Homology modeling and a single-point mutation based on rational design delivered the variant F173A, which displayed greater than 50% conversion in biotransformations for most of the tested substrate panel of 17 substrates that

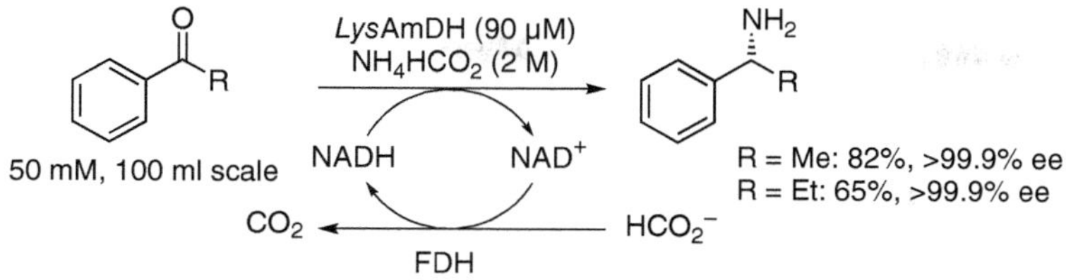

Scheme 7.12 Preparative scale biosynthesis of benzylic amines.

included 2-hexanone, acetophenone, and 4-chromanone. A large section of the substrate scope contained acetophenone derivatives, one substrate class that has been challenging for the earlier described AmDH; this clearly shows that approaches such as this can deliver AmDHs that have a distinct substrate scope and therefore additional synthetic applications. Preparative scale reactions of acetophenone and phenylethylketone delivered the desired amines in 82% and 65% isolated yields, respectively, on 5 mmol scales (Scheme 7.12).

The native AmDH discussed earlier also had a different substrate scope to the engineered AmDH family [51]. Specific activity was determined toward a panel of 22 substrates across the four natAmDH biocatalysts described, with the natAmDH from *Microbacterium sp.* being the most active. It demonstrated >50 mU/mg toward 17 of the substrate panel when $NADP^+$ was used as a cofactor. This was also chosen for one of the preparative scale reactions, with four examples being shown and isolated yields >45% obtained for all.

The new native AmDH families that have been uncovered recently have demonstrated that through new approaches to AmDH discovery and design, the synthetic application of AmDHs can be significantly expanded beyond that of the engineered AmDHs [50, 51, 58]. Section 7.3.3.3 will discuss recent expansions of the substrate scope of AmDHs away from exclusively generating primary alkylamines as products and also how multienzyme cascades can further improve the efficiency of AmDHs.

7.3.3.3 Substrate Promiscuity in AmDH

Reductive Aminase Activity

Some reports have demonstrated that AmDHs may display promiscuous reductive amination activity toward small alkylamine donors under certain conditions [51, 59]. This results in secondary or tertiary amines being produced as products, something that hitherto had only been achieved biocatalytically by a subset of the IRED family [53, 64, 65].

Mutti and coworkers were the first to report reductive amination activity of AmDHs with alkylamines [59]. The study screened the ChiAmDH [48], and the PheAmDH from *Rhodococcus* sp. M4 originally described by the Li group [61]. A substantial substrate scope revealed a limited number of hits, with methylamine, ethylamine, propargylamine, and propylamine all providing conversions of around 40% with at least one of the carbonyl compounds. Nevertheless, these reactions were primarily run with 100-fold excess of amine donor in most cases (1 M amine and 10 mM carbonyl) with a necessity to increase to 600-fold for methylamine (6 M amine and 10 mM carbonyl) to obtain synthetically useful conversions. An unexpected result of the study was the formation of primary amines when using the alkylamine donors. The authors posed several mechanistic rationalizations to explain how this could be happening, with the reversible nature of AmDH thought to play a key role

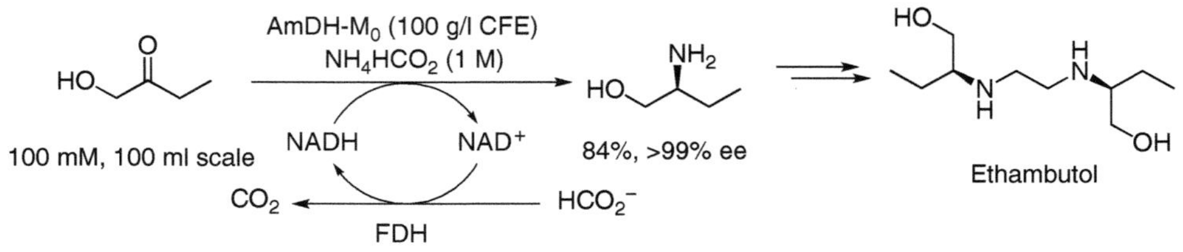

Scheme 7.13 AmDH synthesis of intermediate for ethambutol.

in dealkylating the products by oxidizing the product on the opposite bond of the secondary amine being supported by experimental data. Nonetheless, no preparative scale reactions were undertaken, raising the question of stability of the enzyme in high concentrations of alkylamine solution.

Grogan and coworkers also demonstrated the natAmDH could catalyze similar transformations with methylamine as an amine source [51]. However, only cyclohexanone and isobutyraldehyde were tested, and no conversions were determined, just specific activities. Furthermore, they were also tested with a 100-fold excess of methylamine (0.5 M methylamine and 5 mM carbonyl), showing that this was also just a proof-of-concept demonstration of primary amine reductive amination activity.

With such high requirements for concentration of organic amine donor, these reactions are likely unfeasible for industrial application; however, potential for improvement remains through the use of directed evolution methodologies.

Amino Alcohol Synthesis

Amino alcohols are important motifs in many bioactive molecules, as such methods for the synthesis of such moieties are necessary. As with chiral amines, synthesis using transition metal catalysis generates toxic waste and often proceeds through inefficient routes. A recent report from Turner and coworkers showed an engineered variant of an earlier described LeuAmDH from *L. fusiformis* that could perform the reductive amination of several hydroxyketone substrates to yield enantiopure vicinal amino alcohols [66]. A small investigation on substrate scope revealed that several amino alcohols could be generated from the respective hydroxy ketone substrates with perfect stereoselectivity and generally high conversions. A preparative scale reaction was applied to the synthesis of (*S*)-2-amino-1-butanol, a key intermediate in the synthesis of tuberculosis drug ethambutol, with an 84% isolated yield, and >99% ee on a 10 mmol scale highlighting the reaction utility (Scheme 7.13).

7.3.3.4 Cascade Reactions that Use AmDH

Cascade reactions are chemical reactions in which several catalysts or reagents generate a substrate that can be used immediately in the same vessel, thereby increasing the efficiency as intermediates do not need to be isolated. Biocatalysts possess an intrinsic ability to perform these reactions, as most operate in aqueous media at a consistent temperature range, with high chemoselectivity that allows for reactions with only substrates that are specific to them.

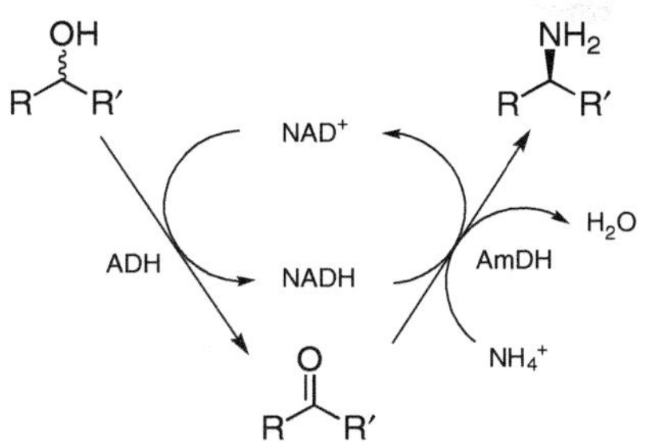

Scheme 7.14 Biocatalytic hydrogen borrowing cascade for chiral amine synthesis from racemic alcohols. Source: Based on Chen et al. [55].

Biocatalytic Hydrogen Borrowing

There have been several reports of AmDH acting within cascade reactions, with one of the earliest being the biocatalytic borrowing hydrogen cascades reported by Turner and coworkers [57]. The original report described the conversion of racemic alcohols into chiral amines through the combined use of stereocomplementary alcohol dehydrogenases (ADHs) and AmDH, specifically the PheAmDH from *B. badius*. The cascade proceeded with catalytic quantities of NAD^+ as with other AmDH reactions; however, the ADH was dependent on NAD^+ and generated NADH as a by-product, which was used in the AmDH reaction, thereby delivering the product and regenerating the cofactor for the alcohol oxidation reaction, without the need of a sacrificial substrate. This elegant cascade only generated water as a by-product and delivered the products in high conversion and high enantioselectivity [57]. This was followed up by a similar study from Chen et al., who identified a nonselective ADH from *Streptomyces coelicolor*, thereby avoiding the need for two complementary ADHs in the alcohol oxidation step (Scheme 7.14) [55].

There have been other advancements of this cascade since, including an expansion of the substrate scope [67], investigating immobilization of biocatalyst for increased stability [68], and in vivo expression of biocatalysts alleviating the need for protein purification and exogenous addition of nicotinamide cofactor [69, 70].

Other Multienzyme Cascades

Xu and coworkers extended the cascade described previously one step further, taking unfunctionalized alkanes and converting them to the prerequisite alcohol using a P450-mediated hydroxylation [71]. Despite the impressive transformation, this was only achieved using cyclohexane to afford cyclohexylamine, and because of the incompatibility of the different systems, a large amount of reaction engineering was required to ensure that the reaction could proceed in acceptable levels. Nevertheless, this did represent a formal C–H amination transformation.

The same group has also utilized AmDH as a recycling enzyme to drive transaminase-catalyzed reductive amination [72]. TA enzymes catalyze the transfer of an amino group via the cofactor pyridoxal phosphate; however, the reaction is reversible, so methods are required to drive the reaction equilibrium to the desired product [1]. By using AmDH to convert the ketone by-product back to the amine donor for the reaction, this not only increased the efficiency of the TA reaction but also meant that the only stoichiometric amino source

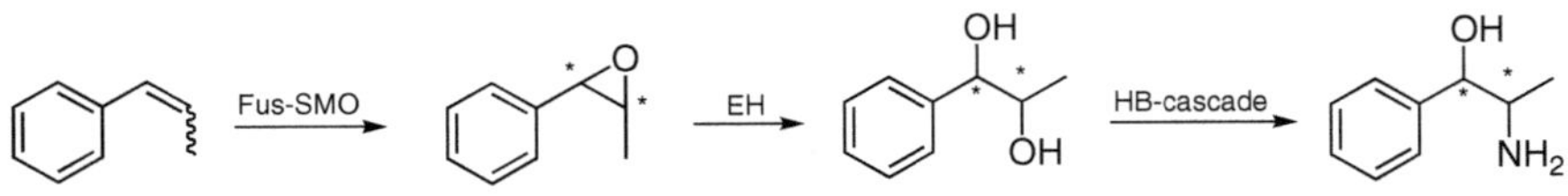

Scheme 7.15 Sequential dihydroxylation/hydrogen borrowing amino-hydroxylation of methylstyren. Fus-SMO, fused styrene monooxygenase; EH, epoxide hydrolase; and HB, hydrogen borrowing.

was the ammonium buffer, a benign and cheaper source of amine than is often used in TA reactions (alanine, isopropylamine, etc.) [72].

Mutti and coworkers revealed a four-enzyme cascade that could convert methylstyrene to 1,2-amino alcohols in sequential reactions [73]. The first sequence witnessed a styrene monooxygenase-catalyzed (SMO) epoxidation of the double bond, followed by a stereoselective ring opening mediated by an epoxide hydrolase (EH). Careful choice of EH could deliver the 1,2-diols in any of the four stereoisomers, which were in turn delivered to the hydrogen borrowing cascade to convert, selectively, the remote alcohol to a chiral amine moiety [73]. The reaction was performed on a preparative scale (up to 100 mg), and the authors calculated a simple *E*-factor (no solvent contribution), which was around five to eight times lower than an equivalent chemical route to make the same compound (Scheme 7.15).

Chemoenzymatic Cascades

The combination of chemo- and biocatalysis offers much promise to access new areas of chemical space [74]. Gröger and coworker recently demonstrated a one-pot, two-step formal hydroamination of styrene to afford α-methylbenzylamine [75]. The cascade was initiated with a $PdCl_2/CuCl$-catalyzed Wacker oxidation to afford acetophenone, followed by an in situ AmDH-catalyzed reductive amination. Initial tests showed that the AmDH was not tolerant of Cu(I) salts necessary for the oxidative step, so the use of a polydimethylsiloxane (PDMS) membrane allowed internal compartmentalization of the transition metal reaction, thus shielding the enzyme and permitting a one-pot transformation. Although only one simple substrate was demonstrated, this technology holds promise for further chemoenzymatic synthesis, especially with regard to AmDH, which requires highly coordinating ammonia as a buffer.

7.4 Imine Reductases

7.4.1 From Biosynthesis to Biocatalysis

With the ubiquity of cyclic amine motifs within not just secondary metabolites but also molecules instrumental in all life, it is unsurprising that imine reduction is utilized by nature to install chirality within many biosynthetic pathways. Present within all organisms, dihydrofolate reductase (DHFR) catalyzes the reduction of dihydrofolate (DHF), a dihydropteridone substrate, to yield 5,6,7,8-tetrahydrofolate (Scheme 7.16a), a cofactor involved in the synthesis of purine bases. Further examples include enzymes performing imine reduction as a subunit of a larger polyketide synthase (PKS), capable of forming

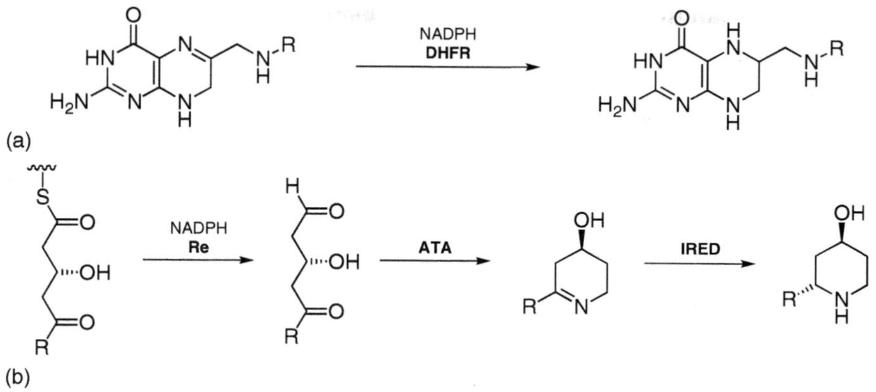

Scheme 7.16 Imine reductases in biosynthesis. Source: (b) Based on Peng et al. [76].

piperidine rings through sequential thioester reduction, transamination, cyclization, and reduction, such as in the biosynthesis of indolizomycin [76] (Scheme 7.16b) and agrimycin [77] alkaloids. However, despite the wealth of natural examples and the clear synthetic importance of enzymes capable of stereoselective imine reduction, application in the form of a biocatalysis platform was limited by these highly specialized enzymes demonstrating an extremely narrow substrate scope.

The initial breakthrough in IRED biocatalysis came from Mitsukura et al., who screened a large collection of microorganisms for reduction activity toward 2-methyl-1-pyrroline (2-MPN), an imine that is not readily hydrolyzed in water. Two species of *Streptomyces* bacteria were identified as possessing enantiocomplementary reduction activity [78]. This was followed by the initial characterization of the (*R*)-selective IRED from *Streptomyces* sp. GF3587 (Scheme 7.17), which showed high specificity toward 2-MPN as a substrate [79]. The (*S*)-selective IRED from *Streptomyces* sp. GF3546 (Scheme 7.17) was also characterized and was observed to act on further substrates such as 3,4-dihydroisoquinolines, initially suggesting that this IRED was more suited to biocatalytic application.

Despite the two enzymes possessing a sequence alignment of only 37%, many similarities were noted such as NADPH dependence, dimer formation in solution, and an approximate molecular mass of 30 kDa. The enzymes both displayed no reduction activity toward DHF and crucially toward a number of ketone substrates, indicating that IREDs were indeed likely chemoselective toward the reduction of imines. This assumption was challenged by Nestl and coworkers, who demonstrated that IREDs are capable of the reduction of the highly activated ketone 2,2,2-trifluoroacetophenone [80]. However, this result was paired with in silico studies, suggesting that the energy barrier for NADPH hydride transfer to

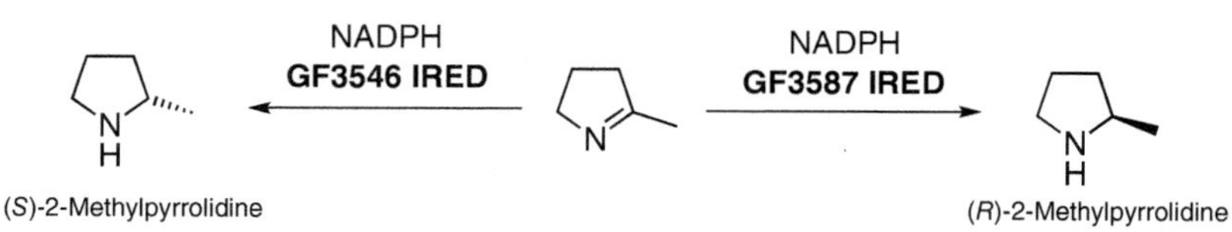

Scheme 7.17 Enantioselective reduction of 2-methylpyrroline.

this substrate falls within the same range as that of iminium substrates. As imines are not typically protonated at the operating pH range of IREDs, it is widely speculated that chemoselectivity arises from selective protonation of imine substrates within the active site. Despite this, the full mechanism of IRED imine reduction remains unknown. Although IREDs are the only known class of biocatalysts capable of chemoselective imine reduction, promiscuous activity has been observed in GDH enzymes, which may reduce iminium substrates stereoselectively [81].

7.4.2 Biocatalytic Application of Imine Reductases

Following the foundational work by Mitsukura and coworkers, Turner and coworkers sought to expand on promising results observed for the (*S*)-selective IRED from *Streptomyces* sp. GF3546. Following codon optimization, the enzyme was expressed within *E. coli* to provide a recombinant whole cell biocatalyst for activity and biotransformation studies [82]. Further probing of the substrate scope revealed that (*S*)-IRED displayed a higher activity for 2-methyl-1-piperidiene than the previously reported 2-MPN and was also capable of reducing seven-membered rings. The biocatalyst was observed to perform best with bulky polycyclic substrates in which the imine formed part of a six-membered ring, with (*S*)-1-methyltetrahydroisoquinoline (Scheme 7.18a) and (*S*)-1-methyltryptoline both synthesized in full conversion from their respective imines with quantitative enantioselectivity after 1 hour.

Despite the narrow substrate scope previously reported, the (*R*)-selective IRED from *Streptomyces* sp. GF3587 was also studied for biocatalytic application, again through the use of heterologous *E. coli* expression [83]. Once again, a preference for piperidine substrates was observed, tolerating a large range of 2-substituents while maintaining high enantioselectivity (Scheme 7.18b). The enzyme was able to catalyze the synthesis of azepanes as well as a small number of THIQs. The enzyme however did not prove to be enantiocomplementary to (*S*)-IRED, yielding for many products the (*S*)-enantiomer.

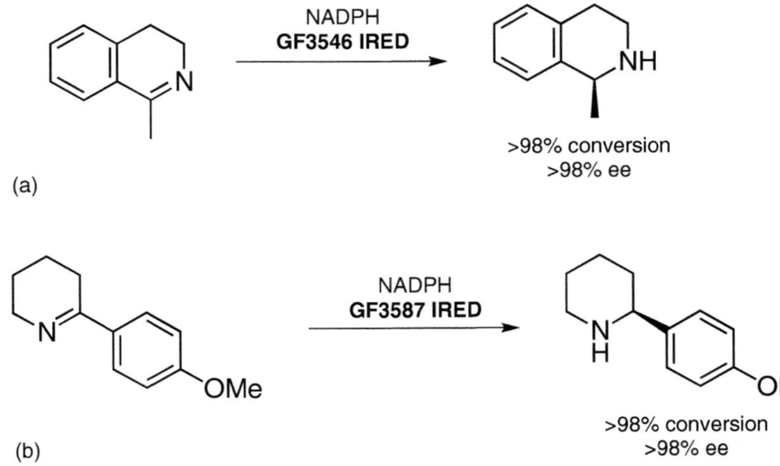

Scheme 7.18 Further Mitsukura IRED transformations.

With the identification and characterization of the Mitsukura IREDs, a starting point had been provided for the identification of related sequences through genomics, facilitating the initial development of a larger biocatalytic toolbox for imine reduction. Multiple groups reported the characterization of new IREDs [84, 85], including researchers from Roche, who marked the entrance of industry into the field with a panel of 20 IREDs [86]. These enzymes demonstrated a similar substrate scope to those of the Mitsukura IREDs reported by the Turner group, including five- to seven-membered rings and bicyclics but crucially provided enantiocomplementary options for all substrates tested, as well as an ability to tolerate aryl-brominated THIQs suitable for cross-coupling. The Turner group reported an IRED from *Amycolatopsis orientalis*, which while displaying high activity across a range of substrates often yielded opposite stereoselectivity for analogous substrates, as seen in the "(*R*)-IRED" from *Streptomyces* sp. GF3587 [87]. Further IREDs suitable for the reduction of bicyclic substrates were subsequently reported, with Zheng and coworkers describing an IRED from *Paenibacillus lactis* capable of synthesizing a range of chiral indolines [88], while the groups of Zheng and Qu expanded the IRED toolbox for THIQ synthesis by reporting an IRED and an IRED panel suitable for the synthesis of bulkier alkyl- and aryl-substituted THIQs, respectively [88, 89]. Finally, Zhu and coworkers demonstrated that a panel of IREDs could be used to synthesize a range of 1-benzyloctahydroquinolines, including examples with aromatic substitution on the aryl motif [90].

The potential for IREDs to reduce cyclic imine substrates carrying an additional heteroatom within the ring has also been widely studied by the Gröger group, who have demonstrated the ability of certain IREDs to tolerate both sulfur and oxygen heterocycles. Although thiazolinyl IREDs are known in nature for the biosynthesis of siderophores [91], biocatalytic application is once again limited by specificity. Gröger and coworker first opted to screen known IREDs for reducing activity toward 3-thiazolines and 2*H*-1,4-benzothiazines, finding enantioselective reduction ability for a range of substrates, including spiro-compounds [92]. This was later expanded to include products monosubstituted at the 2-position, although no enantioselectivity was observed [93]. An investigation into the ability of known IREDs to catalyze the reduction of 2*H*-1,4-benzoxazines revealed both an ability to tolerate steric bulk and achieve high enantioselectivity [94]. These demonstrations of IRED capability to tolerate more complex and polar ring systems and set stereocenters highlight their potential for use in building medicinally relevant architectures.

7.4.2.1 IREDs in Cascade and Chemoenzymatic Synthesis

Often existing as part of a larger PKS in nature, and utilizing substrates that may have limited stability within water, it is unsurprising that IREDs have found application within biocatalytic cascade reactions. The idea of forming a cyclic imine through an initial generation of an amino-carbonyl substrate that may spontaneously cyclize was exploited by the Turner group through the transamination of either diketones or keto-aldehydes (Scheme 7.19a) generated by the prior enzymatic reduction of a keto-acid [96]. Prior enzymatic steps allowed for the Mitsukura IREDs to be applied in the synthesis of a range of disubstituted pyrrolidine and piperidine products with diastereoselectivity. Interestingly, the piperidine forming PKS found in indolizidine biosynthesis that mirrors this cascade was first reported after this manuscript [76], demonstrating that biocatalysis researchers

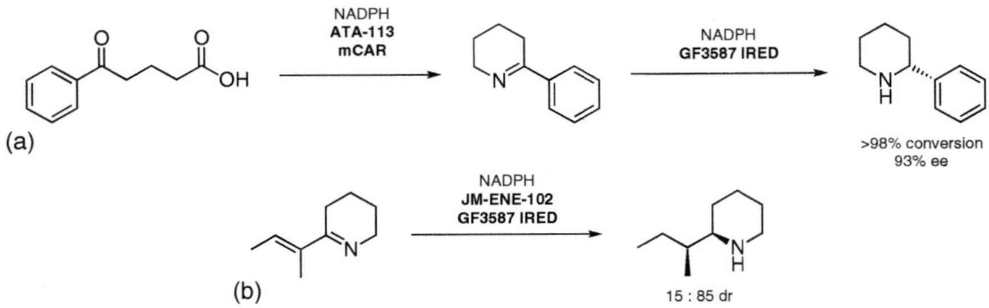

Scheme 7.19 IREDs in biocatalytic cascade synthesis. Source: (b) Thorpe et al. [95].

and nature had arrived at the same retrosynthetic conclusion. This work was later expanded to include the creation of a single whole cell biocatalyst capable of achieving the keto-acid to piperidine transformation through expression of multiple enzymes [97] and the use of longer chain ketones to form long-chain 2,5-disubstitued pyrrolidines [98]. The Turner group later demonstrated how the combination of an IRED with an ene-reductase (ERED) could be used to perform the sequential reduction of α,β-unsaturated cyclic imines, affording a range of diastereomerically enriched heterocyclic products (Scheme 7.19b) [95]. Nestl and coworkers have also contributed widely to the field of IRED cascades, demonstrating the ability of IREDs to perform reduction twice in a single-enzyme multiple reaction cascade in which 1,2-dicarbonyls and 1,2-diamines were combined to form substitute piperazines stereoselectively [99], as well as a cascade in which pyrrolidines and piperidines are formed from diamines through conversion to amino-aldehydes by the enzyme putrescine oxidase [100].

Although cyclic imines may be readily formed by cyclization chemistry, the most established biocatalytic method for their formation is the use of amine oxidases discussed earlier in this chapter. Turner and coworkers speculated that the rate of amine oxidase deracemization could be increased by replacing nonselective chemical reducing agents with an IRED and demonstrated this concept through the one-pot combination of 6-HDNO with the Mitsukura IRED from *Streptomyces sp.* GF3587 to deracemize a number of pyrrolidine and piperidine substrates [101].

In addition to their application in biocatalytic cascades, the ability of IREDs to set stereocenters within amine heterocycles with excellent selectivity has been utilized in chemoenzymatic methodologies to deliver many interesting products. Gröger and coworker expanded their 3-thiazoline reduction work to include the preceding multicomponent Asinger synthesis in one pot through the use of a PDMS thimble, in a similar manner to that discussed earlier (see Section 7.3.3.4.3) (Scheme 7.20a) [102]. Turner and coworkers demonstrated a route to enantiomerically enriched dibenz[*c*,*e*]azepines through the initial Suzuki–Miyaura cross-coupling of a protected benzylamine and benzylketone, followed by deprotection to yield the imine through condensation. This imine could then be reduced using an IRED to install chirality along the biaryl axis [104]. The use of a transaminase in the synthesis of the protected benzylamine was also demonstrated, allowing for a prior vicinal stereocenter to be set in the preceding imine. The Turner group also developed methodology for the asymmetric synthesis of 2,2-disubstituted azepanes through the initial IRED-catalyzed generation of a 2-arylsubstituted azepane, which was

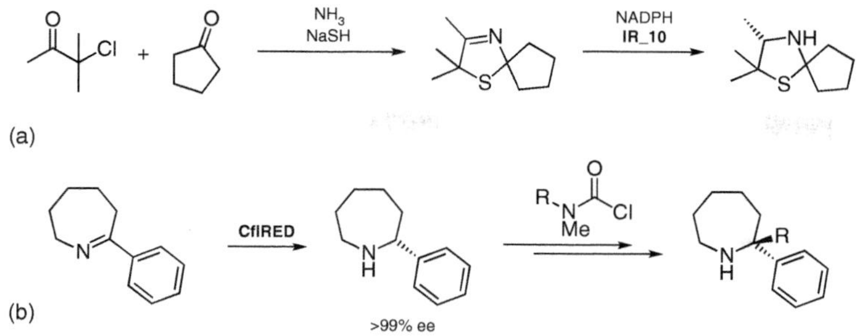

Scheme 7.20 IREDs in chemoenzymatic synthesis. Source: (a) Based on Zumbrägel and Gröger [102]. (b) Based on Zawodny et al. [103].

then subjected to urealation and a lithium-mediated stereospecific transfer to yield the final product (Scheme 7.20b) [103].

7.4.3 IRED Engineering

Although widely established biocatalysts, the engineering of IREDs through directed evolution has been hindered by limitations such as the lack of robust screening methods and mechanistic unknowns. Although there are limited examples of the application of natural IRED mutants in synthesis, one area in which directed evolution has been instrumental is the design of artificial IREDs. The Ward group has been focused on the creation of artificial transfer hydrogenases (ATHase) capable of imine reduction by the anchoring of iridium piano stool complexes within an enzyme-active site through biotin–streptavidin technology. Using wild-type and single-point mutation, streptavidin (Sav) imine reduction was demonstrated with high conversion and enantioselectivity [105, 106]. This was followed by work in which the secondary coordination sphere provided by the active site was subjected to mutation by selection of residues that fell within 5 Å of the iridium center, allowing for the creation of metalloenzymes that rival natural IREDs for combined conversion and enantioselectivity.

7.4.4 Imine Reductases Catalyzing Reductive Amination

Although amine synthesis is perhaps the most developed reaction set in the biocatalytic toolbox, the most established options exist for primary amination and installing chirality within cyclic amines, with options for amine alkylation long eluding researchers. Enzymes capable of performing reductive amination with primary or secondary amines had been limited in the literature beyond examples such as a patent from Codexis detailing the use of engineered opine dehydrogenases for reductive amination and the ketimine reductase "μ-crystallin (CRYM)," capable of performing reductive methylamination of a range of amino acids [107, 108]. With an ability to selectively reduce imine substrates in the presence of carbonyl compounds, IREDs had long been speculated as potential biocatalysts for asymmetric reductive amination, with the difficulty in forming an exocyclic imine in aqueous media the major limitation. In 2014, Müller and coworkers reported the first IRED-mediated reductive aminations, successfully reacting a panel of aromatic ketones with methylamine, albeit in low yield with high enzyme loading and amine equivalence [109]. This was followed by the Nestl group who demonstrated that the reductive amination of aldehydes could be achieved in good conversion under conditions, which promoted the formation of stabilized exocyclic imines in water [110]. Researchers from Roche screened their IRED panel for reductive amination activity, noting particularly good conversions for the combination of cyclohexanone derivatives and methylamine and achieving preparative scale synthesis [111]. The Höhne group demonstrated the synthesis of both enantiomers of the API rasagiline in good yield and enantiomeric excess [112], as well as an ability of IREDs to provide β-chiral amine products through the dynamic kinetic resolution and reductive amination of α-branched aldehyde substrates [113].

Despite a wide range of examples appearing in the literature, high conversions to product were consistently only observed through the application of a large excess of amine donor, limiting the synthetic applicability of IRED-mediated asymmetric reductive amination. A major breakthrough came in 2017 when the Turner group reported an IRED capable of achieving conversions of up to 94% with a 1 : 1 ratio of carbonyl to amine (Scheme 7.21a) alongside further examples requiring a small amine excess [64]. While all previously reported IREDs had been isolated from bacteria, this enzyme constituted the first homolog to be cloned from a eukaryotic source, the fungus *Aspergillus oryzae*. The ability of this biocatalyst to perform asymmetric reductive amination with such low amine equivalence in water suggested a mechanism separate to those seen in previously studied IREDs, in which the initial imine formation step was also catalyzed, leading to the designation of a subclass of IREDs known as "reductive aminases" (RedAms) and the enzyme being termed "*Asp*RedAm." Investigations into this mechanism revealed that the productive conformation for catalysis is only achieved upon the binding of both carbonyl and amine substrate, allowing domain closure to form a much smaller active site than in previously examined IREDs, providing the hydrophobicity required to form an exocyclic imine for reduction [65].

With the capability of IREDs to fully catalyze asymmetric reductive amination demonstrated, bioinformatic efforts were focused on identifying more enzymes of the RedAm subclass. Work from the Turner group determined that this reactivity was not unique to eukaryotic IREDs, with a manuscript that reported a number of novel bacterial enzymes capable of performing low amine equivalence reductive amination utilizing both primary and secondary amine donors, such as pyrrolidine (Scheme 7.21b) [52]. The potential for scaled reductive amination sparked further industrial interest, with researchers from GlaxoSmithKline also reporting a panel of IREDs with reductive amination activity. These biocatalysts significantly expanded the scope of amine donor substrates, with a range of arylamine substrates demonstrated in high conversion. The combination of cyclohexanone and benzene-1,4-diamine was also studied and demonstrated the outstanding chemoselectivity of these biocatalysts, with certain IREDs from the panel showing complete selectivity for the single reductive amination product (Scheme 7.21c) [114].

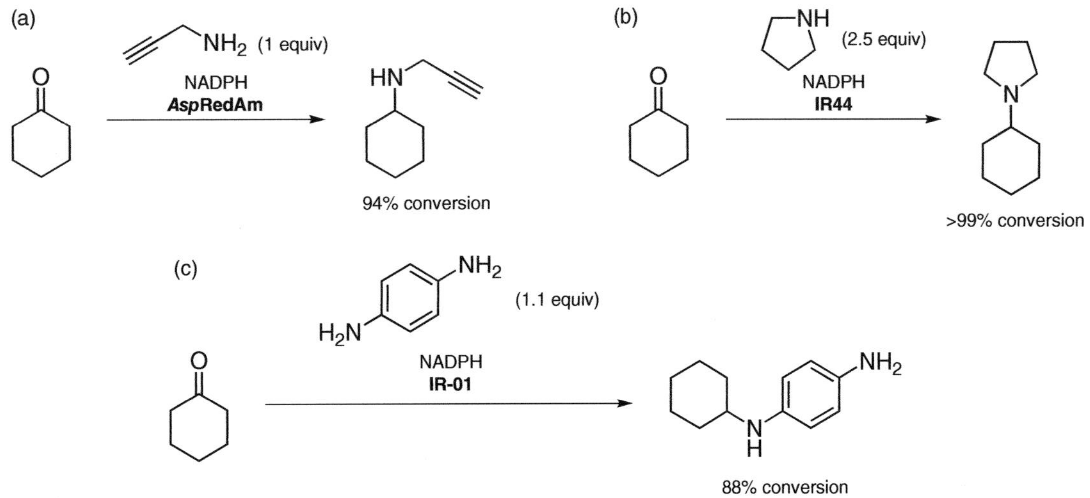

Scheme 7.21 IRED-catalyzed reductive amination reactions. Source: (b) Based on France et al. [52] and (c) Based on Roiban et al. [114].

7.4.5 Imine Reductase-Catalyzed Amine Alkylation Cascades

In recent years, the development of more sustainable methodology for amine alkylation has been identified by researchers as a key step toward more renewable synthesis, with equivalents to reductive amination that instead employ redox surrogate reagents to carbonyl compounds receiving particular focus. The ability of enzymes to mirror borrowing hydrogen catalysis has been previously discussed in this chapter and with the addition of RedAms to the biocatalytic toolbox could be expanded to include a much broader range of amines than ammonia. By combining *Asp*RedAm with a range of single $NADP^+$-dependent ADHs, the Turner group demonstrated an ability to alkylate a range of amines using either primary or secondary alcohols, with excellent enantioselectivity achieved in the latter case (Scheme 7.22a) [115]. Furthermore, this cascade was extended with the addition of a cytochrome P450 monooxygenase to affect amine alkylation with nonfunctionalized cycloalkanes, an unprecedented transformation in the synthetic literature (Scheme 7.22b) [116]. The Turner group have also demonstrated the combination of *Asp*RedAm with an engineered alcohol oxidase [118] and wild-type carboxylic acid reductase (CAR) to alkylate amines with either alcohol or carboxylic acid substrates (Scheme 7.22c) [117]. Although this methodology overcomes the thermodynamic limitations of biocatalytic hydrogen borrowing, it is limited to the in situ generation of aldehydes.

7.4.6 Engineering of Reductive Aminases

With the potential impact of a green platform for asymmetric amine alkylation, it is unsurprising that IREDs with reductive amination activity have begun to find application within chemical industry. Pfizer reported prior art in which an IRED catalyzing asymmetric reductive amination of the preceding tetralone substrate with methylamine may afford the blockbuster drug (*S*,*S*)-sertraline [119]. It is industrial interest in the field that has yielded the first successful directed evolution campaign of IREDs. Although limitations in screening have restricted academic efforts, researchers at GSK successfully applied a robotics platform to develop a biocatalyst capable of catalyzing a key reductive amination step in the process synthesis of the lysine-specific demethylase-1 inhibitor GSK2879552, eliminating the requirement for a prior resolution (Scheme 7.23) [120]. Through three rounds of evolution researchers produced an engineered IRED with a >38 000-fold improvement over the wild-type enzyme, capable of producing the product in near-perfect enantioselectivity under acidic conditions for product stability, an acidophilic enzyme having been previously unreported in IRED catalysis. The researchers also discussed an investigation into replacing the preceding alcohol oxidation step with a hydrogen borrowing cascade but reported that thermodynamic limitations prevented economic viability.

7.5 Engineered Cytochrome P450s

One of the most recent classes of biocatalysts that have been applied in the synthesis of chiral amines is the cytochrome P450. More commonly associated with hydroxylation and the insertion of oxygen atoms, recent efforts from several groups have demonstrated

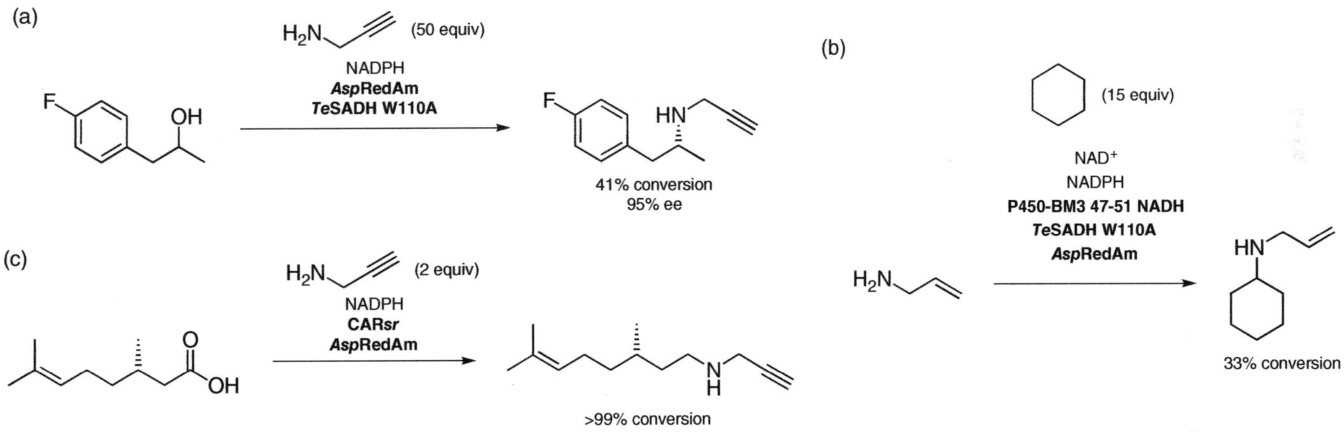

Scheme 7.22 Applications of reductive aminases in amine alkylation cascades. Source: (a) Based on Montgomery et al. [115], (b) based on Tavanti et al. [116], and (c) based on Ramsden et al. [117].

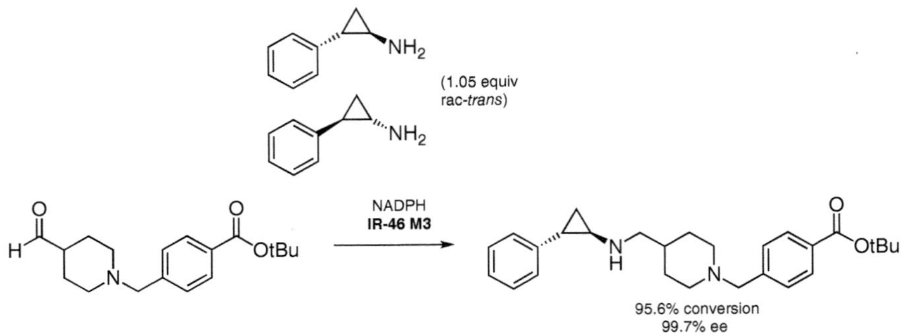

Scheme 7.23 Reductive aminase-catalyzed step in the synthesis of GSK2879552. Source: Based on Schober et al. [120].

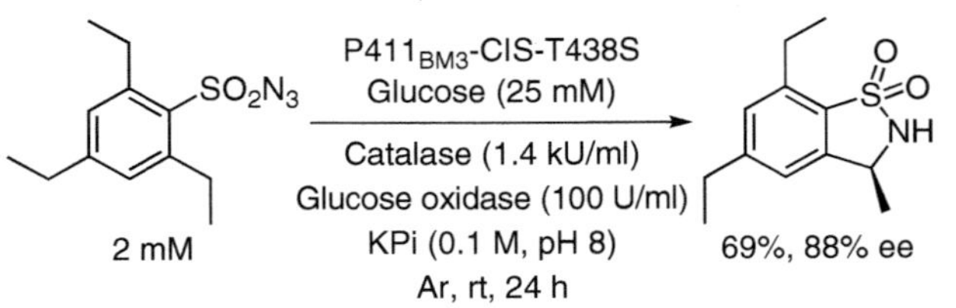

Scheme 7.24 P411-catalyzed intramolecular C–H amination. Source: Based on McIntosh et al. [123].

that these enzymes can be "retasked" to perform non-natural chemistry, including the direct amination of a range of substrates [121]. Inspired by early work from Dawson and coworkers [122], Arnold and coworkers reported that using sulfonyl azides intramolecular C–H amination could be effected by engineered variants of $P450_{BM3}$ to afford cyclic sulfonamides (Scheme 7.24) [123]. The engineering of the $P450_{BM3}$ variant, undertaken previously by the Arnold group, had shown that the key mutation C400S significantly altered the activity of $P450_{BM3}$ [124]. The single-point mutation resulted in the abolition of mono-oxygenation activity, allowing other transformations to be mediated, including carbene and later nitrene transfer. Changing the cysteine for a serine also caused a shift in the characteristic CO-bound complex from 450 to 411 nm, leading to these enzymes to be called P411s. For effective catalysis for these biocatalysts, an oxygen depletion system is essential (glucose, glucose oxidase, and catalase), and rigorous degassing of the reaction mixture alongside an inert atmosphere must also be implemented.

The same transformation has also been demonstrated with myoglobin heme-enzymes and HRP [125].

There has since been several additional reports of heme-containing proteins that utilize azides in both intra- and intermolecular amination reactions [121]. Elegant examples producing chiral amines include a sigmatropic rearrangement of allylic sulfides to afford chiral allylic amines [126], aziridination of benzylic alkenes [127], a system that permits selective amination of primary, secondary, or tertiary C—H bonds [128], and impressively the intermolecular amination of benzylic C—H bonds [129] (Scheme 7.25).

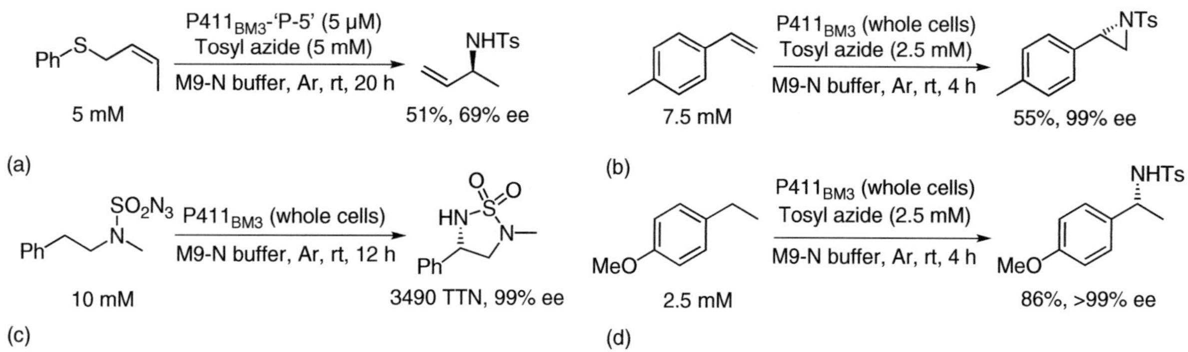

Scheme 7.25 Applications of engineered cytochrome P411 biocatalysts for chiral amine formation.

7.6 Conclusions and Perspectives

AOs, AmDHs, and IREDs show a broad application for the manufacture of chiral amines and constitute an exciting field of study in the coming decade. Nevertheless, several limitations to be overcome remain in this field. For instance, the lack of a (*R*)-selective amine oxidase with a broad substrate scope and high activity to access (*S*)-amines limits the general use of these enzymes. In addition, the number of wild-type enzymes is scarce, and they present a low sequence similarity among them. Tools such as metagenomics and the development of faster and more effective high-throughput screening will greatly help in discovering new amine oxidases to use in synthetic chemistry.

As a class of enzyme, AmDHs have emerged as a promising biocatalyst for the synthesis of primary amines in an efficient and cost-effective manner. The plethora of AADH sequences that are available indicate that many substrates could be accessed, and with the new "native" AmDHs that have been described recently, more chemical space has the potential to be explored. Challenges associated with AmDH adoption include poor kinetics for non-natural substrates, and a low substrate tolerance, both of which could be enhanced through enzyme evolution studies. Additionally, further development of reaction conditions could enable better integration into multicatalyst cascade reactions, with advances such as the PDMS membrane a promising development.

Over the previous decade, IRED biocatalysis has seen to be developed from the enantiocomplementary reduction of 2-MPN to a platform utilized by industry to perform tailored asymmetric reductive amination on a process scale. Through the genomics efforts of both academic researchers and industry alike, a wide range of biocatalysts now populate the imine reduction toolbox, allowing for both imine reduction and reductive amination with a wide range of substrates in both excellent conversion and enantioselectivity. Through application in both cascade and chemoenzymatic synthesis, transformations that are either challenging or unknown to traditional synthesis have been achieved and biocatalytic equivalents to established chemical amine alkylation methodology are emerging as robust strategies. With the intrinsic selectivity of enzymes affording an advantage in the key challenges of developing a platform for asymmetric reductive amination such as the avoidance of over-alkylation and chemoselective reduction, further development of IREDs may deliver a leading role in synthesis from lab to process scale.

References

1 Slabu, I., Galman, J.L., Lloyd, R.C., and Turner, N.J. (2017). Discovery, engineering, and synthetic application of transaminase biocatalysts. *ACS Catalysis* 7 (12): 8263–8284.

2 Busto, E., Gotor-Fernández, V., and Gotor, V. (2011). Hydrolases in the stereoselective synthesis of N-heterocyclic amines and amino acid derivatives. *Chemical Reviews* 111 (7): 3998–4035.

3 Ghislieri, D. and Turner, N.J. (2014). Biocatalytic approaches to the synthesis of enantiomerically pure chiral amines. *Topics in Catalysis* 57 (5): 284–300.

4 Batista, V.F., Galman, J., Pinto, D.C.G.A. et al. (2018). Monoamine oxidase: tunable activity for amine resolution and functionalization. *ACS Catalysis* 8: 11889–11907.

5 MacHeroux, P., Kappes, B., and Ealick, S.E. (2011). Flavogenomics – a genomic and structural view of flavin-dependent proteins. *The FEBS Journal* 278 (15): 2625–2634.

6 Schilling, B. and Lerch, K. (1995). Amine oxidases from *Aspergihs*: identification of a novel flavin-dependent enzyme. *Biochimica et Biophysica Acta* 1243: 529–537.

7 Schilling, B. and Lerch, K. (1995). Cloning, sequencing and heterologous expression of the monoamine oxidase gene from *Aspergillus niger*. *MGG Molecular and General Genetics* 247 (4): 430–438.

8 Pleiss, J. (2012). Rational design of enzymes. In: *Enzyme Catalysis in Organic Synthesis*, 3e, vol. 1 (eds. K. Drauz, H. Gröger and O. May), 89–117. Wiley-VCH Verlag GmbH.

9 Alexeeva, M., Enright, A., Dawson, M.J. et al. (2002). Deracemization of α-methylbenzylamine using an enzyme obtained by in vitro evolution. *Angewandte Chemie International Edition* 41 (17): 3177–3180.

10 Carr, R., Alexeeva, M., Enright, A. et al. (2003). Directed evolution of an amine oxidase possessing both broad substrate specificity and high enantioselectivity. *Angewandte Chemie International Edition* 42 (39): 4807–4810.

11 Carr, R., Alexeeva, M., Dawson, M.J. et al. (2005). Directed evolution of an amine oxidase for the preparative deracemisation of cyclic secondary amines. *Chembiochem* 6 (4): 637–639.

12 Dunsmore, C.J., Carr, R., Fleming, T., and Turner, N.J. (2006). A chemo-enzymatic route to enantiomerically pure cyclic tertiary amines. *Journal of the American Chemical Society* 128 (7): 2224–2225.

13 Eve, T.S.C., Wells, A., and Turner, N.J. (2007). Enantioselective oxidation of O-methyl-N-hydroxylamines using monoamine oxidase N as catalyst. *Chemical Communications* 2 (15): 1530–1531.

14 Köhler, V., Bailey, K.R., Znabet, A. et al. (2010). Enantioselective biocatalytic oxidative desymmetrization of substituted pyrrolidines. *Angewandte Chemie International Edition* 49 (12): 2182–2184.

15 Atkin, K.E., Reiss, R., Koehler, V. et al. (2008). The structure of monoamine oxidase from *Aspergillus niger* provides a molecular context for improvements in activity obtained by directed evolution. *Journal of Molecular Biology* 384 (5): 1218–1231.

16 Rowles, I., Malone, K.J., Etchells, L.L. et al. (2012). Directed evolution of the enzyme monoamine oxidase (MAO-N): highly efficient chemo-enzymatic deracemisation of the alkaloid (±)-crispinea. *ChemCatChem* 4 (9): 1259–1261.

17 Ghislieri, D., Green, A.P., Pontini, M. et al. (2013). Engineering an enantioselective amine oxidase for the synthesis of pharmaceutical building blocks and alkaloid natural products. *Journal of the American Chemical Society* 135 (29): 10863–10869.

18 Chen, Z., Ma, Y., He, M. et al. (2015). Semi-rational directed evolution of monoamine oxidase for kinetic resolution of rac-mexiletine. *Applied Biochemistry and Biotechnology* 176 (8): 2267–2278.

19 Li, G., Yao, P., Gong, R. et al. (2017). Simultaneous engineering of an enzyme's entrance tunnel and active site: the case of monoamine oxidase MAO-N. *Chemical Science* 8 (5): 4093–4099.

20 Li, T., Liang, J., Ambrogelly, A. et al. (2012). Efficient, chemoenzymatic process for manufacture of the boceprevir bicyclic [3.1.0]proline intermediate based on amine oxidase-catalyzed desymmetrization. *Journal of the American Chemical Society* 134 (14): 6467–6472.

21 Currin, A., Kwok, J., Sadler, J.C. et al. (2019). GeneORator: an effective strategy for navigating protein sequence space more efficiently through boolean OR-type DNA libraries. *ACS Synthetic Biology* 8 (6): 1371–1378.

22 Foulkes, J.M., Malone, K.J., Coker, V.S. et al. (2011). Engineering a biometallic whole cell catalyst for enantioselective deracemization reactions. *ACS Catalysis* 1 (11): 1589–1594.

23 Köhler, V., Wilson, Y.M., Dürrenberger, M. et al. (2013). Synthetic cascades are enabled by combining biocatalysts with artificial metalloenzymes. *Nature Chemistry* 5 (2): 93–99.

24 O'Reilly, E., Iglesias, C., Ghislieri, D. et al. (2014). A regio- and stereoselective ω-transaminase/monoamine oxidase cascade for the synthesis of chiral 2,5-disubstituted pyrrolidines. *Angewandte Chemie International Edition* 53 (9): 2447–2450.

25 Bechi, B., Herter, S., McKenna, S. et al. (2014). Catalytic bio-chemo and bio-bio tandem oxidation reactions for amide and carboxylic acid synthesis. *Green Chemistry* 16 (10): 4524–4529.

26 Toscani, A., Risi, C., Black, G.W. et al. (2018). Monoamine oxidase (MAO-N) whole cell biocatalyzed aromatization of 1,2,5,6-tetrahydropyridines into pyridines. *ACS Catalysis* 8 (9): 8781–8787.

27 Scalacci, N., Black, G.W., Mattedi, G. et al. (2017). Unveiling the biocatalytic aromatizing activity of monoamine oxidases MAO-N and 6-HDNO: development of chemoenzymatic cascades for the synthesis of pyrroles. *ACS Catalysis* 7 (2): 1295–1300.

28 Znabet, A., Polak, M.M., Janssen, E. et al. (2010). A highly efficient synthesis of telaprevir by strategic use of biocatalysis and multicomponent reactions. *Chemical Communications* 46 (42): 7918–7920.

29 de Graaff, C., Oppelaar, B., Péruch, O. et al. (2016). Stereoselective monoamine oxidase-catalyzed oxidative aza-Friedel–Crafts reactions of *meso*-pyrrolidines in aqueous buffer. *Advanced Synthesis and Catalysis* 358 (10): 1555–1560.

30 Odachowski, M., Greaney, M.F., and Turner, N.J. (2018). Concurrent biocatalytic oxidation and C—C bond formation via gold catalysis: one-pot alkynylation of N-alkyl tetrahydroisoquinolines. *ACS Catalysis* 8 (11): 10032–10035.

31 Qin, L., Zheng, D., Cui, B. et al. (2016). Bio-mediated oxidative resolution of racemic 2-substituted 1,2,3,4-tetrahydroquinolines. *Tetrahedron Letters* 57 (22): 2403–2405.

32 Deng, G., Wan, N., Qin, L. et al. (2018). Deracemisation of phenyl substituted 2-methyl-1,2,3,4-tetrahydroquinolines by a recombinant monoamine oxidase from Pseudomonas monteilii ZMU-T01. *ChemCatChem* 10: 2374–2377.

33 Iwaki, H., Shimizu, M., Tokuyama, T., and Hasegawa, Y. (1999). Biodegradation of cyclohexylamine by *Brevibacterium oxydans* IH-35A. *Applied and Environmental Microbiology* 65 (5): 2232–2234.

34 Iwaki, H., Shimizu, M., Tokuyama, T., and Hasegawa, Y. (1999). Purification and characterization of a novel cyclohexylamine oxidase from the cyclohexylamine-degrading

Brevibacterium oxydans IH-35A. *Journal of Bioscience and Bioengineering* 88 (3): 264–268.

35 Leisch, H., Grosse, S., Iwaki, H. et al. (2012). Cyclohexylamine oxidase as a useful biocatalyst for the kinetic resolution and dereacemization of amines. *Canadian Journal of Chemistry* 90 (1): 39–45.

36 Li, G., Ren, J., Yao, P. et al. (2014). Deracemization of 2-methyl-1,2,3,4-tetrahydroquinoline using mutant cyclohexylamine oxidase obtained by iterative saturation mutagenesis. *ACS Catalysis* 4 (3): 903–908.

37 Li, G., Yao, P., Cong, P. et al. (2016). New recombinant cyclohexylamine oxidase variants for deracemization of secondary amines by orthogonally assaying designed mutants with structurally diverse substrates. *Scientific Reports* 6 (1): 3–11.

38 Yao, P., Cong, P., Gong, R. et al. (2018). Biocatalytic route to chiral 2-substituted-1,2,3,4-tetrahydroquinolines using cyclohexylamine oxidase muteins. *ACS Catalysis* 8 (3): 1648–1652.

39 Cosgrove, S.C., Hussain, S., Turner, N.J., and Marsden, S.P. (2018). Synergistic chemo/biocatalytic synthesis of alkaloidal tetrahydroquinolines. *ACS Catalysis* 8 (6): 5570–5573.

40 Debon, A., Pott, M., Obexer, R. et al. (2019). Ultrahigh-throughput screening enables efficient single-round oxidase remodelling. *Nature Catalysis* 2 (9): 740–747.

41 Yasukawa, K., Nakano, S., and Asano, Y. (2014). Tailoring D-amino acid oxidase from the pig kidney to R-stereoselective amine oxidase and its use in the deracemization of α-methylbenzylamine. *Angewandte Chemie International Edition* 53 (17): 4428–4431.

42 Yasukawa, K., Motojima, F., Ono, A., and Asano, Y. (2018). Expansion of the substrate specificity of porcine kidney D-amino acid oxidase for S-stereoselective oxidation of 4-Cl-benzhydrylamine. *ChemCatChem* 10 (16): 3500–3505.

43 Brühmüller, M., Möhler, H., and Decker, K. (1972). Covalently bound flavin in D-6-hydroxynicotine oxidase from *Arthrobacter Oxidans* purification and properties of D-6-hydroxynicotine oxidase. *European Journal of Biochemistry* 29 (1): 143–151.

44 Koetter, J.W.A. and Schulz, G.E. (2005). Crystal structure of 6-hydroxy-D-nicotine oxidase from *Arthrobacter nicotinovorans*. *Journal of Molecular Biology* 352 (2): 418–428.

45 Heath, R.S., Pontini, M., Bechi, B., and Turner, N.J. (2014). Development of an R-selective amine oxidase with broad substrate specificity and high enantioselectivity. *ChemCatChem* 6 (4): 996–1002.

46 Abrahamson, M.J., Vázquez-Figueroa, E., Woodall, N.B. et al. (2012). Development of an amine dehydrogenase for synthesis of chiral amines. *Angewandte Chemie International Edition* 51 (16): 3969–3972.

47 Abrahamson, M.J., Wong, J.W., and Bommarius, A.S. (2013). The evolution of an amine dehydrogenase biocatalyst for the asymmetric production of chiral amines. *Advanced Synthesis and Catalysis* 355 (9): 1780–1786.

48 Bommarius, B.R., Schürmann, M., and Bommarius, A.S. (2014). A novel chimeric amine dehydrogenase shows altered substrate specificity compared to its parent enzymes. *Chemical Communications* 50 (95): 14953–14955.

49 Itoh, N., Yachi, C., and Kudome, T. (2000). Determining a novel NAD+-dependent amine dehydrogenase with a broad substrate range from Streptomyces virginiae IFO

12827: purification and characterization. *Journal of Molecular Catalysis B: Enzymatic* 10 (1–3): 281–290.

50 Mayol, O., David, S., Darii, E. et al. (2016). Asymmetric reductive amination by a wild-type amine dehydrogenase from the thermophilic bacteria: *Petrotoga mobilis*. *Catalysis Science & Technology* 6 (20): 7421–7428.

51 Mayol, O., Bastard, K., Beloti, L. et al. (2019). A family of native amine dehydrogenases for the asymmetric reductive amination of ketones. *Nature Catalysis* 2 (4): 324–333.

52 France, S.P., Howard, R.M., Steflik, J. et al. (2017). Identification of novel bacterial members of the imine reductase enzyme family that perform reductive amination. *ChemCatChem* 10 (3): 510–514.

53 Mangas-Sanchez, J., France, S.P., Montgomery, S.L. et al. (2017). Imine reductases (IREDs). *Current Opinion in Chemical Biology* 37: 19–25.

54 Chen, F.F., Zheng, G.W., Liu, L. et al. (2018). Reshaping the active pocket of amine dehydrogenases for asymmetric synthesis of bulky aliphatic amines. *ACS Catalysis* 8 (3): 2622–2628.

55 Chen, F.-F., Liu, Y.-Y.Y., Zheng, G.-W.W., and Xu, J.-H.H. (2015). Asymmetric amination of secondary alcohols by using a redox-neutral two-enzyme cascade. *ChemCatChem* 7 (23): 3838–3841.

56 Knaus, T., Böhmer, W., and Mutti, F.G. (2017). Amine dehydrogenases: efficient biocatalysts for the reductive amination of carbonyl compounds. *Green Chemistry* 19 (2): 453–463.

57 Mutti, F.G., Knaus, T., Scrutton, N.S. et al. (2015). Conversion of alcohols to enantiopure amines through dual-enzyme hydrogen-borrowing cascades. *Science* 349 (6255): 1525–1529.

58 Tseliou, V., Knaus, T., Masman, M.F. et al. (2019). Generation of amine dehydrogenases with increased catalytic performance and substrate scope from ε-deaminating L-Lysine dehydrogenase. *Nature Communications* 10 (1): 1–11.

59 Tseliou, V., Masman, M.F., Böhmer, W. et al. (2019). Mechanistic insight into the catalytic promiscuity of amine dehydrogenases: asymmetric synthesis of secondary and primary amines. *Chembiochem* 20 (6): 800–812.

60 Wang, X., Saba, T., Yiu, H.H.P. et al. (2017). Cofactor NAD(P)H regeneration inspired by heterogeneous pathways. *Chem* 2 (5): 621–654.

61 Ye, L.J., Toh, H.H., Yang, Y. et al. (2015). Engineering of amine dehydrogenase for asymmetric reductive amination of ketone by evolving *Rhodococcus* phenylalanine dehydrogenase. *ACS Catalysis* 5 (2): 1119–1122.

62 Au, S.K., Bommarius, B.R., and Bommarius, A.S. (2014). Biphasic reaction system allows for conversion of hydrophobic substrates by amine dehydrogenases. *ACS Catalysis* 4 (11): 4021–4026.

63 Pushpanath, A., Siirola, E., Bornadel, A. et al. (2017). Understanding and overcoming the limitations of *Bacillus badius* and *Caldalkalibacillus thermarum* amine dehydrogenases for biocatalytic reductive amination. *ACS Catalysis* 7 (5): 3204–3209.

64 Aleku, G.A., France, S.P., Man, H. et al. (2017). A reductive aminase from *Aspergillus oryzae*. *Nature Chemistry* 9 (10): 961–969.

65 Sharma, M., Mangas-Sanchez, J., France, S.P. et al. (2018). A mechanism for reductive amination catalyzed by fungal reductive aminases. *ACS Catalysis* 8 (12): 11534–11541.

66 Chen, F.-F., Cosgrove, S.C., Birmingham, W.R. et al. (2019). Enantioselective synthesis of chiral vicinal amino alcohols using amine dehydrogenases. *ACS Catalysis* 9 (12): 11813–11818.

67 Thompson, M.P. and Turner, N.J. (2017). Two-enzyme hydrogen-borrowing amination of alcohols enabled by a cofactor-switched alcohol dehydrogenase. *ChemCatChem* 9 (20): 3833–3836.

68 Böhmer, W., Knaus, T., and Mutti, F.G. (2018). Hydrogen-borrowing alcohol bioamination with coimmobilized dehydrogenases. *ChemCatChem* 10 (4): 731–735.

69 Houwman, J.A., Knaus, T., Costa, M., and Mutti, F.G. (2019). Efficient synthesis of enantiopure amines from alcohols using resting: *E. coli* cells and ammonia. *Green Chemistry* 21 (14): 3846–3857.

70 Liu, J. and Li, Z. (2019). Enhancing cofactor recycling in the bioconversion of racemic alcohols to chiral amines with alcohol dehydrogenase and amine dehydrogenase by coupling cells and cell-free system. *Biotechnology and Bioengineering* 116 (3): 536–542.

71 Yu, H.L., Li, T., Chen, F.F. et al. (2018). Bioamination of alkane with ammonium by an artificially designed multienzyme cascade. *Metabolic Engineering* 47: 184–189.

72 Chen, F.F., Zhang, Y.H., Zhang, Z.J. et al. (2019). An ammonium-formate-driven trienzymatic cascade for ω-transaminase-catalyzed (R)-selective amination. *The Journal of Organic Chemistry* 84 (22): 14987–14993.

73 Corrado, M.L., Knaus, T., and Mutti, F.G. (2019). Regio- and stereoselective multi-enzymatic aminohydroxylation of β-methylstyrene using dioxygen, ammonia and formate. *Green Chemistry* 21 (23): 6246–6251.

74 Sperl, J.M. and Sieber, V. (2018). Multi-enzyme cascade reactions – status and recent advances. *ACS Catalysis* 8 (3): 2385–2396.

75 Uthoff, F. and Gröger, H. (2018). Asymmetric synthesis of 1-phenylethylamine from styrene via combined Wacker oxidation and enzymatic reductive amination. *The Journal of Organic Chemistry* 83 (16): 9517–9521.

76 Peng, H., Wei, E., Wang, J. et al. (2016). Deciphering piperidine formation in polyketide-derived indolizidines reveals a thioester reduction, transamination, and unusual imine reduction process. *ACS Chemical Biology* 11 (12): 3278–3283.

77 Ye, S., Braña, A.F., González-Sabín, J. et al. (2018). New insights into the biosynthesis pathway of polyketide alkaloid argimycins P in *Streptomyces argillaceus. Frontiers in Microbiology* 9: 1–12.

78 Mitsukura, K., Suzuki, M., Tada, K. et al. (2010). Asymmetric synthesis of chiral cyclic amine from cyclic imine by bacterial whole-cell catalyst of enantioselective imine reductase. *Organic & Biomolecular Chemistry* 8 (20): 4533–4535.

79 Mitsukura, K., Suzuki, M., Shinoda, S. et al. (2011). Purification and characterization of a novel (R)-imine reductase from *Streptomyces* sp. GF3587. *Bioscience, Biotechnology, and Biochemistry* 75 (9): 1778–1782.

80 Lenz, M., Meisner, J., Quertinmont, L. et al. (2017). Asymmetric ketone reduction by imine reductases. *Chembiochem* 18 (3): 253–256.

81 Roth, S., Präg, A., Wechsler, C. et al. (2017). Extended catalytic scope of a well-known enzyme: asymmetric reduction of iminium substrates by glucose dehydrogenase. *Chembiochem* 18 (17): 1703–1706.

82 Leipold, F., Hussain, S., Ghislieri, D., and Turner, N.J. (2013). Asymmetric reduction of cyclic imines catalyzed by a whole-cell biocatalyst containing an (S)-imine reductase. *ChemCatChem* 5 (12): 3505–3508.

83 Hussain, S., Leipold, F., Man, H. et al. (2015). An (R)-imine reductase biocatalyst for the asymmetric reduction of cyclic imines. *ChemCatChem* 7 (4): 579–583.

84 Gand, M., Müller, H., Wardenga, R., and Höhne, M. (2014). Characterization of three novel enzymes with imine reductase activity. *Journal of Molecular Catalysis B: Enzymatic* 110: 126–132.

85 Scheller, P.N., Fademrecht, S., Hofelzer, S. et al. (2014). Enzyme toolbox: novel enantiocomplementary imine reductases. *Chembiochem* 15 (15): 2201–2204.

86 Wetzl, D., Berrera, M., Sandon, N. et al. (2015). Expanding the imine reductase toolbox by exploring the bacterial protein-sequence space. *Chembiochem* 16 (12): 1749–1756.

87 Aleku, G.A., Man, H., France, S.P. et al. (2016). Stereoselectivity and structural characterization of an imine reductase (IRED) from *Amycolatopsis orientalis*. *ACS Catalysis* 6: 3380–3889.

88 Li, H., Tian, P., Xu, J.H., and Zheng, G.W. (2017). Identification of an imine reductase for asymmetric reduction of bulky dihydroisoquinolines. *Organic Letters* 19 (12): 3151–3154.

89 Zhu, J., Tan, H., Yang, L. et al. (2017). Enantioselective synthesis of 1-aryl-substituted tetrahydroisoquinolines employing imine reductase. *ACS Catalysis* 7: 7003–7007.

90 Yao, P., Xu, Z., Yu, S. et al. (2019). Imine reductase-catalyzed enantioselective reduction of bulky α,β-unsaturated imines en route to a pharmaceutically important morphinan skeleton. *Advanced Synthesis and Catalysis* 361 (3): 556–561.

91 Meneely, K.M., Ronnebaum, T.A., Riley, A.P. et al. (2016). Holo structure and steady state kinetics of the thiazolinyl imine reductases for siderophore biosynthesis. *Biochemistry* 55 (38): 5423–5433.

92 Zumbrägel, N. and Gröger, H. (2019). One-pot synthesis of a 3-thiazolidine through combination of an Asinger-type multi-component-condensation reaction with an enzymatic imine reduction. *Journal of Biotechnology* 291: 35–40.

93 Zumbrägel, N., Wagner, K., Weißing, N., and Gröger, H. (2019). Biocatalytic reduction of 2-monosubstituted 3-thiazolines using imine reductases. *Journal of Heterocyclic Chemistry* 56 (3): 788–794.

94 Zumbrägel, N., Machui, P., Nonnhoff, J., and Gröger, H. (2019). Enantioselective biocatalytic reduction of 2H-1,4-benzoxazines using imine reductases. *The Journal of Organic Chemistry* 84 (3): 1440–1447.

95 Thorpe, T.W., France, S.P., Hussain, S. et al. (2019). One-pot biocatalytic cascade reduction of cyclic enimines for the preparation of diastereomerically enriched N-heterocycles. *Journal of the American Chemical Society* 141 (49): 19208–19213.

96 France, S.P., Hussain, S., Hill, A.M. et al. (2016). One-pot cascade synthesis of mono- and disubstituted piperidines and pyrrolidines using carboxylic acid reductase (CAR), ω-transaminase (ω-TA), and imine reductase (IRED) biocatalysts. *ACS Catalysis* 6 (6): 3753–3759.

97 Hepworth, L.J., France, S.P., Hussain, S. et al. (2017). Enzyme cascades in whole cells for the synthesis of chiral cyclic amines. *ACS Catalysis* 7 (4): 2920–2925.

98 Costa, B.Z., Galman, J.L., Slabu, I. et al. (2018). Synthesis of 2,5-disubstituted pyrrolidine alkaloids via a one-pot cascade using transaminase and reductive aminase biocatalysts. *ChemCatChem* 10 (20): 4733–4738.

99 Borlinghaus, N., Gergel, S., and Nestl, B.M. (2018). Biocatalytic access to piperazines from diamines and dicarbonyls. *ACS Catalysis* 8 (4): 3727–3732.

100 Al-Shameri, A., Borlinghaus, N., Weinmann, L. et al. (2019). Synthesis of: N-heterocycles from diamines via H_2-driven NADPH recycling in the presence of O_2. *Green Chemistry* 21 (6): 1396–1400.

101 Heath, R.S., Pontini, M., Hussain, S., and Turner, N.J. (2016). Combined imine reductase and amine oxidase catalyzed deracemization of nitrogen heterocycles. *ChemCatChem* 8 (1): 117–120.

102 Zumbrägel, N. and Gröger, H. (2018). Merging heterocyclic chemistry and biocatalysis in one-pot processes through compartmentalization of the reaction steps. *Bioengineering* 5 (3): 60.

103 Zawodny, W., Montgomery, S.L., Marshall, J.R. et al. (2018). Chemoenzymatic synthesis of substituted azepanes by sequential biocatalytic reduction and organolithium-mediated rearrangement. *Journal of the American Chemical Society* 140 (51): 17872–17877.

104 France, S.P., Aleku, G.A., Sharma, M. et al. (2017). Biocatalytic routes to enantiomerically enriched dibenz[c,e]azepines. *Angewandte Chemie International Edition* 56 (49): 15589–15593.

105 Dürrenberger, M., Heinisch, T., Wilson, Y.M. et al. (2011). Artificial transfer hydrogenases for the enantioselective reduction of cyclic imines. *Angewandte Chemie International Edition* 50 (13): 3026–3029.

106 Quinto, T., Schwizer, F., Zimbron, J.M. et al. (2014). Expanding the chemical diversity in artificial imine reductases based on the biotin-streptavidin technology. *ChemCatChem* 6 (4): 1010–1014.

107 Chen, H., Collier, S.J., Nazor, J. et al. (2016) Engineered imine reductases and methods for the reductive amination of ketone and amine compounds. US Patent 9, 487, 760 B2, issued 2016.

108 Hallen, A., Cooper, A.J.L., Smith, J.R. et al. (2015). Ketimine reductase/CRYM catalyzes reductive alkylamination of α-keto acids, confirming its function as an imine reductase. *Amino Acids* 47 (11): 2457–2461.

109 Huber, T., Schneider, L., Präg, A. et al. (2014). Direct reductive amination of ketones: structure and activity of S-selective imine reductases from streptomyces. *ChemCatChem* 6 (8): 2248–2252.

110 Scheller, P.N., Lenz, M., Hammer, S.C. et al. (2015). Imine reductase-catalyzed intermolecular reductive amination of aldehydes and ketones. *ChemCatChem* 7 (20): 3239–3242.

111 Wetzl, D., Gand, M., Ross, A. et al. (2016). Asymmetric reductive amination of ketones catalyzed by imine reductases. *ChemCatChem* 8 (12): 2023–2026.

112 Matzel, P., Gand, M., and Höhne, M. (2017). One-step asymmetric synthesis of (R)- and (S)-rasagiline by reductive amination applying imine reductases. *Green Chemistry* 19: 385–389.

113 Matzel, P., Wenske, S., Merdivan, S. et al. (2019). Synthesis of β-chiral amines by dynamic kinetic resolution of α-branched aldehydes applying imine reductases. *ChemCatChem* 11 (17): 4281–4285.

114 Roiban, G.D., Kern, M., Liu, Z. et al. (2017). Efficient biocatalytic reductive aminations by extending the imine reductase toolbox. *ChemCatChem* 9 (24): 4475–4479.

115 Montgomery, S.L., Mangas-Sanchez, J., Thompson, M.P. et al. (2017). Direct alkylation of amines with primary and secondary alcohols through biocatalytic hydrogen borrowing. *Angewandte Chemie International Edition* 56 (35): 10491–10494.

116 Tavanti, M., Mangas-Sanchez, J., Montgomery, S.L. et al. (2017). A biocatalytic cascade for the amination of unfunctionalised cycloalkanes. *Organic & Biomolecular Chemistry* 15 (46): 9790–9793.

117 Ramsden, J.I., Heath, R.S., Derrington, S.R. et al. (2019). Biocatalytic N-alkylation of amines using either primary alcohols or carboxylic acids via reductive aminase cascades. *Journal of the American Chemical Society* 141 (3): 1201–1206.

118 Heath, R.S., Birmingham, W.R., Thompson, M.P. et al. (2019). An engineered alcohol oxidase for the oxidation of primary alcohols. *Chembiochem* 20 (2): 276–281.

119 Prior Art Database (2017). Imine Reductase Mediated Synthesis of (S, S)-Sertraline. IPCOM000251929D. https://priorart.ip.com/IPCOM/000251929 (accessed 18 June 2020).

120 Schober, M., MacDermaid, C., Ollis, A.A. et al. (2019). Chiral synthesis of LSD1 inhibitor GSK2879552 enabled by directed evolution of an imine reductase. *Nature Catalysis* 2 (10): 909–915.

121 Brandenberg, O.F., Fasan, R., and Arnold, F.H. (2017). Exploiting and engineering hemoproteins for abiological carbene and nitrene transfer reactions. *Current Opinion in Biotechnology* 47: 102–111.

122 Svastits, E.W., Dawson, J.H., Breslow, R., and Gellman, S.H. (1985). Functionalized nitrogen atom transfer catalyzed by cytochrome P-450. *Journal of the American Chemical Society* 107 (22): 6427–6428.

123 McIntosh, J.A., Coelho, P.S., Farwell, C.C. et al. (2013). Enantioselective intramolecular C—H amination catalyzed by engineered cytochrome P450 enzymes in vitro and in vivo. *Angewandte Chemie International Edition* 52 (35): 9309–9312.

124 Coelho, P.S., Wang, Z.J., Ener, M.E. et al. (2013). A serine-substituted P450 catalyzes highly efficient carbene transfer to olefins in vivo. *Nature Chemical Biology* 9 (8): 485–487.

125 Bordeaux, M., Singh, R., and Fasan, R. (2014). Intramolecular C(sp^3)H amination of arylsulfonyl azides with engineered and artificial myoglobin-based catalysts. *Bioorganic & Medicinal Chemistry* 22 (20): 5697–5704.

126 Prier, C.K., Hyster, T.K., Farwell, C.C. et al. (2016). Asymmetric enzymatic synthesis of allylic amines: a sigmatropic rearrangement strategy. *Angewandte Chemie International Edition* 55 (15): 4711–4715.

127 Farwell, C.C., Zhang, R.K., McIntosh, J.A. et al. (2015). Enantioselective enzyme-catalyzed aziridination enabled by active-site evolution of a cytochrome P450. *ACS Central Science* 1 (2): 89–93.

128 Yang, Y., Cho, I., Qi, X. et al. (2019). An enzymatic platform for the asymmetric amination of primary, secondary and tertiary C(sp^3)—H bonds. *Nature Chemistry* 11 (11): 987–993.

129 Prier, C.K., Zhang, R.K., Buller, A.R. et al. (2017). Enantioselective, intermolecular benzylic C—H amination catalysed by an engineered iron–haem enzyme. *Nature Chemistry* 9 (7): 629–634.

8

Engineering Functional Nanomaterials Through the Amino Group

Giacomo Filippini[1], Paolo Pengo[1], Susanna Bosi[1], Giulio Ragazzon[1], Lucia Pasquato[1] and Maurizio Prato[1,2,3]

[1] *Università degli Studi di Trieste, Dipartimento di Scienze Chimiche e Farmaceutiche, via L. Giorgieri 1, 34127 Trieste, Italy*
[2] *Center for Cooperative Research in Biomaterials (CIC biomaGUNE), Carbon Bionanotechnology Laboratory, CIC biomaGUNE, Paseo Miramòn, 182, 20014 Donostia San Sebastiàn, Guipuzcoa, Spain*
[3] *Ikerbasque Basque Foundation for Science, Maria Diaz de Haro 3, 6 Solairua 48013, Bilbao, Spain*

Abbreviations

AFG	amino-functionalized graphene
AFM	atomic force microscopy
APTS	(aminopropyl)-triethoxysilane
ATR-IR	attenuated total reflectance infrared
BOC	*t*-butyl carbamate
CDs	carbon dots
CND	carbon nanodiamond
CNHs	carbon nanohorns
CNOs	carbon nano-onions
CNT	carbon nanotube
CPT	camptothecin
CVD	chemical vapor deposition
CVD-G	chemical vapor deposition graphene
DFT	density functional theory
DMAP	4-(dimethylamino)pyridine
DMF	dimethylformamide
DNA	deoxyribonucleic acid
Dod	dodecylamine
ECD	electronical circular dichroism
ECL	electrochemiluminescence
EDC	*N*-ethyl-*N′*-(3-(dimethylamino)propyl) carbodiimide hydrochloride
EGFP	enhanced green fluorescent protein

Methodologies in Amine Synthesis: Challenges and Applications, First Edition.
Edited by Alfredo Ricci and Luca Bernardi.

ESR	electron spin resonance
FITC	fluorescein isothiocyanate
FLG	few layers graphene
f-SWCNT	functionalized single-walled carbon nanotube
GO	graphene oxide
HIV	human immunodeficiency viruses
HoBt	1-hydroxy benzotriazole
HRTEM	high-resolution transmission electron microscopy
IPCA	5-oxo-1,2,3,5-tetrahydroimidazo[1,2-α]pyridine-7-carboxylic acid
IR	infrared
KT	Kaiser test
MPS	3-mercaptopropyltrimethoxysilane
MW	microwave
MWCNHs	multiwalled carbon nanohorns
MWCNT	multiwalled carbon nanotube
NCD	nitrogen-doped carbon dot
ND	nanodiamonds
NMP	*N*-methylpyrrolidone
NMR	nuclear magnetic resonance
NP	nanoparticle
NR	nanorods
NTA	nitrilotriacetic acid
Oct	octylamine
ODE	1-octadecene
OEAs	oxygen evolving anodes
OLE	oleylamine
ORR	oxygen reduction reaction
OTS	octadecyltrichlorosilane
PAMAM	poly(amidoamine)
PC	prostate cancer
PEG	polyethylene glycol
PEI	polyethyleneimine
PLQY	photoluminescence quantum yield
POM	polyoxometalate
PSMA	prostate-specific membrane antigen
PVP	polyvinylpyrrolidone
QDs	quantum dots
SAM	self-assembled monolayers
SCE	saturated calomel electrode
siRNA	small-interfering ribonucleic acid
SP	supraparticle
SWCNHs	single-walled carbon nanohorns
SWCNT	single-walled carbon nanotube
SWNT	single-walled nanotube
TEM	transmission electron microscopy

TFA	trifluoroacetic acid
TGA	thermogravimetric analysis
TIPS	triisopropylsilane
TLC	thin-layer chromatography
TMEDA	tetramethylethylendiamine
TOP	tri-*n*-octylphosphine
TOPO	tri-*n*-octylphosphine oxide
VCD	vibrational circular dichroism
XPS	X-ray photoelectron spectroscopy

8.1 Introduction

In the past two decades, the fast scientific and practical evolution of nanomaterials has led to applications that can improve our everyday life. The design, preparation, characterization, and study of nanomaterial properties and features are now moving toward a convergence of technologies developed in different fields such as nanotechnology, biotechnology, information technology, and cognitive sciences. The evolution so far observed and the new outcome are guided by the heavy work of scientists, who develop (nano)materials with new properties, proposing environmentally sustainable procedures for their preparation, introducing innovation for reproducible and large-scale production.

Chemistry and in particular organic chemistry offers a variety of tools to design, synthetize, or assemble building blocks combining organic and inorganic materials for the construction of architectures with nanometer dimension, functional nanomaterials, and integrate biomaterials with nanomaterials. Moreover, organic chemistry enables an infinite variety in the design of electronic, optical, and catalytic properties of (nano)materials.

In this context, the chemistry of the amino group is of paramount importance. Indeed, amines have been widely employed to design and produce a large variety of novel smart nanomaterials addressing many of the targets for innovation and sustainability.

In particular, in this chapter, we describe and analyze the most common strategies that use amines in nanomaterial preparation or that enable efficient ways for introducing amino groups in nanomaterials for further functionalization.

To this end, we selected two main types of nanomaterials, which have become very popular in the past decades: carbon-based nanomaterials and hybrid organic–inorganic nanoparticles (NPs), which are our main research fields. In particular, we mainly focus on fullerenes, carbon nanotubes (CNTs), graphene, and on less common carbon nanomaterials, namely, carbon nano-onions (CNOs) and nanodiamonds (NDs). We also take into consideration the emerging members of this family, i.e. carbon nanodots. Then, we present the use of amines in the synthesis, stabilization, and functionalization of organic–inorganic hybrid nanomaterials, such as two dimensional self-assembled monolayers (2D SAMs), semiconductor quantum dots (QDs), and metal or metal oxide nanoparticles. These apparently very different types of materials heavily rely on similar wet chemistry approaches for their preparation, stabilization, functionalization, and processing. These are rooted down to very fundamental properties of amines such as their Lewis basicity, nucleophilic character, and hydrogen bond acceptor ability. Beside their role in the synthesis or modification of the materials, amines are also keys to unlock numerous practical nanomaterial applications in the fields of biology, medicine, sensor development, and energy production. Some proof-of-principle examples are briefly discussed to further highlight the scope of amines in producing novel ad hoc nanomaterials and how amines can affect and address their final applications.

8.2 Quantification of Nanomaterial-Bound Amino Groups

In an overview of the amino group in nanomaterials, it seems adequate to recall how amino groups can be identified and quantified. Several possible strategies have been used, and often, they can be considered complementary with one another. Here, we recall those which have been most commonly used in the analysis of nanomaterials.

Ideally, spectroscopic techniques that do not require chemical reactivity of the amino group are preferred (Figure 8.1a). However, gaining direct spectroscopic evidence of the presence of amino groups may not be straightforward. Nuclear magnetic resonance (NMR) techniques are certainly very powerful. Simple ^{1}H-NMR can be used if the amine structure is known because the protons of the amino group may not be very diagnostic. However, solution NMR techniques often suffer from the limited mobility of molecular fragments attached to the nanomaterial [3]. ^{15}N-NMR has also been used for this purpose, having the evident advantage of detecting directly the N-nucleus [4]. However, this technique requires the use of ^{15}N-labeled compounds, which is often not practical. X-ray photoelectron spectroscopy (XPS), infrared (IR) spectroscopy, and thermogravimetric analysis (TGA) – among others – can provide qualitative evidence in support of other employed strategies [5]. In some cases, XPS can be particularly useful, although it has two important limitations that should be recalled. The first limitation is that, as it detects the core of the nitrogen atom, in the presence of several types of nitrogen atoms in the nanomaterial, the technique may not be sufficiently selective to identify amino groups. To some extent, the multicomponent fitting of the acquired nitrogen peak data can limit this shortcoming, albeit not completely. The second limitation is that it is a surface technique; therefore, it is inappropriate to test bulk materials, albeit this can also be a strength when the surface properties of the nanomaterial are of particular interest.

A different class of strategies rely on the detection of amino groups exploiting its reactivity. Intrinsically, these methods are strongly dependent on the type of amine to be determined. In the simplest case, acid–base titrations can be used for the quantification [6]. Absorption measurements can also be used, e.g. upon interaction with suitable dyes, and also as fluorogenic precursors such as fluorescamine [7].

In many cases, the chemical detection of amines relies on the formation of imine derivatives (Figure 8.1b). Examples of reaction partners suitable for NMR quantification are 4-nitrobenzaldehyde and 3,5-bistrifluoromethyl-benzaldehyde, which were used for spectrophotometric or ^{19}F-NMR detection, respectively [1]. One of the most widely used tests for the identification of amines is the reaction with ninhydrin, affording a colored product. This colorimetric analysis is based on the same principle of the commonly used thin-layer chromatography (TLC) stain. In particular, a variant of the ninhydrin reaction, performed under basic conditions in the presence of KCN, is known as the Kaiser test (KT),

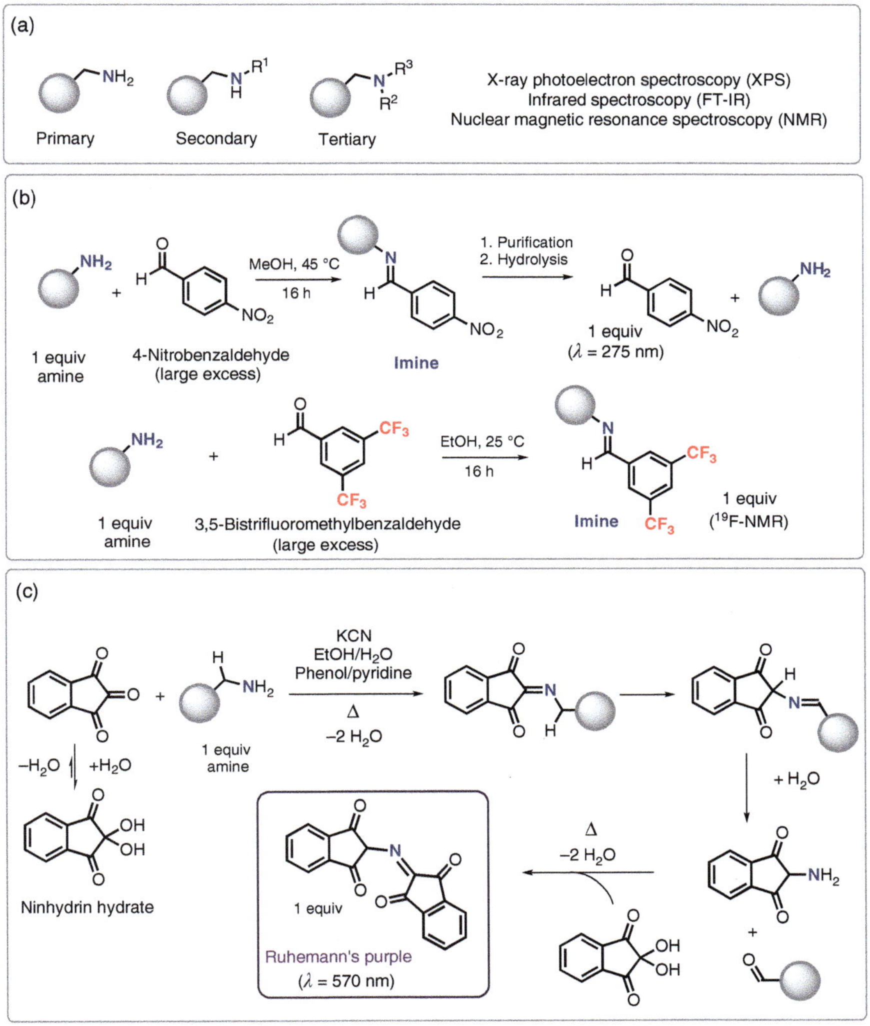

Figure 8.1 Overview of the various methods to identify and quantify amino groups present on nanomaterials. (a) Prominent spectroscopic techniques employed for the detection of primary, secondary, and tertiary amines. (b) Strategies based on imine formation with aromatic aldehydes, for the identification of primary amines. Source: Based on Sun et al. [1]. (c) Reaction sequence associated with the Kaiser test that can disclose primary aliphatic amines having one proton attached to the α-carbon. Source: Based on Sarin et al. [2c].

that is commonly employed for the quantitative identification of amines, especially during peptide synthesis (Figure 8.1c) [2]. This colorimetric test has also been applied for amines on nanomaterials, including CNTs, nanoparticles, and surfaces [8]. In the past decades, we had the opportunity to appreciate the simplicity of the Kaiser test that has been frequently used in our laboratories to test different amino-functionalized nanomaterials.

Importantly, the Kaiser test is selective for the detection of primary amines, with at least one proton on the α carbon atom. Following imine formation, a multistep mechanism affords a colored product known as Ruhemann's purple. Importantly, the molecular structure of this intensely colored product is independent on the starting amine, allowing a reliable colorimetric quantification.

It should be noted, however, that not all primary amines afford a quantitative test, as evidenced by studies performed to optimize the test conditions. Moreover, the colored product suffers from limited stability under the test conditions [2]. For these reasons, the Kaiser test can be considered a quantitative test that however underestimates the total number of amines. This was indeed confirmed by studies on silica matrices rich in amino groups that were tested with different methods for amine detection [1]. Finally, limited accessibility of amino groups has also been described as a possible reason for an incomplete quantification of the reactive groups. Indeed, matrix effects may hamper the reactivity, in particular at a high surface coverage of functional groups.

8.3 Exploiting Amino Compounds for the Functionalization of Carbon-Based Nanomaterials

8.3.1 Historical Backgrounds: Allotropes of Carbon

The discovery of fullerenes as the first molecular allotropes of carbon in 1985 [9] and their further preparation in macroscopic amounts by arc vaporization of graphite in 1990 [10] have profoundly influenced many aspects of contemporary chemistry. Since then, the family of carbon nanoforms has grown significantly, first with the closely related CNTs [11] and, more recently, with the advent of graphene in 2004 [12]. Nowadays, carbon-based nanomaterials are fully recognized as very useful carbon building blocks with potential applications in different scientific fields [13]. In fact, all carbon allotropes may undergo chemical transformation that enhance both their solubility and processability, with the aim of integrating them into inorganic, organic, and biological systems. The reactivity of these nanostructures toward addition reactions was found to be strongly dependent on the curvature of their carbon frameworks and thus on the pyramidalization degrees of their sp^2-hybridized carbon atoms [14]. Elevated pyramidalization levels of the sp^2-hybridized carbon atoms make, typically, the materials more susceptible toward functionalization reactions, and a general reactivity series may be indicated as follows: fullerenes > CNTs > graphene [15]. Moreover, in graphene, which can ideally be considered as a flat polycyclic aromatic hydrocarbon, the sheet edges are generally considered the most reactive sites [16]. This section of the chapter focuses, through the analysis of some explicative examples, on the most common protocols used for the derivatization of carbon-based nanostructures using amino compounds, and on the applications of the resulting functionalized materials.

8.3.2 Use of Amines for the Functionalization of Carbon Nanostructures

Amino compounds have been largely capitalized to covalently functionalize carbon-based nanostructures over recent years. In Figure 8.2a,b, two historical and relevant examples of this type of chemical processes are shown, the 1,3-dipolar cycloaddition of azomethine ylides and the radical arylation reaction. The former reaction, originally performed on C_{60} (fullerene) is a particular version of the well-known cycloaddition to olefins [20]. The amino group in sarcosine reacts with paraformaldehyde to give an iminium salt, which undergoes fast decarboxylation when heated, thus forming the corresponding azomethine ylide intermediate (**Ia**, Figure 8.2a). This species then reacts with C_{60} via a 1,3-dipolar cycloaddition pathway to afford an *N*-methylpyrrolidine derivative (**1a**) in moderate isolated yield (41%) [17]. Remarkably, a wide variety of fullerene derivatives (**1**) can be easily obtained through the rational choice of the amino acids and carbonyl compounds (aldehydes or ketones), showing both the versatility and robustness of this approach (Figure 8.2a, light blue box) [21]. The cycloaddition of azomethine ylides is still an easy and very general approach, commonly followed nowadays, for the production of fullerene derivatives with the structure of **1**. Other convenient methods for the synthesis of pyrrolidinofullerenes (**1**) are based on the thermal ring opening of aziridines, acid-catalyzed or thermal desilylation of trimethylsilylamino derivatives, or photochemical treatment of tertiary amines with C_{60} [22]. In 2006, the reversibility of the reaction on C_{60} was also demonstrated [23]. Indeed, the thermal cycloelimination of the pyrrolidine moiety within **1** could take place in the presence of a strong dipolarophile, such as maleic anhydride, and a metal complex (e.g. Wilkinson's catalyst or copper triflate), thus regenerating the pristine C_{60} fullerene.

Later in 2009, the first asymmetric metal-catalyzed 1,3-dipolar cycloaddition reaction onto C_{60} fullerene was reported, which gives rise to chiral pyrrolidinofullerenes (**1**) with complete control of the stereochemical outcome [24]. This approach has paved the way to the development of numerous catalytic protocols for the enantioselective functionalization of fullerenes, therefore leading to the production of a wide variety of chiral fullerene derivatives with high site and regioselectivity [25].

These optically active materials have found applications in fields where chirality is a key issue, such as medicinal chemistry and nanobiotechnology [26].

CNTs ideally consist of graphitic sheets, which have been rolled up into a cylindrical shape. The length of CNT is in the size of micrometers with diameters up to 100 nm. CNTs form bundles, which are entangled together in the solid state giving rise to complex networks. Depending on the arrangement of the hexagon rings along the tubular surface, CNTs can be metallic or semiconducting [27].

Because of their extraordinary properties, CNTs can be considered as attractive candidates in diverse nanotechnological applications, such as fillers in polymer matrixes, molecular tanks, (bio)sensors, and many others. Since their discovery in 1991, there has been a great deal of interest in the organic derivatization of single-walled carbon nanotubes (SWCNTs) to facilitate their manipulation, enhance their solubility, and make them more amenable to composite formation [28].

In 2001, Tour and coworkers developed a novel and efficient radical approach for the functionalization of small-diameter (c. 0.7 nm) SWCNTs (**2**) [29] that exploits the reactivity of aryl diazonium salts (Figure 8.2b) [18]. As the diameter of these nanotubes (**2**) is approximately the same as that of C_{60}, they were expected to display enhanced reactivity with respect to larger diameter CNTs. The diazonium salts used to derivatize the pristine SWCNTs (**2**) were freshly prepared from the corresponding aniline derivatives using nitrosonium tetrafluoroborate ($NOBF_4$) at low temperature (−30 °C). The

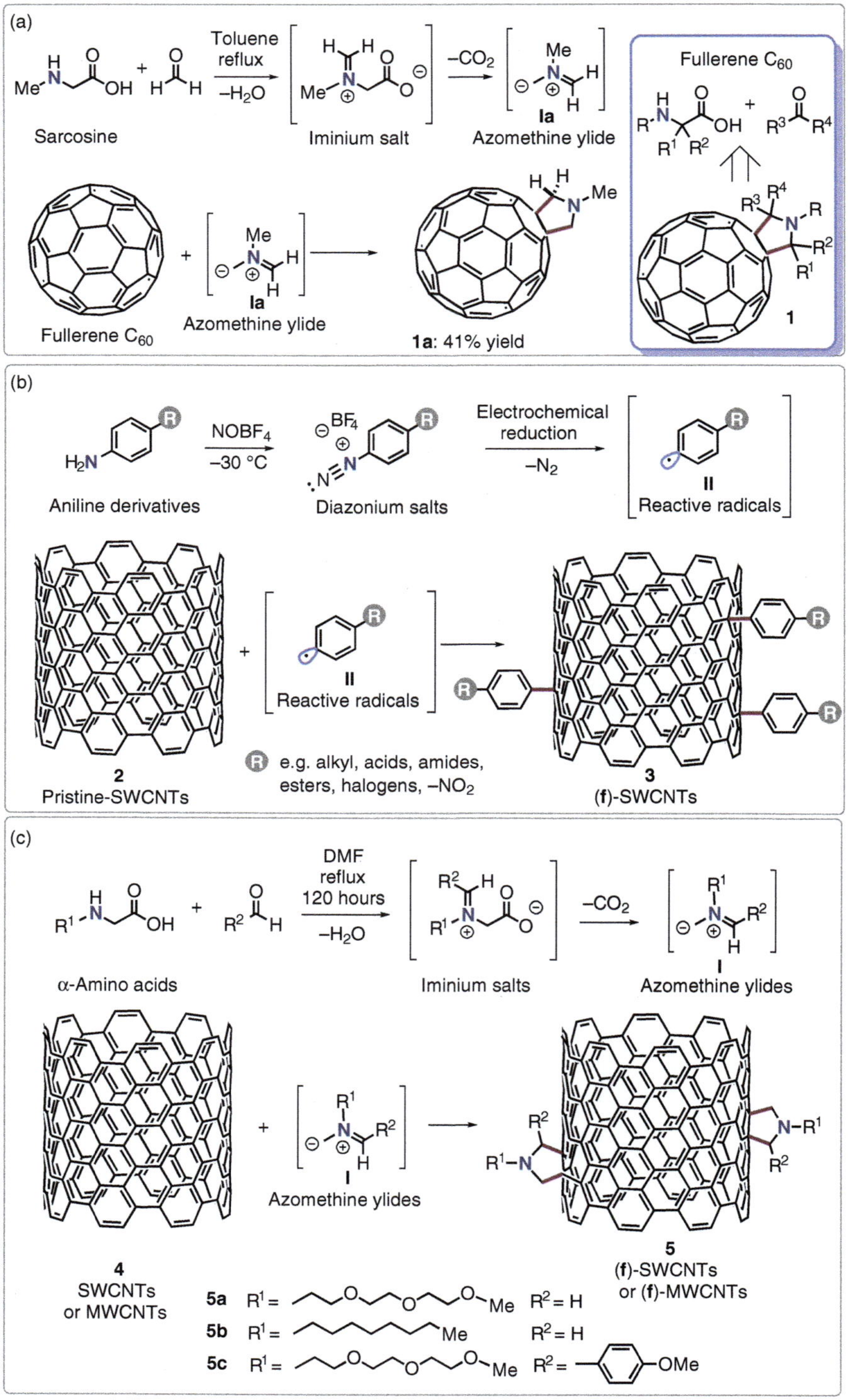

Figure 8.2 (a) Cycloaddition of azomethine ylides to C_{60}. Source: Based on M. Maggini et al. [17]. (b) Arylation reaction on SWCNTs. Source: Based on J. L. Bahr et al. [18]. (c) Cycloaddition of azomethine ylides to CNTs. Source: Based on Georgakilas et al. [19].

subsequent electrochemical reduction of these diazonium salts provided the corresponding reactive aryl radicals (**II**) that covalently bind to the carbon surface, yielding the functionalized single-walled carbon nanotubes (***f***-SWCNTs) (**3**). The functionalization degrees depend on the nature of the para-substituent (–R) within **II** and have been evaluated by combining Raman and TGA. In 2002, a versatile methodology for the organic functionalization of CNTs by using the 1,3-dipolar cycloaddition of azomethine ylides was reported (Figure 8.2c) [19]. The pristine materials (**4**) were suspended in dimethylformamide (DMF), together with an excess aldehyde and the suitable α-amino acid. The heterogeneous reaction mixture was then heated at 130 °C for five days. After workup, the functionalized nanotubes (e.g. **5a–c**) were obtained, which were soluble in chloroform, dichloromethane, acetone, methanol, and ethanol; less soluble in toluene and tetrahydrofuran; and practically insoluble in diethyl ether and hexane. Remarkably, any moiety could be, in principle, attached to the tubular network of CNTs by employing these two derivatization approaches (Figure 8.2b,c) [30]. Moreover, both cycloaddition and arylation reactions, typically, work efficiently with both SWCNTs and multiwalled carbon nanotubes (MWCNTs). The main problem of insolubility of CNTs in aqueous media has been solved by developing a two-step synthetic protocol that allows installing numerous superficial positively charged hydrophilic functionalities at their surfaces (Figure 8.3a). The resultant amino-functionalized CNTs (e.g. **5e–f**) show high aqueous solubility and were particularly suitable for the covalent immobilization of molecules or for further functionalization of the surface based on multiple positive/negative charge interactions (Figure 8.3) [33]. Indeed, various biomolecules have been attached on these amino-CNTs (**5e,f**), such as amino acids, peptides (shown in Figure 8.3b), or nucleic acids [31].

Several applications in the field of medicinal chemistry have been pursued through this approach, including vaccine and drug delivery, gene transfer, and immunopotentiation [34]. Afterward, it was demonstrated that materials of the type of **5e,f** can be subjected to a series of alkylation reactions with glycidyltrimethylammonium chloride, and/or with a mixture of acrylate and ethylenediamine, to obtain a novel family of dendron-functionalized MWCNT derivatives, which are characterized by the presence of numerous positively charged tetra-alkyl ammonium salts at the periphery of the dendrons [35]. These functionalized and charged MWCNTs showed excellent chemical stability and aqueous solubility that increases with the number of the tetra-alkyl ammonium units present at their surfaces and thus with the functionalization degree of the materials (aqueous solubility up to 3 mg/ml). In 2010, the use of these water-soluble dendron MWCNTs for the production of new electrodes suitable for the water-splitting process was exploited (Figure 8.3c) [32]. Indeed, a very efficient and stable nanostructured oxygen-evolving anode can be obtained by the assembly of an oxygen-evolving polyoxometalate (POM) cluster [36], that is, a totally inorganic ruthenium catalyst, with a conducting bed of positively charged dendron MWCNTs. The electrostatic capture of the POMs by the polycationic MWCNTs surface thus allowed for the fabrication of a stable and efficient electrode, which addresses the major challenge of artificial photosynthesis, namely, efficient water oxidation.

Single functionalization protocols of carbon-based nanostructures, namely, C_{60} and CNTs, have been discussed up to this point of the chapter. In addition to single transformations, multiple functionalization reactions represent powerful tools to decorate with more than one functional group of the nanomaterial surfaces (e.g. CNTs) [37]. In 2008, a

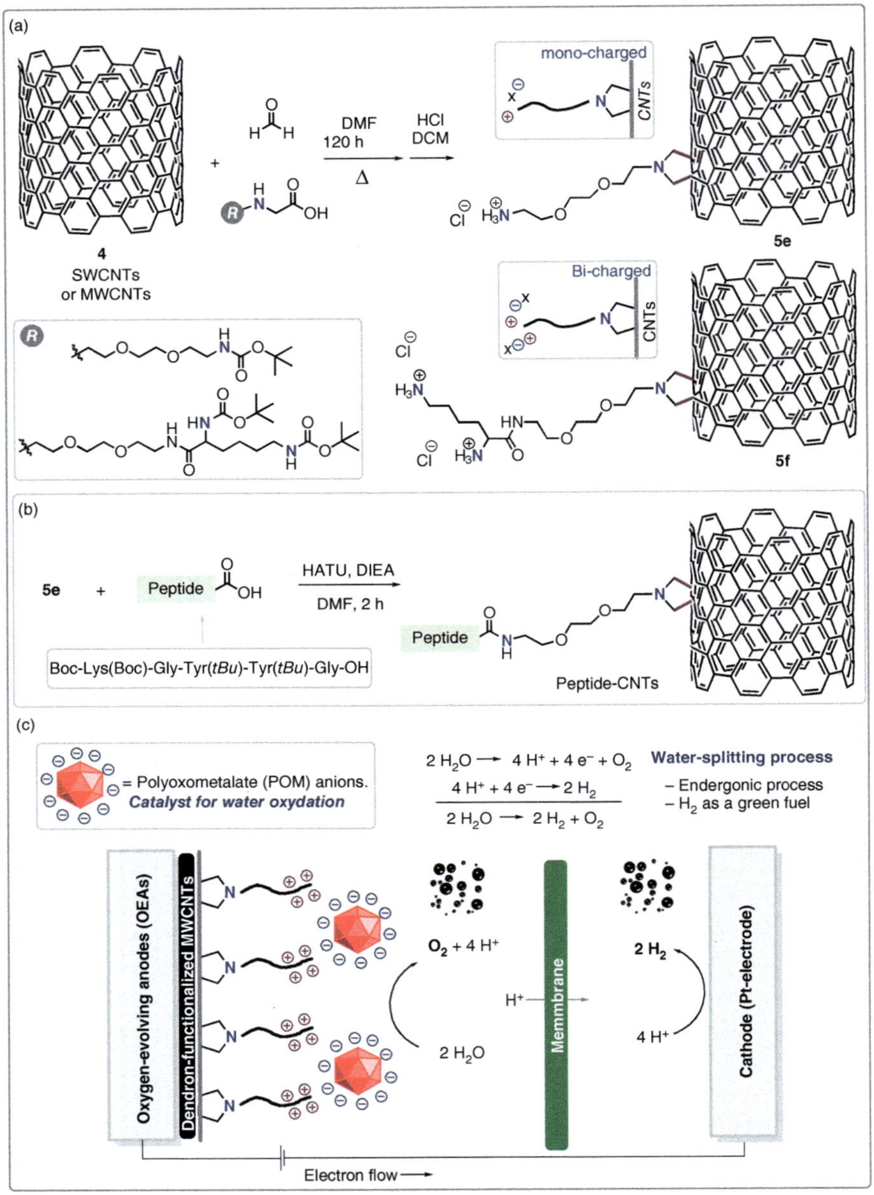

Figure 8.3 (a) Reaction pathway for obtaining water-soluble ammonium-modified nanotubes. (b) Covalent functionalization of superficial amines of CNTs with a biologically active peptide. Source: Based on Pantarotto et al. [31a]. (c) Electrostatic capture of negatively charged polyanionic ruthenium-containing clusters and general scheme for a water-splitting electrocatalytic cell with the integrated nanostructured oxygen evolving anodes (OEAs). Source: Based on Toma et al. [32].

novel, synthetic, two-step strategy to produce multifunctionalized CNTs was described, by employing a combination of two different transformations, the 1,3-dipolar cycloaddition of azomethine ylides (**I**) and the radical addition of diazonium salts, both via a simple and fast microwave (MW)-induced method (Figure 8.4a) [39]. A solvent-free cycloaddition reaction combined with a MW irradiation afforded functionalized nanotubes (**5g–j**) in just one hour. This method significantly improves the productivity of the derivatization process (reaction performed over 120 hours in the primal approach, see Figure 8.2c) and thus paves the way to large-scale functionalization [40]. The resultant materials were then subjected to diazonium salt reaction to introduce other functional groups on the same single-walled nanotubes (SWNTs). Therefore, ***f***-SWNTs (**5g–j**) were dispersed in water with *p*-toluidine and isoamyl nitrite, and this mixture was heated for 90 minutes at 80 °C in a microwave reactor. Mechanistically, the aniline derivative could attack the nitrite group giving the corresponding diazonium and alkoxide species (Figure 8.4a, light blue box). This anionic species (alkoxide) could then add to the electrophilic diazonium salt, thus yielding a diazoether intermediate, which might eventually decompose with release of nitrogen and generation of an aryl radical (**IIa**). This radical could react with the CNT, thus producing **6a–d**. This proposed mechanism is still being discussed at present [38]. In 2011, a novel methodology to covalently functionalized SWCNTs with three different active groups, in one step, by employing the arylation reaction was developed (Figure 8.4b) [8b]. The CNTs (**7a**) are functionalized with numerous benzylamine moieties blocked with three orthogonal protecting groups that can be selectively removed under specific conditions. The sequential removal of the protecting groups of the amine functions might allow for a precisely controlled grafting of molecules of interest onto the nanotube surfaces.

Graphene is a single atomic layer of sp^2-hybridized carbon atoms arranged in a honeycomb lattice with dimensions ranging from a few hundred nanometers to tens of micrometers [41]. Many different types of graphenes can also be defined depending on the number of layers, layer dimension, and amount of oxygen present in the carbon structure [42]. The family of graphene-based materials exhibits a unique set of electronic, mechanical, and thermal properties [43]. Consequently, a wide variety of applications have been designed for graphene materials in sensing, energy storage, catalyst support, supercapacitors, and optoelectronic devices [44]. In most of these applications, graphene is supported on a substrate because the production and handling of graphene in solution is technically challenging. This is due to the metastable nature of graphene and consequently to its tendency toward restacking. The chemical functionalization of single- and few-layer graphene (FLG), both on a support and in solution, is a topic of paramount importance because it allows for the fine-tuning of the materials' chemical and physical properties [45].

The following two examples concern the covalent functionalization of graphene with amines either in solution for the production of a gold-containing nanocomposites or on a substrate to anchor a biologically active protein (Figure 8.5). A mixture of single- and FLG (**8**) can be produced by dispersion and exfoliation of graphite in *N*-methylpyrrolidone (NMP) [47]. The use of this technique allows to effectively produce single- and FLG without the need for intercalants, polymers, or surfactants, which might interfere with the subsequent material manipulations. Functionalized graphene was then prepared by condensation of paraformaldehyde with a modified α-amino acid, followed by the deprotection of the *t*-butyl carbamate (BOC) group to afford **9** (Figure 8.5a) [8f]. To assess

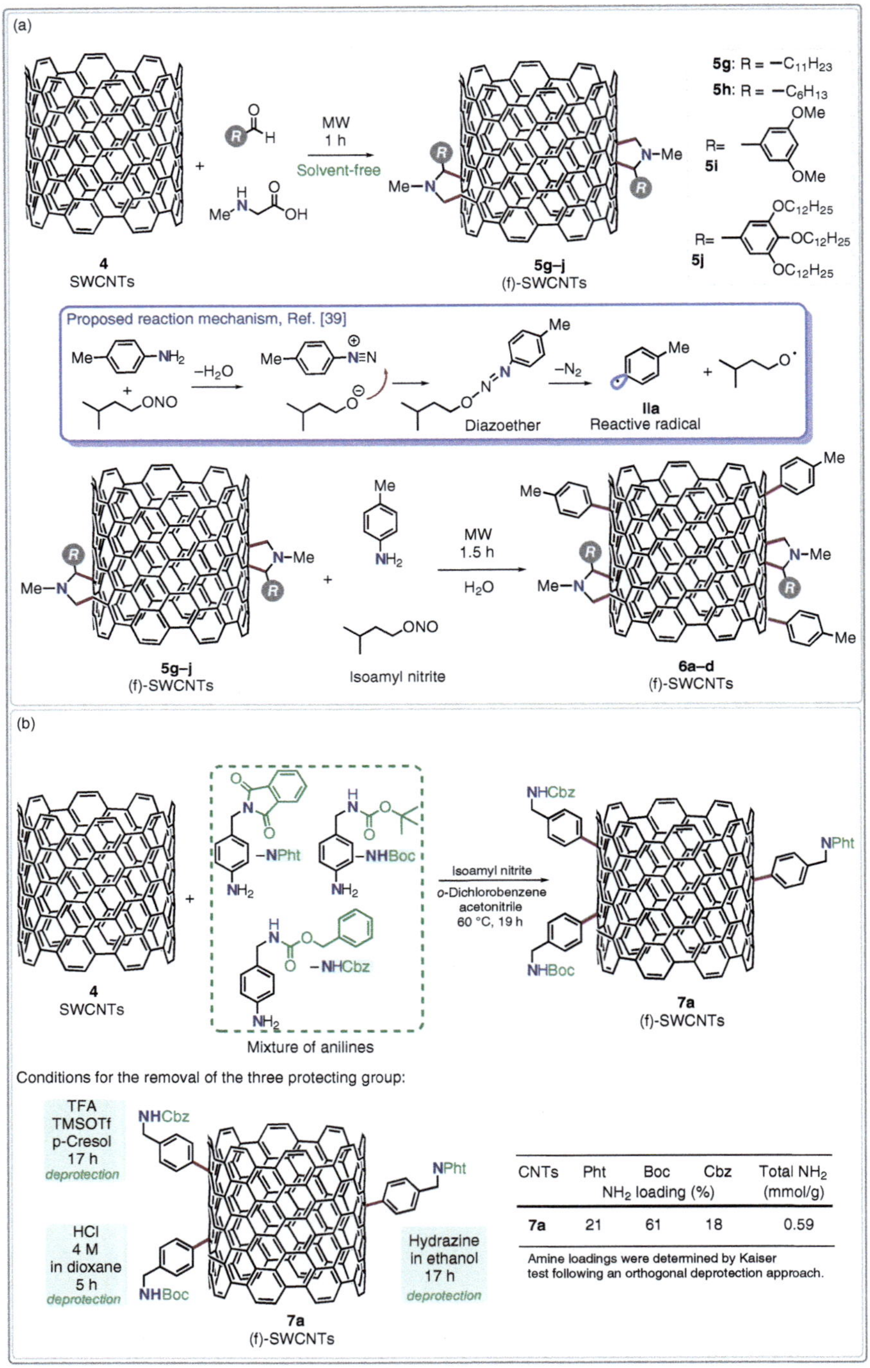

CNTs	Pht	Boc	Cbz	Total NH_2
	NH_2 loading (%)			(mmol/g)
7a	21	61	18	0.59

Figure 8.4 (a) Synthetic protocol for the MW-induced double functionalization of CNTs. Source: Based on Lipinsk et al. [38]. (b) One-pot triple functionalization of CNTs [39]. Source: Based on Ménard-Moyon et al. [8].

the presence of free amino groups in the resultant product (**9**), the authors performed a quantitative Kaiser test [2c]. The amount of amine functions in **9** was therefore measured to be in the range of 0.64 ± 0.07 mmol/g of material. This is a high value if one takes into account only the functionalization of the edges of graphene sheets, which are considered to be the most reactive sites.

This is an indication that 1,3-dipolar cycloaddition must also take place in the less-reactive central carbon–carbon bonds of graphene layers. The presence of all these superficial amines was utilized for producing a nanocomposite (**10**) with gold nanorods (Au-NRs). The sheets of this hybrid material were characterized by transmission electron microscopy (TEM), which demonstrated the presence of Au-NRs distributed uniformly over the graphene surface. This second observation also confirmed that the reaction took place not just at the edges but also at the internal C—C bonds of graphene.

Another and very popular approach for the functionalization of graphene surfaces is the addition of aryl radicals generated by the reduction of diazonium salts [48]. This strategy, typically, leads to high functionalization degrees and also has the advantage of being tolerant to a variety of conditions (e.g. the solvent used).

The chemical properties of graphene, when it is placed on a substrate, depend strongly on the graphene shape, number of layers, and nature of the substrate. Naturally, this dependence is also reflected in the reactivity with diazonium species [49]. As an example, graphene grown via chemical vapor deposition-graphene (CVD-G) was transferred through reactivity imprint lithography onto a patterned substrate displaying both hydrophobic (OTS, octadecyltrichlorosilane) and hydrophilic (SiO_2) areas (Figure 8.5b) [46]. Then, the supported graphene **11** was functionalized with 4-carboxyphenyldiazonium tetrafluoroborate. The efficient introduction of the carboxyphenyl moieties in **12** has been confirmed by attenuated total reflectance infrared (ATR-IR) analysis. Attachment of these organic units on graphene allowed to establish amide bonds with *N*α,*N*α-bis(carboxymethyl)-L-lysine hydrate, thus leading to the formation of **13** with superficial nitrilotriacetic acid (NTA) units that could be employed to chelate metals. In this way, upon addition of $NiCl_2$, functionalized graphene displayed NTA–Ni complexes that were used as anchors for polyhistidine (His)-tagged enhanced green fluorescent protein (EGFP), allowing for detection of resulting **14** by confocal fluorescence microscopy.

8.3.3 Other Functionalization Procedures of Common Carbon Nanostructures

In this section, we discuss other notable examples related to the derivatization of common carbon-based nanostructures, namely, fullerenes, CNTs, and graphene, which may lead to the formation of various different superficial amino-containing functionalities.

The extraordinary chemical and physical properties of C_{60} and its derivatives might confer activity of these compounds as potential therapeutic agents against human immunodeficiency viruses (HIV) and neurodegenerative diseases and as agents for deoxyribonucleic acid (DNA) cleavage [50].

However, the insolubility of fullerenes in aqueous media has dramatically slowed down the realization of these biological applications. In 2000, the Schuster and Wilson groups reported a methodology for the production of a novel family of amino-substituted methanofullerenes (**15**, $n = 1, 3, 6$), which exhibit pronounced water solubility

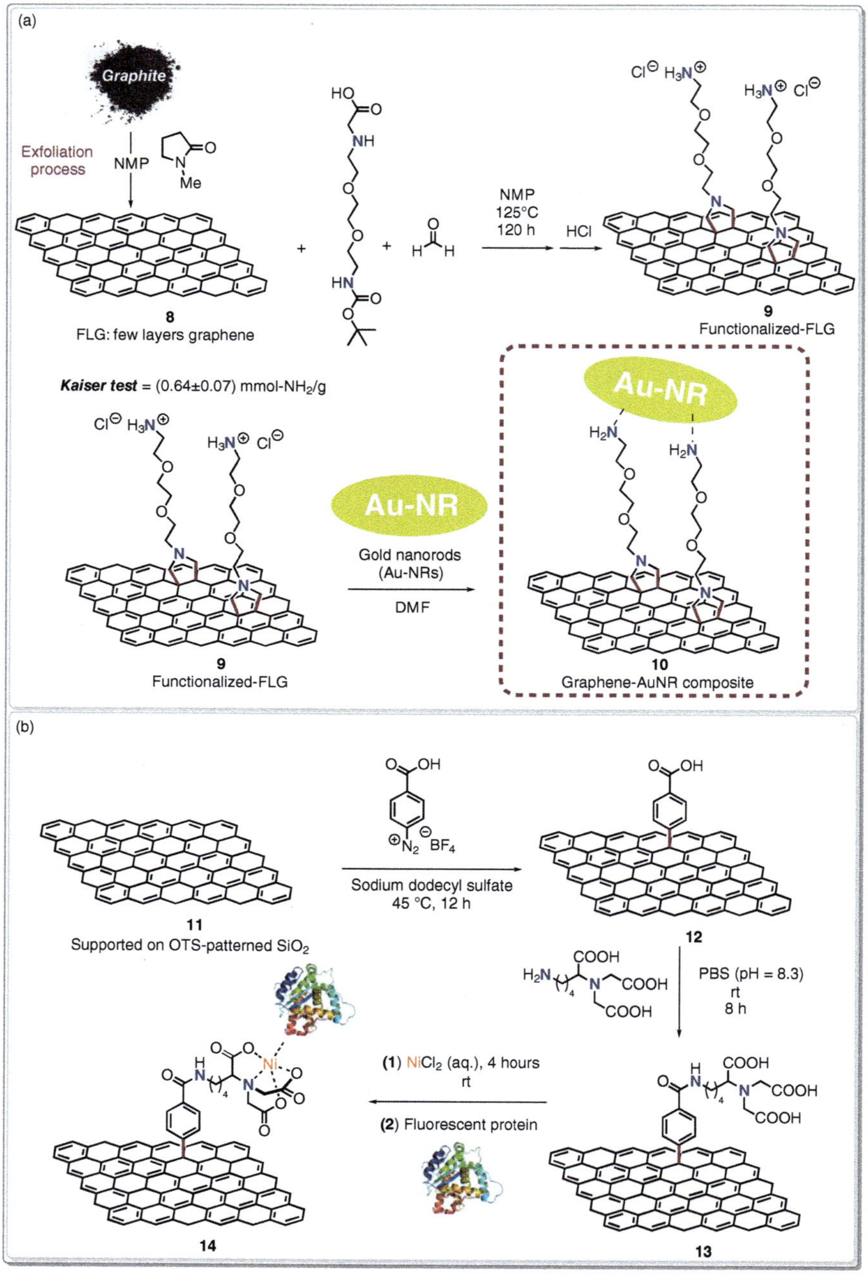

Figure 8.5 (a) Functionalization of graphene via 1,3-dipolar cycloaddition and production of a graphene–Au nanorod composite. Source: Reprinted with permission from Quintana et al. [8]. Copyright 2010, American Chemical Society. (b) Synthetic procedure to anchor a fluorescent protein on graphene . Source: Based on Wang et al. [46].

(Figure 8.6a) [51]. These amino-functionalized fullerenes (**15**) have been obtained through a Bingel–Hirsch [54] reaction between an N-protected malonic derivative and C_{60}, followed by the rapid and effective removal of amine Boc-protecting groups with trifluoroacetic acid (TFA). After the following workup, the crude reaction mixture containing all products **15** was separated on a silica gel column. Interestingly, the exa-amino derivative ($n = 6$), which bears 12 positive charges, showed excellent aqueous solubility (418 mg/ml).

Supercritical fluids (e.g. supercritical carbon dioxide or supercritical water) are regarded as eco-friendly alternatives to organic solvents. There have been many examples of their use in chemical synthesis, usually under homogeneous conditions without the need for other solvents [55]. In contrast to ionic liquids or fluorinated solvents, supercritical fluids represent a different kind of alternative solvent because they are not in the liquid state. In fact, these fluids show low viscosity and no surface tension, and the interactions between solvent molecules and the solute are weak, so high diffusivities are reached, close to those in the gas phase. Recently, supercritical ammonia has been utilized in the chemical modification of MWCNTs (Figure 8.6b) [52]. Such a fluid was able to introduce several amino groups in the pristine nanotubes **16**, giving **17**, while it is more likely that the conversion of carboxylic groups into amides (**19**) took place when oxidized-MWCNTs **18** were used as starting materials.

Recently, it was demonstrated that amino-functionalized graphene (AFG) is a potentially safe material suitable for in vivo biomedical applications [53]. AFG can be produced following the methodology described in Figure 8.6c. Graphene oxide (GO, **20**) sheets can be prepared by refluxing the commercially available graphite powder with a strongly acidic mixture of sulfuric and nitric acid [56]. Material **21** is then obtained by a rapid thermal treatment (1100 °C) of GO (**20**) under argon atmosphere that reduces functional epoxy and hydroxyl groups from the surface of GO-sheets, upon evolution of CO_2, thus leading to the formation of vacancies in the nanostructure. Afterward, the carboxylic acids within **21** are reacted in a mixture of $SOCl_2$ and DMF to produce the corresponding graphene-COCl derivative. This material can be treated first with sodium azide and then with concentrated hydrochloric acid to yield the desired amino-functionalized graphene product (**22**). Subsequently, in 2014, amine-functionalized graphene has also been used as a metal-free catalyst for the oxygen reduction reaction (ORR), a redox process that may produce either water or hydrogen peroxide [57]. The electrochemical results show that AFG is an active catalyst in oxygen reduction reaction and exhibits an electrocatalytic activity much higher than graphene. The significantly high electrochemical performance of AFG could be attributed to its specific structure, which possesses electron-donating groups (e.g. amines), a large number of holes in its sheet plate and the porous structure of the bulk material, arising from the randomly stacked AFG layers.

8.3.4 Exfoliation of Graphite with Melamine

As mentioned above, graphene and its derivatives are novel and extraordinary carbon-based materials expected to revolutionize the electronic industry [58]. Although considerable scientific progress has been achieved in this field of research, the key to commercializing graphene lies in the development of a cost-effective and scalable production method. The biggest problem concerning the exfoliation of graphite is associated with the high energy

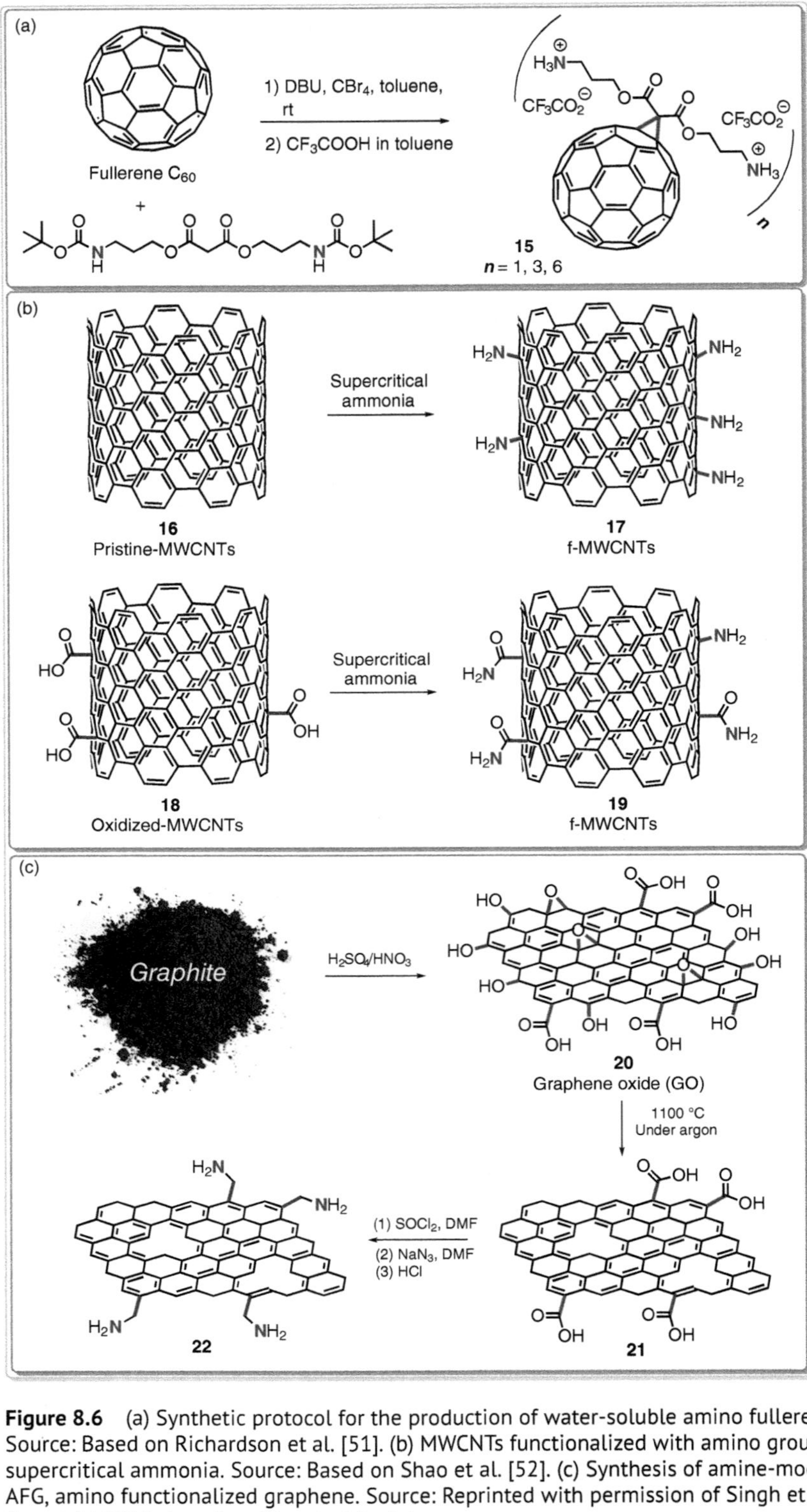

Figure 8.6 (a) Synthetic protocol for the production of water-soluble amino fullerene derivative. Source: Based on Richardson et al. [51]. (b) MWCNTs functionalized with amino groups by the use of supercritical ammonia. Source: Based on Shao et al. [52]. (c) Synthesis of amine-modified graphene. AFG, amino functionalized graphene. Source: Reprinted with permission of Singh et al. 2012 [53]. Copyright 2012, American Chemical Society.

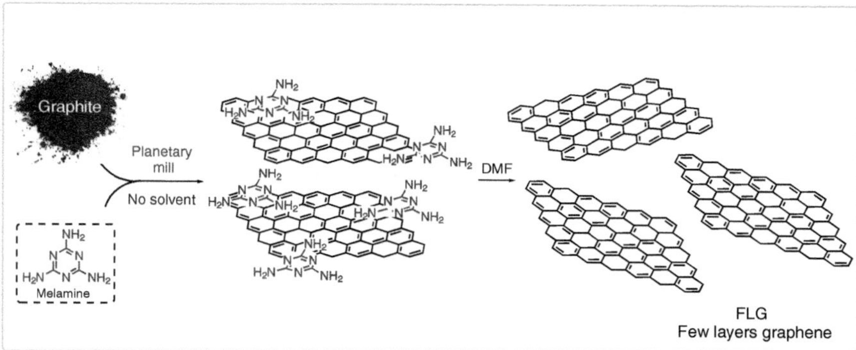

Figure 8.7 Schematic illustration for the exfoliation of graphite through ball milling approach. Source: Based on León et al. [62].

required to break the enormous van der Waals-like forces between the graphite layers [59]. Current techniques produce graphene in both *bottom-up* and *top-down* approaches [60]. *Bottom-up* methods include chemical vapor deposition (CVD) or epitaxial growth on a substrate and tend to produce high-quality graphene films.

In contrast, *top-down* approaches have been explored to exfoliate graphene nanosheets from graphitic precursors. Within this class, the use of an aromatic amine, along with the mechanical activation achieved by milling process, has been recently applied to the exfoliation of graphite [61]. Specifically, it was shown that large quantities of inexpensive materials, namely, graphite and melamine (2,4,6-triamino-1,3,5-triazine), can be used for the massive and fast production of concentrated dispersions of FLG with a low amount of defects (Figure 8.7) [62]. In a typical procedure, a solid mixture of melamine and graphite is ball-milled under air atmosphere. Thus, the adsorption of melamine on graphene is the key step to compensate the huge van der Waals attractive interactions between the layers, thereby allowing the exfoliation process [63]. Then, the resulting solid mixture was dispersed in DMF (or water) producing a black concentrated suspension, which contained the desired FLG.

8.3.5 Other Carbon Nanomaterials

In the family of carbon allotropes, other "minor" members have their own identity and their potential applications. Among them, carbon nanohorns (CNHs), nanodiamonds and CNOs must be mentioned. In general, their reactivity is quite similar to that of the most famous siblings fullerenes, CNTs, and graphene. For this reason, they have been functionalized either covalently or noncovalently using many of the synthetic strategies developed for CNTs and graphene, as we illustrate in the following paragraphs.

8.3.5.1 Carbon Nanohorns

CNHs are conical-shaped carbon nanostructures sometimes also called carbon nanocones constituted by one or more sp^2 carbon sheets typically 2–5 nm in diameter and 40–50 nm in length (see Figure 8.8 for an example of transmission electron micrographs). CNHs belong

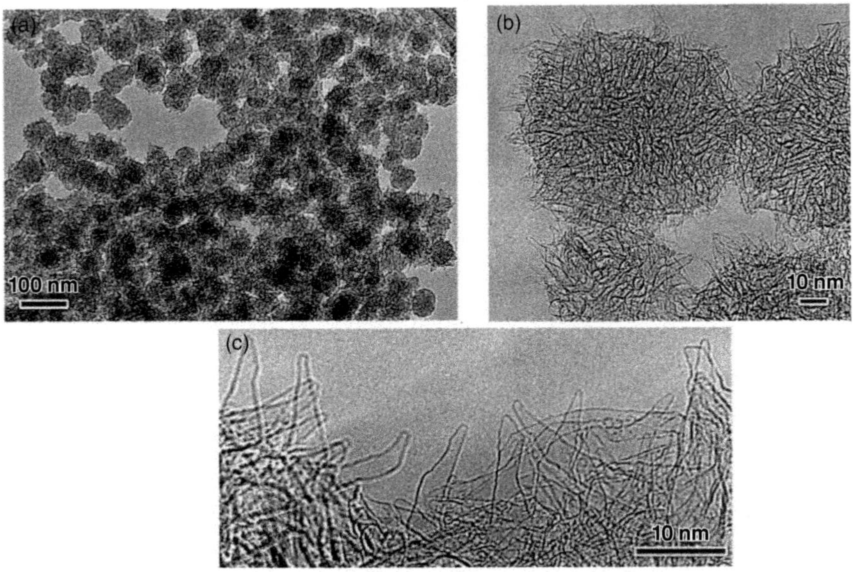

Figure 8.8 Transmission electron micrographs of carbon nanohorns: (a) close-up of individual nanohorns, (b) aggregates of nanohorn clusters, and (c) zoom-in on a few aggregates of nanohorns, also known as dahlias, produced by laser ablation. Source: Reprinted from Iijima et al. 1999 [64]. Copyright 1999, Elsevier.

to the group of one-dimensional carbon nanostructures, but during the synthesis, they can aggregate into spherical clusters ~100 nm in diameter called "dahlia-like" clusters. CNHs have been first observed by Harris et al. from the waste product of soot formed during the production of fullerenes via arc discharge process [65]. After heating to 2500–3500 K with positive-hearth electron gun for approximately four hours, the soot is converted into two types of nanohorns, single-walled carbon nanohorns (SWCNHs) and multiwalled carbon nanohorns (MWCNHs).

"Dahlias" of nanohorns were first synthesized via laser ablation process of graphite in an argon atmosphere by Kasuya and coworkers. CNHs are apt to industrial take-up. Notably, they can be produced cheaply, in large quantities, without the need for subsequent purification steps and without the presence of metal catalysts [66]. CNHs are considered as a good replacement of CNTs because of their large specific surface area, high production yield, high chemical stability, high purity, low toxicities, exceptional catalytic properties, superior porosity, and good conductivity. Because of all these properties, CNHs found applications in the fields of electrochemical sensing or biosensing, biofuel cell development, as an electrode material, supercapacitor, gas storage devices, and biomedical application [67]. CNHs have been modified with amino groups similarly to the previously described carbon-based nanomaterials: oxidized CNHs were conjugated to the fifth generation of polyamidoamine (PAMAM) dendrimer and used as an essential substrate for the incorporation of Pd^{2+} ions, which were subsequently chemically reduced to Pd nanoparticles.

The latter dendrimer-encapsulated Pd nanoparticles onto oxidized CNHs were efficiently employed as nonenzymatic tracing tag for the electrocatalytic reduction of dissolved oxygen

[68]. Oxidized CNHs, after activation with *N*-ethyl-*N*′-(3-(dimethylamino)propyl) carbodiimide hydrochloride (EDC) and 1-hydroxy benzotriazole (HOBt), were treated with *tert*-butyl 2-(2-(2-aminoethoxy)ethoxy)ethylcarbamate for the preparation of modified CNHs carrying amines masked by BOC protecting groups [69]. Acidic deprotection of the BOC moieties liberated free amines on the modified CNHs, which were subsequently conjugated with 5,10,15-tris(3,5-di-tert-butylphenyl)-20-[4-(carbonyloxy)phenyl] porphyrin, providing a donor–acceptor CNH-based hybrid, with significant redox and photophysical properties. Oxidized CNHs carrying ammonium groups were assembled to a benzo-18-crown-6-ether moiety linked to a pyrrolidine-functionalized C_{60} [70]. The so-formed C_{60}-CNH nanohybrid similarly showed high dispersibility in organic solvents, which allowed its spectroscopic characterization. In addition, femtosecond laser flash photolysis disclosed efficient electron transfer events from the singlet excited state of the C_{60} unit to CNHs, thus indicating the action of the CNH moiety as an electron donor and C_{60} as an electron acceptor, effectively facilitating charge-separation while suppressing the recombination of photoexcited electron–hole pairs.

Many studies are dedicated to the covalent functionalization of CNHs sidewalls with amino groups. Large aggregates of CNHs were successfully deagglomerated and solubilized in aqueous media after treatment with $NaNH_2$ in refluxing liquid NH_3, which resulted in the introduction of amino groups onto their outer surface [71]. The so-formed amino-modified CNHs were further derivatized with the fluorescent dye Oregon Green and incubated with growing mammalian cells (3T3 and HeLa cells), revealing low cytotoxicity. Furthermore, the covalent attachment of a biotinylated diamide onto amino-CNHs through an amidation reaction led to a biotinylated CNH intermediate, which proved to be an excellent specimen for the conformational study of such modified CNHs by TEM imaging (Figure 8.9) [72].

Acidic deprotection of BOC-protected amino-modified CNHs allowed their labeling with fluorescein isothiocyanate (FITC), which made them useful for monitoring their interaction with primary murine macrophages. In a similar approach, 2,2′-(ethylenedioxy)bis(ethylamine) was directly cross-linked with intact CNHs using a simple bath sonication method for 10 hours [73]. Subsequently, such amino-functionalized CNHs were conjugated with the fluorescent IR Dye 800 CW via an amidation reaction, thus constructing a light-driven nanomodulator that generates heat and reactive oxygen species under biologically transparent near-infrared laser irradiation.

1,3-Dipolar cycloaddition of azomethine ylides, in situ generated upon thermal condensation of aldehydes and α-amino acids, was also employed for the modification of the sidewalls of CNHs with amino groups. With this particular cycloaddition reaction, a large number of pyrrolidine units fused to C—C bonds on the CNH sidewalls were introduced, affording highly dispersible functionalized materials. Most importantly, a significant variety of pyrrolidino-modified CNHs with diverse substituents on the pyrrolidine ring can be prepared, depending on the nature of the utilized aldehyde and/or α-amino acid. Thus, different aldehydes give rise to functionalized CNHs carrying diverse substituents on the α-carbon atom of the pyrrolidine ring, while N-modified α-amino acids lead to different N-substituted pyrrolidines on the skeleton of CNHs [74]. One example of this kind of derivatives was reported as a new selective antibody–drug nanosystem based on CNHs for the treatment of prostate cancer (PC) [75]. In particular, cisplatin in a prodrug form

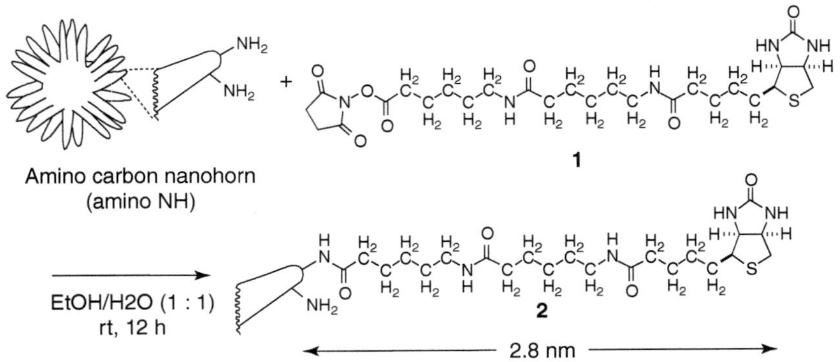

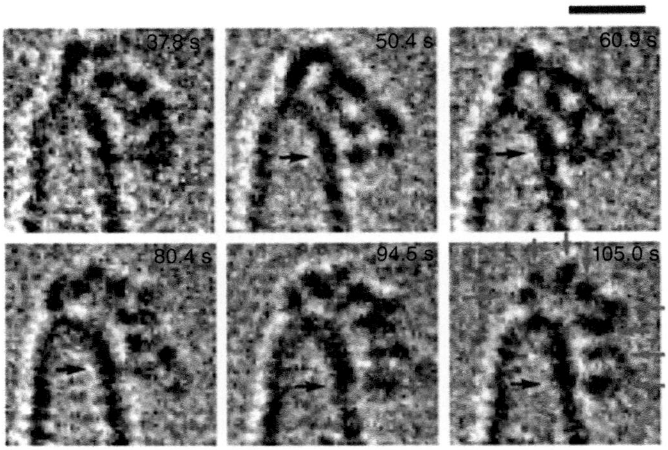

Figure 8.9 TEM images of triamide–NH conjugate in a series of bent conformations. The figure captions refer to the time (seconds) after initiation of the observation, i.e. initiation of electron irradiation. Source: Reprinted with permission from Nakamura et al. [72a] and Lacotte et al. [72b].

and the monoclonal antibody (Ab) D2B, selective for prostate-specific membrane antigen (PSMA) + cancer cells, have been attached to CNHs because of the current application of this antigen in PC therapy. The hybrids Ab–CNHs, cisplatin–CNHs, and functionalized CNHs have also been synthesized to be used as control systems. The efficacy and specificity of the D2B–cisplatin–CNH conjugate to selectively target and kill PSMA$^+$ prostate cancer cells compare very well with other derivatives.

The radical addition of in situ generated aryl diazonium salts was applied to covalently modify the outer skeleton of CNHs. By this method, the thermally induced reaction of pristine CNHs with aryl diazonium intermediates, generated in situ from aniline derivatives after treatment with isoamyl nitrite, resulted in the covalent grafting of a wide range of aryl units onto the outer walls of CNHs, thus enhancing their solubility in common solvents. Following this procedure, either simple anilines, such as 4-nitroaniline, 4-aminobenzoic acid, and 4-butylaniline, or more complex aniline derivatives, such as *tert*-butyl (2-(2-(2-(4-aminobenzamido)ethoxy)ethoxy)ethyl)carbamate, were reacted with CNHs in the presence of isoamyl nitrite in a mixture of *o*-dichlorobenzene/acetonitrile as solvent to afford novel highly functionalized hybrid materials [76].

Recently, microwave-assisted chemistry proved to be a powerful tool toward chemical functionalization of carbon nanostructured materials, presenting significant advantages by shortening considerably the reaction times and offering higher yields and/or fewer by-products. Furthermore, taking advantage of the strong microwave absorption that carbon nanostructures exhibit, the presence of toxic and irritant solvents can be reduced, thus establishing microwave-assisted chemical functionalization as a general concept in green chemistry. Therefore, without surprise, most of the already known reactions that conventionally take place for the chemical modification of CNHs were also tested with the aid of microwave irradiation. In this line, the 1,3-dipolar cycloaddition of in situ generated azomethine ylides onto the outer surface of CNHs was achieved by using microwave irradiation in the absence of any solvent and within one hour of reaction time. The latter result was a significant improvement compared with the corresponding conventional methodology, which takes place in DMF as a solvent and needs thermal treatment for more than 100 hours [77]. In the same report, the introduction of various aryl units onto the CNH-surface via the radical addition of in situ generated aryl diazonium salts was achieved with the aid of microwave irradiation and in environmentally friendly aqueous media. In a further step, taking advantage of the fact that the cycloaddition reaction was more selective and slower than the radical addition, the double functionalization of CNHs, namely, the 1,3-cycloaddition under solvent-free conditions, followed by the radical addition of aryl diazonium salts in water, led to multifunctionalized hybrid materials with great potential in diverse applications.

8.3.5.2 Carbon Nanodiamonds

Carbon nanodiamonds (CNDs) are in the group of zero-dimensional carbon nanoallotropes, consisting primarily of tetrahedral sp^3 carbon atoms. They have a diamond-like monocrystalline topology with crystal domains and particle size less than 10 nm. Nanodiamonds consist of complex structures with an inner diamond core and an outer amorphous carbon shell. Because of this type of structure, they display unique properties such as high surface area, hardness, high electrochemical stability, mechanical robustness, high thermal conductivity, environmental inertness, high insulating capability, modifiable color centers, nanotribology, thermal expansion, chemical corrosiveness protection, nontoxicity, and biocompatibility [78].

Nanocrystalline diamonds are characterized by an irregular shape and, depending on the synthesis and purifications, different populations of nanometer-sized materials can be recovered. The shape of ultrananocrystalline particles is almost spherical, with an average size equal to 5 nm. By high resolution transmission electron microscopy (HRTEM), the diamond core is easily recognized. HRTEM can be also helpful to pinpoint structural defects inside the nanoparticles (Figure 8.10) [80].

Nanodiamonds can be produced by the detonation method, laser ablation, CVD, high impact on diamond microcrystals, ball milling, meteoritic impacts process, jet milling, or abrasion of microdiamonds. Nanodiamonds offer a potential novel template for drug delivery and targeting because of their small primary particle size, purity, excellent

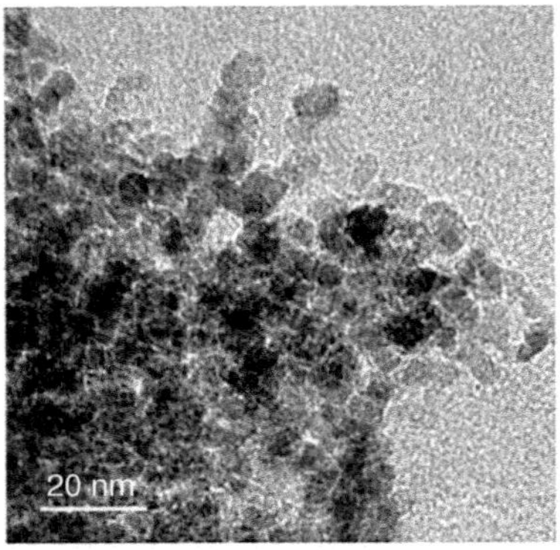

Figure 8.10 TEM micrograph of 10 nm diameter nanodiamond. Source: Reprinted from Valey et al. [79].

properties, facile surface functionalization, high biocompatibility, and inexpensive large-scale synthesis.

However, nanodiamond properties such as aggregation state, surface chemistry, and presence of impurities, as well as the localization and accumulation behavior at the organism level, must be controlled to maintain safety and retain efficacy of conjugated therapeutic agents.

Nanodiamonds functionalized with different organic moieties carrying terminal amino groups have been synthesized. Fluorinated nanodiamonds can further react with ethylenediamine, resulting in amino-nanodiamond [81]. Alternatively, carboxylated NDs were linked to ethylenediamine via acyl chloride intermediate [82]. Reaction of aminated NDs with epoxy resin results in covalent incorporation of the ND into a polymer network. Covalently bonded nanodiamond-epoxy composites showed a three times higher hardness, 50% higher Young's modulus, and two times lower creep compared to the composites in which the nanodiamond was not chemically linked to the matrix.

For the quantification of primary amino groups, a modified photometric assay based on the Kaiser test has been developed and validated for different types of aminated nanodiamonds. The results match well with the values obtained by thermogravimetry analysis. The method represents an alternative wet-chemical quantification method in cases where other techniques, such as elemental analysis, fail because of unfavorable combustion behavior of the analyte or other impediments [8d].

Another method to obtain more homogeneous nanodiamond surfaces is the reduction of surface carbonyl groups. Kruger et al. reported on the reaction with borane, leading to hydroxylated diamond samples [83]. This material can be modified in a variety of ways. The grafting of trialkoxysilanes easily yields functionalized nanodiamond particles. The reaction can be carried out with different silanes, for example, trimethoxy- and triethoxy-3-amino-propylsilane. The amino groups of the material are accessible for further surface modification. Besides the build-up of a short model peptide on the diamond solid phase, biotin was covalently immobilized on the particle surface by coupling with the amino groups of the silane. Its activity in the binding to streptavidin was demonstrated, making it a model for bioactive moieties covalently grafted onto the diamond surface. An interesting application of amino-functionalized ND was reported by Yu et al. To align the effective nanocarrier features of the supraparticles (SPs) with alkyl amine molecules,

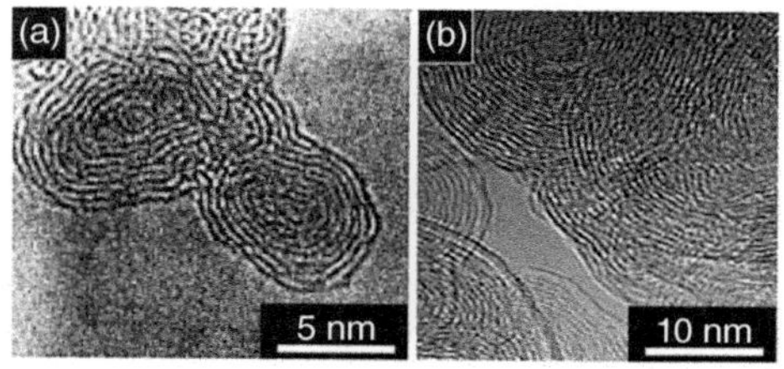

Figure 8.11 TEM images of carbon nano-onions. Source: Reprinted with permission from Plonska-Brzezinska [86].

carboxylic ND (ND-ori)-based supraparticles (ND-SPs) were designed and synthesized via a condensation reaction between the carboxylic group on the surface of ND-ori and the amino group of the primary alkyl amines, *n*-octylamine (Oct), dodecylamine (Dod), and oleylamine (Ole) [84]. Alkyl amine molecules have indeed a relatively lower bioaccumulation compared to fluorine compounds usually employed to achieve this kind of entities [84a]. Oct-ND-SPs and Dod-ND-SPs spontaneously self-assembled in SP and were used for further drug efficacy tests because of their low cytotoxicity. A model anticancer drug camptothecin (CPT) was incorporated into these ND-SPs and conventional polyethylene glycol (PEG)-modified polymer micelles, and the ND-SPs showed good drug efficacy against U_2OS bone osteosarcoma cells. In particular, CPT-loaded ND-SPs displayed the highest anticancer therapeutic effect. In vivo antitumor assays showed that CPT delivered by ND-SPs caused stronger tumor suppression of HT-29 colorectal adenocarcinoma xenografts.

8.3.5.3 Carbon Nano-onions

CNOs are essentially multilayer fullerenes, which were first observed by the Ugarte group in 1992 [85]. CNOs consist of spherical closed carbon shells, multilayered, quasi-spherical, and polyhedral shaped shells with structure resembling that of an onion (Figure 8.11). CNOs are totally insoluble in organic and inorganic solvents, similar to SWNTs. This lack of solubility hinders the study of their chemical and physical properties.

Currently, the most common method used for the production of spherical CNOs is the method based on annealing ND particles (average diameter of 5 nm) at high temperatures in an inert atmosphere under a high vacuum, frequently called the Kuznetsov method (named after the creator of this method) [84b].

The first covalent functionalization of CNOs was reported in 2003 by Georgakilas et al. using the 1,3-dipolar cycloaddition of azomethine ylides [87]. A few years later, an azomethine ylide addition reaction was used for nanodiamond-derived CNOs [88]. *N*-Boc-protected amino-CNO-pyrrolidine was synthesized using the procedure proposed by Prato and coworkers for the synthesis of fullerene derivatives (Figure 8.12a) [21b]. Functionalization at defect sites is also possible exploiting the carboxylic acid moieties present therein to link polymeric and oligomeric functional groups. For further successful reactions with amino-terminated compounds, the presence of these carboxyl groups is very crucial because a considerable difference between nanotubes and CNOs is, of course, the fact that the latters do not have easily accessible open ends, which facilitate oxidation. Under conditions similar to those reported for the direct functionalization and solubilization of single-walled CNTs by diamine-terminated PEG [89], CNOs could be

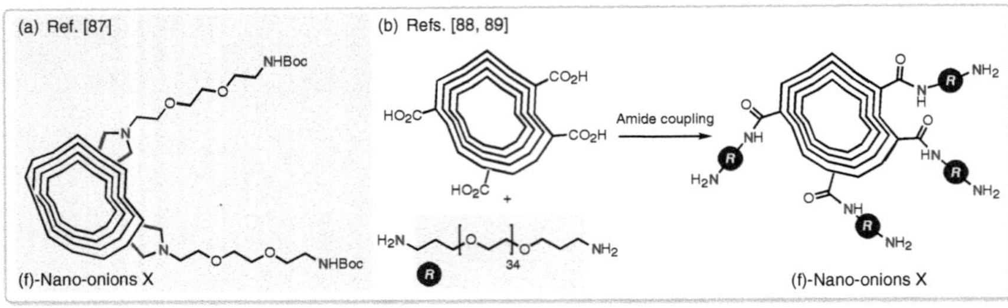

Figure 8.12 Amino functionalization of CNO by (a) cycloaddition reaction on nano-onions. Source: Based on Georgakilas et al. [87]. (b) Amide coupling on nano-onions. Source: Cioffi et al. [88] and Huang et al. [89].

made water soluble [90]. The CNOs were allowed to react at 140 °C for reaction times up to 19 days (Figure 8.12b).

8.3.6 Amino-Functionalized Carbon-Based Nanomaterials for Analytical Applications

The sensitivity of the electronic properties of CNT toward molecules adsorbed on their surface, and the high surface area providing for this high sensitivity, make CNT a promising starting material for the development of superminiaturized chemical and biological sensors [91]. The operation principle of these sensors is based on changes in the V–I curves of nanotubes as a result of adsorption of specific molecules on their surface. The use of CNT in sensor devices is one of the most promising applications in electronics and the functionalization with amino groups has often provided a good improvement in the performances.

The nitrogen dioxide (NO_2) gas sensitivity of monolayer CNT functionalized by the amino group $-NH_2$ was studied [92]. The amino group acts as a charge transfer agent of the semiconducting CNT that increases the number of electrons transferred from the nanotube to the NO_2 molecule.

A CNT-PAMAM hybrid obtained by covalently linking PAMAM onto CNTs has demonstrated to be useful for preparing multiple nanocomposites and for carrying biomacromolecules for different purposes. Zeng et al showed that they can be very effective as platform components of biosensors. The GO_x and horseradish peroxidase (HRP) immobilized CNTs-PAMAM show excellent performance for glucose sensing detection. The dispersion of the hybrid possesses excellent stability in water because of the large number of amino groups on the surface of the CNT-PAMAM. The amino groups on the CNT-PAMAM can be easily regulated by using different generations of PAMAM [93].

Graphene and its derivatives have generated a great deal of interest in the field of analytical sensors, mainly thanks to their unique physicochemical features [12, 94]. Indeed, both the large surface area and planar geometry of these carbon-based materials may generally enhance the loading of targeted molecules (gas and liquid), thus resulting in improved detection abilities [95]. As an example, the superficial functionalization of graphene with numerous amino groups has been recently used to produce a gas sensor for the detection of formaldehyde, which is the most common indoor pollutant gas and harmful to human health [96]. The superficial amines on this graphene-based sensor may therefore react with the gaseous formaldehyde to provide the corresponding imine adducts, thus allowing the detection of the target analyte.

8.4 Amines in the Synthesis and Functionalization of Carbon Dots

In recent years, carbon dots (CDs) have emerged as a new class of carbon-based nanomaterials [97]. CDs are quasi-spherical nanoparticles with dimensions below 10 nm. They are typically composed mostly of carbon and oxygen atoms, but doping with other types of atoms – in particular nitrogen – is commonly encountered, affording improved properties.

CDs have been isolated serendipitously from the purification of SWCNTs obtained from arc discharge [98]. Their synthesis is indeed possible with the so-called top-down methods, such as laser ablation, arc discharge, and electrochemical oxidation of larger carbon structures. On the other hand, it has become clear that similar materials can also be obtained from the thermal treatment of small molecular precursors [99]. These synthetic routes include pyrolysis and hydro- or solvothermal synthesis. In contrast with top-down methods, the bottom-up strategies offer more precise control over the final material. In our experience, microwave hydrothermal treatments provided quick and highly reproducible syntheses, with the possibility to include additional molecular doping agents directly during the synthesis. This allowed to obtain materials tailored for specific applications [100]. Indeed, some properties such as luminescence, and the functional group present on CDs, can be controlled at least to some extent with a proper choice of the molecular precursors. Common precursors are citric acid, amino acids, urea, simple aliphatic amines, and small aromatic molecules, such as aniline and phenol derivatives.

For what concerns the synthetic conditions, the use of high temperature is a consistent trend, with the employed temperatures in most cases being above 140 °C. High temperature promotes the formation of a carbonaceous core, following cross-linking and aromatization. Under hydrothermal conditions, 300 °C may be considered as a boundary temperature: lower temperatures are associated with the formation of amorphous, sp^3 carbon-rich CDs, while higher temperatures promote the formation of more graphitic materials. The time also affects the properties of the final material. In this case, longer reaction times lead to more graphitic CDs, while shorter times are associated with amorphous materials. Common synthesis times range from a few hours up to a few days. In this regard, microwave-assisted reactions have demonstrated their potential, allowing to shorten considerably the required time, down to a few minutes, providing a quick access route to this class of nanomaterials.

At present, the formation mechanism of CDs remains largely unknown and is currently under investigation [101]. As a consequence, the control on the synthesis is also limited, albeit some general trends have emerged. One consistent element is the presence of polar superficial groups, such as acids, alcohols, and amines, which provide solubility in water and polar solvents. This is typically inferred from IR data, combined with XPS analysis. The second element that emerges rather consistently from the literature is that some of the moieties present in the molecular precursors are retained during the synthesis [97]. This is a first element that allows tuning the properties of the final material depending on the starting mixture of precursors.

The possibility to tune the properties of the final materials directly during the synthetic process is one of the reasons why CDs have stimulated a growing interest among scientists. Indeed, in most nanomaterials, the functionalization has to be performed at later stages, as discussed also in the initial part of this chapter. For this aim, as well as for the engineering of specific functional properties, the amino group has proven extremely versatile.

8.4.1 Amines as CD Constituents

Soon in the development of CDs, the use of a molecular precursor bearing acidic functionalities, such as citric acid, has been combined with the use of amino groups. Under the commonly employed hydrothermal conditions, the formation of amide bonds can occur, leading to oligomeric structures. Despite the detailed mechanism of formation of CDs is unknown and is currently the subject of debate, it seems reasonable to envision a polymerization, followed by aromatization reactions that lead to graphitic structures. Experimentally, the degree of graphitization is inferred from XPS data, TGA, and TEM images [102]. The latter technique has allowed imaging nanoparticles with layered graphitic structures.

The concept of polymerization followed by aromatization is also consistent with extensive screening of synthetic conditions that indicated multifunctional small molecule units as precursors of election, bearing complementary units, such as acid and amine, or amine precursors, such as urea, that is known to undergo thermal decomposition to give ammonia and isocyanic acid under hydrothermal conditions. Indeed, two of the most studied combinations of CDs precursors are citric acid with either ethylenediamine or urea [103]. In these cases, the nitrogen-containing precursor is considered a nitrogen doping agent because it introduces the heteroatom in the CD structure. The combination of ethylenediamine with citric acid is a particularly significant case. Indeed, these two precursors have been frequently used jointly, and their reaction has been studied in considerable more detail than other systems. In this case, amide formation occurs readily, followed by a cyclization reaction, ultimately leading to the formation of 5-oxo-1,2,3,5-tetrahydroimidazo[1,2-α]pyridine-7-carboxylic acid (IPCA) [103]. This small molecule is highly fluorescent and has been identified by several authors as a product formed under the employed synthetic condition of CDs from citric acid and ethylenediamine [104]. In particular, IPCA was observed after purification of the reaction products by column chromatography, and analysis of the prominent fluorescent fraction by NMR and mass spectrometry [105]. IPCA has therefore been identified unambiguously as the principal fluorescent unit of that type of carbon dots. Whether this small fluorophore is present as a single entity, it is entrapped or is covalently bound to a carbon-based polymeric matrix remains unclear. Fluorescence correlation spectroscopy studies suggest that the fluorophore is present as an independent unit [106]. Similar conclusions have been reported for CDs obtained from citric acid and urea. Indeed, it has been observed that a thorough purification protocol is required in the preparation of these nanomaterials because of the "ubiquitous" presence of fluorescent impurities that originate during the synthesis [107].

The use of amino acids as precursors seems to partially overcome this issue. Several amino acids have been used to prepare CDs. These include histidine, aspartic acid, arginine, lysine, and cysteine, among others [108]. In the case of amino acid, an obvious role of CDs is to serve as polymerization units via amide bond formation. However, additional roles have

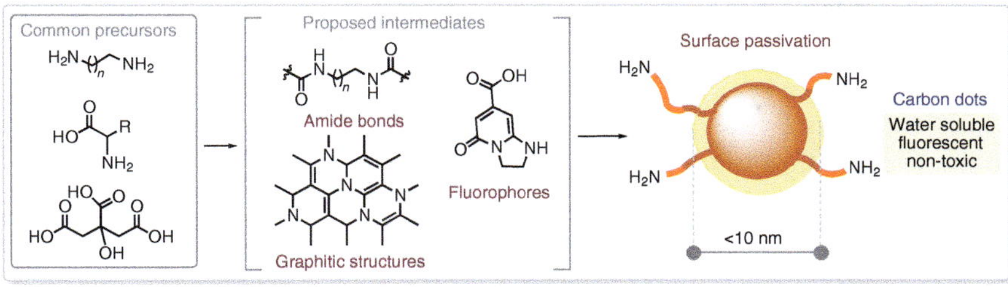

Figure 8.13 Role of amine precursors in the synthesis of CD. Depending on the synthetic conditions and the chosen precursors, intermediate structures can form, such as oligomers, molecular fluorophores, and graphitized N-doped structures. When the synthesis is performed at relatively low temperature (i.e. below 300 °C), the amines may have the role of passivating agents.

also been attributed to amine precursors, in particular in relation to the luminescence of CDs (Figure 8.13).

A consistent observation is that heteroatom doping, and in particular nitrogen doping, affords better luminescence properties, i.e. higher quantum yields. In amorphous CDs, this effect is related to the formation of fluorophores, whereas in graphitic systems, it has been associated with the emission of N-centered defects in the structure [109]. Even beyond the simple analysis of quantum yields, a clear example of the useful transformation of amino groups into graphitic structures has been reported by Reisner and coworkers, who studied graphitic carbon dots obtained from a prolonged thermal treatment above 300 °C, using either citric acid or aspartic acid as single-source precursors [108]. In both cases, CDs with a comparable size of c. 3 nm were obtained, with those obtained from aspartic acid displaying clear lattice fringes. XPS data revealed that nitrogen was included in the structure only as pyridinic or pyrrolic units. These N-doped CDs were used as photosensitizers for photocatalytic hydrogen evolution and displayed enhanced absorption in the visible region of the electromagnetic spectrum, as well as better photosensitizer abilities.

A complementary role is attributed to nitrogen (mostly amine)-doping in the synthesis carried out at lower temperatures. In these cases, amines are described as passivating agents that protect the photoactive core of CDs from the surrounding environment. Indeed, several studies have demonstrated the importance of passivation to enhance the CD properties, and amines emerged as one of the best performing passivating agents [110]. In a comparative study, citric-acid-based CDs were synthesized at temperatures close to 200 °C, adding ammonium hydroxide, ethylenediamine, or diethylenetriamine as doping agents [111]. The authors observed that the use of primary amine could boost the luminescence quantum yield, which increased from below 1% in the case of nonpassivated CDs, up to above 50% when the triamine was used. Interestingly, increasing the amount of triamine increased the quantum yield almost linearly. On the other hand, amine passivation decreased the photostability of the material. In the same study, electrophoretic experiments revealed negatively charged CDs and CDs with both positive and negative charge, indicating that what is described as surface passivation, or doping, most likely reflects a broad range of chemical reactions, leading to a multitude of different structures. Indeed, for simple surface passivation, one would expect a net positive charge, uniformly present on CDs. To this aim, CDs obtained solely from the thermal treatment of polyamines, which could be successfully used in biological application such as gene delivery, owing to the charge complementarity with nucleic acids, have been prepared [112].

8.4.2 Amine-Rich CDs from Arginine and Ethylenediamine (NCDs)

Considering the role of amino groups in amide bond formation, graphitization, and passivation, our group focused on the engineering of nitrogen-doped CDs obtained from the microwave-assisted hydrothermal synthesis of arginine and ethylenediamine (nitrogen-doped carbon dots, NCDs) [100, 113]. A short treatment (three minutes) at 240 °C of the precursors, mixed with a minimal amount of water, turns the light yellow mixture into a brown solution, which is purified by filtration and dialysis. The resulting NCDs were imaged with atomic force microscopy (AFM), revealing small particles with a size of 2.5 ± 0.8 nm, having good solubility in water, and sufficient solubility in polar organic solvents such as methanol and DMF. These NCDs displayed a good quantum yield of 0.17, which could be increased up to 0.46 upon chromatographic separation of different families of NCDs. Additional structural information was obtained by NMR experiments performed with ^{13}C-labeled precursors. These experiments revealed that the aromatic domains of NCDs mostly originated from arginine, whereas the carbon atoms of ethylenediamine were associated with aliphatic portions of the material. XPS analysis revealed the presence of nitrogen in different forms. Importantly, this material provided a positive Kaiser test, leading to the quantification of 1350 μmol/g of primary aliphatic amines. The oxidation potential of NCDs is consistent with the presence of primary amines. A combination of NMR and electrochemical studies indicated that these amines are somehow similar to benzylamine or phenethylamine [114]. Despite the importance of this rationalization, it is stressed that it should be taken with great caution, considering the complexity of the system under study, and the extremely limited knowledge of its structural features. Having established the presence of amines even in the final material, it became possible to exploit the amino group for different purposes, during the synthesis, or in postfunctionalization [100].

8.4.2.1 One-Pot Functionalization of NCDs

A particularly attractive strategy to tailor the properties of NCDs is to exploit the reactivity of the amino group to incorporate the desired molecular units directly during the synthesis (Figure 8.14).

A significant example of the exploitation of amino groups of NCDs during the synthesis has been their reaction with naphthalene dianhydride (either bare or substituted) [115]. Reaction of ethylenediamine to form naphthalene diimides was expected under the synthetic conditions employed, and the resulting imides could further participate in NCD formation, leading to the inclusion of the naphthalene-based chromophore in the final nanomaterial. The envisioned reactivity was indeed observed, leading to NCDs emissive in the blue or orange region of the spectrum, depending on the substitution pattern of the starting anhydride. Computational insights suggested that when a naphthalene dianhydride precursor bearing bromide substituents in the bay region was employed, amine could participate in nucleophilic aromatic substitution reactions. By adjusting the ratio between two different anhydride precursors, white-light-emitting NCDs were obtained. Overall, this proved to be

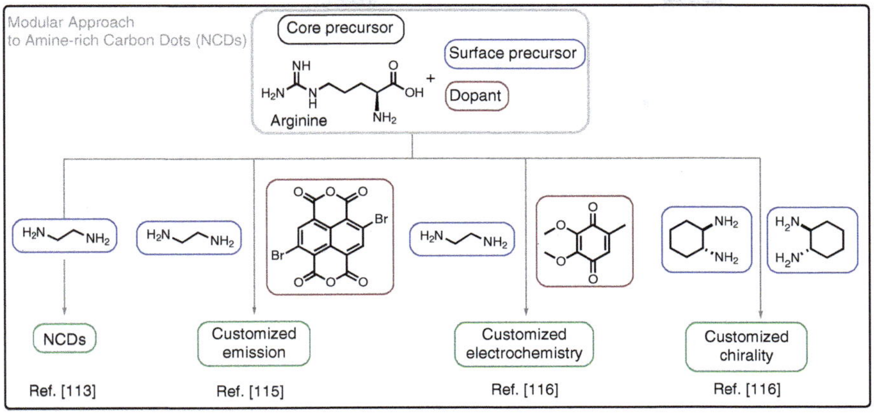

Figure 8.14 Examples of doping strategies to tailor the properties of NCDs during the synthesis, exploiting the amino group reactivity, with acids, anhydrides, electron-poor aromatic groups, and quinones.

an effective strategy to exploit the reactivity of the amino precursors tailoring the absorption and emission properties of the nanomaterial.

Using a conceptually analogous strategy, the redox properties of NCDs could also be modulated [116]. In this case, quinones were used as doping agents, which could react with amines to afford the corresponding imine derivatives. Introduction of quinones in the reaction mixture afforded NCDs with less negative reduction potentials, which could be modulated from −1.52 to −2.04 V vs. saturated calomel electrode (SCE). The oxidation potentials were also affected by the doping procedure. This is coherent with the participation of the amino groups in the formation of adducts with quinones during the synthesis, thus limiting their availability to undergo oxidation as such.

As a final example of modification during the synthesis, chirality was controlled using a chiral diamine, i.e. (*R*,*R*)- or (*S*,*S*)-cyclohexanediamine, instead of ethylenediamine [117]. Albeit conceptually simple, the inclusion of chiral precursors into CDs is not straightforward because of the harsh synthetic conditions that are typically required for the synthesis of CDs. Indeed, even if in our standard protocol (*S*)-arginine is employed, racemization occurs during the synthesis, despite its short time (three minutes). In fact, most of the chiral CDs reported rely on postsynthetic modifications. On the contrary, (*R*,*R*)- and (*S*,*S*)-cyclohexanediamine retained their chirality at least to an extent that was appreciable spectroscopically, by both electronical circular dichroism (ECD) and vibrational circular dichroism (VCD). In addition, the obtained chiral nanomaterials could be effectively employed to transfer their chirality to a supramolecular polymer.

Overall, from these examples, it should emerge the versatility of the amino group in the synthesis of NCDs as a mean to tailor the functionality of the final material via small-molecule doping.

8.4.2.2 Postfunctionalization of NCDs

Besides in-synthesis doping, postfunctionalization strategies are also extremely useful to embed specific functionalities to the nanomaterial (Figure 8.15) [108, 118]. This route is

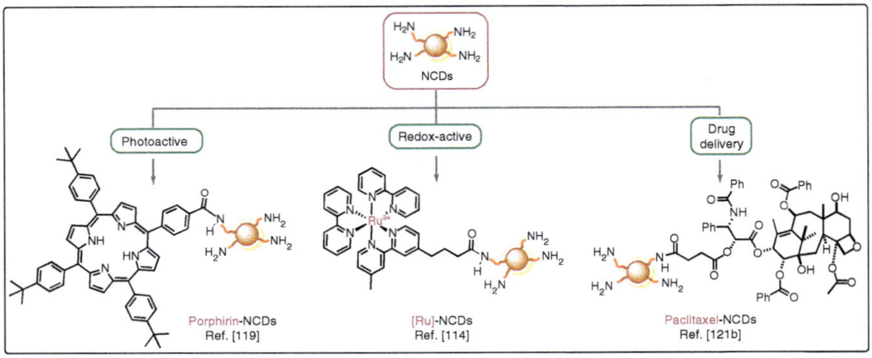

Figure 8.15 Examples of nanoconjugates obtained from the coupling of functional units to NCDs via amide bond formation.

particularly attractive to introduce residues that may degrade during the synthesis or interfere with it, as well as units that benefit from the interaction with the CD platform. Moreover, this strategy offers a much better control on the molecular structures present in the final material because the fragment introduced at the postfunctionalization stage retains its defined molecular structure. One of the first development envisioned, in these regards, has been the use of NCDs in a covalently bound photoactive nanoconjugate for charge separation purposes [119]. Porphyrin was selected as a coupling partner because of its proven ability to be engaged in electron transfer processes [120].

Covalent functionalization was achieved in a straightforward manner by coupling a porphyrin derivative bearing a pendant carboxylic acid via amide bond formation promoted under classical coupling conditions.

In the resulting nanoconjugate, detailed spectroscopic analysis revealed that upon light excitation, a charge-separated state formed. In this state, the positive charge is localized on the porphyrin, while the negative charge is localized on the NCD. This observation may seem counterintuitive for an amine-rich nanoparticle because the pendant groups are prone to oxidation. This partially unexpected behavior further highlights the complexity of NCDs as nanomaterials.

The capability of NCDs to undergo oxidation was exploited in a different nanohybrid, in which a derivative of ruthenium(II)tris(2,2′-bipyridyl) ($Ru(bpy)_3^{2+}$) bearing a pendant carboxylic functionality was covalently attached to NCDs [114]. In this case, the nanohybrid was exploited in the efficient generation of electrochemiluminescence (ECL). Indeed, the metal complex is a well-known luminophore in this context, which usually operates in the presence of an oxidizable coreactant. Alkyl amines are the most representative class of coreactants but have some drawbacks, e.g. the commonly employed tripropylamine is toxic, corrosive, and volatile, while it has to be used in high concentration. On the contrary, use of the nanohybrid as an ECL platform allowed the system to operate via an intramolecular electron transfer, leading to an enhanced ECL signal compared to the separated components, without the need to employ toxic amines in high concentration.

Besides the engineering of photophysical and electrochemical processes, the postfunctionalization of superficial amines can also be employed for the conjugation of drugs, leading to the formation of a prodrug [121]. The use of NCDs as a drug delivery system is particularly attractive, given their water solubility, low toxicity, and fluorescence, which in principle may allow to monitor the localization of the conjugate. In particular, studies in our group focused on the delivery of paclitaxel because it is a drug with poor water solubility, which could thus benefit from the conjugation to a highly water-soluble platform [121b]. Upon derivatization of paclitaxel with succinic anhydride at the hydroxyl C2′ position, the residual pendant acidic moiety was again used to conjugate the nanohybrid via amide bond formation. Successful conjugation was confirmed via a number of techniques, including diffusion-ordered NMR spectroscopy, matrix-assisted laser desorption ionization, and AFM. The hybrid nanomaterial-induced apoptosis in cancer cells, displaying a slightly better anticancer activity compared to the drug alone, evaluated in breast (MCF-7 and MDA-MB-231), lung (A-549), prostate (PC-3), and cervix (HeLa and C33-A) cancer cell lines. Because the drug is known to be inactive when the 2′ hydroxyl moiety is derivatized, the hybrid is likely active only upon ester hydrolysis.

8.4.2.3 Use of CD-Supported Amines in Organocatalysis

Amino groups are prominent in organocatalysis [122], as also reported elsewhere in this book. Their support on CDs as aminocatalytic platforms would be an advantage, in particular aiming at their uses as organocatalysts in biological media. One tantalizing opportunity enabled by this process is in situ drug synthesis [123]. The exploitation of amino groups in the context of CD-promoted organocatalysis is still in its infancy, albeit some applications related to catalysis started to emerge [124]. In our group, amine-rich NCDs have been employed for the photocatalytic perfluoroalkylation of electron-rich substrates, including phenols, olefins, and caffeine [125]. In this case, the important role played by the superficial amino groups is related with its ability to engage in halogen bond interactions with iodoperfluoroalkyl substrates, as supported by ^{19}F-NMR experiments. The proximity of the substrate facilitates electron transfer from the excited state of NCDs to the iodinated substrate, generating a highly reactive perfluoroalkyl radical. Overall, this strategy allowed the formation of C—C bonds under mild and operationally simple conditions.

In relation to catalytic applications, a different type of amine-bearing carbon dots was used in combination with CO_2, which was employed to recycle the system [126]. In this case, the CDs used were obtained from citric acid and alkyl diamines including primary, secondary, and tertiary amines bearing short alkyl chain with up to eight carbon atoms. Changes in the amine precursor did not affect the final material in terms of morphology, and all samples displayed a high number of amines, up to c. 4 mmol/g, as evaluated from a combination of XPS and TGA experiments. The carbon dots obtained using 1,4-diaminobutane as an amine precursor were used to promote the Knoevenagel condensation affording 2-(4-bromobenzylidene)malononitrile in 1-octanol. After product filtration, the CDs were recovered by CO_2 bubbling in a biphasic system. This operation led to the transfer of CDs to the aqueous phase, likely as a consequence of carbonate ion formation at the surface of CDs, as revealed by ^{13}C-NMR experiments. This catalyst recycling procedure allowed to use the same batch of CDs to perform the reaction up to six times.

Clearly, additional work remains to be done regarding the exploitation of amino groups as catalytic units supported onto CDs.

8.5 Amines for the Engineering of Hybrid Organic–Inorganic Nanomaterials

The role of amines in the engineering of hybrid organic–inorganic nanomaterials is manifold; these valuable compounds find applications in several stages of nanomaterial development, from the synthesis and functionalization to their practical implementation. Within the context of developing synthetic strategies for organic–inorganic hybrids, amines have been applied to a very broad range of nanosized materials encompassing metal chalcogenide and perovskite quantum dots, noble metal nanoparticles, and metal oxide nanoparticles, including transition or rare-earth metals. Another class of organic–inorganic hybrids for which amines can provide tailored properties is that of SAMs on flat surfaces. Nanomaterial surface functionalization is a relevant feature that controls their interaction with other nanoscale or mesoscale objects and therefore underpins their applications. In this respect, amines and amino groups have been exploited in various manners. Indeed, positively charged nanoparticles or SAMs displaying amino groups on their surface can effectively interact with other polyelectrolytes, either natural or synthetic, by charge complementarity or hydrogen bond formation. Alternatively, surface-accessible amines can be used for developing complex, yet well-defined, higher order systems by, for instance, complexation of metal ions or covalent bond formation.

The reliability, safety, and scalability of the nanomaterial preparation attained by using amines are of outmost importance and have nowadays set a benchmark. Besides, the excellent control over nanomaterial properties allowed by these approaches is witnessed by the large body of literature in this field and the impact on everyday life considering the increasing role of nanotechnology products. This makes an exhaustive description of the amino group chemistry in the context of hybrid organic–inorganic nanomaterial development out of reach. This notwithstanding, some relevant aspects and approaches are quite general and will be presented in Section 8.5.1, with specific reference at SAMs on flat surfaces, semiconductor quantum dots, and metal and metal oxide nanoparticles.

8.5.1 Amines as Head Groups or End Groups on Self-assembled Monolayers on Flat Surfaces

SAMs on flat surfaces are assemblies of small molecules anchored by either chemisorption or physisorption on a substrate. A variety of species can form SAMs, but the most studied are alkanethiols that represent the preferred species for SAMs formation on metal surfaces [127]. SAMs formed by using alkylamines on flat metal surfaces are usually less stable than those formed by thiols.

The amine head groups interact weakly with the metal surfaces (Figure 8.16a) [128], and this is consistent with the mismatch between the relatively hard nature of amines and the soft acid nature of metals in the zero oxidation state. Density functional theory (DFT) calculations pointed out that the interaction energy of amines with gold surfaces is about seven times smaller than the gold–thiol interaction [129]. Therefore, if the latter is about 40 kcal/mol [130], the nitrogen–gold interaction is about 6 kcal/mol, comparable to those of

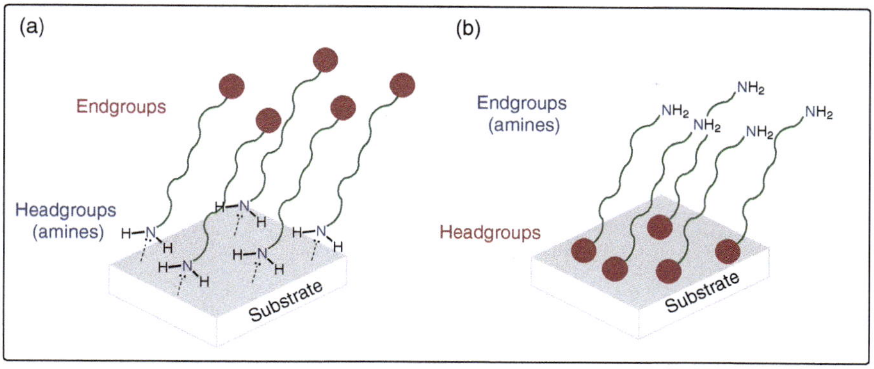

Figure 8.16 Schematic representations of 2D-SAM, (a) amine head groups anchored to the substrate surface and (b) amine end groups exposed to the outer surface of the SAM.

hydrogen bonds. A notable exception is monolayer formation on iron surfaces. In this case, a series of anilines [131] and alkylamines [132] were reported to effectively inhibit surface corrosion by monolayer formation. Most likely, such a strong binding is a consequence of the amino group interacting with iron in the pseudo-fourfold hollow sites of the Fe(110) surface. The adsorption of amines remains weak even on surfaces exposing hydroxyl groups such as mica, quartz, or glass [133]. On such materials, the formation of a monolayer is mainly driven by hydrogen bonding to the substrate and strongly depends on the nature of the solvent used. The obtained SAM is often incomplete and mechanically unstable, as probed by AFM [134]. More frequently, amines are used as capping agents for noble metal nanoparticles and their use in this domain will be analyzed with more detail in Section 8.5.2.

Amino groups are very often introduced on the outer surface of the 2D SAMs (Figure 8.16b), where they serve as reactive moieties for further functionalization or for the formation of adhesive surfaces, e.g. for cells, proteins, and biomolecules [135]. In this context, the amine-terminated SAMs proved to be more efficient than hydroxyl-terminated SAMs and are essentially as active as carboxyl group-terminated SAMs, albeit a large degree of nonspecific adsorption was also pinpointed for the amine-terminated monolayers [135c, 136]. One of the most exploited reactions has been the functionalization of SAMs formed by aminopropylsilane, on Si/SiO_2 or glass. Although 2D SAMs are mainly formed by reaction of surface available hydroxy groups with the aminopropylsilane, SAMs can also be formed by CVD [137]. The reason for the popularity of such a simple SAM has been the possibility to interface it with DNA, either covalently or noncovalently, retaining the hybridization ability of the employed strands [138]. The reliability of multiple noncovalent interactions has allowed their use in the formation of elaborated architectures. As an example, multilayer films containing DNA and polyamines were easily prepared, by simply dipping the substrate into solutions of polyamine and DNA, sequentially [139]. In the resulting multilayer architecture, the change complementarity between protonated amines and deprotonated phosphate groups, together with hydrogen bonding interactions, provided the observed stability. Interestingly, these interactions occurred at the expense of

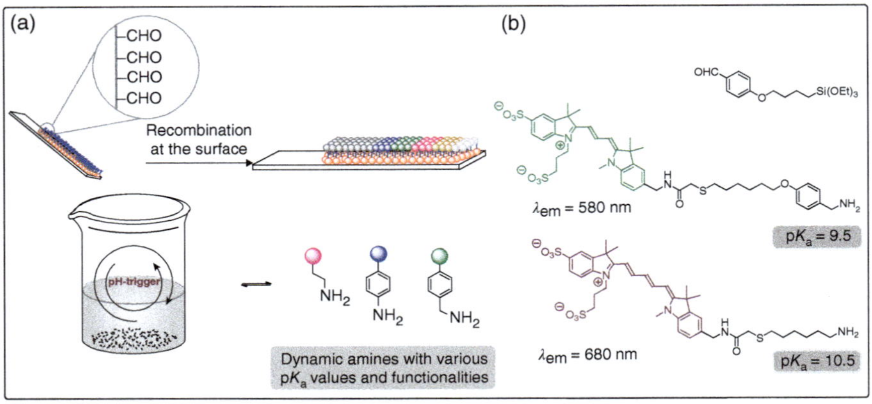

Figure 8.17 Schematic representation of the strategy described in Ref. [147] to prepare surface gradients via imine bond formation. Source: Based on Tauk et al. [147].

hydration of the two charged groups, leading to a weakly hydrated multilayer structure. To complete the overview on noncovalent exploitation of amino groups, the use of polylysine should also be mentioned because it is a prominent polycationic amine-rich polymer, frequently used for surface functionalization [140].

Besides noncovalent strategies, the covalent reactivity of amine-terminated SAMs has been extensively studied and proved to be extremely versatile in the context of SAM functionalization [141]. Covalent functionalization included (i) amide bond formation upon reaction with an activated ester or an acid chloride [142]; (ii) nucleophilic acyl substitutions, e.g. upon reaction with di-(*N*,*N*-succinimidyl) carbonate, which allowed further reaction with the hydroxy group of β-cyclodextrin forming carbamate bonds [143]; or nucleophilic acyl substitutions affording (9-fluorenyl)methoxycarbonyl-protected amines, thus changing the hydrophobicity of the monolayer [144]. The reactivity of amino groups was also exploited in surface-patterning applications, e.g. via microcontact printing or photodeprotection reactions involving amines as reagent or – more frequently – products [145].

One strategy that has proven particularly versatile for the functionalization of surfaces is imine bond formation, which can potentially merge the advantages of covalent chemistry with the benefits originating from its dynamic nature [146]. An elegant example of the versatility of imine formation reactions at surfaces has been presented by Giuseppone and coworkers [147]. In this work, a substrate was covered via silanization with an appropriate aldehyde-terminating unit, and the obtained monolayer was used to prepare surface gradients via imine bond formation. To achieve this target, the monolayer with terminal aldehydes was immersed in a solution containing both aliphatic and benzylic amines, having different pK_a values, thus displaying a different reactivity toward imine formation at a given pH (Figure 8.17). While imine formation was taking place at the surface, the substrate was slowly removed from the solution. Simultaneously, the pH was modified with the addition of triethylamine or trichloroacetic acid.

Under these conditions, as a consequence of amine protonation, the preferential partner for imine formation changed over time inducing the formation of surface gradients. The solidity of this approach was also demonstrated by implementing this strategy in an organic solvent or for the functionalization with biomolecules.

8.5.2 Alkylamines in the Preparation of Semiconductor Quantum Dots

II–VI semiconductor QDs such as CdSe, CdS, ZnS, ZnSe, etc., are very often prepared under conditions resembling those described in the landmark work by Bawendi and coworkers [148]. These required the use of tri-*n*-octylphosphine (TOP) and tri-*n*-octylphosphine oxide (TOPO) acting both as a solvent system and capping agents. However, TOP and TOPO are far from ideal, suffering the disadvantage of high toxicity. These issues prompted the development of synthetic conditions replacing the need of phosphine species in part or altogether. Long-chain alkylamines can be used in the temperature range, above 250 or 300 °C, required for the synthesis of quantum dots and are safer and more environmentally benign than TOP and TOPO. Indeed, several alkylamines, and especially oleylamine (OLE), are commonly used in the syntheses of these nanomaterials, often in association with fatty acids, high boiling point alkenes, 1-octadecene (ODE) being the most common, and TOP/TOPO mixtures [149].

Although suitable as solvents for high-temperature reactions, long-chain alkylamines are solid at room temperature (Dod has a melting point of 27–28 °C and a boiling point of 250 °C; tetradecylamine has a melting point of 40 °C and a boiling point of 290 °C), which hinders the work-up procedures. Accordingly, among fatty amines, the choice of election for the preparation of quantum dots and other nanomaterials is OLE because at room temperature, it is in the liquid state (melting point 18–26 °C), has a boiling point of 350 °C, and is commercially available, although of technical grade. Besides serving as suitable solvents, alkylamines also act as reagents, mainly in the activation of the chalcogenide precursors and capping agents for the nanocrystals. Activation of metal carboxylate precursors by amines has also been reported in the preparation of ZnSe and ZnS QDs [150].

8.5.2.1 Sulfur–Amine and Selenium–Amine Systems

Although the complex nature of the solutions obtained from elemental sulfur in certain amines is known since 1930 [151], and despite the use of these compounds in the preparation of metal chalcogenide semiconductor NPs is well established, it is only recently that a clear understanding of these systems is emerging. Elemental sulfur dissolves in certain amines to give colored solutions, which were thought to contain *N,N'*-polythiobisamines, formed by scission of S—S bonds in the S_8 molecules with evolution of hydrogen sulfide. Electron spin resonance (ESR) studies of these solutions pointed out the presence of sulfur-centered free radicals formed by homolytic scission of the S—S bonds of the *N,N'*-polythiobisamines [152]; Raman spectroscopy analyses were in line with this suggestion [153]. More recently 1D-, 2D-, and DOSY NMR analyses of the S_8-octylamine model system pointed out that, at low temperature, S_8-alkylamine solutions contain mainly alkylammonium polysulfides [154], and a small amount of *N,N'*-polythiobisamine, which is however sufficient to produce enough hydrogen sulfide necessary to the formation of alkylammonium polysulfides. Under the high-temperature conditions typically used for the nanocrystal preparation, the polysulfide ion reacting with excess amine produces hydrogen sulfide and other by-products, which likely comprise

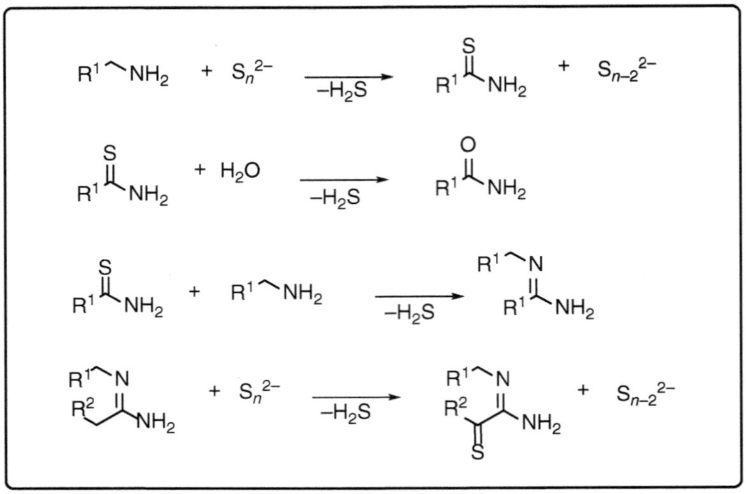

Figure 8.18 Reaction pathways in sulfur–octylamine solution at 130 °C. Source: From Thomson et al. [154].

amides and amidines that arise from a common intermediate identified in octylthioamide. Possibly, α-thioketoamidines are also formed by reaction of amidines with polysulfides. Overall, the authors propose the mechanistic scheme depicted in Figure 8.18 and suggest alkylammonium hydrogen sulfide as the sulfur precursor in the preparation of metal sulfide nanocrystals.

As a compelling demonstration of the importance of thioamide intermediates in the synthesis of quantum dots, the authors report the synthesis of PbS nanocrystals at 25 °C using commercially available thioacetamide dissolved in OLE as a sulfur precursor [154].

In contrast with elemental sulfur, selenium dissolves poorly in neat alkylamines. However, it dissolves at high concentration in the simultaneous presence of OLE and of reducing agents such as sodium borohydride [155], or alkanethiols [156]. In the latter case, the mechanism of dissolution was found to be dependent on the reduction potential of the thiol, and the rate of dissolution on the nature of the amine that is faster in the order primary > secondary > tertiary. Using low molecular weight alkanethiols such as ethanethiol, which could be removed as well as the corresponding disulfide under vacuum, leads to highly concentrated selenium solutions in OLE. In the reaction system made of OLE/ODE/sodium borohydride, elemental selenium can be dissolved in 10 hours at room temperature. Conversely, by replacing the unsaturated OLE with hexylamine, no reaction occurs after 10 hours at 50 °C. Octadecene itself is known to dissolve selenium, but reaction takes place at temperatures exceeding 200 °C [157]. This indicates that the simultaneous presence of an unsaturation and amino group are essential for reaction with selenium at low temperature [158], while at higher temperature, amines are not necessary because in both ODE and saturated alkanes, selenium powder generates the hydrogen selenide or other reactive selenium hydride species involved in the QD nucleation [154, 159].

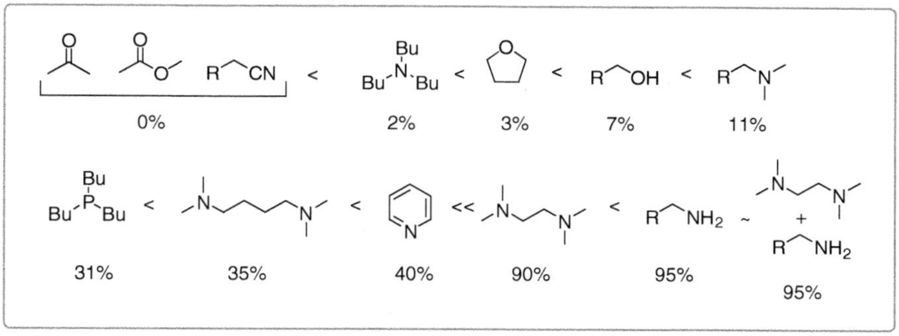

Figure 8.19 Relative displacement potency labeled with the percentage of Lewis base-Cd(carboxylate)$_2$ displaced in a 2.0 M solution of the of Lewis base-type ligand. Source: From Anderson et al. [163].

8.5.2.2 Capping Ligands for Quantum Dots and Ligand Exchange by Amines

Primary alkylamines are able to displace other weak ligands from the surface of quantum dots, in particular, they can readily displace TOPO [160]. However, amines used as a cosurfactant during CdSe clusters or NP synthesis cannot compete with fatty acids for QD passivation if fatty acids are present as in the surfactant mixture [161]. Conversely, if fatty acids are not used in excess, i.e. not used as cosurfactants, the coating by amines dominates even with TOP and TOPO in solution [162]. On the other hand, Owen demonstrated that interaction of tetramethylethylendiamine (TMEDA) or mixtures of primary amines and TMEDA with carboxylate-protected CdSe QD results in the displacement of cadmium carboxylate from cadmium selenide nanocrystals, and not in the displacement of the carboxylate as such [163]. Different Lewis bases exert the same effect albeit to a very different extent (Figure 8.19).

By using TMEDA, the displacement of cadmium carboxylate complexes is nearly quantitative in few minutes and results in a dramatic drop in photoluminescence quantum yield (PLQY). However, samples exposed to solely primary amines display significantly increased PLQY, despite losing a large proportion of surface-bound cadmium carboxylates. Recent works by Peng and coworkers point out that the coordination of fatty acids on CdSe nanocrystals is strongly facet dependent [164], with the polar {100} and {111} facets being involved in chelating and bridging binding modes with the carboxylate ligands. On the other hand, the nonpolar {110} facets are substantially reconstructed with yet very weak coordination bonds to the surface lattice. Considering CdSe nanoplates, Peng and coworkers displayed that exposure to amines maintains the carboxylate ligands on the polar {100} basal planes, while the mostly nonpolar side facets are weakly passivated with amines [165].

Amine ligands are important stabilizers for QD, and among the ligands usually encountered in the passivation of CdSe quantum dots, amines give more dynamic passivation than carboxylates and less than thiols [166]. Therefore, the ease of displacement depends on the relative affinity of the ligands originally present on the nanocrystals surface. Desorption rate constants reported in the literature are variable, ranging from 0.01 to 1.5×10^{-4} s^{-1}.

These values were generally obtained by analysis of the photoluminescence of the QD upon exposure to amines, but an NMR analysis by Hens and coworkers using octylamine- and pyridine-capped CdSe QD gave a significantly lower limit, which sets the desorption rate constant to values higher than $50\,s^{-1}$ [167]. The very same system studied by Peng and coworkers monitoring the photoluminescence increase upon exposure to octylamine gave a desorption rate constant of $0.01\,s^{-1}$ [166].

Because the CdSe pyridine was shown to offer a labile and reversible bonding with the surface cadmium ions, refluxing TOPO-capped CdSe QD in pyridine allows complete removal of the original ligand without altering the structural properties of the QD [168]. The large difference in Lewis basicity between pyridine and primary amines or thiols leads to consider pyridine-treated CdSe nanocrystals essentially "naked" once used in the presence of stronger ligands, such as the above-mentioned amines and thiols.

Displacement of weak ligands such as TOPO from the surface of CdSe QD by alkylamines affords materials with higher PLQY. Mulvaney and coworker displayed that PLQY enhancement by amines decreases in the order primary ≫ secondary > tertiary [160]. This enhancement is correlated with the passivation of surface trap states that enable non radiative deactivation of the excited QD reducing PLQY. Fatty amines are electron-donating ligands and should thus passivate Cd sites on the surface [169]. However, the exact nature of these surface trap states is elusive, and only recently, experimental and theoretical work is shedding light on the proper ligands needed to increase the PLQY of semiconductor of different compositions [170].

Polymeric amines also interact with QD. Indeed, polyallylamine has been used to transfer TOPO-capped CdSe QD in water [171], and CdSe/ZnS QDs protected by PAMAM dendrimers, soluble in water, have been proposed as QD for cancer cell targeting [172]. Polyethyleneimine (PEI) has also been used similarly, representing the "handle" for QD functionalization in a copolymer ligand consisting of PEG-grafted PEI reported by Nie and coworker [173].

Because of their high PLQY, semiconductor QDs are interesting alternatives to organic fluorophores for a range of applications in the chemical, biochemical, and biomedical field. All these applications rely on the anchoring to the NP surface suitable moieties to endow the QD with solubility in aqueous media. Amine-functionalized ligands also represent convenient species for the introduction of functional groups on the outer surface of the QD. However, given the dynamic nature of amine passivation, it has to rely on multivalent interaction with the nanocrystals. In this perspective, Parak and coworker demonstrated that the interaction of pyridine-modified amphiphilic polymers can be used to functionalize TOPO-capped QD by displacing part of the native ligands, while establishing hydrophobic interactions between the hydrophobic moieties in the polymer and the remaining TOPO on the QD surface [174].

8.5.3 Alkylamines as Reagents for the Synthesis and Passivation of Metal Nanoparticles

8.5.3.1 Alkylamines as Capping Agents for Metal Nanoparticles

Amines are commonly used in the preparation of noble metal and metal oxide NPs. In this context, amines may act as a cosolvent, reducing agent, and surfactant or may react with the metal precursor forming the actual species involved in the nanostructure preparation. The amino group anchors on the metal surface while the alkyl chain provides stabilization by Van der Waals interactions and steric hindrance, preventing irreversible particle aggregation. Different kinds of amines have been used in the preparation of metal NPs, acting according to one or more of the above-mentioned roles. These range from simple long-chain alkylamines and dialkylamines to 4-(dimethylamino)pyridine (DMAP) and amino-functionalized dendrimers [175]. In addition, the availability of chiral enantiopure amines, either synthetic or from the chiral pool, enables the development of NPs, which may be used for enantioselective catalysis [176].

Amines have been reported to promote the formation of gold and platinum NPs from Au(III) or Pt(II) precursors, without the need of any additional reducing agent [177]. In other cases, the reactions are carried out in the presence of stronger reducing agents such as borane–amine complexes or sodium borohydride [175d, 178]. Usually, the particles obtained by using amines as the capping agent can be recovered by precipitation from the reaction mixture and centrifugation. In addition, amines can be completely removed, if necessary, leaving "naked" particles by treatment with acetic acid and/or heating [179].

Among amines, OLE is, very often, the choice of election [180]. Nearly monodisperse amine-capped Ag and Au NPs in the 6–21-nm (Au) and 8–32-nm (Ag) size ranges and polydispersites below 7% could be prepared in large scale by refluxing hydrogen tetrachloroaurate or silver acetate and OLE in toluene for 120 minutes. The metal NP sizes can be tuned by varying the concentrations of the metal precursor and of the amine. The surfactant concentration has only a small influence on NPs size, but it influences polydispersity, which can be reduced below 10% when, for example, at least 65 equiv of the amine with respect to the gold precursor are used [181]. Xia and coworkers reported the preparation of silver multiply twinned particles by reaction of silver trifluoroacetate in *o*-dichlorobenzene in the presence of OLE at 180 °C [182]. Upon reaction of these silver NPs with hydrogen tetrachloroaurate, and excess OLE, galvanic replacement of silver takes place with the formation of hollow particles. The extra OLE was found to be critical for the formation of well-defined, uniform hollow structures free of AgCl contamination because of the formation of a soluble complex between AgCl and OLE during the replacement reaction.

Complexation of AuCl and AuBr with OLE makes possible to prepare Au NPs by thermolysis under mild conditions (60 °C in $CHCl_3$) without using an additional reducing agent [183]. In this procedure, particles with 12.7 nm diameter and size distribution of 8% could be obtained and the authors suggest a putative reaction mechanism involving the formation of dioleylamine as a by-product. Besides its role in Au(I) complexation, OLE acts as surface stabilizer and the authors point out that in the presence of octadecylamine instead of OLE, the particles displayed an average diameter of 100 nm, a clear evidence that even though the *Z*-configuration of OLE double bond does not play a role in the synthesis, the steric requirements of the OLE backbone strongly drive the formation of smaller NPs. The use of OLE as a surfactant and aminoborane complexes as reducing agents has been studied as a convenient method for the fine-tuning of NPs size from 2 to about 7 nm, using a mixture of OLE/octane 1/1 and by running the reaction at different temperatures [184].

Similar results were achieved using OLE and tetralin as a solvent (Figure 8.20) [178b]. The critical role of straight chain alkane solvents in the polydispersity of these materials in these syntheses was pointed out by Zheng and coworkers [185].

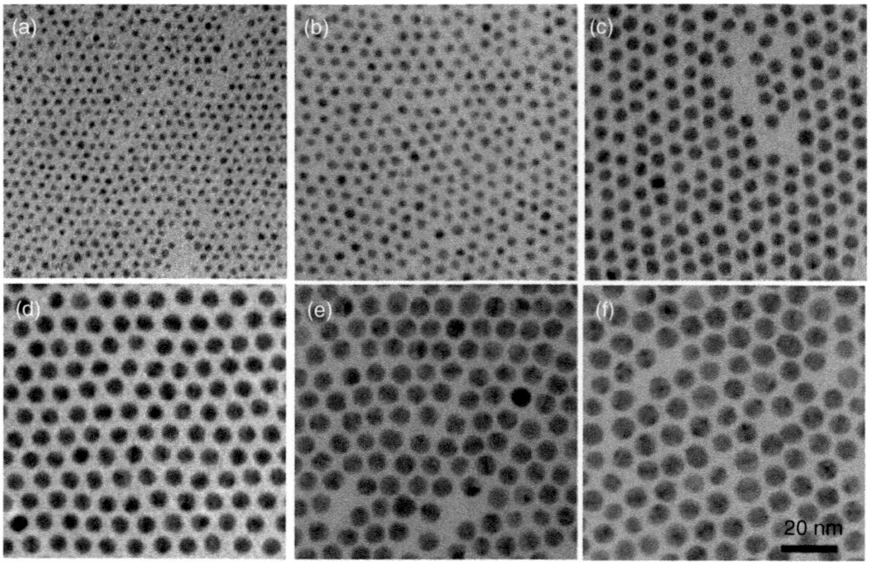

Figure 8.20 Representative TEM images for oleylamine-based synthesis of gold NPs run at temperatures. (a) 40 °C, (b) 35 °C, (c) 25 °C, (d) 15 °C, (e) 10 °C, and (f) 2 °C. Source: Reproduced from Peng et al. [178b]. Copyright 2008, Springer.

In association with fatty acids, amines have provided a way to produce anisotropic nanostructures. OLE has been used for the preparation of gold nanowires starting from AuCl as a precursor. In this case, the amine forms a complex with AuCl and these complexes interact among themselves via aurophilic interactions spontaneously forming extended 1D superstructures that upon reduction give the ultrathin nanowires [186]. Xing reported a method for preparing ultrathin single-crystalline gold nanowires in solution using $HAuCl_4$, OLE as a surfactant, and triisopropylsilane (TIPS) as a reducing agent, the reaction being complete within a few hours at room temperature [187]. Strongly anisotropic nanocrystals of transition metals and rare-earth oxide are obtained by using oleic acid and OLE in octadecene [188]. OLE, in association with oleic acid, has been used as a stabilizer for FePt NPs [179] and Pt nanocubes [189] where the surfactant mixture also served for shape control. Concave Pt nanocrystals with high-index {411} facets (Figure 8.21) were prepared solvothermally by reduction of hexachloroplatinic acid using poly(vinylpyrrolidone) (PVP) as the surfactant in the presence of methylamine [190]. Tuning of the concavity could be achieved by changing the amount of methylamine used in the synthesis.

After the pioneering work by Crooks and coworkers [191], dendrimers containing amino groups have been widely used as templates for the preparation of encapsulated metal NPs using sodium borohydride as a reducing agent in these syntheses [192]. Amino dendrimers are particularly effective because they are able to preconcentrate metal ions by complexation and a wide range of metal NPs has been prepared by this method, including Cu, Au, Pt, Pd, and Ru NPs. In these systems, the limited stabilization provided by the amine moiety is reinforced by the multiple contacts provided by the dendrimeric structure. Dendrimeric

Figure 8.21 (a) TEM images of the concave Pt nanocrystals. (b) SEM image of the as-prepared concave Pt nanocrystals. (c) High-magnification SEM image of a single concave Pt nanocrystal. The top-right inset shows an ideal geometrical model of the concave Pt nanocrystal with the same orientation as the nanocrystal in the SEM image. Source: Reproduced with permission from Huang et al. [190]. Copyright 2011, American Chemical Society.

amines grafted on solid supports or NPs can also serve for the syntheses of metal NPs, which results in an assembled composite material [193].

8.5.3.2 Displacement of Amines from the Surface of Metal Nanoparticles

Qualitatively, metals in the zero oxidation state behave as soft acids and will interact strongly with soft bases, it is therefore reasonable that a hard base such as an amine will interact with metal particles only weakly and in a reversible manner. Indeed, gold NPs protected by amines have been considered to be kinetically stable instead of thermodynamically stable [194]. Therefore, metal NPs prepared in the presence of amines are amenable of an easy ligand displacement reaction with ligands that bind more strongly to the metal surfaces. The weakly absorbed OLE on gold NPs can be readily displaced with aliphatic thiols by adding a toluene solution of the OLE-capped gold NPs to a boiling solution of 5–10 equiv (based on gold) of the thiol in the same solvent [181]. OLE-capped metal NPs have been used in exchange reactions with different thiols giving essentially clean substitution products. This procedure was found to be effective even using binary blends of thiols, which give rise to NP displaying in the monolayer the same composition of the feeding mixture [184]. On the other hand, exchange of OLE by octadecanethiol has been reported to be incomplete on batches of NPs containing excess amine surfactant even after prolonged time and repeated exchanges in the presence of excess thiol [195].

8.5.4 Amines on the Outer Surface of Organic–inorganic Hybrid Nanoparticles

Introduction of amino groups at the outer nanoparticle surface can be used to provide hybrid NP systems with specific ways to interact with other nanosized objects or a reactive handle for further functionalization. To this end, bifunctional ligands are required. One functional moiety binds the metal NP surface while a distal amino group is displayed on the outer surface of the nanoparticle. Typically, for transition metal oxide or silica NPs, these bifunctional ligands are aminosilanes such as (aminopropyl)-triethoxysilane (APTS) [196], while for noble metal NPs or semiconductor quantum dots, the preferred ligands are ω-aminoalkanethiols. In the case of iron oxide NPs, a cathecolate ligand such as dopamine has also been used, which was also reported to increase the colloidal stability of these systems [197]. However, the silane chemistry can also be used to introduce amino groups on the surface of gold NPs and QD by first functionalizing the metal surface with silane thiol, such as 3-mercaptopropyltrimethoxysilane (MPS), and afterward reacting the alkoxysilane groups with aminosilanes [198].

Alternatively, NPs can be coated with polycationic polymers, either natural or synthetic, such as chitosan [199], polyallylamine, or PEI [200]. In this case, the polymers provide both stabilization and reactive groups that can be used for further functionalization. For this approach to be effective, the surface of the metal particles has to be negatively charged or protected by labile ligands. In some cases, negative charge on the particle surface may come from the synthesis such as in the case of colloidal gold prepared by using the citrate reduction method [201]. Alternatively, a negatively charged surface can be obtained by exposing the particles to negatively charged polymers to form a first layer, which is covered afterward by an amine-functionalized polymer. Lennox and coworkers displayed the viability of this approach by reporting a series of Au/DMAP/anionic polymer/cationic polymer systems [202] and, with little variation, this strategy is still largely pursued. Quantification of reactive amino groups accessible on the surface of gold NPs is possible by labeling the amino groups with suitable reporter moieties, used for conventional amine assay, that can be quantified after disruption of the nanostructure [8]. Instead, quantification of the total number of amino groups on the nanostructure surface can be achieved by assessing the amount of amine containing ligands left behind in the solution after incubation with the Aunanostructures [8].

The large majority of applications of these nanosystems concerns biomedical applications. Indeed, the protonated amines on the outer NP surface can interact with negatively charged biomolecules, thanks to charge complementarity or hydrogen bonding, such as proteins [203] or nucleic acids [204]. In this latter context, amine-functionalized nanostructures and the related nanostructures decorated with ammonium ions can be efficiently used for gene transfection, or drug delivery [205], or to target the bacterial cell wall [206]. Amine derived from lysine or dendronized lysine have been introduced on the outer surface of NPs, providing a way to modulate the ability of the nanoparticle system to condense DNA and to deliver this cargo to cells [207]. The same approach has been applied to the delivery of small-interfering ribonucleic acid (siRNA) either by using metal NPs [208] or polymeric core–shell NPs [209].

Polyallylamine and its derivatives are often used as polymeric coating in passivating the nanoparticle surface but also to provide surface-accessible amino groups. Very often, successive layers of polymers with opposite charges are used. This is also a strategy to endow particles with improved gene delivery properties by reducing the synthetic efforts to a minimum. In this respect, gold NPs stabilized by chitosan, coated with a layer of charge-reversible poly(allylamine hydrochloride)-citraconic anhydride, and PEI could be loaded with siRNA protecting the payload against enzymatic degradation and displaying excellent transfection properties (Figure 8.22) [210]. PEI can also be used in the same way for siRNA delivery [211].

Besides gold NPs, which are interesting because of their low toxicity, other materials such as silica have been used to develop amine-functionalized scaffolds that can be loaded with nucleic acids for gene delivery [212]. Amine-functionalized hollow mesoporous silica NPs

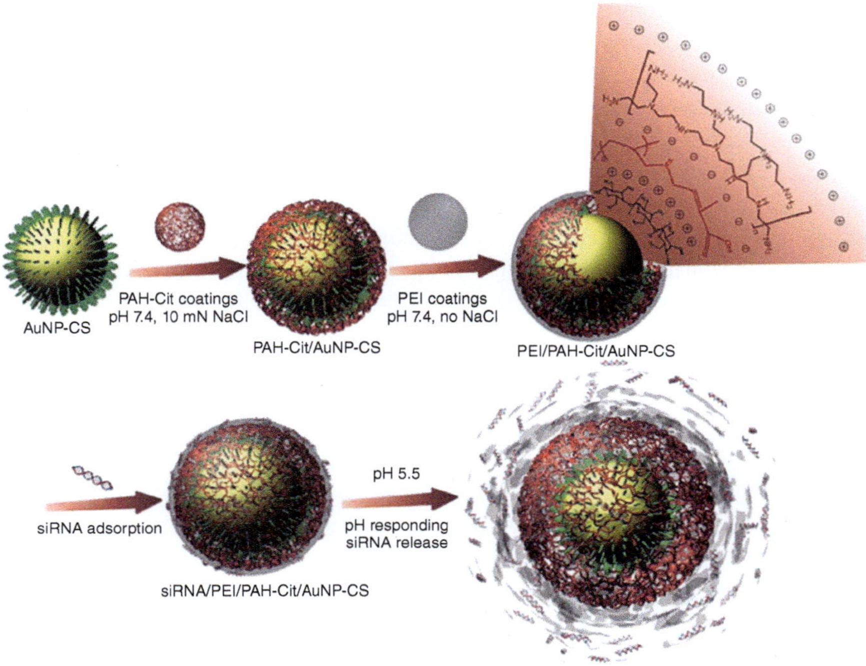

Figure 8.22 Assembly steps for siRNA/PEI/PAH-Cit/AuNP–CS complexes and pH-responsive release of siRNA. Source: Reproduced with permission from Han et al. [210]. Copyright 2011, American Chemical Society.

could be efficiently used as vectors for drug delivery with high efficiency, thanks to their large amine surface giving enhanced interactions with cells [213].

The advantage of these systems is the relatively easy introduction of amino groups on silica surface by using readily available aminosilanes. Nevertheless, more engineered particles such as polylysine-functionalized mesostructured silica were also proposed for gene delivery [214]. The extent of surface coverage in amino-functionalized silica has also an impact on the toxicity of this material [215]. Although these applications rely on electrostatic interactions between a positively charged nanoparticle surface and negatively charged biopolymers or biosystems, surface available amines can be exploited as ligands for the complexation of transition metal ions. Indeed, macrocyclic secondary amines such as triazacyclononane or chelating amines have been extensively used to anchor metal ions on the outer surface of gold NPs. Some of these constructs display remarkable catalytic activity in hydrolytic processes [216]. However, NPs decorated with macrocyclic, chelating, and ligands have found application in the complexation of metal ions such as Gd(III) that have a relevant role in diagnostic techniques based on magnetic resonance imaging [217].

The ability of amine-functionalized NPs to complex metal ions has also an impact on water detoxification. In this respect, by associating the magnetic properties of iron oxide NPs with surface-exposed amines, it has been possible to devise systems capable of capturing toxic metal cations from wastewater or contaminated solutions with the benefit of an easy recovery by applying magnetic fields [218]. The same concept has been proposed for the magnetic capture of bacterial pathogens [219].

Ligands on the surface of metal NPs can be designed to include short-peptide sequences displaying amino groups from lysine or histidine residues. These constructs were originally developed to mimic natural enzymes, and indeed, NPs functionalized with peptides containing histidine display catalytic activity in the hydrolysis of active esters with cooperative mechanisms and substrate recognition [220]. The imidazole moiety has also been exploited in the development of nanoparticles-based multivalent exo-receptors for porphyrin arrays [221]. Peptides have been studied as suitable surface capping agents for gold NPs, identifying the minimal requirement, in terms of sequence length and sequence composition to endow the particles with high stability [222]. Peptides have also been applied as coating molecules on silica-based nanomaterials, where lysine-rich sequences have been exploited to boost cell penetration and drug delivery properties [223]. This strategy is also useful when lysine-rich peptides have therapeutic properties themselves [224].

Amino groups at the outer surface of NPs can also be covalently coupled with other species by standard amide bond formation with reactive esters [225]. Nanoparticles for this kind of applications are commercially available from many suppliers. Amino groups can be reacted with activated linkers to further enlarge the variety of surface-reactive groups such as maleimide units suitable for click chemistry reactions [226].

Amino groups on the surface of NPs can also be used to covalently link NPs among themselves, or NPs with polymers carrying carboxylic groups spatially localized along the backbone, forming well-defined arrays [227]. Covalent conjugation with proteins is also possible [228]. Bawendi and coworkers displayed the viability of this strategy reporting an early compelling example of biocompatible QD for cellular imaging [229]. In this system, the protecting layer was engineered by using heterobifunctional ligands incorporating dihydrolipoic acid for anchoring to the nanocrystal, a short PEG spacer for solubility purposes, and an amine or carboxylate end group for further functionalization.

8.5.5 Postfunctionalization of Amine-Terminated Organic–Inorganic Hybrid Nanoparticles

Amine-functionalized surfaces have been exploited to set up approaches in which a solid support carrying amino groups is used for the immobilization of NPs. Immobilization relies on either charge complementarity between the particles and the support or the affinity of metal surfaces for amine ligands. Given the reversible nature of the binding between amines and metal surfaces, the immobilization can be temporary or permanent. In the case of temporary immobilization, this strategy can be used to mask part of the NP surface while the exposed surface can be conveniently functionalized. Overall, this process yields Janus particles. Shumaker-Parry demonstrated the viability of this strategy by immobilizing citrate-capped Au-NPs on amine-functionalized glass slides and replacing the exposed citrate ligands with α,ω-hydroxyundecanethiol. Removal of the particles from the glass slide by sonication and their treatment with amine- or acid-terminated thiols gave two sets of Janus Au-NPs with hydroxyl/amine and hydroxyl/acid functional hemispheres. Coupling of these particles with amide bond formation gave rise to nanoparticle dimers in high yield [230]. On similar lines, amine-functionalized Janus gold NPs were prepared by grafting gold NPs on silica particles coated with a ligand comprising an imine moiety. After functionalization of the exposed second half of the nanoparticle surface, hydrolysis of the imine released the Janus particles in solution [231].

Permanent immobilization on amine-functionalized solid support is also possible, and this approach has been used to anchor NPs for the preparation of composite materials and catalysts, which display improved performance with respect to model system. Indeed, amine-functionalized zeolites, obtained by using 3-aminopropyltrimethoxysilane, were found to bind Pd and Pt NPs giving materials useful in hydrogenation and Heck reactions [232], and Pd NPs anchored on amino group-terminated SiO_2 were proposed as highly effective catalysts in the model reduction of 4-nitrophenol [233]. Pd NPs supported on amino-functionalized silica was reported as a catalyst for the reduction of Cr(VI) [234]. Nitrogen-rich polymers have also been used to entrap Pd NPs, and the system has been tested in model processes for biofuel upgrading [235]. Citrate-capped gold NPs could also be anchored to poly(allylamine hydrochloride)-modified polymer poly(glycidyl) methacrylate submicrospheres to yield raspberry-like composites with catalytic activity in the reduction of 4-nitrophenol [236].

DMAP-functionalized gold NPs could be grafted, with ligand displacement, on the surface of amine-functionalized boron nitride nanotubes as nanoscale templates for assembly and integration with other nanoscale materials [237]. Nanoparticle assembly on carbon nanostructures gives rise to an important class of composites materials for which amine functionalization can provide strong interaction and coupling [238]. Indeed, amino silica layers formed on the surface of CNTs has been used for the assembly of gold NPs on the surface obtaining materials useful as electrocatalysts for oxygen reduction [239]. Accelerated catalytic activity of Pd NPs in the hydrogenation of quinoline was observed upon supporting these materials on amine-rich silica hollow nanospheres [240].

References

1 Sun, Y., Kunc, F., Balhara, V. et al. (2019). *Nanoscale Advances* 1: 1598–1607.

2 (a) Troll, W. and Cannan, R.K. (1953). *The Journal of Biological Chemistry* 200: 803–811. (b) Kaiser, E., Colescott, R.L., Bossinger, C.D., and Cook, P.I. (1970). *Analytical Biochemistry* 34: 595–598. (c) Sarin, V.K., Kent, S.B.H., Tam, J.P., and Merrifield, R.B. (1981). *Analytical Biochemistry* 117: 147–157.

3 Kunc, F., Balhara, V., Brinkmann, A. et al. (2018). *Analytical Chemistry* 90: 13322–13330.

4 Zhao, L., Baccile, N., Gross, S. et al. (2010). *Carbon* 48 (13): 3778–3787.

5 Baer, D.R., Engelhard, M.H., Johnson, G.E. et al. (2013). *Journal of Vacuum Science and Technology A* 31 (5): 050820.

6 Kralj, S., Drofenik, M., and Makovec, D. (2011). *Journal of Nanoparticle Research* 13: 2829–2841.

7 (a) Ederer, J., Janoš, P., Ecorchard, P. et al. (2017). *RSC Advances* 7: 12464–12473. (b) Chen, Y. and Zhang, Y. (2011). *Analytical and Bioanalytical Chemistry* 399: 2503–2509.

8 (a) Battigelli, A., Wang, J.T.-W., Russier, J. et al. (2013). *Small* 9 (21): 3610–3619. (b) Ménard-Moyon, C., Fabbro, C., Prato, M., and Bianco, A. (2011). *Chemistry* 17

(11): 3222–3227. (c) Xia, X., Yang, M., Wang, Y. et al. (2012). *ACS Nano* 6 (1): 512–522. (d) Jarre, G., Heyer, S., Memmel, E. et al. (2014). *Beilstein Journal of Organic Chemistry* 10: 2729–2737. (e) Poli, E., Chaleix, V., Damia, C. et al. (2014). *Analytical Methods* 6: 9622–9627. (f) Quintana, M., Spyrou, K., Grzelczak, M. et al. (2010). *ACS Nano* 4 (6): 3527–3533.

9 Kroto, H.W., Heath, J.R., O'Brien, S.C. et al. (1985). *Nature* 318: 162–164.

10 Krätschmer, W., Lamb, L.D., Fostiropoulos, K., and Huffman, D.R. (1990). *Nature* 347: 354–358.

11 Iijima, S. (1991). *Nature* 354: 56–58.

12 Novoselov, K.S., Geim, A.K., Morozov, S.V. et al. (2004). *Science* 306: 666–669.

13 (a) Martin, N. and Bodwell, G. (2019). *Accounts of Chemical Research* 52: 2757–2759. (b) Delgado, J.L., Ángeles Herranza, M., and Martín, N. (2008). *Journal of Materials Chemistry* 18: 1417–1426.

14 (a) Hirsch, A. (1998). *Topics in Current Chemistry* 199: 1–65. (b) Chen, Z., Thiel, W., and Hirsch, A. (2003). *ChemPhysChem* 1: 93–97. (c) Quintana, M., Vázquez, E., and Prato, M. (2013). *Accounts of Chemical Research* 46: 138–148.

15 Lin, T., Zhang, W.D., Huang, J., and He, C. (2005). *The Journal of Physical Chemistry. B* 109: 13755–13760.

16 Jiang, D., Sumpter, B.G., and Dai, S. (2007). *The Journal of Chemical Physics* 126: 134701–134706.

17 Maggini, M., Scorrano, G., and Prato, M. (1993). *Journal of the American Chemical Society* 115: 9798–9799.

18 Bahr, J.L., Yang, J., Kosynkin, D.V. et al. (2001). *Journal of the American Chemical Society* 123: 6536–6542.

19 Georgakilas, V., Kordatos, K., Prato, M. et al. (2002). *Journal of the American Chemical Society* 124: 760–761.

20 Hashimoto, T. and Maruoka, K. (2015). *Chemical Reviews* 115: 5366–5412.

21 (a) Prato, M. and Maggini, M. (1998). *Accounts of Chemical Research* 31 (9): 519–526. (b) Kordatos, K., Da Ros, T., Bosi, S. et al. (2001). *The Journal of Organic Chemistry* 66: 4915–4920.

22 Maggini, M. and Menna, E. (2002). *Fullerenes: From Synthesis to Optoelectronic Properties*, Chapter 1 (eds. D.M. Guldi and N. Martin), 1–50. Dordrech: Kluwer Academic Publishers.

23 Martín, N., Altable, M., Filippone, S. et al. (2006). *Angewandte Chemie, International Edition* 45: 110–114.

24 Filippone, S., Maroto, E.E., Martín-Domenech, A. et al. (2009). *Nature Chemistry* 1: 578–582.

25 (a) Maroto, E.E., de Cózar, A., Filippone, S. et al. (2011). *Angewandte Chemie, International Edition* 50: 6060–6064. (b) Sawai, K., Takano, Y., Izquierdo, M. et al. (2011). *Journal of the American Chemical Society* 133: 17746–17752. (c) Maroto, E.E., Filippone, S., Martín-Domenech, A. et al. (2012). *Journal of the American Chemical Society* 134: 12936–12938. (d) Marco-Martínez, J., Marcos, V., Reboredo, S. et al. (2013). *Angewandte Chemie, International Edition* 52: 5115–5119. (e) Marco-Martínez, J., Reboredo, S., Izquierdo, M. et al. (2014). *Journal of the American Chemical Society* 136: 2897–2904.

26 Fernández-García, J.M., Evans, P.J., Filippone, S. et al. (2019). *Accounts of Chemical Research* 52: 1565–1574.

27 Ménard-Moyon, C., Izard, N., Doris, E., and Mioskowski, C. (2006). *Journal of the American Chemical Society* 128: 6552–6553.

28 Harris, P.J.F. (2001). *Carbon Nanotubes and Related Structures: New Materials for the Twenty-First Century*. Cambridge: Cambridge University Press.

29 Nikolaev, P., Bronikowski, M.J., Bradley, R.K. et al. (1999). *Chemical Physics Letters* 313: 91–97.

30 Tasis, D., Tagmatarchis, N., Bianco, A., and Prato, M. (2006). *Chemical Reviews* 106: 1105–1136.

31 (a) Pantarotto, D., Partidos, C.D., Graff, R. et al. (2003). *Journal of the American Chemical Society* 125: 6160–6164. (b) Pantarotto, D., Singh, R., McCarthy, D. et al. (2004). *Angewandte Chemie, International Edition* 43: 5242–5246. (c) Bianco, A., Hoebeke, J., Godefroy, S. et al. (2005). *Journal of the American Chemical Society* 127: 58–59. (d) Singh, R., Pantarotto, D., McCarthy, D. et al. (2005). *Journal of the American Chemical Society* 127: 4388–4396. (e) Wu, W., Wieckowski, S., Pastorin, G. et al. (2005). *Angewandte Chemie, International Edition* 44: 6358–6362.

32 Toma, F.M., Sartorel, A., Iurlo, M. et al. (2010). *Nature Chemistry* 2: 826–831.

33 (a) Bianco, A. and Prato, M. (2003). *Advanced Materials* 15: 1765–1768. (b) Bianco, A. (2004). *Expert Opinion on Drug Delivery* 1: 57–65. (c) Bianco, A., Kostarelos, K., Partidos, C.D., and Prato, M. (2005). *Chemical Communications*: 571–577.

34 (a) Prato, M., Kostarelos, K., and Bianco, A. (2008). *Accounts of Chemical Research* 41: 60–68. (b) Kostarelos, K., Bianco, A., and Prato, M. (2009). *Nature Nanotechnology* 4: 627–633.

35 Herrero, M.A., Toma, F.M., Al-Jamal, K.T. et al. (2009). *Journal of the American Chemical Society* 131: 9843–9848.

36 Sartorel, A., Carraro, M., Scorrano, G. et al. (2008). *Journal of the American Chemical Society* 130: 5006–5007.

37 (a) Lee, K.M., Li, L., and Dai, L. (2005). *Journal of the American Chemical Society* 127: 4122–4123. (b) Stephenson, J.J., Hudson, J.L., Leonard, A.D. et al. (2007). *Chemistry of Materials* 19: 3491–3498.

38 Lipinsk, M.E., Rebelo, S.L.H., Pereira, M.F.R. et al. (2012). *Carbon* 50: 3280–3294.

39 Brunetti, F.G., Herrero, M.A., Muñoz, J.d.M. et al. (2008). *Journal of the American Chemical Society* 130: 8094–8100.

40 Brunetti, F.G., Herrero, M.A., Muñoz, J.d.M. et al. (2007). *Journal of the American Chemical Society* 129: 14580–14581.

41 (a) Geim, A.K. and Novoselov, K.S. (2007). *Nature Materials* 6: 183–191. (b) Chen, L., Hernandez, Y., Feng, X., and Müllen, K. (2012). *Angewandte Chemie, International Edition* 51: 7640–7654.

42 Wick, P., Louw-Gaume, A.E., Kucki, M. et al. (2014). *Angewandte Chemie, International Edition* 53: 7714–7718.

43 (a) Geim, A.K. (2009). *Science* 324: 1530–1534. (b) Castro Neto, A.H., Guinea, F., Peres, N.M.R. et al. (2009). *Reviews of Modern Physics* 81: 109–162. (c) Balandin, A. (2011). *Nature Materials* 10: 569–581.

44 (a) Sheng Su, D., Perathoner, S., and Centi, G. (2013). *Chemical Reviews* 113: 5782–5816. (b) Wu, S., He, Q., Tan, C. et al. (2013). *Small* 9: 1160–1172. (c) Bonaccorso, F., Sun, Z., Hasan, T., and Ferrari, A.C. (2010). *Nature Photonics* 4: 611–622. (d) Pumera, M. (2011). *Energy & Environmental Science* 4: 668–674.
45 Criado, A., Melchionna, M., Marchesan, S., and Prato, M. (2015). *Angewandte Chemie, International Edition* 54: 10734–10750.
46 Wang, Q.H., Jin, Z., Kim, K.K. et al. (2012). *Nature Chemistry* 4: 724–732.
47 Hernandez, Y., Nicolosi, V., Lotya, M. et al. (2008). *Nature Nanotechnology* 3: 563–568.
48 Hossain, Z., Walsh, M.A., and Hersam, M.C. (2010). *Journal of the American Chemical Society* 132: 15399–15403.
49 Fan, X., Nouchi, R., and Tanigaki, K. (2011). *Journal of Physical Chemistry C* 115: 12960–12964.
50 (a) Montellano, A., Da Ros, T., Bianco, A., and Prato, M. (2011). *Nanoscale* 3: 4035–4041. (b) Schinazi, R.F., Sijbesma, R., Srdanov, G. et al. (1993). *Antimicrobial Agents and Chemotherapy* 37: 1707–1710. (c) Ikeda, A., Hatano, T., Kawaguchi, M. et al. (1999). *Chemical Communications*: 1403–1404.
51 Richardson, C.F., Schuster, D.I., and Wilson, S.R. (2000). *Organic Letters* 2: 1011–1014.
52 Shao, L., Bai, Y., Huang, X. et al. (2009). *Chemical Physics* 116: 323–326.
53 Singh, S.K., Singh, M.K., Kulkarni, P.P. et al. (2012). *ACS Nano* 6: 2731–2740.
54 Biglova, Y.N. and Mustafin, A.G. (2019). *RSC Advances* 9: 22428–22498.
55 Nakamura, T., Ishihara, M., Ohana, T. et al. (2004). *Chemical Communications*: 1336–1337.
56 Singh, M.K., Titus, E., Gonçalves, G. et al. (2010). *Nanoscale* 2: 700–708.
57 Jiang, Z., Jiang, Z.J., Tian, X., and Chen, W. (2014). *Journal of Materials Chemistry A* 2: 441–450.
58 (a) Guo, S.J. and Dong, S.J. (2011). *Chemical Society Reviews* 40: 2644–2672. (b) Ferrari, A.C. et al. (2015). *Nanoscale* 7: 4598–4810.
59 Ciesielski, A. and Samorì, P. (2014). *Chemical Society Reviews* 43: 381–398.
60 Parviz, D., Irin, F., Shah, S.A. et al. (2016). *Advanced Materials* 28: 8796–8818.
61 Vázquez, E., Giacalone, F., and Prato, M. (2014). *Chemical Society Reviews* 43: 58–69.
62 León, V., Quintana, M., Herrero, M.A. et al. (2011). *Chemical Communications* 47: 10936–10938.
63 Backes, C., Hauke, F., and Hirsch, A. (2011). *Advanced Materials* 23: 2588–2601.
64 Iijima, S., Yudasaka, M., Yamada, R. et al. (1999). *Chemical Physics Letters* 309: 165–170.
65 Harris, P.J.F., Tsang, S.C., Claridge, J.B., and Green, M.L.H. (1994). *Journal of the Chemical Society, Faraday Transactions* 90: 2799–2802.
66 Azami, T., Kasuya, D., Yuge, R. et al. (2008). *Journal of Physical Chemistry C* 112 (5): 1330–1334.
67 (a) Murata, K., Kaneko, K., Kokai, F. et al. (2000). *Chemical Physics Letters* 331 (1): 14–20. (b) Zhu, S. and Xu, G. (2010). *Nanoscale* 2: 2538–2549. (c) Ajima, K., Yudasaka, M., Murakami, T. et al. (2005). *Molecular Pharmaceutics* 2 (6): 475–480.
68 Deng, S.Y., Lei, J.P., Huang, Y. et al. (2012). *Chemical Communications* 48: 9159–9161.
69 Cioffi, C., Campidelli, S., Sooambar, C. et al. (2007). *Journal of the American Chemical Society* 129: 3938–3945.

70 Vizuete, M., Gomez-Escalonilla, M.J., Fierro, J.L.G. et al. (2014). *Chemical Science* 5: 2072–2080.

71 Isobe, H., Tanaka, T., Maeda, R. et al. (2006). *Angewandte Chemie, International Edition* 45: 6676–6680.

72 (a) Nakamura, E., Koshino, M., Tanaka, T. et al. (2008). *Journal of the American Chemical Society* 130: 7808–7809. (b) Lacotte, L., Garcia, A., Decossas, M. et al. (2008). *Advanced Materials* 20: 2421–2426.

73 Miyako, E., Russier, J., Mauro, M. et al. (2014). *Angewandte Chemie, International Edition* 53: 13121–13125.

74 Tagmatarchis, N., Maigne, A., Yudasaka, M., and Iijima, S. (2006). *Small* 2: 490–494.

75 Lucío, M.I., Opri, R., Pinto, M. et al. (2017). *Journal of Materials Chemistry B* 5: 8821–8832.

76 Pagona, G., Karousis, N., and Tagmatarchis, N. (2008). *Carbon* 46: 604–610.

77 Cioffi, C., Campidelli, S., Brunetti, F.G. et al. (2006). *Chemical Communications*: 2129–2131.

78 Karfa, P., De, S., Majhi, K. et al. (2019). Functionalization of carbon nanostructures. In: *Comprehensive Nanoscience and Nanotechnology*, 2e, vol. 2, 123–144.

79 Valey, T.S., Hirani, M., Harrison, G., and Holt, K.B. (2014). *Faraday Discussions* 172: 349.

80 Popov, V.A., Egorovn, A.V., Savilov, S.V. et al. (2013). *Journal of Surface Investigation: X-Ray, Synchrotron and Neutron Techniques* 7: 1034–1043.

81 Liu, Y., Gu, Z., Margrave, J.L., and Khabashesku, V.N. (2004). *Chemistry of Materials* 16: 3924–3930.

82 Mochalin, V.N., Neitzel, I., Etzold, B.J.M. et al. (2011). *ACS Nano* 5 (9): 7494–7502.

83 Krüger, A., Liang, Y., Jarre, G., and Stegk, J. (2006). *Journal of Materials Chemistry* 16: 2322–2328.

84 (a) Yu, Y., Yang, X., Liu, M. et al. (2019). *Nanoscale Adv.* 1: 3406–3412; (b) Yu, Y., Nishikawa, M., Liu, M. et al. (2018). *Nanoscale* 10: 8969–8978.

85 Ugarte, D. (1992). *Nature* 359: 707–709.

86 Plonska-Brzezinska, M.E. (2019). *Chemistry of Nanomaterials* 5: 568–580.

87 Georgakilas, V., Guldi, D.M., Signorini, R. et al. (2003). *Journal of the American Chemical Society* 125: 14268–14269.

88 Cioffi, C.T., Palkar, A., Melin, F. et al. (2009). *Chemistry - A European Journal* 15: 4419–4427.

89 Huang, W., Fernando, S., Allard, L.F., and Sun, Y.-P. (2003). *Nano Letters* 3: 565–568.

90 Rettenbacher, A.S., Elliott, B., Hudson, J.S. et al. (2006). *Chemistry - A European Journal* 12: 376–387.

91 Wei-De, Z. and Wen-Hui, Z. (2009). *Journal of Sensors* 2009: 160698.

92 Tran, T.H., Lee, J.-W., Lee, K. et al. (2008). *Sensors and Actuators B: Chemical* 129: 67–71.

93 Zeng, Y.-L., Huang, Y.-F., Jiang, J.-H. et al. (2007). *Electrochemistry Communications* 9 (1): 185–190.

94 (a) Schedin, F., Geim, A., Morozov, S. et al. (2007). *Nature Materials* 6: 652. (b) Wehling, T., Novoselov, K., Morozov, S. et al. (2008). *Nano Letters* 8: 173–177.

95 Tung, T.T., Nine, M.J., Krebsz, M. et al. (2017). *Advanced Functional Materials* 27: 1702891–1702947.

96 Tang, X., Mager, N., Vanhorenbeke, B. et al. (2017). *Nanotechnology* 28: 1–11.

97 (a) Baker, S.N. and Baker, G.A. (2010). *Angewandte Chemie, International Edition* 49: 6726–6744. (b) Georgakilas, V., Perman, J.A., Tucek, J., and Zboril, R. (2015). *Chemical Reviews* 115: 4744–4822. c Lim, S.Y., Shen, W., and Gao, Z. (2015). *Chemical Society Reviews* 44: 362–381.

98 Xu, X., Ray, R., Gu, Y. et al. (2004). *Journal of the American Chemical Society* 126: 12736–12737.

99 (a) Roy, P., Chen, P.-C., Periasamy, A.P. et al. (2015). *Materials Today* 18: 447–458. (b) Sun, Y.-P., Zhou, B., Lin, Y. et al. (2006). *Journal of the American Chemical Society* 128: 7756–7757. (c) Hutton, G.A.M., Martindale, B.C.M., and Reisner, E. (2017). *Chemical Society Reviews* 46: 6111–6123. (d) Li, H., Kang, Z., Liu, Y., and Lee, S.-T. (2012). *Journal of Materials Chemistry* 22: 24230–24253.

100 Arcudi, F., Đorđević, L., and Prato, M. (2019). *Accounts of Chemical Research* 52 (8): 2070–2079.

101 (a) Hu, Y., Yang, J., Tian, J., and Yu, J.-S. (2015). *Journal of Materials Chemistry B* 3 (27): 5608–5614. (b) Shi, Q.-Q., Li, Y.-H., Xu, Y. et al. (2014). *RSC Advances* 4 (4): 1563–1566. (c) Krysmann, M.J., Kelarakis, A., Dallas, P., and Giannelis, E.P. (2012). *Journal of the American Chemical Society* 134 (2): 747–750.

102 See e.g. Verma, N.C., Yadav, A., and Nandi, C.K. (2019). *Nature Communications* 10: 2391.

103 (a) Schneider, J., Reckmeier, C.J., Xiong, Y. et al. (2017). *Journal of Physical Chemistry C* 121: 2014–2022. (b) Li, D., Jing, P., Sun, L. et al. (2018). *Advanced Materials* 30: 1705913.

104 Fang, Q., Dong, Y., Chen, Y. et al. (2017). *Carbon* 118: 319–326.

105 Song, Y., Zhu, S., Zhang, S. et al. (2015). *Journal of Materials Chemistry C* 3: 5976–5984.

106 Righetto, M., Privitera, A., Fortunati, I. et al. (2017). *Journal of Physical Chemistry Letters* 8: 2236–2242.

107 Essner, J.B., Kist, J.A., Polo-Parada, L., and Baker, G.A. (2018). *Chemistry of Materials* 30 (6): 1878–1887.

108 (a) Jiang, J., He, Y., Li, S., and Cui, H. (2012). *Chemical Communications* 48: 9634–9636. (b) Martindale, B.C.M., Hutton, G.A.M., Caputo, C.A. et al. (2017). *Angewandte Chemie, International Edition* 56: 6459–6463. (c) Mazzier, D., Favaro, M., Agnoli, S. et al. (2014). *Chemical Communications* 50: 6592–6595. (d) Choi, Y., Thongsai, N., Chae, A. et al. (2017). *Journal of Industrial and Engineering Chemistry* 47: 329–335. (e) Li, F., Li, Y., Yang, X. et al. (2018). *Angewandte Chemie, International Edition* 130 (9): 2401–2406.

109 (a) Strauss, V., Margraf, J.T., Dolle, C. et al. (2014). *Journal of the American Chemical Society* 136 (49): 17308–17316. (b) Sciortino, A., Cannizzo, A., and Messina, F. (2018). *Carbon* 4: 67.

110 (a) Tong, G., Wang, J., Wang, R. et al. (2015). *Journal of Materials Chemistry B* 3: 700–706. (b) Konstantinos, D. (2016). *Current Organic Chemistry* 20 (6): 682–695.

111 Manioudakis, J., Victoria, F., Thompson, C.A. et al. (2019). *Journal of Materials Chemistry C* 7: 853–862.

112 (a) Liu, C., Zhang, P., Zhai, X. et al. (2012). *Biomaterials* 33 (13): 3604–3613. (b) Hu, L., Sun, Y., Li, S. et al. (2014). *Carbon* 67: 508–513. (c) Dou, Q., Fang, X., Jiang, S. et al. (2015). *RSC Advances* 5: 46817–46822. (d) Pierrat, P., Wang, R., Kereselidze, D. et al. (2015). *Biomaterials* 51: 290–302.

113 (a) Arcudi, F., Đorđević, L., and Prato, M. (2016). *Angewandte Chemie, International Edition* 55: 2107–2112. (b) Đorđević, L., Arcudi, F., and Prato, M. (2019). *Nature Protocols* 14: 2931–2953.

114 Carrara, S., Arcudi, F., Prato, M., and De Cola, L. (2017). *Angewandte Chemie, International Edition* 56: 4757–4761.

115 Arcudi, F., Đorđević, L., and Prato, M. (2017). *Angewandte Chemie, International Edition* 56: 4170–4173.

116 Rigodanza, F., Đorđević, L., Arcudi, F., and Prato, M. (2018). *Angewandte Chemie, International Edition* 57: 5062–5067.

117 Đorđević, L., Arcudi, F., D'Urso, A. et al. (2018). *Nature Communications* 9: 3442.

118 Liu, W., Li, C., Ren, Y. et al. (2016). *Journal of Materials Chemistry B* 4: 5772–5788.

119 Arcudi, F., Strauss, V., Đorđević, L. et al. (2017). *Angewandte Chemie, International Edition* 56: 12097–12101.

120 D'Souza, F. and Ito, O. (2009). *Chemical Communications*: 4913–4928.

121 (a) Zheng, M., Liu, S., Li, J. et al. (2014). *Advanced Materials* 26 (21): 3554–3560. (b) Gomez, I.J., Arnaiz, B., Cacioppo, M. et al. (2018). *Journal of Materials Chemistry B* 6: 5540–5548.

122 (a) Bertelsen, S. and Jørgensen, K.A. (2009). *Chemical Communications*: 2178–2189. (b) Melchiorre, P., Marigo, M., Carlone, A., and Bartoli, G. (2008). *Angewandte Chemie, International Edition* 47: 6138–6171. (c) Melchiorre, P. (2012). *Angewandte Chemie, International Edition* 51: 9748–9770. (d) Lelais, G. and MacMillan, D.W.C. (2006). *Aldrichimica Acta* 39: 79–87. (e) Nielsen, M., Worgull, D., Zweifel, T. et al. (2011). *Chemical Communications* 47: 632–649. (f) List, B. (2006). *Chemical Communications*: 819–824. (g) Seebach, D., Groselj, U., Badine, D.M. et al. (2008). *Helvetica Chimica Acta* 91: 1999–2034. (h) Seebach, D., Gilmour, R., Groselj, U. et al. (2010). *Helvetica Chimica Acta* 93: 603–634. (i) Schmid, M.B., Zeitler, K., and Gschwind, R.M. (2010). *Angewandte Chemie, International Edition* 49: 4997–5003. (j) Schmid, M.B., Zeitler, K., and Gschwind, R.M. (2011). *Chemical Science* 2: 1793–1803.

123 van der Helm, M.P., Klemm, B., and Eelkema, R. (2019). *Nature Reviews Chemistry* 3: 491–508.

124 Testa, C., Zammataro, A., Pappalardo, A., and Trusso Sfrazzetto, G. (2019). *RSC Advances* 9: 27659–27664.

125 Rosso, C., Filippini, G., and Prato, M. (2019). *Chemistry - A European Journal* 25: 16032–16036.

126 Pei, X., Xiong, D., Wang, H. et al. (2018). *Angewandte Chemie, International Edition* 57: 3687–3691.

127 Love, J.C., Estroff, L.A., Kriebel, J.K. et al. (2005). *Chemical Reviews* 105: 1103–1170.

128 de la Llave, E., Clarenc, R., Schiffrin, D.J., and Williams, F.J. (2014). *Journal of Physical Chemistry C* 118: 468–475.

129 Hoft, R.C., Ford, M.J., McDonagh, A.M., and Cortie, M.B. (2007). *Journal of Physical Chemistry C* 111: 13886–13891.

130 Ulman, A. (1996). *Chemical Reviews* 96: 1533–1554.

131 (a) Jeyaprabha, C., Sathiyanarayanan, S., and Venkatachari, G. (2005). *Applied Surface Science* 246: 108. (b) Khaled, K.F. and Hackerman, N. (2003). *Materials Chemistry and Physics* 82: 949–960.

132 Feng, Y., Chen, S., You, J., and Guo, W. (2007). *Electrochimica Acta* 53: 1743–1753.

133 Pujari, S.P., Scheres, L., Marcelis, A.T.M., and Zuilhof, H. (2014). *Angewandte Chemie, International Edition* 53: 6322–6356.

134 Bentez, J.J., Kopta, S., Ogletree, D.F., and Salmeron, M. (2002). *Langmuir* 18: 6096–6100.

135 (a) Faucheux, N., Schweiss, R., Lützow, K. et al. (2004). *Biomaterials* 25: 2721–2730. (b) Arima, Y. and Iwata, H. (2007). *Journal of Materials Chemistry* 17: 4079–4087. (c) Arima, Y. and Iwata, H. (2015). *Acta Biomaterialia* 26: 72–81. (d) Hasan, A., Pattanayek, S.K., and Pandey, L.M. (2018). *ACS Biomaterials Science & Engineering* 4: 3224–3233.

136 Leea, M.H., Brass, D.A., Morris, R. et al. (2005). *Biomaterials* 26: 1721–1730.

137 (a) Feng Zhang, F., Sautter, K., Larsen, A.M. et al. (2010). *Langmuir* 26: 14648–14654. (b) Yadav, A.R., Sriram, R., Carter, J.A., and Miller, B.L. (2014). *Materials Science and Engineering: C* 35: 283–290.

138 (a) Chriseyl, L.A., Lee, G.U., and O'Ferrall, C.E. (1996). *Nucleic Acids Research* 24 (15): 3031–3039. (b) Caviani-Pease, A., Solas, D., Sullivan, E.J. et al. (1994). *Proceedings of the National Academy of Sciences of the United States of America* 91: 5022–5026.

139 Sukhorukov, G.B., Montrel, M.M., Petrov, A.I. et al. (1996). *Biosensors & Bioelectronics* 11 (9): 913–922.

140 (a) Jordon, C.E., Frey, B.L., Kornguth, S., and Corn, R.M. (1994). *Langmuir* 10 (10): 3642–3648. (b) Frey, B.L. and Corn, R.M. (1996). *Analytical Chemistry* 68 (18): 3187–3193. (c) Ruiz-Taylor, L.A., Martin, T.L., Zaugg, F.G. et al. (2001). *Proceedings of the National Academy of Sciences of the United States of America* 98 (3): 852–857.

141 (a)Chechik, V., Crooks, R.M., and Stirling, C.J.M. (2000). *Advanced Materials* 12: 1161–1171. (b) Sullivan, T.P. and Huck, W.T.S. (2003). *European Journal of Organic Chemistry*: 17–29.

142 Fox, M.A., Whitesell, J.K., and McKerrow, A.J. (1998). *Langmuir* 14: 816–820.

143 Campiña, J.M., Martins, A., and Silva, F. (2009). *Electrochimica Acta* 55: 90–103.

144 Frutos, A.G., Brockman, J.M., and Corn, R.M. (2000). *Langmuir* 16: 2192–2197.

145 (a) Han, S.W., Lee, I., and Kim, K. (2002). *Langmuir* 18: 182–187. (b) Alonso, J.M., Reichel, A., Piehler, J., and del Campo, A. (2008). *Langmuir* 24: 448–457. (c) Whitesell, J.K. and Chang, H.K. (1993). *Science* 261: 73–76. (d) Kratzmuller, T., Appelhans, D., and Braun, H.G. (1999). *Advanced Materials* 11 (7): 555–558.

146 (a) Moon, J.H., Shin, W.J., Kim, S.Y., and Park, J.W. (1996). *Langmuir* 12: 4621–4624. (b) Jiang, D., Tang, J., Liu, B. et al. (2003). *Biosensors & Bioelectronics* 18: 1183–1191.

147 Tauk, L., Schröder, A.P., Decher, G., and Giuseppone, N. (2009). *Nature Chemistry* 1: 649–656.

148 Murray, C.B., Norris, D.J., and Bawendi, M.G. (1993). *Journal of the American Chemical Society* 115: 8706–8715.

149 Talapin, D.V., Rogach, A.L., Kornowski, A. et al. (2001). *Nano Letters* 1: 207–211.
150 Li, L.S., Pradhan, N., Wang, Y., and Peng, X. (2004). *Nano Letters* 4: 2261–2264.
151 (a) Levi, T.G. (1930). *Gazzetta Chimica Italiana* 60: 975–987. (b) Levi, T.G. (1931). *Gazzetta Chimica Italiana* 61: 286–293.
152 Hodgson, W.G., Buckler, S.A., and Peters, G.J. (1963). *Journal of the American Chemical Society* 85: 543–546.
153 Sullivan, P.D. (1973). *The Journal of Physical Chemistry* 77: 1859–1861.
154 Thomson, J.W., Nagashima, K., Macdonald, P.M., and Ozin, G.A. (2011). *Journal of the American Chemical Society* 133: 5036–5041.
155 (a) Riha, S., Parkinson, B.A., and Prieto, A.L. (2011). *Journal of the American Chemical Society* 133: 15272–15275. (b) Wang, W., Geng, Y., Yan, P. et al. (1999). *Journal of the American Chemical Society* 121: 4062–4063. (c) Wei, Y., Yang, J., Lin, A.W.H., and Ying, J.Y. (2010). *Chemistry of Materials* 22: 5672–5677.
156 (a) Liu, Y., Yao, D., Shen, L. et al. (2012). *Journal of the American Chemical Society* 134: 7207–7210. (b) Walker, B.C. and Agrawal, R. (2014). *Chemical Communications* 50: 8331–8334.
157 Bullen, C., van Embden, J., Jasieniak, J. et al. (2010). *Chemistry of Materials* 22: 4135–4143.
158 Zimdars, J. and Bredol, M. (2016). *New Journal of Chemistry* 40: 1137–1142.
159 Deng, Z., Cao, L., Tang, F., and Zou, B. (2005). *The Journal of Physical Chemistry. B* 109: 16671–16675.
160 (a) Bullen, C. and Mulvaney, P. (2006). *Langmuir* 22: 3007–3013. (b) Green, M. (2010). *Journal of Materials Chemistry* 20: 5797–5809.
161 Fisher, A.A.E., Osborne, M.A., Day, I.J., and Lucena Alcalde, G. (2019). *Communications Chemistry* 2: 63.
162 Cooper, J.K., Franco, A.M., Gul, S. et al. (2011). *Langmuir* 27: 8486–8493.
163 Anderson, N.C., Hendricks, M.P., Choi, J.J., and Owen, J.S. (2013). *Journal of the American Chemical Society* 135: 18536–18548.
164 Zhang, J., Zhang, H., Cao, W. et al. (2019). *Journal of the American Chemical Society* 141: 15675–15683.
165 Zhu, C., Chen, D., Cao, W. et al. (2019). *Angewandte Chemie, International Edition* 58: 17764–17770.
166 Ji, X., Copenhaver, D., Sichmeller, C., and Peng, X. (2008). *Journal of the American Chemical Society* 130: 5726–5735.
167 Hassinen, A., Moreels, I., de Mello Donegá, C. et al. (2010). *Journal of Physical Chemistry Letters* 1: 2577–2581.
168 Peng, X., Schlamp, M.C., Kadavanich, A.V., and Alivisatos, A.P. (1997). *Journal of the American Chemical Society* 119: 7019–7029.
169 (a) Gao, Y. and Peng, X. (2015). *Journal of the American Chemical Society* 137: 4230–4235. (b) Pu, C. and Peng, X. (2016). *Journal of the American Chemical Society* 138: 8134–8142.
170 (a) Kirkwood, N., Monchen, J.O.V., Crisp, R.W. et al. (2018). *Journal of the American Chemical Society* 140: 15712–15723. (b) du Fossé, I., ten Brinck, S., Infante, I., and Houtepen, A.J. (2019). *Chemistry of Materials* 31: 4575–4583.
171 Zhang, T., Ge, J., Hu, Y., and Yin, Y. (2007). *Nano Letters* 7: 3203–3207.

172 Akin, M., Bongartz, R., Walter, J.G. et al. (2012). *Journal of Materials Chemistry* 22: 11529–11536.

173 Duan, H. and Nie, S. (2007). *Journal of the American Chemical Society* 129: 3333–3338.

174 Carrillo-Carrion, C. and Parak, W.J. (2016). *Journal of Colloid and Interface Science* 478: 88–96.

175 (a) Gandubert, V.J. and Lennox, R.B. (2005). *Langmuir* 21: 6532–6539. (b) Rucareanu, S., Gandubert, V.J., and Lennox, R.B. (2006). *Chemistry of Materials* 18: 4674–4680. (c) Kumar, A., Mandal, S., Selvakannan, P.R. et al. (2003). *Langmuir* 19: 15, 6277–6282. (d) Metin, Ö., Sun, X., and Sun, S. (2013). *Nanoscale* 5: 910–912. (e) Crooks, R.M., Zhao, M., Sun, L. et al. (2001). *Accounts of Chemical Research* 34: 181–190.

176 Zaera, F. (2017). *Chemical Society Reviews* 46: 7374–7398.

177 (a) Liu, X., Atwater, M., Jinhai, W. et al. (2007). *Journal of Nanoscience and Nanotechnology* 7: 3126–3133. (b) Wang, C., Yongjie Hu, Y., Lieber, C.M., and Sun, S. (2008). *Journal of the American Chemical Society* 130: 8902–8903. (c) Wang, C., Hou, Y., Kim, J., and Sun, S. (2007). *Angewandte Chemie, International Edition* 46: 6333–6335. (d) Liu, X., Atwater, M., Wang, J., and Huo, Q. (2007). *Colloids and Surfaces, B: Biointerfaces* 58: 3–7.

178 (a) Mazumder, V. and Sun, S. (2009). *Journal of the American Chemical Society* 131: 4588–4589. (b) Peng, S., Lee, Y., Wang, C. et al. (2008). *Nano Research* 1: 229–234. (c) Manea, F., Bindoli, C., Polizzi, S. et al. (2008). *Langmuir* 24: 4120–4124.

179 Sun, S., Murray, C.B., Weller, D. et al. (2000). *Science* 287: 1989–1992.

180 Mourdikoudis, S. and Liz-Marzán, L.M. (2013). *Chemistry of Materials* 25: 1465–1476.

181 Hiramatsu, H. and Osterloh, F.E. (2004). *Chemistry of Materials* 13: 2509–2511.

182 Lu, X., Tuan, H.-Y., Chen, J. et al. (2007). *Journal of the American Chemical Society* 129: 1733–1742.

183 Lu, X., Tuan, H.-Y., Korgel, B.A., and Xia, Y. (2008). *Chemistry - A European Journal* 14: 1584–1591.

184 Yang, Y., Serrano, L.A., and Guldin, S. (2018). *Langmuir* 34: 6820–6826.

185 Wu, B.-H., Yang, H.-Y., Huang, H.-Q. et al. (2013). *Chinese Chemical Letters* 24: 457–462.

186 Lu, X., Yavuz, M.S., Tuan, H.-Y. et al. (2008). *Journal of the American Chemical Society* 130: 8900–8901.

187 Feng, H., Yang, Y., You, Y. et al. (1984-1986). *Chemical Communications* 2009.

188 Huo, Z., Tsung, C.-K., Huang, W. et al. (2009). *Nano Letters* 9: 1260–1264.

189 Wang, C., Daimon, H., Lee, Y. et al. (2007). *Journal of the American Chemical Society* 129: 6974–6975.

190 Huang, X., Zhao, Z., Fan, J. et al. (2011). *Journal of the American Chemical Society* 133: 4718–4721.

191 Zhao, M., Sun, L., and Crooks, R.M. (1998). *Journal of the American Chemical Society* 120: 4877–4878.

192 (a) Esumi, K., Suzuki, A., Yamahira, A., and Torigoe, K. (2000). *Langmuir* 16: 2604–2608. (b) Pan, C., Pelzer, K., Philippot, K. et al. (2001). *Journal of the American Chemical Society* 123: 7584–7593.

193 (a) Sun, L. and Crooks, R.M. (2002). *Langmuir* 18: 8231–8236. (b) Dhiman, M., Chalke, B., and Polshettiwar, V. (2015). *ACS Sustainable Chemistry & Engineering* 3: 3224–3230.

194 Leff, D.V., Brandt, L., and Heath, J.R. (1996). *Langmuir* 12: 4723–4730.
195 Kluenker, M., Mondeshki, M., Tahir, M.N., and Tremel, W. (2018). *Langmuir* 34: 1700–1710.
196 (a) Bruce, I.J. and Sen, T. (2005). *Langmuir* 21: 7029–7035. (b) Bagwe, R.P., Hilliard, L.R., and Tan, W. (2006). *Langmuir* 22: 4357–4362.
197 Ye, Q., Zhou, F., and Liu, W. (2011). *Chemical Society Reviews* 40: 4244–4258.
198 Jana, N.R., Earhart, C., and Ying, J.Y. (2007). *Chemistry of Materials* 19: 5074–5082.
199 Ferreira, H., Martins, A., Alves da Silva, M.L. et al. (2018). *Journal of Materials Chemistry B* 6: 2104–2115.
200 Zheng, Z., Saar, J., Zhi, B. et al. (2018). *Langmuir* 34: 4614–4625.
201 (a) Turkevich, J., Stevenson, P.C., and Hillier, P. (1953). *The Journal of Physical Chemistry* 57: 670–673. (b) Frens, G. (1973). *Nature Physical Science* 241: 20–22. (c) Kimling, J., Maier, M., Okenve, B. et al. (2006). *The Journal of Physical Chemistry. B* 110: 15700–15707.
202 Dorris, A., Rucareanu, S., Reven, L. et al. (2008). *Langmuir* 24: 2532–2538.
203 Fleischer, C.C. and Payne, C.K. (2014). *Accounts of Chemical Research* 47: 2651–2659.
204 Ghosh, P., Han, G., De, M. et al. (2008). *Advanced Drug Delivery Reviews* 60: 1307–1315.
205 (a) Tiwari, P.M., Vig, K., Dennis, V.A., and Singh, S.R. (2011). *Nanomaterials* 1: 31–63. (b) Daraee, H., Eatemadi, A., Abbasi, E. et al. (2016). *Artificial Cells, Nanomedicine, and Biotechnology* 44: 410–422. (c) Giljohann, D.A., Seferos, D.S., Daniel, W.L. et al. (2010). *Angewandte Chemie, International Edition* 49: 3280–3294. (d) Yang, X., Yang, M., Pang, B. et al. (2015). *Chemical Reviews* 115: 10410–10488. (e) Sandhu, K.K., McIntosh, C.M., Simard, J.M. et al. (2002). *Bioconjugate Chemistry* 13: 3–6.
206 (a) Scaletti, F., Hardie, J., Lee, Y.-W. et al. (2018). *Chemical Society Reviews* 47: 3421–3432. (b) Chen, J., Andler, S.M., Goddard, J.M. et al. (2017). *Chemical Society Reviews* 46: 1272–1283.
207 Ghosh, P.S., Kim, C.-K., Han, G. et al. (2008). *ACS Nano* 2: 2213–2218.
208 Lee, S.H., Bae, K.H., Kim, S.H. et al. (2008). *International Journal of Pharmaceutics* 364: 94–101.
209 Siegwart, D.J., Whitehead, K.A., Nuhn, L. et al. (2011). *Proceedings of the National Academy of Sciences of the United States of America* 108: 12996–13001.
210 Han, L., Zhao, J., Zhang, X. et al. (2012). *ACS Nano* 6: 7340–7351.
211 Lee, Y., Lee, S.H., Kim, J.S. et al. (2011). *Journal of Controlled Release* 155: 3–10.
212 (a) Zelikin, A.N., Becker, A.L., Johnston, A.P.R. et al. (2007). *ACS Nano* 1: 63–69. (b) Chang, J.H., Mou, K.Y., and Mou, C.Y. (2019). *Scientific Reports* 9: 11457.
213 Hao, N., Jayawardana, K.W., Chen, X., and Yan, M. (2015). *ACS Applied Materials & Interfaces* 7: 1040–1045.
214 Hartono, S.B., Gu, W., Kleitz, F. et al. (2012). *ACS Nano* 6: 2104–2117.
215 Hsiao, I.L., Fritsch-Decker, S., Leidner, A. et al. (2019). *Small* 15: 1805400.
216 (a) Bonomi, R., Selvestrel, F., Lombardo, V. et al. (2008). *Journal of the American Chemical Society* 130: 15744–15745. b Bonomi, R., Scrimin, P., and Mancin, F. (2010). *Organic & Biomolecular Chemistry* 8: 2622–2626. (c) Manea, F., Bodar Houillon, F., Pasquato, L., and Scrimin, P. (2004). *Angewandte Chemie, International Edition* 43: 6165–6169.

217 (a) Kreyling, W.G., Abdelmonem, A.M., Ali, Z. et al. (2015). *Nature Nanotechnology* 10: 619–623. (b) Nicholls, F.J., Rotz, M.W., Ghuman, H. et al. (2016). *Biomaterials* 77: 291–306. (c) Song, Y., Xu, X., MacRenaris, K.W. et al. (2009). *Angewandte Chemie, International Edition* 48: 9143–9147. (d) Moriggi, L., Cannizzo, C., Dumas, E. et al. (2009). *Journal of the American Chemical Society* 131: 10828–10829. (e) Şologan, M., Padelli, F., Giachetti, I. et al. (2019). *Nanomaterials* 9: 879. (f) Rotz, M.W., Culver, K.S.B., Parigi, G. et al. (2015). *ACS Nano* 9: 3385–3396.
218 Hao, Y.M., Man, C., and Hu, Z.B. (2010). *Journal of Hazardous Materials* 184: 392–399.
219 Huang, Y.-F., Wang, Y.-F., and Yan, X.-P. (2010). *Environmental Science & Technology* 44: 7908–7913.
220 (a) Pengo, P., Baltzer, L., Pasquato, L., and Scrimin, P. (2007). *Angewandte Chemie (International Ed. in English)* 46: 400–404. (b) Pengo, P., Polizzi, S., Pasquato, L., and Scrimin, P. (2005). *Journal of the American Chemical Society* 127: 1616–1617. (c) Pengo, P., Broxterman, Q.B., Kaptein, B. et al. (2003). *Langmuir* 19: 2521–2524.
221 Fantuzzi, G., Pengo, P., Gomila, R. et al. (2003). *Chemical Communications*: 1004–1005.
222 Christopher Doty, R., Hussain, I., Nichols, R.J. et al. (2004). *Journal of the American Chemical Society* 126: 10076–10084.
223 (a) Zhang, D., Wang, J., and Xua, D. (2016). *Journal of Controlled Release* 229: 130–139. (b) Hu, J.-J., Xiao, D., and Zhang, X.-Z. (2016). *Small* 12: 3344–3359.
224 Luo, G.-F., Chen, W.-H., Liu, Y. et al. (2014). *Scientific Reports* 4: 6064.
225 (a) Mikhaylova, M., Kim, D.K., Berry, C.C. et al. (2004). *Chemistry of Materials* 16: 2344–2354. (b) Gu, Y.J., Cheng, J., Lin, C.C. et al. (2009). *Toxicology and Applied Pharmacology* 237: 196–204.
226 Hristov, D.R., Rocks, L., Kelly, P.M. et al. (2015). *Scientific Reports* 5: 17040.
227 Sardar, R. and Shumaker-Parry, J.S. (2008). *ACS Nano* 8: 731–736.
228 Clarke, S., Pinaud, F., Beutel, O. et al. (2012). *ACS Nano* 6: 11080–11087.
229 Liu, W., Howarth, M., Greytak, A.B. et al. (2008). *Journal of the American Chemical Society* 130: 1274–1284.
230 Sardar, R., Heap, T.B., and Shumaker-Parry, J.S. (2007). *Journal of the American Chemical Society* 129: 5356–5357.
231 Wu, S., Tan, S.-Y., Ang, C.Y. et al. (2015). *Chemical Communications* 51: 11622–11625.
232 Mandal, S., Roy, D., Chaudhari, R.V., and Sastry, M. (2004). *Chemistry of Materials* 16: 3714–3724.
233 Liu, J., Hao, J., Hu, C. et al. (2018). *Journal of Physical Chemistry C* 122: 2696–2703.
234 Celebi, M., Yurderi, M., Bulut, A. et al. (2016). *Applied Catalysis, B: Environmental* 180: 53–64.
235 Singuru, R., Dhanalaxmi, K., Shit, S.C. et al. (2017). *ChemCatChem* 9: 2550–2564.
236 Li, M. and Chen, G. (2013). *Nanoscale* 5: 11919–11192.
237 Sainsbury, T., Ikuno, T., Okawa, D. et al. (2007). *Journal of Physical Chemistry C* 111: 12992–12999.
238 Georgakilas, V., Gournis, D., Tzitzios, V. et al. (2007). *Journal of Materials Chemistry* 17: 2679–2694.
239 Guo, S., Dong, S., and Wang, E. (2008). *Journal of Physical Chemistry C* 112: 2389–2393.
240 Guo, M., Li, C., and Yang, Q. (2017). *Catalysis Science & Technology* 7: 2221–2227.

9

Recent Advances in the Synthesis of Nitrogen Compounds from Biomass Derivatives

Ana C. Fernandes

Centro de Química Estrutural, Departamento de Química, Instituto Superior Técnico, Universidade de Lisboa, Avenida Rovisco Pais, 1049-001 Lisboa, Portugal

9.1 Introduction

Organic nitrogen-containing compounds display an enormous structural diversity and play an important role in many aspects of life and in industry. For example, they are important building blocks for the production of polymers, dyes, lubricants, surfactants, food additives, pharmaceuticals, agrochemicals, and biologically active compounds.

It is estimated that the global amines market in 2020 will be US$ 19.90 billion, and in 2025, it will reach US$ 29.3 billion. Unfortunately, most of the industrially relevant nitrogen compounds are currently manufactured from fossil-based feedstocks (nonrenewable). Nowadays, the importance of a sustainable economy requires the synthesis of nitrogen compounds by transformation of renewable resources. These compounds can be prepared from bioresources that already incorporate the amino group, such as chitin or amino acids, or by modifying bio-based compounds such as carbohydrates, fatty oils, etc.

In the literature, there are already some interesting reviews discussing the synthesis of bio-based amines [1–5]. This chapter will highlight some selected examples of recent advances in the synthesis of nitrogen compounds from biomass derivatives.

9.2 Synthesis of Nitrogen Compounds from Chitin and Its Derivatives

Chitin, the second most abundant biopolymer on Earth after cellulose, is a linear polymer composed of *N*-acetylglucosamine (NAG) units linked by β-1,4-glycosidic bonds (Figure 9.1). This polymer exists in crustacean shells (e.g. crabs and shrimps), insect exoskeletons (e.g. ants), and fungi and can be readily obtained from the millions of tons of crustacean shell waste generated by fishing industry. Chitin has a structure very similar to that of cellulose but contains an acetamido group at the C2 position within the monomer unit (Figure 9.1). This functional group makes chitin an attractive feedstock for the production of bio-based small molecules containing one or more nitrogen atoms [6–8] and

Methodologies in Amine Synthesis: Challenges and Applications, First Edition.
Edited by Alfredo Ricci and Luca Bernardi.

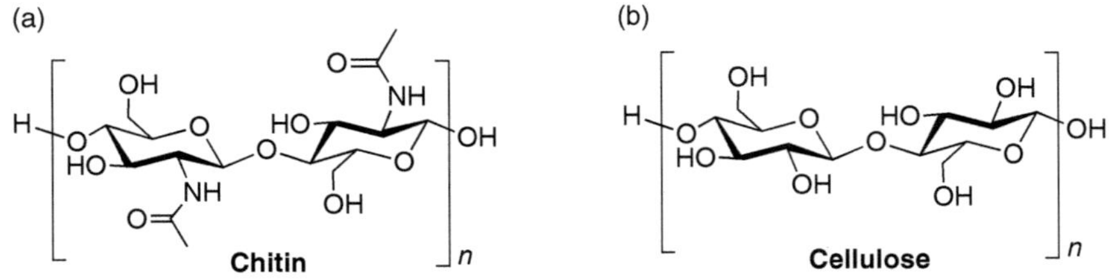

Figure 9.1 (a) Chitin and (b) cellulose structures.

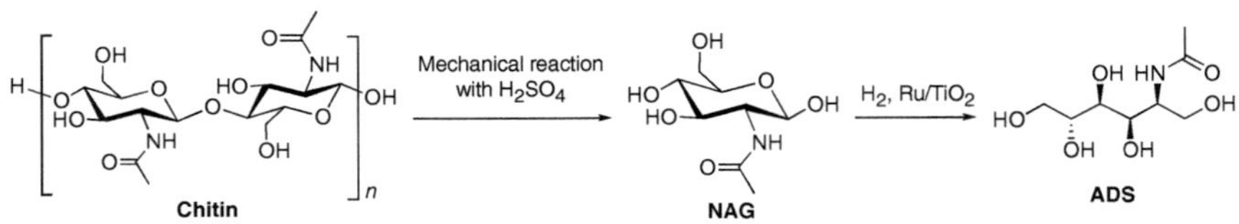

Figure 9.2 One-pot synthesis of ADS from chitin. Source: Based on Kobayashi et al. [9].

that are not directly attainable from lignocellulosic biomass, which only contains C, H, and O atoms.

One of the nitrogen compounds synthesized from chitin was 2-acetamido-2-deoxysorbitol (ADS), a valuable feedstock for the production of polyester amides, nylon resins, and commodity chemicals such as alkylamines and acetamides [9]. This compound was obtained, with 52% overall yield, by a one-pot, two-reaction process combining the mechanocatalytic hydrolysis of chitin to NAG and oligomers with 83% yield in the presence of H_2SO_4, with hydrogenation catalyzed by Ru/TiO_2 (Figure 9.2).

These authors [10] also observed the formation of erythritol and *N*-acetylethanolamine, as side products, by retro-aldol reaction when the hydrogenation was carried out with Ru catalyst in the presence of HCO_3^-. Based on these results, this group explored the synthesis of the protected amino acid acetylglycine (AcGly) from NAG in two steps using a heterogeneous catalyst (Figure 9.3). The synthesis of this amino acid from biomass resources is very important as it is widely used in the production of foods, agricultural chemicals, chelating agents, and cosmetics. In the first step, NAG was converted into *N*-acetylmonoethanolamine (AMEA) through retro-aldol reaction and hydrogenation reaction using a Ru/C catalyst in a $NaHCO_3$ aqueous solution by introduction of high-pressure H_2 after reaching 393 K. The second step consisted in the oxidation of the hydroxyl group of AMEA to carboxylic acid producing AcGly. This process has the advantage that the same Ru/C catalyst promotes both NAG-to-AMEA and AMEA-to-AcGly

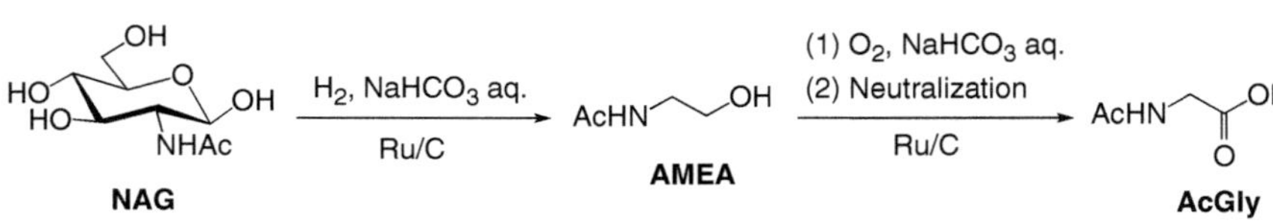

Figure 9.3 Conversion of NAG to AcGly catalyzed by Ru/C. Source: Based on Techikawara et al. [10].

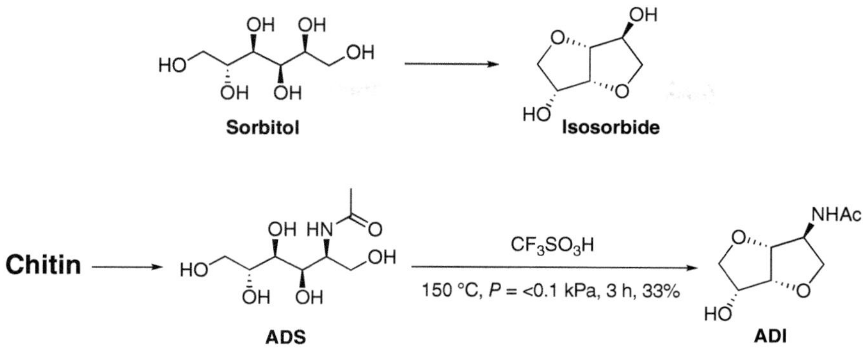

Figure 9.4 Synthesis of ADI from ADS. Source: Based on Sagawa et al. [11].

conversions using the same concentration of $NaHCO_3$ aqueous solution by just changing the atmosphere of H_2 for O_2.

Inspired by the synthesis of isosorbide from the dehydrative condensation of sorbitol, this research group [11] also investigated the conversion of ADS, obtained from chitin, into 2-acetamide-2-deoxyisosorbide (ADI) (Figure 9.4), which is a potential starting material for the preparation of a wide variety of nitrogen compounds such as polymers. This reaction required significantly more severe reaction conditions than that of sorbitol dehydration. ADI was obtained in 33% yield using a large amount of a superstrong acid, CF_3SO_3H, under a decreased pressure of less than <0.1 kPa.

Several methodologies have been reported in the literature for the synthesis of 3-acetamido-5-acetylfuran (3A5AF) from biomass resources [12–14]. For example, Kerton and coworkers [14] described the conversion of NAG into 3A5AF, with 58% yield, using boric acid ($B(OH)_3$) and sodium chloride under microwave irradiation at 220 °C for 15 minutes (Figure 9.5).

Encouraged by these results, Yan and coworkers [15] studied, for the first time, the direct conversion of chitin into 3A5AF. This process combined the chitin hydrolysis with NAG dehydration and was carried out by using boric acid and alkaline chlorides as additives and *N*-methyl-2-pyrrolidone (NMP) as a solvent, leading to a 50% chitin conversion and 7.5% 3A5AF yield (Figure 9.6).

The chitin-derived N-compound 3A5AF is a valuable bio-based starting material for the synthesis of a wide variety of heteroaromatic building blocks and important biologically active compounds (Figure 9.7). For instance, alkylation of 3A5AF allowed the preparation

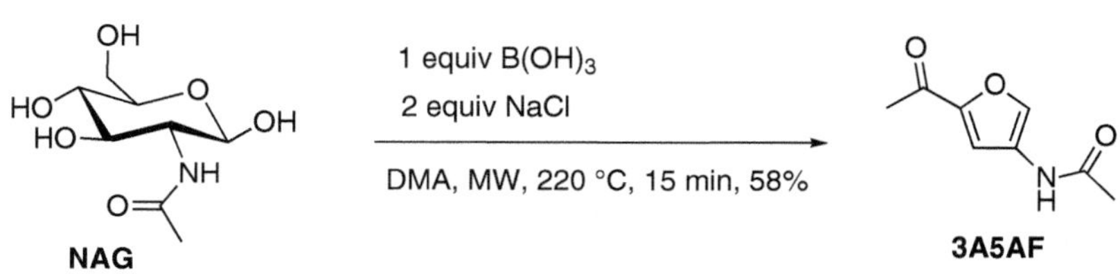

Figure 9.5 Synthesis of 3A5AF from NAG. Source: Based on Omari et al. [14].

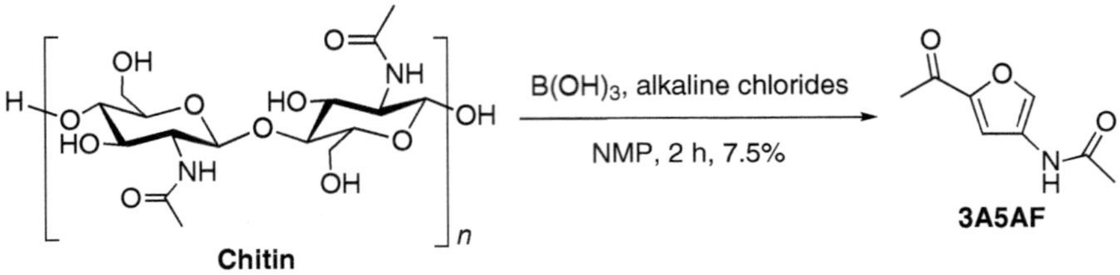

Figure 9.6 Direct conversion of chitin into 3A5AF. Source: Based on Chen et al. [15].

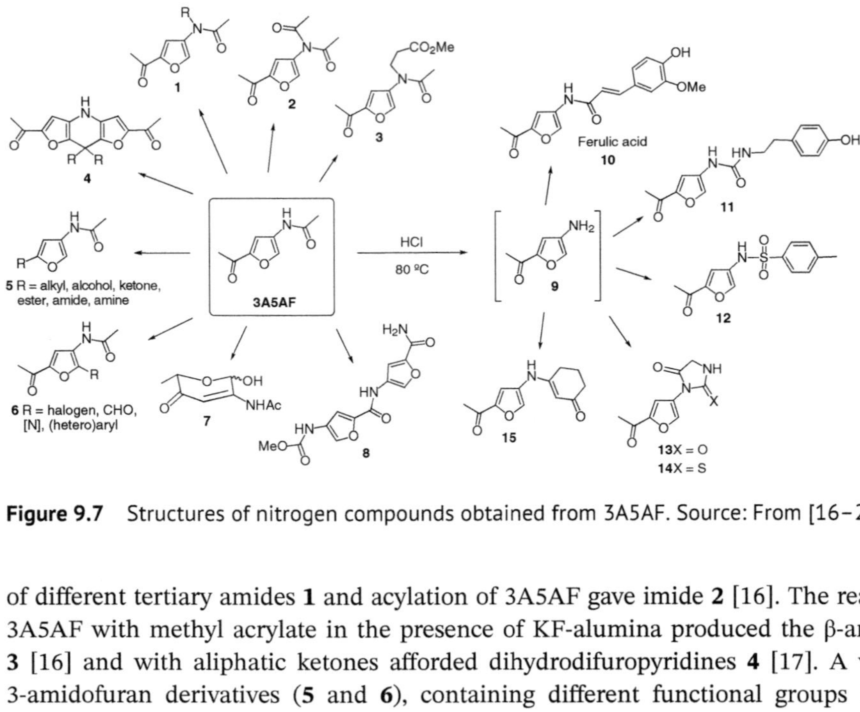

Figure 9.7 Structures of nitrogen compounds obtained from 3A5AF. Source: From [16–22].

of different tertiary amides **1** and acylation of 3A5AF gave imide **2** [16]. The reactions of 3A5AF with methyl acrylate in the presence of KF-alumina produced the β-amidoester **3** [16] and with aliphatic ketones afforded dihydrodifuropyridines **4** [17]. A variety of 3-amidofuran derivatives (**5** and **6**), containing different functional groups including carbonyl, halogen, alcohol, ester, and amide in position 2 or 5, was prepared from 3A5AF [18]. *N*-Acetyl-L-rednose **7**, a 2-aminosugar, was synthesized from 3A5AF by a Noyori reduction–Achmatowicz rearrangement sequence in excellent yield and enantiopurity [19]. Subsequent acetamide hydrolysis gave L-rednose, a rare 2-aminosugar found in natural anthracycline and angucycline antibiotics. The synthesis of the anticancer alkaloid proximicin A (**8**) from 3A5AF is another very important example of the incorporation of the biorenewable nitrogen present in the chitin into biologically active compounds [20]. Proximicin A is an aminofuran natural product isolated from marine strain MG-37 of the actinomycete *Verrucosispora* [21], which exhibits a powerful cytostatic effect in a wide variety of human tumor cell lines [22]. Compound 3A5AF can also be deacetylated producing 3-aminofuran **9**, which can then be converted into a variety of N-functional groups including acrylamine **10**, urea **11**, sulfonamide **12**, hydantoin **13**, thiohydantoin **14**, and enamines **15** [16].

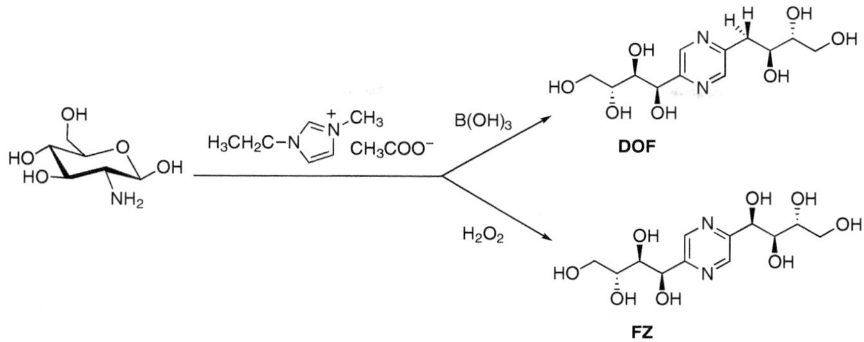

Figure 9.8 Selective synthesis of deoxyfructosazine and fructosazine from D-glucosamine. Source: Based on Lingyu Jia et al. [23].

The development of synthetic methodologies for the production of N-heterocyclic chemicals including pyrazines from biomass resources is also currently an important topic. For example, the selective synthesis of deoxyfructosazine (DOF) and fructosazine (FZ), compounds with many applications in food and pharmaceutical industries, was achieved from D-glucosamine, the building monomer unit of the polysaccharide chitosan catalyzed by the basic ionic liquid 1-ethyl-3-methylimidazolium acetate (Figure 9.8) [23]. DOF was obtained with 40.2% yield via a dehydration process using $B(OH)_3$ as an additive and FZ was formed as the main product in 25.3% yield employing H_2O_2 as an oxidant, which facilitated the dehydrogenation process.

Wang and coworkers [24] also reported the synthesis of DOF from glucosamine using chloride-based deep eutectic solvents (DESs) as catalysts and solvents. This method involved the use of choline chloride-urea DES and led to the formation of DOF with 13.5% yield at a temperature of 100 °C after only 150 minutes. The DOF yield can be increased by using an amino acid as a cocatalyst. For example, in the presence of arginine, DOF was obtained with 30.1% yield.

9.3 Synthesis of Amines and Formamides from α-Amino Acids

α-Amino acids are also a valuable renewable resource of nitrogen compounds that can be used as an excellent alternative to produce amines derivatives from fossil resources [25–29]. As an example, here is highlighted the recent work developed by De Vos and coworkers [28, 29] on the conversion of amino acids into amine derivatives. This group reported a method for the sustainable synthesis of primary amines by indirect decarboxylation of amino acids (via hydrogenation–decarbonylation) catalyzed by ruthenium complexes [28]. This reaction was performed in water at 150 °C using 40 bar H_2, 5 mol% of Ru, and 1.2 equiv of H_3PO_4 for 1 hour 20 minutes (Figure 9.9). Amino acids having an aliphatic side chain gave the corresponding primary amines with yields up to 84%, along with a small amount of 2-amino-alkane as a side product. The reaction of L-proline produced pyrrolidine in 40% yield, which was further hydrogenolyzed giving 27% yield of *n*-butylamine. The hydroxyl groups present in amino acids were not very stable under these reaction conditions. For

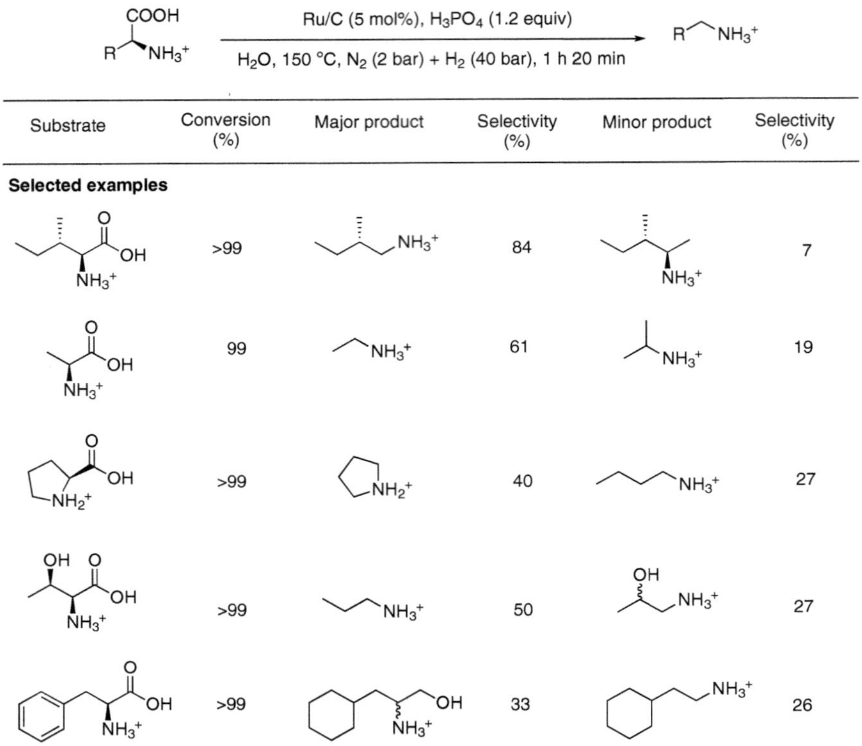

Figure 9.9 Hydrogenation–decarbonylation of natural amino acids. Source: Based on Verduyckt et al. [28].

example, L-threonine lost the secondary hydroxyl group by hydrogenolysis and gave mainly *n*-propylamine. In the reaction of L-phenylalanine, the carboxylic group and the aromatic ring were hydrogenated, producing mainly cyclohexylalaninol.

Very recently, the same research group reported a metal-free methodology for the decarboxylation of amino acids, leading to the production of bio-based primary amines and amides [29]. This approach involves the reaction of the amino acid with a carbonyl compound, which is temporarily trapped into a Schiff base, from which the elimination of CO_2 occurs more easily (Figure 9.10). Subsequent protonation of the azomethine ylide intermediate **16** and hydrolysis of the Schiff base adduct **17** lead to the release of the amine and the regeneration of the catalyst.

These authors found that isophorone, a cheap and commercial available ketone, was the best organocatalyst for this reaction, allowing the synthesis of a wide variety of primary amines in 2-propanol at 150 °C (Figure 9.11). For instance, the reactions of glycine and amino acids with an aliphatic side chain including alanine, valine, leucine, isoleucine, and norleucine gave the corresponding primary amines in excellent yields after 24 hours. Pyrrolidine and tryptamine were also produced in excellent yields from proline and tryptophan, respectively, after only 4 hours.

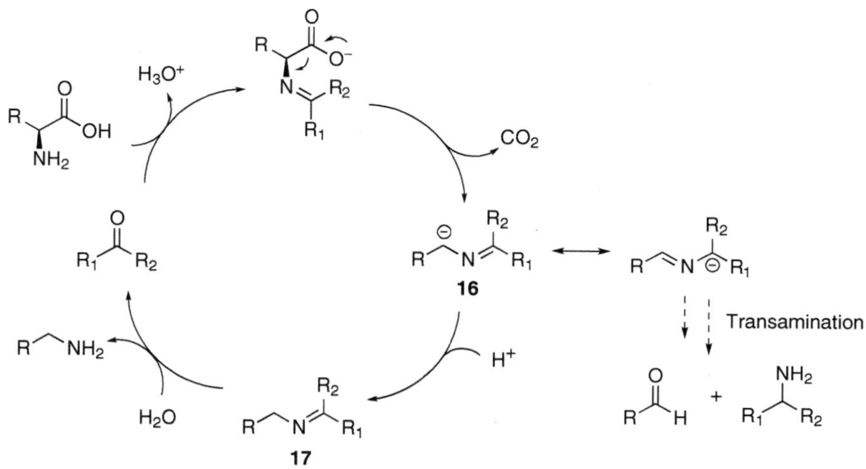

Figure 9.10 Proposed mechanism for the metal-free decarboxylation of amino acids. Source: Based on Claes et al. [29].

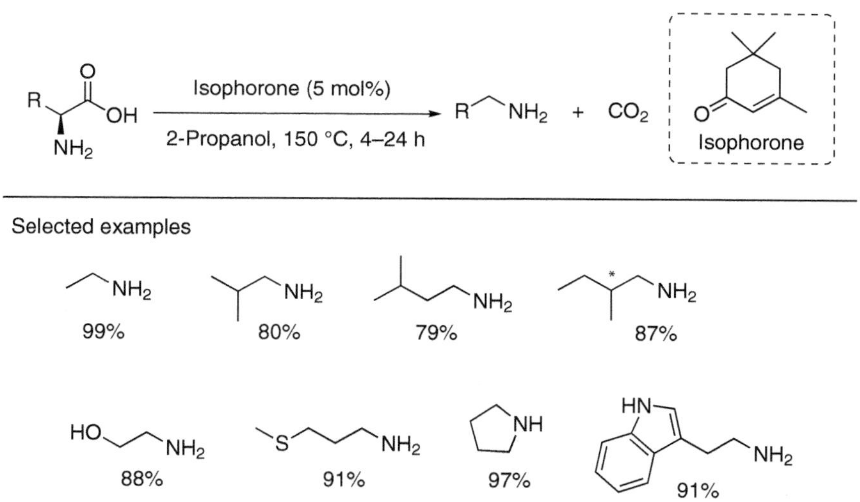

Figure 9.11 Synthesis of primary amines by decarboxylation of amino acids. Source: Based on Claes et al. [29].

When this reaction is performed in DMF, the amine is susceptible to a subsequent acid-catalyzed N-acylation, as result of the nucleophilic attack of the amine on the DMF carbonyl group, leading to the formation of the corresponding formamides. This reaction is catalyzed by the acidity of the amino acid reactant itself. Formamides are important intermediates in the production of pharmaceuticals, agrochemicals, and isocyanides. This one-pot tandem decarboxylation-N-formylation of amino acids process was used to synthesize a wide variety of formamides, in excellent yields, from different amino acids such as alanine, valine, leucine, isoleucine, norleucine, and proline in the presence of

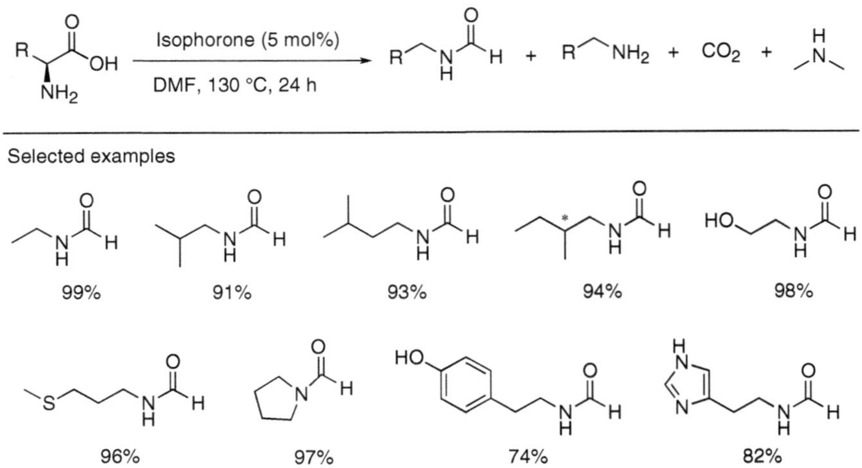

Figure 9.12 Synthesis of formamides by tandem decarboxylation-*N*-formylation of amino acids. Source: Based on Claes et al. [29].

5 mol% of isophorone at 130 °C for 24 hours (Figure 9.12). Furthermore, this reaction was successfully applied to tyrosine and histidine, in contrast to the low yields obtained on decarboxylation of these amino acids.

9.4 Synthesis of Nitrogen Compounds from Cellulosic Biomass Derivatives

Nowadays, there is a growing interest in the production of renewable chemicals containing heteroatoms including N from the cellulosic biomass, but the lack of nitrogen-containing functionalities in the cellulose-derived molecules limits the production of these compounds. For this reason, in the recent years, great research efforts have been devoted to the development of new synthetic strategies for the conversion of renewable O-containing chemicals into their nitrogen analogs.

Among the biomass resources, carbohydrates form by far the largest natural source of carbon and are considered the ideal feedstock for the development of cost-effective methods to access valuable nitrogen compounds. For example, Sels and coworkers [30] developed a low-temperature, one-step reductive aminolysis methodology for the synthesis of bio-based short amines and aminoalcohols by reacting glucose with primary and secondary alkyl- or alkanolamines using commercial Ru and Ni catalysts in the absence of a solvent (Figure 9.13). Short-chain amines including ethanolamines and ethylenediamines are very important compounds with diverse applications in the industry or commonly encountered moieties in many bioactive natural products and pharmaceuticals. This one-pot process involves a retro-aldol C—C bond scission of the sugar molecule, followed by a reductive amination of the short carbonyl intermediate, without release of CO_2.

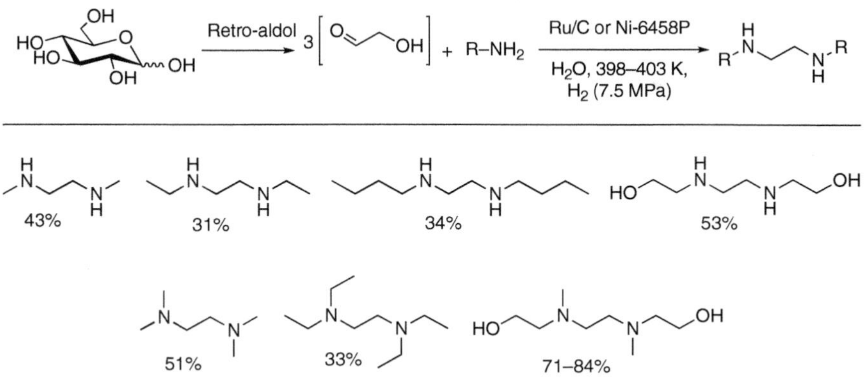

Figure 9.13 Reductive aminolysis of glucose into C2-diamines. Source: Based on Pelckmans et al. [30].

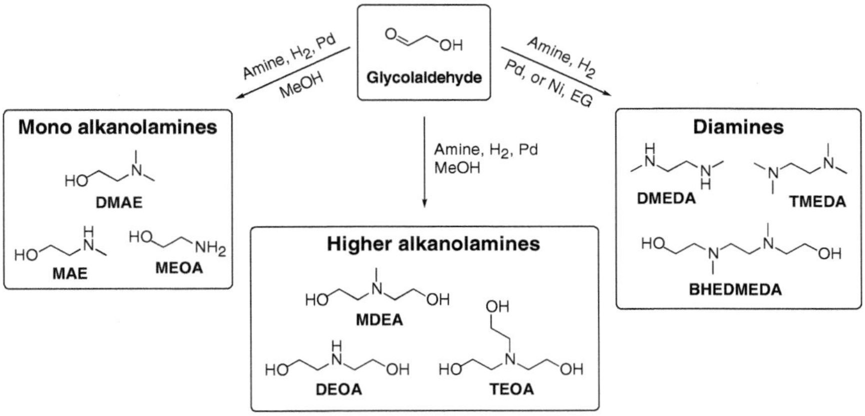

Figure 9.14 Synthesis of alkanolamines and ethylene diamines from glycolaldehyde. Source: Based on Faveere et al. [31].

This reductive aminolysis process also allowed the successful conversion of an industrial starch-derived sugar syrup into *N*,*N*′-bis(2-hydroxyethyl)-*N*,*N*′-dimethylethylenediamine (BHEDMEDA), with a yield similar to the one obtained from crystalline glucose.

Glycolaldehyde, the simplest of all common aldoses, is another versatile renewable feedstock for the production of amines. Very recently, Sels and coworkers [31] have also reported the synthesis of several alkanolamines and ethylene diamines by reductive aminolysis of glycolaldehyde using Ni and Pd catalysts. The choice of the solvent was crucial to obtain high yields in the reductive aminolysis of glycolaldehyde. For instance, protic solvents such as methanol led to excellent yields of alkanolamines such as 2-(methylamino)ethanol (MAE) from monomethylamine and 2-(dimethylamino)ethanol (DMAE) from dimethyl amine, without requiring an excess of amine as observed in other methodologies. Only the reaction of glycolaldehyde with NH_3 benefited of an excess amine reagent to produce monoethanolamine (MEOA) in moderate yield (Figure 9.14).

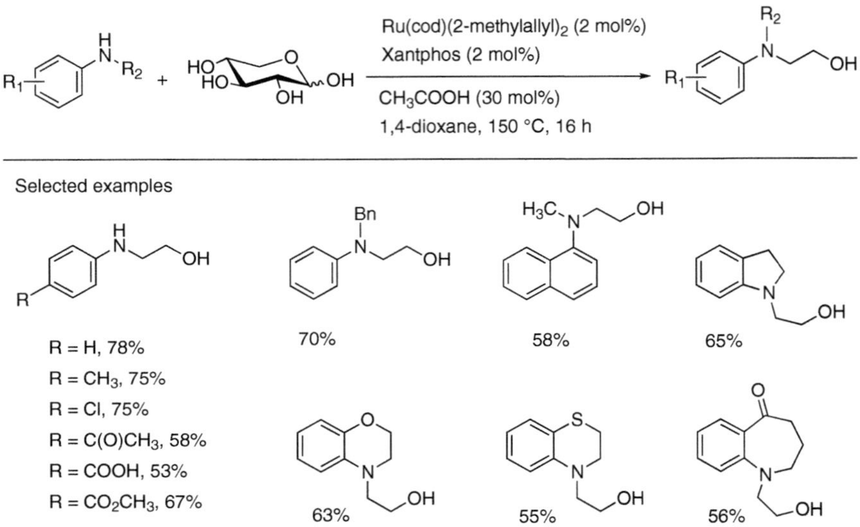

Figure 9.15 Direct hydroxyethylation of amines by carbohydrates. Source: Based on Jia et al. [32].

This methodology also allowed the synthesis of valuable higher secondary and tertiary alkanolamines in good yields including *N*-methyldiethanolamine (MDEA, 91% C), diethanolamine (DEOA, 70% C), and triethanolamine (TEOA, 77% C), using a substoichiometric amount of amine compared to glycolaldehyde in methanol.

The diamines *N*,*N*′-dimethylethylenediamine (DMEDA), *N*,*N*,*N*′,*N*′-tetramethylethylenediamine (TMEDA), and BHEDMEDA were also prepared selectively in high yields using ethylene glycol as a solvent. This two-step, one-pot process involved the synthesis of an unsaturated C_2-ene-diamine precursor, followed by its hydrogenation at elevated temperature.

Recently, Li and coworkers [32] reported the first protocol for the synthesis of β-amino alcohols from arylamines employing commercially available carbohydrates as alkylating reagents in the presence of the ruthenium complex Ru(cod)(2-methylallyl)$_2$ and the ligand Xantphos under acid conditions (Figure 9.15). Moderate to good yields of β-amino alcohols were obtained in the reactions of *N*-methylaniline with both C6- and C5-sugars, including D-glucose (54%), D-fructose (84%), D-mannose (70%), D-galactose (51%), D-xylose (87%), D-ribose (64%), xylitol (67%), and erythritol (66%). Furthermore, the disaccharides cellobiose and sucrose were also tested in this reaction, leading to the formation of β-amino alcohol with yields of 8% and 18%, respectively.

This direct hydroxyethylation methodology was applied to a variety of *N*-alkyl-anilines containing electron-donating and electron-withdrawing groups on the phenyl ring and to anilines bearing various alkyl groups (Et, *i*-Pr, *n*-Bu, $C_{12}H_{25}$, Cy, and benzyl) using xylose as the alkylating reagent, producing the corresponding β-amino alcohols with moderate to good yields (Figure 9.15). This reaction was also performed with heterocyclic compounds including benzomorpholine, benzothiazine, and 5*H*-1-benzazepin-5-one with moderate yields.

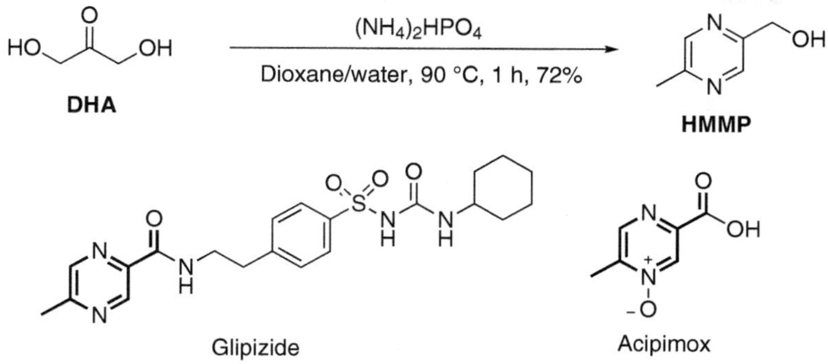

Figure 9.16 Synthesis of HMMP from DHA. Source: Based on Song et al. [33].

The synthesis of valuable N-heterocyclic compounds including pyrazines has also been investigated from cellulose-derived molecules using a nitrogen source. For example, 2-hydroxymethyl-5-methylpyrazine (HMMP) is used as an intermediate in the production of commercial medicines including the antidiabetic glipizide and the hyperlipidemic acipimox. Zhang and coworkers [33] studied the synthesis of HMMP from biomass-derived C3 hydroxyl aldehydes using diammonium phosphate as nitrogen source under optimized conditions of pH = 8.0–9.1 at 90 °C for 1 hour in a dioxane and water mixture. Glyceraldehyde and 1,3-dihydroxyacetone (DHA) were the best substrates for this reaction, leading to the formation of HMMP with 72% yield from 1,3-DHA (Figure 9.16).

The proposed mechanism involved DHA isomerization to glyceraldehyde, followed by ketimine condensation, cyclization, and dehydration reactions (Figure 9.17). The isomerization of DHA to glyceraldehyde observed during the reaction suggests that glyceraldehyde is also a suitable raw material for this reaction. The hydroxyl group present in this pyrazine allows further chemical modifications and the synthesis of pharmaceuticals and high-value added chemicals.

2-Methyl pyrazine (MP) was also prepared from glucose in a one-pot process with 25.6% yield, along with 15% yield of 4 and 5-methyl imidazoles as by-products, within 15 minutes using an ammonia solution and a tungsten-based catalyst (Figure 9.18) [34]. The reaction

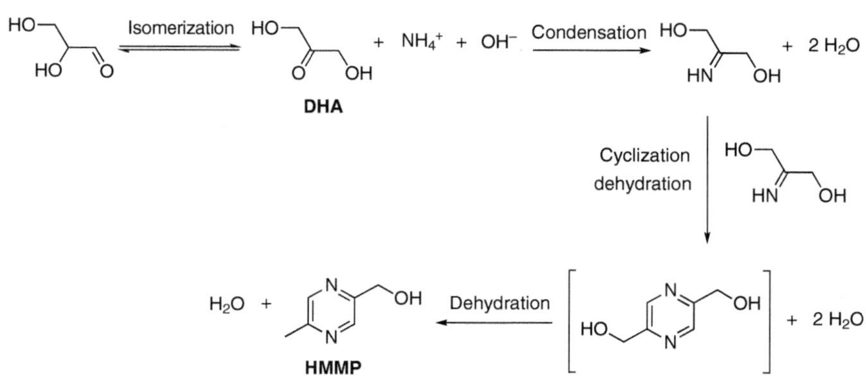

Figure 9.17 Proposed mechanism for the synthesis of HMMP. Source: Based on Song et al. [33].

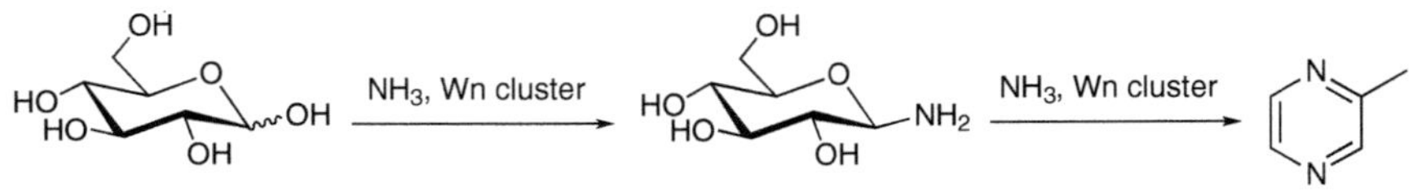

Figure 9.18 One-pot synthesis of 2-methyl pyrazine from glucose. Source: Based on Chen et al. [34].

involved the rapid formation of β-D-glucopyranosylamine as an intermediate catalyzed by tungsten clusters such as $[HW_2O_7]^-$ and $[W_4O_{13}]^{2-}$, which also promoted the chain fragmentation and the subsequent cyclization reaction to generate the pyrazine ring.

This pyrazine was also synthesized from other monosaccharides and disaccharides. For instance, 2-MP was obtained from xylose in 23.3% yield and from the disaccharides cellobiose and maltose in 10.3% and 10.5% yield, respectively.

Furan derivatives are a very important class of compounds, usually employed as building blocks in the production of numerous drugs possessing a wide variety of biological activities such as antimicrobial, anti-inflammatory, antitubercular, antitumor, anti-HIV, antimalarial, antiviral, antifugal, etc. [35] Direct connection of the furan ring to biomass resources provides an efficient and ecological approach for the sustainable long-term production of these pharmaceutical substances. In this context, the synthesis of furan compounds from the hemicellulose-derived furfural and 5-hydroxymethylfurfural has attracted considerable attention.

Among furan derivatives, furfurylamines and 5-hydroxymethylfurfurylamines have diverse applications in different research areas, including the preparation of polymers, biologically active compounds, and also as intermediates in the synthesis of pharmaceuticals such as antiseptic agents, antihypertensives, and diuretics (e.g. furosemide).

The synthesis of furfurylamines from carbohydrates was reported by Caetano and Fernandes [36]. This novel one-pot, three-reaction process involved the initial conversion of xylose into furfural catalyzed by an aqueous solution of $HReO_4$, followed by the in situ reductive amination of furfural with different anilines using silane as a reducing agent (Figure 9.19).

A large variety of secondary and tertiary amines were prepared by this methodology with good overall yields and good chemoselectivity (Figure 9.20). The best yields were obtained in the reactions with anilines containing electron-withdrawing groups. The reaction with the secondary amine *N*-methylaniline led to the synthesis of the tertiary amine with an overall yield of 75%. This one-pot, multireaction process also has the advantage of minimizing chemical wastes and reducing cost, time, and the separation of intermediate compounds. In

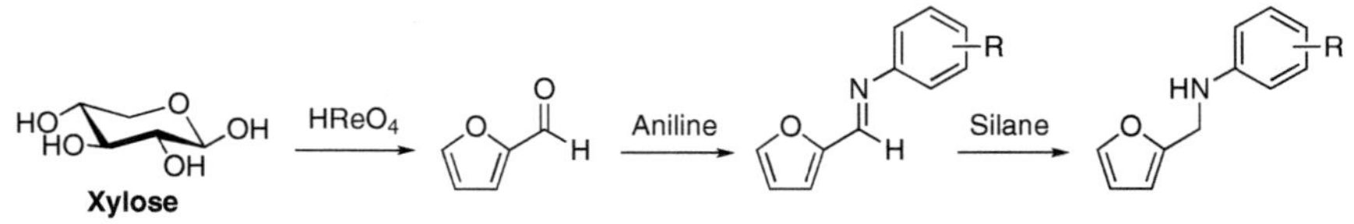

Figure 9.19 One-pot synthesis of furfurylamines from xylose. Source: Based on Caetano and Fernandes [36].

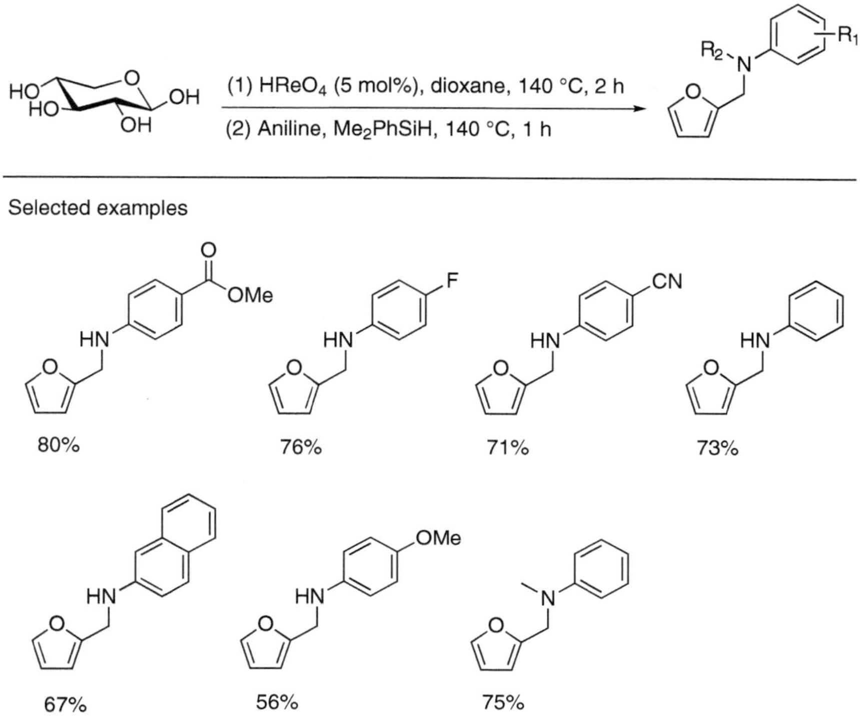

Figure 9.20 One-pot synthesis of furfurylamines from xylose catalyzed by $HReO_4$. Source: Based on Caetano and Fernandes [36].

contrast, the synthesis of furfurylamines from ribose and arabinose, the second most abundant pentose in biomass after xylose, gave only 36% and 29% yields, respectively. These low yields are due to the low conversion of these pentoses to furfural.

Using this methodology, several furfurylamines were also prepared from the polysaccharide xylan, which is made from units of xylose and is found in plant cell walls and in some algae. This chemical transformation involved one-pot, four reactions, including the cleavage of xylan into xylose, promoted by the aqueous solution of $HReO_4$, followed by dehydration of xylose to furfural and reductive amination (Figure 9.21).

The reaction of xylan with different anilines catalyzed by $HReO_4$ using dimethylphenylsilane as a reducing agent led to the synthesis of secondary and tertiary furfurylamines with 53–62% overall yields (Figure 9.22).

Very recently, Kim and coworkers [37] also reported the one-pot synthesis of a wide variety of furfurylamines, with 64–72% yields, by a metal-free catalytic reductive amination of xylose using the pyridine-based ionic liquid $[Pro(Etpy)_3][Lac]_3$ and $(EtO)_3SiH$ as a reducing agent.

N-substituted 5-hydroxymethyfurfurylamines are also very important compounds in the pharmaceutical industry because of their wide range of pharmaceutical activities such as calcium antagonistic, cholinergic or antimuscarinic agents, and carcinogenesis inhibitors. The synthesis of N-substituted 5-hydroxymethyfurfurylamines from the biomass derivative

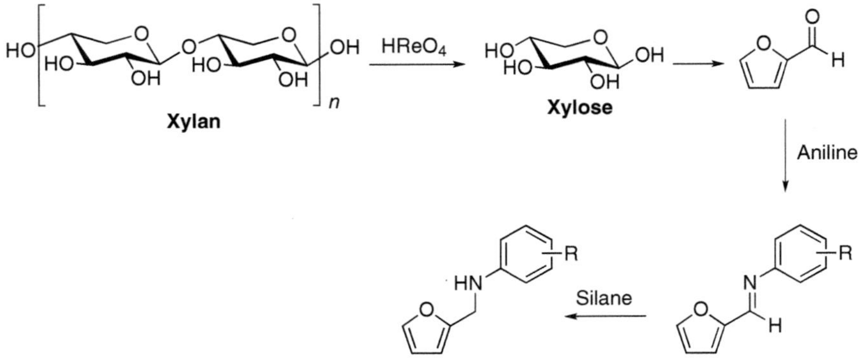

Figure 9.21 One-pot conversion of xylan into furfurylamines. Source: Based on Caetano and Fernandes [36].

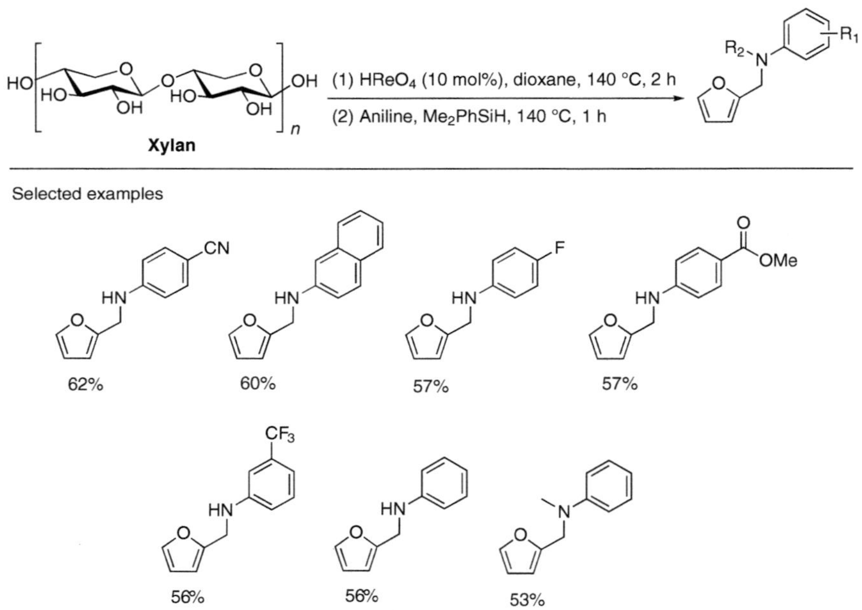

Figure 9.22 One-pot synthesis of furfurylamines from xylan. Source: Based on Caetano and Fernandes [36].

5-hydroxmethylfurfural by reductive amination has attracted considerable attention in recent years. For example, Zhang and coworkers [38] investigated the synthesis of these amines by reaction of 5-HMF with primary and secondary amines catalyzed by the ruthenium complex dichlorobis(2,9-dimethyl-1,10-phenanthroline)ruthenium(II) ($Ru(DMP)_2Cl_2$) in the presence of H_2 (Figure 9.23). The best results were obtained with aromatic amines bearing electron-donating groups. The amines containing electron-withdrawing groups on aromatic rings required much longer reaction times or higher temperatures. In addition, the reaction with cyclic aliphatic morpholine could

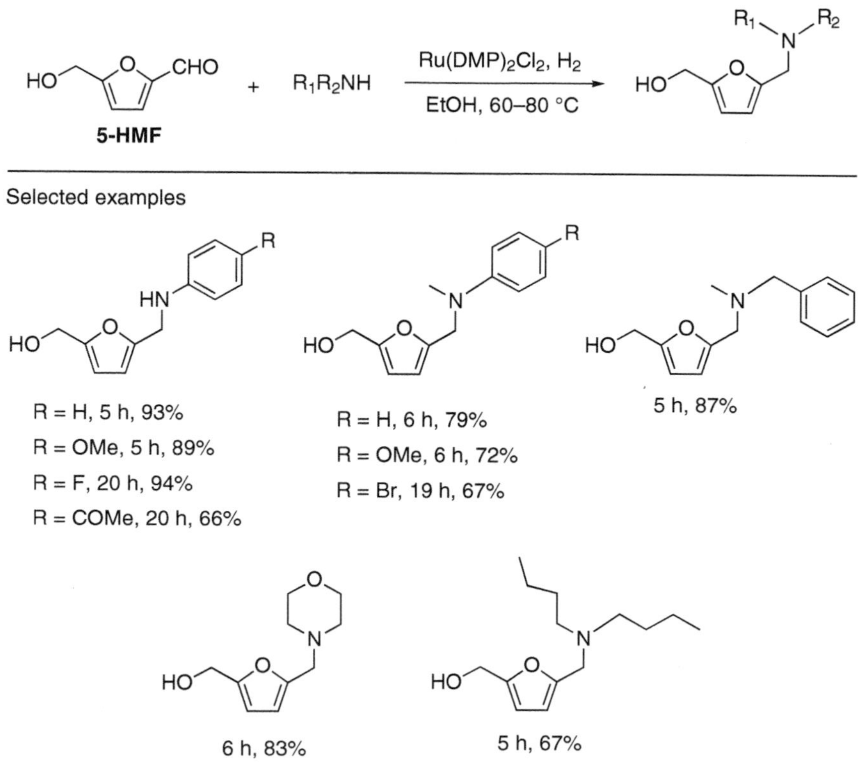

Figure 9.23 Reductive amination of 5-HMF catalyzed by $Ru(DMP)_2Cl_2$. Source: Based on Xu et al. [38].

be performed at 30 °C, producing the corresponding amine with 83% yield. Moderate yield (67%) was observed in the reaction of dibutylamine, and no product was formed with the primary alkylamine butylamine.

Later, Zhang and coworkers [39] extended this methodology to the preparation of a variety of bis(hydroxylmethylfurfuryl)amines (BHMFA), a furan-based monomer with great potential for the production of functional biopolymers. The reaction of 5-HMF with several benzyl and aliphatic amines was performed in the presence of the catalyst $Ru(DMP)_2Cl_2$ (Figure 9.24). Good yields were obtained in the reactions with straight-chain amines including *n*-hexyl and *n*-dodecyl amines. The reaction carried out with *sec*-butylamine gave the product in 75.1% yield, whereas no product was obtained using *tert*-butylamine. The bis(hydroxylmethylfurfuryl)amines were also prepared in good yields using benzylamines containing electron-donating or electron-withdrawing groups.

Many other homogeneous and heterogeneous catalysts containing different metals including Pd [40, 41], Au [42], Ni [43–45], Ru/ZrO_2 [46], and Co [47] have also been applied to the synthesis of furfurylamines by reductive amination of furfural or 5-hydroxymethylfurfural, giving generally good yields.

The production of renewable polyamides is another topic of interest that has stimulated the development of various methods for the preparation of diamines from

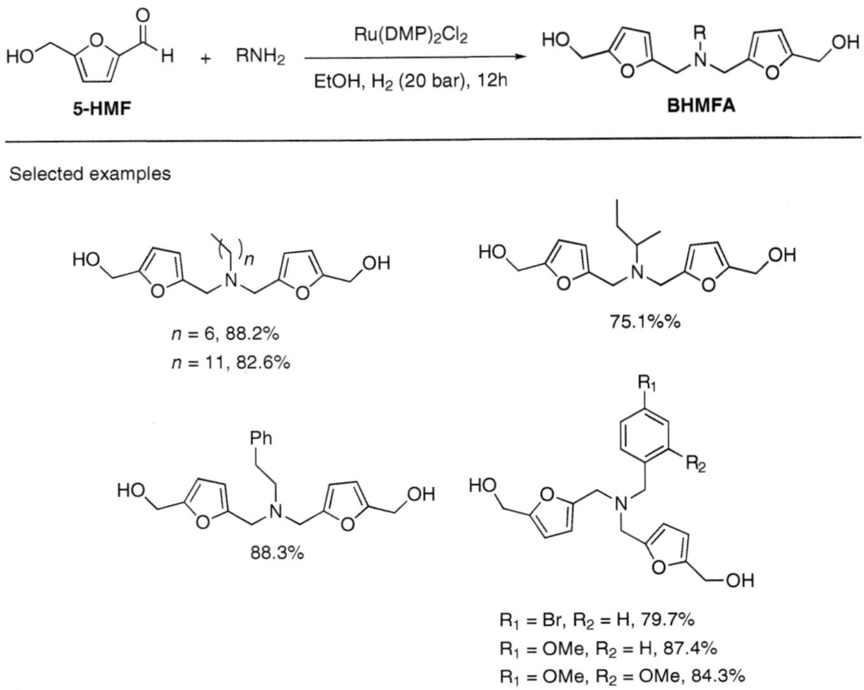

Figure 9.24 Synthesis of BHMFA from 5-HMF. Source: Based on Xu et al. [39].

furan derivatives [48–53]. For example, Mecking and coworkers [51] reported the synthesis of 2,5-bis(aminomethyl)furan (BAMF) from 5-HMF by one-pot direct reductive amination-alcohol amination, with 85% yield, using ammonia and the catalyst [RuHCl(acridine-^{i}Pr-PNP)(CO)] at 140 °C after 11 hours (Figure 9.25). The reaction profile showed a very fast conversion of 5-HMF to the intermediate aminoalcohol by reductive amination of the aldehyde with ammonia. In contrast, the subsequent conversion of the aminoalcohol to BAMF was much slower.

The synthesis of BAMF was also achieved in 65% yield from 2,5-bis(hydroxymethyl)furan (BHMF) in the presence of ammonia catalyzed by [RuHCl(acridine-^{i}Pr-PNP)(CO)] after 7 hours at 150 °C (Figure 9.25).

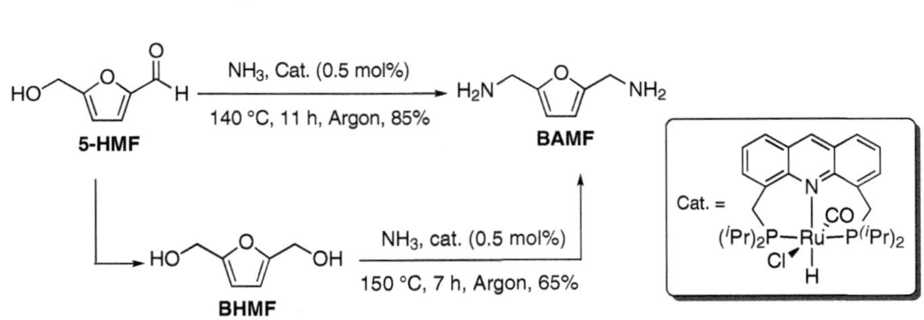

Figure 9.25 Synthesis of BAMF from 5-HMF. Source: Based on Pingen et al. [51].

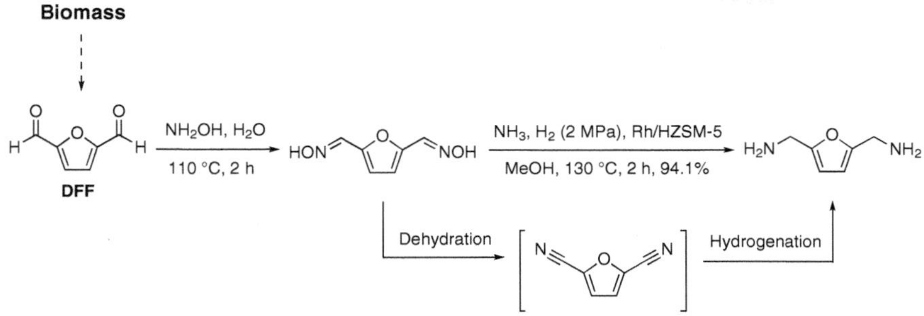

Figure 9.26 Synthesis of BAMF from DFF catalyzed by Rh/HZSM-5. Source: Based on Xu et al. [53].

Recently, Shi and coworkers [52] have also developed a methodology for the production of BAMF, with 85.9% yield, from 5-HMF by reductive amination with NH_3 over bifunctional $CuNiAlO_x$ at 90 °C, followed by hydrogen-borrowing reaction of hydroxyl group with NH_3 at 210 °C.

The synthesis of BAMF was also investigated from 2,5-diformylfuran (DFF), which can be readily obtained from 5-HMF or directly from carbohydrates (Figure 9.26). This process involved the oximation of DFF producing 2,5-diformylfuran dioxime, followed by dehydration of aldoxime to nitrile and hydrogenation of cyano group catalyzed by Rh/HZSM-5 [53]. This method avoids the formation of diimine intermediates, which are very susceptible to irreversible condensation, resulting in the formation of oligomer by-products, limiting the yield of BAMF to 50%.

α-Aminophosphonates and α-aminophosphonic acids are structurally analogous to α-amino acids and constitute an important class of compounds with diverse biological activities and potential to be employed as enzyme inhibitors [54], antibiotics [55], antitumor [56], antifungal [57], or herbicidal [58] agents.

Using a synthetic strategy similar to that developed for the one-pot preparation of furfurylamines from carbohydrates [36], Isca and Fernandes [59] investigated the one-pot synthesis of α-aminophosphonates from xylose and xylan by in situ reaction of furfural with several anilines and *H*-phosphonates (Figure 9.27).

In general, the best yields of α-aminophosphonates were obtained in the reactions with anilines containing electron-withdrawing groups (Figure 9.28). No reaction occurred using benzylamine, cyclohexylamine, and cyclopropylaniline.

Almost at the same time, Popowycz and coworkers also reported the synthesis of a variety of α-aminophosphonates from 5-HMF employing different amines and I_2 as a catalyst (Figure 9.29) [60]. Excellent to good yields were obtained in the reactions with anilines containing electron-donating or electron-withdrawing para substituents. Aliphatic amines were less reactive, producing the α-aminophosphonates in moderate yields. A mixture of two isomers was obtained with 72% overall yield from the reaction of 5-HMF with the chiral amine (*R*)-α-methylbenzylamine. The presence of the hydroxymethyl group in these products allows further chemical modifications and the synthesis of new derivatives.

The Biginelli reaction is a three-component reaction to prepare dihydropyrimidinones, which display a wide variety of biological activities such as antiviral, antitumor,

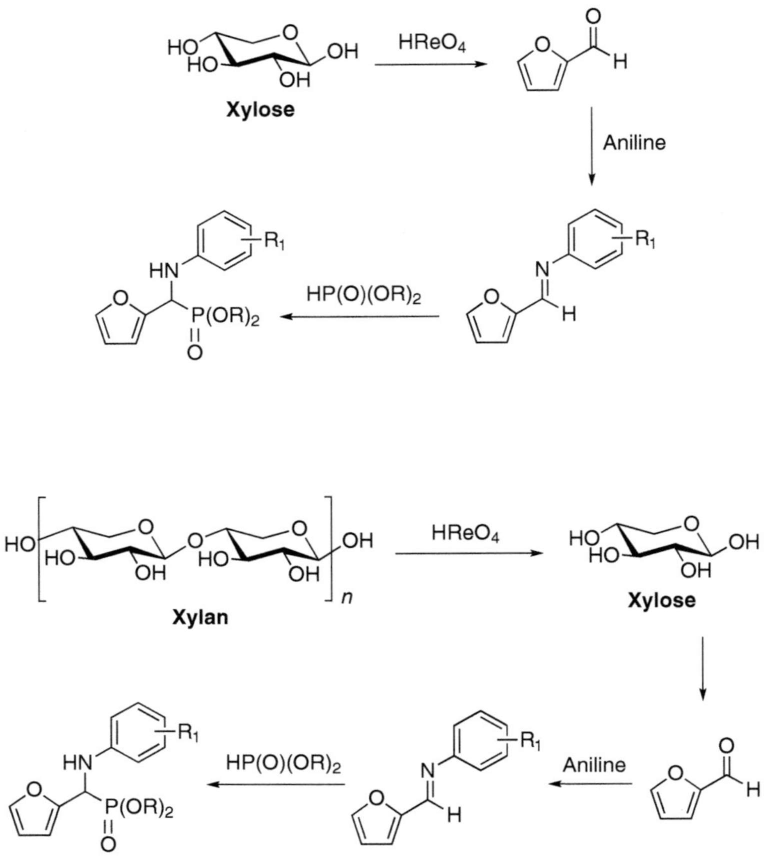

Figure 9.27 One-pot conversion of xylose and xylan into α-aminophosphonates. Source: Based on Isca et al. [59].

anti-inflammatory, antibacterial, antifungal, antiepileptic, antimalarial, and antileishmanial [61–63].

The synthesis of 3,4-dihydropyrimidin-2-ones has been investigated from the biomass derivatives 5-HMF and furfural [64, 65]. For example, Popowycz and coworkers [66] reported the preparation of a large diversity of 3,4-dihydropyrimidin-2-ones from the reaction of 5-HMF with several 1,3-diketones or β-keto esters and ureas catalyzed by $ZnCl_2$ under solvent-free conditions (Figure 9.30). A variety of acetoacetates were tested in this reaction, leading to the production of dihydropyrimidinones with moderate to good yields. The Biginelli reaction was also investigated with *N*-methylurea, thiourea, and heterocyclic urea derivatives. Low yields were obtained from the reactions with *N*-methylurea and thiourea, whereas using 3-amino-1,2,4-triazole, the corresponding product was prepared with 83% yield.

Pyrrole is considered one of the main units in many natural products and synthetic pharmaceuticals and is also a valuable building block for the preparation of biologically active molecules and functional materials [67–69]. The synthesis of a wide variety of

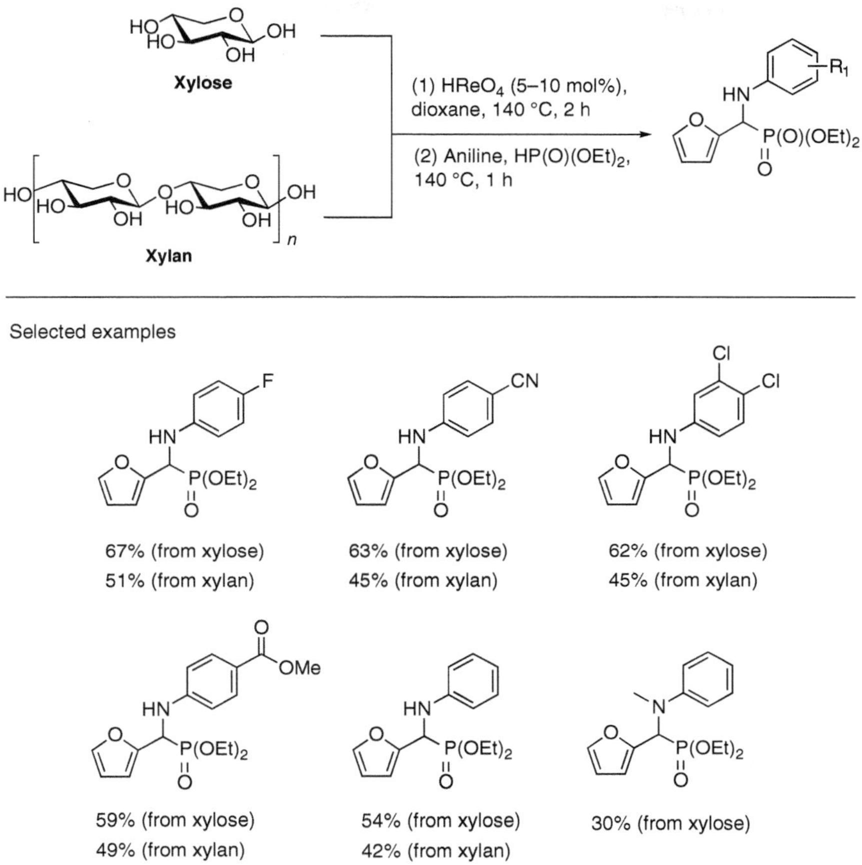

Figure 9.28 One-pot synthesis of α-aminophosphonates from xylose and xylan. Source: Based on Isca et al. [59].

N-substituted 2-hydroxymethyl-5-methylpyrroles from 5-HMF, in two steps, was reported by de Vries and coworkers (Figure 9.31) [70]. This strategy involved, in the first step, the hydrogenation of 5-HMF into 1-hydroxyhexane-2,5-dione (HHD) catalyzed by a ruthenium complex. In the second step, a Paal–Knorr reaction using several alkylamines gave *N*-alkyl-substituted 2-hydroxymethyl-5-methylpyrroles with excellent yields in a few minutes. In contrast, the reaction with the sterically hindered cyclopentylamine needed 48 hours, and the reaction with *tert*-butylamine did not lead to the synthesis of pyrrole. Anilines required prolonged reaction times and gave lower yields. Several benzylamines were also tested, producing excellent yields. This methodology has also been successfully applied to the preparation of pyrrole-based compounds containing donor atoms, such as sulfur, oxygen, and nitrogen that could serve as ligand precursors, in combination with the 2-hydroxymethyl-5-methyl-pyrrole moiety, for the synthesis of metal complexes. The N-substituted 2-hydroxymethyl-5-methylpyrroles prepared have the advantage of containing an alcohol functionality, which can be further modified in order to obtain novel bioactive derivatives.

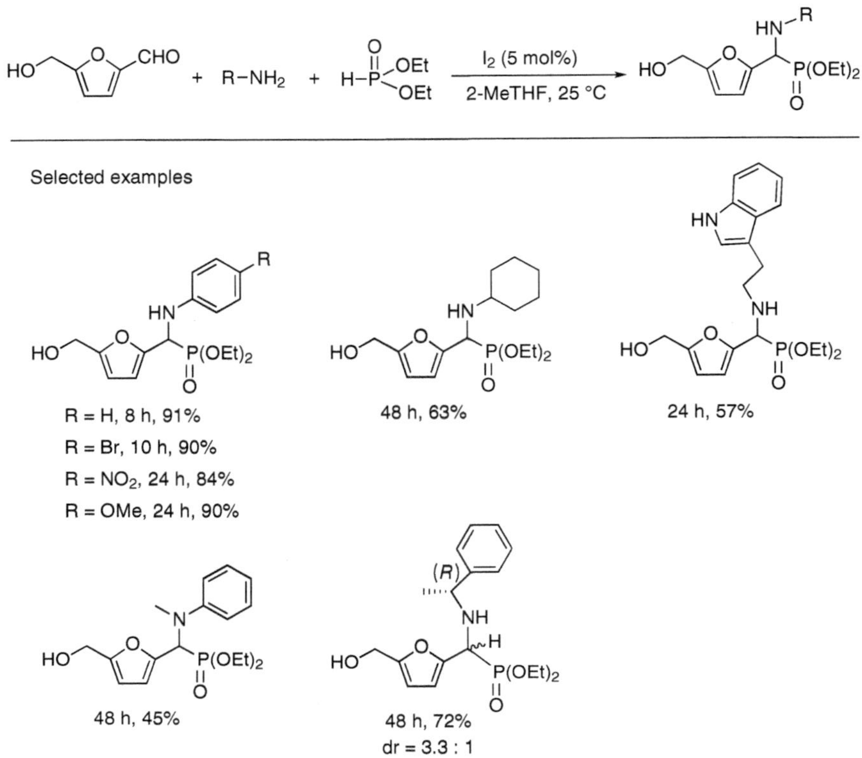

Figure 9.29 Synthesis of α-aminophosphonates from 5-HMF. Source: Based on Fan et al. [60].

Levulinic acid is generally obtained via the acid-catalyzed solvolysis of sugars such as glucose or fructose and has been recognized as one of the top 10 most promising biomass-derived platform molecules by the U.S. Department of Energy. From this versatile feedstock, several valuable nitrogen compounds [71] have been synthesized including pyrrolidones and pyrrolidines, which have many applications in diverse areas including the synthesis of agrochemicals and pharmaceuticals. The sustainable synthesis of pyrrolidones and pyrrolidines from levulinic acid has been investigated using several catalytic systems containing different metals such as Pt, Ru, Ir, In, and Al and reducing agents including H_2, formic acid, and silanes [72–76]. Very recently, Darcel and coworkers [77] reported the selective synthesis of pyrrolidines and pyrrolidinones from levulinic acid using iron complexes as catalysts and phenylsilane as a reducing agent. The reaction of levulinic acid with anilines containing different functional groups catalyzed by [CpFe(CO)$_2$(*IMes*)][I] (*IMes* = 1,3-bis(2,4,6-trimethylphenyl)imidazol-2-ylidene), upon visible light irradiation under neat conditions, led to the preparation of several pyrrolidinones with good yields and chemoselectivity (Figure 9.32).

These authors [77] also demonstrated that using the catalyst [Fe(CO)$_4$(*IMes*)], it is possible to switch the selectivity of the reaction and produce pyrrolidines with good yields (Figure 9.32). Under these reactions conditions, several functional groups, including a boronate ester or a trifluoromethyl, were well tolerated.

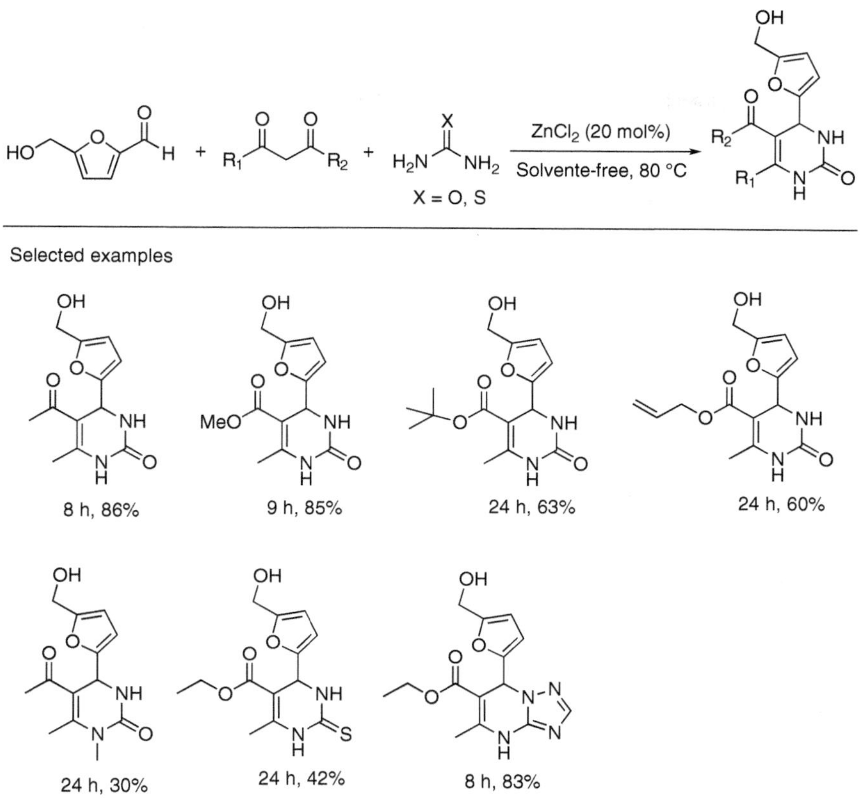

Figure 9.30 Synthesis of 3,4-dihydropyrimidin-2-ones from 5-HMF. Source: Based on Fan et al. [66].

The synthesis of pyrrolidones using glucose as biomass feedstock was also reported by Fischmeister and coworkers [78]. The process, catalyzed by an iridium complex and employing H_2 as a hydrogen source, gives the product in 31% overall yield (Figure 9.33).

Levulinic acid has recently been used as a renewable raw material, along with amines and amino acid-based isocyanides, for the synthesis of several peptide-like heterocyclic molecules containing 2-pyrrolidone (γ-lactam), amide, and ester functional groups, with potential pharmaceutical applications (Figure 9.34) [79]. In this process, levulinic acid is employed as a bifunctional chemical, as aldehyde and carboxylic acid, allowing the conversion of the traditional Ugi four-component condensation reaction into an Ugi three-component condensation reaction. Several aliphatic and aromatic amines such as 4-fluoroaniline and tryptamine and racemic isocyanides, prepared from D,L-tryptophane, D,L-leucine and D,L-phenylalanine, were used in this reaction, leading to the synthesis of pseudopeptides, obtained as a mixture of diastereomers, with good to excellent yields.

The synthesis of amines and their derivatives from cellulosic biomass-derived alcohols has also attracted increasing interest. Among them, glycerol is an excellent renewable feedstock for the production of valuable nitrogen compounds, and nowadays, it is also obtained in large quantities as a by-product of biodiesel production.

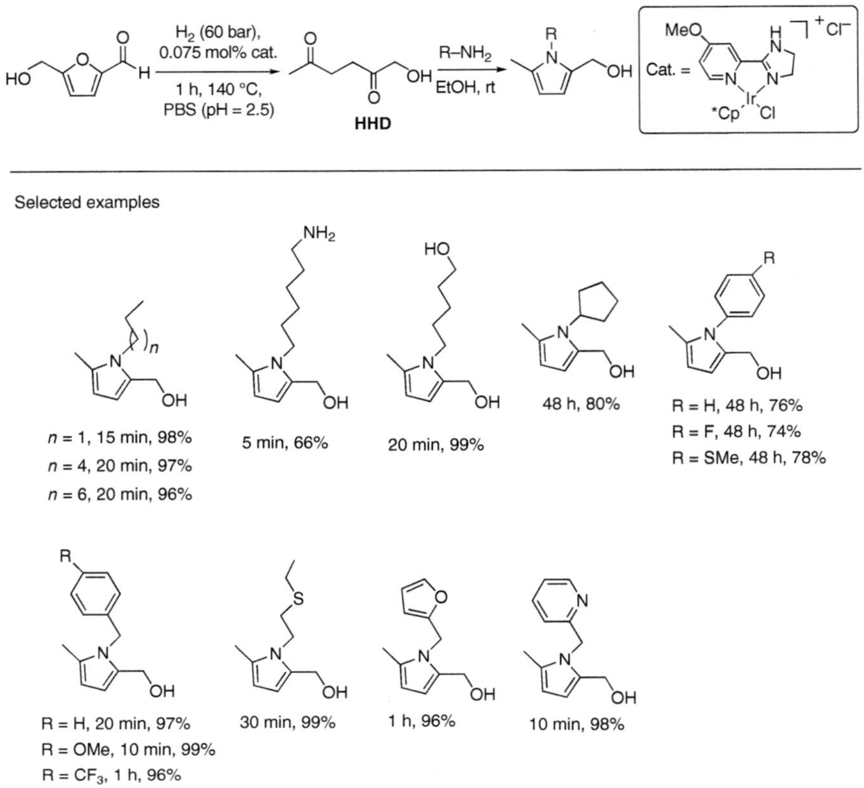

Figure 9.31 Synthesis of 2-hydroxymethyl-5-methylpyrroles from 5-HMF. Source: Based on Wozniak et al. [70].

Shi and coworkers [80] reported the conversion of glycerol into a variety of prochiral aminoketones, *N*-formamides, and *N*-methyl amines using $CuNiAlO_x$ as a catalyst (Figure 9.35). Initially, these authors investigated the synthesis of *N*-acetonyl amines by reaction of glycerol with different amines. Good to excellent yields (78–94%) were obtained in the reactions with morpholine, piperidine, and piperazine derivatives. *N*-Acetonyl amines were also prepared in good yields (70–82%) from common secondary amines, including dipropylamine, dibutylamine, dioctylamine, and dibenzylamine. In contrast, low yields were observed in the reactions with aliphatic primary and aromatic amines.

Several formamides were also prepared from glycerol involving the production of formic acid through oxidative C—C bond cleavage of glycerol, followed by a quick reaction with amine (Figure 9.35) [80]. Good yields of formamides were obtained from the reactions of glycerol with cyclic and linear secondary amines and moderate yields from the reactions with aliphatic primary amines.

As glycerol can be used as a hydrogen source in metal-catalyzed transfer hydrogenation of carbonyl compounds, this group also prepared *N*-methylamines following the generation of *N*-formyl amine (Figure 9.35) [80]. Several *N*-methylamines were obtained in good yields

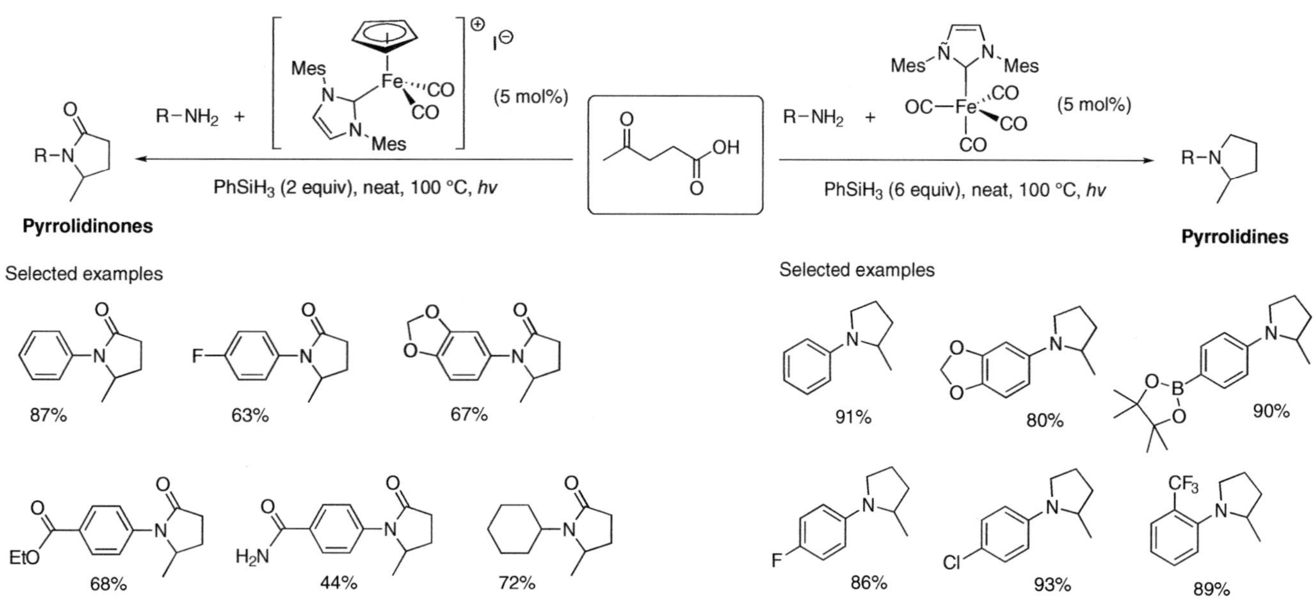

Figure 9.32 Synthesis of pyrrolidinones and pyrrolidines from levulinic acid. Source: Based on Wei et al. [77].

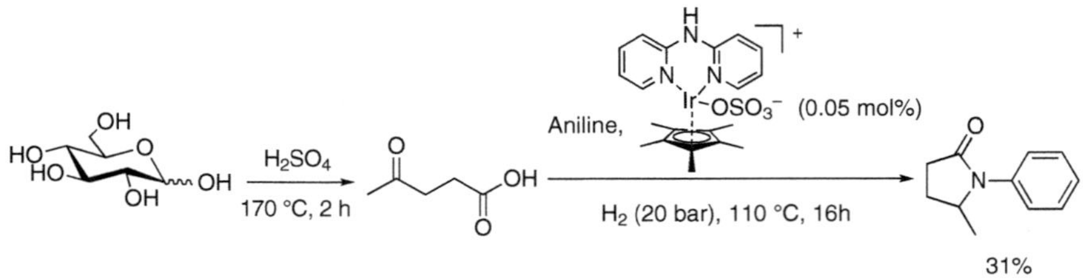

Figure 9.33 Synthesis of a pyrrolidine from glucose catalyzed by an iridium catalyst. Source: Based on Wang et al. [78].

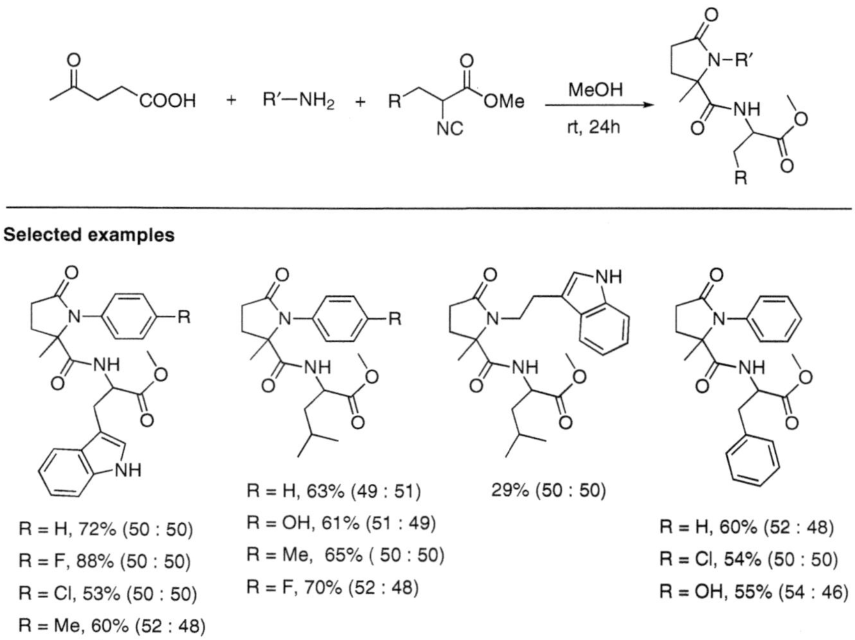

Figure 9.34 Synthesis of pseudopeptides from levulinic acid. Source: Based on Khalesi et al. [79].

from the reaction of glycerol with cyclic and linear amines in a mixed reaction medium (water +1,4-dioxane).

The production of amino acids from biomass resources has also been explored. For example, very recently, Yan and coworkers [81] developed a one-step protocol for the synthesis of alanine, with 43% yield, from glycerol in aqueous ammonia catalyzed by Ru_1Ni_7/MgO (Figure 9.36).

The authors proposed a mechanism involving sequential dehydrogenation, dehydration, and rearrangement of glycerol, leading to the formation of lactic acid. Then, lactic acid reacts with ammonia, producing alanine by a dehydrogenation–imination–hydrogenation pathway (Figure 9.37). This novel method can contribute to the valorization of waste of biodiesel production into value-added amino acids.

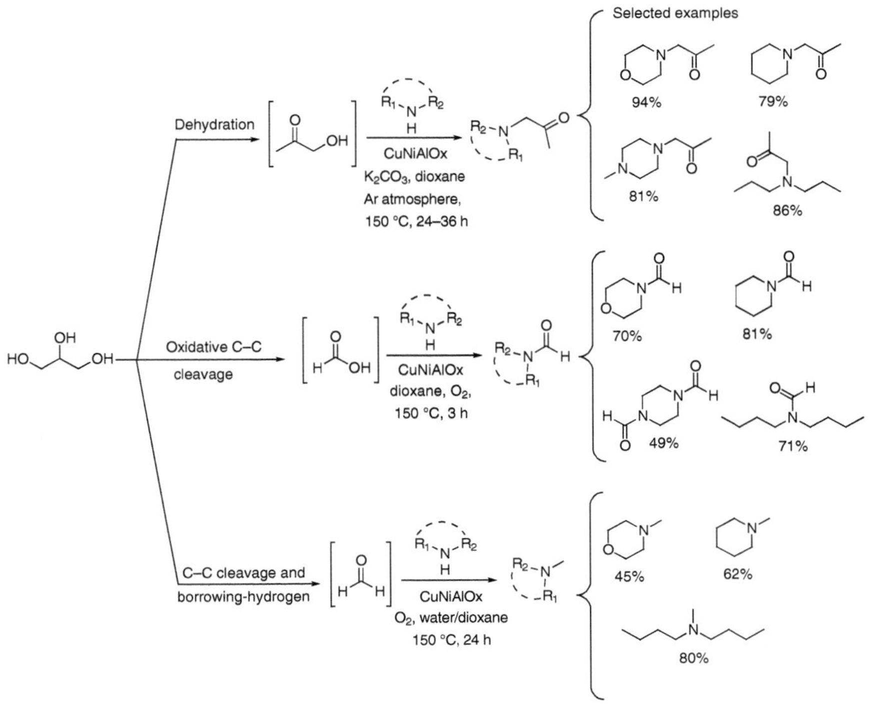

Figure 9.35 Synthesis of aminoketones, *N*-formamides, and *N*-methyl amines from glycerol. Source: Based on Dai et al. [80].

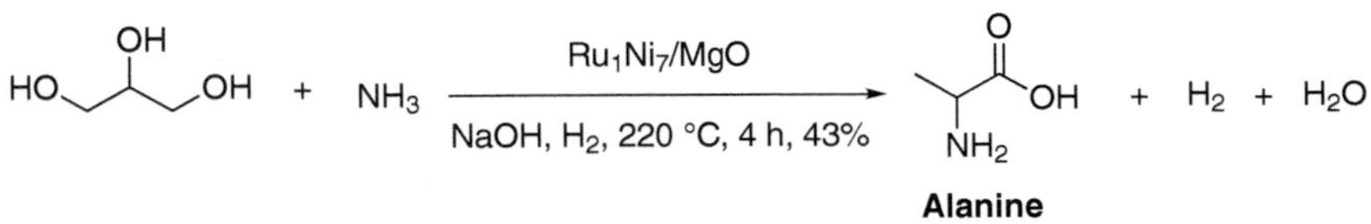

Figure 9.36 Synthesis of alanine from glycerol. Source: Based on Wang et al. [81].

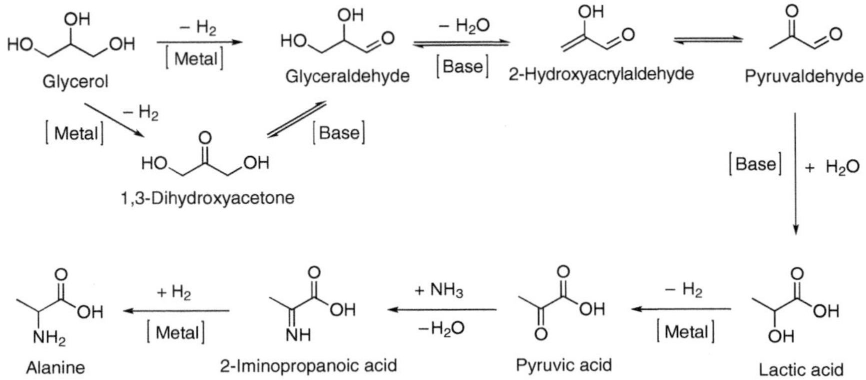

Figure 9.37 Proposed mechanism for the synthesis of alanine from glycerol. Source: Based on Wang et al. [81]..

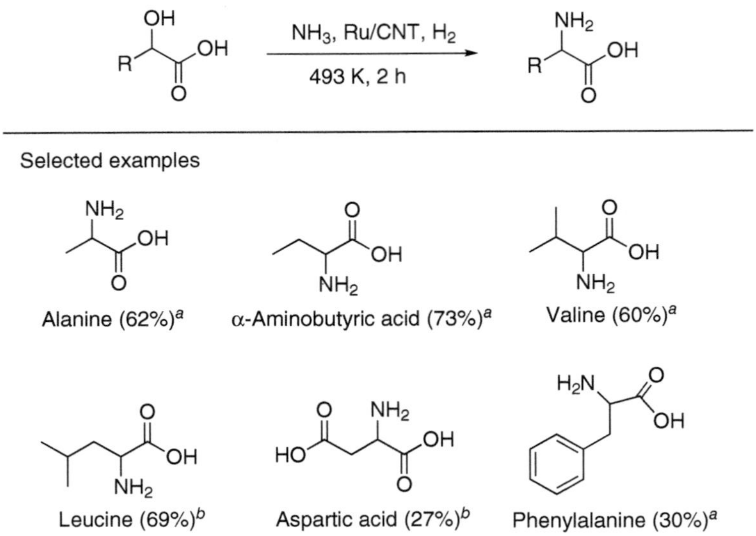

[a] Ru/CNT was modified with 10 wt% Ni.
[b] KOH or NaOH was added.

Figure 9.38 Synthesis of α-amino acids from α-hydroxy acids. Source: Based on Deng et al. [82].

Lignocellulosic biomass-derived α-hydroxy acids are also useful feedstocks for the production of amino acids. These compounds were applied by Wang and coworkers [82] for the synthesis of α-amino acids including alanine, leucine, valine, aspartic acid, and phenylalanine by a dehydrogenation-reductive amination using ammonia catalyzed by ruthenium nanoparticles supported on carbon nanotubes (Ru/CNT) (Figure 9.38). In some reactions, the catalyst was modified with 10 wt% nickel or a small amount of base (NaOH or KOH) was added to promote dehydrogenation and improve amino acid yields. The best results were obtained in the preparation of α-aminobutyric acid, leucine, and valine with 73%, 69%, and 60% yields, respectively. These authors also reported the production of alanine directly from glucose in a two-step process with an overall yield of 43%. In the first step, glucose was converted into lactic acid with 75% yield catalyzed by $Ba(OH)_2$, and then, the resulting solution was acidified, freeze-dried, and directly used in the amination. This work hints to the future production of amino acids from lignocellulosic biomass or agricultural waste.

9.5 Synthesis of Nitrogen Compounds from Lignin Derivatives

Lignin is considered as the largest source of biorenewable aromatic monomers comprising valuable structural moieties present in a wide variety of nitrogen compounds used as building blocks in the synthesis of pharmaceutical and biologically active molecules. Among the lignin derivatives, vanillin has attracted considerable attention as a substrate for the synthesis of nitrogen compounds with great potential to be used as antioxidant compounds [83] and scaffold ligands against Alzheimer's disease [84].

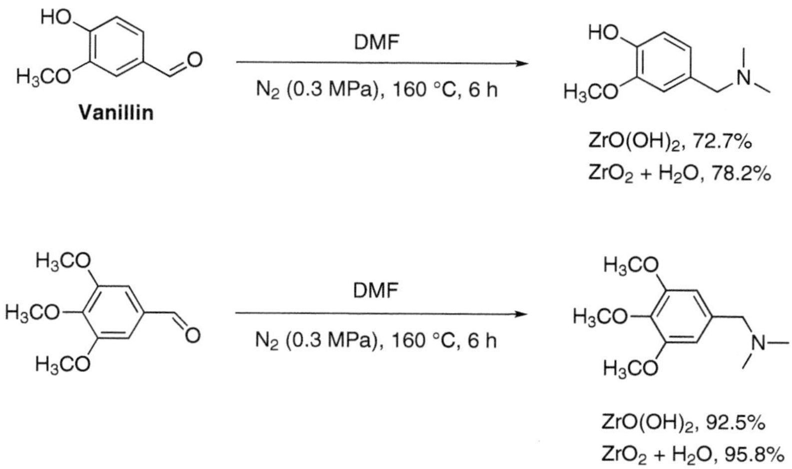

Figure 9.39 Synthesis of aromatic *N,N*-dimethyl tertiary amines from lignin-derived aldehydes. Source: Based on Zhang et al. [85].

Zhang and coworkers reported a novel method for the synthesis of aromatic *N,N*-dimethyl tertiary amines from lignin-derived aromatic aldehydes by a Leuckart reaction using heterogeneous zirconium-based catalysts and *N,N*-dimethyl formamide as a solvent, low-molecular-weight amine source and reductant [85]. From the reaction of vanillin in the presence of $ZrO(OH)_2$ or $ZrO_2 + H_2O$, the corresponding *N,N*-dimethylamine was obtained in 72.7% and 78.2% yields, respectively (Figure 9.39). *N,N*-Dimethyl-1-(3,4,5-trimethoxyphenyl)methanamine was also synthesized from the lignin derivative 3,4,5-trimethoxybenzaldehyde catalyzed by $ZrO(OH)_2$ or $ZrO_2 + H_2O$, with 92.5% and 95.8% yields, respectively (Figure 9.39). The addition of water contributed to the more rapid decomposition of DMF, accelerating the catalytic process.

Very recently, Maes and coworkers [86] developed a synthetic strategy to access 3,4-dialkoxyanilines from 4-propylguaiacol and 4-propylcatechol involving the alkylation of the –OH group, followed by a benzylic oxidation, a Beckmann rearrangement, and finally an amide alcoholysis (Figure 9.40). Propionate esters were also obtained as by-products in this process, which are valuable compounds for the industry.

Using this four-step procedure, several 3,4-dialkoxyanilines were prepared in moderate overall yields, without involving any purification after each individual synthetic step (Figure 9.41). Some of these bio-based 3,4-dialkoxyanilines are important intermediates in the synthesis of anticancer drugs (prazosin, alfuzosin, doxazosin, terazosin, gefitinib, and erlotinib), agrochemicals (diethofencarb), and dyes.

Very recently, Barta and coworkers [87] developed a new synthetic strategy for the preparation of aminoalkylphenol derivatives and benzazepines, important pharmaceutical intermediates, or biologically active molecules, directly from wood in three waste-free steps (Figure 9.42). This process involved, in the step 1, the reductive catalytic fractionalization of lignocelluloses using copper-doped porous metal oxides (Cu20-PMO) and 40 bar H_2, producing aromatic monomers, predominantly dihydroconiferyl alcohol **18** (obtained from pine with selectivity >90%), along with small amounts of 4-ethylguaiacol, 4-propylguaiacol,

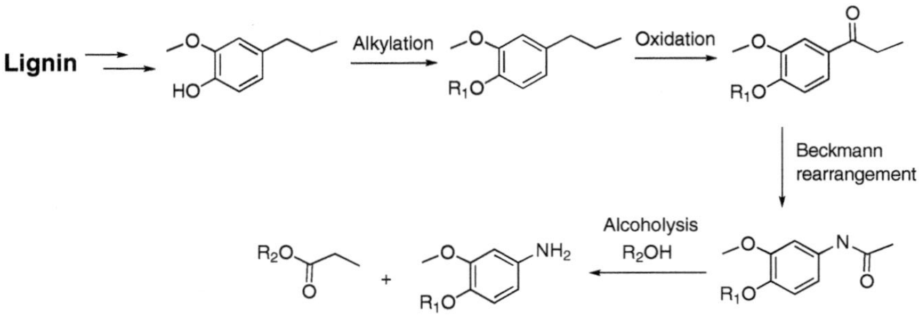

Figure 9.40 Synthesis 3,4-dialkoxyanilines from 4-propylguaiacol. Source: Based on Blondiaux et al. [86].

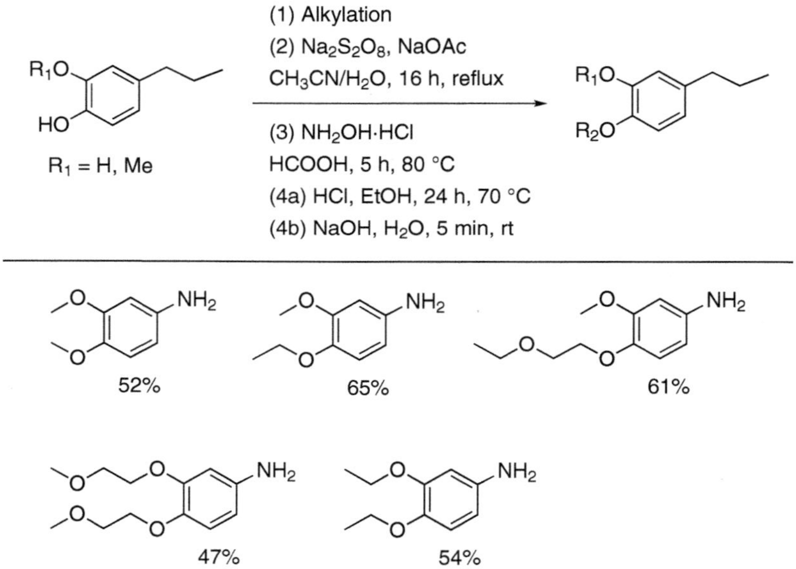

Figure 9.41 Synthesis of 3,4-dialkoxyanilines from 4-propylguaiacol and 4-propylcatechol. Source: Based on Blondiaux et al. [86].

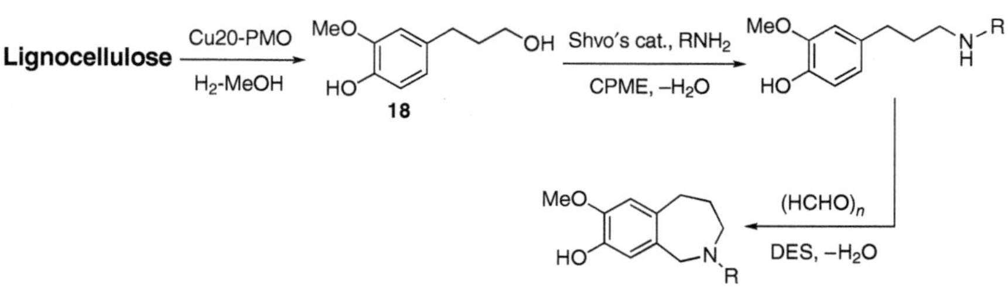

Figure 9.42 Synthesis of aminoalkylphenol derivatives and benzazepines from lignocellulose. Source: Based on Elangovan et al. [87].

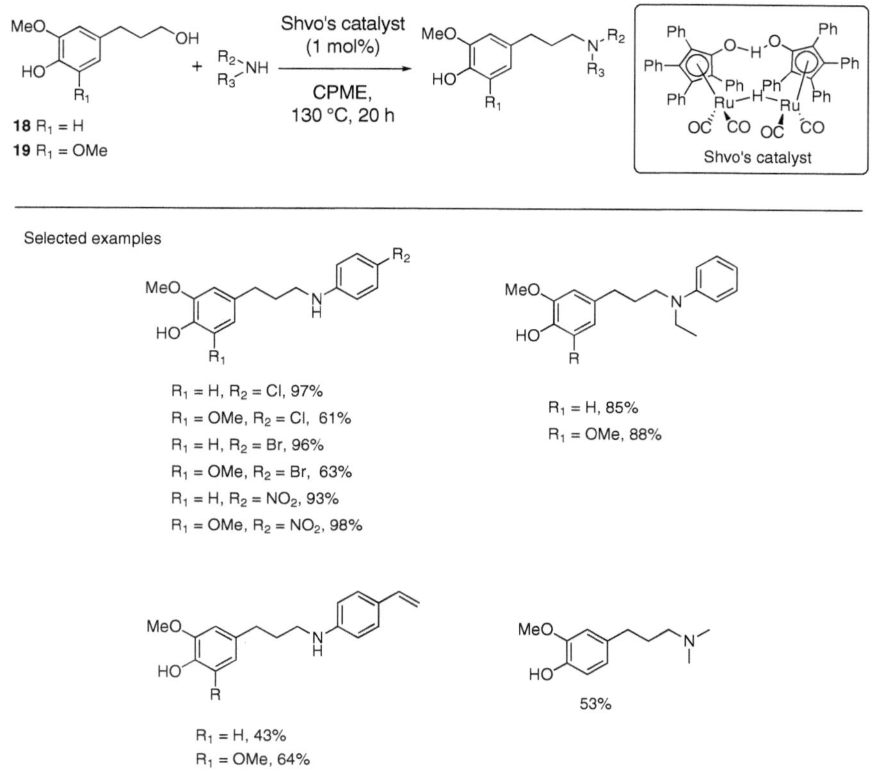

Figure 9.43 Catalytic amination of dihydroconiferyl and dihydrosinapyl alcohols. Source: Based on Elangovan et al. [87].

and residual amounts of sugars and lignin oligomers. After extraction with ethyl acetate, in the step 2, this organic phase containing the dihydroconiferyl alcohol and other components was subjected to the amination reaction in the presence of Shvo's catalyst using cyclopentyl methyl ether (CPME) as a solvent. Finally, this methodology allowed the preparation of several benzazepines by a Pictet–Spengler cyclization of the secondary amines in DESs. This sustainable catalytic process used benign solvents and produced no by-products except water.

These authors also extended the amination method to the production of a variety of lignin-based aminoalkyl- and aryl-guaiacols by reacting dihydroconiferyl alcohol **18** and dihydrosinapyl alcohol **19** with different (hetero)aromatic, aliphatic primary, and secondary amines using Shvo's catalyst (Figure 9.43) [87]. Good to excellent yields were obtained in the reactions with anilines containing both electron-withdrawing and electron-donating groups and also with secondary amines.

A library of benzazepines was also prepared in good to excellent yields by Pictet–Spengler cyclization of lignin-based aminoalkyl- and aryl-guaiacols prepared before (Figure 9.43), using DESs comprising choline chloride/oxalic acid (ChCl/OA) (Figure 9.44) [87]. Some of these benzazepines exhibited interesting antibacterial or anticancer activities.

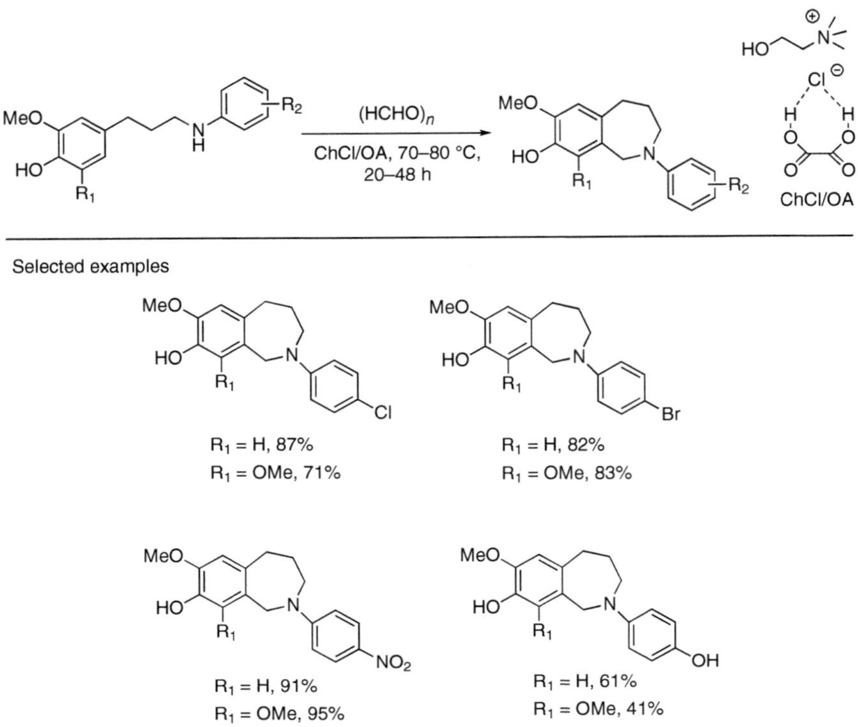

Figure 9.44 Synthesis of lignin-derived tetrahydro-2-benzazepines in deep eutectic solvent. Source: Based on Elangovan et al. [87].

9.6 Synthesis of Nitrogen Compounds from Triglycerides and Fatty Alcohols

Triglycerides are also a valuable biomass feedstock for the production of fatty nitrogen compounds [88–90]. For example, they were used as an alkylating agent for the *N*-mono-alkylation of primary and secondary amines by Beller and coworkers [91]. This method has the advantage of using a benign alkylating agent, in contrast to the traditional procedures that generally employ toxic reagents. The reactions were performed with the catalytic system [Ru(acac)$_3$]/Triphos (Triphos = 1,1,1-tris(diphenylphosphinomethyl)-ethane) in the presence of $HNTf_2$ and 60 bar of H_2 at 130 °C. Under these reaction conditions, a variety of aromatic and aliphatic amines were efficiently alkylated (Figure 9.45). Good yields were obtained in the alkylation of amines bearing both electron-donating and electron-withdrawing substituents. The reactions with electron-rich primary amines were more efficient and required lower amounts of triglycerides to avoid over alkylation. This method was also successfully applied to the N-alkylation of secondary amines using an excess of triglycerides (3–4 equiv). Furthermore, both benzylic and aliphatic amines, as well as amino alcohols, were also alkylated.

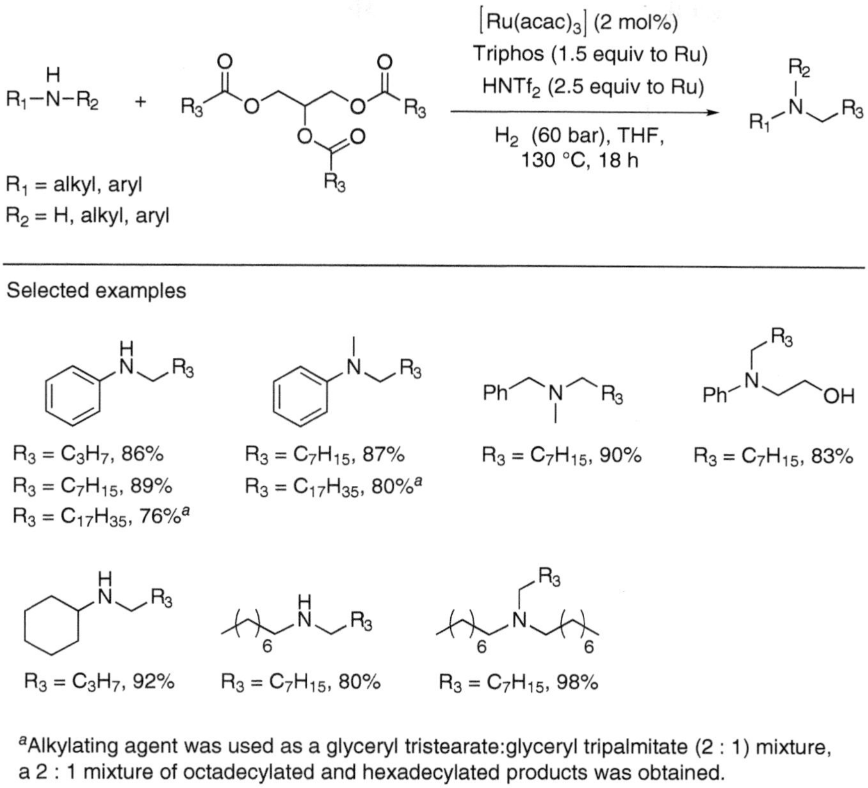

Figure 9.45 N-mono-alkylation of aromatic and aliphatic amines with triglycerides and H_2. Source: Based on Adam et al. [91].

This reaction was also efficiently applied to the alkylation of aniline using sunflower oil as an alkylating agent, producing a mixture of *N*-octadecylaniline and *N*-hexadecylaniline in 95% isolated yield (9.5 : 0.5 ratio of octadecylated: hexadecylated products).

More recently, Shimizu and coworkers [92] developed a one-pot methodology for the synthesis of fatty amines by reductive amination of triglycerides using aqueous NH_3 under an atmosphere of H_2 catalyzed by ZrO_2-supported Pt metal nanoparticles at 220 °C for 36 hours (Figure 9.46). A wide variety of triglycerides with different chain lengths (C4–C18) was tested under these reaction conditions, producing the corresponding primary and secondary amines in moderate to good yields (48–59%). Kinetic studies demonstrated that triglycerides were initially converted into the corresponding amides by direct amination, followed by hydrogenation, producing primary amines, which can also undergo a self-coupling reaction, leading to the formation of secondary amines.

These authors also investigated the synthesis of fatty amides and nitriles from triglycerides [92]. After testing different catalysts using trilaurin as a model substrate, the zeolite Hβ-75 was identified as the best catalyst for these reactions. At 180 °C, various triglycerides were transformed into the corresponding fatty amides, with high yields (82–95%),

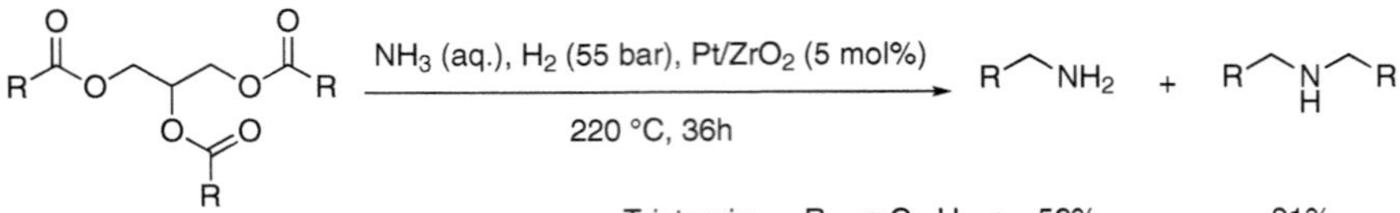

Selected examples

Tristearin,	$R = n\text{-}C_{17}H_{35}$:	56%	21%
Tripalmitin,	$R = n\text{-}C_{15}H_{21}$:	54%	19%
Trimyristin,	$R = n\text{-}C_{13}H_{27}$:	53%	18%
Trilaurin,	$R = n\text{-}C_{11}H_{23}$:	59%	28%
Tricaprin,	$R = n\text{-}C_{9}H_{19}$:	51%	30%

Figure 9.46 Synthesis of fatty amines from triglycerides. Source: Based on Jamil et al. [92].

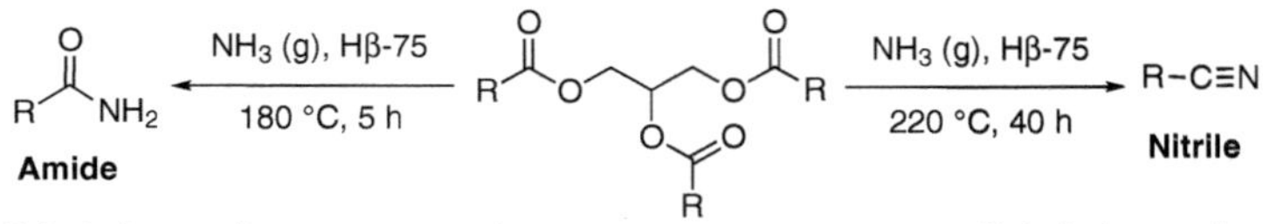

Selected examples (Amide)

Tristearin,	$R = n\text{-}C_{17}H_{35}$:	91%
Tripalmitin,	$R = n\text{-}C_{15}H_{21}$:	89%
Trimyristin,	$R = n\text{-}C_{13}H_{27}$:	90%
Trilaurin,	$R = n\text{-}C_{11}H_{23}$:	95%
Tricaprin,	$R = n\text{-}C_{9}H_{19}$:	86%

Selected examples (Nitrile)

Tristearin,	$R = n\text{-}C_{17}H_{35}$:	98%
Tripalmitin,	$R = n\text{-}C_{15}H_{21}$:	94%
Trimyristin,	$R = n\text{-}C_{13}H_{27}$:	91%
Trilaurin,	$R = n\text{-}C_{11}H_{23}$:	96%
Tricaprin,	$R = n\text{-}C_{9}H_{19}$:	89%

Figure 9.47 Synthesis of fatty amides and nitriles from triglycerides. Source: Based on Jamil et al. [92]..

by reaction with gaseous NH_3 after 5 hours (Figure 9.47). Long-chain triglycerides (C13, C15, and C17) required a slightly longer reaction time (6 hours).

The reaction of triglycerides with gaseous NH_3 catalyzed by zeolite Hβ-75 is affected by the temperature and the reaction time. At 220 °C and after prolonged reaction time (40 hours), several C4–C18 triglycerides were converted into the corresponding fatty nitriles with good yields (Figure 9.47). This method has the advantage of allowing the direct conversion of triglycerides to fatty nitriles, in contrast to the traditional process involving methyl esterification, followed by amidation with ammonia and dehydration of amides into the corresponding fatty nitriles.

Long-chain diols derived from plant oils have also been used as a biomass raw material for the synthesis of long-chain diamines by catalytic direct amination. Mecking and coworkers [51] reported the production of these amines using NH_3 and the catalyst [RuHCl(acridine-^{i}Pr-PNP)(CO)] at 150 °C for 65 hours (Figure 9.48). In all reactions, a large excess of ammonia (120 equiv) was used to obtain the primary amines and avoid the formation of secondary and cyclic amines.

These authors also reported the polycondensation of these diamines with long-chain diacids, obtaining higher molecular weight polyamides.

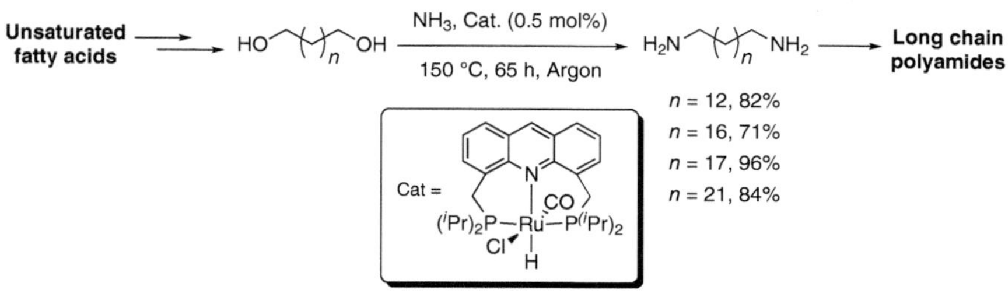

Figure 9.48 Synthesis of long-chain diamines from long-chain diols by direct amination. Source: Based on Pingen et al. [51].

9.7 Conclusion

This chapter provides selected examples of the most relevant methodologies recently developed for the synthesis of nitrogen compounds from biomass derivatives. These examples clearly demonstrated the importance of these new reactions for the production of a wide variety of valuable nitrogen compounds from biomass resources, contributing to the reduction of fossil resources.

Despite great achievements over the past few years, there are still many challenges in this area such as the development of new and more efficient processes for the conversion of chitin into value-added nitrogen compounds. These processes will allow the valorization of million tons of crustacean shell waste generated by the global fishing industry and will bring significant economic and environmental benefits.

Certainly, in the near future, other economic, efficient, and sustainable methodologies for the synthesis of nitrogen compounds from biomass resources using green reagents and mild conditions will be developed, contributing to the development of more sustainable chemical and pharmaceutical industries.

References

1 Froidevaux, V., Negrell, C., Caillol, S. et al. (2016). *Chemical Reviews* 116: 14181–14224.
2 Pelckmans, M., Renders, T., Vyver, S.V., and Sels, B.F. (2017). *Green Chemistry* 19: 5303–5331.
3 Simakova, I.L., Simakov, A.V., and Murzin, D.Y. (2018). *Catalysts* 8: 365.
4 Hülsey, M.J., Yang, H., and Yan, N. (2018). *ACS Sustainable Chemistry & Engineering* 6: 5694–5707.
5 Wang, Y., Furukawa, S., Fu, X., and Yan, N. (2020). *ACS Catalysis* 10: 311–335.
6 Gao, X., Chen, X., Zhang, J. et al. (2016). *ACS Sustainable Chemistry & Engineering* 4: 3912–3920.

7 Pierson, Y., Chen, X., Bobbink, F.D. et al. (2014). *ACS Sustainable Chemistry & Engineering* 2: 2081–2089.

8 Osada, M., Kikuta, K., Yoshida, K. et al. (2013). *Green Chemistry* 15: 2960–2966.

9 Kobayashi, H., Techikawara, K., and Fukuoka, A. (2017). *Green Chemistry* 19: 3350–3356.

10 Techikawara, K., Kobayashi, H., and Fukuoka, A. (2018). *ACS Sustainable Chemistry & Engineering* 6: 12411–12418.

11 Sagawa, T., Kobayashi, H., Murata, C. et al. (2019). *ACS Sustainable Chemistry & Engineering* 7: 14883–14888.

12 Drover, M.W., Omari, K.W., Murphy, J.N., and Kerton, F.M. (2012). *RSC Advances* 2: 4642–4644.

13 Chen, X., Liu, Y., Kerton, F.M., and Yan, N. (2015). *RSC Advances* 5: 20073–20080.

14 Omari, K.W., Dodot, L., and Kerton, F.M. (2012). *ChemSusChem* 5: 1767–1772.

15 Chen, X., Chew, S.L., Kerton, F.M., and Yan, N. (2014). *Green Chemistry* 16: 2204–2212.

16 Pham, T.T., Lindsay, A.C., Chen, X. et al. (2019). *Sustainable Chemistry and Pharmacy* 13: 100143.

17 Pham, T.T., Chen, X., Yan, N., and Sperry, J. (2018). *Monatshefte fuer Chemie* 149: 857–861.

18 Pham, T.T., Lindsay, A.C., Kim, S.-W. et al. (2019). *ChemistrySelect* 4: 10097–10099.

19 Pham, T.T., Gözaydın, G., Söhnel, T. et al. (2019). *European Journal of Organic Chemistry*: 1355–1360.

20 Sadiq, A.D., Chen, X., Yan, N., and Sperry, J. (2018). *ChemSusChem* 1: 532–535.

21 Schneider, K., Keller, S., Wolter, F.E. et al. (2008). *Angewandte Chemie, International Edition* 47: 3258–3261.

22 Fiedler, H.-P., Bruntner, C., Riedlinger, J. et al. (2008). *The Journal of Antibiotics* 61: 158–163.

23 Lingyu Jia, L., Zhang, Z., Qiao, Y. et al. (2017). *Industrial and Engineering Chemistry Research* 56: 2925–2934.

24 Wu, M., Ma, H., Ma, Z. et al. (2018). *ACS Sustainable Chemistry & Engineering* 6: 9434–9441.

25 Kirchhecker, S., Tröger-Müller, S., Bake, S. et al. (2015). *Green Chemistry* 17: 4151–4156.

26 Kirchhecker, S., Antonietti, M., and Esposito, D. (2014). *Green Chemistry* 16: 3705–3709.

27 De Schouwer, F., Claes, L., Claes, N. et al. (2015). *Green Chemistry* 17: 2263–2270.

28 Verduyckt, J., Coeck, R., and De Vos, D.E. (2017). *ACS Sustainable Chemistry & Engineering* 5: 3290–3295.

29 Claes, L., Janssen, M., and De Vos, D.E. (2019). *ChemCatChem* 11: 4297–4306.

30 Pelckmans, M., Vermandel, W., Waes, F.V. et al. (2017). *Angewandte Chemie, International Edition* 56: 14540–14544.

31 Faveere, W., Mihaylov, T., Pelckmans, M. et al. (2020). *ACS Catalysis* 10: 391–404.

32 Jia, L., Makha, M., Du, C.-X. et al. (2019). *Green Chemistry* 21: 3127–3132.

33 Song, L., Zheng, M., Pang, J. et al. (2017). *Green Chemistry* 19: 3515–3519.

34 Chen, X., Yang, H., Hülsey, M.J., and Yan, N. (2017). *ACS Sustainable Chemistry & Engineering* 5: 11096–11104.

35 Banerjee, R., HKS, K., and Banerjee, M. (2012). *International Journal of Review Life Science* 2: 7–16.

36 Caetano, J.A.T. and Fernandes, A.C. (2018). *Green Chemistry* 20: 2494–2498.
37 Rathod, P.V., Deonikar, V.G., Puguan, J.M.C., and Kim, H. (2020). *Fuel* 264: 116822.
38 Xu, Z., Yan, P., Xu, W. et al. (2014). *RSC Advances* 4: 59083–59087.
39 Xu, Z., Yan, P., Liu, L. et al. (2016). *ChemSusChem* 9: 1255–1258.
40 Xiao, J., Jin, Q., Yang, J. et al. (2019). *Asian Journal of Organic Chemistry* 8: 328–334.
41 García-Ortiz, A., Vidal, J.D., Climent, M.J. et al. (2019). *ACS Sustainable Chemistry & Engineering* 7: 6243–6250.
42 Zhu, M.-M., Tao, L., Zhang, Q. et al. (2017). *Green Chemistry* 19: 3880–3887.
43 Chieffi, G., Braun, M., and Esposito, D. (2015). *ChemSusChem* 8: 3590–3594.
44 Yuan, H., Li, J.-P., Su, F. et al. (2019). *ACS Omega* 4: 2510–2516.
45 Chen, W., Sun, Y., Du, J. et al. (2018). *Journal of Chemical Technology and Biotechnology* 93: 3028–3034.
46 Liang, G., Wang, A., Li, L. et al. (2017). *Angewandte Chemie, International Edition* 56: 3050–3054.
47 Zhou, K., Chen, B., Zhou, X. et al. (2019). *ChemCatChem* 11: 5562–5569.
48 Zhou, K., Liu, H., Shu, H. et al. (2019). *ChemCatChem* 11: 2649–2656.
49 Komanoya, T., Kinemura, T., Kita, Y. et al. (2017). *Journal of the American Chemical Society* 139: 11493–11499.
50 Le, N.-T., Byun, A., Han, Y. et al. (2015). *Green and Sustainable Chemistry* 5: 115–127.
51 Pingen, D., Schwaderer, J.B., Walter, J. et al. (2018). *ChemCatChem* 10: 3027–3033.
52 Yuan, H., Kusema, B.T., Yan, Z. et al. (2019). *RSC Advances* 9: 38877–38881.
53 Xu, Y., Jia, X., Ma, J. et al. (2018). *Green Chemistry* 20: 2697–2701.
54 Alonso, E., Alonso, E., Solis, A., and del Pozo, C. (2000). *Synlett*: 698–700.
55 Gokha, A.A.E., Ghanim, I.M.S., Megeed, A.E.S.A. et al. (2016). *International Journal of Pharmaceutical Sciences and Research* 7: 181–189.
56 Gu, L. and Jinb, C. (2012). *Organic & Biomolecular Chemistry* 10: 7098–7102.
57 Rezaei, Z., Khabnadideh, S., Zomorodian, K. et al. (2011). *International Journal of Medicinal Chemistry*: 678101.
58 Rogacz, D., Lewkowski, J., Malinowski, Z. et al. (2018). *Molecules* 23: 3173.
59 Isca, V.M.S. and Fernandes, A.C. (2018). *Green Chemistry* 20: 3242–3245.
60 Fan, W., Queneau, Y., and Popowycz, F. (2018). *RSC Advances* 8: 31496–31501.
61 Crespo, A., Maatougui, A.E., Biagini, P. et al. (2013). *ACS Medicinal Chemistry Letters* 4: 1031–1036.
62 Fátima, A., Braga, T.C., Neto, L.S. et al. (2015). *Journal of Advanced Research* 6: 363–373.
63 Kaur, R., Chaudhary, S., Kumar, K. et al. (2017). *European Journal of Medicinal Chemistry* 132: 108–134.
64 Portilla-Zuñiga, O.M., Sathicq, A.G., Martínez, J.J. et al. (2018). *Sustainable Chemistry and Pharmacy* 10: 50–55.
65 Martínez, J.J., Nope, E., Rojas, H. et al. (2014). *Catalysis Letters* 144: 1322–1331.
66 Fan, W., Queneau, Y., and Popowycz, F. (2018). *Green Chemistry* 20: 485–492.
67 Arikawa, Y., Nishida, H., Kurasawa, O. et al. (2012). *Journal of Medicinal Chemistry* 55: 4446–4456.
68 Luňák, S. Jr., Vala, M., Vyňuchal, J. et al. (2011). *Dyes and Pigments* 91: 269–278.
69 Takase, M., Yoshida, N., Narita, T. et al. (2012). *RSC Advances* 2: 3221–3224.

70 Wozniak, B., Li, Y., Hinze, S. et al. (2018). *European Journal of Organic Chemistry*: 2009–2012.

71 Xue, Z., Yu, D., Zhaob, X., and Mu, T. (2019). *Green Chemistry* 21: 5449–5468.

72 Wei, Y., Wang, C., Jiang, X. et al. (2013). *Chemical Communications* 49: 5408–5410.

73 Vidal, J.D., Climent, M.J., Concepcion, P. et al. (2015). *ACS Catalysis* 5: 5812–5821.

74 Ogiwara, Y., Uchiyama, T., and Sakai, N. (2016). *Angewandte Chemie, International Edition* 55: 1864–1867.

75 Wu, C., Luo, X., Zhang, H. et al. (2017). *Green Chemistry* 19: 3525–3529.

76 Xu, Z., Yan, P., Jiang, H. et al. (2017). *Chinese Journal of Chemistry* 35: 581–585.

77 Wei, D., Netkaew, C., and Darcel, C. (2019). *Advanced Synthesis and Catalysis* 361: 1781–1786.

78 Wang, S., Huang, H., Bruneau, C., and Fischmeister, C. (2017). *ChemSusChem* 10: 4150–4154.

79 Khalesi, M., Halimehjani, A.Z., and Martens, J. (2019). *Beilstein Journal of Organic Chemistry* 15: 852–857.

80 Dai, X., Rabeah, J., Yuan, H. et al. (2016). *ChemSusChem* 9: 3133–3138.

81 Wang, Y., Furukawa, S., Song, S. et al. (2020). *Angewandte Chemie, International Edition* 59: 2289–2293.

82 Deng, W., Wang, Y., Zhang, S. et al. (2018). *PNAS* 115: 5093–5098.

83 Scipioni, M., Kay, G., Megson, I., and Lin, P.K.T. (2018). *European Journal of Medicinal Chemistry* 143: 745–754.

84 Scipioni, M., Kay, G., Megson, I.L., and Lin, P.K.T. (2019). *Medicinal Chemistry Communications* 10: 764–777.

85 Zhang, H., Tong, X., Liu, Z. et al. (2018). *Catalysis Science & Technology* 8: 5396–5400.

86 Blondiaux, E., Bomon, J., Smoleń, M. et al. (2019). *ACS Sustainable Chemistry & Engineering* 7: 6906–6916.

87 Elangovan, S., Afanasenko, A., Haupenthal, J. et al. (2019). *ACS Central Science* 5: 1707–1716.

88 Araújo, P.H.F., Barata, P.H.S., Araújo, I.F. et al. (2018). *Catalysis Letters* 148: 843–851.

89 Shirazi, Y., Tafazolian, H., Viamajala, S. et al. (2017). *ACS Omega* 2: 9013–9020.

90 Kumar, D., Park, C.H., and Kim, C.S. (2019). *Catalysts* 9: 774.

91 Adam, R., Cabrero-Antonino, J.R., Junge, K. et al. (2016). *Angewandte Chemie, International Edition* 55: 11049–11053.

92 Jamil, A.R., Siddiki, S.M.A.H., Touchy, A.S. et al. (2019). *ChemSusChem* 12: 3115–3125.

10

Recent Advances in the Synthesis of Arylamines in the Light of Application in Pharmaceutical and Chemical Industry

Dino Berthold[1], *Alexander M. Haydl*[2], *Joyce C. Leung*[3], *Ulrich Scholz*[4], *Qing Xiao*[5] *and Zhibin Zhu*[5]

[1] *Albert-Ludwigs-Universität Freiburg, Institut für Organische Chemie, Albertstraße 21, 79104 Freiburg, Germany*
[2] *Department for Chemical Intermediates, Amine Chemistry, BASF SE, Carl-Bosch Str. 38, 67056 Ludwigshafen, Germany*
[3] *Boehringer Ingelheim Pharmaceutical, Inc., Chemical Development, 900 Ridgebury Road, Ridgefield, CT 06877, USA*
[4] *Boehringer Ingelheim Pharma GmbH & Co.KG, Chemical Process Development, Binger Straße 173, 55216 Ingelheim am Rhein, Germany*
[5] *Boehringer Ingelheim Shanghai Pharmaceutical Co., Ltd., Center of Chemistry, 1010 Longdong Avenue, 201203 Shanghai, China*

10.1 Modern Approaches to Transition-Metal-Catalyzed C–N Coupling in Industry

10.1.1 Introduction

Amines bearing C(sp^2)—N bonds are fundamental products and building blocks in chemistry, and for this reason, their formation has become a fundamental target in organic synthesis. The ubiquitous nature of arylamines and N-heterocycles in natural products, pharmaceuticals, agrochemicals, and organic molecules relevant for materials science has encouraged countless chemists for the development of general methodologies, which today find applications in both academic research and industrial processes [1].

Historically, the synthesis of aromatic amines (and N-heterocycles) mainly relied on two different strategies: nitration followed by a reduction step and nucleophilic aromatic substitution of activated substrates (Scheme 10.1).

Although being very useful, certain disadvantages are concomitant to these two transformations, such as limited substrate scope or functional group tolerance. In particular, nitration and subsequent reduction is an economical and well-established technology but suffers from poor step economy in the overall synthesis of aromatic amines. Nucleophilic aromatic substitution requires strong activating electron-withdrawing groups on the aromatic substrates in order to achieve sufficient activity, and therefore, a call for new synthetic methodologies was drawn in the recent past.

10.1.2 Transition-Metal-Catalyzed C–N-Bond Formation

In this context, the synthesis of substituted aromatic amines by transition metal catalysis directly from amines and aryl (pseudo)halides is of great interest. Therefore, immense

Methodologies in Amine Synthesis: Challenges and Applications, First Edition.
Edited by Alfredo Ricci and Luca Bernardi.

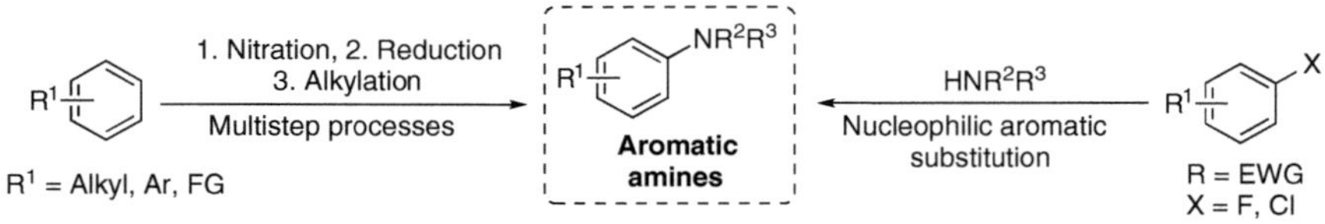

Scheme 10.1 Historic strategies for the synthesis of aromatic amines.

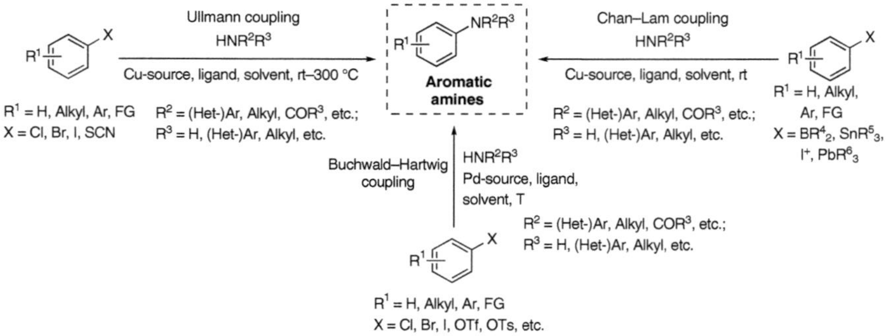

Scheme 10.2 Transition-metal-based methodologies for the preparation of aromatic amines.

progress has been made since the first pioneering work by Ullmann [2] and Goldberg [3] at the beginning of the twentieth century. Besides the development of the so-called Ullmann-type couplings, new methodologies for C—N bond formation have been established in the context of the related Chan–Lam coupling [4] and the palladium-catalyzed synthesis of aromatic amines, now commonly known as the Buchwald–Hartwig amination (Scheme 10.2) [5].

This section presents an overview of these three approaches for transition-metal-catalyzed C—N bond formation, with particular emphasis on their application in industry. The presentation of different methodologies is organized primarily according to their chronological development. Consequently, only small information will be given concerning the underlying mechanisms and the development of certain reaction conditions because several reviews and book chapters have already covered these aspects [6].

10.1.2.1 Ullmann-Type Amination

Ullmann and Goldberg found in their seminal work starting in 1903 that the aromatic substitution of aryl halides with phenols and amines was significantly improved in terms of yield and reactivity in the presence of copper powder (Scheme 10.3). Ever since these early findings, this reaction generally suffers from harsh conditions, such as high reaction temperatures (typically >200 °C) and high copper catalyst loadings. An increased understanding of the mechanism of different transition-metal-catalyzed coupling reactions has led to a wide application of this methodology in organic synthesis for the preparation of potential drugs, natural products, and other useful compounds, not only in academic research but also in industrial-scale synthesis.

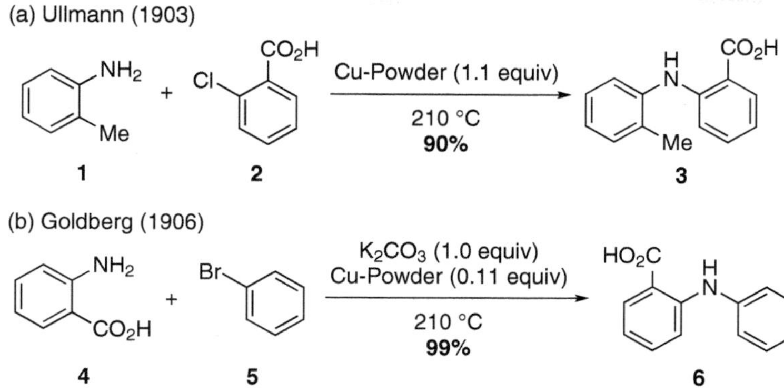

Scheme 10.3 (a) Ullmann's (1903) and (b) Goldberg's (1906) initial reports of a copper-catalyzed C–N coupling of aryl halides and anilines.

Mechanistic Studies

Since these early findings by Ullmann and Goldberg, many different copper sources with regard to the oxidation state were utilized to catalyze Ullmann couplings. Because pure copper as well as many different copper salts proved to work well for the arylation of different nucleophiles, a common understanding implied only one single catalytic active copper species being formed during the reaction. Much work has been carried out to identify reaction intermediates and to provide a general understanding for the reaction mechanism. Unfortunately, to date, the exact pathway remains rather unclear. However, a possible and anticipated mechanism involves the copper catalyst in a redox path with a change over the oxidation states from **I** to **III**, as shown in Scheme 10.4 [7].

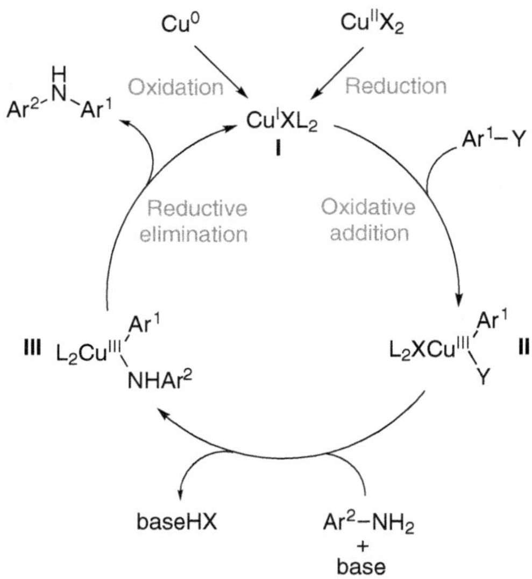

Scheme 10.4 Anticipated mechanism of the copper-catalyzed C–N coupling reaction referred as the Ullmann coupling.

It has been shown that under the reaction conditions, especially in the presence of a nucleophilic aniline derivative, both reaction pathways for the formation of the Cu^{I}-species **I**, either by oxidation of a Cu^{0}-source or by reduction of a Cu^{II}-source, are likely. This Cu^{I}-intermediate **I** can undergo oxidative insertion into the aryl halide moiety affording Cu^{III}-species **II**. In the next step, substrate coordination and deprotonation by a non-nucleophilic base is assumed to form the key intermediate **III**, which finally undergoes reductive elimination to give the arylated amine and regenerate the Cu^{I}-catalyst **I**.

Catalysts and Variants of the Ullmann Coupling for the Formation of C—N Bonds

Since the first observations made more than 100 years ago, an overwhelming number of different catalysts and conditions for the coupling of amines and aryl halides have been reported. Mostly applying (sub-)stoichiometric amounts of a Cu^{I}-source and weak, non-nucleophilic bases in polar, aprotic solvents have become a standard strategy for the efficient and cheap synthesis of arylamines. The choice of ligand mainly influences the copper loading, which is why different O,O-, N,N-, and N,O-based ligands have been utilized for the stabilization of the catalyst. Despite a multitude of described reaction protocols, the ligand itself needs to be optimized for every single transformation, even if quite similar coupling partners are predescribed in the literature. A small overview is given below (Scheme 10.5) [6, 8].

Alongside to this manifold of previously reported conditions, a variety of modifications of this reaction have been developed in the past to overcome certain disadvantages, especially the harsh reaction conditions and the low yields of the primal Ullmann coupling. Therefore, new coupling partners other than aryl halides were utilized, including aryl boronic acids and esters, aryl silanes, and aryl stannanes, aryl iodonium salts, aryl lead, and aryl bismuth compounds. In particular, the first two mentioned modifications, commonly referred as the Chan–Lam coupling, were applied frequently in organic synthesis (Scheme 10.6).

Unfortunately, the exact mechanism of the Chan–Lam coupling remains, as for the Ullmann coupling, unknown because of the liability of the copper reagents and the multicomponent nature of the reaction. However, for the formation of the C—N bond, a reductive elimination starting from a Cu^{II}- or Cu^{III}-species is assumed to take place in analogy to the Ullmann coupling. The presence of oxygen or a stoichiometric amount of oxidant is not only necessary for the successful outcome of the reaction by oxidizing the Cu catalyst but also facilitates the reaction. This fact and the low reaction temperatures give certain advantages over the Ullmann coupling and the Buchwald–Hartwig amination [9].

Recent Applications on Industrial Scale of Ullmann and Chan–Lam Coupling

Being the Ullmann and the Chan–Lam couplings, the first versatile tools for the preparation of aromatic amines, through the years an extensive research activity, was performed in academia, and many industrial processes have been established. In the following section, a small selection of their application in the large-scale production of pharmaceuticals and agrochemicals is provided [10].

In 2007, researchers from Abott Laboratories reported on an efficient synthesis of the naphthalenoid H_3 antagonist **9** by a Ullmann coupling of naphthalene bromide **7** and pyridazinone (**8**) on a 14.5 kg scale [11]. The authors first focused on a Buchwald–Hartwig coupling for this key step but were not successful, although several catalysts and bases were

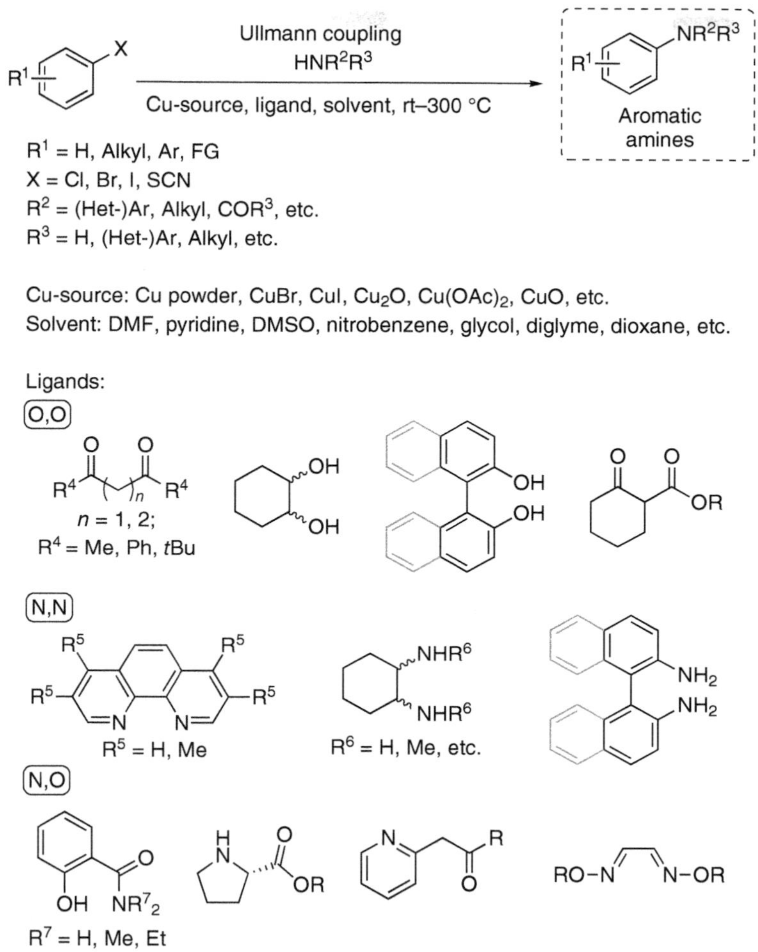

Scheme 10.5 Examples of different conditions for the copper-catalyzed C–N bond formation (Ullmann coupling).

applied. These discouraging results prompted them to consider a Cu-catalyzed C—N bond forming approach as an alternative, viable method. After extensive optimization of the reaction conditions, the desired product **9** could be obtained in a yield of 85% by applying a catalyst system based on CuCl and 8-hydroxyquinoline in DMF at 140 °C. It is worthwhile to highlight that most of the copper catalyst was efficiently removed by washing the organic phase with aqueous ammonia and Na_2EDTA solutions (Scheme 10.7) [11].

In 2012, within a collaboration between Amri, Serenex, and Pfizer, an Ullmann coupling provided straightforward access to SNX-5422, a member of a novel class of heat-shock protein 90 (Hsp90) inhibitors [12]. In the key step, a quite complex substituted and annulated pyrazole **10** was coupled with the aryl bromide **11** in 73% yield, giving only minor impurities in terms of undesired N^2-isomer. By optimizing these reaction conditions, in particular the Cu-catalyst species and the solvent, it was found that only 20 mol% of CuI

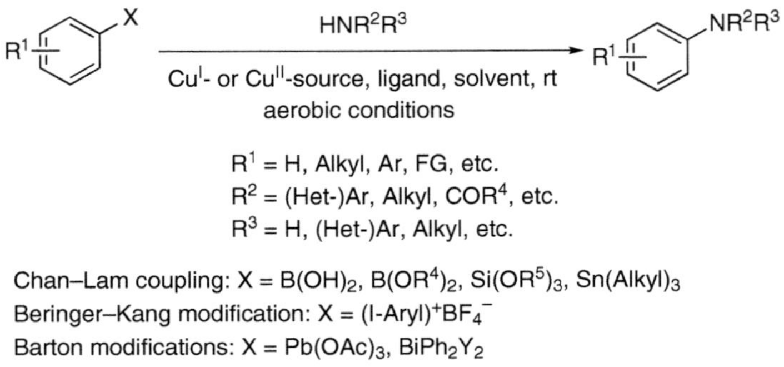

Scheme 10.6 Examples of different modifications of the Ullmann C—N bond formation.

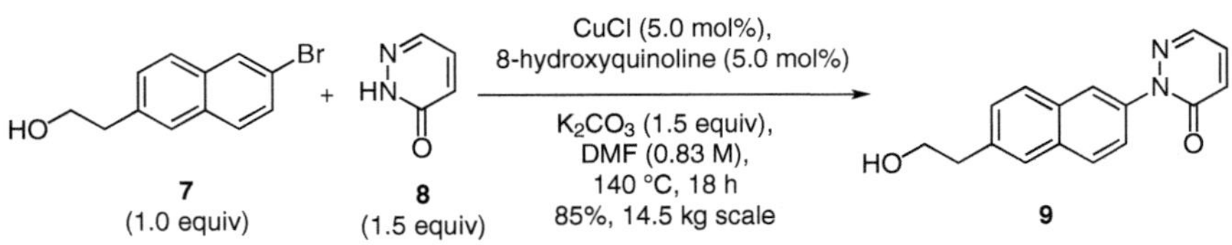

Scheme 10.7 Preparation of synthetic naphthalenoid H_3 antagonist **9** on a 14.5 kg scale.

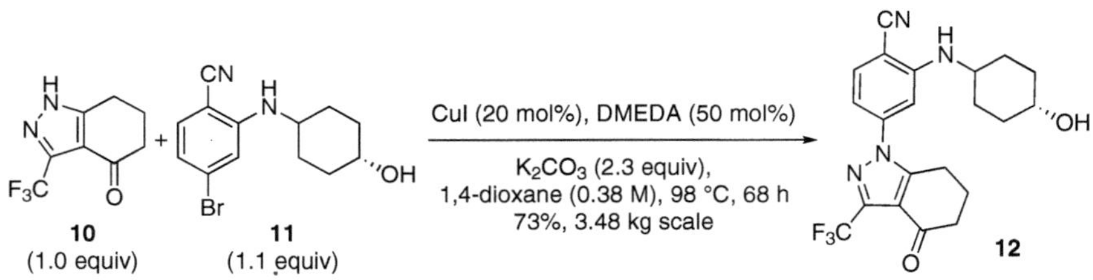

Scheme 10.8 Synthesis of pyrazole **12** by large-scale Ullmann coupling.

and 50 mol% DMEDA in dioxane are appropriate to catalyze this reaction on a 3.48 kg scale (Scheme 10.8). Minor impurities, as a result of regioselectivity issues, as well as the catalyst, could be separated by reslurrying the crude product in EtOAc, after washing the organic phases with an ammonium hydroxide solution [12].

More recently, a group from the Shenyang Pharmaceutical University developed a new approach toward tamibarotene, a retinoic acid receptor alpha antagonist, by employing and refining a coupling of ammonium chloride as an ammonia source. Utilizing a catalytic mixture of CuI (20 mol%) and L-proline (40 mol%), they obtained aniline **14**, a synthetic intermediate toward tamibarotene in 75% yield on a 0.56 kg scale (Scheme 10.9) [13].

Besides the described disadvantages for aromatic amination via Ullmann coupling, these selected examples also reveal an advantage, which relies on the nature of the copper salts to be complexed by ammonia or other suitable ligands. These types of complexes are mostly water soluble and may therefore be easily separated via a simple washing procedure, in

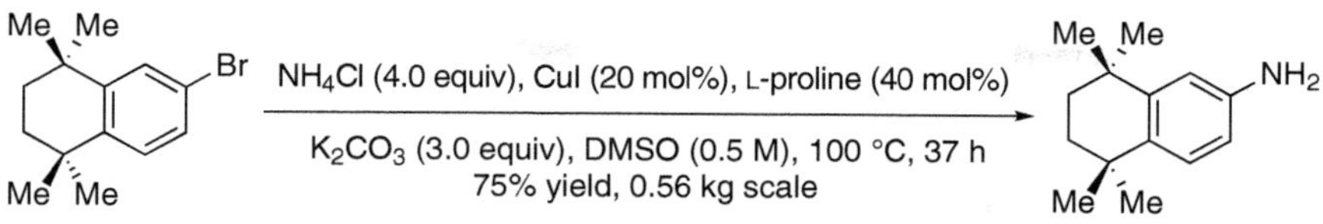

Scheme 10.9 Preparation of annulated aniline derivative **14** as an intermediate structure toward tamibarotene.

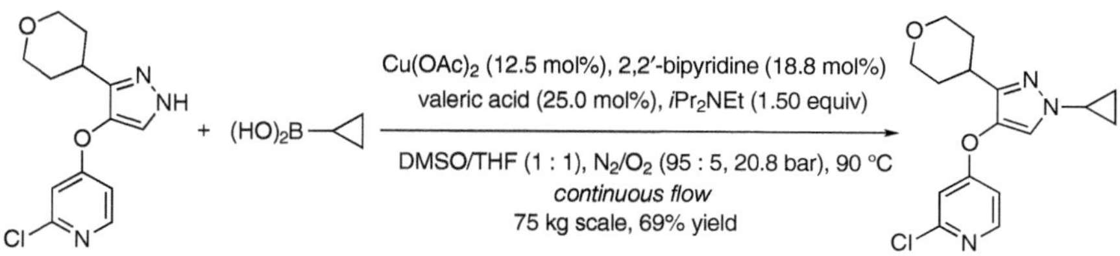

Scheme 10.10 Industrial-scale synthesis of *N*-cyclopropyl pyrazole derivative **17** via Chan–Lam coupling by use of a continuous-flow setup.

contrast to palladium-based catalysts. This advantage and the low cost of copper in comparison to palladium encouraged many more scientists to apply Ullmann-type C–N coupling reactions on industrial scale [14].

The ability and necessity of the Chan–Lam coupling to be run successfully under aerobic conditions also bears – besides the strong economic and environmental advantages of O_2 – significant safety and practical concerns surrounding the use of oxygen in the presence of organic solvents. The transition from multipurpose batch reactors, typically used to carry out these types of reactions, to continuous-flow reactors as safer and more practical alternatives enabled a more safely operation [15]. In this context, scientists from Eli Lilly developed a continuous aerobic oxidative Chan–Lam coupling for the penultimate (**17**) of an active pharmaceutical ingredient (API, cp. also Section 10.3.2.4). In their approach, the key to a safe and reliable procedure relied on using diluted air as the oxygen source and identification of homogeneous reaction conditions, which allowed for carrying out the process in a continuous vapor–liquid pipe-in series reactor. In particular, applying a catalyst formed by combining $Cu(OAc)_2$ and 2,2′-bipyridine (bipy), the desired compound **17** was obtained in 69% yield on a 75 kg scale (Scheme 10.10) [16].

To conclude, since the seminal observations by Ullmann and Goldberg more than 100 years ago, the scope of methodologies for C–N coupling reactions mainly relied on the Ullmann coupling, until it was revolutionized by Buchwald and Hartwig in the year 1995. Thus, many different conditions and a variety of protocols have been developed, with the Chan–Lam coupling being one of the most prominent and representative.

10.1.2.2 Buchwald–Hartwig Amination

The use of palladium complexes as catalysts for the formation of $C(sp^2)$—N bonds has a long story and was pioneered in 1983 by Migita and coworkers, who reported on the Pd-catalyzed

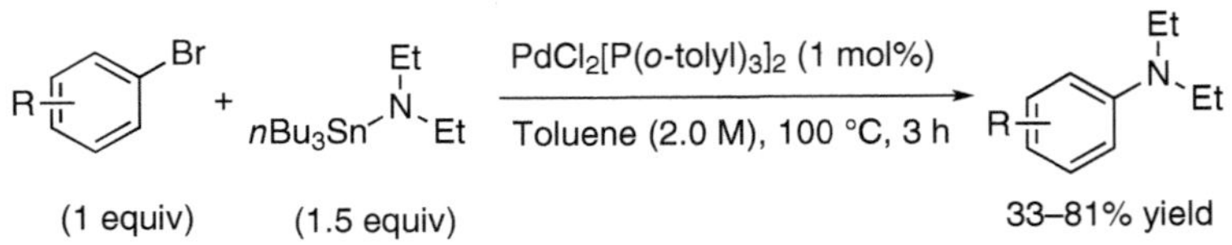

Scheme 10.11 Migita's early report on a palladium-catalyzed C–N coupling reaction of aryl bromides with organotin compounds.

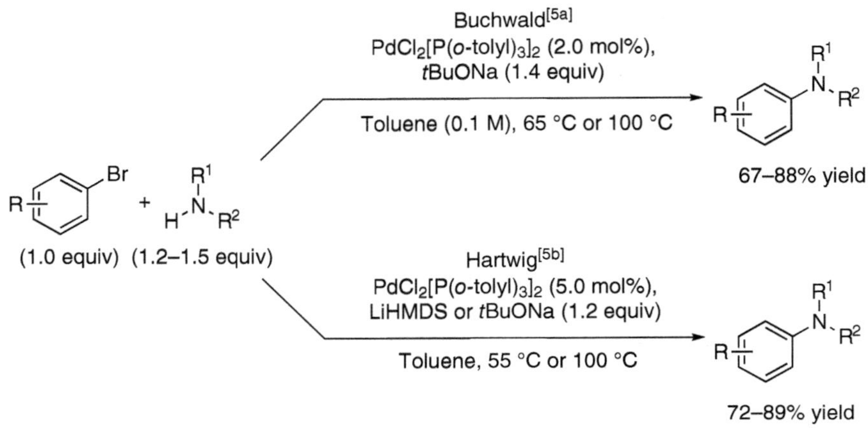

Scheme 10.12 Buchwald's and Hartwig's initial reports of a palladium-catalyzed C–N bond forming reaction of aryl bromides with amines.

formation of arylamines by coupling aryl bromides with amino tin reagents (Scheme 10.11) [17].

The limitations described in this seminal work were later circumvented by Buchwald and Hartwig, who independently reported on a tin-free Pd-catalyzed coupling of aryl bromides with amines in 1995 (Scheme 10.12) [5]. This breakthrough, accompanied by subsequent efforts from both groups toward a general and efficient methodology for C—N bond formation and an in-depth mechanistic understanding of the process, led to the established general reaction, which is nowadays known as the Buchwald–Hartwig amination [18].

After these initial reports, the reaction quickly took off and was adopted in a variety of academic and industrial applications. About 10% of all publications in the fields of medicinal chemistry of the year 2014 relied at least once on a Buchwald–Hartwig-type C—N bond forming reaction, pointing out the importance of this transformation in the pharmaceutical industry at early-stage drug discovery [19]. Within this context, the scope of the Buchwald–Hartwig amination has been widely illustrated for a variety of arene and amine coupling partners. Nonetheless, the choice of the catalyst is highly dependent on the geometric and electronic features of the utilized substrates.

Mechanistic Studies

Early studies by Hartwig established the following general mechanism using $P(Ar)_3$-type ligands [5]. Oxidative insertion of a Pd^0-species **I** into the aryl halide substrate affords a

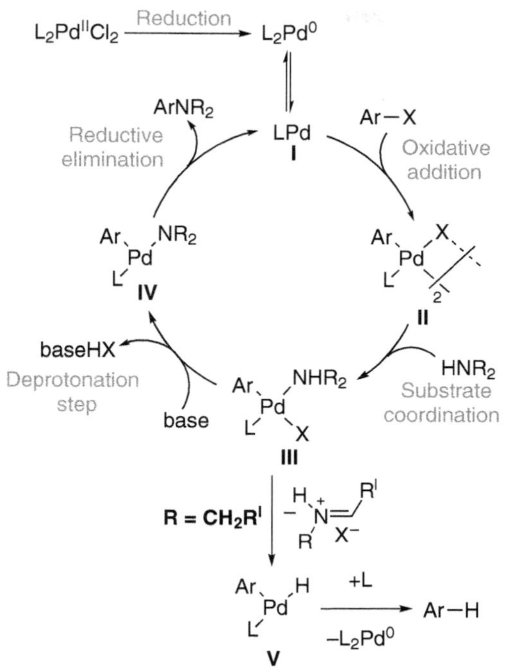

Scheme 10.13 Mechanistic cycle of the palladium-catalyzed C–N coupling reaction. The competing side reaction takes place out of complex **III** via β-hydride elimination.

dimeric species of type **II**. Coordination of the amine significantly increases its acidity, allowing for deprotonation by a strong but hindered and non-nucleophilic base, such as *t*BuONa or LiHMDS, to form the palladium amine complex **IV**, which finally undergoes reductive elimination to lead to the arylated amine and regenerate the Pd^0-catalyst **I** (Scheme 10.13).

Further and extensive investigations concerning the reaction mechanism conducted by Hartwig as well as Buchwald led to an excellent understanding of the factors influencing the efficiency within the course of the overall reaction from the oxidative addition to the reductive elimination process, including the competing side reaction such as the β-hydride elimination [20].

Catalyst Design

Understanding the above-mentioned factors allowed for rational development of a large variety of catalytic systems, mainly improving the reactivity and applicability of the amination reaction. By using varieties of different bidentate ligands including families of biaryl monophosphines and also *N*-heterocyclic carbene (NHC) ligands, the scope of the Buchwald–Hartwig amination could be broadened and moreover expanded from aryl bromides and iodides to triflates, chlorides, and tosylates (Scheme 10.14). Furthermore, primary amines, prone to undergo the side reaction of β-hydride elimination, are now suitable substrates by adapting the system with different optimized reaction conditions, accelerating the deprotonation and reductive amination steps [5, 21–30].

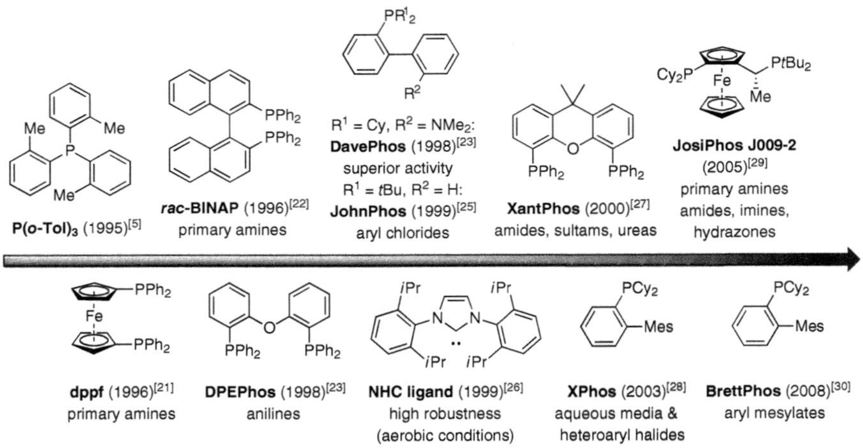

Scheme 10.14 Evolution of the ligands for the Buchwald–Hartwig amination reaction: examples and advantages thereof.

Notably, Buchwald published a user's guide of the numerous biaryl monophosphine ligands developed in his group and their preferred applications in order to aid the selection of reaction conditions and further optimization [18c]. To conclude, this ongoing development of new ligands and the general optimization of the coupling conditions enabled a large variety of (hetero-)aryl (pseudo-)halides to react with amines, amides, and even N–H heterocycles, often at relatively low temperatures and catalyst loadings, even on the order of parts per million [29, 31]. As an additional proof of the applicability and robustness of this methodology, it is also possible to use more challenging sources of nitrogen, such as fluoroalkylamines [32], or even ammonia [33] as a direct coupling partner.

Recent Applications in Industry

In addition of being an indispensable tool in academic research, a variety of industrial processes using the Buchwald–Hartwig amination on industrial scale have been reported. In this section, a selection of its application in large-scale production of pharmaceuticals and agrochemicals developed over the course of the past years is discussed.

In 2004, a team from Rhodia reported on a hydrazonation, applying a $Pd(OAc)_2$- and MePhos-based catalyst system for the synthesis of *N*-(*p*-tolyl)benzophenone hydrazone (**20**) on a 3.4 kg scale [34] (Scheme 10.15).

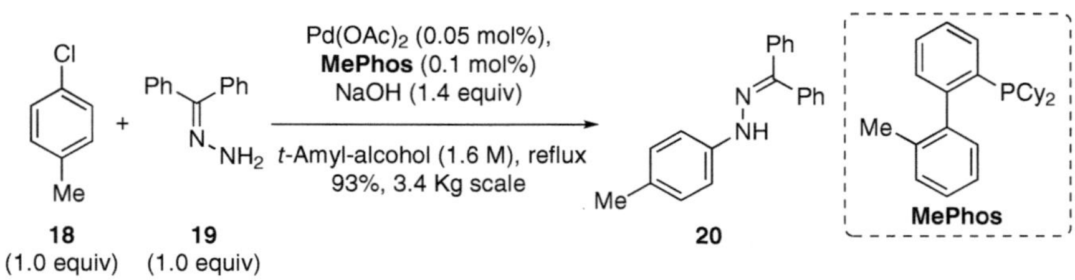

Scheme 10.15 Preparation of *N*-*p*-tolyl-substituted benzophenone hydrazine derivative (**20**) on a 3.4 kg scale by a Buchwald–Hartwig amination.

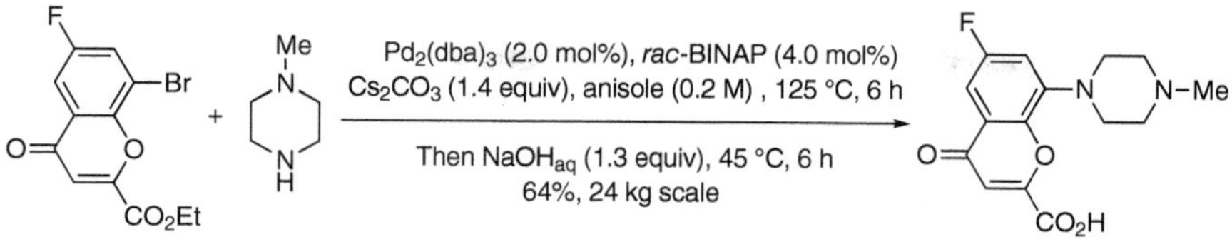

21 (1.0 equiv) **22** (1.1 equiv) **23**

Scheme 10.16 Preparation of intermediate **23** for the synthesis of a 5-HT receptor antagonist on 24 kg scale.

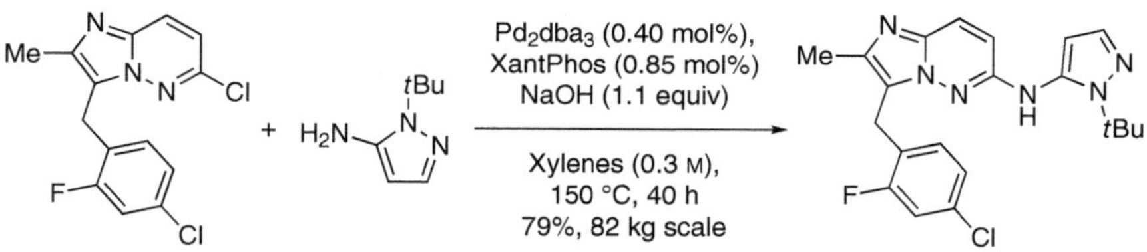

24 (1.00 equiv) **25** (1.05 equiv) **26**

Scheme 10.17 C–N coupling of pyridazinium chloride **24** and amino pyrazole **25**.

In the same year, chemists from Astra Zeneca were capable of producing a 5-HT receptor antagonist, making use of a Pd_2dba_3 catalyst and racemic BINAP. The application of this system together with cesium carbonate as the base and anisole as the solvent allowed for an efficient and economical coupling, proceeding within only six hours and providing 24 kg of the desired compound **23** (Scheme 10.16) [35].

A team from Eli Lilly developed a catalyst system, employed at only 0.4 mol% loading, for the efficient, regioselective preparation of compound **26**. Remarkably, this Buchwald–Hartwig amination catalyzed by Pd_2dba_3 and XantPhos was carried out after extensive refinement of the process conditions on a 82 kg scale (Scheme 10.17) [36]. Notably, the vast efforts for refinement were due to an extreme sensitivity of the reaction to oxygen.

A high functional group tolerance even on a 2.5 kg scale was demonstrated by chemists from Pfizer, being able to couple an isoquinoline bromide **27** with the chiral amino alcohol **28** to deliver the desired N-arylated product **29** in 80% yield. By applying *rac*-BINAP, the chiral product **29** was obtained in the absence of erosion with regards to the enantiomeric purity and without the need of protection of the primary alcohol functionality (Scheme 10.18) [37].

A more recent example was reported by a team from Amgen, accomplishing a synthesis of the precursor (**32**) for AMG 925, a drug candidate required for early phase clinical studies. The application of the biaryl based ligand BrettPhos in addition to *t*BuONa as a suitable base delivered **32** in 90% yield, at low temperature of 60 °C, and in a short reaction time of only three hours (Scheme 10.19) [38].

Twenty-five years ago, the scope of methodologies for C–N coupling reactions was revolutionized by Buchwald and Hartwig by their development of the Pd-catalyzed coupling of amines and aromatic (pseudo-)halides. This transformation has become one of the main

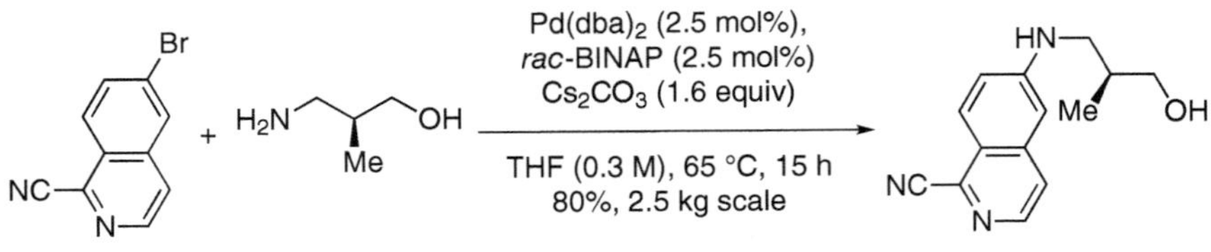

27 (1.0 equiv) **28** (1.4 equiv) **29**

Scheme 10.18 Regioselective and enantiospecific C–N coupling of isoquinoline bromide **27** and chiral amino alcohol **28**.

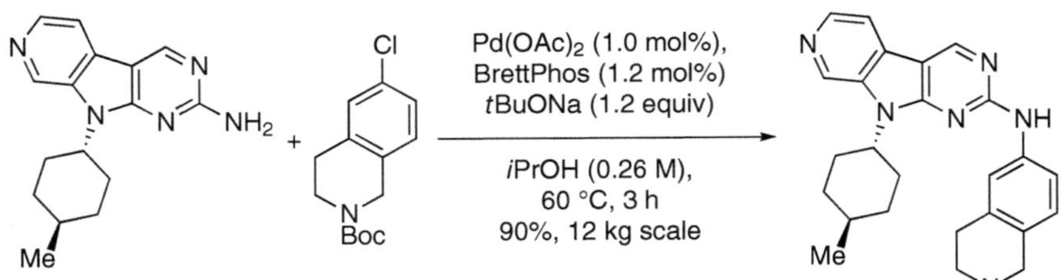

30 (1.0 equiv) **31** (1.1 equiv) **32**

Scheme 10.19 Preparation of the precursor **32** for AMG 925 by Amgen.

tools for organic synthesis. It has found countless applications in academia as well as in industry, illustrated by selected examples within in this section.

10.2 New Methodologies in the Synthesis of Arylamines on the Brink of Industrial Application

10.2.1 Introduction

In recent years, research efforts in transition metal catalysis have heavily focused on embracing the principles of atom economy to achieve efficient C—N bond formation. Aside from the aforementioned classical aryl amination methods that have been widely developed on industrial scale to tackle syntheses of complex molecules in pharmaceuticals, agrochemicals, and fine chemicals, modern methods that utilize C–H functionalization technology have emerged, albeit still in their infancy to be applicable in industrial settings. From industrial perspectives, transition metal catalysis still poses some challenges regarding its scaling up. To date, the industrial feasibility of C–H amination remains under study, owing to the typically required chelation assistance that reduces the synthetic benefits of this methodology.

This section focuses primarily on developments in arylamine synthesis by means of catalytic C—N bond formation appeared in the last decade. Because of vast literature reports in the area of C—N bond forming methodologies, only a small selection of topics and examples will be covered. Detailed mechanistic studies and related catalytic systems design can be found in the cited reviews.

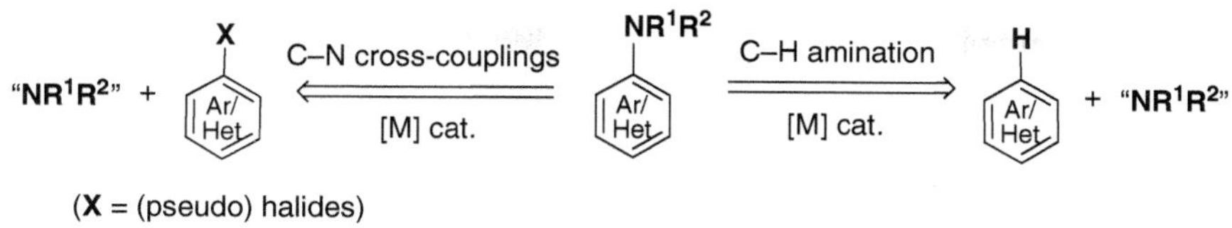

Scheme 10.20 Aromatic C–N bond formation approaches.

10.2.2 Catalytic C–H Amination

C—N bond formation by way of transition metal catalysis provided a platform for rapid and robust synthesis of nitrogen-containing aryl/heteroaryl compounds in materials science and agrochemical and pharmaceutical industries. A new paradigm in this research area has grown tremendously to suit the increasing demand of safe and economical methodologies to synthesize nitrogen-containing molecules (Scheme 10.20). Lack of practical amination protocols inspired organic synthetic chemists to develop novel approaches to address the shortcomings of existing methods that require preinstallation of aryl (pseudo)halides.

Catalytic C–H functionalization processes offer many advantages to the classical cross-coupling reactions. This step-economical approach bypasses the preactivation typically required for aryl C—C or C—X bond formations (where X is any heteroatom). In fact, as early as the late 1960s, researchers have already discovered the efficiency of C–H amination by the way of metal-catalyzed azide decomposition [39]. It was not until early 1980s when Breslow and Mansuy paved the way for intensive investigations in this field, by demonstrating the viability of metal catalysts to effect intramolecular and intermolecular C–H amination, respectively [40]. Later on, Daugulis and coworkers pioneered the work of using bidentate directing groups in metal-catalyzed directed C–H functionalization, enhancing the site selectivity of the transformation [41].

10.2.2.1 Catalytic C–H Amination under Standard Conditions

Driven by the aforementioned and appealing opportunities offered by this strategy, numerous catalytic systems have been studied and developed. Despite the economical feature that this methodology presents, there are inevitably some intrinsic challenges in achieving C–H functionalization. For instance, a well-designed and elegant catalytic system is necessary to induce activation of a high-energy C—H bond in a selective manner. Moreover, use of directing groups is typically a prerequisite to achieve a high degree of selectivity, which poses limitations on substrate scope. Among commonly used directing groups, such as 2-aminophenyl oxazoline, 2-aminophenyl pyrazole, and 2-aminopyridine-1-oxide, use of 8-aminoquinoline has dominated the field, providing an entry to novel constructions of C—N bond by way of C–H activation [42].

Several review articles featuring the advancements in the field of C–H amination were published within the past decade [43], including a comprehensive review published by Chang and coworkers [43], highlighting both inter- and intramolecular variations of this transformation in great detail. A small selection of the recent development of intermolecular arene C–H amination methodologies will be presented in this section.

Among the C—N bond forming reactions developed by adopting the C–H activation strategy, arylamine syntheses via this approach continue to be an active area of research.

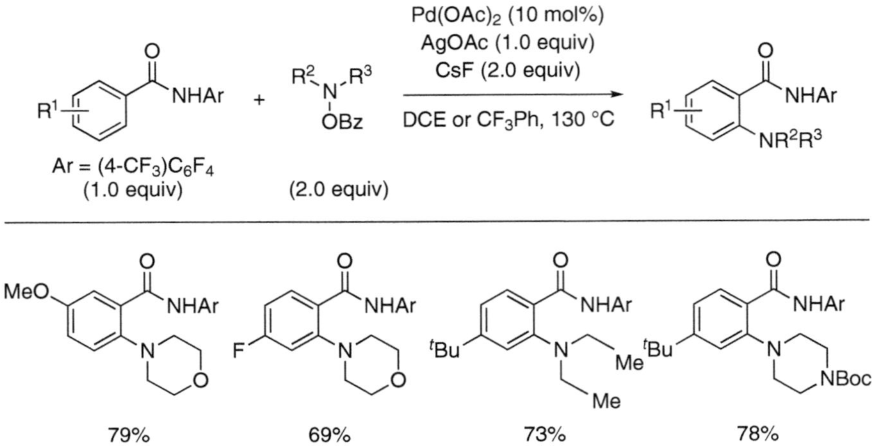

Scheme 10.21 Palladium-catalyzed aryl C–H amination.

A diverse range of metal-based catalytic systems have been shown to be effective in promoting aryl C–H activation. Inter- and intramolecular palladium-catalyzed C—H bond functionalization is well established. In-depth studies of this methodology have provided great insights into plausible reaction mechanisms and palladium redox chemistry. Early work on C—N bond formation via palladium catalysis focused on intramolecular coupling reactions to form nitrogen-containing heterocycles [44]. An intermolecular version of this transformation was later discovered, albeit with amides as suitable coupling partners [45]. In 2011, Yu and coworkers reported the first example of Pd-catalyzed intermolecular C–H amination by developing a new auxiliary capable of directing *ortho*-C–H amination [46]. The author described the use of *O*-benzoyl hydroxylamines as electrophilic amine partners, which could be readily accessed by treating the amine of interest with benzoyl peroxide. Using the developed conditions, a broad scope of arene nucleophiles and amines were tolerated, resulting in aminated arenes in good yields (Scheme 10.21). Although this approach provided an entry for further development in palladium-catalyzed C–H amination reactions, removal of the requisite auxiliary remains challenging and poses limitations in industrial applications.

As previously mentioned, intermolecular C–H amination of arenes remains a continuing challenge. Research progress has been made to identify a practical protocol that obviates the chelation assistance or generation of stoichiometric amount of by-product as a result of external oxidants. Chang reported a rhodium-catalyzed C–H amination method using aryl azide as a nitrogen source for C—N bond formation, leading to a wide diversity of diarylamines [47]. The author reasoned that the in situ generated cationic Rh(III) catalyst was responsible for the success of this transformation. Notably, under this condition, N_2 was the sole stoichiometric by-product, rendering this reaction free of oxidants and halide salts by-products. The author also highlighted the high functional group tolerance and broad applications of the synthesized diarylamines from C–H amination of carboxamide and ketoxime derivatives, wherein facile transformations could elaborate diarylamines to synthetically useful building blocks (Schemes 10.22 and 10.23).

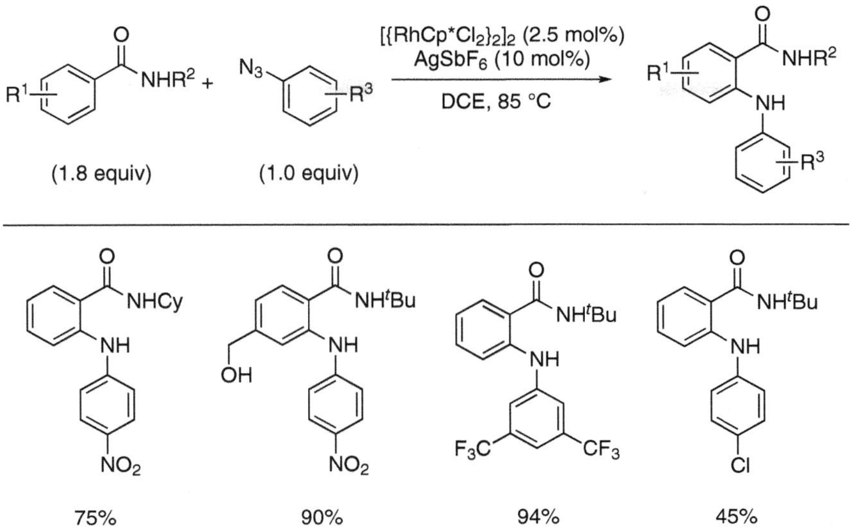

Scheme 10.22 Rhodium-catalyzed direct C–H amination with aryl azides.

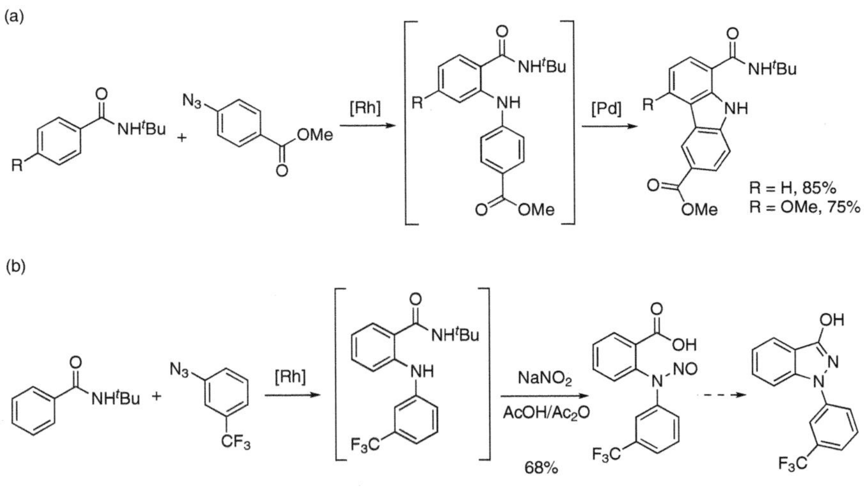

Scheme 10.23 Synthetic versatility of diarylamine products from C–H amination of benzamides. (a) Synthesis of carbazoles and (b) synthesis of 1-aryl indazolol precursor.

Similarly, Glorius and coworkers reported the use of the same Rh(III) complex to effect C—N bond formation using *N*-pivaloyloxy amide as directing group and *N*-chloromorpholine as the aminating partner [48]. The author described a mild set of conditions, wherein CsOAc was employed as a base, and the reaction proceeded smoothly at room temperature. Notably, by exploiting *N*-pivaloyloxy amide as the directing group, arylamine products could be readily diversified into the corresponding benzamide and 1-aminoaniline derivatives, outlining the applications of this methodology (Scheme 10.24).

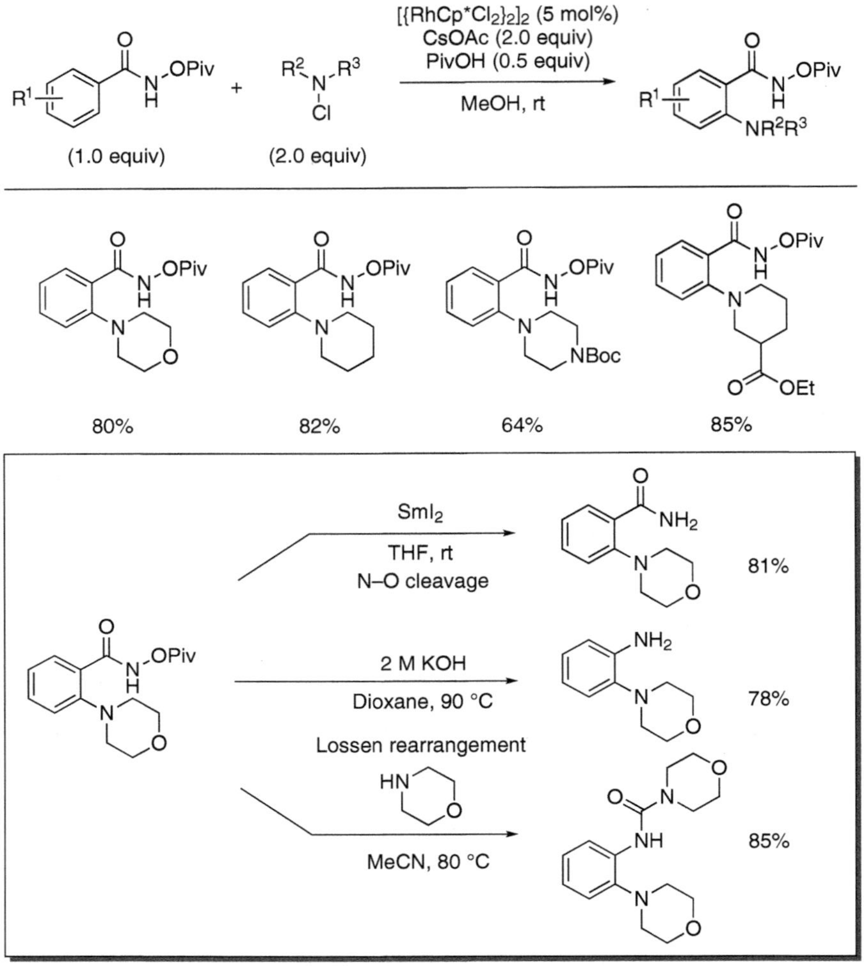

Scheme 10.24 Rhodium-catalyzed C–H amination and synthetic utilities.

Recently, Falck disclosed a dirhodium catalyst, $Rh_2(esp)_2$, previously developed by Du Bois and coworkers [49], to promote aryl C–H amination inter- and intramolecularly with high levels of regioselectivity (Scheme 10.25) [50]. Differing from a direct metal catalyzed C–H activation pathway, the author proposed that generation of a metal-nitrenoid species was accountable for the subsequent C—N bond formation. Remarkably, the use of *N*-methyl-*O*-tosylhydroxylamine as the aminating agent allowed for in situ generation of an internal oxidant upon N—O bond cleavage, circumventing the need for an external oxidant, which in turn enhanced the practicality of this transformation. As a result, an array of products of direct arene amination was furnished in high chemo- and regioselectivity, encompassing structurally diverse *N*-methyl-aniline derivatives by way of late-stage functionalization of complex products.

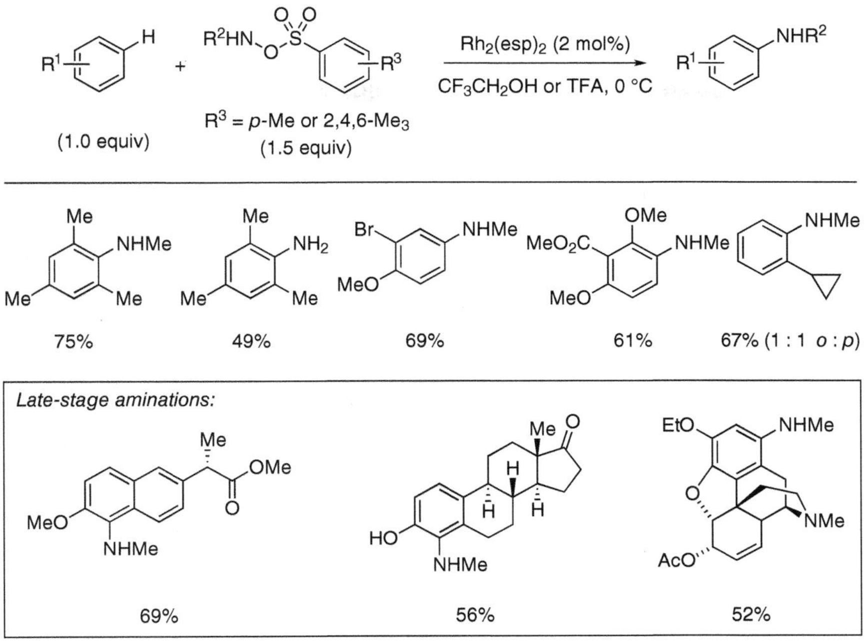

Scheme 10.25 C–H amination enabled by a dirhodium catalyst.

Cyclometalated iridium(III) complexes have been shown to be capable of promoting C–H functionalization when used in stoichiometric quantities [51]. In 2014, Chang and coworkers developed a directed C–H amination to synthesize diarylamines using an iridium(III) catalyst under very mild conditions [52]. In this work, the author described that the use of sterically encumbered *N*-adamatylbenzamide as a directing group could facilitate the formation of an iridacyclic intermediate, required for the subsequent C—N bond formation. Additionally, the choice of acetate additive and an external oxidant, in this case $Cu(OAc)_2$ and $AgNTf_2$, respectively, are required to produce reasonable reactivity (Scheme 10.26).

Modern methods for constructing C—N bonds by means of precious metal-based catalysis are of paramount importance in organic synthesis. However, nonprecious metals have gained considerable attention from researchers owing to their low cost, environmental friendliness, and natural abundance. Nickel has been shown to be a competent metal to promote aryl C–H amination. In 2011, Chatani and coworkers reported the first example of nickel-catalyzed transformations by chelation-assisted C–H activation [53]. Intrigued by this work, Zhang and coworkers published the first example of direct arene amination using Ni-catalysis [54]. By exploiting the chelation assistance of 8-aminoquinoline benzamide, this approach exhibited a broad substrate scope of amines and carboxamide counterparts (Scheme 10.27). Radical trapping and deuterium labeling experiments suggested that this process might proceed through Ni^{I}/Ni^{III} catalytic pathway.

Very recently, Tan and coworkers reported the first example of primary aniline synthesis enabled by nickel in the presence of ammonia [55]. Encouraged by the previous work developed in the group, wherein stoichiometric nickel promoted C–H amination using NaN_3 as

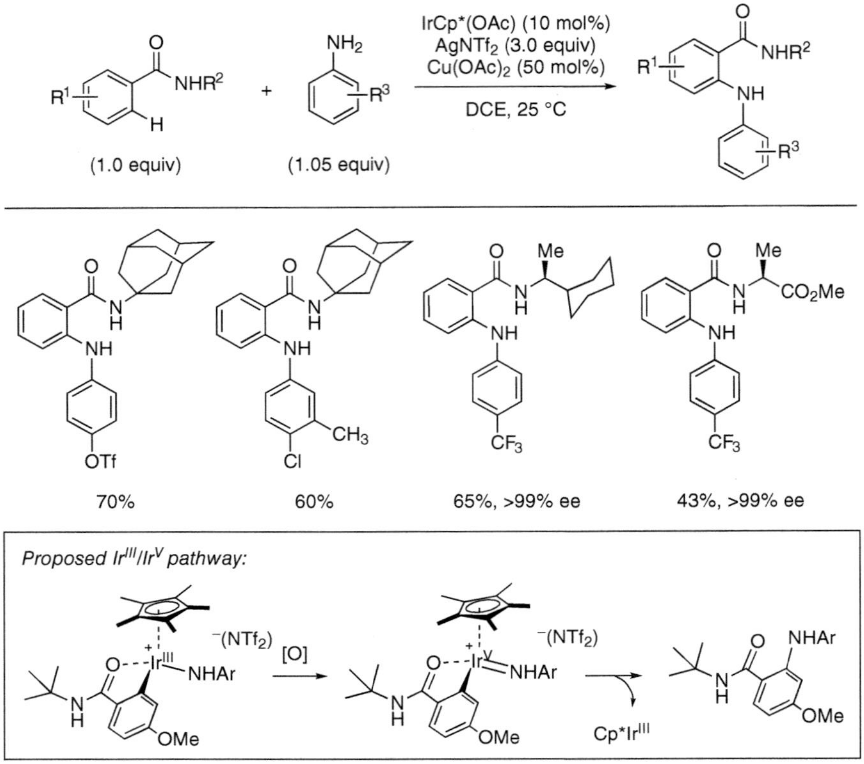

Scheme 10.26 Diarylamine synthesis via iridium-nitrenoid catalyzed C–H amination.

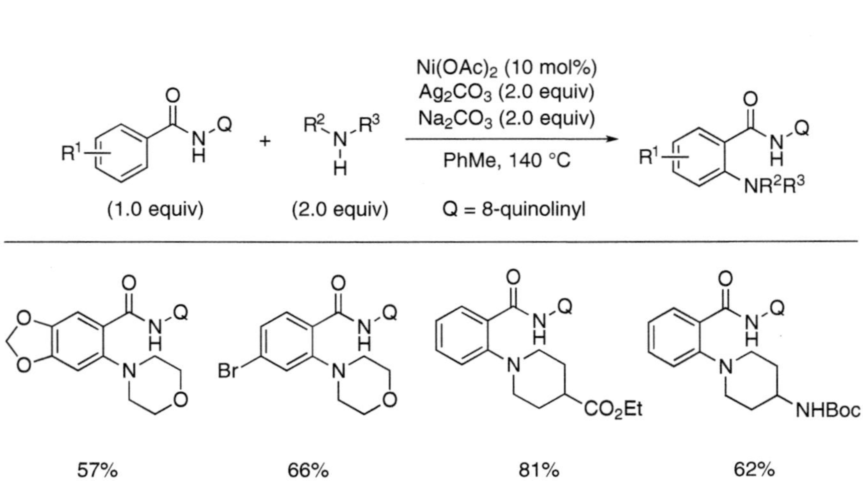

Scheme 10.27 Nickel-catalyzed aryl C–H amination with alkyl amines.

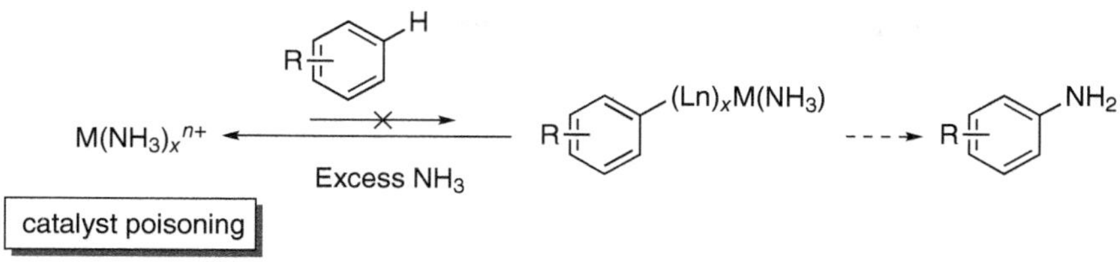

Scheme 10.28 Long-standing challenge of using ammonia as an aminating agent (Ln = directing group).

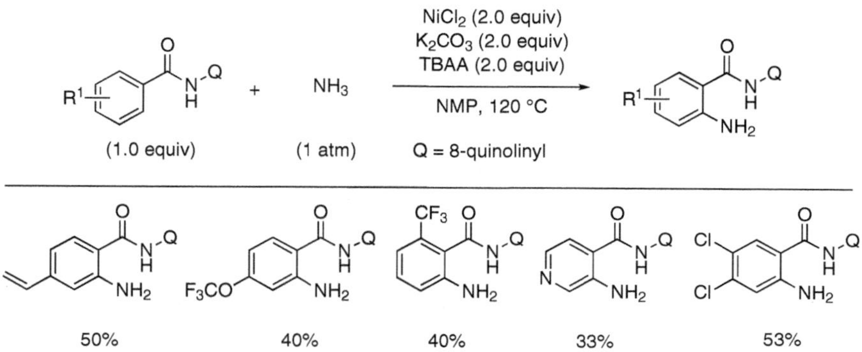

Scheme 10.29 Nickel-mediated C–H amination with ammonia.

a nitrogen source [56], the author sought to tackle the use of ammonia as an amino source, which is challenging because of its propensity to strongly coordinate metal complexes and its capability to reduce high-valent organometallic intermediates (Scheme 10.28).

Use of 8-aminoquinoline as the directing group and of a stoichiometric amount of $NiCl_2$ constituted the key to successful promotion of C–H amination with ammonia, albeit in only fair to moderate yields (Scheme 10.29).

Cobalt-catalyzed C–H amination has recently emerged. In particular, it has been shown that a Co(III) complex could catalyze heteroarenes using azide as preactivated aminating agent [57]. Encouraged by these reports, Song and coworkers developed a new amination protocol, wherein simple amines were compatible aminating reagents under Co(II) catalysis [58]. To achieve this transformation, a bidentate N,O-directing group, i.e. 2-aminopyridine-1-oxide, played a crucial role and could be readily removed upon basic hydrolysis (Scheme 10.30). In the same year, Zhang and coworkers independently developed a similar transformation employing 8-aminoquinolinebenzamide as the directing group with $Co(acac)_2$ as the catalyst [59]. A similar range of alkyl amines and carboxamide substrates were demonstrated to be compatible under the developed reaction conditions. Preliminary mechanistic studies implicated that single electron transfer (SET) processes might be operative.

The pioneering work in aryl C–H amination mediated by copper using pyridine as a directing group was developed in 2006 [60]. Following these reports, immense efforts have focused on rendering this transformation catalytic and expanding the scope of the aryl and heteroaryl coupling partners [61]. In 2013, Daugulis and coworkers discovered

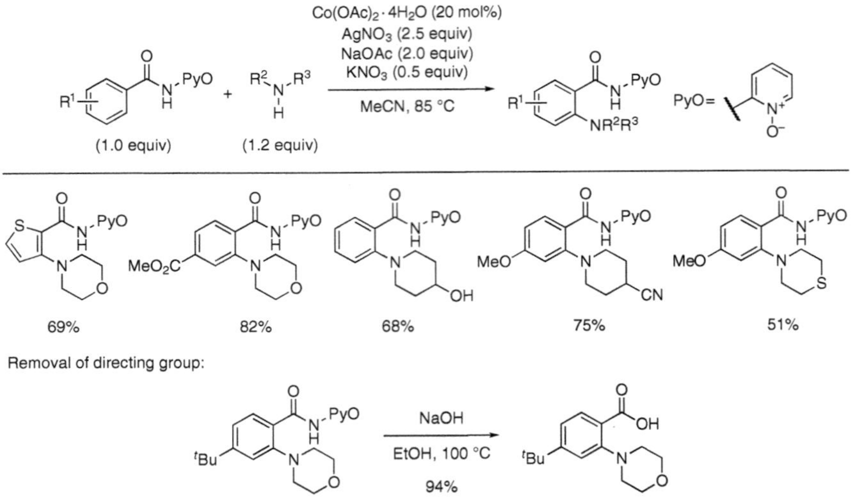

Scheme 10.30 Cobalt-catalyzed aryl C–H amination and directing group hydrolysis.

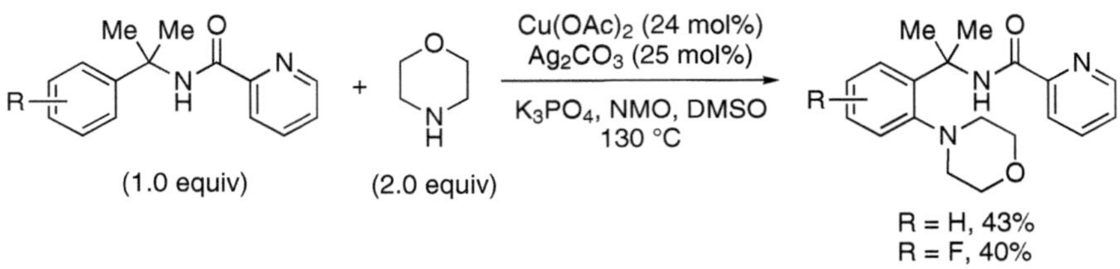

Scheme 10.31 Amination of benzylamine picolinamide.

a dual catalytic system in which the use of a Cu(II) catalyst in combination with a silver-based catalyst allowed for directed *ortho*-amination of aryl C—H bonds [62]. Using 8-aminoquinoline as the directing group, a broad scope of *ortho*-aminobenzoic acid derivatives could be achieved. Interestingly, under the developed conditions, α,α-disubstituted benzylamine picolinamides could also undergo C–H amination, affording aminated products in moderate yields (Scheme 10.31).

Many approaches using copper to promote aromatic C—N bond formation have demonstrated the viability of copper in the area of C–H functionalization. However, to date, one drawback of this application in arylamine synthesis remains the stoichiometric use of copper to render these transformations synthetically useful, in addition to the need for an appropriate directing group. Furthermore, synthesis of primary anilines is particularly challenging because of the intrinsic nature of ammonia, which is an excellent ligand for metals and can inhibit subsequent C—H bond activation events. Thus, a compatible nitrogen source to serve as an ammonia surrogate continues to be the subject of intense investigations [63]. In 2017, Bolm and coworkers discovered the use of dibenzothiophene sulfoximine **33** as an ammonia surrogate that could undergo in situ hydrolysis after aryl C—N bond formation. This two-step, one-pot method took advantage of the facile radical S—N bond cleavage to allow for deamination of sulfoximine and subsequent liberation of aminated arylamine

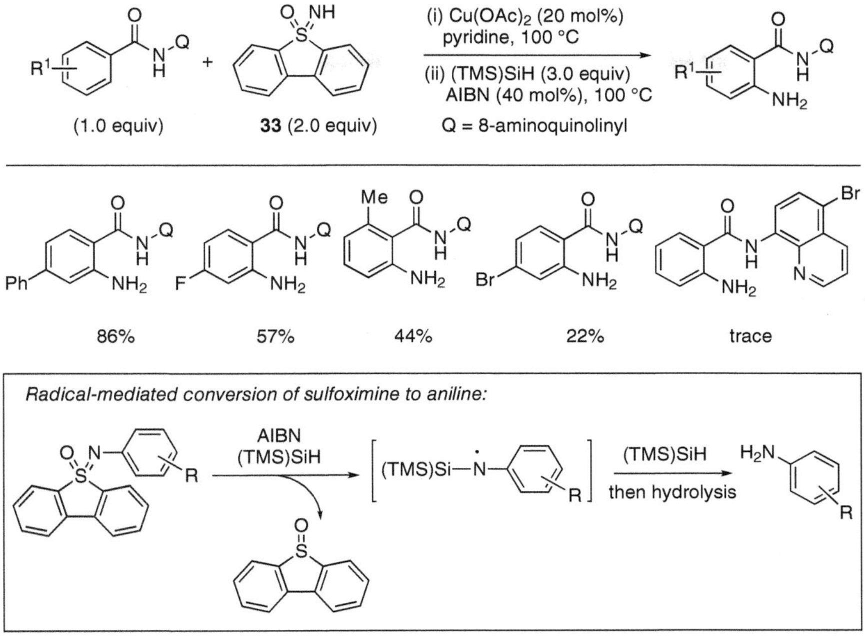

Scheme 10.32 Copper-catalyzed C–N bond formation using dibenzothiophene sulfoximine **33** as an ammonia surrogate.

product [64]. Although this strategy bypassed a stepwise operation, substrates with halogen substituents suffered from dehalogenation because of the radical-mediated hydrolysis process, posing limitations on the reaction scope and potential subsequent transformations that could exploit halogen functional handles (Scheme 10.32).

Later, Chang and coworkers identified a catalytic system that allowed for direct use of aqueous ammonia to access products of arene amination [65]. The key feature of this strategy includes the in situ generation of Cu(III) intermediates by disproportionation of a Cu(I) complex, circumventing the strong coordination of ammonia. Taking advantage of the low-valent Cu(I) complexes and the principle of subtle disproportionation of copper, deprotonation and subsequent reductive amination of ammonia were made viable without metal poisoning of the catalyst. Density functional theory (DFT) calculations were conducted to gain insights into the feasible redox behavior of active copper species as well as the energy profile of a plausible reaction pathway. The author also described the comparable results using stoichiometric and substoichiometric copper to achieve C–H amination, enabling late-stage functionalization of medicinally relevant molecules (Scheme 10.33).

In contrast to many literature precedents in metal-catalyzed C–H activation wherein a chelator is the requisite for directing C—N bond formation, Morandi and coworkers developed an operationally simple and practical transformation affording primary anilines by way of iron-catalyzed C–H functionalization [66]. Unlike typical C—H bond forming reactions that require an additional deprotection step to prepare anilines, the author described a direct synthesis of anilines using $FeSO_4$ as a catalyst. Noteworthily, the author also disclosed a new benchtop-stable aminating reagent (**34**) that together with the iron catalyst resulted in

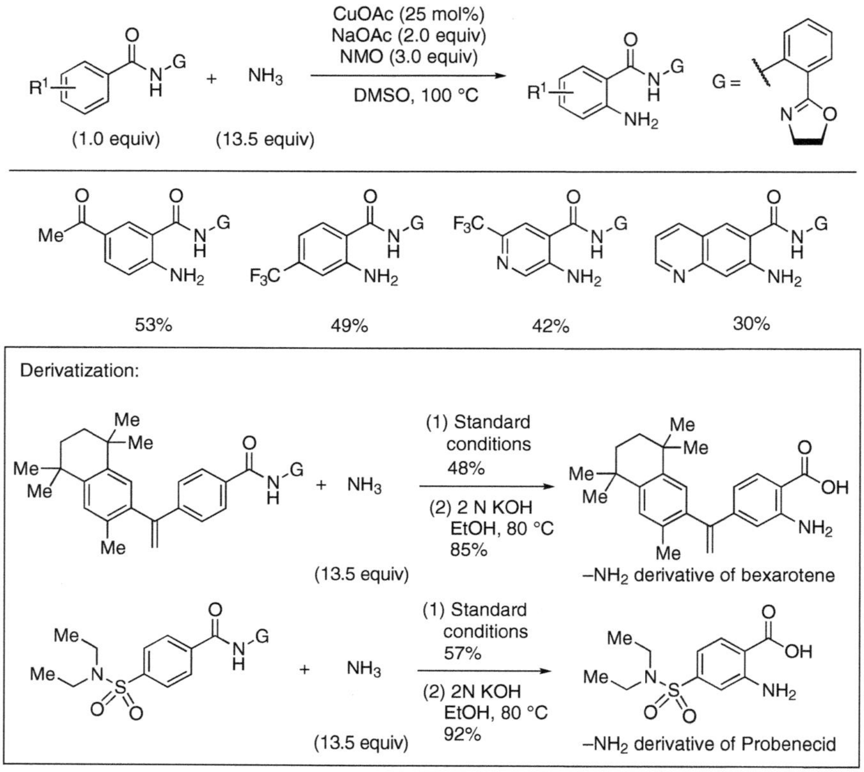

Scheme 10.33 Copper-catalyzed primary aniline synthesis.

C—N bond formation in acetonitrile/water mixture at room temperature (Scheme 10.34). Late-stage C–H functionalization of architecturally complex molecules to access primary anilines in one operation step demonstrated the utility of this method. Despite many attractive features of this approach, selectivity remained a challenge because of the directing effects of the functional groups on the arene.

Apart from C—N bond forming amination via metal catalysis, metal-free protocols to construct *N*-arylamines have also attracted considerable attention. Chang and DeBoef independently reported C–H oxidative amidation using hypervalent iodine reagent in the absence of metal catalysts [67]. Inspired by this metal-free direct amination methodology, Chupakhin and coworkers later published the discovery of arene amination with amino heterocycles using a hypervalent iodine(III) reagent [68]. Presumably, the hypervalent iodine(III) activates the amino group toward electrophilic aromatic substitution with the arene of choice. Further investigation led to subsequent development of organocatalytic conditions that utilized catalytic iodobenzene in the presence of peracetic acid as an oxidant to effect aryl C–H amination, enabling efficient access to amino heterocycles with broad scope (Scheme 10.35).

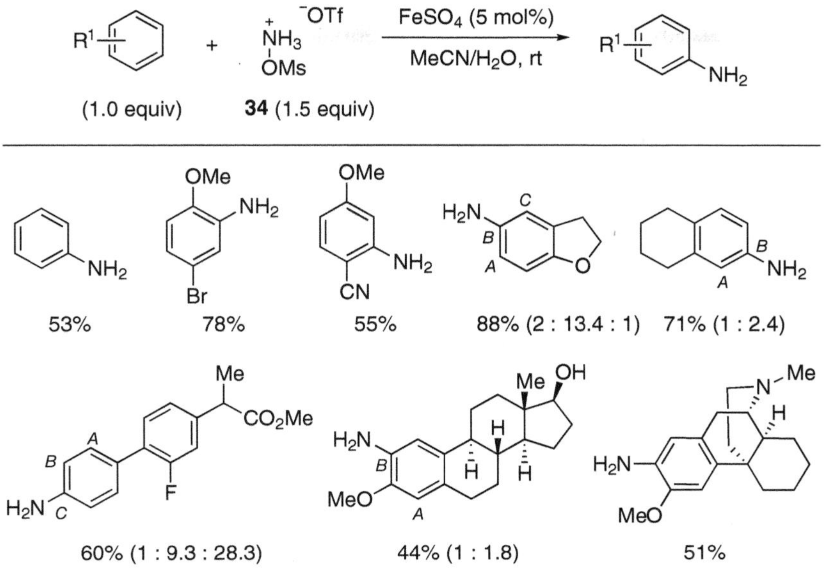

Scheme 10.34 Iron-catalyzed C–H functionalization using ammonia surrogate **34**.

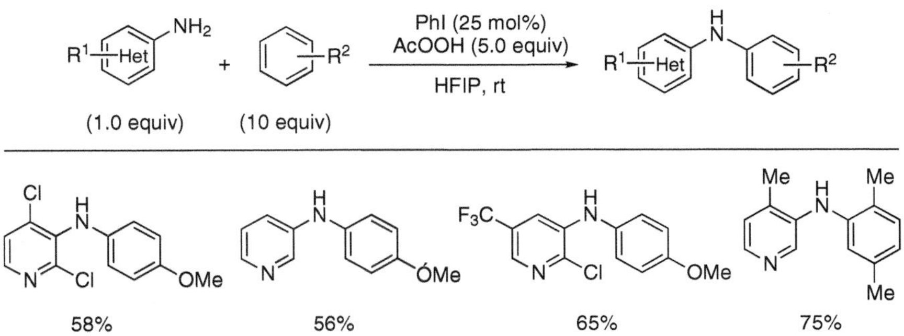

Scheme 10.35 Organocatalyic C–H oxidative amination.

To date, directed C–H amination remains an area of great interest in the synthetic chemistry community. Chelation assistance to guide site-selective C—N bond formation continues to be one of the challenges that hampers the extension of the aminated products to synthetically useful compounds. Although certain directing groups are readily susceptible to hydrolysis, the additional removal step compromises the step economy for this bond forming strategy. Vigorous efforts are devoted to designing new systems that are more practical to enable large-scale applications.

10.2.2.2 Photoredox Catalysis

In the realms of C–H activation chemistry, site selectivity and limited scope of amine coupling partners remain formidable challenges for achieving atom-economical aryl C–H amination. Ongoing research seeks to develop new methodologies, including designing new

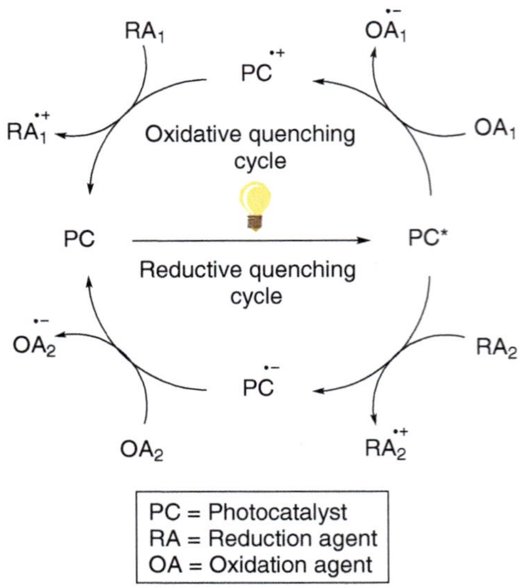

Scheme 10.36 Typical modes of action of photocatalysis.

catalytic systems, to address these issues and provide a powerful toolbox for C—N bond forming reactions.

In the past decade, electron transfer photocatalysis played a vital role in organic synthesis as a complementary alternative to traditional methods of C—C bond constructions. Thermodynamically prohibited processes under transition metal catalysis were made possible by exploiting a photocatalyst to harness visible light to induce chemical transformations. Upon irradiation, an excited photocatalyst accepts or donates an electron to initiate a SET event that enables the subsequent oxidative or reductive quenching cycles (Scheme 10.36). Although vast literature reports have been published on its applications in C—C bond formations, aryl C—N bond construction using this technology remains a less-explored research area [69]. Readers can also refer to the chapters of this book by Leonori and Xia for a complementary description of the application of photocatalysis to the synthesis of nitrogen-containing compounds.

In 2015, Nicewicz and coworkers developed an elegant system that circumvents many of the limitations in direct arene amination protocols, namely excess of arene, directing group assistance, preactivation of amine counterparts, etc. [70]. This methodology relied on the use of a photoredox acridinium catalyst capable of oxidizing arenes, which was employed in combination with nitroxyl radical as a cocatalyst to achieve regioselective C—N bond formation. Despite the incomplete mechanistic studies, the author attributed the observed regioselectivity to the stabilized radical formed upon nucleophilic addition to arene cation radical. Upon screening an array of nitrogen sources, synthesis of aniline derivatives applying this methodology was achieved by using benchtop stable ammonium carbamate **35** ($H_4N^+O_2CNH_2^-$) as an ammonia surrogate (Scheme 10.37).

Following this seminal work, the same group designed a strategic approach to construct C—N bond with aliphatic amines by means of photoredox catalysis [71]. Through generation of an aryl cation radical intermediate in the presence of an acridinium

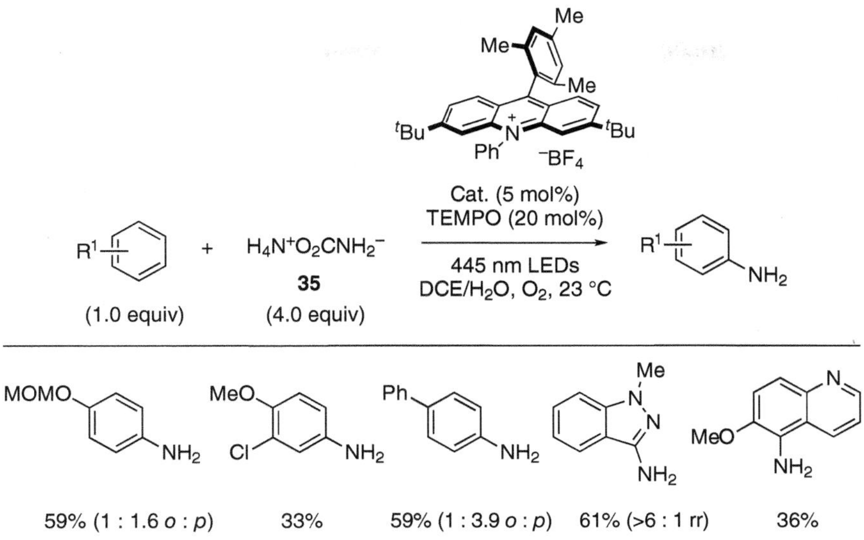

Scheme 10.37 Synthesis of anilines enabled by a photoredox acridinium catalyst using ammonia surrogate **35**.

photocatalyst, the reaction with an amine coupling partner proceeded smoothly to furnish aminated arene products. It is worth noting that the excited acridinium photocatalyst could be quenched by arene and/or amines. The mechanistic pathways leading to products of amination depended on the electronic properties of the arenes. To highlight the practicality and generalization of this approach, a wide variety of architecturally complex amines were used in this coupling reaction (Scheme 10.38). Amino acids were evaluated to demonstrate the absence of racemization under the mild reaction conditions. The author also demonstrated the viability of this methodology as a late-stage functionalization tool by derivatizing complex arene-containing pharmaceuticals.

As photoredox-catalyzed C—N bond forming processes became a more appealing strategy for constructions of arylamines, Leonori developed another direct aryl C–H amination method featuring the formation of aminium radicals from *O*-aryl hydroxylamines using a ruthenium-based photocatalyst [72]. Highlighting the strategic generation of aminium radicals to surmount the shortcoming of aminyl radicals, which typically suffer from their intrinsic nucleophilic properties, the author showed the broad scope of arenes that are susceptible to amination. Amination of marketed pharmaceuticals and other functional complex molecules using this methodology further established the utility of this strategy in late-stage amination (Scheme 10.39).

More recently, Togni and coworkers published a novel approach to construct anilines through a C–H pyridination-aminolysis sequence [73]. Using a ruthenium-based photocatalyst, generation of N-centered pyridyl radical cation from triflyoxy pyridinium **36** mediated by visible-light afforded *N*-aryl pyridinium intermediates, which could undergo different transformations to access functionally diverse products. The authors demonstrated the viability of C–H pyridination using photoredox catalysis and subsequently highlighted the utility of these products through an in situ aminolysis to furnish aniline

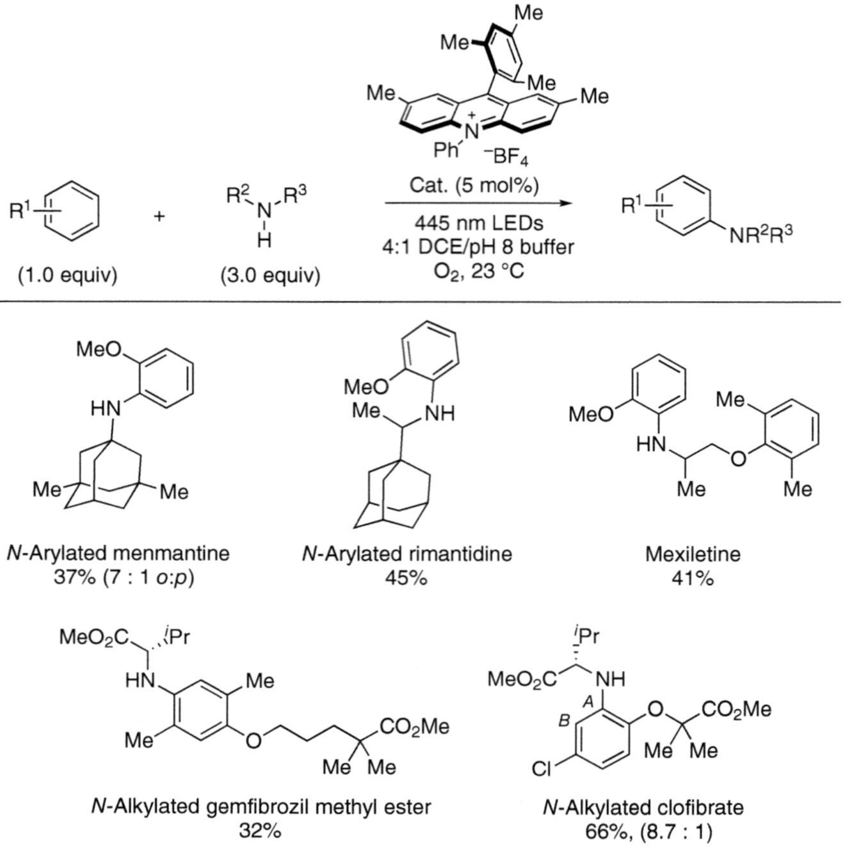

Scheme 10.38 Photoredox-catalyzed C–H amination: substrate scope.

products (Scheme 10.40). Notably, the products of C–H pyridination could be hydrogenated to the corresponding piperidines through a multistep process.

In the past years, the intriguing concept of dual catalysis emerged. For example, photocatalysis has been coupled with nonphotochemical catalysis to modulate the reactivity of intermediates generated by photoinduced electron transfer [74]. This strategic, cooperative approach opened new avenues for modulating redox properties of metal catalysts to achieve high chemoselectivity with exceptional functional group tolerance under mild conditions.

In the context of C–H functionalization, Wu and coworkers adopted a dual catalyst system by using photocatalysis in combination with cobalt catalysis to access aniline products. Benzene was coupled with ammonia or Boc-protected amine **37** to achieve dehydrogenative C–N couplings under dual catalysis, generating H_2 gas as the sole by-product (Scheme 10.41) [75].

10.2.2.3 Electrochemical Approaches

In recent decades, construction of arylamines via C–H activation utilizing electrochemical methods, bypassing the use of prefunctionalized aryl halides and production of

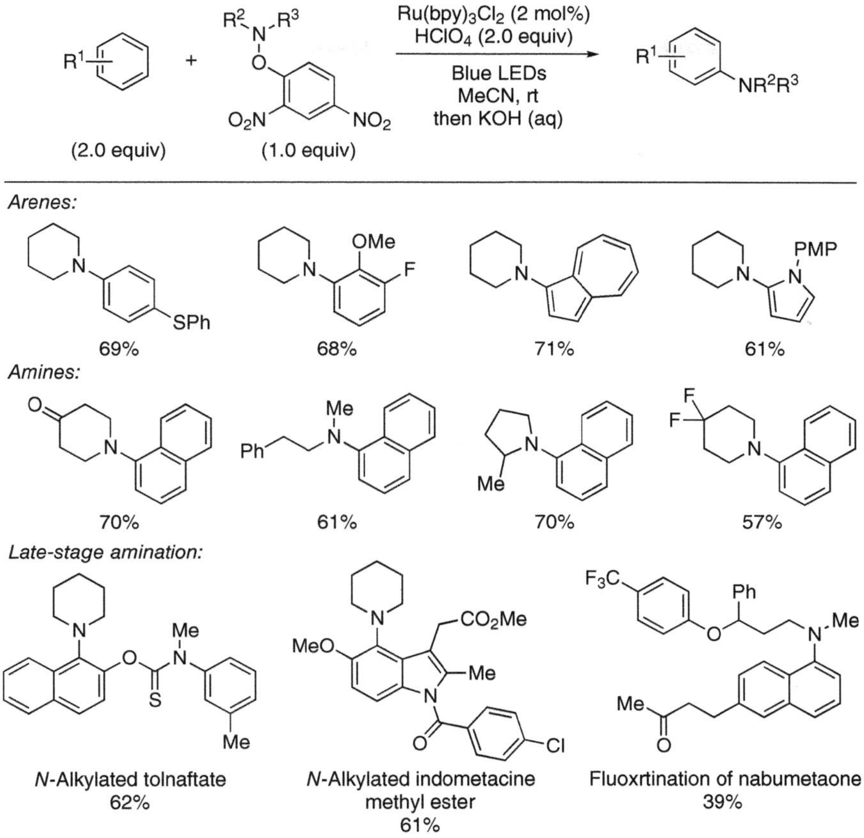

Scheme 10.39 Reaction scope of ruthenium-catalyzed C–H amination.

undesired organometallic by-products, has been disclosed. Direct use of electric current avoids the need for a photosensitizer to induce a chemical transformation, rendering this approach more desirable. Early development of electrochemical methodology demonstrated its simple setup, tunability, sustainability, and efficiency in comparison to existing metal-catalyzed cross-coupling protocols, offering many advantages unique to electrochemistry [76]. The ability to exploit redox behavior by fine-tuning electric current and modulating redox potentials renders these electrochemical processes a greener and more attractive alternative, especially when selectivity is a concern [76b, 77].

In 2013, Yoshida and coworkers developed a powerful electrochemical C—H bond amination method via *N*-arylpyridinium ions [78]. By exploiting the high oxidation potential and high nucleophilicity of pyridine, trapping of the initially formed arene radical cation produced *N*-arylpyridinium under anodic conditions, which underwent ring opening to the corresponding neutral nitrogen-containing products under nonelectrolytic conditions. This metal-free approach offered an efficient and chemoselective construction of a variety of aniline derivatives and was demonstrated in the synthesis of a key intermediate for the synthesis of a VLA-4 antagonist (Scheme 10.42).

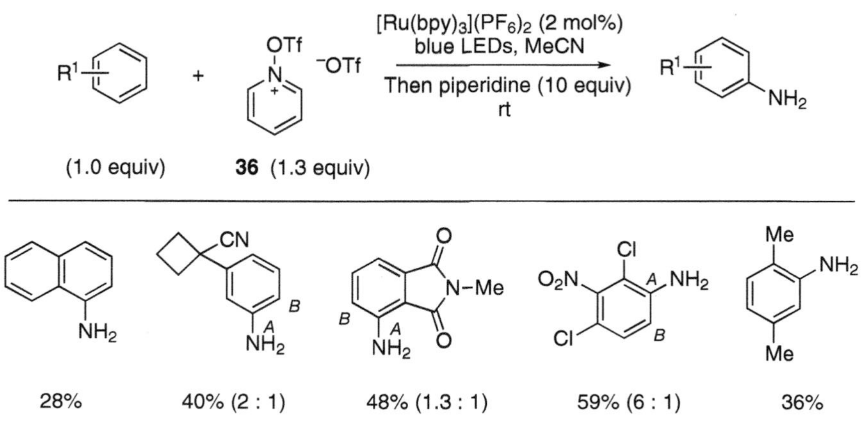

Scheme 10.40 Synthesis of aniline derivatives via a sequence of C–H pyridination and aminolysis of **36**.

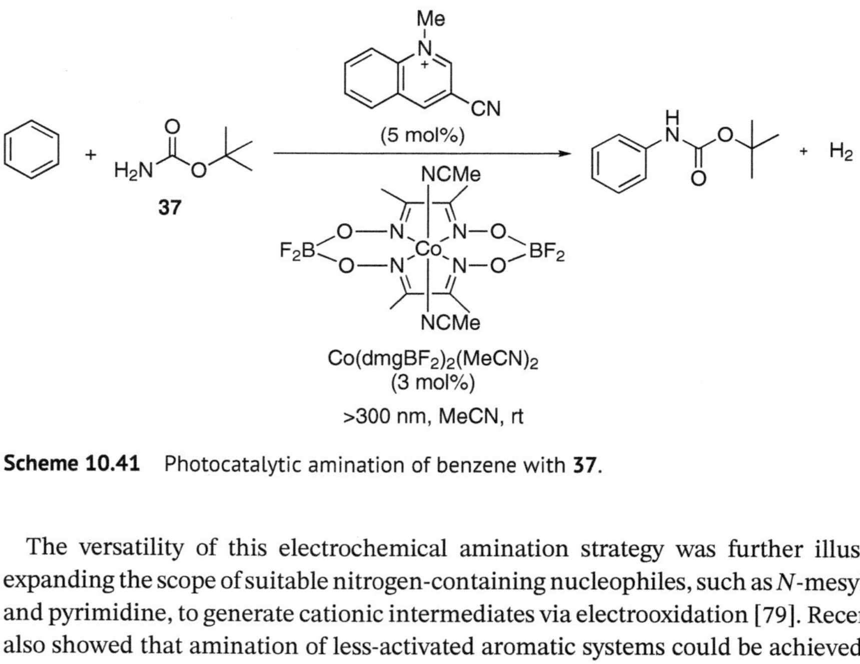

Scheme 10.41 Photocatalytic amination of benzene with **37**.

The versatility of this electrochemical amination strategy was further illustrated by expanding the scope of suitable nitrogen-containing nucleophiles, such as *N*-mesylimidazole and pyrimidine, to generate cationic intermediates via electrooxidation [79]. Recent studies also showed that amination of less-activated aromatic systems could be achieved by using boron-doped diamond (BDD) anodes, surmounting the limitation of electron-rich arenes typical of previous reports [80].

More recently, Lei and Ackermann independently reported versatile C–H amination by way of electrocatalysis under mild conditions [81]. Both authors highlighted the crucial role of carboxylate additives in assisting C–H activation. Interestingly, under nearly identical conditions, Lei's use of 8-aminoquinolinyl as a directing group was not compatible under the conditions developed by Ackermann. Nonetheless, this cooperative approach took advantage of cocatalytic systems to functionalize arene C—H bonds and of electricity

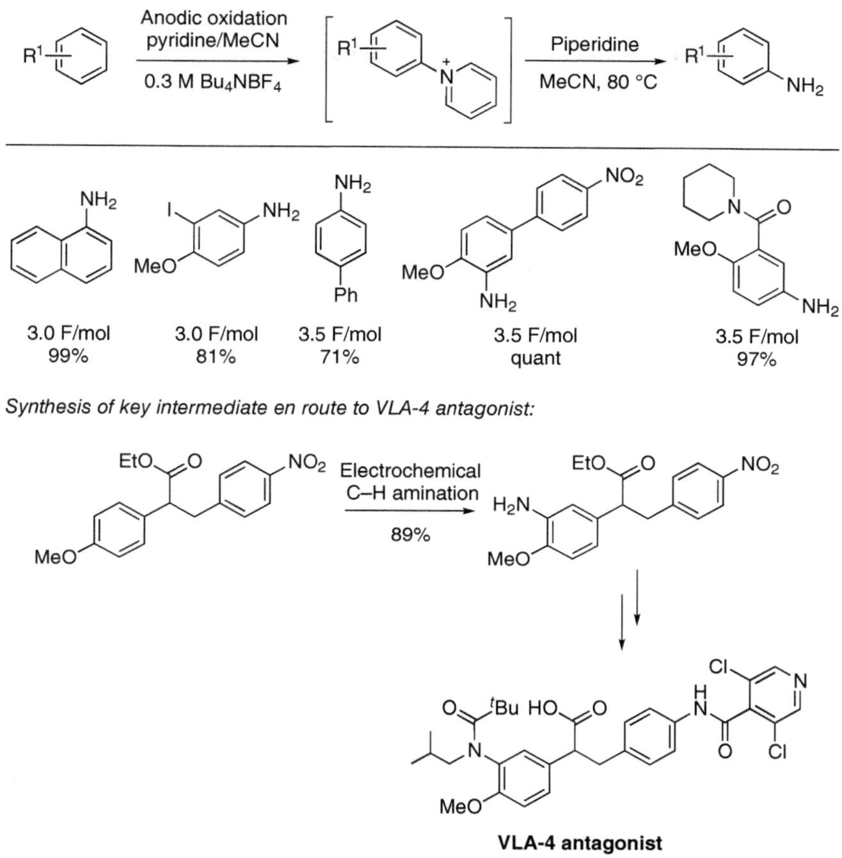

Scheme 10.42 Aniline synthesis via electrochemical C–H amination.

as the sole oxidant to render this transformation environmentally friendly and efficient (Scheme 10.43).

10.2.3 Decarboxylative Aryl Amination

In recent years, transition-metal-catalyzed C—C bond formation has taken the advantage of ubiquitous carboxylic acid moieties embedded in readily available commodity chemicals to effect decarboxylative cross-couplings. This strategic approach utilizes decarboxylative metalation, followed by reductive elimination, to enable direct access to a broad range of synthetically useful C–C coupling products. Additionally, this transformation offers many synthetic benefits, including nontoxic CO_2 as the only stoichiometric by-product, neutral reaction conditions, and site-specific cross-couplings [82]. Adopting this principle, researchers ventured into developing decarboxylative C—N bond forming methods to further complement the synthetic applications and expand the scope of C–N cross-coupling reactions [83].

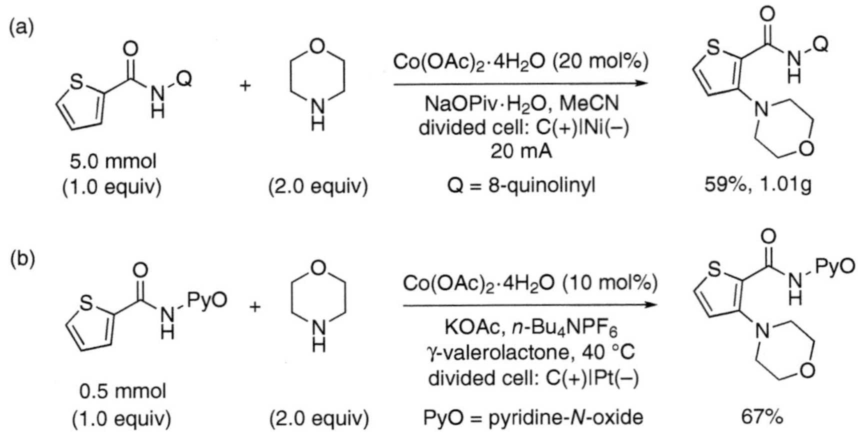

Scheme 10.43 Electrochemical C−H amination by cobalt catalysis. (a) Lei's system and (b) Ackermann's system.

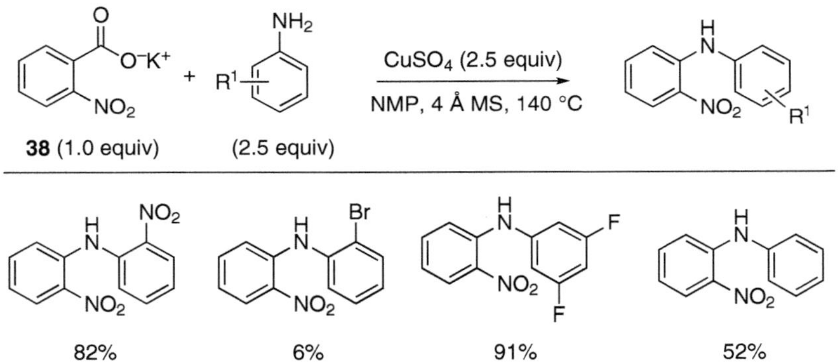

Scheme 10.44 Copper-mediated decarboxylative amination of **38**.

Mainolfi and coworkers reported the first example of decarboxylative C—N bond formation utilizing aryl carboxylic acid as a carbon partner, enabling further advancement of this methodology as an alternative to conventional aryl C—N bond forming transformations [84]. Jia and coworkers subsequently showed that $CuSO_4$ was able to mediate decarboxylative coupling with potassium benzoate **38** with a variety of anilines to furnish diarylamine products in modest to good yields (Scheme 10.44) [85].

Shortly thereafter, another strategy evolved using palladium to catalyze an intramolecular decarboxylative coupling by exploiting the carboxylate and amine moieties both embedded in acroyloxycarbamates [86]. To probe the applicability of this approach, *t*-butyl[(1,1′-biphenyl)-2-carbonyl]oxycarbamate **39** was subjected to coupling conditions on 10 mmol scale, affording Boc-protected biphenyl amine **40** in good yield (Scheme 10.45).

Having established the possibility of introducing nitrogen-containing functional groups into arenes by way of transition-metal-catalyzed decarboxylative couplings, Gooßen and coworkers sought to design a Pd/Cu bimetallic catalytic system to circumvent the potential

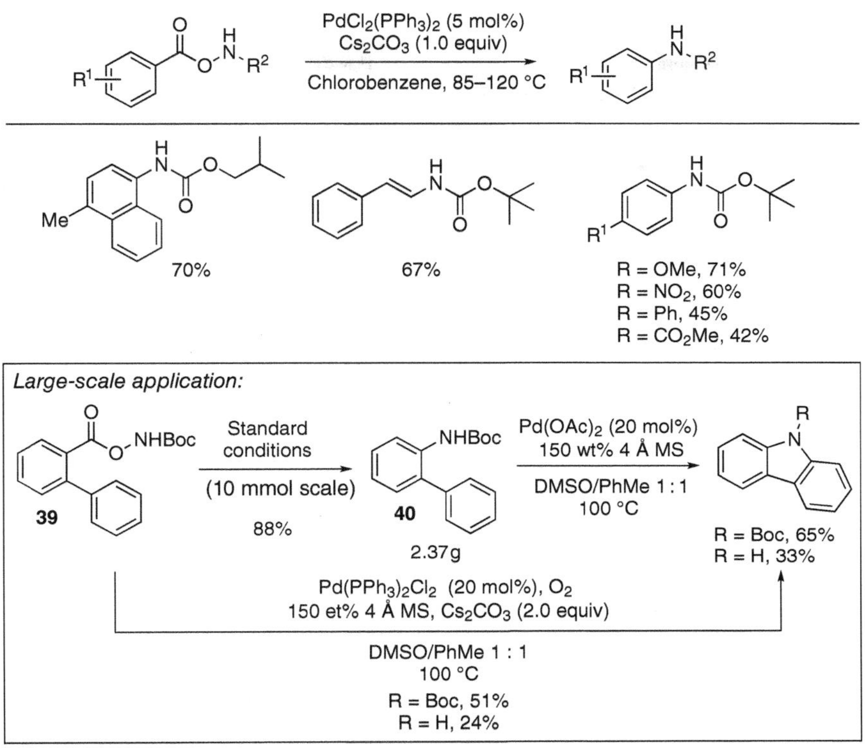

Scheme 10.45 Palladium-catalyzed decarboxylative couplings.

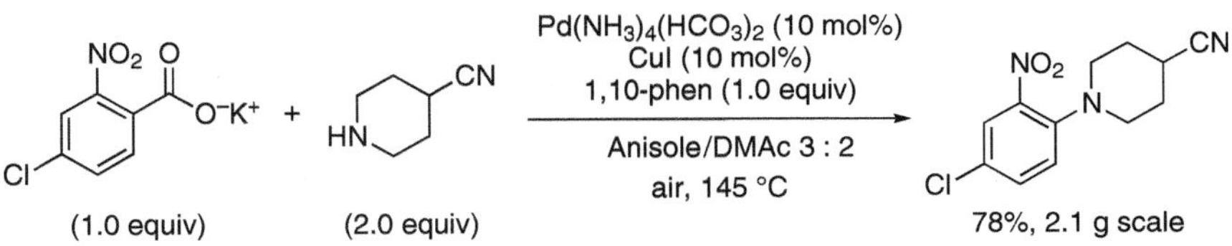

Scheme 10.46 C–H amination under Pd/Cu dual catalysis.

side reactions associated with use of unactivated aliphatic amines as coupling partners under oxidative decarboxylative coupling conditions [87]. Extensive optimization indicated that air could serve as a terminal oxidant in this system, without causing detrimental effects on halogens or esters functional groups. Extensive kinetic studies suggested that a Pd(II)/Pd(IV) catalytic cycle was operative, delivering the target aromatic amine products (Scheme 10.46).

10.2.4 Nickel-Catalyzed C–N Coupling

As noted in Section 10.1.2.2, Pd-catalyzed methods in C—N bond couplings have grown exponentially, offering high level of catalytic activity by modulating and fine-tuning ligands' structural and electronic properties. Despite the notable accomplishments

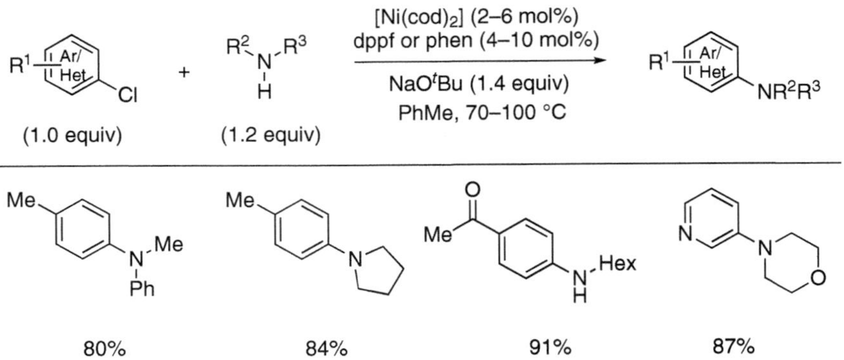

Scheme 10.47 Nickel-catalyzed amination of aryl/heteroaryl chlorides.

in Pd-catalyzed amination, cost of palladium, inaccessibility of tailored ligands, and subsequent removal of metals prompted the quest for alternatives circumventing these shortcomings in large-scale applications [88]. Thus, joint research efforts from academia and industry have been devoted to methodology development to replace precious metals with nonprecious metals in catalysis.

In recent years, interests in employing nickel-based catalysts in cross-coupling reactions have grown extensively because of the abundance and low cost of nickel. Aside from being an inexpensive alternative to palladium, nickel also exhibits exceptional chemical properties, namely, its electropositive characteristics that allow for a high level of reactivity toward less reactive electrophiles such as aryl chlorides. In 2018, Stradiotto and coworker published a review providing a detailed overview of ligand design for applications in Ni-catalyzed C–N cross-couplings. [89] In the arena of Ni-catalyzed aryl amination chemistry, the first example of amination using $NiCl_2$ as a catalyst was reported by the Standard Oil Company in 1950 [90]. Following this report, further studies have shown that air-sensitive Ni(0) complexes could be replaced by a Ni(II) catalyst in amination of bromobenzene, addressing the problem of air sensitivity [91]. During the investigation of the amination of bromobenzene with primary and secondary amines, Cristau and Desmurs discovered that the use of $NiBr_2(bipy)_2$ as a catalyst resulted in the exclusive formation of C–N coupling products, in contrast to the predominant protodehalogenation pathway observed when Pd-catalysts were employed [92].

With these preliminary observations revealing the feasibility of amination chemistry under nickel catalysis, more robust and efficient catalytic systems emerged. In 1997, Wolfe and Buchwald began to optimize the reaction conditions to expand the substrate scope in Ni-catalyzed amination [93]. In this report, amination of aryl chlorides was achieved in good yields by using a Ni(0) precatalyst in the presence of NaO^{t}Bu at high temperature (Scheme 10.47). In some cases, catalyst loading could be as low as 2 mol% to achieve high yields of coupling products, which demonstrated the efficiency of this methodology.

Several research groups (Fort, Nolan, Yang, Masubara, etc.) subsequently modified the catalytic system to bypass the use of discrete, unstable Ni(0) complexes, attempting to generate active Ni(0) species in situ from air- and moisture-stable Ni(II) precatalysts [94]. As described in these reports, NHC as ligands for Ni(0) offer an advantage in couplings with

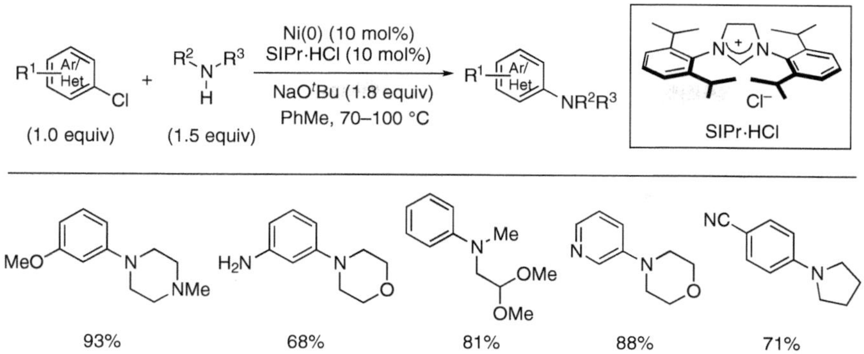

Scheme 10.48 Ni(NHC)-catalyzed amination of aryl/heteroaryl chlorides.

aryl/heteroaryl chlorides without the aid of an external reductant (Scheme 10.48). A review detailing the design and applications of NHC ligand in Ni-catalyzed C–N cross-couplings was published by Chetcuti and coworkers [95].

In addition, Chen and Yang have shown that a Grignard reagent could play a dual role in the reduction of Ni(II) precatalyst and formation of amido-Mg nucleophilic species [96]. Unfortunately, this system was limited to aryl bromides and aryl iodides and suffered from narrow substrate scope because of functional group incompatibility.

Overall, to exploit in situ generation of active Ni(0) species in catalyzing C—N bond formation, assistance of an inherent or external reductant remained a prerequisite. In many cases, the choice of reductant was crucial to accommodate a wide variety of functional groups embedded in the substrates. In general, the necessity of a strongly basic reductant to activate Ni(II) precursors led to poor functional group compatibility. To surmount this, Knochel developed a protocol in which polymethylhydrosiloxane (PMHS) was employed as a reductant in conjunction with the use of Ni(II)-precursors in Ni-catalyzed amination (Scheme 10.49) [97].

Early research efforts were heavily devoted to identify an effective nickel-based system employed in catalytic amination reactions, although high reaction temperatures associated with these conditions remained a limitation. In 2010, Nicasio and coworkers

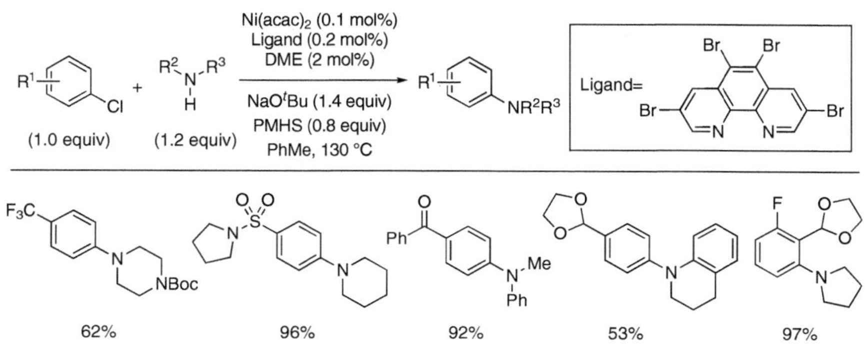

Scheme 10.49 Nickel-catalyzed amination of aryl chlorides using silane reductant.

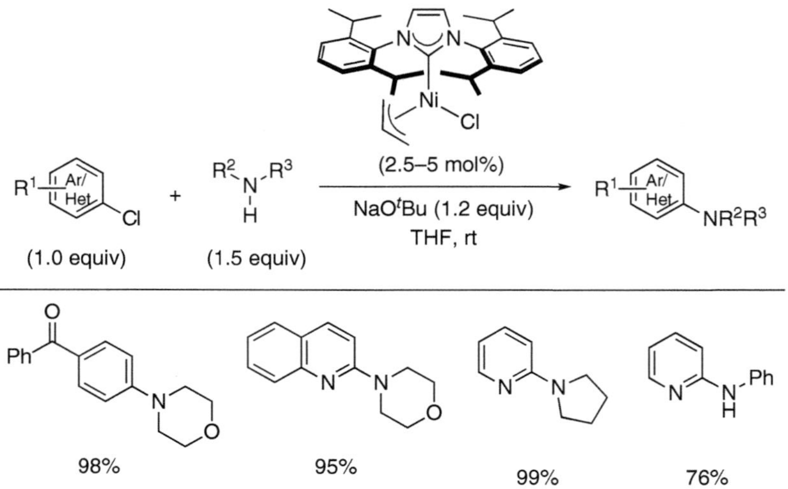

Scheme 10.50 First example of room temperature nickel-catalyzed amination of aryl/heteroaryl chlorides.

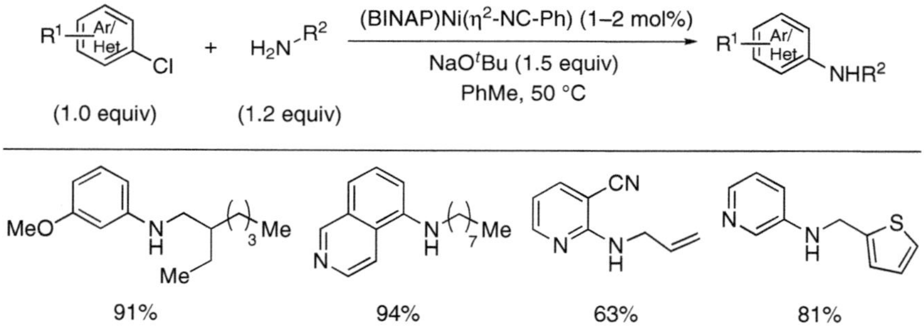

Scheme 10.51 Nickel-catalyzed amination of aryl/heteroaryl chlorides with primary amines.

reported the first example of Ni-catalyzed amination conducted at room temperature using Ni(allyl)(IPr)Cl (Scheme 10.50) [98]. Remarkably, under these mild conditions, heteroaryl chlorides were suitable coupling partners, enabling access to aryl- and heteroaryl amines. Following this report, Yang and Matsubara also demonstrated the feasibility of room temperature nickel-catalyzed amination [99].

In 2014, Hartwig and coworkers described the first amination of unactivated aryl/hetereoaryl chlorides with primary aliphatic amines catalyzed by [Ni(η^2-NCPh)(BINAP)] as Ni(0) source [100]. With very low catalyst loading (1–2 mol%) and mild conditions, a wide range of electronically varied aryl/heteroaryl chlorides were successfully coupled with primary alkyl amines. In this report, detailed kinetic studies were conducted to elucidate the mechanism and to provide a better understanding of the catalytic cycle, in order to further broaden the scope of this transformation, and to provide insights into future catalyst design (Scheme 10.51).

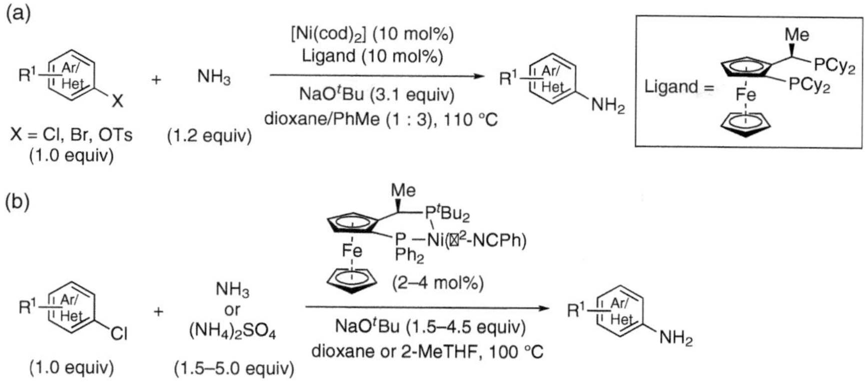

Scheme 10.52 Comparison of (a) Stradiotto's and (b) Hartwig's nickel-catalyzed amination.

In the following year, Stradiotto and Hartwig independently investigated the viability of employing ammonia as amine nucleophile (Scheme 10.52) [101]. Coupling with ammonia poses many challenges because of its inherent affinity to transition metals and its strong N—H bond that renders the oxidative addition of N—H bond unfavorable. Both groups arrived at nearly identical conclusions, the conditions differing in the discrete formation of ferrocene-derived nickel complex developed by Hartwig. Additionally, Hartwig and coworkers showed the compatibility of ammonium salts in their catalytic system. Thus, this breakthrough of using ammonia or an ammonium salt as an amine source in nickel-catalyzed amination demonstrated the potential of nickel-based catalysts as a comparable alternative to modern palladium complexes.

More recently, Fout and coworkers identified the use of $LiN(SiMe_3)_2$ (lithium bis[trimethylsilyl]amide) as a practical ammonia surrogate to effect C—N bond formation using nickel [102]. This method displayed good functional group tolerance and was applicable to amination of heterocycles. Notably, coupling of 2-bromonapthalene with $LiN(SiMe_3)_2$ was performed on gram scale and at low catalyst loading, rendering this process amenable to large-scale production (Scheme 10.53).

In the past decade, significant advances were made to expand and enhance the synthetic utility of this methodology. Particular effort has been devoted to investigations of synthetic equivalents alternative to aryl halides as electrophilic coupling partners (Scheme 10.54). In 2008, Gao and Yang reported the first example of the use of aryl tosylates derived from readily accessible phenols in Ni-catalyzed amination [103]. Later, Chatani and coworkers published the first example of amination of aryl/heteroaryl methyl ether under nickel catalysis, albeit with limited substrate scope with respect to amine coupling partners and

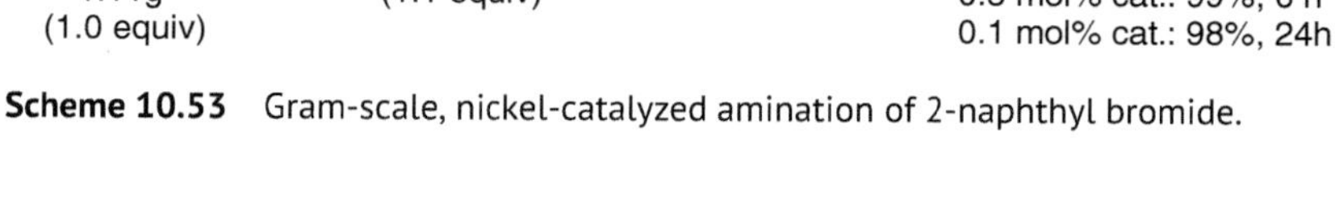

Scheme 10.53 Gram-scale, nickel-catalyzed amination of 2-naphthyl bromide.

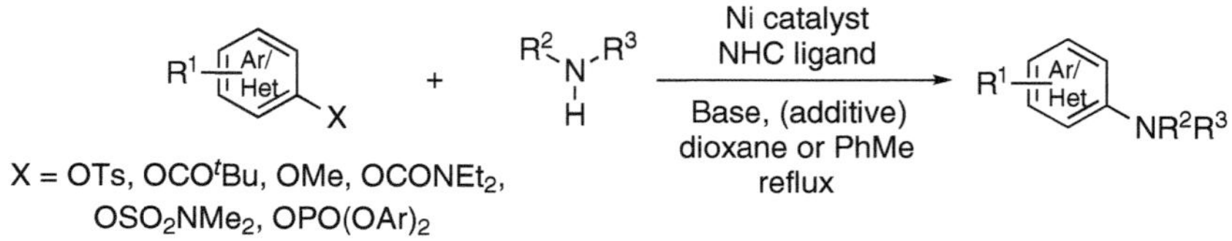

Scheme 10.54 Nickel-catalyzed amination of electrophilic coupling partners other than aryl halides.

high loadings of Ni(0) catalyst and ligands [104]. The author also subsequently discovered the use of aryl carboxylates as aryl electrophiles in nickel-catalyzed amination [105]. The type of phenolic derivatives as coupling partners in amination was further extended to aryl phosphates by Yang and coworker [106]. Following these reports, Garg and Ackermann independently demonstrated the utility of aryl sulfamates as phenol-derived electrophilic counterparts in amination [107, 108a, c–d]. Garg and coworkers also showed that aryl carbamates could participate in nickel-catalyzed amination [108b]. Later, they reported that by employing $NiCl_2$(DME) as a Ni(II) precatalyst in combination with phenylboronic acid pinacol ester as mild reductant, amination of both aryl carbamates could be achieved [108c].

Apart from phenol-derived electrophilic arene coupling partners, Wang and coworker published the first example of nickel-catalyzed amination of aryl fluoride with secondary and benzyl amines [109]. Following this work, Iwai and coworkers expanded the scope of amine nucleophile to primary amines in the presence of a bisphosphine-ligated nickel complex. Application of this method was demonstrated with the synthesis of *N*-PMB-protected paroxetine, a bioactive molecule, and of a liquid crystal material (Scheme 10.55) [110].

Buchwald and coworkers later reported a new air-stable DPPF-ligated nickel precatalyst that allows for efficient couplings with arene substrates bearing base-sensitive functional groups. This robust catalyst demonstrated its ability to effect C–N couplings between a wide range of amine nucleophiles and aryl/heteroaryl electrophiles, namely, aryl/heteraoaryl chlorides, triflates, mesylates, and sulfamates (Scheme 10.56) [111].

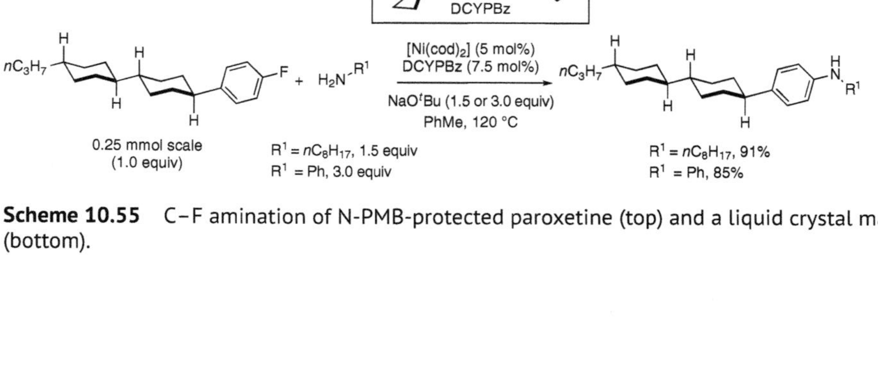

Scheme 10.55 C–F amination of N-PMB-protected paroxetine (top) and a liquid crystal material (bottom).

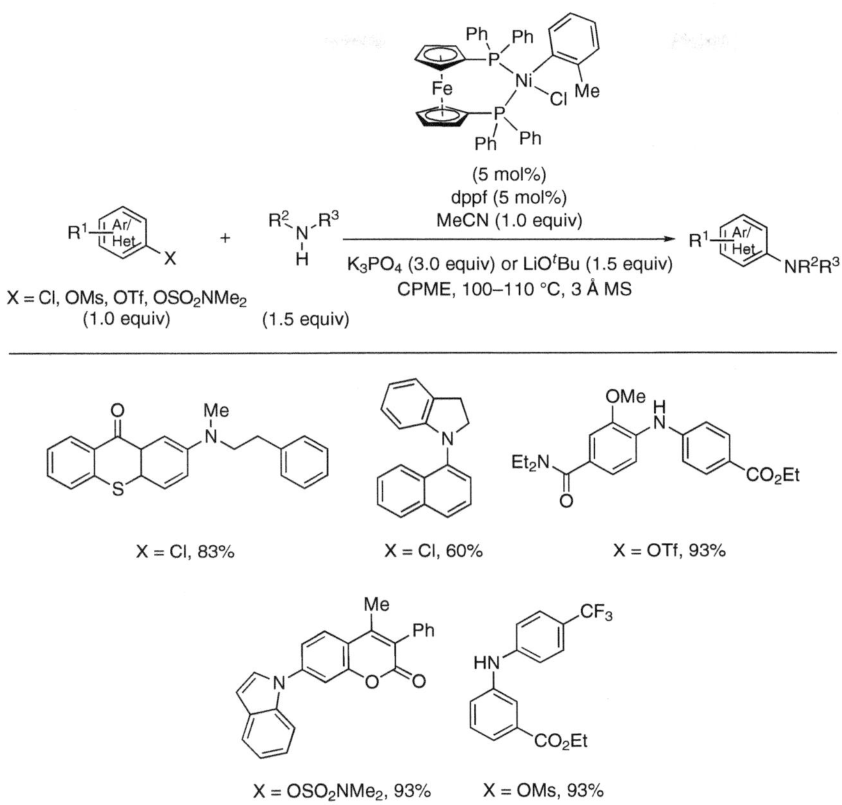

Scheme 10.56 Nickel-catalyzed C–N couplings developed by Buchwald.

Panahi and coworker reported nickel-catalyzed amination of phenol by way of in situ activation of C—O bond in the presence of catalytic 2,4,6-trichloro-1,3,5-triazine (TCT) as an activating reagent [112]. This strategy took advantage of catalytic activating reagent to enable direct C–O amination with a wide range of coupling partners with respect to phenolic electrophiles and amine nucleophiles. Moreover, functional group compatibility of this methodology was demonstrated by using estrone as a phenol coupling partner. The corresponding arylamine product was obtained in good yield (Scheme 10.57).

In 2018, the Dow Chemical Company reported the use of aryl fluorosulfonates as an inexpensive alternative to existing aryl halides in nickel-catalyzed amination with aromatic and alkyl amines [113]. By taking advantage of the simple preparation of aryl fluorosulfonates from phenol and sulfuryl fluoride, a wide range of phenolic derivatives could undergo nickel-catalyzed amination (Scheme 10.58). However, one-pot operation exploiting the in situ formation of aryl fluorosulfonate followed by subsequent amination has not yet been demonstrated with Ni-catalytic system.

Although some drawbacks of nickel-based catalysts hamper their widespread use in construction of C—N bond, the economic benefits of nickel as first-row transition metal is continuing to draw attention from academia and industries. Recent advances in

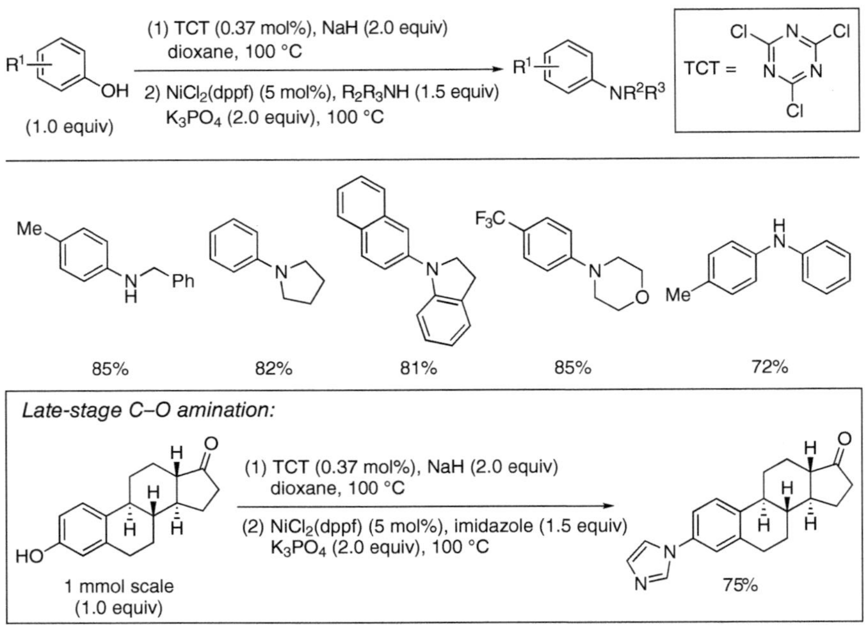

Scheme 10.57 Nickel-catalyzed amination of phenols via C–O activation.

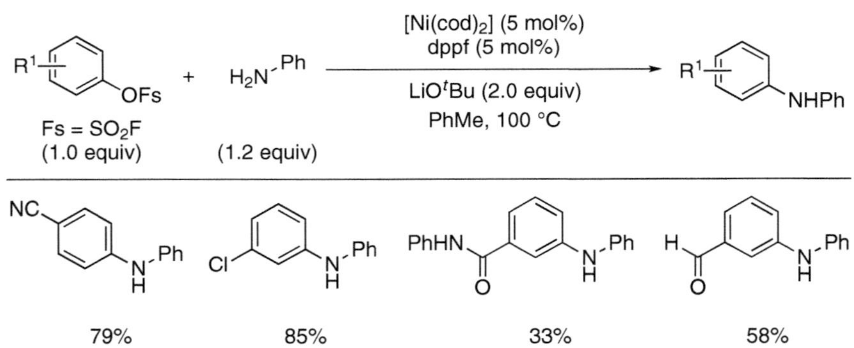

Scheme 10.58 Nickel-catalyzed amination of aryl fluorosulfonates.

nickel-catalyzed C–N coupling have collectively highlighted its potential to deliver comparable outcomes as those obtainable with the state-of-the-art palladium and copper-based catalysts. Furthermore, understanding and designing nickel–ligand complexes would allow for innovative discovery of future applications in C–N cross-couplings.

10.2.5 Other Metal-Catalyzed Cross-Couplings

The growing interest in identifying metal catalysts for amination that are of low cost and environmentally friendly has led to the disclosure of several protocols based on metals other

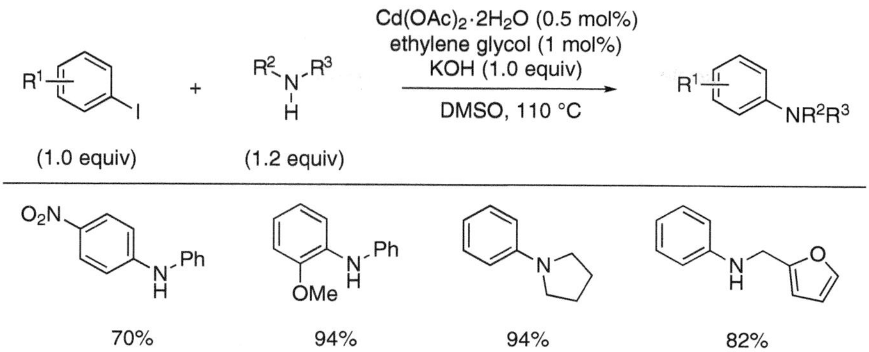

Scheme 10.59 Cadmium-catalyzed amination of iodobenzene.

than palladium, copper, or nickel. Punniyamurthy and coworkers showed that the combined use of a cadmium(II) precatalyst and ethylene glycol was able to effect the coupling of iodobenzene with a variety of amines (Scheme 10.59). Lack of regioselectivity observed in substituted aryl iodides suggested the in situ formation of a benzyne intermediate [114].

In the context of the exploration of sustainable, inexpensive, and nontoxic catalytic systems complementary to the existing aryl amination protocols, iron-based catalysts are of considerable interests because of their high abundance in nature, low cost, and environmentally benign character. Early developments of iron-catalyzed *N*-arylation of various nitrogen nucleophiles demonstrated the viability of iron in promoting C—N bond formation [115]. These reports sparked the interest in expanding the scope of amine nucleophile in coupling with aryl halides. Liu and coworkers identified Fe_2O_3 as a suitable catalyst, in combination with L-proline as a ligand, to effect C—N bond formation using both aliphatic and aromatic amines as coupling partners (Scheme 10.60). Remarkably, halopyridines were tolerated in couplings with morpholine, albeit poor regioselectivity was observed when substituted aryl halides were used.

Following this work, a new heterogeneous catalytic system was developed using a recyclable graphite-supported iron(III) complex in the absence of a ligand [116]. Aryl iodides successfully coupled with various amines, delivering products of C—N bond formation in

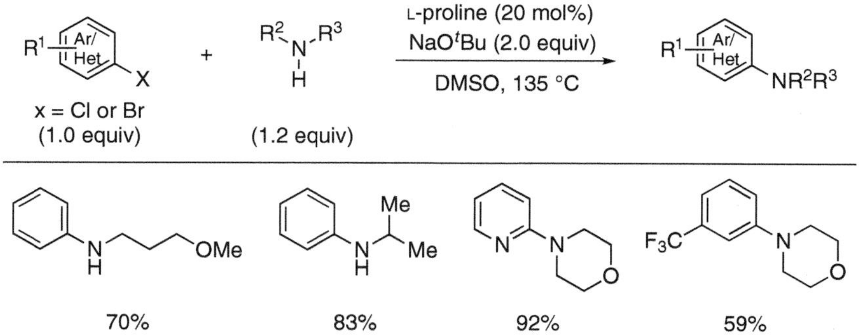

Scheme 10.60 Iron-catalyzed amination of aryl halides.

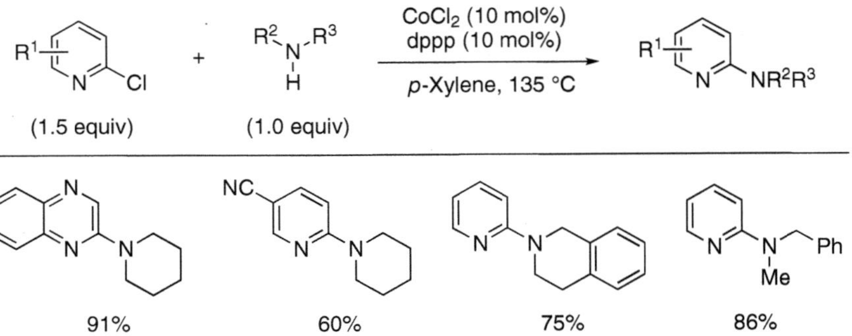

Scheme 10.61 Cobalt-catalyzed amination of *N*-aromatic-2-chlorides.

modest to good yields. The reusability of this catalyst was tested, and no substantial loss of reactivity was observed for up to five cycles, rendering this protocol robust and attractive.

Besides iron-based catalysts, cobalt-based catalytic systems in aryl C—N bond formation have been explored [117]. In the case of couplings with aryl iodides, Co(II) catalyst in combination with *N*,*N*′-dimethylethylenediamine (DMEDA) was found to be effective, albeit with very limited scope regarding the nitrogen nucleophiles. Unfortunately, under these conditions, aromatic and alkyl amines did not produce useful yields. Later, a new catalyst was reported, in which 1,3-bis(diphenylphosphanyl)propane (DPPP) was employed as a ligand (Scheme 10.61). This catalytic system was specific for amination of 2-chloroazines with secondary amines, affording a variety of heteroaryl amines [118].

Inspired by the capability of rhodium-based catalysts in constructing C—N bonds in the area of hydroamination and other transformations, Chang and coworker reported the first example of using NHC-bound rhodium complex in amination of aryl halides [119]. Initial conditions utilized NaO^{t}Bu as a base and afforded aminated product in high to excellent yields. In order to achieve higher levels of functional group tolerance, Cs_2CO_3 was employed as a milder base to accommodate base-labile groups. This modification resulted in facile access to a diverse range of arylamines (Scheme 10.62). Remarkably, this protocol was compatible with amino acid esters, further expanding the scope of this chemistry.

Although Chan–Lam amination of arylboronates mediated by copper is well established, as detailed in Section 10.3.2.4, it is worth noting that Hartwig demonstrated that arylsilanes could also participate in C—N bond formation mediated by copper. Following the report by Lam wherein formation of hypervalent aryltrimethoxysilanes assisted by a fluoride source could effect C—N bond formation [120], Hartwig and coworkers reported the use of arylsilanes readily prepared by way of C–H functionalization as coupling partners [121]. This approach exploited the highly selective C–H silylation methodology developed in the group that surmounts multiple stages of prefunctionalizations to access arylsiloxanes. A wide range of arylsilanes were prepared and employed in the copper-mediated amination reaction with a variety of nitrogen nucleophiles to afford products of C–Si amination (Scheme 10.63).

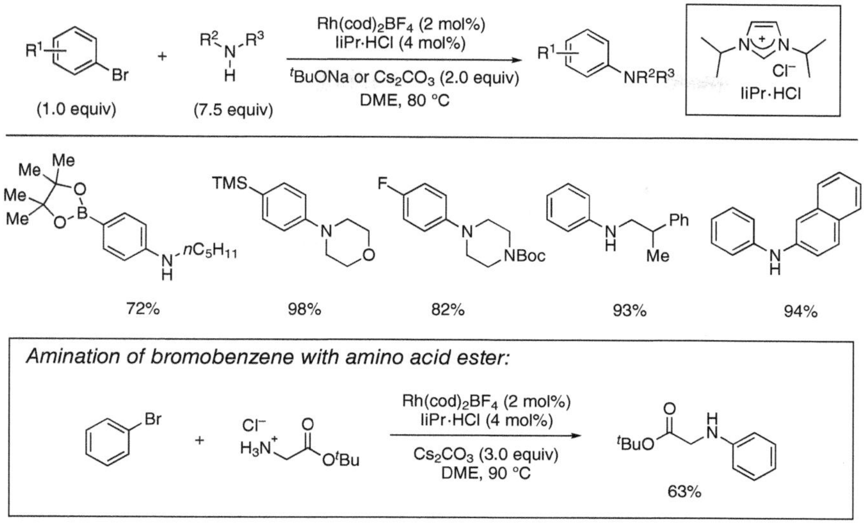

Scheme 10.62 Rhodium-catalyzed amination of bromobenzene derivatives.

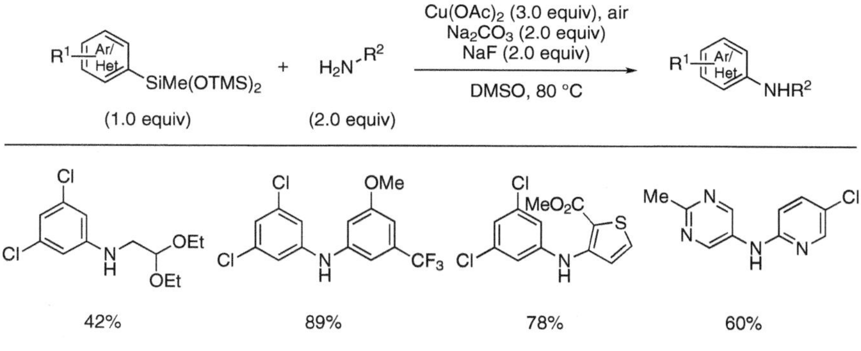

Scheme 10.63 Copper-mediated amination of arylsilanes.

10.2.6 Reductive Amination

Among C—N bond forming methods performed in the pharmaceutical industry, the reductive amination reaction plays an utmost important role primarily because of its simple operation, mild reaction conditions, and well-established records in synthetic chemistry community [122]. A recently published review highlighted the applications of the reductive amination reaction in the syntheses of commercial drug substances and compiled the corresponding reductants used in large-scale reductive amination and relevant green chemistry metrics [123]. Current effort involves fine-tuning the conditions of the classical protocol by implementing new methodologies and technologies to render this transformation stereoselective.

In the context of arylamine synthesis, reductive amination is employed to further functionalize simple arylamines with carbonyl counterparts, resulting, in other words, in the

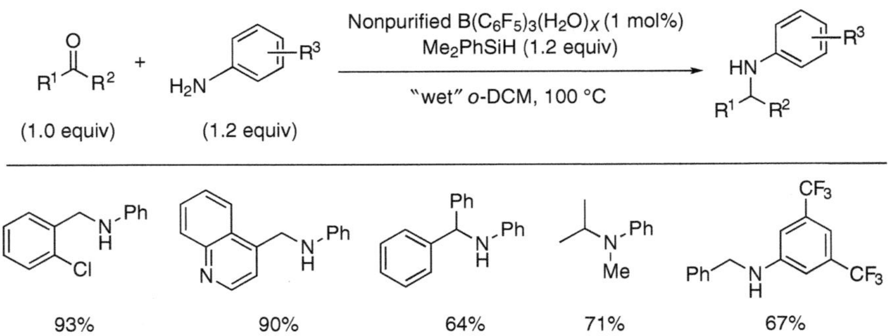

Scheme 10.64 $B(C_6F_5)_3-H_2O$-catalyzed reductive amination using silane reductant.

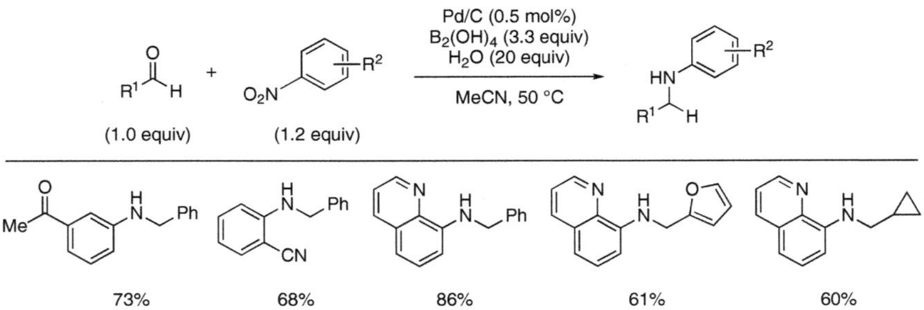

Scheme 10.65 Palladium-catalyzed reduction and reductive amination of nitroarenes using water as a hydrogen source.

mono- and/or bis-alkylation of aniline derivatives. Traditional methods involve formation of arylamine-derived imine from an aldehyde or a ketone, followed by imine reduction in the presence of a hydride donor [124]. This approach can serve as a useful alternative to C–N cross-coupling reactions, in which conditions are typically substrate dependent and relatively more challenging with primary amines under traditional metal catalysis. In 2016, Ingleson and coworkers reported the use of $B(C_6F_5)_3$ as a catalyst for the reductive amination of primary and secondary arylamines in the presence of a silane reductant (Scheme 10.64) [125]. Taking advantage of the water tolerability of $B(C_6F_5)_3$, the author attributed the success of this transformation to the ability of arylamines and their derived imines to (i) reversibly deprotonate $H_2O–B(C_6F_5)_3$ and (ii) reversibly bind to $B(C_6F_5)_3$. In the absence of these reversible events, this catalytic system would not be effective.

In the following year, Zhou and coworkers discovered the use of water as the hydride source to effect reductive amination of nitroarenes under palladium catalysis, mediated by diboronic acid (Scheme 10.65) [126]. In this report, the author proposed a nontypical reductive amination mechanism, in which a boronic acid-bound palladium hydride initially reduced the nitro group to the corresponding aniline that subsequently underwent reductive amination with the aldehyde.

More recently, metal-free reductive aminations to synthesize *N*-alkylated anilines were accomplished by exploiting the unique properties of 2-aminoquinazolin-4(3*H*)-one, a hydrogen bonding organocatalyst that possesses Brønsted acidic and Lewis basic sites

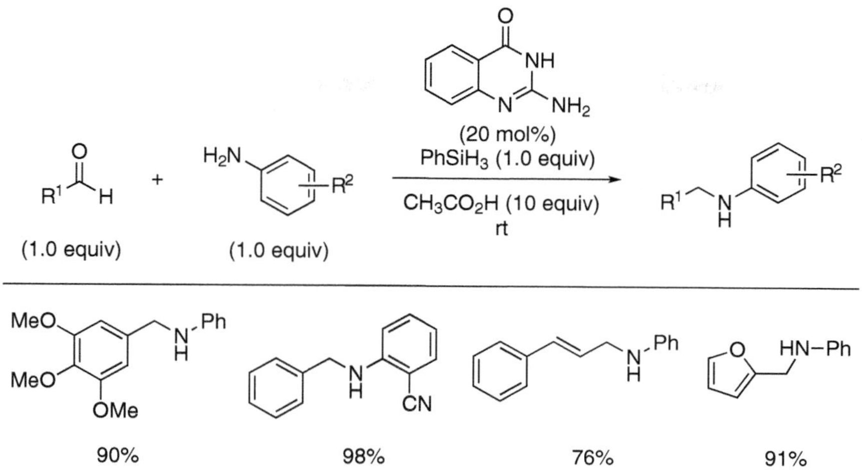

Scheme 10.66 Metal-free reductive amination using hydrosilanes.

[127]. In the presence of a silane reductant, the author describes that the use of acetic acid as a dehydrating agent lowered the reaction temperature to room temperature, enabling a simple and efficient protocol to access a diverse range of anilines (Scheme 10.66).

A great potential of C—N bond construction via reductive amination is its amenability to synthesize chiral amines. Asymmetric reductive amination has drawn considerable attention [128]. In particular, enzymatic reductive amination brings several key advantages. For instance, highly selective processes would circumvent the carbonyl reduction as a side reaction and minimize side products formation. The hallmark of this method is the synthesis of chiral amines from prochiral ketones with an excellent level of selectivity in one enzymatic step. In a recent report, a fungal reductive aminase, *Asp*RedAm, was identified to effect the asymmetric reductive amination with a variety of simple alkyl amines [129]. Unfortunately, use of aniline remained an unsolved problem. Although mutation of this enzyme resulted in improved reactivity, further optimization is still required to enhance specificity. Nonetheless, this work highlights the potentials of engineering RedAm to accommodate a larger class of ketones and amines to achieve stereoselective enzyme-catalyzed reductive amination.

10.2.7 Hydroamination

Another class of C—N bond forming reactions to access arylamines is hydroamination of unsaturated C—C bonds. The significance of this methodology is displayed by the enormous body of previous research, including multiple review articles published on this topic [130]. This atom-economical approach accounts for its long-standing history, establishing a strong foundation for understanding the limitations and drawbacks inherent to this transformation. Particularly, hydroamination of unactivated substrates is not a kinetically viable process; hence, an appropriate catalyst is typically required. This in turn enables opportunities to design new catalytic systems to address the existing shortcomings, namely,

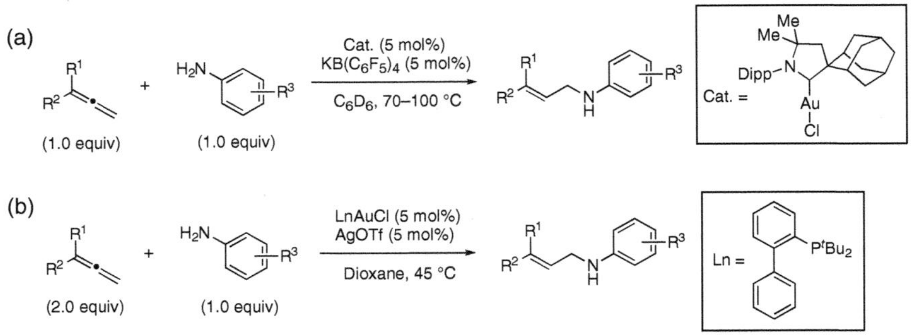

Scheme 10.67 Gold-catalyzed intermolecular hydroamination of allenes with arylamines. (a) Bertrand's methodology and (b) Wildenhoefer's methodology.

the electronic properties of the C—C multiple bonds and the amine counterpart. Additionally, chiral amine synthesis by way of this methodology is of great value to the synthetic organic chemistry community. A review article highlighting the achievements of nonprecious metal-catalyzed hydroamination was published by Parvulescu and coworker [131]. This section focuses on the more recent examples of hydroamination, reported in the past decade.

Intermolecular hydroamination of allenes with arylamines has been reported using palladium catalysis, albeit with limited substrate scope [132]. Notably, Yamamoto and coworker reported the use of a gold(III) catalytic system to access arylamines via hydroamination of allenes accompanied by allene chirality transfer, thus enabling preparation of chiral amines [133]. This method, however, was restricted to monosubstituted and 1,3-disubstituted allenes. Further studies by Bertrand and Widenhoefer independently led to the development of a more defined gold(I) catalyst associated with complex ligands to address the limitations in allene scope [134]. Both systems delivered allylic amines in good yields, demonstrating the applicability of the methodology (Scheme 10.67).

One worth noting feature of hydroamination is the regiochemical outcome of the transformation. Good control of regio- and stereoselectivity is crucial for productive functionalization of the π-system to deliver either branched or linear products. Unlike the observed predominantly linear allylic amine product under the aforementioned conditions, Breit and coworkers reported an exclusive formation of branched product under rhodium catalysis [135]. Even more remarkably, in combination with a chiral ligand (Josiphos), good enantioselectivity was achieved, rendering this method the first example of enantioselective intermolecular hydroamination of terminal allenes with aniline derivatives (Scheme 10.68). Mechanistically speaking, in contrast to gold-catalyzed hydroamination which involves amino-auration followed by protonation to afford the allylic amine product, this catalytic system potentially underwent hydrometalation of the terminal double bond to generate a π-allyl-Rh intermediate, an event that dictated the regioselectivity outcome.

In 2016, Taillefer and coworkers disclosed the use of a copper-based catalyst to effect hydroamination of terminal allenes [136]. A diverse range of allenes and anilines were evaluated, and exclusively linear products were obtained under the developed conditions (Scheme 10.69).

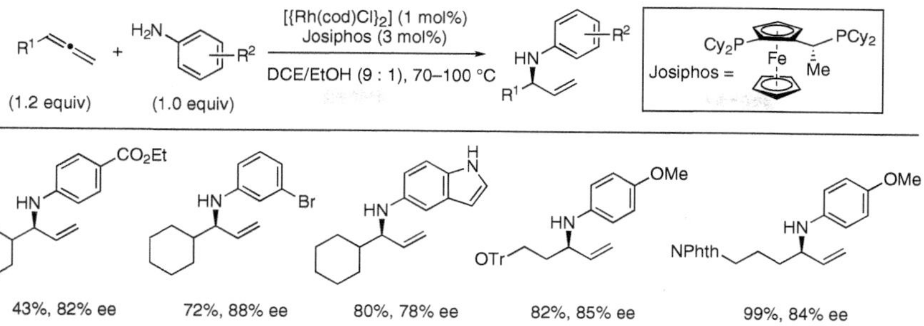

Scheme 10.68 Enantioselective rhodium-catalyzed hydroamination of monosubstituted allenes.

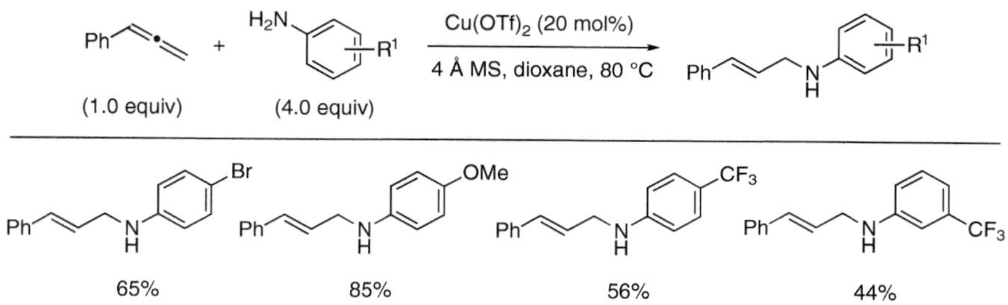

Scheme 10.69 Copper-catalyzed hydroamination of phenylallene.

In the arena of hydroamination, a variety of π-electrophiles have been intensively studied. Aside from allenes with different substitution patterns, alkenes and alkynes are suitable electrophilic partners for this transformation. Similarly, the regio-, chemo-, and stereoselectivity of the reaction rely heavily on the design of the catalytic system. In cases where alkynes are the electrophilic counterparts, a hydrogenation step is necessary to provide the amine product. By taking advantage of a dual catalysis approach, Che and coworker developed a tandem intermolecular hydroamination/transfer hydrogenation of alkynes to access enantiomerically enriched chiral secondary amines (Scheme 10.70) [137]. The author described the use of a Au(I) complex in combination with a chiral Brønsted acid catalyst to accomplish the efficient synthesis of a broad range of chiral amines. One drawback for this transformation was the stoichiometric amount of Hantzsch ester **41** required to reduce the ketimine intermediate formed in situ, resulting in a pyridine as a coproduct.

Later in 2012, Beller and coworkers reported an alternative three-component catalytic system consisting in a Au(I) catalyst, a Knölker's iron complex, and a chiral Brønsted acid to achieve reductive hydroamination enantioselectively [138]. This stepwise process commenced with gold(I)-catalyzed hydroamination, followed by a cooperative hydrogenation catalyzed by chiral phosphoric acid in the presence of the iron catalyst and H_2. A broad scope of alkynes and aniline derivatives demonstrated the applicability of this approach, enabling facile synthesis of chiral amines with high levels of enantioselectivity.

Copper(I) hydride catalysis is of immense importance in hydroamination reaction. In 2013, two research groups independently developed the use of copper hydride to catalyze hydroamination of alkenes with *N*,*N*-dialkylbenzoate [139]. Over the past decade, Buchwald has published notable work to show the versatile applications of this copper(I)

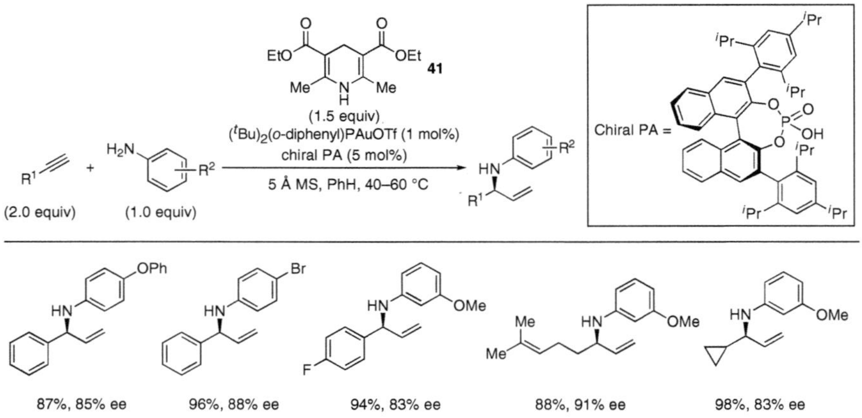

Scheme 10.70 Enantioselective hydroamination of alkynes by Au(I)/chiral Brønsted acid dual catalysis.

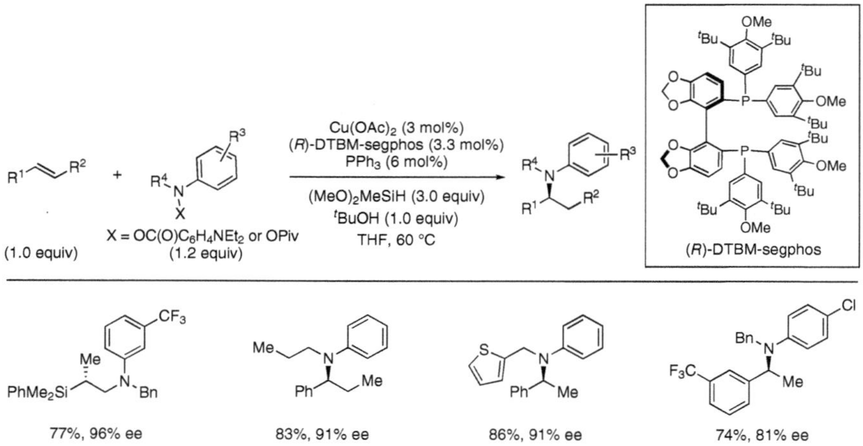

Scheme 10.71 Copper-catalyzed asymmetric hydroamination.

hydride chemistry. In 2016, a review article highlighted the achievements of copper hydride-catalyzed hydroamination of alkenes and alkynes [140]. Recently, Buchwald and coworkers disclosed the synthesis of chiral *N*-arylamine by way of copper catalysis using *N*-arylhydroxylamine esters as electrophilic aminating agents [141]. Under the conditions previously developed in the group for hydroamination using *N*,*N*-dialkylbenzoate, a major by-product indicative of the unproductive reductive cleavage of N—O bond of aminating agent was observed. Screening of reaction conditions identified PPh_3 as the crucial additive to prevent the unproductive pathway, which was hampering the reaction. Adopting this newly established protocol, a variety of α- and β-chiral arylamines were prepared from a wide range of olefins bearing a diversity of functional groups (Scheme 10.71).

In the same year, Li applied copper hydride and nitrene chemistry to access enantioenriched α- and β-chiral arylamines through hydroamination of styrene derivatives with anthranil (Scheme 10.72) [142]. By exploiting the alcohol functional handle

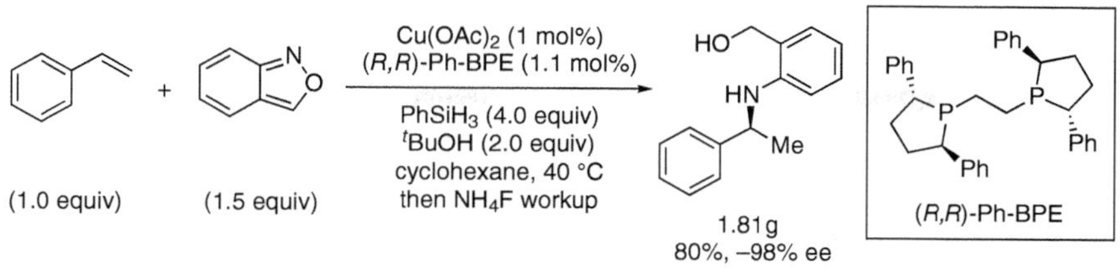

Scheme 10.72 Enantioselective copper-catalyzed hydroamination of styrene with anthranil.

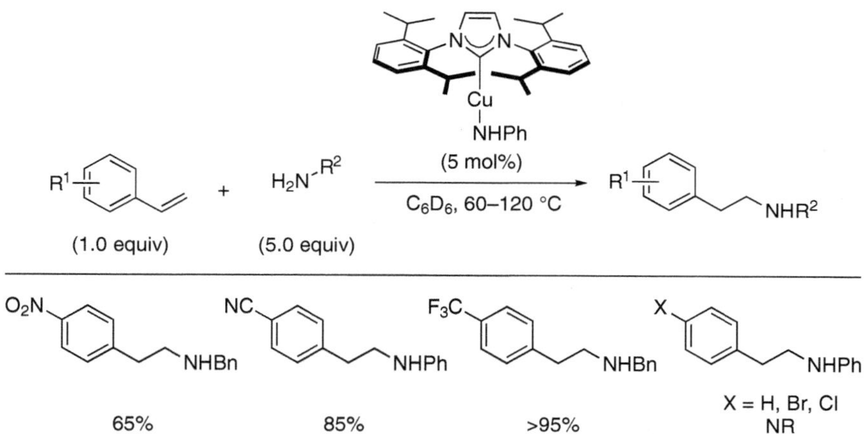

Scheme 10.73 Copper-catalyzed anti-Markovnikov hydroamination of vinylarenes.

resulting from the reaction with anthranils, final bifunctional arylamine product could be further elaborated to a variety of synthetically useful chiral nitrogen-containing compounds.

The first report of anti-Markovnikov hydroamination in arylamine synthesis dates back to 2008 when Gunnoe and coworkers utilized a well-defined copper-amido catalyst complexed with a NHC ligand to effect hydroamination of styrene with aniline derivatives. Although the reaction was highly regioselective, the method suffered from lengthy synthesis of the catalyst and restriction of the scope to terminal styrenes (Scheme 10.73) [143]. The observed regioselectivity was governed by the electronics of the catalytic system. The NHC-bound copper-amido intermediate selectively inserts into the styrene olefin to form a C—N bond at the terminal carbon of the alkene, differing from the typical hydrocupration mechanism that is operative in copper hydride catalysis [144].

To surmount these limitations, Lee developed a readily accessible NHC–CuCl catalyst in combination with KO^{t}Bu to promote hydroamination of a variety of mono- and 1,2-disubstituted styrenes, leading to arylamine products with remarkable anti-Markovnikov selectivity (Scheme 10.74) [145]. Moreover, to demonstrate the utility of this methodology, arylamine **42** subsequently underwent palladium-catalyzed cyclization to the corresponding indoline **43**.

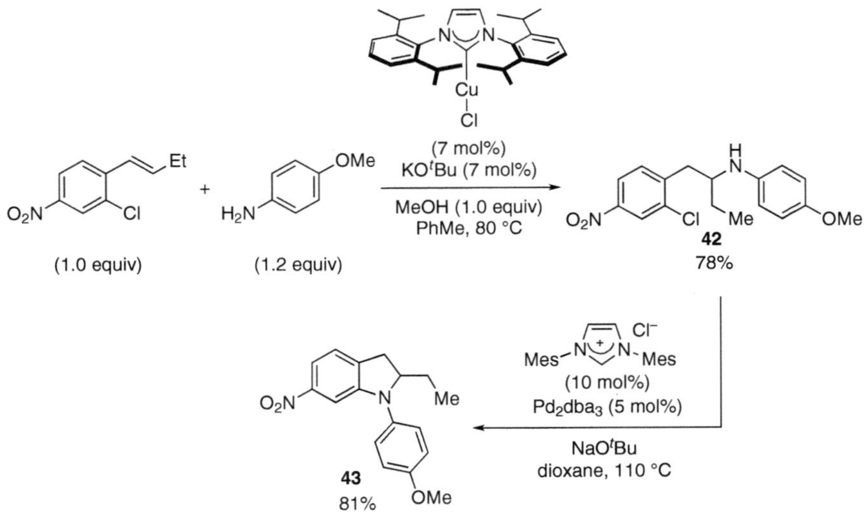

Scheme 10.74 Synthesis of indoline via Cu-catalyzed hydroamination/Pd-catalyzed cyclization.

10.2.8 Summary and Conclusions

Amination technologies have evolved in the past decade to accommodate a large demand in industrial applications. The C–H functionalization approach to incorporate amino functional groups is appealing because of its atom- and step-economical values. However, this arena still requires in-depth studies and investigations to render this set of methodologies available for industrial use.

On the other hand, replacing noble metals with nonprecious metals has been an inspiring area of research. Nickel, in particular, is very attractive in part because of its wide availability and low cost. Additionally, certain nickel catalysts have the unique ability to activate a wide range of electrophilic coupling partners, well beyond the scope of traditional cross-couplings under palladium catalysis.

Although these novel approaches of C—N bond formation remain far away from being widely applicable on industrial scale, these synthetic efforts provide new perspectives and strategies for bond disconnections, which pave the way for future fine-tuning of the newly established processes with the ultimate goal of enhancing efficiency and scalability.

10.3 Advances to Arylamine Formation Using Intensified and More Sustainable Process Technologies

10.3.1 Introduction

In industrialized countries, the intensity of air pollution has attracted broad attention. In China, it triggered in 2013 the promulgation of the strictest-ever clean air policy in the country [146]. Since 2018, environmental protection is one of the key points of the latest “Bluesky Initiative” and has an important impact on the development of chemical manufacturing industry. The major aspects are stricter regulations, enforcement, and environmental health and safety (EHS) policies. According to the new regulations, the laws on waste emission permits will become more stringent. Waste treatment costs are expected to increase significantly (http://www.gov.cn/zhengce/content/2018-07/03/content_5303158.htm). As an example, relevant to amino chemistry, traditional aromatic nitration reactions, such as mixing nitric acid and sulfuric acid, as well as the related downstream reduction and Sandmeyer reactions, are strictly evaluated in fine chemical factories and contract manufacturing organizations (CMOs). This leads to supply chain challenges and often boosts the prices of raw materials that were formerly easily available, especially aniline and nitrobenzene derivatives.

In this evolving scenario, a real event put this subject even more in the focus. Large-scale production of nitro compounds, or even their storage, possesses the potential to cause serious accidents. On 21 March 2019, a major explosion occurred at the Jiangsu Tianjiayi Chemical Co., Ltd. in the Xiangshui Chemical Industrial Park in Jiangsu Province, China. The serious explosion led to 78 deaths, 76 serious injuries, and 640 people were treated in the hospital. The economic loss was estimated to approximately 2 billion ¥/ US$ 286 million. The long-term illegal storage of solid nitration waste was identified as the root cause of the accident. Heat accumulation and spontaneous combustion resulted eventually in explosion (http://www.xinhuanet.com/politics/2019-11/15/c_1125237644.htm). Because of these and many other reasons, advances in C–N coupling with higher compliance to safety concerns, greener processes with less waste, and process intensification attracted the attention of researchers. Flow chemistry, including tubular and microreactors, nanofiltration for metal removal, recovery and recycling, immobilization of catalysts, and supported catalysts are major tools to mitigate the environmental and safety impact.

10.3.2 Flow Chemistry

There have been many advances in flow chemistry/continuous processing in recent years, rendering this technology more feasible and attractive to the pharmaceutical industry. There are many benefits in continuous process approaches to the manufacturing of small-molecule pharmaceutical ingredients, such as process safety, level of control, throughput, higher selectivity, and economic factors [147]. Flow/continuous processes can be useful alternatives to traditional batch procedures, and this has been demonstrated by both industrial and academic chemists [148]. In this section, it will be discussed how flow technology has been used for C—N bond forming reactions, mainly focusing on formation of arylamines. In the past 15 years, this has been one of the most important reactions for the production of APIs.

10.3.2.1 Pd-Catalyzed C—N Bond Forming Reaction

Publications on Pd-catalyzed C–N coupling to form arylamine performed under continuous-flow conditions in academic settings have increased in recent years. In 2016, Buchwald and coworkers [149] published a general and efficient method for C–N cross-coupling with *N*,*N*-dimethyloctanamide as a catalytic cosolvent for biphasic, continuous-flow application. The method was used to synthesize a variety of biarylamines and further applied to a three-step synthesis of imatinib **46**, the API of medical

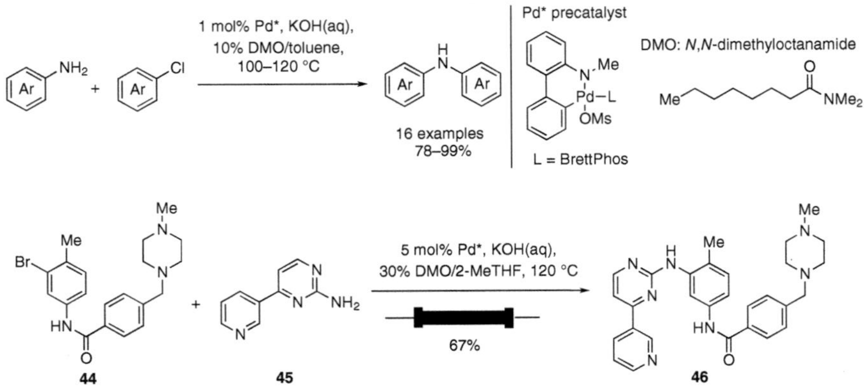

Scheme 10.75 Buchwald's flow synthesis of biarylamines and imatinib **46**.

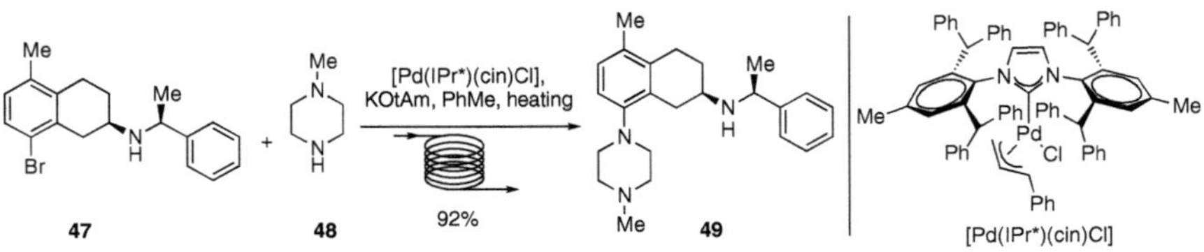

Scheme 10.76 Meadows' continuous synthesis of arylamine **49** with bulky NHC precatalyst [Pd(IPr*)(cin)Cl].

product Gleevec, in good yield without the need of solvent switches. The bromide **44** was mixed with a stream of 2-aminopyrimidine hydrochloride **45** in water, and C–N cross-coupling occurred in the presence of 5 mol% of BrettPhos G4 precatalyst (in 30% *N,N*-dimethyloctanamide/2-MeTHF) at 120 °C in a second packed bed column for an increased mixing efficiency (Scheme 10.75).

Nowadays, the continuous Pd-catalyzed C–N coupling is only rarely used in small- to medium-scale chemical or pharmaceutical production. AR-A2 is a candidate drug previously under development at Astra Zeneca for central nervous system (CNS) disorders. Meadows and coworkers [150] reported a continuous Pd-catalyzed C–N coupling process using a bulky NHC precatalyst for the synthesis of a key intermediate in the synthesis of AR-A2. The stability of the NHC-based catalyst toward acidic conditions played an important role not only for reaction performance but also for allowing a simpler continuous acidic extraction of the product and on-stream recycling of the catalyst in the organic phase. Finally, the flow process demonstrated the amination reaction of **47** and **48** at medium scale in 92% yield of **49** with 92.2% purity (Scheme 10.76). This novel reactor concept for running heterogeneous reactions in flow combines the flexibility of continuously stirred tank reactors (CSTRs) with the smooth operation, low residence time distribution, and excellent heat transfer capability of a conventional flow reactor.

10.3.2.2 Nucleophilic Aromatic Substitution

Bogdan and coworkers [151] reported a high-temperature and high-pressure continuous-flow protocol to carry out nucleophilic aromatic substitution (S_NAr) of heterocycles

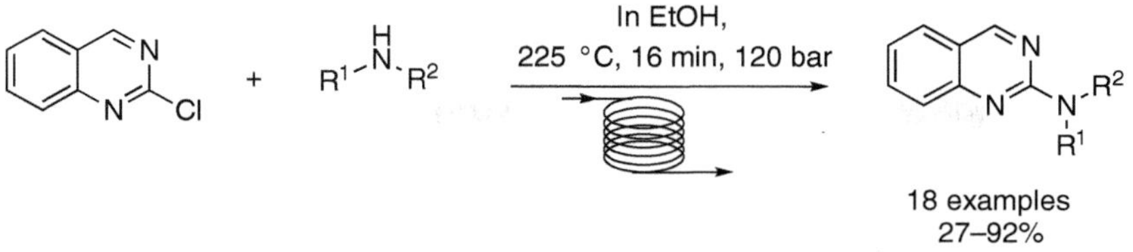

Scheme 10.77 Continuous-flow S_NAr of heterocycles with nitrogen nucleophiles.

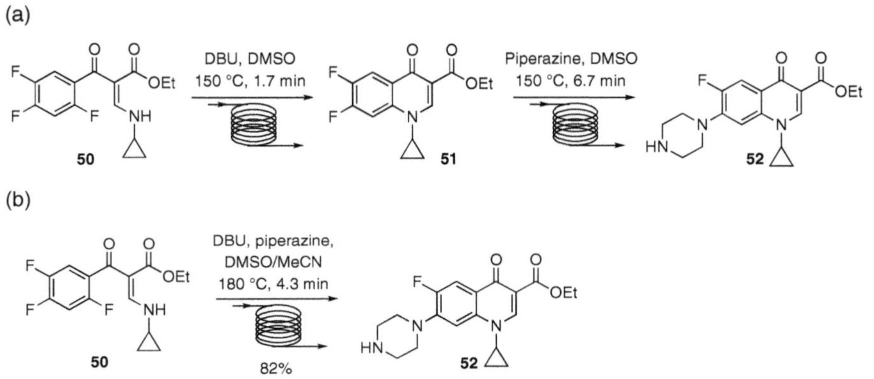

Scheme 10.78 (a) Stepwise and (b) one-pot synthesis with two C—N bond formation in flow.

with nitrogen nucleophiles to synthesize 2-aminoquinazoline analogs in modest to excellent yields. This straightforward procedure used ethanol as the solvent at 225 °C, providing a greener alternative to traditional S_NAr routes. Furthermore, no additional base was needed in the reaction. Product scope was expanded to 2-chloroquinoxalines and 2-chlorobenzimidazoles (Scheme 10.77).

To increase the synthetic efficiency of the manufacturing process of antibiotic ciprofloxacin, Jamison and coworkers [152] developed its rapid total synthesis in continuous flow from simple building blocks via a sequence of six chemical reactions. Two C—N bonds were formed in the cyclization and S_NAr substitution in the flow system. Intermediate **50** was first deprotonated triggering cyclization giving **51**. Aromatic nucleophilic substitution with piperazine on this substrate could be carried out either in a stepwise or one-pot mode. The telescoped process afforded the ester **52** in 82% overall yield in flow under optimized conditions (Scheme 10.78).

A S_NAr reaction in flow between aminopyrazole **53** and chloropyrazine **54** was used by researchers at Eli Lilly [153] in the continuous-flow synthesis of prexasertib monohydrate on an industrial scale. The S_NAr was conducted at 85 °C for a residence time of 180 minutes with good yield and selectivity in a low-pressure plug flow reactor (PFR) constructed from inexpensive perfluoroalkoxy (PFA) tubing. At production scale (2.9 kg/day of product), the PFR volume was 2.8 l and consisted of a PFA tube (6.35 mm ID by 91 m) immersed in a heated bath (Scheme 10.79).

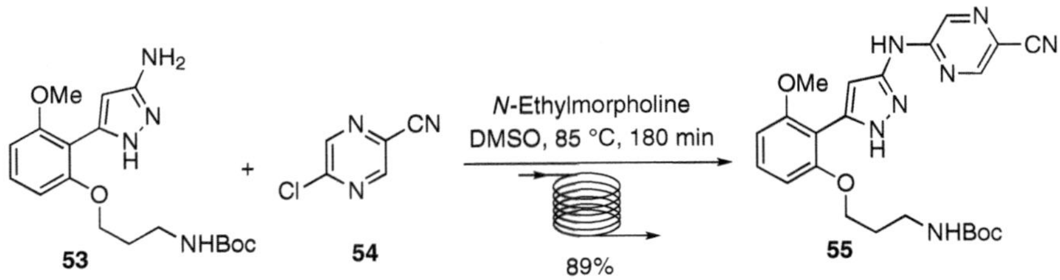

Scheme 10.79 Continuous synthesis of pyrazole **55** with a S_NAr reaction.

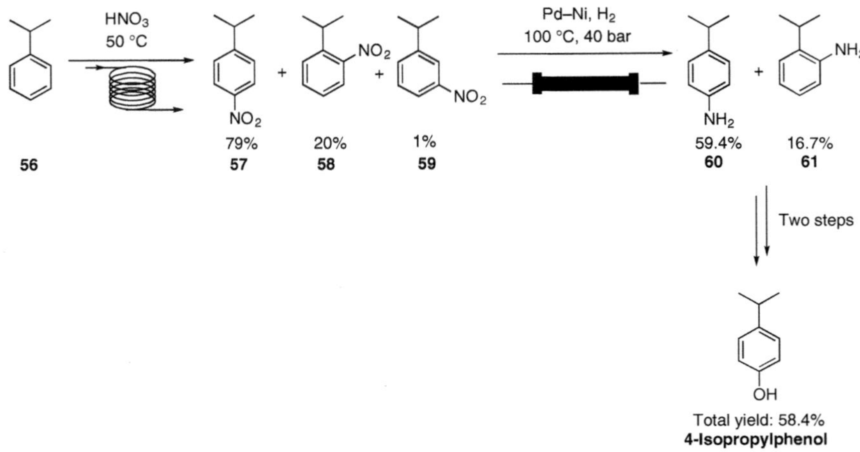

Scheme 10.80 Continuous synthesis of cumidine via nitration and hydrogenation.

10.3.2.3 Telescoped Sequence of Nitration and Hydrogenation in Flow Synthesis

The continuous-flow process for the synthesis of arylamine using a sequence of nitration and reduction is a common approach and numerous examples have been published recently. For these two reactions, usually a good yield is obtained with rapid mixing, high interfacial mass transfer rate, good temperature control via efficient heat transfer, and adequate residence time. Recently, Kulkarni and coworkers [154] reported a sequential flow synthesis of isopropyl phenol involving four-step exothermic, endothermic, and temperature-sensitive reactions such as nitration, reduction, diazotization, and high-temperature hydrolysis. The intermediate nitroarene of cumidine **60** was synthesized by nitration of cumene with fuming nitric acid producing 2- and 4-nitrocumene **57** and **58**, which converted into the corresponding cumidines **60** and **61** by hydrogenation using Pd/Ni catalyst in an H-cube reactor (Scheme 10.80). The sequential flow synthesis was applied to synthesize 4-isopropyl phenol with a total yield of 58%.

10.3.2.4 Chan–Lam Coupling

In 2009, the groups of Stevens and van der Eychen's [155] reported the copper(II)-mediated Chan–Lam reaction in a commercial microreactor. More recently, the group of Baxendale and coworkers [156] developed a catalytic process of Chan–Lam coupling reactions of various amines and arylboronic acids with molecular oxygen as an oxidant in a flow system

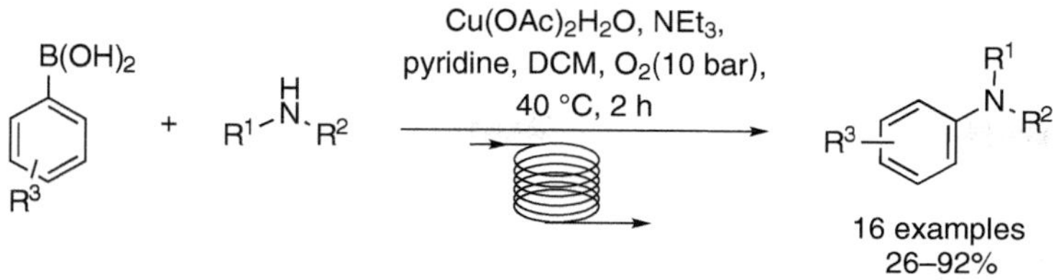

Scheme 10.81 Catalytic Chan–Lam reaction in continuous flow developed by Baxendale.

using a tube-in-tube gas reactor. Moderate to good yields (26–92%) were obtained for a small library of heteroaromatic products. The process has been shown to be robust over extended reaction times (Scheme 10.81).

10.3.3 Immobilization of Catalysts/Supported Catalysts

Transition-metal-catalyzed C—N bond formation, more particularly with palladium catalysis, has emerged as a common technology for the preparation of aromatic amines in fine chemicals and pharmaceuticals [6c]. The recycling of the catalyst has been performed in several ways, including the use of nanofiltration membranes, column chromatography separation, aqueous biphasic systems, and specially designed NHC-Pd catalyst system [157]. The high cost and the difficulty to remove palladium and ligand remains in general one of the major challenges for homogeneous catalysis. The need for more environmentally benign process and the requirement to develop simplified protocols has been clearly identified. One special example is the immobilization of catalysts with little metal leaching and catalyst decomposition. Recently, remarkable progress has been summarized regarding the activity and recyclability of heterogeneous supported palladium-based catalytic systems, mostly for C–C cross-coupling reactions [158]. Immobilization of catalysts for C–N formation draws attention to academic and industry study especially because of the high price of palladium (see also Section 10.4.3).

A study of the perspectives offered by heterogeneous catalysts in a Buchwald–Hartwig amination reaction has been reported by Kiil and coworkers [159]. The applicability of fiber-supported palladium dichloride/triphenylphosphine catalyst, FibreCat 1026, has been evaluated for the Buchwald–Hartwig amination reaction between *p*-bromotoluene and piperazine. With the aim of having a reusable catalyst, the authors found that the separation of the catalyst from the reaction mixture led to a change in the activity of the catalyst. Because of an irreversible leach of both palladium and triphenylphosphine from the support material, the recovered catalyst delivered a slower conversion of *p*-bromotoluene, and the primary product had changed from the amination products to the debromination side product toluene.

A soluble polyethylene glycol-anchored phosphine ligand was reported for C–N coupling by García and coworkers in 2007. Silica-supported SiO_2–SPhos, non-cross-linked polystyrene (solPS)-supported SPhos (solPS-SPhos), and SPhos phosphine ligand anchored on PEG (PEG–SPhos) have been employed for Buchwald–Hartwig coupling of chloro-*p*-xylene and morpholine (Scheme 10.82). PEG–SPhos was found to be the most effective ligand, whereas SiO_2–SPhos gave only traces of product [160]. In contrast,

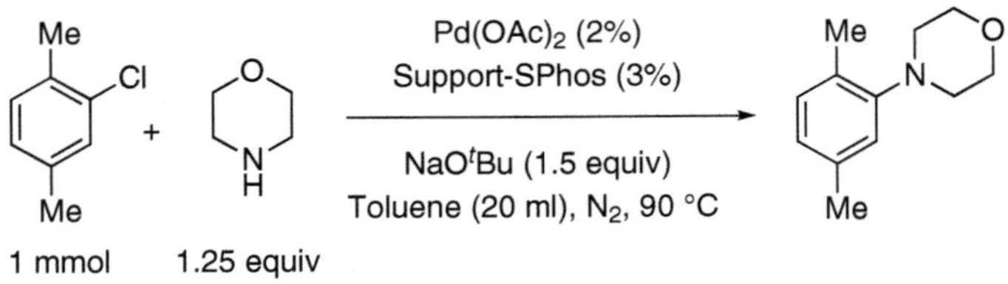

Scheme 10.82 C–N coupling of secondary amine with supported SPhos.

recently, Panahi et al. reported immobilized Pd nanoparticles on silica–starch substrate (PNP-SSS). This silica-based catalyst gave high catalytic activity and was readily reused for the coupling of bromobenzene and morpholine [161].

Other applications of nanostructured palladium-catalyzed C–N couplings have been recently reported. Following the previous use of Merrifield peptide resin-supported dialkylphosphinobiphenyl ligand systems reported by Buchwald and coworker [162], Fareghi-Alamdari et al. found in 2016 that diphenylphosphine grafted to graphene and palladium nanoparticle catalysts (named PFG-Pd) showed high catalytic activity in Buchwald–Hartwig coupling, with at least five times recyclability [163]. Sarvestani and Azadi reported the application of heterogeneous graphene oxide-functionalized chitosan-palladium (GO-Chit-Pd) for Buchwald–Hartwig reaction of aryl halides with different secondary amines in 2018. The stabilizing palladium nanoparticle (NP) catalyst system is reusable at least five times without any significant decrease in its catalytic activities [164]. As reported by Veisi et al. in 2019, a multiwalled carbon nanotube (MWCNT/CC-SH/Pd) nanocatalyst can be effectively applied in Buchwald–Hartwig coupling reactions with a wide variety of functionalized substrates. The catalyst could be reused six times without change of catalytic activity [165]. Immobilizing Pd-NPs on the surface of magnetic cysteine material and its application in N-arylation reactions have been reported by Hajipour et al. in 2019. In addition to high catalytic activity in green media, and mild reaction conditions, the system also exhibited excellent reusability [166].

Several supported NHC-Pd catalysts have been reported for Buchwald–Hartwig amination (Scheme 10.83). With polyvinylpyridine (PVP)-supported catalyst, amination of aryl chlorides under mild conditions and with low catalyst loadings was reported by Kirschning and Mennecke [167]. In 2017, Balogh et al. reported polyisobutylene (PIB)-supported palladium system for aromatic chloride amination with cyclic dialkylamines and aniline derivatives in heptane [168]. More recently, a Wang resin-immobilized NHC with silyl linker catalyst, $ClMe_2Si$–IPr–Pd complex, was developed by Choi and coworkers. The desired amination products were obtained in high yields within 10 minutes in the presence of only 0.2 mol% Pd; however, the catalyst could not be used again because of the formation and leaching of Pd NPs [169].

In conclusion, the current advances of immobilization/supported catalysts allow the use of recyclable catalysts for aromatic amination. However, the need to avoid large excesses of reagents, the metal leaching, and the necessity of a further improvement of the catalytic activity coupled with the high cost for the preparation of the supported catalytic systems demand further investigations. For the time being, the use of this methodology on a large scale is limited by these shortcomings.

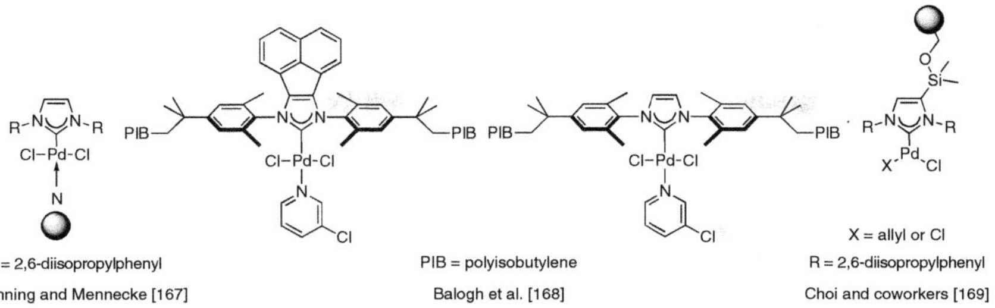

Scheme 10.83 Selected examples of supported NHC-Pd catalysts.

10.3.4 Personal Accounts from Contract Manufacturing Companies on Utility of Modern Flow Amination Methods

CMO and contract development and manufacturing organization (CDMO) have become strategic partners or key suppliers for pharmaceutical companies, both during the development and the commercialization of APIs. Regarding the synthesis of aromatic amines, most CDMO/CMO organizations state that the H_2SO_4/HNO_3 nitration combined with catalytic hydrogenation is still a very useful and economic route for simple raw materials or API starting materials. However, they also agree that transition-metal-catalyzed C–N coupling offers multiple advantages compared to the nitration/hydrogenation strategy because of the lower waste and safety concerns. Especially with the inexpensive copper- or nickel-based catalysts, or alternatively palladium catalysis with turnover number (TON) > 2000, such catalytic strategy is deemed to be economical. At the same time, an increasing number of CMOs/CDMOs make significant investments in continuous-flow technology. It has to be noticed that the safety evaluation of nitration is critical in all settings, even for milliliter-scale flow reaction. For instance, the compatibility of cleaning solvents needs to be considered, i.e. could remaining solvents in the equipment react with nitration mixtures? Methanol for example cannot be used for nitration flow reaction system because of the formation of explosive methyl nitrate.

10.4 Miscellaneous Aspects of Aromatic Amination Reactions in the World of Active Pharmaceutical Ingredients

10.4.1 Cohort of Concerns and its Regulatory Impact on Amine-Based Active Pharmaceutical Compounds

The pharmaceutical industry is one of the most strictly regulated business sectors and critical events sometimes very quickly turn the tides on new technologies. An example is given by the devastating and unexpected finding of carcinogenic *N*-nitrosodimethylamine impurity in Valsartan during a Good Manufacturing Practice (GMP) inspection in early 2018, the so-called "Valsartan-Case." The working hypothesis by the inspector in 2018 was that because of a change in the process to produce Valsartan, trace amounts of *N*,*N*-dimethylamine – derived from a chemical step in which *N*,*N*-dimethylformamide was used as a solvent – reacted with a nitrite in a subsequent step, resulting in its N-nitrosation.

The chemical process of the originator was in fact slightly changed during a technical transfer for a better fit to the process requirements of the local Chinese manufacturer. However, no assessment of the risk that carcinogenic *N*-nitrosodimethylamine could form in the revised process was done (https://www.ema.europa.eu/en/news/ema-reviewing-medicines-containing-valsartan-zhejiang-huahai-following-detection-impurity-some) [170]. When analytical proof that no *N*-nitrosamines had formed was requested, subsequent analytics showed that at least trace amounts were actually present. Furthermore, trace amounts of highly carcinogenic *N*-nitrosamine impurities were, shortly after, found in other drugs of the Sartan family, as well in other APIs, triggering a huge recall activity (https://www.dailymail.co.uk/health/article-6554477/Thousands-blood-pressure-pills-recalled.html).

Aliphatic *N*-nitrosamines are genotoxic impurities and part of the so-called "Cohort of Concern" (CoC) (https://jordilabs.com/wp-content/uploads/2017/07/Whitepaper-Cohort-of-Concern.pdf), which also includes alkylazo compounds, i.e. intermediates that may form during a nitro group reduction, and aflatoxins, a common mycotoxin produced by *Aspergillus flavus* and *Aspergillus parasiticus*. Because of their high carcinogenic and/or toxicological potential, the traditional threshold of toxicological concern (TTC) approach to impurity control cannot be applied to compounds belonging to the CoC (https://www.ema.europa.eu/en/ich-m7-assessment-control-dna-reactive-mutagenic-impurities-pharmaceuticals-limit-potential). Although the TTC considers an intake of a potential genotoxic impurity (PGI) of 1.5 μg/day of a patient treated with the API as acceptable, the levels of control for the CoC are much more stringent because of their potentially even higher carcinogenicity. For Human Pharma, companies have to carefully assess the risk of formation of these compounds, when submitting a new drug application, including analytical prove.

The failure of predicting and controlling *N*-nitrosodimethylamine formation, in Valsartan and other APIs, triggered multiple requests by the regulatory authorities to revise existing and under development processes in light of this failure. Thus, the risk of formation of such impurities has to be reassessed, with very short timelines, considering for example if there is potential contact of secondary amines with nitrite sources, and, if so, if the resulting compounds can be controlled, i.e. removed or their formation avoided.

In the field of aromatic amination, we deal with amines – compounds that could be transformed into *N*-nitrosamines. Thus, the risk of *N*-nitrosamine formation has to be carefully assessed even for this compound class, in light of the new findings, directions, and guidelines.

In a broader perspective, if a new chemical technology may cause problems with control of any of these impurities, its application in human pharma is very unlikely, even if it offers great advantages in yield, quality, or green chemistry.

10.4.2 Role of Control of Elemental Impurities in Human Pharma Applications

Not on the same level of alertness, but still an issue that has to be considered, every time a typical metal catalyst is used for a human pharma application, the marketing authorization holder, i.e. the owner of a new active pharmaceutical compound, needs to be fully aware if and to which extent elemental impurities may be present in the final API. How this has to be assessed is summarized in the so-called ICH Guideline Q3d (https://www.fda.gov/regulatory-information/search-fda-guidance-documents/elemental-impurities-drug-products-guidance-industry), a framework provided by the International Conference of Harmonization (ICH). Next to elemental impurities that may be present due to abrasion from steel or glass lined equipment, the addition of a homogeneous or hetereogeneous catalyst obviously raises the question on how much of it resides in the final product. The elemental impurities are divided in three classes, where the typically used catalyst metals are found in the so-called Class 2b. As they are intentionally added during the manufacturing process, the chemist has to assess if traces of it may end up in the final product. However, as for this kind of compounds a very accurate analytical set of methods is readily available, the assessment can easily be supported by analytical data.

10.4.3 Transition Metal Accounting

As most modern approaches to aromatic amination require the use of transition metal catalysts, we have to consider – next to the usual supply aspects – the problem of costly transition metal catalyst precursors, as well as that of the ligands that lead to the expected catalyst activity and selectivity. Just by looking at the three very prominent metals such as iridium, copper, and palladium over the course of past 20 years, prices have been very volatile, especially in the case of palladium. For the sake of exemplification, any new press release on the next generation of battery technology may cause the price to rise or fall dramatically. Compared to the nineties, we currently see a more than 20-fold increase of the price in the case of palladium. The iridium price seems to be less variable. Metals seen more as commodities as copper resides on a much lower price level, however, are still affected by considerable cost volatility over the time. On these grounds, vendors of transition metal catalyst precursors have established partnership agreements with their customers to keep track of the volume of precious metals that are currently used by the customer partners. Once the amount of the fresh precious metal payed by the customer is used to generate the catalytic species and employed in the chemical process, the spent catalyst goes back to the supplier company. Therefore, the actual loss of the precious metal payed by the customers will be related only to the amount that cannot be reisolated after the reaction, plus the amount lost during the recycling process performed by the metal vendor. In this way, once an industrial process has been established, the cost of goods for the entire process becomes less susceptible to the volatility of the price of the precious metals. However, this procedure requires a very accurate quantitative analysis of the precious metals in the spent catalysts using advanced analytical techniques, on which both the vendor and the customer need to fully agree. In general, the vendor is in charge of developing the analytic methodology. In the frame of the established partnership, after formal validation of the analytical method, it will be transferred to the customer partner. A regular cross-check by both partners follows. The rentability of recycling of various metals has been thoroughly assessed by Graedel et al. [171], however clearly stating that the actual recovery rate of most recycled metals is far below the theoretical amounts, especially for the case of pharma applications, wherein the recycling of precious metal containing wastes is only done for large-scale applications.

10.4.4 Recycling of Metals, Ligands, and Other Cost Drivers of Aromatic Amination

Next to the recycling of the precious metal from precipitates and distillation pits of the reaction, numerous other methods exist to keep the metal in a more localized form – be it as a metal alone or even better in combination with the ligand. These include various immobilization technologies (also see Section 10.3.3), catch and release methodologies [172], micellar catalysis, biphasic reaction modes, but also nanofiltration of homogeneous catalysts [173], or the immobilization in flow setups [174]. However, it is not always clear if the recycling of the catalyst actually reduces the catalyst cost or improves the reproducibility of the catalytic transformation.

As can be seen in Sections 10.2 and 10.3, most catalytic methods require a ligating additive that electronically tunes the transition metal precursor, stable enough to remain in the catalytic cycle. These additives, being phosphorous based, NHCs, or amines, are complex fine chemicals that significantly increase the cost of a chemical transformation, and, at the same time, may contribute to a major part of the environmental footprint of the synthetic step. Ligand recycling should thus become a field of interest. However, very few examples can be found in the literature where the ligands can be fully recovered after the chemical transformation.

Last but not least, the solvents that usually have been applied in typical cross-coupling or aromatic amination reactions, such as THF, Me-THF, dioxane, MTBE, DMSO, or acetonitrile, are classified as "problematic" [175] in the recent work toward the use of greener solvents in organic chemistry inside the CHEM21 Initiative [176]. "Recommended" solvents, such as anisole, *tert*-amyl methyl ether, some alcohols, ketones, or esters are only rarely used.

10.4.5 Regulatory Requirements for the Submission of a Catalytic Reaction in a New Drug Application

High stability, reproducibility, and purity are all attributes that go along quite well with the world of GMP (https://www.fda.gov/drugs/pharmaceutical-quality-resources/current-good-manufacturing-practice-cgmp-regulations) and have thus led to the frequent use of aromatic amination reactions even in the GMP-regulated part of new drug applications. However, the same requirements prevent the process development departments of pharmaceutical companies to extensively apply concepts to reuse solvent, catalyst, or ligands, as recycling may lead to deactivation of catalysts, accumulation of impurities, or degradation of ligands. Additionally, modern pharma supply strategies require an alternative supply source for each component of a new chemical step, which leads the chemists to discard sometimes the best performing, but at the same time also complex and patented, catalysts. A prominent example is given by homogeneous asymmetric hydrogenation. To gain freedom to operate for pharmaceutical and other applications, dozens of BINAP clones have been developed in the 1990s by academia and industry, thus prohibiting the development of a common, standardized protocol for this transformation. Today, even the complex synthetic route to make the ligand is cross-checked by process development to ensure that freedom to operate is kept during the lifecycle of a new drug. Once a chemical step is registered, it is extremely difficult to change it moving to another catalyst, solvent or modifying other reaction details. As far as the Buchwald–Hartwig amination is concerned, the key patents of 1996 [177] have expired, and even the major patents covering the exceedingly active dialkylbiarylphosphine ligand class [178] will become of public domain very soon, opening this technology to further spread. The Buchwald–Hartwig amination is nowadays in the top five reactions most commonly applied by medicinal chemists [179]. Because of the distinct modularity of the approach with its Lego-like building blocks, this will lead to modern drugs even further promoting the disruptive approach that was firstly anticipated with Buchwald's and Hartwig's papers in the middle of the 1990s.

References

1 Vitaku, E., Smith, D.T., and Njardarson, J.T. (2014). *Journal of Medicinal Chemistry* 57: 10257–10274.

2 Ullmann, F. (1903). *Chemische Berichte* 36: 2382–2384.

3 Goldberg, I. (1906). *Chemische Berichte* 39: 1691–1692.

4 (a) Chan, D.M.T., Monaco, K.L., Wang, R.-P. et al. (1998). *Tetrahedron Letters* 39: 2933–2936. (b) Lam, P.Y.S., Clark, C.G., Saubern, S. et al. (1998). *Tetrahedron Letters* 39: 2941–2944.

5 (a) Guram, A.S., Rennels, R.A., and Buchwald, S.L. (1995). *Angewandte Chemie International Edition in English* 34: 1348–1350. (b) Louie, J. and Hartwig, J.F. (1995). *Tetrahedron Letters* 36: 3609–3612.

6 (a) Okano, K., Tokuyama, H., and Fukuyama, T. (2014). *Chemical Communications* 50: 13650–13663. (b) Bhunia, S., Pawar, G.G., Kumar, S.V. et al. (2017). *Angewandte Chemie International Edition* 56: 16136–16179. (c) Ruiz-Castillo, P. and Buchwald, S.L. (2016). *Chemical Reviews* 116: 12564–12649. (d) Ma, D. and Cai, Q. (2008). *Accounts of Chemical Research* 41: 1450–1460. (e) Sambiagio, C., Marsden, S.P., Blacker, A.J. et al. (2014). *Chemical Society Reviews* 43: 3525–3550.

7 (a) Ley, S.V. and Thomas, A.W. (2003). *Angewandte Chemie International Edition* 42: 5400–5449. (b) Litvak, V.V. and Shein, S.M. (1975). *Zhurnal Organicheskoi Khimii* 11: 92–96. (c) Paine, A.J. (1987). *Journal of the American Chemical Society* 109: 1496–1502. (d) Collman, J.P. and Zhong, M. (2000). *Organic Letters* 2: 1233–1236.

8 Lin, H. and Sun, D. (2013). *Organic Preparations and Procedures International* 45: 341–394.

9 (a) Munir, I., Zahoor, A.F., Rasool, N. et al. (2019). *Molecular Diversity* 23: 215–259. (b) Qiao, J.X. and Lam, P.Y.S. (2013). *Boronic Acids: Preparation and Applications in Organic Synthesis*, vol. 2 (ed. D.G. Hall), 315–361. Weinheim: Wiley-VCH.

10 Evano, G., Blanchard, N., and Toumi, M. (2008). *Chemical Reviews* 108: 3054–3131.

11 Pu, Y.-M., Ku, Y.-Y., Grieme, T. et al. (2007). *Organic Process Research and Development* 11: 1004–1009.

12 Duan, S., Venkatraman, S., Hong, X. et al. (2012). *Organic Process Research and Development* 16: 1787–1793.

13 Bao, X., Qiao, X., Bao, C. et al. (2017). *Organic Process Research and Development* 21: 748–753.

14 (a) Graham, J.P., Langlade, N., Northall, J.M. et al. (2011). *Organic Process Research and Development* 15: 44–48. (b) Li, W., Ma, X.Q., and Yu, J.H. (2011). *Advanced Materials Research* 236–238: 2594–2597. (c) Kalhapure, R.S., Patil, B.P., Jadhav, M.N. et al. (2011). *E-Journal of Chemistry* 8: 1747–1749.

15 (a) Mudryk, B., Zheng, B., Chen, K. et al. (2014). *Organic Process Research and Development* 18: 520–527. (b) Newman, S.G. and Jensen, K.F. (2013). *Green Chemistry* 15: 1456–1472. (c) Baumann, M., Baxendale, I.R., and Beilstein, J. (2015). *Organic Chemistry* 11: 1194–1219.

16 Campbell Brewer, A., Hoffman, P.C., Martinelli, J.R. et al. (2019). *Organic Process Research and Development* 23: 1484–1498.

17 (a) Kosugi, M., Kameyama, M., and Migita, T. (1983). *Chemistry Letters* 12: 927–928. (b) Kosugi, M., Kameyama, M., Sano, H. et al. (1985). *Nippon Kagaku Kaishi*: 547–551.

18 (a) Hartwig, J.F. (1998). *Angewandte Chemie International Edition in English* 37: 2046–2067. (b) Yang, B.H. and Buchwald, S.L. (1999). *Journal of Organometallic Chemistry* 576: 125–146. (c) Surry, D.S. and Buchwald, S.L. (2011). *Chemical Science* 2: 27–50. (d) Wolfe, J.P., Wagaw, S., Marcoux, J.-F. et al. (1998). *Accounts of Chemical Research* 31: 805–818. (e) Hartwig, J.F. (2002). *Handbook of Organopalladium Chemistry for Organic Synthesis*, 1051–1096. New York: Wiley-Interscience, USA. (f) Heravi, M.M., Kheilkordi, Z., Zadsirjan, V. et al. (2018). *Journal of Organometallic Chemistry* 861: 17–104.

19 Brown, D.G. and Bostrom, J. (2016). *Journal of Medicinal Chemistry* 59: 4443–4458.

20 (a) Driver, M.S. and Hartwig, J.F. (1995). *Journal of the American Chemical Society* 117: 4708–4709. (b) Hartwig, J.F., Richards, S., Barañano, D. et al. (1996). *Journal of the American Chemical Society* 118: 3626–3633. (c) Widenhoefer, R.A. and Buchwald, S.L. (1996). *Organometallics* 15: 2755–2763. (d) Driver, M.S. and Hartwig, J.F. (1997). *Journal of the American Chemical Society* 119: 8232–8245.

21 Driver, M.S. and Hartwig, J.F. (1996). *Journal of the American Chemical Society* 118: 7217–7218.

22 (a) Wolfe, J.P., Wagaw, S., and Buchwald, S.L. (1996). *Journal of the American Chemical Society* 118: 7215–7216. (b) Wolfe, J.P. and Buchwald, S.L. (2000). *The Journal of Organic Chemistry* 65: 1144–1157.

23 Sadighi, J.P., Harris, M.C., and Buchwald, S.L. (1998). *Tetrahedron Letters* 39: 5327–5330.

24 Old, D.W., Wolfe, J.P., and Buchwald, S.L. (1998). *Journal of the American Chemical Society* 120: 9722–9723.

25 (a) Wolfe, J.P. and Buchwald, S.L. (1999). *Angewandte Chemie International Edition* 38: 2413–2416. (b) Wolfe, J.P., Tomori, H., Sadighi, J.P. et al. (2000). *The Journal of Organic Chemistry* 65: 1158–1174.

26 (a) Huang, J., Grasa, G., and Nolan, S.P. (1999). *Organic Letters* 1: 1307–1309. (b) Stauffer, S.R., Lee, S., Stambuli, J.P. et al. (2000). *Organic Letters* 2: 1423–1426. (c) Viciu, M.S., Kissling, R.M., Stevens, E.D. et al. (2002). *Organic Letters* 4: 2229–2231. (d) Viciu, M.S., Kelly, R.A. III, Stevens, E.D. et al. (2003). *Organic Letters* 5: 1479–1482.

27 (a) Yin, J. and Buchwald, S.L. (2000). *Organic Letters* 2: 1101–1104. (b) Yin, J. and Buchwald, S.L. (2002). *Journal of the American Chemical Society* 124: 6043–6048. (c) Artamkina, G.A., Sergeev, A.G., and Beletskaya, I.P. (2001). *Tetrahedron Letters* 42:

4381–4384. (d) Steinhuebel, D., Palucki, M., Askin, D. et al. (2004). *Tetrahedron Letters* 45: 3305–3307.

28 Huang, X., Anderson, K.W., Zim, D. et al. (2003). *Journal of the American Chemical Society* 125: 6653–6655.

29 Shen, Q., Shekhar, S., Stambuli, J.P. et al. (2005). *Angewandte Chemie International Edition* 44: 1371–1375.

30 Fors, B.P., Watson, D.A., Biscoe, M.R. et al. (2008). *Journal of the American Chemical Society* 130: 13552–13554.

31 Shen, Q., Ogata, T., and Hartwig, J.F. (2008). *Journal of the American Chemical Society* 130: 6586–6596.

32 Brusoe, A.T. and Hartwig, J.F. (2015). *Journal of the American Chemical Society* 137: 8460–8468.

33 (a) Shen, Q. and Hartwig, J.F. (2006). *Journal of the American Chemical Society* 128: 10028–10029. (b) Lundgren, R.J., Peters, B.D., Alsabeh, P.G. et al. (2010). *Angewandte Chemie International Edition* 49: 4071–4074.

34 Mauger, C.C. and Mignani, G.A. (2004). *Organic Process Research and Development* 8: 1065–1071.

35 Robinson, G.E., Cunningham, O.R., Dekhane, M. et al. (2004). *Organic Process Research and Development* 8: 925–930.

36 (a) Mitchell, D., Cole, K.P., Pollock, P.M. et al. (2011). *Organic Process Research and Development* 16: 70–81. (b) Merritt, J.M., Buser, J.Y., Campbell, A.N. et al. (2013). *Organic Process Research and Development* 18: 246–256.

37 Sperry, J.B., Price Wiglesworth, K.E., Edmonds, I. et al. (2014). *Organic Process Research and Development* 18: 1752–1758.

38 Affouard, C., Crockett, R.D., Diker, K. et al. (2015). *Organic Process Research and Development* 19: 476–485.

39 (a) Kwart, H. and Khan, A.A. (1967). *Journal of the American Chemical Society* 89: 1951–1953. (b) Breslow, D.S. and Sloan, M.F. (1968). *Tetrahedron Letters* 9: 5349–5352. (c) Carr, D., Seden, T.P., and Turner, R.W. (1969). *Tetrahedron Letters* 10: 477–478.

40 (a) Breslow, R. and Gellman, S.H. (1982). *Journal of the Chemical Society, Chemical Communications*: 1400–1401. (b) Breslow, R. and Gellman, S.H. (1983). *Journal of the American Chemical Society* 105: 6728–6729. (c) Mahy, J.-P., Bedi, G., Battioni, P. et al. (1989). *New Journal of Chemistry* 13: 651–657.

41 Zaitsev, V.G., Shabashov, D., and Daugulis, O. (2005). *Journal of the American Chemical Society* 127: 13154–13155.

42 (a) Sambiagio, C., Schonbauer, D., Blieck, R. et al. (2018). *Chemical Society Reviews* 47: 6603–6743. (b) Shang, M., Sun, S.Z., Dai, H.X. et al. (2014). *Journal of the American Chemical Society* 136: 3354–3357. (c) Lee, W.C., Shen, Y., Gutierrez, D.A. et al. (2016). *Organic Letters* 18: 2660–2663. (d) Liu, J., Chen, G., and Tan, Z. (2016). *Advanced Synthesis and Catalysis* 358: 1174–1194. (e) Rouquet, G. and Chatani, N. (2013). *Angewandte Chemie International Edition* 52: 11726–11743. (f) Rao, W.-H. and Shi, B.-F. (2016). *Org. Chem. Front.* 3: 1028–1047. (g) Reddy, V.P., Qiu, R., Iwasaki, T. et al. (2015). *Organic & Biomolecular Chemistry* 13: 6803–6813. (h) Wang, X., Xie, P., Qiu, R. et al. (2017). *Chemical Communications* 53: 8316–8319. (i) Liu, J., Xue, Z., Zeng, Z.

et al. (2016). *Advanced Synthesis and Catalysis* 358: 3694–3699. (j) Chen, G., Zhang, X., Zeng, Z. et al. (2017). *ChemistrySelect* 2: 1979–1982.

43 (a) Collet, F., Dodd, R.H., and Dauban, P. (2009). *Chemical Communications*: 5061–5074. (b) Thansandote, P. and Lautens, M. (2009). *Chemistry* 15: 5874–5883. (c) Louillat, M.L. and Patureau, F.W. (2014). *Chemical Society Reviews* 43: 901–910. (d) Jiao, J., Murakami, K., and Itami, K. (2015). *ACS Catalysis* 6: 610–633. (e) Henry, M., Mostafa, M., and Sutherland, A. (2017). *Synthesis* 49: 4586–4598. (f) Park, Y., Kim, Y., and Chang, S. (2017). *Chemical Reviews* 117: 9247–9301.

44 (a) Tsang, W.C., Zheng, N., and Buchwald, S.L. (2005). *Journal of the American Chemical Society* 127: 14560–14561. (b) Inamoto, K., Saito, T., Katsuno, M. et al. (2007). *Organic Letters* 9: 2931–2934. (c) Yamamoto, M. and Matsubara, S. (2007). *Chemistry Letters* 36: 172–173. (d) Wasa, M. and Yu, J.Q. (2008). *Journal of the American Chemical Society* 130: 14058–14059. (e) Jordan-Hore, J.A., Johansson, C.C., Beck, E.M. et al. (2008). *Journal of the American Chemical Society* 130: 16184–16186. (f) Mei, T.S., Wang, X., and Yu, J.Q. (2009). *Journal of the American Chemical Society* 131: 10806–10807. (g) Tan, Y. and Hartwig, J.F. (2010). *Journal of the American Chemical Society* 132: 3676–3677.

45 (a) Thu, H.Y., Yu, W.Y., and Che, C.M. (2006). *Journal of the American Chemical Society* 128: 9048–9049. (b) Ng, K.H., Chan, A.S., and Yu, W.Y. (2010). *Journal of the American Chemical Society* 132: 12862–12864. (c) Xiao, B., Gong, T.J., Xu, J. et al. (2011). *Journal of the American Chemical Society* 133: 1466–1474. (d) Sun, K., Li, Y., Xiong, T. et al. (2011). *Journal of the American Chemical Society* 133: 1694–1697.

46 Yoo, E.J., Ma, S., Mei, T.S. et al. (2011). *Journal of the American Chemical Society* 133: 7652–7655.

47 Ryu, J., Shin, K., Park, S.H. et al. (2012). *Angewandte Chemie International Edition* 51: 9904–9908.

48 Grohmann, C., Wang, H., and Glorius, F. (2012). *Organic Letters* 14: 656–659.

49 Espino, C.G., Fiori, K.W., Kim, M. et al. (2004). *Journal of the American Chemical Society* 126: 15378–15379.

50 Paudyal, M.P., Adebesin, A.M., Burt, S.R. et al. (2016). *Science* 353: 1144–1147.

51 (a) Arndtsen, B.A. and Bergman, R.G. (1995). *Science* 270: 1970–1973. (b) Lohrenz, J.C.W. and Jacobsen, H. (1996). *Angewandte Chemie International Edition in English* 35: 1305–1307. (c) Hoyano, J.K., McMaster, A.D., and Graham, W.A.G. (1983). *Journal of the American Chemical Society* 105: 7190–7191. (d) Young, K.J.H., Mironov, O.A., and Periana, R.A. (2007). *Organometallics* 26: 2137–2140. (e) Turlington, C.R., White, P.S., Brookhart, M. et al. (2014). *Journal of the American Chemical Society* 136: 3981–3994.

52 Kim, H., Shin, K., and Chang, S. (2014). *Journal of the American Chemical Society* 136: 5904–5907.

53 Shiota, H., Ano, Y., Aihara, Y. et al. (2011). *Journal of the American Chemical Society* 133: 14952–14955.

54 Yan, Q., Chen, Z., Yu, W. et al. (2015). *Organic Letters* 17: 2482–2485.

55 Yu, L., Yang, C., Yu, Y. et al. (2019). *Organic Letters* 21: 5634–5638.

56 Yu, L., Chen, X., Liu, D. et al. (2018). *Advanced Synthesis and Catalysis* 360: 1346–1351.

57 (a) Sun, B., Yoshino, T., Matsunaga, S. et al. (2014). *Advanced Synthesis and Catalysis* 356: 1491–1495. (b) Park, J. and Chang, S. (2015). *Angewandte Chemie International Edition* 54: 14103–14107. (c) Liang, Y., Liang, Y.F., Tang, C. et al. (2015). *Chemistry* 21: 16395–16399.

58 Zhang, L.B., Zhang, S.K., Wei, D. et al. (2016). *Organic Letters* 18: 1318–1321.

59 Yan, Q., Xiao, T., Liu, Z. et al. (2016). *Advanced Synthesis and Catalysis* 358: 2707–2711.

60 (a) Chen, X., Hao, X.S., Goodhue, C.E. et al. (2006). *Journal of the American Chemical Society* 128: 6790–6791. (b) Uemura, T., Imoto, S., and Chatani, N. (2006). *Chemistry Letters* 35: 842–843.

61 (a) Monguchi, D., Fujiwara, T., Furukawa, H. et al. (2009). *Organic Letters* 11: 1607–1610. (b) Kienle, M., Dunst, C., and Knochel, P. (2009). *Organic Letters* 11: 5158–5161. (c) Armstrong, A. and Collins, J.C. (2010). *Angewandte Chemie International Edition* 49: 2282–2285. (d) Zhao, H., Wang, M., Su, W. et al. (2010). *Advanced Synthesis and Catalysis* 352: 1301–1306. (e) Cho, S.H., Yoon, J., and Chang, S. (2011). *Journal of the American Chemical Society* 133: 5996–6005.

62 Tran, L.D., Roane, J., and Daugulis, O. (2013). *Angewandte Chemie International Edition* 52: 6043–6046.

63 (a) Peng, J., Chen, M., Xie, Z. et al. (2014). *Organic Chemistry Frontiers* 1: 777–781. (b) Grohmann, C., Wang, H., and Glorius, F. (2013). *Organic Letters* 15: 3014–3017. (c) Tezuka, N., Shimojo, K., Hirano, K. et al. (2016). *Journal of the American Chemical Society* 138: 9166–9171. (d) Kim, H., Park, G., Park, J. et al. (2016). *ACS Catalysis* 6: 5922–5929. (e) Xu, L.L., Wang, X., Ma, B. et al. (2018). *Chemical Science* 9: 5160–5164.

64 Li, Z., Yu, H., and Bolm, C. (2017). *Angewandte Chemie International Edition* 56: 9532–9535.

65 Kim, H., Heo, J., Kim, J. et al. (2018). *Journal of the American Chemical Society* 140: 14350–14356.

66 Legnani, L., Prina Cerai, G., and Morandi, B. (2016). *ACS Catalysis* 6: 8162–8165.

67 (a) Kantak, A.A., Potavathri, S., Barham, R.A. et al. (2011). *Journal of the American Chemical Society* 133: 19960–19965. (b) Kim, H.J., Kim, J., Cho, S.H. et al. (2011). *Journal of the American Chemical Society* 133: 16382–16385.

68 Manna, S., Serebrennikova, P.O., Utepova, I.A. et al. (2015). *Organic Letters* 17: 4588–4591.

69 (a) Prier, C.K., Rankic, D.A., and MacMillan, D.W. (2013). *Chemical Reviews* 113: 5322–5363. (b) Narayanam, J.M. and Stephenson, C.R. (2011). *Chemical Society Reviews* 40: 102–113. (c) Nicewicz, D.A. and Nguyen, T.M. (2013). *ACS Catalysis* 4: 355–360. (d) Qin, Q. and Yu, S. (2014). *Organic Letters* 16: 3504–3507. (e) Brachet, E., Ghosh, T., Ghosh, I. et al. (2015). *Chemical Science* 6: 987–992.

70 Romero, N.A., Margrey, K.A., Tay, N.E. et al. (2015). *Science* 349: 1326–1330.

71 Margrey, K.A., Levens, A., and Nicewicz, D.A. (2017). *Angewandte Chemie International Edition* 56: 15644–15648.

72 Svejstrup, T.D., Ruffoni, A., Julia, F. et al. (2017). *Angewandte Chemie International Edition* 56: 14948–14952.

73 Rossler, S.L., Jelier, B.J., Tripet, P.F. et al. (2019). *Angewandte Chemie International Edition* 58: 526–531.

74 (a) Hopkinson, M.N., Sahoo, B., Li, J.L. et al. (2014). *Chemistry* 20: 3874–3886. (b) Jahn, E. and Jahn, U. (2014). *Angewandte Chemie International Edition* 53: 13326–13328. (c) Vila, C. (2015). *ChemCatChem* 7: 1790–1793. (d) Gui, Y.-Y., Sun, L., Lu, Z.-P. et al. (2016). *Org. Chem. Front.* 3: 522–526. (e) Skubi, K.L., Blum, T.R., and Yoon, T.P. (2016). *Chemical Reviews* 116: 10035–10074.

75 Zheng, Y.W., Chen, B., Ye, P. et al. (2016). *Journal of the American Chemical Society* 138: 10080–10083.

76 (a) Horn, E.J., Rosen, B.R., and Baran, P.S. (2016). *ACS Central Science* 2: 302–308. (b) Moeller, K.D. (2000). *Tetrahedron* 56: 9527–9554. (c) Sperry, J.B. and Wright, D.L. (2006). *Chemical Society Reviews* 35: 605–621. (d) Yoshida, J., Kataoka, K., Horcajada, R. et al. (2008). *Chemical Reviews* 108: 2265–2299. (e) Francke, R. and Little, R.D. (2014). *Chemical Society Reviews* 43: 2492–2521.

77 Frontana-Uribe, B.A., Little, R.D., Ibanez, J.G. et al. (2010). *Green Chemistry* 12: 2099–2119.

78 Morofuji, T., Shimizu, A., and Yoshida, J. (2013). *Journal of the American Chemical Society* 135: 5000–5003.

79 (a) Morofuji, T., Shimizu, A., and Yoshida, J. (2014). *Journal of the American Chemical Society* 136: 4496–4499. (b) Morofuji, T., Shimizu, A., and Yoshida, J. (2015). *Chemistry* 21: 3211–3214. (c) Waldvogel, S.R. and Mohle, S. (2015). *Angewandte Chemie International Edition* 54: 6398–6399.

80 Herold, S., Möhle, S., Zirbes, M. et al. (2016). *European Journal of Organic Chemistry* 2016: 1274–1278.

81 (a) Gao, X., Wang, P., Zeng, L. et al. (2018). *Journal of the American Chemical Society* 140: 4195–4199. (b) Sauermann, N., Mei, R., and Ackermann, L. (2018). *Angewandte Chemie International Edition* 57: 5090–5094.

82 (a) Rodriguez, N. and Goossen, L.J. (2011). *Chemical Society Reviews* 40: 5030–5048. (b) Weaver, J.D., Recio, A. 3rd, Grenning, A.J. et al. (2011). *Chemical Reviews* 111: 1846–1913. (c) Shang, R. and Liu, L. (2011). *Science China Chemistry* 54: 1670–1687.

83 (a) Cho, C.C., Liu, J.N., Chien, C.H. et al. (2009). *The Journal of Organic Chemistry* 74: 1549–1556. (b) Jia, W. and Jiao, N. (2010). *Organic Letters* 12: 2000–2003. (c) Ju, L., Bode, J.W., Toma, T. et al. (2010). *Organic Syntheses* 87: 218–225.

84 Zhang, Y., Patel, S., and Mainolfi, N. (2012). *Chemical Science* 3: 3196–3199.

85 Sheng, W.-J., Ye, Q., Yu, W.-B. et al. (2015). *Tetrahedron Letters* 56: 599–601.

86 Dai, Q., Li, P., Ma, N. et al. (2016). *Organic Letters* 18: 5560–5563.

87 Drapeau, M.P., Bahri, J., Lichte, D. et al. (2019). *Angewandte Chemie International Edition* 58: 892–896.

88 Marin, M., Rama, R.J., and Nicasio, M.C. (2016). *Chemical Record* 16: 1819–1832.

89 Lavoie, C.M. and Stradiotto, M. (2018). *ACS Catalysis* 8: 7228–7250.

90 Hughes, E.C., Veatch, F., and Elersich, V. (1950). *Industrial and Engineering Chemistry Research* 42: 787–790.

91 Cramer, R. and Coulson, D.R. (1975). *The Journal of Organic Chemistry* 40: 2267–2273.

92 Cristau, H.-J. and Desmurs, J.-R. (1995). *Industrial Chemistry Library* 7: 240–263.

93 Wolfe, J.P. and Buchwald, S.L. (1997). *Journal of the American Chemical Society* 119: 6054–6058.

94 (a) Brenner, E. and Fort, Y. (1998). *Tetrahedron Letters* 39: 5359–5362. (b) Brenner, E., Schneider, R., and Fort, Y. (1999). *Tetrahedron* 55: 12829–12842. (c) Gradel, B.T., Brenner, E., Schneider, R. et al. (2001). *Tetrahedron Letters* 42: 5689–5692. (d) Desmarets, C., Schneider, R., and Fort, Y. (2002). *The Journal of Organic Chemistry* 67: 3029–3036. (e) Kelly, R.A., Scott, N.M., Díez-González, S. et al. (2005). *Organometallics* 24: 3442–3447.

95 Ritleng, V., Henrion, M., and Chetcuti, M.J. (2016). *ACS Catalysis* 6: 890–906.

96 Chen, C. and Yang, L.M. (2005). *Organic Letters* 7: 2209–2211.

97 Manolikakes, G., Gavryushin, A., and Knochel, P. (2008). *The Journal of Organic Chemistry* 73: 1429–1434.

98 Iglesias, M.J., Prieto, A., and Nicasio, M.C. (2010). *Advanced Synthesis and Catalysis* 352: 1949–1954.

99 (a) Fan, X.-H., Li, G., and Yang, L.-M. (2011). *Journal of Organometallic Chemistry* 696: 2482–2484. (b) Nagao, S., Matsumoto, T., Koga, Y. et al. (2011). *Chemistry Letters* 40: 1036–1038.

100 Ge, S., Green, R.A., and Hartwig, J.F. (2014). *Journal of the American Chemical Society* 136: 1617–1627.

101 (a) Borzenko, A., Rotta-Loria, N.L., MacQueen, P.M. et al. (2015). *Angewandte Chemie International Edition* 54: 3773–3777. (b) Green, R.A. and Hartwig, J.F. (2015). *Angewandte Chemie International Edition* 54: 3768–3772.

102 Martinez, G.E., Nugent, J.W., and Fout, A.R. (2018). *Organometallics* 37: 2941–2944.

103 Gao, C.Y. and Yang, L.M. (2008). *The Journal of Organic Chemistry* 73: 1624–1627.

104 (a) Tobisu, M., Shimasaki, T., and Chatani, N. (2009). *Chemistry Letters* 38: 710–711. (b) Tobisu, M., Yasutome, A., Yamakawa, K. et al. (2012). *Tetrahedron* 68: 5157–5161.

105 Shimasaki, T., Tobisu, M., and Chatani, N. (2010). *Angewandte Chemie International Edition* 49: 2929–2932.

106 Huang, J.H. and Yang, L.M. (2011). *Organic Letters* 13: 3750–3753.

107 Ackermann, L., Sandmann, R., and Song, W. (2011). *Organic Letters* 13: 1784–1786.

108 (a) Ramgren, S.D., Silberstein, A.L., Yang, Y. et al. (2011). *Angewandte Chemie International Edition* 50: 2171–2173. (b) Mesganaw, T., Silberstein, A.L., Ramgren, S.D. et al. (2011). *Chemical Science* 2: 1766–1771. (c) Hie, L., Ramgren, S.D., Mesganaw, T. et al. (2012). *Organic Letters* 14: 4182–4185. (d) Fine Nathel, N.F., Kim, J., Hie, L. et al. (2014). *ACS Catalysis* 4: 3289–3293.

109 Zhu, F. and Wang, Z.X. (2013). *Advanced Synthesis and Catalysis* 355: 3694–3702.

110 Harada, T., Ueda, Y., Iwai, T. et al. (2018). *Chemical Communications* 54: 1718–1721.

111 Park, N.H., Teverovskiy, G., and Buchwald, S.L. (2014). *Organic Letters* 16: 220–223.

112 Iranpoor, N. and Panahi, F. (2014). *Advanced Synthesis and Catalysis* 356: 3067–3073.

113 Hanley, P.S., Clark, T.P., Krasovskiy, A.L. et al. (2016). *ACS Catalysis* 6: 3515–3519.

114 Rout, L., Saha, P., Jammi, S. et al. (2008). *Advanced Synthesis and Catalysis* 350: 395–398.

115 (a) Guo, D., Huang, H., Xu, J. et al. (2008). *Organic Letters* 10: 4513–4516. (b) Correa, A. and Bolm, C. (2007). *Angewandte Chemie International Edition* 46: 8862–8865. (c) Correa, A., Elmore, S., and Bolm, C. (2008). *Chemistry* 14: 3527–3529. (d) Correa, A. and Bolm, C. (2008). *Advanced Synthesis and Catalysis* 350: 391–394.

116 Swapna, K., Vijay Kumar, A., Prakash Reddy, V. et al. (2009). *The Journal of Organic Chemistry* 74: 7514–7517.

117 Cahiez, G. and Moyeux, A. (2010). *Chemical Reviews* 110: 1435–1462.

118 (a) Teo, Y.C. and Chua, G.L. (2009). *Chemistry* 15: 3072–3075. (b) Toma, G. and Fujita, K.-i. and Yamaguchi, R. (2009). *European Journal of Organic Chemistry* 2009: 4586–4588.

119 Kim, M. and Chang, S. (2010). *Organic Letters* 12: 1640–1643.

120 Lam, P.Y.S., Deudon, S., Averill, K.M. et al. (2000). *Journal of the American Chemical Society* 122: 7600–7601.

121 Morstein, J., Kalkman, E.D., Bold, C. et al. (2016). *Organic Letters* 18: 5244–5247.

122 Roughley, S.D. and Jordan, A.M. (2011). *Journal of Medicinal Chemistry* 54: 3451–3479.

123 Afanasyev, O.I., Kuchuk, E., Usanov, D.L. et al. (2019). *Chemical Reviews* 119: 11857–11911.

124 Abdel-Magid, A.F., Carson, K.G., Harris, B.D. et al. (1996). *The Journal of Organic Chemistry* 61: 3849–3862.

125 Fasano, V., Radcliffe, J.E., and Ingleson, M.J. (2016). *ACS Catalysis* 6: 1793–1798.

126 Zhou, Y., Zhou, H., Liu, S. et al. (2017). *Tetrahedron* 73: 3898–3904.

127 (a) Thakur, M.S., Nayal, O.S., Upadhyay, R. et al. (2018). *Organic Letters* 20: 1359–1362. (b) Thakur, M.S., Nayal, O.S., Sharma, A. et al. (2018). *European Journal of Organic Chemistry* 2018: 6729–6732.

128 (a) Wang, C. and Xiao, J. (2014). *Stereoselective Formation of Amines*, vol. 343 (eds. W. Li and X. Zhang), 261–282. Berlin, Heidelberg: Springer-Veralg. (b) Fleury-Brégeot, N., de la Fuente, V., Castillón, S. et al. (2010). *ChemCatChem* 2: 1346–1371. (c) Huang, H., Liu, X., Zhou, L. et al. (2016). *Angewandte Chemie International Edition* 55: 5309–5312. (d) Natte, K., Neumann, H., Jagadeesh, R.V. et al. (2017). *Nature Communications* 8: 1344–1348. (e) Stemmler, T., Westerhaus, F.A., Surkus, A.-E. et al. (2014). *Green Chemistry* 16: 4535–4540.

129 Sharma, M., Mangas-Sanchez, J., France, S.P. et al. (2018). *ACS Catalysis* 8: 11534–11541.

130 (a) Muller, T.E. and Beller, M. (1998). *Chemical Reviews* 98: 675–704. (b) Brunet, J.J. and Neibecker, D. (2001). *Catalytic Heterofunctionalization* (eds. A. Togni and H. Grützmacher), 91–141. Weinheim, Germany: Wiley-VCH. (c) Pohlki, F. and Doye, S. (2003). *Chemical Society Reviews* 32: 104–114. (d) Alonso, F., Beletskaya, I.P., and Yus, M. (2004). *Chemical Reviews* 104: 3079–3159. (e) Hultzsch, K.C. (2005). *Advanced Synthesis and Catalysis* 347: 367–391. (f) Muller, T.E., Hultzsch, K.C., Yus, M. et al. (2008). *Chemical Reviews* 108: 3795–3892. (g) Severin, R. and Doye, S. (2007). *Chemical Society Reviews* 36: 1407–1420.

131 Coman, S.M. and Parvulescu, V.I. (2015). *Organic Process Research and Development* 19: 1327–1355.

132 (a) Al-Masum, M., Meguro, M., and Yamamoto, Y. (1997). *Tetrahedron Letters* 38: 6071–6074. (b) Besson, L., Goré, J., and Cazes, B. (1995). *Tetrahedron Letters* 36: 3857–3860.

133 Nishina, N. and Yamamoto, Y. (2006). *Angewandte Chemie International Edition* 45: 3314–3317.

134 (a) Zeng, X., Soleilhavoup, M., and Bertrand, G. (2009). *Organic Letters* 11: 3166–3169. (b) Duncan, A.N. and Widenhoefer, R.A. (2010). *Synlett* 2010: 419–422.

135 Cooke, M.L., Xu, K., and Breit, B. (2012). *Angewandte Chemie International Edition* 51: 10876–10879.

136 Blieck, R., Bahri, J., Taillefer, M. et al. (2016). *Organic Letters* 18: 1482–1485.

137 Liu, X.Y. and Che, C.M. (2009). *Organic Letters* 11: 4204–4207.

138 Fleischer, S., Werkmeister, S., Zhou, S. et al. (2012). *Chemistry A European Journal* 18: 9005–9010.

139 (a) Miki, Y., Hirano, K., Satoh, T. et al. (2013). *Angewandte Chemie International Edition* 52: 10830–10834. (b) Zhu, S., Niljianskul, N., and Buchwald, S.L. (2013). *Journal of the American Chemical Society* 135: 15746–15749.

140 Pirnot, M.T., Wang, Y.M., and Buchwald, S.L. (2016). *Angewandte Chemie International Edition* 55: 48–57.

141 Ichikawa, S., Zhu, S., and Buchwald, S.L. (2018). *Angewandte Chemie International Edition* 57: 8714–8718.

142 Xie, F., Shen, B., and Li, X. (2018). *Organic Letters* 20: 7154–7157.

143 Munro-Leighton, C., Delp, S.A., Alsop, N.M. et al. (2008). *Chemical Communications*: 111–113.

144 Munro-Leighton, C., Delp, S.A., Blue, E.D. et al. (2007). *Organometallics* 26: 1483–1493.

145 Park, S., Kang, S., and Lee, Y. (2019). *Advanced Synthesis and Catalysis* 361: 1071–1083.

146 Zhang, Q., Zheng, Y., Tong, D. et al. (2019). *Proceedings of the National Academy of Sciences of the United States of America* 116: 24463–24469.

147 Cole, K.P. and Johnson, M.D. (2018). *Expert Review of Clinical Pharmacology* 11: 5–13.

148 (a) Malet-Sanz, L. and Susanne, F. (2012). *Journal of Medicinal Chemistry* 55: 4062–4098. (b) Britton, J. and Jamison, T.F. (2017). *Nature Protocols* 12: 2423–2446. (c) Plutschack, M.B., Pieber, B., Gilmore, K. et al. (2017). *Chemical Reviews* 117: 11796–11893.

149 Yang, J.C., Niu, D., Karsten, B.P. et al. (2016). *Angewandte Chemie International Edition* 55: 2531–2535.

150 Falß, S., Tomaiuolo, G., Perazzo, A. et al. (2016). *Organic Process Research and Development* 20: 558–567.

151 Charaschanya, M., Bogdan, A.R., Wang, Y. et al. (2016). *Tetrahedron Letters* 57: 1035–1039.

152 Lin, H., Dai, C., Jamison, T.F. et al. (2017). *Angewandte Chemie International Edition* 56: 8870–8873.

153 Cole, K.P., Groh, J.M., Johnson, M.D. et al. (2017). *Science* 356: 1144.

154 Sharma, Y., Nikam, A.V., and Kulkarni, A.A. (2019). *Organic Process Research and Development* 23: 170–176.

155 Singh, B.K., Stevens, C.V., Acke, D.R.J. et al. (2009). *Tetrahedron Letters* 50: 15–18.

156 Mallia, C.J., Burton, P.M., Smith, A.M.R. et al. (2016). *Beilstein J. Organic Chemistry*: 1598–1607.

157 (a) Janssen, M., Müller, C., and Vogt, D. (2011). *Green Chemistry* 13: 2247–2257. (b) Molnár, Á. and Papp, A. (2017). *Coordination Chemistry Reviews* 349: 1–65. (c) Shende, V.S., Saptal, V.B., and Bhanage, B.M. (2019). *Chemical Record* 19: 2022–2043.

(d) Chartoire, A., Claver, C., Corpet, M. et al. (2016). *Organic Process Research and Development* 20: 551–557.

158 (a) Mpungose, P.P., Vundla, Z.P., Maguire, G.E.M. et al. (2018). *Molecules* 23: 1676–1700. (b) Polshettiwar, V., Len, C., and Fihri, A. (2009). *Coordination Chemistry Reviews* 253: 2599–2626. (c) Shakil Hussain, S.M., Kamal, M.S., and Hossain, M.K. (2019). *Journal of Nanomaterials* 2019: 1–17.

159 Christensen, H., Kiil, S., Dam-Johansen, K. et al. (2007). *Organic Process Research and Development* 11: 956–965.

160 Leyva, A., García, H., and Corma, A. (2007). *Tetrahedron* 63: 7097–7111.

161 Panahi, F., Daneshgar, F., Haghighi, F. et al. (2017). *Journal of Organometallic Chemistry* 851: 210–217.

162 Parrish, C.A. and Buchwald, S.L. (2001). *The Journal of Organic Chemistry* 66: 3820–3827.

163 Fareghi-Alamdari, R., Haqiqi, M.G., and Zekri, N. (2016). *New J. Chem.* 40: 1287–1296.

164 Sarvestani, M. and Azadi, R. (2018). *Applied Organometallic Chemistry* 32: e3906.

165 Veisi, H., Safarimehr, P., and Hemmati, S. (2019). *Materials Science and Engineering C* 96: 310–318.

166 Hajipour, A.R., Khorsandi, Z., and Metkazini, S.F.M. (2019). *Journal of Organometallic Chemistry* 899: 120793.

167 Kirschning, A. and Mennecke, K. (2008). *Synthesis* 2008: 3267–3272.

168 Balogh, J., Hlil, A.R., El-Zoghbi, I. et al. (2017). *Macromolecular Rapid Communications* 38: 1700214.

169 Mizusaki, T., Matsumoto, K., Takeuchi, K. et al. (2019). *Organometallics* 38: 1872–1876.

170 Farrukh, M.J., Tariq, M.H., Malik, O. et al. (2019). *Ther. Adv. Drug Saf.* 10: 2042098618823458.

171 Graedel, T.E., Allwood, J., Birat, J.-P. et al. (2011). *Journal of Industrial Ecology* 15: 355–366.

172 Hübner, S., de Vries, J.G., and Farina, V. (2016). *Advanced Synthesis and Catalysis* 358: 3–25.

173 Dreimann, J.M., Skiborowski, M., Behr, A. et al. (2016). *ChemCatChem* 8: 3330–3333.

174 Mandoli, A. (2019). *Catalyst Immobilization* (eds. M. Benaglia and A. Puglisi), 257–306. Wiley-VCH.

175 Ashcroft, C.P., Dunn, P.J., Hayler, J.D. et al. (2015). *Organic Process Research and Development* 19: 740–747.

176 Prat, D., Pardigon, O., Flemming, H.-W. et al. (2013). *Organic Process Research and Development* 17: 1517–1525.

177 Buchwald, S.L. and Guram, A. (1994). Preparation of arylamines. US Patent 5, 576, 460 A.

178 Buchwald, S., Old, D.W., Wolfe, J.P. et al. (1998). Ligands for metals and metal-catalyzed processes. WO2000002887A2.

179 Hayler, J.D., Leahy, D.K., and Simmons, E.M. (2018). *Organometallics* 38: 36–46.

Index

a

AADH 252
AADH *vs.* AmDH 254
2-acetamide-2-deoxyisosorbide (ADI) 343
3-acetamido-5-acetylfuran (3A5AF) 343
2-acetamido-2-deoxysorbitol (ADS) 342
acrylamine 344
active pharmaceutical ingredients (APIs) 12, 25, 158, 161, 188, 246, 399, 431–435
alanine 127, 257, 262, 361, 364–366
alanine dehydrogenase (AlaDH) 127
alcohol dehydrogenases (ADH) 128, 129, 261
alfuzosin 367
aliphatic amines 67, 74, 131, 133, 244, 289, 312, 355, 357
aliphatic *N*-nitrosamines 432
alkanolamines 348–350
alkene-alkyne (enyne) metathesis 139
alkyl amines 306, 307
  capping agents 322–325
  displacement, of amines 325
alkyl aryl ketones 160
alkylazo compounds 432
α,α-disubstituted benzylamine picolinamides 396
α-alkyl acrylates 175, 176
α-aminobutyric acid 366
α-aminophosphonates 357
α-aminophosphonic acids 357
α,β-unsaturated 2-acylimidazoles 177
AmDH (amine dehydrogenase) 128, 129, 252–262
amine dehydrogenase (AmDH) 128, 129, 252–262
  cascade reactions
    biocatalytic hydrogen borrowing 261
    chemoenzymatic cascades 262
    multi-enzyme cascades 261–262
  leucine dehydrogenase 253–254
  native amine dehydrogenase 254–256
  phenylalanine dehydrogenase 254
  substrate promiscuity
    amino alcohol synthesis 260
    reductive aminase activity 259–260
  synthetic applications
    primary amine synthesis 256–258
amidyl radical addition 55–62
amination via N-atom nucleophilic addition
  aromatic $C(sp^2)$-H bonds amination 77–82

Methodologies in Amine Synthesis: Challenges and Applications, First Edition.
Edited by Alfredo Ricci and Luca Bernardi.

amination via N-atom nucleophilic addition (*contd.*)
  benzylic $C(sp^3)$-H bonds amination 85–87
  N-α-C(sp3)-H bonds amination 87–90
  olefinic $C(sp^2)$-H bonds amination 82–85
amination via radicals cross-coupling
  aryl $C(sp^2)$-N bonds formation via radicals cross-coupling 94–96
  C–N bonds formation 96–98
  using diarylamines 91
amine-functionalized graphene (AFG) 299
amine oxidases (AOs) 103, 155, 214, 243–252, 266, 274
amines
  and formamides synthesis 345–348
aminium radical cation addition 64–71
amino α-carbanion 204
aminoalkylphenol derivatives 367, 368
amino-functionalization, of CNO 308–309
amino functionalized carbon-based nanomaterials 308–309
3-aminofuran 344
aminoketones 362, 364, 365
2-aminophenyl oxazoline 389
2-aminophenyl pyrazole 389
amino pyrazole 387, 427
2-aminopyridine-1-oxide 389, 395
2-aminopyrimidine hydrochloride 426
2-aminoquinazolin-4(3*H*)-one 418
8-aminoquinoline 389, 393, 395, 396
  benzamide 393, 395
8-aminoquinolinyl 404
2-aminosugar 344
3-amino-1,2,4-triazole 358
2-and 4-nitrocumene 428
aniline 382
  synthesis 403
annulated aniline derivative 382
anti-cancer alkaloid proximicin A 344
anti-cancer drugs 367
antihypertensives 352
antiseptic agents 352
arene amination 392, 393, 397, 398, 400
arene nucleophiles 390
arginine and ethylenediamine (NCDs) 312
  CD-supported amines 315
  one-pot functionalization 312, 313
  post-functionalization 313–315
aromatic amination reactions
  human pharma application 432–433
  recycling of metals, ligands and other cost drivers of 434
  regulated part of new drug applications 434–435
  transition metal accounting 433
aromatic amines 68, 249, 354, 361, 362, 377, 378, 415, 429, 431
aromatic $C(sp^2)$-H bonds amination 77
aromatic C–N bond formation approaches 389
aromatic *N,N*-dimethyl tertiary amines 367
aromatization reactions 310
Arthrobacter nicotinovorans 252
artificial imine reductase (ATHase) 247
artificial transfer hydrogenases (ATHase) 267
arylamine formation
  Chan-Lam coupling 428–429

flow amination methods 431
immobilization of catalysts/supported catalysts 429–431
nucleophilic aromatic substitution 426–428
Pd-catalyzed C–N bond forming reaction 425–426
telescoped sequence of nitration and hydrogenation 428
arylamines 423, 426
catalytic C–H amination 389–399
decarboxylative arylamination 405–407
electrochemical approach 402–405
hydroamination 419–424
metal catalyzed cross couplings 414–417
nickel catalyzed C–N coupling 407–414
photoredox catalysis 399–402
reductive amination 417–419
syntheses 389
arylboronates 6, 7, 416
aryl bromide 167, 381, 384, 385, 409
aryl C–H amination 74, 390, 392–396, 398, 417
aryl C(sp^2)-N bonds formation via radicals cross-coupling
other C–N bonds formation 96–98
using diarylamines 91
aryl-diazonium salts 291, 304, 305
aryl fluorosulfonates 413, 414
aryl/heteraoaryl chlorides 412
O-aryl hydroxylamines 401
aspartic acid 310, 311, 366
Aspergillus Niger 214, 244, 245, 247, 255, 257
Aspergillus oryzae (AspRedAm) 128, 129, 268, 419
asymmetric hydroamination 162
alkene 162
asymmetric hydroaminoalkylation 168–170
asymmetric hydroaminomethylation 171–176
asymmetric hydrogenation (AH) 156, 171, 172, 434
asymmetric reductive amination 156–161, 267, 268, 274, 419
atomic force microscopy (AFM) 312
attenuated total reflectance infrared (ATR-IR) 297
Au(I)/chiral Brønsted acid dual catalysis 421
aza-Michael reaction 194, 196
azomethine ylides 291, 292, 295, 305, 307, 346
intermediate 291, 346

b

Bacillus badius 254, 256, 257
Bacillus sphaericus 256
Bacillus steareothermophilus 253
benchtop stable aminating reagent 398
benchtop stable ammonium carbamate 400
benzazepines 367–369
benzomorpholine 350
benzothiazine 350
O-benzoyl hydroxylamines 390
benzylamine 87, 157, 246, 249, 252, 254, 295, 357
benzylic amines 259, 318
benzylic C(sp^3)-H bonds amination 85–87
β-amidoester 344
β-amino alcohols 350
β-D-glucopyranosylamine 352
β-1,4-glycosidic bonds 341
β-*N*-glycosidic bond 196
$B(C_6F_5)_3$-H_2O catalyzed reductive amination 418

biaryl based ligand BrettPhos 387
biaryl monophosphine ligands 386
bicyclic α-amino acid 214
Biginelli reaction 357, 358
Bingel–Hirsch reaction 299
bio-based 3,4-dialkoxyanilines 367
biocatalytic cascade synthesis 266
biocatalytic hydrogen borrowing 261, 270
biodiesel production 361, 364
biorenewable aromatic monomers 366
bis(hydroxylmethylfurfuryl)amines (BHMFA) 355
3,5-bis(trifluoromethyl)benzyl catalyst 225
2,5-bis(aminomethyl)furan (BAMF) 356
2,5-bis(hydroxymethyl)furan (BHMF) 356
N,N′-bis(2-hydroxyethyl)-*N,N′*-dimethyl-ethylenediamine (BHEDMEDA) 349, 350
1,3-bis(diphenylphosphanyl)propane 416
3,5-bistrifluoromethyl-benzaldehyde 288
boron-doped diamond anodes (BDD) 404
2-bromonapthalene 411
bromotrichloromethane ($BrCCl_3$) 87
Brønsted acid (BH) 90, 134, 135, 160, 161, 165, 418, 421
Buchwald–Hartwig amination 378
  applications 386–388
  catalyst design 385–386
  limitations 384
  mechanistic studies 384–385
Buchwald–Hartwig reactions 16, 430
Buchwald's flow synthesis 426
bulky N-heterocyclic carbene (NHC) precatalyst 426

c
cadmium-catalyzed amination of iodobenzene 415
camptothecin (CPT) 307
Candida antarctica lipase B (CAL-B) 126
carbon-based nanomaterials 287
  carbon, allotropes of 290
  carbon nanostructures
    Au-NR 297
    azomethine ylides, cycloaddition 291
    chiral fullerene derivatives 291
    glycidyltrimethylammonium chloride 293
    nanotechnological applications 291
    POM cluster 293
    radical arylation reaction 291
  functionalization procedures 297–299
carbon-based nanostructures 290, 295, 297
carbon dots (CDs) 309
  amines, as constituent 310–312
carbon nanocones 301
carbon nanodiamonds (CNDs) 305–307
carbon nanoHorns (CNHs) 301–305
carbon nanomaterials
  CND 305–307
  CNH 301–305
  CNO 307–308
carbon nano-onions (CNOs) 287, 307–308
carbon nanotubes (CNTs) 287, 290, 291
1,1′-carbonyldiimidazole 190
carboxylic ND (ND-ori)-based supraparticles (ND-SPs) 307
catalytic C–H amination 389–405
CD-promoted organocatalysis 315

cellulosic biomass derivatives 348–366
Chan–Lam coupling 378, 380–383, 428–429
chemical vapor deposition (CVD) 301
chemoenzymatic cascades 249, 262
chimeric amine dehydrogenase 254
chiral amine frameworks 155
chiral amine *(R)*-α-methylbenzylamine 357
chiral amine synthesis, biocatalytic strategies 244
chiral amino alcohol 387, 388
chiral bisphosphine-thiourea ligand 158
chiral Brønsted acid 134, 160, 421, 422
chiral metal center 176–177
chiral *N*-arylamine 422
chiral pyrrolidinofullerenes 291
chiral switch 155
chitin 341–345
chitin-derived N-compound 343
chloropyrazine 427
C–H methyl bond 168
choline chloride-urea DES 345
C–H pyridination-aminolysis sequence 401
ciprofloxacin 427
cis-N-(2-methylcyclohexyl) 161
ClMe2Si-IPr-Pd complex 430
Clostridium sticklandii 258
C–N bond formation
  amination via N-atom nucleophilic addition
    activated C(sp^3)-H bonds amination 85–90
    aromatic C(sp^2)-H bonds amination 77–82
    olefinic C(sp^2)-H bonds amination 82–85
  amination via radicals cross-coupling
    aryl C(sp^2)-N bonds formation via radicals cross-coupling 90–96
  radical addition to C-C double/triple bonds
    amidyl radical addition 55–62
    aminium radical cation addition 64–71
  radical species addition to aromatic rings 71–76
  via N-radical species addition 54–76
cobalt-catalyzed C–H amination 395
colorimetric test 289
confocal fluorescence microscopy 297
continuous Pd-catalyzed C–N coupling process 426
copper-catalysed heterocyclisation 191
copper-catalyzed anti-Markovnikov hydroamination 423
copper-catalyzed asymmetric hydroamination 422
copper-catalyzed C–N coupling reaction
  anticipated mechanism of 379
  different conditions 380
copper-catalyzed hydroamination 421, 423
copper-catalyzed primary aniline synthesis 397
copper-doped porous metal oxides (Cu20-PMO) 367
copper(I) hydride catalysis 421
copper-mediated decarboxylative amination 406
copper triflate 108, 291
covalent functionalization 294, 295, 303, 307, 314
cross-dehydrogenative coupling (CDC) 53

Cu-catalyzed amination reagents 3
aminoboration of styrenes 9
aryl boronates 7
aryl silanes 8
hydroamination of styrenes 10
organolithium reagents 5
organozinc reagents 3
silyl enol ethers 9
via aryne intermedate 4
Cu-catalyzed annulative electrophilic amination 12
Cu-catalyzed C–H Amination of heterocycles 15
Cu-catalyzed C–N bond-forming approach 381
Cu-catalyzed electrophilic amination of alkynes 11
Cu-catalyzed electrophilic amino-lactonization 14
Cu-catalyzed electrophilic diamination 13
Cu-catalyzed ring-opening amination 14
cumidine 428
Curtin–Hammett control 199
cyclic voltammetry (CV) 58, 61
1,4-cyclohexadiene (1,4-CHD) 59
cyclohexylamine 249, 357
cyclohexylamine oxidase (CHAO) 249
cyclometalated iridium(III) complexes 393
cyclopropylaniline 357

d

dahlia-like clusters 302
dahlias 302
D-amino acid oxidase (pkDAO) 252
decarboxylation 131, 210, 211, 220, 291, 345–348
decarboxylative arylamination 405–407
deep eutectic solvents (DES) 345, 369, 370
dehydrogenative C–N couplings 91, 402
density functional theory (DFT) 60, 170, 316, 397
deoxyfructosazine (DOF) 345
deoxyribonucleic acid (DNA) 297
D-fructose 350
D-galactose 350
D-glucose 350
3,4-dialkoxyanilines 367, 368
dialkylphosphinobiphenyl ligand systems 430
diarylamines 64, 91–94, 390, 391, 393
dibenzothiophene sulfoximine 396, 397
dibenzylamine 108, 109, 215, 362
dibutylamine 355, 362
dichlorobis(2,9-dimethyl-1,10-phenanthroline)ruthenium (II) ($Ru(DMP)_2Cl_2$) 354
(1,2-DCE) 74
9,10-dicyanoanthracene (DCA) 85
2,3-dicyano-5,6-dichlorobenzoquinone (DDQ) 81
diethanolamine (DEOA) 350
diethofencarb 367
diethoxy(methyl)silane (DEMS) 167
2,5-diformylfuran (DFF) 357
2,5-diformylfuran dioxime 357
dihydroconiferyl alcohol 367, 369
dihydrofolate (DHF) 262
dihydrofolate reductase (DHFR) 262
dihydrolipoic acid 328
dihydropyrimidinones 357, 358
3,4-dihydropyrimidin-2-ones 358, 361
dihydrosinapyl alcohol 369
1,3-dihydroxyacetone (DHA) 351
2-(dimethylamino)ethanol (DMAE) 349

4-(*N,N*-dimethylamino)pyridine (DMAP) 198
4-(dimethylamino)pyridine (DMAP) 323
N,N′-dimethylethylenediamine (DMEDA) 350, 416
di-(*N,N*-succinimidyl) carbonate 318
dioctylamine 362
diphenylmethanamine 160
1,3-dipolar cycloaddition pathway 291
dipolarophile 291
dipropylamine 362
diuretics 352
D,L-leucine 361
D,L-phenylalanine 361
D,L-tryptophane 361
D-mannose 350
Doebner–Knovenagel reaction 189
doxazosin 367
D-ribose 350
D5 variant 246, 247
D-xylose 350
dynamic kinetic asymmetric trasformation (DYKAT) 199

e

electrochemical approach 75, 402–405
electrochemiluminescence (ECL) 314
electron donor-acceptor (EDA) 34
electron paramagnetic resonance (EPR) 89
electron spin resonance (ESR) 319
electrooxidation cationic intermediates 404
electrophilic amination
  arylcuprates, NH-oxaziridine 6
  via C–H cupration 7
electrophilic amination reactions, substitution-type 1
electrophilic amination reagents 1
  Cu-catalyzed reactions 2–15
  hydroxylamine-derived metallanitrenes 18–22
  structure of 2
  transition metal free 22–27
  transition metals 15–18
  types of 2
electrophilic diazonium salt 295
enamines 8, 118–120, 344
enantioenriched α-and β-chiral arylamines 422
enantiomerically enriched chiral secondary amines 421
enantioselective A^3 coupling 104–108
  primary amines 106–108
  secondary amines 108–112
enantioselective C–H activation 123–125
enantioselective hydroamination 162, 164
enantioselective hydrogenation, photoredox synthesis 130–133
enantioselective rhodium-catalyzed hydroamination 420
ene-reductase (ERED) 266
Engineered cytochrome P450s 270–273
engineering functional nanomaterials
  carbon-based nanomaterials, functionalization of
    carbon, allotropes of 290
    carbon nanostructures 290–297
    functionalization procedures 297–299
  nanomaterial-bound amino groups 288–290
environmental health and safety (EHS) policies 425
enzymatic reductive amination 419
enzymatic synthesis, propargylamines 126–130
epoxide hydrolase (EH) 262

erlotinib 367
erythritol 342, 350
ethanolamines 348
ethylene diamine 125, 310, 348, 349
4-ethylguaiacol 367
1-ethyl-3-methylimidazolium acetate 345
ethynylbenziodoxolone (EBX) 130, 132
evolution of, MAO-N 244–247
Exiguobacterium sibiricum 256

f
fatty alcohols 370–373
fatty amines 319, 322, 371
few layers graphene (FLG) 301
fiber-supported palladium dichloride/triphenylphosphine catalyst 429
FibreCat 1026 429
fluorescein isothiocyanate (FITC) 303
fluorescence correlation spectroscopy 310
fluoroalkylamines 386
4-fluoroaniline 361
^{19}F-NMR detection 288
N-formamides 362, 364, 365
formamides 345–348
formate dehydrogenase (FDH) 256
N-formyl amine 362
fossil-based feedstocks 341
Friedel-Crafts-type arylation 135, 138
fructosazine (FZ) 345
fullerene derivatives 30, 291, 307
furan derivatives 352
furfurylamines 352–354
furosemide 352

g
gefitinib 367
Geobacillus stearothermophilus 258
glucose dehydrogenase (GDH) 130, 256, 264
glutaric anhydride 201–203
glyceraldehyde 351
glycerol 361, 362, 364, 365
glycolaldehyde 349, 350
gold(I) catalyzed hydroamination 421
gold-catalyzed intermolecular hydroamination 420
gold nanorods (Au-NRs) 297
gram scale nickel-catalyzed amination of 2-naphthyl bromide 411
graphene 290, 295, 297–299, 308
graphene oxide 299, 430
graphite exfoliation, with melamine 299–301

h
Hajos–Parrish–Eder–Sauer–Wiechert reactions 188
N-haloamine 32
halopyridines 415
Hantzsch ester (HEH) 172, 421
1,5-H-atom transfer (1,5-HAT) 31
- enthalpic effects 34
- geometrical factors 32
- intramolecular 31–34
- oxidative strategies
  - via amidyl and sulfamidyl radicals 40–43
  - via iminyl radicals 39–40
- photoinduced bond homolysis 43–46
- polar effects 34
- reductive strategies
  - via amidyl and sulfamidyl radicals 37–39
  - via iminyl radicals 34
- thermal strategies 46–49

5*H*-1-benzazepin-5-one 350
heat shock protein 90 (Hsp90) inhibitors 381

hemicellulose-derived furfural 352
(hetero-)aryl (pseudo-)halides 386
heteroatom doping 311
heterogeneous zirconium-based catalysts 367
hexafluoroisopropanol (HFIP) 21, 84
horseradish-peroxidase (HRP) 308
H_2SO_4/HNO_3 nitration 431
human immunodeficiency viruses (HIV) 297
hybrid organic-inorganic nanomaterials, amines 316
  alkylamines, semiconductor QD 319
    capping ligands 321–322
    selenium-amine system 319–321
    sulfur-amine system 319, 320
  SAMs 316–319
hydantoin 344
hydroamination 8, 10, 11, 17, 54–56, 58, 62, 66–69, 83, 162–164, 419–424
hydrogen-atom transfer (HAT) 31–50, 55
hydrogen bond acceptor 287
hydro-or solvo-thermal synthesis 309
1-hydroxy benzotriazole (HOBt) 303
6-hydroxy-D-nicotine oxidase (6-HDNO) 252
1-hydroxyhexane-2,5-dione (HHD) 359
hydroxylamine-*O*-sulfonic acid (HOSA) 20
hydroxyl moiety 315
5-hydroxymethylfurfural 352, 355
5-hydroxymethylfurfurylamines 352
2-hydroxymethyl-5-methylpyrazine (HMMP) 351
2-hydroxymethyl-5-methyl-pyrrole moiety 359
hypervalent iodine(III) 398

i
imatinib 425, 426
imine formation 160, 247, 268, 290, 318
imine reductases (IREDs) 129, 244, 255, 262–270
imine reductases catalysing reductive amination 267
immobilization of catalysts/supported catalysts 429–431
indoline 59, 66, 115, 265, 423
infrared spectroscopy (IR) 288
initial breakthrough in imine reductase (IRED) 263, 268
  biocatalytic cascade synthesis 266
  catalysed reductive amination reactions 269
  chemoenzymatic synthesis 265–267
  engineering 267
inorganic ruthenium catalyst 293
in single-electron transfer (SET) processes 130
in situ generated cationic Rh(III) catalyst 390
inter-and intramolecular palladium catalyzed C–H bond functionalization 390
intermolecular hydroamination 68, 163, 420, 421
International Council of Harmonization (ICH) 433
iridium-catalyzed hydroamination 165
iridium-nitrenoid catalyzed C–H amination 393, 394
iron-catalyzed amination of aryl halides 415
iron-catalyzed C–H functionalization 397, 399
isophorone 346–348

4-isopropyl phenol 428
isoquinoline bromide coupling 387, 388
isosorbide 343

j
Jørgensen–Hayashi catalyst 189

k
Kaiser test 288–290, 306, 312
ketimine reductases (KIREDs) 129, 267
kinetic isotope effect (KIE) 77
Knoevenagel condensation 315
Knölker's iron complex 421
Kuznetsov method 307

l
laser ablation 302, 305, 309
leucine 253, 346, 366
leucine dehydrogenase (LeuDH) 253–254
levulinic acid 360, 361, 364
Lewis basicity 287, 322
lignin 366–370
lignin-based aminoalkyl-and aryl-guaiacols 369
lignocellulosic biomass-derived α-hydroxy acids 366
$LiN(SiMe_3)_2$ (lithium bis(trimethylsilyl)amide) 411
long-chain diols 372, 373
luminescence 309
luminophore 314
Lysinibacillus fusiformis 256, 260

m
macrocyclic secondary amines 327
maleic anhydride 291
Meadows' continuous synthesis 426
3-mercaptopropyltrimethoxysilane (MPS) 326
Merrifield peptide resin 430
meta-chloroperbenzoic acid (*m*CPBA) 5
meta-hydroxyacetophenone 159
metal catalysed asymmetric synthesis
  enantioselective A^3 coupling 104–112
  enantioselective C–H activation 123–125
  Rh/Ru-catalyzed enantioselective hydrogenation 120–122
  terminal alkynes to enamines/enamides 118–120
metal catalyzed cross couplings 8, 403, 414–417
metal or metal oxide nanoparticles 287
2-(methylamino)ethanol (MAE) 349
2-methyl cyclopentanone 161
methyl isobutyl ketone (MIBK) 253
2-methyl pyrazine (MP) 351, 352
microcontact printing 318
microwave (MW)-induced method 295
monoamine oxidase 155, 214, 244, 249
monoamine oxidase A (MAO-A) 103
monoamine oxidase from *Aspergillus Niger* (MAO-N) 214, 244
monoethanolamine (MEOA) 349
mono-substituted allenes 420
morpholine 354, 362, 429
multi-enzyme cascades 261
MW-induced double functionalization 296

n
N-acetonyl amines 362
N-acetylethanolamine 342
N-acetyl-L-rednose 344
N-α-$C(sp^3)$-H bonds amination 87–90
N-acyl-propargylamines 145

N-alkyl-substituted 2-hydroxymethyl-5-methylpyrroles 359
N-alkyl tetrahydroisoquinolines (THIQs) 248
naphthalene bromide coupling 380
nanomaterial-bound amino groups 288–290
naphthalenoid H_3 antagonist 380
N-arylhydroxylamine esters 422
N-aryl-propargylamines 142, 143, 145
N-arylpyridinium ions 403
native AmDHs (natAmDH) 254, 259
native amine dehydrogenase 254–256
N-Boc-protected amino-CNO-pyrrolidine 307
N-ethyl-*N*′-(3-(dimethylamino)propyl) carbodiimide hydrochloride (EDC) 303
neurodegenerative diseases 297
N-heterocyclic carbenes (NHC) 385, 408
N-heterocyclic chemicals 345
N-heterocyclic compounds 351
N-hexadecylaniline 371
Ni(NHC)-catalyzed amination of aryl/heteroaryl chlorides 409
Ni-catalyzed aryl amination chemistry 408
Ni-catalyzed C–N cross-couplings 408, 409
nickel-catalyzed amination 408–414
nickel catalyzed C–N coupling 407–414
nickel-mediated C–H amination 395
Ni(allyl)(IPr)Cl 410
nitrene 18, 125
nitrilotriacetic acid (NTA) 297
nitrogen compound synthesis
  from α-amino acids 345–348
  from cellulosic biomass derivatives 348–366
  from chitin and its derivatives 341–345
  lignin derivatives 366–370
  from triglycerides and fatty alcohols 370–373
nitrogen dioxide (NO_2) 308
nitrogen (mostly amine)-doping 311
nitrogen radicals 31–50
nitrosonium tetrafluoroborate ($NOBF_4$) 291
^{15}N-labeled compounds 288
N-mesylimidazole 404
N-methylamines 362, 364
N-methylamino acid dehydrogenases (NMAADHs) 129
N-methyldiethanolamine (MDEA) 350
N-methyl-*O*-tosylhydroxylamine 392
N-methylpyrrolidine derivative 291
N-modified α-amino acids 303
N,N-dimethyl-1-(3,4,5-trimethoxy-phenyl)methanamine 367
N-octadecylaniline 371
non-cross-linked polystyrene (solPS) supported SPhos (solPS-SPhos) 429
N-(*p*-tolyl)benzophenone hydrazone 386
N-substituted 5-hydroxymethyfurfurylamines 353
N-substituted 2-hydroxymethyl-5-methylpyrroles 359
nuclear magnetic resonance (NMR) techniques 172, 288, 315
nucleophilic aromatic substitution 312, 377, 426–428

o

1-octadecene (ODE) 319
O-(2,4-dinitrophenyl)hydroxylamine (DPH) 20

olefinic C(sp^2)-H bonds amination 82
organic-inorganic hybrid nanomaterials 287
organic-inorganic hybrid nanoparticles 328
amines 325–328
post functionalization of amine 328–329
organocatalyic C–H oxidative amination 398
organocatalysis 187–231
organocatalytic stereoselective synthesis 134
organolithium reagents 4, 5
organozinc reagents 2, 3, 15
ortho-aminobenzoic acid derivatives 396
oxazoles synthesis 145
oxazoles thiazoles 148
oxygen reduction reaction (ORR) 299

p

Paal-Knorr reaction 359
Paenibacillus lactis 265
palladium amine complex 385
palladium catalyzed aryl C–H amination 390
paloxetine 412
Pd-catalyzed C–N bond forming reaction 425–429
Pd/Cu bimetallic catalytic system 406
Pd nanoparticles on silica-starch substrate (PNP-SSS) 430
penicillin G amidohydrolase (PGA) 127
peptide-like heterocyclic molecules 361
Petrotogamobilis 254
phase-transfer catalysed (PTC) 187
phenol-derived electrophilic arene coupling partners 412
phenylalanine 253, 254, 366
phenylalanine dehydrogenase (PheDH) 253, 254
phenylallene 421
(S)-1-phenylethylamine 202
photocatalysis 131, 178, 400, 416, 418
photocatalytic perfluoroalkylation 315
photodeprotection reactions 318
photoinduced electron transfer 402
photoinduced strategies
oxidative strategies
1,5-HAT via amidyl and sulfamidyl radicals 40–43
1,5-HAT via iminyl radicals 39–40
photoinduced bond homolysis 43–46
reductive strategies
1,5-HAT via amidyl and sulfamidyl radicals 37–39
1,5-HAT via iminyl radicals 34–37
thermal strategies 46–49
photoredox catalysis 37, 54, 73, 77, 78, 81, 130, 150, 399–402
photoredox catalysts 69, 176
photoredox catalyzed C–N bond forming processes 401
photoredox synthesis, propargylamines 130
Pictet-Spengler cyclization 369
Pinnick oxidation 205
p-methoxyphenyl (PMP) 56
polar effects 34
poly(vinylpyrrolidone) (PVP) 324
polyallylamine 322, 326
polyamidoamine (PAMAM) dendrimer 302
polyethylene glycol-anchored phosphine ligand 429
polyethyleneimine (PEI) 322
polymeric coating 326

polymethylhydrosiloxane (PMHS) 8, 409
polysaccharide xylan 353
polyvinylpyridine (PVP) supported catalyst 430
porphyrin 303, 314, 328
Potential Genotoxic Impurities (PGIs) 211
prazosin 367
pregabalin
  acetate enolate equivalents to nitroalkene 206–214
  amino α-carbanion to Michael acceptors 204–206
  glutaric anhydride 202–204
pristine C_{60} fullerene 291
prochiral aminoketones 362
proline-based catalyst and *p*-nitrobenzoic acid (PNBA) 134
propargylamines 103
  enzymatic synthesis 126–130
  metal catalysed asymmetric synthesis
    alkynyl-substituted epoxides/lactones/carbonates 116–118
    enantioselective A^3 coupling 104–112
    enantioselective C–H activation 123–125
    Rh/Ru-catalyzed enantioselective hydrogenation 120–122
    terminal alkynes to enamines/enamides 118–120
  organocatalytic stereoselective synthesis 134
  oxazoles synthesis 145–147
  oxazoles thiazoles 148–150
  propargylic amination reaction 112
  pyridines synthesis 140–142
  pyrroles synthesis 138–139
  pyrrolines synthesis 139–140
  quinolines synthesis 142–145
  synthetic methods 104
propionate esters 367
4-propylcatechol 367, 368
4-propylguaiacol 367, 368
prostate specific membrane antigen (PSMA)+ cancer cells 304
proton-coupled electron transfer (PCET) 40, 55, 176
proximicin A 344
pseudo-peptides 361
$P(Ar)_3$-type ligands 384
pyrazines 345, 351, 352
pyrazole 77, 80, 381, 387, 389, 427
pyridazinium chloride 387
pyridazinone 380
pyridines synthesis 140
pyridinium chlorochromate (PCC) 174, 215
pyrimidine 166, 196, 404
pyrolysis 309
pyrroles 71, 72, 80, 81, 103, 137–139, 358, 359
moiety 291
pyrrolidine 346, 360
pyrrolidinofullerenes 291
pyrrolidones 12, 360, 361
pyrrolines synthesis 139

r

racemic isocyanides 361
radical addition to C-C double/triple bonds
  amidyl radical addition 55–62
  aminium radical cation addition 64–71
radical arylation reaction 291
radical species addition to aromatic rings 71–76
Raman spectroscopy 319
reductive aminases (RedAms) 268
reductive amination 417

reductive amination (*contd.*)
aldehyde/ketone 156
aldehydes 173
intramolecular asymmetric 158
2-methyl-6-ethyl-aniline 157
reductive amination reaction 128, 156, 158, 417
renewable polyamides 355
Rh-catalysed hydroaminomethylation 171
rhodium-catalyzed C–H amination 390, 392
Rhodococcus sp. 253, 256, 257, 259
(R)-selective amine oxidases
D-amino acid oxidase 252
6-HDNO 252
Ru-catalyzed C–H amination 17
Ruhemann's purple 290
ruthenium-based photocatalyst 401
ruthenium catalyzed C–H amination 403

s
Schiff base adduct 346
self-assembled monolayers on flat surfaces (SAMs) 316–319
semiconductor quantum dots 287, 316, 319–322
Shvo's catalyst 369
silane reductant 409, 418
silica supported SiO_2-SPhos 429
silyl enol ethers 9, 26, 27
single electron transfer (SET) 35, 55, 130, 176, 395
single-walled carbon nanotubes (SWNTs) 291, 293, 307
spectrophotometric detection 288
spectroscopic techniques 288, 289
Sphingobium yanoikuyae (SyADH) 128
sp^2-hybridized carbon atoms 290, 295
(S)-selective amine oxidases
CHAO 249–251
MAO-N 244–247
Pseudomonas monteilii ZMU-T01 249
synthetic applications and cascades 247–249
sterically encumbered *N*-adamatylbenzamide 393
Stradiotto's and Hartwig's nickel-catalyzed amination 411
Streptomyces virginiae 254
styrene monooxygenase-catalysed (SMO) 262
2,5-substituted pyrrolidines 167
sulfamates 412
sulfonamide 45, 56, 71, 72, 80, 81, 91, 99, 166, 167, 344
sunflower oil 371
supercritical fluids 299
superficial amines 294, 297, 308, 314

t
Tamibarotene 382, 383
t-butyl[(1,1′-biphenyl)-2-carbonyl] oxycarbamate 406
telcagepant 189, 190, 221, 231
telescoped sequence of nitration and hydrogenation 428
terazosin 367
terminal alkynes to enamines/enamides 118–120
tert-butyl 2-(2-(2-aminoethoxy)ethoxy) ethylcarbamate 303

2-*tert*-butylanthraquinone (2-*t*-Bu-AQN) [illegible]
tert-butyl (2-(2-(2-(aminobenzamido) ethoxy)ethoxy)ethyl)carbamate 305
tert-butyl hydroperoxide 197
tert-butyl nitrite (TBN) 81
tetrahydroisoquinolines 130, 158, 247, 248, 250
N,N,N′,N′-tetramethylethylenediamine (TMEDA) 350
thermal cycloelimination 291
Thermoanaerobacter ethanolicus (TeSADH) 128
thermogravimetric analysis (TGA) 288
thiohydantoin 344
threshold of toxicological concern (TTC) 432
tin-free Pd-catalyzed coupling 384
TM-catalyzed electrophilic amination 2, 15
TM-free Prilezhaev reaction 26
TM-free Rubottom oxidation 26
Tofacitinib 159
topaquinone (TPQ) 243
top-down methods 309
transaminases (TAs) 243, 244
transition metal accounting 433
transition metal catalyzed C–N coupling
  aromatic amines 377
  Buchwald-Hartwig amination 383–388
  nucleophilic aromatic substitution 377
  substituted aromatic amines 377
  Ullmann-type amination 378–383
transition metal (TM) free amination 22
transmission electron micrographs 301, 302
transmission electron microscopy (TEM) 297
2,4,6-triamino-1,3,5-triazine 301
triazacyclononane or chelating amines 327
trichloroacetic acid 318
triethanolamine (TEOA) 350
trifluoroacetic acid (TFA) 42, 80, 192, 299
trifluoromethanesulfonamide ($TfNH_2$) 83
5-(trifluoromethyl)-2-isoxazolines 223–230
triflyloxy pyridinium 401
triglycerides 370–373
triisopropylsilane (TIPS) 324
2,4,6-triisopropyl-thiophenol (TRIP thiol) 56
trilaurin 371
3,4,5-trimethoxybenzaldehyde 367
trimethylsilylamino derivatives 291
tri-*n*-octylphosphine (TOP) 319
tri-*n*-octylphosphine oxide (TOPO) 319
2,2,5-triphenylpent-4-en-1-amine 166
2,4,6-triphenylpyrylium (TPT) 93
5,10,15-tris(3,5-di-tertbutylphenyl)-20-[4-(carbonyloxy)phenyl] porphyrin 303
tryptamine 346, 361
2-D self-assembled monolayers 287

u

Ugi three-component condensation reaction 361
Ullmann-type amination
  applications 380–383

Ullmann-type amination (*contd.*)
  catalysts and variants 380
  mechanistic studies 379–380
urea 59, 163, 344

V
valine 346, 347, 366
van der Waals-like forces 301
vinylarenes 423
VLA-4 antagonist 403, 405

W
waste treatment 425
Wilkinson's catalyst 291

X
X-ray photoelectron spectroscopy (XPS) 288
xylan 353, 358, 359
xylitol 350